ACCIDENT PREVENTION MANUAL
for Business & Industry

Administration & Programs
Occupational Safety and Health Series

The National Safety Council's OCCUPATIONAL SAFETY AND HEALTH SERIES is composed of four volumes and two study guides written to help readers establish and maintain safety and health programs. The latest information on establishing priorities, collecting and analyzing data to help identify problems, and developing methods and procedures to reduce or eliminate illness and accidents, thus mitigating injury and minimizing economic loss resulting from accidents, is contained in all volumes in the series:

ACCIDENT PREVENTION MANUAL FOR BUSINESS & INDUSTRY (2-volume set)
  Administration & Programs
  Engineering & Technology
STUDY GUIDE: ACCIDENT PREVENTION MANUAL FOR BUSINESS
  & INDUSTRY
FUNDAMENTALS OF INDUSTRIAL HYGIENE
STUDY GUIDE: FUNDAMENTALS OF INDUSTRIAL HYGIENE
OCCUPATIONAL HEALTH AND SAFETY

Other safety and health references published by the Council include:
ACCIDENT FACTS
MOTOR FLEET SAFETY MANUAL
OUT IN FRONT: EFFECTIVE SUPERVISION IN THE WORKPLACE
PRODUCT SAFETY: MANAGEMENT GUIDELINES
PUBLIC EMPLOYEE SAFETY & HEALTH MANGEMENT
SUPERVISORS' SAFETY MANUAL

ACCIDENT PREVENTION MANUAL
for Business & Industry

Administration & Programs   **10th edition**

Project Editor: Patricia M. Laing
Technical Coordinator: Philip E. Schmidt
Composition: Publishing Solutions
Cover Designer: Beth Herman Design Associates

Copyright, Waiver of First Sale Doctrine

©1992 by the National Safety Council
All Rights Reserved
Printed in the United States of America
96 95 94 93 92   5 4 3 2 1

Library of Congress Cataloging in Publication Data
National Safety Council
Accident Prevention Manual for Business & Industry:
Administration & Programs
International Standard Book Number: 0–87912–155–6
Library of Congress Catalog Card Number: 91–066437
10M1091   Product Number: 12144

Contents

Foreword

Emerging Trends in Safety and Health

by
William E. Tarrants, PhD, PE, CSP

The state of occupational safety and health in the United States and throughout the world is, and always has been, dynamic. Although change is to be expected, the rate of change has increased substantially in recent decades and, from all indications, will continue to accelerate during the 1990s and through the year 2000. This trend will have considerable impact on the users of this 10th Edition of the *Accident Prevention Manual for Business & Industry*—the student using it as a textbook, the corporate or company manager searching for solutions to safety and health problems, the new safety specialist who must plan and organize a safety and health program within a company, or the experienced safety professional seeking to improve an operating program and to learn more about advances in the field of safety and health.

This two-volume manual contains information based on the most current knowledge and practices within the safety and health field in 1991. Much of this information is based on practice and research and will remain relevant during the next decade. However, changes are bound to occur that will require managers to find innovative solutions to safety and health problems on and off the job. Thus, occupational safety and health professionals must keep abreast of current and future economic, environmental, regulatory, and technical advances affecting their field.

A good place to begin when planning for the future is to study the projections through the 1990s into the year 2000 made by experts in the field. A consensus of these studies and recommendations will provide a general guide for planning, organizing, and implementing safety and health programs. For example, a study entitled *Workforce 2000—A Survey Report on Corporate Responses to Demographic and Labor Force Trends* conducted within 645 organizations by Towers Perrin for the Hudson Institute identified four key trends that will shape the last years of the 20th century:

- The U.S. economy should grow at a relatively healthy pace, boosted by renewed productivity, market growth, and a strong world economy.
- Manufacturing will account for a much smaller share of the U. S. economy in the year 2000 than it does today. Over the next decade, service industries will create almost all of the new jobs and most of the new wealth.
- The workforce will grow slowly and comprise more older workers, women, and minority workers. Only 15% of the new entrants into the labor force over the next 10 years will be native-born white males, compared with 47% in that category today.
- The new jobs in service industries will demand much higher skill levels than do jobs today. Few new jobs will be available for those who cannot read, follow directions, or use basic mathematics.

    Risk is the individual likelihood of harm (one airplane crash in over nine million airline flights) from some level of exposure (110 million airline flights [1975–1985]) to a particular hazard (windsheer at airports). Risk and exposure are combined to determine how much harm (e.g., 12 crashes and 400 lives lost) is caused. The term "risk" is

also used to represent the sum of all individual risks, particularly when developing population estimates.

In 1985 the Executive Committee of the Industrial Division of the National Safety Council asked its Research Projects Committee to conduct a study of emerging trends in the safety field to assist future safety program planning. The Research Committee ultimately identified 34 emerging trends for evaluation and ranking. A panel of 13 experts from the Research Projects Committee and Industrial Division Group Directors were asked to evaluate each trend on the basis of three criteria:

- *importance*—the trend's impact on safety and health policy and procedures and the degree to which the trend may be associated with deaths and injuries
- *interest*—the degree to which the public may become involved in each trend
- *time frame*—the period of time during which the trend will have its maximum impact.

Most trends were rated "moderate" in importance and interest, and "long term" in their impact. The trends that the panel ranked the highest on the basis of all these criteria were:

- changing composition/nature of the workforce/jobs
- occupational versus nonoccupational illnesses
- long-term effects of toxic materials in the environment
- defining acceptable risks
- development of safety standards

A second study, begun in December, 1989, and summarized in May, 1991, was completed by a 26-member panel of experts from the Research Projects Committee, the Industrial Division Group Directors, the Research Committee of the NSC Board of Directors, and the Chairman of the Labor Division. This group used four criteria to select emerging trends/issues:

- *importance*—the degree to which the trend/issue is or will become a major safety and health concern or factor
- *persistence*—the degree to which the trend/issue will remain throughout the decade
- *significance*—the degree to which the trend/issue is or should be considered a priority by the National Safety Council
- *confidence*—the respondents' overall confidence in their responses.

The group agreed that five trends or issues ranked highest on the basis of these criteria: ergonomics; consideration of safety, health, and ergonomic issues at the concept and design stage; incorporating safety into the management system; repetitive motion trauma, including carpal tunnel syndrome; and back injuries. Only slightly less important in the rankings were trends in emergency planning and global environmental issues.

The final step in the project was to examine all the trends/issues and to form clusters of the trends that were judged to be closely or strongly related. This composite set of groups could provide a more global view of the subject areas covered by the original trends/issues submitted. The panel devised the following five cluster groups, considering them all top priority areas for the future:

- occupational health and wellness
- public safety and health issues
- competence of health and safety professionals
- changing nature of the workforce/jobs

- responses to changing technology.

Based on the results of these two emerging trends studies, it is possible to draw some conclusions and to make specific recommendations to support enhanced safety and health efforts during the coming decade. Specific major problem issues and corrective actions recommended include the following:

1. Employee cutbacks may continue in the near future, which could mean fewer safety and health professionals to perform the work necessary to achieve appropriate safety and health goals within any organization. Managers should make hiring and supporting qualified safety and health professionals a top priority.

2. If there are too few qualified staff members to accomplish the jobs required, then companies will have to pay more attention to time management. Because all safety actions can be completed within a given time, companies will have to set priorities based on the loss potential of identified risks and the cost-benefit impact of available countermeasures. The process will require a reliable, accurate system of hazard identification and evaluation.

3. Many major safety and health problems existing within an organization can be traced to what managers have judged to be "acceptable risk." That is, as long as no major losses occur, mangers assume that the risks they are taking are "acceptable." They need to improve their ability to define and assess risk and "acceptable risk" based on *loss-potential* measurement criteria instead of after-the-fact accident losses. Risk at the general level involves two major components: (1) the existence of a possible unwanted consequence or loss and (2) the probability such a consequence will occur. Managers can reduce risk either by decreasing the probability of occurrence or by educating the risk taker to recognize and to control the hazard.

4. In times of financial crises organizations may decide not to spend money to correct a safety or health problem. If such a choice must be made, only problems with the lowest loss potential should remain uncorrected for a time. This approach assumes that managers are able to measure accurately the potential loss associated with various identified safety and health problems.

5. Personal safety, health, environmental safety, equipment safety, process safety, and product safety are becoming more interrelated and creating an expanded role for the safety professional. Practicing safety and health professionals must upgrade their knowledge, skills, and job performance abilities to handle these new responsibilities. Course work and content within university degree programs in the safety field should be examined to insure adequate coverage of those subjects needed to prepare the safety professional properly.

6. Physical hazards and unsafe behaviors, as well as high-risk human-machine-environment interactions, must be included among the measurement criteria used to evaluate safety performance. The key to proper safety and health program evaluation is measurement criteria based on determining the loss potential of certain hazards and behaviors rather than on after-the-fact analysis of actual ac-

cident losses. Two examples of evaluation methodologies based on no-loss measurement criteria include behavior-environment sampling and the Critical Incident Technique.

7. Accident prevention within an organization is not the exclusive responsibility of the safety and health professional. Managers and supervisors at all levels must be involved in day-to-day safety and health activities and decisions. Managers should be properly trained to identify and assess potential accident situations before they result in actual losses (injury, death, property damage, production stoppage, etc.). Unfortunately, many managers believe that no safety or health problem exists if no accidents occur that cause immediately measurable losses. Safety and health training must be included in the education of future managers and in upgrading the qualifications and abilities of present managers.

8. Ergonomics needs more attention. Fifty percent of OSHA incidents at present are ergonomics related. The key points addressed in the U.S. Occupational Safety and Health Administration Ergonomic Guidelines include management commitment and employee involvement in safety; worksite analysis requiring a thorough evaluation of workplace ergonomic hazards; hazard prevention and control, including the requirement to follow established procedures to control ergonomic hazards; and education and training of managers, supervisors, and employees to help them understand the ergonomic hazards associated with their jobs.

9. Managements need to place greater emphasis on the use of engineering techniques in accident prevention. For example, they can apply engineering concepts and design to make manufacturing processes safer and to minimize waste. Instead of focusing primarily on correcting workers' unsafe practices, management should examine and assess the total human-machine-environment system to appraise its loss potential. They can then apply the principles of engineering science to minimize future accidents and their subsequent losses.

10. Companies need a better, more accurate method for determining direct and indirect costs of employer accident losses on a regular, sustained basis and for assessing the impact of accident prevention programs in dollar units that can be translated into profits. Employers need a cost accounting/cost investigation procedure to identify both direct and indirect (uninsured) costs of employee accidents occurring on and off the job. Because of employers' vested interest in loss reduction, a true determination of dollar losses resulting from accidents should strongly motivate management to initiate or expand various worker safety and health programs.

11. With an increase in vehicular traffic, both on and off the job, highway traffic safety is becoming more important to employers. Although management should pay particular attention to on-the-job motor vehicle safety, off-the-job vehicle accidents are also costly to employers. As a result, managers need to focus on a systems approach when developing a comprehensive company highway traffic safety program.

12. The demand for well-qualified, competent safety and health professionals will increase significantly during the 1990s. Companies will need not only more staff but more sophisticated, well-educated, technically competent individuals able to respond to the emerging trends/issues identified in the safety and health field. Safety and health professionals of the future can help to prepare themselves by graduating from an accredited safety degree program and acquiring the appropriate certificates and degrees.

13. The next decade will witness the emergence of a world market and a greater need for international standards in accident recordkeeping, labeling and transporting hazardous materials, and occupational health and safety issues. For example, formation of the European Economic Community (EEC) has brought about changes in worker safety, commercial and industrial products, and other goods and services. The Framework Directive of the EEC (passed in 1989), encourages the formation of on-the-job safety and health programs for workers and management and applies to all private and public sectors of business. As more nations join the world economic community, solving occupational health and safety problems and preserving the environment must become top international priorities for all countries.

Preface

The 10th edition of the ACCIDENT PREVENTION MANUAL was written with an eye on the future—the decade of the nineties and the 21st century. To recognize the expanding role of the safety and health professional, we have chosen a new name—ACCIDENT PREVENTION MANUAL FOR BUSINESS & INDUSTRY—and added new chapters and new topics. We have also totally revised the organization of the two volumes.

Covering a broad spectrum of subjects, this *Manual* pinpoints specific hazards and directs the reader to the appropriate sources of help. The *Administration & Programs* volume includes a chapter on management techniques, government regulations, and safety and health program development. The *Engineering & Technology* volume covers more technical information related to specific businesses and industries. The National Safety Council's *Fundamentals of Industrial Hygiene*, 3rd edition, and *Occupational Health and Safety*, 2nd edition, are additional volumes vital to complete your safety and health library. Contact the National Safety Council for more information. (Note that as of June, 1992, the National Safety Council will be located at 1121 Spring Lake Drive, Itasca, IL 60143-3201.)

New Material

In the *Administration & Programs* volume, we have added Chapter 5, Environmental Management, because the safety and health professional needs to know the *what* and *how* of controlling emissions to the environment. We have also added Chapter 7, Employee Assistance Programs, to give the reader a background in how to assess the need for an EAP, how to set it up, and how to evaluate outside service providers. In the *Engineering & Technology* volume, we have added a Chapter 23, Automated Lines, Systems, or Processes, to give the safety and health professional vital information about safety issues related to the use of automation in business and industry.

Revised Material

We have totally rewritten Chapter 14, Computers and Information Management, Chapter 16, Safety Training, and Chapter 17, Audiovisual Media, in the *Administration & Programs* volume. We have revised many topics in the *Engineering & Technology* volume as well, including lockout/tagout, confined space, lifting recommendations, and ergonomics.

New Organization

To help the safety and health professional, we have reorganized each volume into four parts. In the *Administration & Programs* volume, Part 1, Introduction to Safety and Health deals with the history of the safety and health movement and government regulations. Part 2, Program Organization, includes chapters on the development or evaluation of specific programs. Part 3, Hazard Information and Analysis, covers how to acquire and analyze hazard information and costs, and how to manage the data acquired. Part 4, Program Implementation, gives important pointers on motivation, safety training, using audiovisual media, promoting safety in-

side and outside the organization, and providing safe, healthful environments for office personnel, workers with disabilities, and nonemployees.

In the *Engineering & Technology* volume, Part 1, Facilities and Workstations, covers safety issues related to the design, construction, and maintenance of facilities, boilers, pressure vessels, fire protection, and personnel facilities. It also includes an important chapter on how to design jobs and workstations to better fit the physiological needs of workers. Part 2, Material Handling, discusses safety issues related to use of manual and powered equipment to move and store material. Part 3, Workplace Exposures and Protection, covers safeguards for protecting workers from specific exposures. Part 4, Production Operations, covers safety issues related to power tools; the woodworking, welding, and metalworking industries; and automation and robotics.

Contributors

The ACCIDENT PREVENTION MANUAL FOR BUSINESS & INDUSTRY is unique—the compilation of the experience and expertise of contributors from all major occupations and industries. Each of the reviewers and contributors is a practicing expert. To assure uniformity and accuracy, the final version of the text was reviewed by Raymond P. Boylston, CSP, Sr. Vice President, ELB and Associates, Inc.; William J. Larson, CSP, PE; and Floyd G. Miller, PhD, University of Illinois at Chicago. The National Safety Council and the editors wish to express their appreciation and gratitude to these reviewers of the final text and to the many contributors who devoted hours to updating and checking the accuracy of this publication. These contributors to the *Administration & Programs* volume include: Cendrella Abdallah, William W. Allison, Jan I. Amundson, Ernest V. Anderson, Raymond E. Arntson, Matilda A. Babbitz, Robert L. Baldwin, Wesley H. Beitl, Jean E. Benton, Patrick H. Boyer, Raymond P. Boylston, Susan Byrd, David E. Carothers, Allen W. Carpenter, Janice Cave, William M. Clardy, Ronald W. Clark, David F. Coble, Dene Corless, Robert W. Crolins, John A. Davenport, Georgiann DeCenzo, Ruth Deutsch, Kenneth C. Eck, J. Nigel Ellis, Fred Kohloff, Michael J. Fagel, Mark A. Friend, Wilfred G. Gallardo, Sr., Daniel F. Garde, Bahador Ghahramani, Patricia A. Gleason, Harold M. Gordon, Warren D. Graef, Jr., Maurice L. Greiner, Jamie E. Haines, Earl E. Hansen, Jay B. Hart, Merrie L. Healy, C. Nelson, W. R. Hooper, Gerald C. Hurley, Gary M. Hutter, Bobby J. Jackson, Dieter W. Jahns, Ann Johnson, C.W. Jordan, Donald W. Kase, Christine B. Kelly, James G. Kendzel, Marvin E. Kennebeck, Jr., Alpheus R. Klashak, Fred Kohloff, Abraham L. Kooiman, Richard S. Kraus, Helen M. Kutska, William J. Larson, Gordon L. Leach, Richard Lewis, Ray Lehr, William R. Mairson, J. Malley, Jerome Mansfield, Alvin V. Marks, Craig G. May, Katherine S. McCarter, Jeffrey D. Meddin, Barrett C. Miller, William M. Montante, Paul Myers, Michael E. Nave, PhD, Thomas E. Nidiffer, Lawrence O'Brien, Lynn O'Donnell, N. Olson, Richard D. Olson, Michael K. Orn, Theodore Oslay, Anthony Parente, Robert Puttoff, P.L. Rives, Treft W. Schultetus, Jack L. Skidmore, V.J. Siminski, F.L. Smith, Dan Stockwell, Wain W. Stowe, George Swartz, John J. Szwarc, William E. Tarrants, Lori Traweek, Todd W. Turriff, Eric L. Van Fleet, PhD, Diane L. Viera, Alfred B. Vimont, Robert L. Vines, Steve Vojtaski, Jr., Diana Wilbur, Lester H. Wittenberg, and Susan J. Younkin.

The National Safety Council also wishes to thank members of its staff who contributed significantly to this *Manual:* Linda Bennett, Thomas Danko, Barbra Jean Dembski, Gary Fisher, John Fogel, Steve Jackson, Dan Jones, Patrina Jones, Alan Hoskin, Ronald Koziol, Joseph Lasek, Randy Logsdon, Robert LoMastro, Elizabeth Lucas, Robert J. Marecek, John McHale, Jill Niland, Robert O'Brien, Michael Peltier, Carl Piepho, Douglas Poncelow, Gilbert Riley, and Philip E. Schmidt.

PART 1

INTRODUCTION TO SAFETY AND HEALTH

Refinery at Grangemouth, Scotland. (Courtesy UOP.)

1

Historical Perspectives

THE MISSION OF THE NATIONAL SAFETY COUNCIL

The mission of the National Safety Council is to educate and influence society to adopt safety and health policies, practices, and procedures that prevent and mitigate human and economic losses arising from accidental causes and adverse occupational and environmental health exposures. (Approved by the Board of Directors, October 18, 1983.)

Protecting life and promoting health is the basic mission of the National Safety Council. The Council's hallmarks over the years have been flexibility and the ability to create new programs to cope with ever-changing challenges. Many of the major safety and health problems that faced safety professionals more than eight decades ago are virtually nonexistent today. However, occupational and environmental health, bloodborne diseases, and drug and alcohol abuse present new challenges to today's safety professionals.

In an imperfect world, there will always be risks, and the National Safety Council will continually strive to reduce the number and severity of those risks—no matter what the cause. In working to protect people from accidental death or injury, the Council seeks ways to ensure that everyone enjoys a safe and healthful environment.

In this, the tenth edition of the *Accident Prevention Manual for Business & Industry*, the National Safety Council presents a compilation of knowledge and experience that form part of the safety movement's general heritage. For specific information regarding occupational health and industrial hygiene, see two other National Safety Council books: *Occupational Health and Safety* and the *Fundamentals of Industrial Hygiene*, respectively.

PHILOSOPHY OF ACCIDENT PREVENTION

In medieval times, the master craftsman instructed apprentices and journeymen to work skillfully and safely because he knew the value of high-quality, uninterrupted production. However, the Industrial Revolution, which began in England during the 18th century, shifted the emphasis to faster and greater production and created the conditions that inspired the development of accident prevention as a specialized field.

The industrial safety philosophy developed because the hazardous work environment of early factories and other production and distribution sites produced an appalling rate of worker injuries and deaths. If these conditions had not been corrected to stop the waste of personnel and resources, the growing number of accidents and injuries would have staggered the imagination.

In the beginning, one way to encourage management to accept responsibility for preventing accidents was to pass workers' compensation laws. This "new" line of thinking held the employer responsible for a share of the economic loss suffered by an employee involved in an accident.

It was a rather short step from this approach to the realization that most accidents could be prevented and that the same industrial knowledge used to develop mass production

3

methods also could be applied to accident prevention. Managers soon discovered that efficient production and safety were closely related. From this beginning grew the safety movement as it is known today.

The progress in reducing the number of accidents and injuries in the relatively short time since this movement began has exceeded the most optimistic expectations of early safety pioneers. The accidental death rate per 100,000 persons in the United States has decreased 55% over the past 78 years (*Accident Facts*, 1991).

Experience has shown that virtually any hazard can be overcome by practical safety measures. To further that belief, the National Safety Council continues its concerted efforts to prevent accidents and occupational illnesses.

In summary, here are six reasons for working hard to prevent accidents and occupational illnesses:

1. Needless destruction of life and health is morally unjustified.
2. Failure to take necessary precautions against predictable accidents and occupational illnesses makes management and workers morally responsible for those accidents and occupational illnesses.
3. Accidents and occupational illnesses severely limit efficiency and productivity.
4. Accidents and occupational illnesses produce far-reaching social harm.
5. The safety movement has demonstrated that its techniques are effective in reducing accident rates and promoting efficiency.
6. Recent state and federal legislation mandates management responsibility to provide a safe, healthful workplace.

THE BEGINNINGS OF SAFETY AND HEALTH AWARENESS

The written history of safety and health began about the time of the building of the Egyptian pyramids. *The Ebers Papyrus* and the *Edwin Smith Papyrus*, both found in 1862 and dating from about 3000 B.C., addressed many of these issues. *The Ebers Papyrus* contains household and medical recipes to cope with various traumatic events such as crocodile bites, burns, and the removal of foreign objects (splinters). The *Edwin Smith Papyrus* served as a textbook of surgery that discussed a variety of injuries and treatments involving splints, dressings, and ointments.

About 2000 B.C., Hammurabi, a Babylonian ruler, revised the old laws of the land and produced a Code of some 280 paragraphs. It covered bodily injury and physicians' fees and probably was the first document that included a beginning of what today is known as workers' compensation laws. Two clauses that would most interest modern safety and health professionals are:

§199. If [a man] has caused the loss of the eye of a gentleman's servant or has shattered the limb of a gentleman's servant, he shall pay half his price.

§206. If a man has struck a man in a quarrel, and has caused him a wound, that man shall swear "I do not strike him knowingly," and shall [be responsible for] the doctor.

Ramses III in about 1500 B.C. hired physicians to care for mine and quarry workers as well as those engaged in the construction of public works such as canals and large temples. His decision was based more on a desire to retain a healthy work force than to be loved by his subjects.

Other ancient healers contributed to the growing body of health and safety knowledge. In 400 B.C., the Greek physician Hippocrates, usually called the father of medicine, wrote the first known description of tetanus. About 200 B.C., the effects of lead poisoning were described by the Greek poet and physician Nicander.

Various Roman writers from 100 B.C. through the second century A.D. reported on the plague of Athens, the ill effects of mining environments on workers, and the health hazards of lead water pipes and containers for blending wine.

As early as the first century A.D., Pliny the Younger mentioned lead poisoning as a disease present among mine slaves. Pliny the Elder wrote about ox bladders being used as primitive respirators by workers who produced vermillion to keep the mercury fumes out of their lungs.

The Middle Ages

Throughout the Middle Ages, workers must have suffered from the ill effects of using pigments, grinding metalware, and silvering mirrors. Unfortunately, their ordinary living conditions so shortened their lives that tuberculosis and various plagues killed or crippled many workers before occupational diseases could take their tolls. Thus, health and safety issues in the workplace were not considered a major problem.

In the seventh century in ancient Lombardy, King Rothari codified existing laws in 388 chapters. This document was probably the origin of the basic principles of compensation for injury. The edict applied to personal injuries received in brawls, fights, and feuds, and established payments for disability and death.

In the eleventh century, King Canute, King of Denmark, Norway, and England, stated the principles of compensation for particular injuries. The importance of losing a thumb was recognized—compensation for its loss was twice that for the loss of the second digit and two-and-one-half times that given for the loss of the third digit.

In 1473, Ulrich Ellenbog, an Austrian physician, wrote a tract directed toward goldsmiths and other handlers of metal. He warned against burning coal in confined spaces and inhaling vapors produced by heating such metals as lead, antimony, silver, and mercury. This tract is considered to be the first writing devoted exclusively to industrial metal poisoning.

Six years after the Saxon physician, George Agricola, died, his book, *De Re Metallica*, was published in 1555. It emphasized the need to ventilate mines and depicted various devices that would force air below ground. Other illustrations showed personal protective devices—gloves, leggings, and masks. The work made such an impact on industry that it endured for centuries.

In 1567, Philippus Aurelous, a.k.a. Theophrastus Bombastus von Hohenheim, who later called himself Paracelsus, published a treatise, *On the Miners' Sickness and*

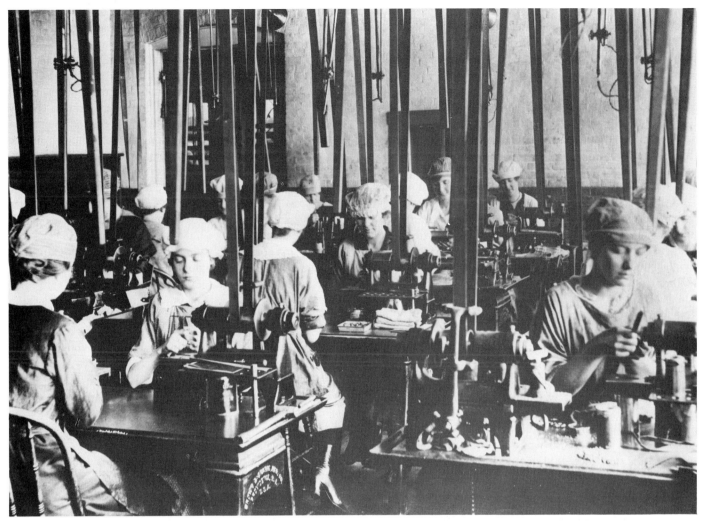

Figure 1–1. Little was known in the early years of industrialization about the effects of work environment on the worker. Poorly lit and ventilated shops and accommodations that show no awareness of ergonomic concerns were the rule rather than the exception.

Other Miners' Diseases. Paracelsus had grown up in Switzerland, studied medicine in Italy, and practiced medicine as an itinerant teacher, visiting mines and workshops. In his treatise, he distinguished between acute and chronic metal poisoning. His monograph was the first to deal with the diseases of a specific occupational group.

In the early part of the 18th century, Bernardino Ramazzini published the classic *Discourse on the Diseases of Workers,* which is still relevant today. Ramazzini pointed out that in addition to the standard questions asked of a patient, one more should be added: "What is your occupation?"

Ramazzini, dubbed the father of occupational medicine, summarized the two causes that he believed were responsible for the occupational diseases of workers of his day:

> The first and most potent is the harmful character of the materials that they handle, for these emit noxious vapors and very fine particles inimical to human beings and induce particular diseases; the second cause I ascribe to certain violent and irregular motions and unnatural postures of the body, by reason of which the natural structure of the vital machine is so impaired that serious diseases gradually develop therefrom.

Mass Production Appears

Until the 1700s, production methods were labor-intensive, with work being done by hand in cottages. Three developments were to change this way of life: In England, inventors developed the spinning jenny in 1764 and perfected the power loom in 1784. In America, Eli Whitney added his invention, the cotton gin, in 1792. These and other innovations ushered in what would later be called the Industrial Revolution. What began in Britain in the 18th century and spread to the European continent and the United States transformed the life of Western culture, not only the nature of society, but traditional relationships between groups of people.

Specifically, the innovations in the processes and organization of production included:

- substitution of mechanical energy for animal sources of power, particularly steam power through the combustion of coal
- substitution of machines for human skills and strength
- invention of new methods for transforming raw materials into finished goods, particularly in iron and steel production and industrial chemicals
- organization of work into large units, such as factories or

forges or mills. This made possible direct supervision of the manufacturing process and an efficient division of labor.

Paralleling these production changes were the altered technologies employed in agriculture and transportation.

Initially, this new way of organizing work was termed the "factory system." Later, however, when it reached larger and more complex scales, it was designated the Industrial Revolution by A. Toynbee (Toynbee, 1884). His nephew, Arnold J. Toynbee, is described as "the first economic historian to think of, and to set out to describe, the Industrial Revolution as a single great historical event, in which all the details come together to make an intelligent and significant picture."

Unfortunately, these changes in production methods with their need for masses of workers also created hazards never before encountered. These conditions greatly affected the history of occupational safety and health. Many health workers and industrial experts recognized the increasing need for hazard control.

The Industrial Revolution Comes to America

The effects of the Industrial Revolution were first felt in the United States about a century after the revolution began in Great Britain. Before the 19th century, most families in the United States lived and worked on farms. Some industries had developed in the new country, namely printing, ship-building, quarrying, cabinetmaking, bookbinding, clockmaking, and the production of paper, chocolate, and cottonseed oil. However, it was the textile industry that introduced the new factory system into the United States, especially in New England where hundreds of spinning mills shot up. As the Industrial Revolution continued its unbounded growth, the toll on workers began to show (Felton, 1986). (See Figure 1–1.)

HISTORY OF U.S. SAFETY AND HEALTH MOVEMENT

During the last half of the 19th century, American factories were expanding their product lines and producing at previously unimagined rates. While the factories were far superior in terms of production to the preceding small handicraft shops, they were often vastly inferior in terms of human values, health, and safety.

In the area of human values alone, the facts make a grim picture. The 1900 census showed 1,750,178 working children between 10 and 15 years of age. Some 25,000 were employed in mines and quarries; 12,000 in making chewing tobacco and cigars; 5,000 in sawmills; 5,000 at or near steam-driven planers and lathes; 7,000 in laundries; 2,000 in bakeries; and 138,000 as servants and waiters in hotels and restaurants. These children often worked 12 to 14 hours a day; there were no health or safety guidelines in effect, even for children under age 10 (Figure 1–2).

The lag between the emergence of new working methods and the creation of health and safety standards was probably inevitable. The tools of mass production had to be invented and applied before anyone could begin to imagine the problems they might create. In turn, the problems had to be known before corrective measures could be considered, tested, and proved. Thus, for some time deaths and injuries were accepted as part of "industrial progress"—one of the costs of doing business.

While this revolution in the work environment was taking place, the thinking of the public, management, and the law still reflected the past, when the worker was an independent craftsman or a member of the family-owned shop. Common law provided the employer with a defense that gave the injured worker little chance for compensation. The three doctrines of common law that favored the employer were:

Fellow servant rule—Employer was not liable for injury to an employee that resulted from negligence of a fellow employee.

Contributory negligence—Employer was not liable if the employee was injured due to his own negligence.

Assumption of risk—Employer was not liable because the employee took the job with full knowledge of the risks and hazards involved.

In 1906, the Pittsburgh Survey, sponsored by the Russell Sage Foundation, marked the first attempt to pinpoint the serious nature of occupational accidents and deaths. The survey team realized they did not have the resources to survey the problem throughout the United States and instead concentrated on Allegheny County in Pennsylvania. They constructed a "death calendar" of the county, showing that industrial accidents accounted for an average of nearly two deaths per day throughout the year. The number of crippling injuries was far higher. If this was the case in only one county, people asked, what must the situation be for the entire United States? The Pittsburgh Survey made it clear that the accident and death rate was serious and gave the safety movement a much-needed boost.

In large industrial centers, the ugly results of industrial accidents and poor occupational health conditions became more and more obvious. Individuals and public and private organizations raised their voices to protest these conditions. Though some employers denied the problem existed, wiser managements began to take action to improve their work environments.

As early as 1867, Massachusetts had begun to use factory inspectors. Ten years later, the state passed a law requiring employers to safeguard hazardous machinery. During 1877, Massachusetts also passed the Employer's Liability Law making employers liable for damages when a worker was injured. However, court decisions based on common law often let the employer escape liability.

From 1898 on, there were additional efforts to make the employer financially liable for accidents. In his presidential message of 1908, Theodore Roosevelt stated: "The number of accidents which result in the death or crippling of wage earners is simply appalling. In a very few years it runs up a total far in excess of the aggregate of the dead in any major war."

Roosevelt's message acquired force when his social legislation passed that year in Congress. Although this first

Figure 1–2. Barefoot children amidst heavy machinery, c. 1900-1910, show what a long way the safety and health movement has come.

workers' compensation law covered only federal employees, it set a precedent for state laws to follow.

The first bill for workers' compensation (the Wainwright Law) was passed in New York in 1910, but it was declared unconstitutional by the New York Court of Appeals. The court ruled that the law violated both the federal and New York State constitutions, "because it took property from the employer and gave it to his employee without due process of law." On the same day the 1910 act was declared unconstitutional, March 25, 1911, a devastating fire in New York City's Triangle clothing factory killed 146 employees. This disaster, called the Triangle Fire, outraged the public and spurred demand for factory legislation and health and safety reform. After an amendment to the state constitution was approved in 1913 at the general election, a compulsory Workmen's Compensation Act finally became effective in mid–1914.

In 1911, Wisconsin passed its first effective Workers' Compensation Act, but it was declared unconstitutional by the Wisconsin Supreme Court within a few months. New Jersey and Washington also passed laws that year.

At first, the courts continued to declare such laws invalid

because they conflicted with the due process of law provisions of the 14th Amendment. However, after the United States Supreme Court in 1916 declared workers' compensation to be constitutional in *New York Central Railroad Co. v White*, 243 U.S. 188, many states passed compulsory workers' compensation laws.

By the late 1800s and early 1900s, the railroads had criss-crossed the East and West but exacted a heavy toll among employees. It was said that a man was killed for each mile of track laid. By 1907, annual railroad employee deaths had reached 4,353.

Industry experts made some progress on the technical side of the problem. The railroads adopted air brakes and the automatic coupler well before the turn of the century. They also worked on guarding and fire prevention. They came to realize, however, that guarding was not the total answer. People's actions were equally important factors in creating accident situations.

At the same time, insurance companies started to relate the cost of premiums for workers' compensation insurance to the cost of accidents. Management began to understand the close relationship between successful production and safe production.

During the first decade of the 20th century, two giant industries, railroads and steel, began the first large-scale organized safety programs (Figure 1–3). From this period comes one of the historic documents of safety. In 1906, Judge Elbert Gary, president of the United States Steel Corporation, wrote:

The United States Steel Corporation expects its subsidiary companies to make every effort practicable to prevent injury to its employees. Expenditures necessary for such purposes will be authorized. Nothing which will add to the protection of the workmen should be neglected.

The Association of Iron and Steel Electrical Engineers, organized soon after this announcement, devoted considerable attention to safety problems.

Birth of the National Safety Council

The year 1912 proved to be a landmark for accident prevention. In the previous year, the Association of Iron and Steel Electrical Engineers (which had been formed in 1907) had called for a general industrial safety conference on a national scale. The result was the First Cooperative Safety Congress, which met in 1912 in Milwaukee. The following year, at a second national meeting in New York, the delegates formed the National Council for Industrial Safety. Shortly afterward, members changed the organization's name to the National Safety Council and broadened its program to include all aspects of accident prevention. The program also includes occupational health. Yet, it must be remembered that the Council was the creation of industry and that its activities have always been heavily concentrated on industrial safety.

The group that met in Milwaukee and New York was composed of a few safety professionals, some management leaders, public officials, and insurance specialists. Their one point in common was a desire to attack a problem which most people considered either unimportant or insoluble. The determination of these safety pioneers helped to create the safety movement as we know it today.

Actually, the members' underlying objective in forming the National Safety Council in 1913 was standardization. Thus, the primary purpose of the Council was to provide an avenue of communication, an exchange of views, and various solutions to common problems in accident prevention.

In 1918, the Council conducted the first national survey of state, federal, and municipal regulations, together with a study of insurance recommendations, technical association recommendations, and the practices of industry. The survey revealed utter chaos in industrial safety, and a clear need for industry-wide methods and practices.

Realizing its own limitations, the Council consulted the National Bureau of Standards, which agreed to call a conference to discuss establishing procedures to standardize safety methods and practices. Meeting in Washington, DC, in 1919, the attendees expressed the belief that uniform industry standards were not only desirable but essential to promote effective worker health and safety. The conference voted to formulate safety standards under the auspices and

procedures of the American Engineering Standards Committee (AESC), which had been formed in 1918 by five engineering societies and three governmental departments.

American Standards Association Beginnings

In 1920, the National Safety Code Program was brought into the AESC. This resulted in the first reorganization of the Committee and marked the beginning of what later became the American Standards Association (ASA). A national code committee was organized to suggest the initial safety code projects. This later became the Safety Codes Correlating Committee, the first of ASA's 18 standards boards. Bringing manufacturing companies and trade associations into AESC membership also initiated a broader program of engineering standards. These steps launched an enlarged national standardization program.

In 1928, recognizing that the extending activities called for a more formal type of organization, the member groups reorganized the AESC as the American Standards Association, now known as the American National Standards Institute (ANSI). ASA continued to be an important partner in the safety movement. This group handled the "things" aspect of safety while the National Safety Council focused on the "people" portion of accident and occupational illness prevention.

Accident Prevention Discoveries

As industry developed some experience in safety, it discovered that engineering could prevent accidents, that employees could be reached through education, and that safety rules could be established and enforced. Thus the "Three E's of Safety"— engineering, education, and enforcement—were developed.

Two of the many breakthroughs that safety groups made during the 1900–1990 era were the identification of occupational diseases such as mercury and lead poisoning, and the efforts to control these hazards. Other discoveries have had equally profound implications. For example, asbestos was found to be a carcinogen that causes lung cancer and another type of cancer, mesothelioma. Health professionals also studied the effects of chromium compounds and beryllium on industrial workers.

These and other discoveries led safety and health professionals to argue that savings in compensation costs and medical expenses would repay safety expenditures many times over. Thoughtful business leaders soon learned that these savings were only a fraction of the financial benefits to be derived from accident prevention work. Newer, more effective techniques have been discovered and are described elsewhere in this volume. See especially Chapters 3 and 11 through 14 in this volume, and Chapter 5 in the *Engineering and Technology* volume.

Acceleration of the Drive for Safety and Health

Industrial safety received wide acceptance in the years between World Wars I and II. During World War II the growth of safety procedures and policies intensified, particularly as the federal government began encouraging its contractors to

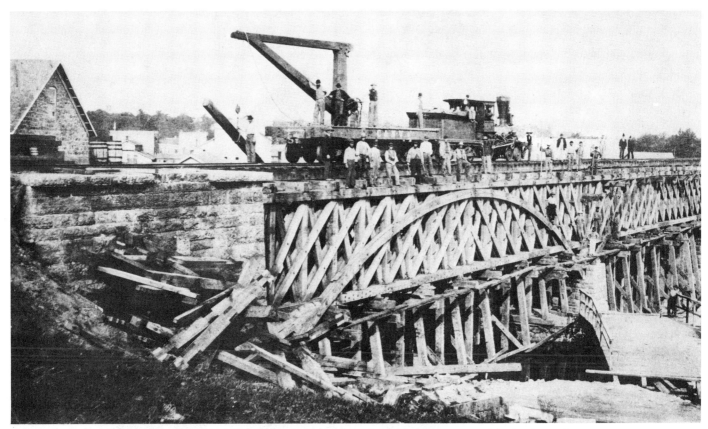

Figure 1–3. Because of hazards encountered in railroad operations, the railroad industry was one of the first industries to develop organized safety programs. This picture shows some of the hazards associated with building a bridge at Rockford, Illinois, in 1869. (Courtesy of Chicago & North Western Railroad.)

adopt safe work practices. As industry expanded to meet the needs of the war effort, additional safety personnel were hastily trained in an effort to keep pace with the new risks and hazards in the workplace. The acceptance of safety activities as part of the industrial picture did not diminish with the end of the war. By then, the importance of safety to quality production was well established. The small handful of people dedicated to safety in 1912 had grown to tens of thousands of trained personnel. In 1948, for example, Admiral Ben Moreell, then president of Jones and Laughlin Steel Corporation, wrote:

> Although safe and healthful working conditions can be justified on a cold dollars-and-cents basis, I prefer to justify them on the basic principle that it is the right thing to do. In discussing safety in industrial operations, I have often heard it stated that the cost of adequate health and safety measures would be prohibitive and that "we can't afford it."
>
> My answer to that is quite simple and quite direct. It is this: "If we can't afford safety, we can't afford to be in business."

A discussion of current United States safety legislation follows later in this chapter under Safety and the Law, and also in Chapter 2, Regulatory History and Compliance.

One by-product of organized safety activities has been a growing interest in safety engineering on the part of colleges and universities. Many schools offer degrees and advanced courses on this subject and are contributing to a higher standard of knowledge among professionals in the field.

In addition, the World War II labor shortage dramatically brought home to management the magnitude and seriousness of the problem of off-the-job accidents to industrial employees. The wartime theme of the National Safety Council, "Save Manpower for Warpower," focused attention on the need to reduce off-the-job accidents in order to maintain efficient, safe production on the job.

An increasing number of employers are including off-the-job safety in their overall safety programs. Companies realize their operating costs and production schedules are affected almost as much when employees are injured away from work as when they are injured on the job. Off-the-job safety generally is an extension of a company's on-the-job safety program and is intended to educate the employee to follow, in outside activities, the safe practices used on the job. Companies have found that on-the-job and off-the-job programs complement each other.

From the earliest days of industrial safety, it has been difficult to make a clear distinction between health and accident hazards. Is dermatitis an accident or a disease? What about hernias, hearing loss, and heart trouble? Inevitably, safety professionals have become interested in many health problems on the borderline between diseases and accidents. In 1939, the American Industrial Hygiene Association was established to promote the recognition, evaluation, and con-

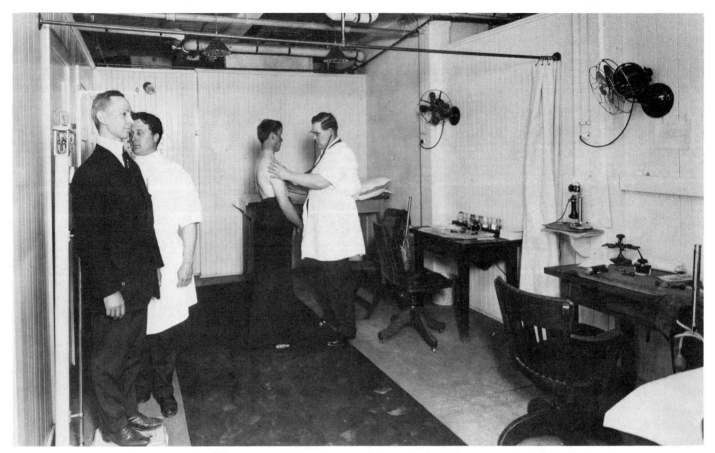

Figure 1–4. As the relationship between health and employment hazards became recognized, employees were given medical examinations.

trol of environmental stresses arising in or from the work-place (Figure 1–4).

EVALUATION OF ACCOMPLISHMENTS

Since the factors are complex, no simple rating scale can indicate all the answers to the question, "What has the safety movement accomplished?" In the absence of such a rating scale, an attempt to answer the question must be made by assembling several kinds of data.

First, the question must be asked, "Has the safety movement, in fact, really helped to prevent accidents?" The answer to that question is a clear "Yes!" (Figure 1–5). If the annual accidental death rate per 100,000 of population recorded in 1912 had continued, more than 3,100,000 additional accidental deaths would have occurred since that time. Instead, the death rate for persons of normal working age—25 to 64 years—has steadily declined by some 59%, while the rate for all ages of the entire population has declined 54%. Medical progress accounts for some of this gain, but the larger part is certainly the product of organized safety work.

Since World War II, the number of work-related deaths per 100,000 population, standardized to the age distribution of the population in 1940, also has decreased steadily (Figure 1–6). This indicates that the risk of on-the-job death has declined for the population as a whole. Part of the progress made in lowering the overall death rate, however, can be

Table 1-A. Work Deaths per 100,000 Workers.

Industry Group	1945	1990	Percent Change
Agriculture, forestry, and fishing	53	42	− 21
Mining and quarrying	187	43	− 77
Construction	126	33	− 74
Manufacturing	19	6	− 68
Transportation and public utilities	52	22	− 58
Wholesale and retail trade	10	4	− 60
Services	20	4	− 80

Source: National Safety Council, *Accident Facts,* 1946 and 1991 editions.

attributed to the rapid growth in recent years of the economy's service sector with its lower death rate, and the decline of some fairly high-risk segments of the manufacturing sector. In 1945, 43% of the nonagricultural workforce was in production-related industries (mining, construction, and manufacturing). By 1990 that proportion had declined to 23%.

A clearer picture emerges by looking at the trends on a more detailed level. Table 1-A shows a significant decline in death rates within the major industry groups. In five of the

Figure 1–5. These women working during World War I are shown risking their lives on unsecured scaffolding. Note also the version of protective footwear of that era. (Courtesy Women's Bureau, National Archives.)

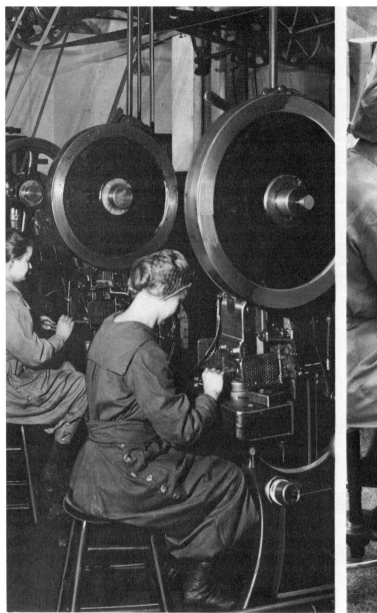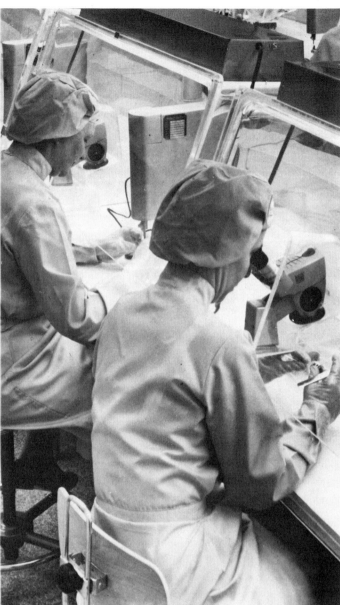

Figure 1–6. In 1919, the protective clothing for women making small parts for telephones was coveralls, the workstations were dark, the chairs were uncomfortable and lacked back support, and the machinery was unguarded. In 1969, employees wore lint-free clothing, gloves, and head coverings, and had more comfortable workstations. Improvements continue today. (Courtesy Western Electric Company.)

seven private sector groups, death rates have been reduced by 50% or more over the past 46 years. This clearly indicates, by one criterion, the effectiveness of the organized safety movement.

Long-term trends in nonfatal occupational injury rates cannot be examined because of a break in continuity of the historical statistical series. Until the early 1970s, injury rates were based on the voluntary American National Standards *Method of Recording and Measuring Work Injury Experience,* ANSI Z16.1. With the passage of the Occupational Safety and Health Act of 1970, it became mandatory for most private-sector employers in the United States to keep occupational injury and illness records in accordance with OSHA recordkeeping requirements.

A clear trend has not yet emerged in the occupational injury and illness incidence rates published by the Bureau of Labor Statistics since 1972. Business cycles and changes in the distribution of the labor force among industries can mask any short-term changes in rates caused by more effective or more intensive safety efforts.

The Dollar Values

It has been estimated that the annual cost of occupational accidents in the United States exceeds $37 billion. If the 1912 accident rate had continued unchanged and if there had been no organized safety movement, this annual cost could easily have been two to three times as great, even in constant dollars.

Against such dollar savings, the relatively small expenditures for safety throughout the United States provide a striking contrast. Each dollar industry spends for safety may be returning a clear profit of up to several hundred percent.

Industry and Nonwork Accidents

Directly and indirectly, industry bears a substantial part of the cost related to nonwork accidents and their prevention. Although the National Safety Council is the creation of industry and largely supported by it, the Council, along with state and local safety organizations, plays a major role in the fight to prevent such accidents. Industry is a major supporter of the efforts to inform the general public on safety issues through the press, radio, and television.

This nonwork accident-prevention campaign is having a definite impact on public safety. From the time records on nonwork accidents were first kept in 1921, both home and public accident death rates have generally declined.

Although industry has been a large contributor to this successful work, it has also been a major beneficiary. The reduction in nonwork injuries, illnesses, and deaths has lessened disruption of the labor force, and has reduced hardship among employees, consumers' loss of purchasing power, and tax burdens required to support hospitals and relief agencies.

RESOURCES FOR SAFETY

Safety *statistics* measure what has been accomplished in this area. Safety *resources* describe the tools, methods, and knowledge developed for safety and health professionals in their efforts to overcome future accident and occupational illness problems.

Knowledge and Experience

This *Accident Prevention Manual for Business & Industry*, for example, is an accumulation of facts and experiences that are a part of the safety and health movement. Its purpose is to present key points of knowledge to people interested in safety, whether they are students new in the field or advanced and experienced practitioners.

An individual using this *Manual* can find better answers to a wider range of industrial safety problems than were available to the wisest and best-trained professional safety practitioner several decades ago. Yet, even this *Manual* cannot contain all of the knowledge available to fight the never-ending war against accidents and occupational illnesses.

Other material may be found in numerous pamphlets, books, and periodicals published by safety and health organizations, government agencies, and insurance companies, and in the studies and directives of individual industrial concerns. The literature of various trades and professions is likewise rich in safety information. A list of handbooks is presented in Safety Tables, an appendix of the *Engineering and Technology* volume. At the end of this chapter is a list of the general safety books used as sources of questions for the Certified Safety Professional examination.

The National Safety Council offers a series of training courses, at both the beginning and advanced levels, for professionals. The Council also offers extensive consulting services, books, and software.

Finally, through conferences, technical seminars, newsletters, and other publications, professional safety engineers, executives, supervisors, and rank-and-file employees regularly exchange safety information. The annual National Safety Congress and Exhibition, held in Chicago or another major U.S. city, is an excellent means of enhancing professional development.

The Heritage of Cooperation

The safety movement would be far less effective if its members had concealed their discoveries from their colleagues who worked in competing companies. It was teamwork that created the safety activities of the Association of Iron and Steel Electrical Engineers. It was broadened teamwork that organized the first Milwaukee Conference, which led to the formation of the National Safety Council and other safety-related organizations.

Effective accident prevention requires cooperation at all levels of industry and government. Through the Council and other safety organizations, safety professionals meet to exchange ideas, develop safety publications, and stimulate one another in friendly competition. The tradition that there should be "no secrets in safety," no denial of help even to a competitor when it involves saving lives, is one of the great strengths in the safety movement.

Goodwill

In its early days, safety did not rank highly among management concerns. Today, a significant part of the safety professional's capital is the prestige and goodwill that have accumulated over the years, making management more receptive to safety proposals and expenditures. Where yesterday the pioneers had to battle every step of the budgetary way, safety professionals today usually obtain a far more sympathetic hearing from management.

Professionalism

Dedicated safety and health professionals continue to be the most valuable asset of accident and occupational illness prevention. Their ranks have grown; in 1990, membership in the American Society of Safety Engineers (ASSE) was 24,000. This organization, dedicated to these professionals' interests and development, has more than 120 chapters in the United States and Canada. Individual membership is worldwide. Other professional societies include the National Safety Management Society, the Board of Certified Hazard Control Management, and the System Safety Society.

In 1968, the ASSE was instrumental in forming the Board of Certified Safety Professionals (BCSP). Its purpose is to certify qualified safety people as professionals once they meet strict educational and experience requirements and pass an examination. Similarly, professional certification of industrial hygienists (CIH) was sponsored by the American Industrial Hygiene Association (AIHA). Both the ASSE and the AIHA are described in the Appendix, Sources of Help.

Advancement of Knowledge

The tremendous increase in scientific knowledge and technological advancement since the 1950s has added to the complexities of safety work (Figure 1–6). Prevention and control measures have oscillated between those that emphasize environmental control or engineering and those that emphasize human factors. From this, several important trends in safety work and the safety professional's development have emerged. All are discussed in subsequent chapters of this volume.

- First, more emphasis is being placed on analyzing the loss potential of any organization or projected activity. Such analysis requires the ability (1) to predict where and how loss- and injury-producing events will occur and (2) to find ways of preventing such events.
- Second, industry is developing more factual, unbiased, and objective information about loss-producing problems and accident causation to help those who are ultimately responsible for worker health and safety make sound decisions.
- Third, management is making greater use of the safety and health professional's knowledge and assistance in developing safe products. The application of the principles of accident causation and control to product manufacturing is assuming more importance because of the rise in product liability cases, the recent legal emphasis on the concept of negligent design, and the obvious impact a safer product would have on the environment.

To identify and evaluate the magnitude of the safety problem, safety professionals must be concerned with all facets of the problem—personal and environmental, transient and permanent. This perspective will help to determine the causes of accidents or to identify loss-producing conditions, practices, or materials. On the basis of this collected and analyzed information, safety professionals propose alternate solutions, together with recommendations founded on their specialized knowledge and experience, to those who have decision-making responsibilities.

Therefore, application of this knowledge—whether to industry, transportation, the home, or in recreation—makes it imperative that those in the safety field be trained to use scientific principles and methods to achieve adequate results. Most important, they need to have the knowledge, skill, and ability to integrate machines, equipment, and environments with people and their capabilities.

In performing these functions, safety and health professionals draw upon specialized knowledge in both the physical and social sciences. They should know how to apply the principles of measurement and analysis to evaluate safety performance and must have a fundamental knowledge of statistics, mathematics, physics, chemistry, and engineering. Students in safety and health degree programs should be certain these subjects are included. Safety and health professionals must know and understand management systems so they can properly advise line managers concerning safety and health management.

They also require training in the field of behavior, motivation, communications, and management principles along with the theory of business and government organization. Their specialized knowledge must include a thorough understanding of the causative factors contributing to accidents and occupational illnesses as well as methods and procedures designed to control such events.

Safety professionals also need a good general education if they are to meet future challenges. The population explosion, problems of urban areas and future transportation systems, weakening of the family structure, decline of respect for authority, and an uncertain economy, coupled with the increasing complexities of everyday life, will create many problems and stretch safety professionals' creativity to its maximum. They will need all their skills if they are to provide knowledge and leadership to conserve life, health, and property.

Training for the safety and health professional of the future can no longer be solely "on-the-job" or one-on-one education. It must include specialized undergraduate courses that lead to a bachelor's or higher degree. Training courses, such as those offered by the National Safety Council, will continue to educate a large number of people who began performing safety functions as part of their jobs and who need initial or advanced training in certain specialized areas.

A large number of four-year U.S. colleges and universities offer courses in safety and health, and several dozen offer a bachelor's or higher degree in safety. Two-year community colleges offer associate degrees or certificates for courses designed for the safety technician or part-time administrator.

ACHIEVEMENTS OF THE SAFETY MOVEMENT

The safety movement has helped save more than 3 million lives. It is saving industry and its employees billions of dollars every year. It faces the future with numerous resources for preventing and controlling accidents and occupational illnesses—resources in know-how, teamwork, goodwill, and education programs that produce trained and dedicated safety workers.

The safety movement has, therefore, done much to meet the double challenge presented to it. The movement has been able to deal with current accidents and occupational illnesses and to build a solid foundation for the long-range attack upon these problems in the future.

To answer the question, "What has the safety movement accomplished?" we look at continued growth in safety awareness and accident reduction. To answer the question, "Where does the safety movement stand?" we must look at what is wrong, as well as what is right, with the present situation. The answer can be found by comparing where the safety movement now stands to where it ought to be. The first point to be considered is simple and grim:

- Accidents still cost this country more than 93,500 lives a year, cause about 9 million disabling injuries, and account for a total financial loss of more than $173.8 billion (*Accident Facts*, 1991). Work accidents destroy more than

Figure 1–7. Ford assembly line, ca. 1910.

10,500 lives a year. Disabling injuries in work accidents affect about 1.8 million persons annually and cost more than $63.8 billion (1990 figures).

In recent years, the ratio of off-the-job deaths to on-the-job deaths has been about 3.5 to 1; more than half of the injuries suffered by employees occurred off the job. In terms of time lost, all injuries to workers, both on and off the job, resulted in a loss of about 100 million staff days of work per year.

Accident rates vary widely from industry to industry and from company to company. The wholesale and retail trade, services, finance, insurance, and real estate industries all have occupational injury and illness incidence rates below the private sector average. On the other hand, rates are above average in construction, agriculture, manufacturing, transportation and public utilities, and mining industries.

Injury incidence rates by size of establishment are lowest for very small businesses (1 to 19 employees) but rise steadily until they reach a maximum in establishments of 100 to 249 employees. The rates then decline steadily as establishment size continues to increase (*U.S. Bureau of Labor Statistics Annual*, 1989). (Figure 1–7.)

Small Establishments

It has been stated that small businesses, those with 100 to 249 employees, have proportionately more work injuries than

large corporations or very small companies, 1 to 19 employees. As a general rule, companies—large or small—that ignore safety and health efforts will have more than their share of accidents and occupational illnesses.

The serious safety and health problems of small enterprises are widely recognized, and the National Safety Council has devoted considerable effort to reducing accident and illness rates in this sector. One way has been to establish a liaison between the National Safety Council and the trade associations representing many small companies.

Certain aspects of the small-company problem can be stated with assurance:

1. The small establishment may not need or cannot employ specialized safety and health personnel to deal with its accident and occupational health problems.
2. The number of accidents or the financial position of many small companies makes it difficult to convince them that spending money for proper equipment, layout, guarding, and other elements is important.
3. Managers of small operations deal with a host of problems and seldom have the expertise or time for the proper study of accidents and occupational illnesses and their causes.
4. In small units, statistical measures of performance are unreliable. As a result, it is difficult to produce clear-cut evidence of the cost of accidents versus the effectiveness of accident-prevention work. In other words, a small operation may have, by luck, a good or bad accident record over a few years, whether or not its safety program is sound.

These aspects present serious obstacles to any progress in worker health and safety. They are not, of course, excuses for failure to prevent accidents and occupational illnesses. The trade association approach offers the best hope for improvement in this sector.

Labor-Management Cooperation

From its inception, one of the prime goals of organized labor has been the safety and health of its members. Many of today's international unions were organized originally to deal with severely hazardous situations in the workplace. They have a sincere desire to work with management on methods to prevent occupational injury and illness.

In 1949, the National Safety Council issued a policy statement declaring the common interest of labor and management in accident prevention. Even before this date, representatives of leading labor organizations served as members of the Council's governing boards. In 1955, a Council Labor Department and a Labor Conference (now known as Labor Division) were formally established. The Labor Division is a vital link between industry management and the nation's labor unions.

Labor Division representatives review products, training materials, and policy statements (Figure 1–8). The division shares information with labor leaders and with more than 550 volunteers from international and local labor unions. These groups have combined their educational efforts to help

Figure 1–8. Mine rescue training was one of the earliest cooperative efforts involving industry and labor. Wearing self-contained breathing devices such as this Gibbs model (approved by the U.S. Bureau of Mines in 1920), teams rescued trapped miners and recovered underground mines. (Courtesy Bureau of Mines, U.S. Department of the Interior.)

the nearly 20 million members of organized labor improve the quality of their lives, both on and off the job. Some unions have done extensive safety work, published printed matter, and released films that promote the safety movement.

The National Safety Council Labor Division and the Industrial Division, along with other affected divisions, often prepare Council position statements for such matters as standards action, oversight testimony, publicity releases, and other areas bearing on occupational safety and health. As a result of these efforts, Council positions are recognized as representative of all elements of society, which gives them even greater impact on administrative agencies and legislative bodies.

As the result of a program started in January 1978 with a symposium of leaders from government, industry, and

organized labor, the National Safety Council has launched an extensive inquiry to determine causal factors of injuries. Because most data in the past have cataloged only types of injuries, the focus of this program is twofold: (1) to change investigatory and reporting methods, and (2) to provide an information exchange bank that lists the factors actually causing injury or occupational illness.

Statistics, Standards, and Research

Statistical data on industrial accidents have been compiled by the National Safety Council for more than 60 years. Analyses are computed annually and published in *Work Injury and Illness Rates* and *Accident Facts*.

Some industries through their trade associations have recorded accident rates for almost 60 years. In most instances, even the divisions of an industry can record specific numbers and types of accidents and compare their experiences with national averages.

A large number of standards by the American National Standards Institute (ANSI) and other professional organizations relate to safety. Continuing research over the years has kept these standards in line with current industrial developments and the creation of new products and materials.

Special research projects, such as those studying walking surfaces and safety belts, can be and have been financed by private sources and coordinated by the National Safety Council. Recently, the Council has conducted research surveys on the organization and safety management system. The primary component of an organization's activities and the value of each component in the total safety structure was analyzed. A value has been established for each item separately. The results of these projects are presented as a summary of findings.

Safety and the Law

Early legal action in industrial safety took the form of laws to regulate and investigate industry working conditions and death and injury rates. The next phase was largely concerned with workers' compensation payments.

In subsequent years, all governments gradually expanded their roles in regulating industry on safety matters. The Walsh-Healey Act, which mandates safety measures in companies having supply contracts with the federal government, is an example of such regulation.

In certain industries—notably mining and transportation—U.S. government regulation and inspection have been extensive. The Construction Safety Act, which Congress passed in 1969, addresses the particular health and safety problems of that industry.

In 1970, the Williams-Steiger Occupational Safety and Health Act (OSHAct) was passed. For the first time, the United States had a comprehensive national safety law. The legislation covers every business affected by interstate commerce and employing one or more persons. Safety took on a new direction and meaning as a result of the Occupational Safety and Health Act.

But change is also happening worldwide. For example, in the United Kingdom the Health and Safety at Work Act was passed in 1974. The Act permitted ministers to make regulations to replace existing piecemeal legislation by regulations and codes of practice requiring improved standards of safety, health, and welfare. It provides for a coordinating enforcement authority and gives inspectors powers to initiate actions. Coverage was extended to anyone employed. The Act addressed discussions on the atmosphere and also dealt with certain building codes and standards. It was a major step involving legislation.

In Australia the Victoria Occupational Health and Safety Act of 1985 provides for mandatory regulation and is supplemented by codes of practice. The Act requires union, industry, and government input, and it intends for most safety and health issues to be resolved at the workplace.

In Canada the Ontario Occupational Health and Safety Act of 1978 for Industrial Establishments, Construction, Mines and Mining Plants requires the recording and posting of accident records, penalties, and inspection of workplaces. The Act's provisions were coupled with the Workers' Compensation program.

Another Canadian act developed out of provisions of OSHA. Working with all Canadian provinces, the Workplace Hazardous Management Information System (WHMIS) was created. WHMIS has three parts: (1) labels, (2) MSDS, and (3) a worker education program.

A third Canadian act, the Workers' Compensation Act, became law on January 1, 1990. It covers only the province of Ontario. The law deals with a number of subjects: injury, employee retention, accommodation for disabled employees, employee benefit continuation after employee injury, and disability employee benefits.

Besides the OSHAct, several other laws have impacted both industry and the safety professional. Largely under the Environmental Protection Acts, they include: Clean Air and Clean Water Acts (CAA, CWA), Toxic Substance Control Act (TSCA), Resource Conservation and Recovery Act (RCRA), Comprehensive Environmental Response Compensation and Liability Act (CERCLA, "Superfund"), Superfund Amendments and Reauthorization Act (SARA or Community Right-to-Know). These acts will not be discussed in detail. (See Chapter 5, Environmental Management.)

Concern with the safety and health of workers is a major priority for management. This concern goes beyond the obvious benefits of less downtime, reduced costs for Workers' Compensation insurance, and lower medical and administrative expenses resulting from disability, death, and impaired productivity. Management also knows there are penalties for not adhering to the law. For example, failure to comply with health and safety requirements can mean citations, which (at the least) create administrative costs but could also lead to serious monetary penalties. The federal government can also institute criminal sanctions against employers and even against individual managers who ignore or disregard the law. The criminal action has not only come from federal and state job safety and health agencies, however. Local prosecutors have successfully convicted individual

managers for murder and aggravated assault in the deaths and injuries of workers on the job.

Management must address serious emerging issues in worker health and safety law. These issues include ways to deal with the special problem of employees who are at risk in the work environment because of physical condition, language problems, or particular susceptibility to injury or disease. Another issue is the burgeoning paperwork required to comply with OSHA and other agency recordkeeping regulations, with the Medical Access Standard, and with the Hazard Communication Standard.

Industry accepts almost without question the concept of financial responsibility for work injuries. Not all of industry, however, is convinced of the effectiveness of government regulation of safety procedures.

Some states have established laws requiring compulsory health and accident insurance to cover employee disabilities from diseases or accidents occurring off the job.

This compulsory insurance might be considered either a drastic extension of the principle of workers' compensation or an extension of social security legislation. It differs from workers' compensation in that it puts a financial burden upon management for diseases and accidents brought about by conditions beyond its control.

Whatever the theory, the result of these laws is to give the employer a direct financial stake in dealing with the off-the-job accident problem. (See Chapter 3, Hazard Control Program.)

European Communities

Regulatory change has not only occurred in the United States, but also in other nations. The formation of the European Economic Community has caused changes affecting (1) worker safety, (2) products, and (3) other goods and services. The nations involved are Belgium, Denmark, France, Germany, Greece, Iceland, Italy, Luxembourg, the Netherlands, Portugal, Spain, and the United Kingdom.

One directive, Framework Directive (89/391/EEC) of June 12, 1989, covers the introduction of measures to encourage improvements in on-the-job safety and health of workers. The range of application for the Framework Directive is very broad indeed. It applies to all sectors of activity, both private and public, with the exception of specific working forces such as the armed forces and the police.

The Framework Directive lays down the obligations of employers relating to the safety and health of workers:

- Taking the measures necessary for their safety and health, bearing in mind technical progress
- Evaluating hazards and instructing workers accordingly
- Setting up protection and prevention services—for example, the precision of safety and health practitioners—within the workplace, possibly by enlisting competent external services or persons
- Organization of first aid
- Evacuation of workers in the event of serious damage.

The Directive also lists workers' obligations in relation to

their own health and safety:

- Correct use of machinery
- Correct use of protective equipment
- The need to report defects in equipment, defects in procedures, and potentially dangerous situations such as near-miss accidents.

To those who export goods and products to the EEC countries, the materials will only have to conform to one standard rather than 12 different ones. Although exporting may be easier, conforming to EEC '92 product standards may be more difficult. Under current regulations testing and certification is required before exporting to EEC countries. Companies not in EEC countries will not be permitted to self-certify. Companies in countries who are EEC members can self-certify.

Safety and Occupational Health

Cooperation between medical and safety personnel in accident prevention activities began during the earliest days of the safety movement. Nevertheless, interest in safety on the part of the medical profession and, conversely, interest in employee and public health on the part of the safety professional is increasing. This trend, brought to the forefront in the 1980s, will assume increasing importance in the 1990s.

Part of the interest in workers' health results from concern with occupational disease, noise, radiation, and other problems that extend beyond the former concepts of occupational accident prevention. Interest in protecting the health of citizens in nearby communities comes partly from such infamous and well-publicized events as the Three-Mile Island accident; the Chernobyl, Ukraine, meltdown; and the Bhopal, India, disaster.

Work environment. Over time, safety and health professionals became aware of the relationship between physical illness and working conditions. They found that workers in certain industries such as mining, chemical manufacturing, and steel exhibited a higher than normal incidence of such problems as dermatitis, musculoskeletal problems, pulmonary disease, mental illness, and cancer. The improved safety and health of today's workers is the result of concerted efforts by a safety, industrial hygiene, and occupational health team working with a management that realizes an organization's primary asset is a safe and healthy workforce (Figure 1–9).

For more details, refer to the Council's *Fundamentals of Industrial Hygiene* and *Occupational Health and Safety,* part of the *Occupational Safety and Health Series.*

Community environments. Increasingly, the public has demanded a larger role in the management of community environmental risks. Both public and private risk managers realize that providing avenues for public participation is a necessary part of their decision-making process. The problem is how to ensure public involvement and at the same time improve the quality of safety and health decisions.

To help fill the gap in credible risk communication on environmental health and safety issues, the National Safety Council established the Environmental Health Center. This

Figure 1–9. Only time, experience, and a concerted effort by all involved disciplines revealed the many relationships between physical illness and work environment. (Courtesy Library of Congress.)

special-purpose organization is led by a Board of Governors and operates mostly through philanthropic funding from concerned corporations, foundations, labor unions, and individuals. Its goals include development of accurate and objective information on environmental and public health risks, improvement of public knowledge about these risks, and dissemination of this information to the public.

Workers with disabilities. The employment of workers with disabilities by progressive companies and the impact of federal and state equal opportunity laws have modified the practice of preemployment examinations to screen out unfit or undesirable prospects. Medical personnel perform a preplacement examination simply to determine what physical or mental restrictions are appropriate to the prospective employee; they do not determine fitness for a specific job. The job description must specify realistic physical and mental requirements that the human resources department can match to medical restrictions. The Americans with Disabilities Act requires that the job be modified to accommodate the disabled worker. (See the discussion in Chapter 20, Workers with Disabilities.)

Psychology and "Accident Proneness"

Safety professionals who are thoughtfully looking for ways to improve their work encounter a great deal of useful information in modern psychological writing—and a great deal of careless and misleading generalizations. For example, concern about the so-called "accident-prone" individual in industry is as old as the safety movement itself. Statistical information suggests that such individuals exist, though clear and sharp data demonstrating this suggestion are remarkably hard to obtain. Too many alleged "proofs" turn out to be statistically deceptive, based on inadequate samples, or the result of highly subjective diagnoses. Justification for use of "proofs" of accident proneness may reflect the common law doctrines in which employers tried to justify accidents as being caused by the employee's action.

Statistical support for the existence of accident-prone individuals is elusive. Some safety professionals believe in accident proneness and feel it may be a passing phase in the individual and not a permanent characteristic. At most, it may be a problem encountered only by an insignificant minority. Realistically, objective analysis might reveal a su-

pervisory deficiency or procedural weakness that can increase the risks of certain operations or interfere with the performance of individuals or groups of workers.

The same observation applies to psychological tests used as screening devices for new employees. So-called experts may make spectacular claims for the ability of these tests to predict accident proneness, but so far no test has proven itself to the general satisfaction of the safety profession. In the past, the work of psychologists like Dunbar and the Menningers aroused great interest among safety professionals. However, the best weapons in the practical, day-to-day fight against accidents tend to come from strong line management safety awareness and from the disciplines of engineering and behavioral psychology, such as human factors engineering, system safety, and risk management or assessment. (Refer to the discussions in Chapters 15 and 16.)

The field of industrial safety is one of progress and improvement, largely through the continued application of techniques and knowledge slowly and systematically acquired through the years. There appears to be no limit to the progress possible through the application of the universally accepted safety techniques of education, engineering, and enforcement.

Yet large and serious problems remain unsolved. A number of industries still have high accident rates. In far too many instances, management and labor either fail to work together or have different goals for the safety program. The efforts in creating effective joint safety and health committees are helping meet common goals in safety and health.

The resources of the safety movement to tackle these problems are impressive: a growing body of knowledge, a corps of able safety and health professionals, a high level of prestige, and strong organizations for cooperation and exchange of information.

CURRENT PROBLEMS

Some problems of the safety movement are directly related to the movement's traditional strengths and weaknesses. A few of these problems are social and political, while others are essentially organizational.

Technology and Public Interest

There is no reason for the safety professional to be alarmed at the public's interest in product safety, a better environment, and general technological trends. An informed, educated public can be a powerful ally in safety work. Emphasis upon automation and more refined instrumentation will probably continue. New problems will arise, but they will generally be those that well-established methods of safety engineering can solve.

The use of new materials and techniques—particularly radioactive materials, automation, and lasers—is likely to present more serious difficulties to the safety professional. However, even here, the professional can draw on the movement's considerable experience to handle these problems.

Finally, safety professionals need to keep up with the rapid developments in communications and computerization as they affect the safety field. (See Chapter 14, Computers and Information Management.)

Political Problems

On the political side remains the age-old problem of industry-union-government relations. The key issue here is what regulatory role should governments play in protecting worker health and safety and what aspects of national life should it regulate? This issue is likely to remain a source of controversy for years to come.

Organizational Problems

On the worldwide scale, a wide variety of organizations are attacking specific aspects of occupational safety and health problems. The National Safety Council is a strong, constructive, nonpolitical, nonprofit leader among these organizations. It has repeatedly sought and often achieved a cooperative division of labor between itself and other organizations in the safety and health field. One of the guiding principles of the Council has been that there is work enough and credit enough for all.

It remains to be seen whether the best organizational forms have been found for participation by all businesses in safety and health work. Safety and health professionals should be ready to consider new ideas and new solutions.

A Look to the Future

The greatest reasons for intensifying the safety effort are humane and moral. The worth of neighbors, friends, and family cannot be measured in dollars or coded into computer records.

During the coming decade, the U.S. population will continue to grow although more slowly than in the past. By 1995 there will be about 260 million Americans. The average life expectancy of these persons will increase due, in part, to health care advances. The "baby boom" generation will be middle-aged adults crowding the workforce.

The shift from extractive and manufacturing to service areas of employment will continue well into the next century. This fact will hopefully contribute to a decline in occupational death rates since the number of people employed in high-risk industry will be lower.

The workforce continues to undergo major changes as minorities and women move into more industries and management levels. Safety issues associated with the growing numbers of women workers are complicated, and many difficult choices regarding working conditions will have to be made.

As social changes continue to take place, including a high divorce rate, more single-parent families, and more two-income households, the effect on the structure and values of family life will be felt more strongly. Many perceive a reduced respect for either parental or social authority, a factor they believe creates considerable conflict and safety problems in the workplace.

One of the current trends is the evolution of a world market. The establishment of a European Economic Community market in 1992 will have a major impact on world economics due to the EEC '92 regulations on products and goods. The complete restructuring of the economies of the Soviet Union and Eastern Europe may have an even greater impact on the world markets during the 1990s and the early 21st century. As a result, there will be a great demand for standardizing accident recordkeeping, so that comparisons can be made between various segments of the global market. Labeling of hazardous material must also be standardized so that all personnel handling chemicals manufactured in one country and shipped to another will understand the warnings on the container.

Accident prevention information will become much more a global commodity than it has been. Although to some extent accident prevention information is shared with worldwide affiliates of large multinational corporations, the vast majority of information developed in one country rarely finds its way to others. There are excellent behavioral techniques in use in Sweden today that are hardly known in the United States. As the world develops a global market, these national barriers will come down and information sharing will also become global.

The United States, Western Europe, and Japan are moving into a post-Industrial Revolution era. Developing countries, however, are just coming out of the early Industrial Revolution as new technologies and manufacturing industries spread to these nations. This trend should create a strong demand for U.S. safety and health expertise in the developing countries.

What will the government's regulatory agencies be doing in the next decade? What kind of standards can be expected from U.S. OSHA, EPA, MSHA in the next few years? Probably the most far reaching will be the "generic standards" regulating exposure to chemicals. In 1989 OSHA promulgated its PEL (Permissible Exposure Limits) standard. Because of rules laid down in the 1970 OSHAct, OSHA was able to adopt at that time an existing federal standard (the 1968 Walsh-Healey Act) to establish permissible exposure levels for about 400 hazardous chemicals.

The next few years should see development of a generic exposure monitoring standard and a generic medical monitoring standard. It will be oriented to performance rather than to specification, that is, it will state an intended result and allow the employer the latitude to find ways to achieve the result. At the same time, OSHA will also promulgate a generic, performance-oriented standard requiring medical monitoring of personnel exposed to hazardous chemicals. This standard will require periodic medical evaluation by blood or urine analysis, as well as other diagnostic techniques.

The current high level of public concern in the U.S. for control of hazardous chemicals will probably translate into new legislation. Another area in which OSHA will be getting involved is setting a standard protecting workers in hospitals and health care institutions from exposure to infectious diseases. The need for this standard is growing because of public

concern over the AIDS epidemic. Further in the future is a generic standard on exposure to biological agents. This is a highly specialized area. Though the pharmaceutical industry has for many years been aware of hazards of biological agents and taken measures to protect its employees, methods of detection and protection are not widely known.

In addition, criminal prosecution of employers and safety and health professionals for negligence resulting in serious injury or accidental death of employees will become more common. Several recent state supreme court decisions in Illinois and Wisconsin are paving the way for continued activity by state district attorneys to press such criminal cases. As the safety profession becomes increasingly recognized, the public may hold some practitioners accountable for any omissions that may result in death or injury to workers in their organizations.

The days of the plant that operated incognito are gone forever. In the past, industry operators and owners felt the less anyone knew about their businesses, the better off they were. That philosophy became obsolete as the public and industry members alike realized the necessity to control accidents, injuries, and work-related illnesses. The trend toward openness and cooperation with the public, the government, and the media has spread to all industries. In fact, the leading companies will be those most successful in convincing the public and its local governments that they are good neighbors. Safety professionals will have to understand the legal and political ramifications of their companies' safety practices in light of the public's changing expectations.

These are issues of major importance and will need the expertise and guidance of everyone in the safety and health field. The future is, as it always has been, most uncertain. By working together, all those in the safety and health community can reduce some of that uncertainty by helping to make workplaces and off-the-job environments safer and healthier.

Opportunities for the safety professional. The 1990s are an exciting time for the safety profession. Professionals in safety management, safety engineering, industrial hygiene, occupational medicine, and the new field of holistic medicine are discovering new and compelling reasons for cooperating more closely and for opening new areas of employment for the safety and health professional.

Other expanding employment opportunities for safety professionals lie in the safety departments of international (and some local) labor unions, on the staffs of a number of trade associations, and, of course, in government service. Safety consulting has expanded rapidly, both in individuals offering their talents on a contract basis, in safety and health consulting firms, and in consulting services offered by nonprofit associations and by a few industrial concerns.

Although advanced education courses and programs for safety professionals increased dramatically in the 1980s, the number could not keep pace with the expanding need. The result was a serious shortage of highly qualified safety professionals. At the same time, the population bulge in the prime working-age group created a surplus of less-skilled

people in management. This situation had two effects on safety positions in companies and in government: increased competition for senior safety positions and increased demand for advanced safety training beyond that offered by technical colleges. In fact, one of the greatest growth opportunities for highly qualified safety professionals lies in teaching college safety and health courses.

These are the challenges of tomorrow—improving performance on the job and in the safety profession, coupled with obtaining the education necessary to compete and work effectively in the safety and health field.

SUMMARY

- The basic mission of the National Safety Council is to protect life and promote health. The Council carries out this mission by providing knowledge, expertise, and assistance to safety professionals in all fields.
- The philosophy of accident prevention holds that society as a whole has a moral responsibility to prevent needless destruction of life and health, particularly in the workplace. The employer is primarily responsible for ensuring a safe, healthy work environment. Employees are held accountable for following prescribed safety standards and guidelines.
- The safety movement had its beginnings in the ancient world. From the early days of the Egyptians, Greeks, and Romans through the Middle Ages, various codes and edicts sought to establish guidelines for working conditions and workers' compensation. Not until the Industrial Revolution, however, did the safety movement become a major force in industry.
- In the United States, new methods of organizing work and new industries appeared quickly, exacting a grim toll on workers' health and safety. Initially, common law favored the employer, but over time, the high rates of occupational deaths and injuries stirred up public opinion and encouraged passage of workers' compensation laws. Management began to realize that productivity and safety were closely related.
- In 1913, the National Council for Industrial Safety, later renamed the National Safety Council, was founded. The initial objective was to provide a forum for members and those in industry to exchange ideas and to develop national safety standards. Other groups soon followed, including the American Engineering Standards Committee and the American National Standards Institute. For the first time, these groups began to keep statistics on accident and injury rates in industry.
- The safety movement soon began to identify occupational diseases as well as safety issues and to develop methods to control hazardous and toxic materials and to protect workers who handled them. In recognition of their effect on the job, off-the-job safety issues have been added to the agenda.
- Since the end of World War II, the safety movement has become an accepted and valued part of industry. It has

not only helped to reduce deaths and injuries but has shown management that every dollar spent on safety measures returns a clear profit of up to several hundred percent.
- Over the past eight decades, safety workers have gradually become safety professionals. Today's safety and health problems require a highly trained, experienced, and dedicated team of professionals to find solutions.
- Organizations within the safety movement will continue to encourage close cooperation among labor, management, and government to upgrade standards and develop more legislation.
- The National Safety Council is seeking to improve investigative methods and to establish an information exchange bank regarding work-related hazards. The Council also supports special research projects to study the effectiveness of safety measures and legislation.
- The safety movement faces serious challenges in the future, including new technologies and their hazards, major changes in the workforce, formidable environmental hazards, bloodborne diseases such as AIDS, and widespread drug and alcohol abuse. As the world witnesses the creation of a global marketplace, the need for worldwide safety and health standards will increase—as will the need for highly skilled and trained safety professionals to protect workers in every country.

GENERAL SAFETY BOOKS

In addition to this *Manual*, the following books cover the basics of occupational safety and health. Note: These resources serve as the basis for the questions on management in the Board of Certified Safety Professional examinations.

Browning RL. *The Loss Rate Concept in Safety Engineering.* New York: Marcel Dekker, 1980.

DeReamer R. *Modern Safety and Health Technology.* New York: John Wiley & Sons, 1981.

Ferry TS. *Modern Accident Investigation and Analysis.* New York: John Wiley & Sons, 1981.

Firenze RJ. *The Process of Hazard Control.* Dubuque, IA: Kendall-Hunt, 1978.

Gilmore CL. *Accident Prevention and Loss Control.* New York: The American Management Association, 1970.

Grimaldi JV. *Safety Management*, 4th ed. Homewood, IL: Richard D. Irwin, Inc., 1984.

Heinrich HW. *Industrial Accident Prevention*, 5th ed. New York: McGraw-Hill Book Co., 1980.

Petersen DC. *Analyzing Safety Performance.* Reprint. River Dale, NJ: Aloray Publishers, 1984.

———. *Safety by Objectives.* River Dale, NJ: Aloray Publishers, 1978.

————. *Techniques of Safety Management.* 3rd ed. River Dale, NJ: Aloray Publishers, 1989.

Tarrents WE. *The Measurement of Safety Performance.* New York: Garland STPM Press, 1980.

REFERENCES

American Engineering Council. *Safety and Production.* New York: Harper & Brothers Publishers, 1928.

American National Standards Institute, 11 West 42nd Street, New York, NY 10036.

Andrews EW. "The Pioneers of 1912." *National Safety News* 66:24-25, 64-65, 1952.

Beyer DS. *Industrial Accident Prevention,* 3rd ed. Boston: Houghton Mifflin Co., 1928.

Campbell RW. "The National Safety Movement." *Proceedings of the Second Safety Congress of the National Council for Industrial Safety,* 188-192, 1913.

DeBlois LA. *Industrial Safety Organization for Executive and Engineer.* New York: McGraw-Hill Book Company, 1926.

Eastman C. *Work Accidents and the Law.* New York: Charities Publication Committee, 1910. Reprint. New York: Arno Press, 1969.

Felton JS. "History of Occupational Health and Safety," Chapter 3 of *Occupational Health and Safety,* ed. by J LaDou. Chicago: National Safety Council, 1992.

Heinrich HW. *Industrial Accident Prevention,* 5th ed. New York: McGraw-Hill Book Company, 1980.

Holbrook SH. *Let Them Live.* New York: The Macmillan Co., 1939.

Menninger KA. *Man Against Himself.* New York: Harcourt, Brace and World, Inc., 1956.

Meyer RL. Series commemorating the Diamond Anniversary of the Council. *Safety and Health,* (January through October 1987).

Mock HE. *Industrial Medicine and Surgery.* Philadelphia: WB Saunders, 1920.

National Safety Council, 444 North Michigan Avenue, Chicago, IL 60611.
    Accident Facts. Issued annually.
    "Golden Anniversary Issue." *National Safety News,* 87:5, 1963.
    Proceedings of the First Co-Operative Safety Congress, 1912.
    Proceedings of the National Safety Congress. Issued annually from 1914-1925.
    Proceedings of the Second Safety Congress of the National Council for Industrial Safety, 1913.
    Safety and Health. Issued monthly.

Ramazzini B. *Diseases of Workers,* trans. from the Latin text *De Morbis Artificum of 1713,* by WC Wright. New York: Hafner Publishing, 1964.

Schaefer VG. *Safety Supervision.* New York: McGraw-Hill Book Company, 1941.

Schulzinger MS. "Accident Syndrome—A Clinical Approach." *Archives of Industrial Health,* 11:66-71, 1955. Chicago, IL, American Medical Assn.

Schwedtman FA. *Accident Prevention and Relief.* New York: National Association of Manufacturers in the United States of America, 1911.

Toynbee A. *The Industrial Revolution.* (First published in 1884.) Boston: Beacon Press, 1956.

2

Regulatory History and Compliance

In the past three decades, Congress has enacted two major pieces of federal legislation impacting the field of occupational safety and health. These laws are the Occupational Safety and Health Act of 1970 (OSHAct) and the U.S. Mine Safety and Health Act of 1977. This chapter focuses only on these two Acts.

PART I
The Occupational Safety and Health Act

A new national policy was established on December 29, 1970, when President Richard M. Nixon signed into law the OSHAct (Public Law No. 91–596 found in 29 *United States Code* (USC) §§651–678). The Congress of the United States declared that the purpose of this piece of legislation was "to assure so far as possible every working man and woman in the Nation safe and healthful working conditions and to preserve our human resources."

The OSHAct took effect April 28, 1971. Coauthored by Senator Harrison A. Williams (Dem.-NJ) and the late Congressman William Steiger (Rep.-WI), the Act is sometimes referred to as the Williams-Steiger Act. It is regarded by many as landmark legislation because it goes beyond the present workplace and considers long-term health hazards in the working environment of the future.

The information provided in Part I of this chapter focuses on federal OSHA programs. State OSHA programs may differ from the federal program in certain areas, but are required to be equal to the federal requirements. However, unless specifically stated to the contrary, the recommendations in this chapter can be followed whether jurisdiction rests at the federal or state level.

LEGISLATIVE HISTORY

Historically, the enactment of safety and health laws has been left to the states. Prior to the 1960s only a few federal laws (such as the Walsh-Healey Public Contracts Act and the Longshoremen's and Harbor Workers' Compensation Act) directed any attention to occupational safety and health. Several pieces of legislation passed by the Congress during the 1960s, including the Service Contract Act of 1965, the National Foundation on Arts and Humanities Act, the Federal Metal and Nonmetallic Mine Safety Act, the Federal Coal Mine Safety and Health Act, and the Contract Workers and Safety Standards Act (Construction Safety Act), focused industry attention on occupational safety and health.

Each of these federal laws was applicable only to a lim-

ited number of employers. The laws were either directed at those who had obtained federal contracts, or they targeted a specific industry. Even collectively, all the federal safety legislation passed prior to 1970 was not applicable to most employers or employees. Up to that time, congressional action on occupational safety and health issues was, at best, sporadic, covering only specific sets of employers and employees. There was little attempt to establish the omnibus coverage that is a central feature of the OSHAct.

Proponents of a more significant federal role in occupational safety and health, mostly represented by organized labor, based their position primarily on the following:

1. With rare exceptions, the states failed to meet their obligation in regard to occupational safety and health. A few had reasonable or adequate safety and health legislation, but most states legislated safety and health only in specific industries. In general, states had inadequate safety and health standards, inadequate enforcement procedures, inadequate staff with respect to quality and quantity, and inadequate budgets.

2. In the late 1960s, approximately 14,300 employees were killed annually on or in connection with their job and more than 2.2 million employees suffered a disabling injury each year as a result of work-related accidents. The injury/death toll was considered by most to be unacceptably high.

3. The nation's work-injury rates in most industries were increasing throughout the 1960s. Because the trend was moving in the wrong direction, proponents of federal intervention felt that national legislation would help to reverse this trend.

The Act evolved amid stormy controversy in both houses of Congress as the legislators debated state versus federal roles, industry versus government control, and the like. Such issues were responsible for sharply drawn lines between political parties and between the business community and organized labor. After three years of political tug-of-war, numerous compromises were made to allow passage of the OSHAct by both houses of Congress.

ADMINISTRATION

Administration and enforcement of the OSHAct are vested primarily with the Secretary of Labor and the Occupational Safety and Health Review Commission (OSHRC), discussed later. With respect to the enforcement process, the Secretary of Labor performs the investigation and prosecution aspects, and the OSHRC performs the adjudication portion.

Research and related functions and certain educational activities are vested in the Secretary of Health and Human Services. These responsibilities, for the most part, are carried out by the National Institute for Occupational Safety and Health (NIOSH) established within the Department of Health and Human Services (DHHS). Compiling injury and illness statistical data is handled by the Bureau of Labor Statistics (BLS), United States Department of Labor (DOL).

To assist the Secretary of Labor, the Act authorizes the appointment of an Assistant Secretary of Labor for Occupational Safety and Health. This position is filled by presidential appointment with the advice and consent of the Senate. The Assistant Secretary is the chief of the Occupational Safety and Health Administration (OSHA) established within the DOL. The Assistant Secretary acts on behalf of the Secretary of Labor. For the purposes of this chapter, OSHA is also synonymous with the term Secretary or Assistant Secretary of Labor.

The primary functions of the four major governmental units assigned to carry out the provisions of the Act are described in this section on administration.

Occupational Safety and Health Administration

The Occupational Safety and Health Administration came into existence officially on April 28, 1971, the date the OSHAct became law. This agency was created by the DOL to discharge the Department's responsibilities assigned by the Act.

Major areas of authority. The Act grants OSHA the authority, among other things, (1) to promulgate, modify, and revoke safety and health standards; (2) to conduct inspections and investigations and to issue citations, including proposed penalties; (3) to require employers to keep records of safety and health data; (4) to petition the courts to restrain imminent danger situations; and (5) to approve or reject state plans for programs under the Act.

The Act also authorizes OSHA (1) to provide training and education to employers and employees; (2) to consult with employers, employees, and organizations regarding prevention of injuries and illnesses; (3) to grant funds to the states for identification of program needs and for plan development, experiments, demonstrations, administration, and operation of programs; and (4) to develop and maintain a statistics program for occupational safety and health.

Major duties delegated. In establishing OSHA, the Secretary of Labor delegated to the Assistant Secretary for Occupational Safety and Health the authority and responsibility for safety and health programs and activities of the DOL, including responsibilities derived from:

1. Occupational Safety and Health Act of 1970
2. Walsh-Healey Public Contracts Act of 1936, as amended
3. Service Contract Act of 1965
4. Public Law 91–54 of 1969 (construction safety amendments)
5. Public Law 85–742 of 1958 (maritime safety amendments)
6. National Foundation on the Arts and Humanities Act of 1965
7. Longshoremen's and Harbor Workers' Compensation Act (Title 33, Chapter 18, §§901, 904, *U.S. Code*; Act of March 4, 1927, Chapter 509, 44 Stat. 1424)
8. Federal safety program under Title 5 *U.S. Code* 7902.

Similarly, the Commissioner of the BLS was delegated the authority and given responsibility for developing and maintaining an effective program for collection, compilation, and analysis of occupational safety and health statistics; for providing grants to the states to assist in developing and administering programs in such statistics; and for coordinat-

ing functions with the Assistant Secretary for Occupational Safety and Health.

The Solicitor of Labor is assigned responsibility for providing legal advice and assistance to the Secretary and all officers of the Department in the administration of statutes and Executive Orders relating to occupational safety and health. In enforcing the Act's requirements, the Solicitor of Labor represents the Secretary in litigation before OSHRC and, subject to the control and direction of the Attorney General, before the federal courts.

To assist in carrying out its responsibilities, OSHA has established 10 regional offices in the cities of Boston, New York, Philadelphia, Atlanta, Chicago, Dallas, Kansas City, Denver, San Francisco, and Seattle. (See Directory of Federal Agencies at the end of this chapter.) The primary mission of the regional office chief, known as the Regional Administrator, is to supervise, coordinate, evaluate, and execute all programs of OSHA in the region. Assisting the Regional Administrator are Assistant Regional Administrators for (1) training, education, consultation, and federal agency programs; (2) technical support; and (3) state and federal operations.

Area offices have been established within each region, each office headed by an Area Director. The mission of the Area Director is to carry out the compliance program of OSHA within designated geographic areas. The area office staff carries out its activities under the general supervision of the Area Director with guidance of the Regional Administrator, using policy instructions received from the national headquarters. The real action for implementing the enforcement portion of the OSHAct is carried out by the area offices in those states that do not have an approved state plan. In states with an approved plan, the area office monitors state activities. (See Federal-State Relationships, later in this chapter.)

Occupational Safety and Health Review Commission

The Occupational Safety and Health Review Commission is a quasi-judicial board of three members appointed by the president and confirmed by the Senate. OSHRC is an independent agency of the executive branch of the U.S. government and is not a part of the DOL. The principal function of the Commission is to adjudicate cases when an enforcement action taken by OSHA against an employer is contested by the employer, the employees, or their representatives.

OSHRC's actions are limited to contested cases. In such instances, OSHA first notifies the Commission of the contested cases. The Commission then hears all appeals on actions taken by OSHA concerning citations, proposed penalties, and abatement periods and determines the appropriateness of such actions. When necessary, the Commission may conduct its own investigation and may affirm, modify, or vacate OSHA's findings.

There are two levels of adjudication within the Commission: (1) the administrative law judge (ALJ), and (2) the three-member Commission. All cases not resolved on informal proceedings are heard and decided by one of the Commission's ALJs. The judge's decision can be changed by a majority vote of the Commission if one of the members, within 30 days of the judge's decision, directs that the decision be reviewed by the Commission members. The Commission is the final administrative authority to rule on a particular case, but its findings and orders can be subject to further review by the courts. (For further information, see Contested Cases later in this chapter.)

The headquarters of OSHRC is located at 1825 K Street, NW, Washington, DC 20006.

National Institute for Occupational Safety and Health

The National Institute for Occupational Safety and Health was established within the DHHS under the provisions of the OSHAct. Administratively, NIOSH is located in DHHS's Centers for Disease Control in Atlanta, GA. NIOSH is the principal federal agency engaged in research, education, and training related to occupational safety and health.

The primary functions of NIOSH are (1) to develop and establish recommended occupational safety and health standards, (2) to conduct research experiments and demonstrations related to occupational safety and health, and (3) to develop educational programs to provide an adequate supply of qualified personnel to carry out the purposes of the OSHAct.

Research and related functions. Under the OSHAct, NIOSH has the responsibility for conducting research for new occupational safety and health standards. NIOSH develops criteria for establishing these standards and transmits the criteria to OSHA. This agency is responsible for the final setting, promulgation, and enforcement of the standards.

The OSHAct also requires NIOSH to publish an annual listing of all known toxic substances and the concentrations at which such toxicity is known to occur. The inclusion of a substance on the list does not necessarily mean that it should be avoided. It does mean that the listed substance has a documented potential of being hazardous if misused and, therefore, care must be exercised to control the substance. Conversely, the absence of a substance from the list does not mean that it is nontoxic. Some hazardous substances may not qualify for the list simply because the dose that causes a toxic effect is not known.

Education and training. NIOSH also has the responsibility to develop (1) education and training programs aimed at providing an adequate supply of qualified personnel to carry out the purpose of the Act and (2) informational programs on the importance and proper use of adequate safety and health equipment. The long-term approach to an adequate supply of training personnel in occupational safety and health is found in the colleges and universities and other institutions in the private sector. NIOSH encourages such institutions, by contracts and grants, to expand their curricula in occupational medicine, occupational health nursing, industrial hygiene, and occupational safety engineering.

Employer and employee services. Of principal interest to individual employers and employees are the technical services offered by NIOSH. The five main services are provided upon request to NIOSH's Division of Technical Services, 4676 Columbia Parkway, Cincinnati, OH 45226 (1–800–356–4674). These services are:

1. Hazard evaluation—Onsite evaluations of potentially toxic substances used or found on the job.
2. Technical information—Detailed technical information concerning health or safety conditions at workplaces, such as the possible hazards of working with specific solvents, and guidelines for use of protective equipment.
3. Accident prevention—Technical assistance for controlling on-the-job injuries, including the evaluation of special problems and recommendations for corrective action.
4. Industrial hygiene—Technical assistance in the areas of engineering and industrial hygiene, including the evaluation of special health-related problems in the workplace and recommendations for control measures.
5. Medical service—Assistance in solving occupational medical and nursing problems in the workplace, including assessment of existing medically related needs and development of recommended means for meeting such needs.

NIOSH and the Mine Safety and Health Administration test and approve personal sampler units for coal-mine dust and respiratory protective devices, including self-contained breathing apparatus, gas masks, supplied-air respirators, chemical-cartridge respirators, and dust, fume, and mist respirators.

NIOSH representatives, although not authorized to enforce the OSHAct, are authorized to make inspections and to question employers and employees in carrying out the duties assigned to the DHHS under the Act. NIOSH has both warrant and subpoena power, if necessary, to obtain the information needed for its investigations. It may also request access to employee records. However, it must obtain the consent of employees or use methods that maintain the employee's right to privacy concerning information in the records.

Injury and Illness Data

The responsibility for conducting statistical surveys and establishing methods used to acquire injury and illness data is placed in the BLS. Questions regarding recordkeeping requirements and reporting procedures can be directed to any of the OSHA regional or area offices. (See the directory at the end of this chapter.)

Advisory Committees

The Act established a 12-member National Advisory Committee on Occupational Safety and Health (NACOSH) to advise, consult with, and make recommendations to the Secretaries of Labor and Health and Human Services with respect to the administration of the Act. Eight members are designated by the Secretary of Labor and four by the Secretary of Health and Human Services. Members include representatives from management, labor, occupational safety

and health professions, and the public.

The Act also authorizes the appointment of 15-member advisory committees to assist OSHA in the development of standards. The Standards Advisory Committees on Construction Safety and Health and on Cutaneous Hazards are the two currently in place.

MAJOR PROVISIONS OF THE OSHAct

Coverage

Except for specific exclusions, the Act applies to every employer who has one or more employees and who is engaged in a business involving interstate commerce. The law applies to all 50 states, the District of Columbia, Puerto Rico, and all U.S. possessions.

Specifically *excluded* from coverage are all federal, state, and local government employees. There are, however, special provisions in the Act for federal employees and potential coverage for state and local government employees. The Act requires each federal agency head to establish and maintain an occupational safety and health program consistent with the standards promulgated by the Secretary of Labor. Executive Orders setting requirements for federal programs have been issued over the years. OSHA regulations implementing the Executive Order are found in 29 *Code of Federal Regulations (CFR)* Part 1960 (1983).

Employees of states and political subdivisions of the states are excluded from the federal OSHAct. However, states with approved state plans are required to provide coverage for these public employees. Public employees in states without approved plans are not covered by the OSHAct in any manner.

However, two states—Connecticut and New York—have state plans covering only public employees. The District of Columbia and the states of Mississippi, New Hampshire, New Jersey, Rhode Island, and Wisconsin have laws specifically providing job safety and health protection to public employees.

The OSHAct also does apply not to those operations in which a federal agency (and state agencies acting under the Atomic Energy Act of 1954), other than the DOL, already has the authority to prescribe or enforce standards or regulations affecting occupational safety or health and is performing that function. An example of this exclusion is specific issues covered by Department of Transportation regulations in the railroad industry.

Also excluded from the OSHAct are operators and miners covered by the U.S. Mine Safety and Health Act of 1977. (See Part II of this chapter.) This Act applies to mines of all types: coal and noncoal, surface and underground.

In its yearly appropriations bills since 1977, Congress has placed restrictions on OSHA enforcement. For example, items exempt from inspection in fiscal year (FY) 1991 include:

- farmers with ten or fewer employees on the day of inspection and the 12 months preceding the day of inspection
- any work activity in any recreational, hunting, fishing, or

shooting area

- employers with 10 or fewer employees in selected Standard Industrial Classification industries with rates below the national average in lost workdays case rate (1989 rate was 3.9 per 100).

However, the exemptions do not apply to situations involving such issues as employee complaints, imminent dangers, health hazards, accidents resulting in a fatality or hospitalization involving five or more employees, or discrimination complaints.

OSHA clarified its interpretation of coverage with respect to certain employees by issuing a regulation. (This policy regarding "Coverage of Employees Under the Williams-Steiger Occupational Safety and Health Act of 1970" is contained in Title 29 *CFR*, Chapter XVII, Part 1975.) OSHA has also stated that churches and religious organizations are not regarded as employers, nor are persons who, in their own residences, employ others to perform domestic household tasks. Further, any person engaged in agriculture who is a member of the farmer's immediate family is not regarded as an employee and hence is not covered by the Act.

Employer and Employee Duties

OSHA clearly delineates employer and employee responsibilities. Each employer covered by the Act:

1. has the general duty to furnish each employee with employment and places of employment free from recognized hazards causing or likely to cause death or serious physical harm (this is commonly known as the "general duty clause")
2. has the specific duty of complying with safety and health standards promulgated under the Act.

Each employee, in turn, has the duty to comply with the safety and health standards and with all rules, regulations, and orders that apply to employee actions and conduct on the job.

For employers, the general duty provision is used only where there are no specific standards applicable to a particular hazard involved. A hazard is "recognized" if it is a condition generally regarded as a hazard in the particular industry in which it occurs and is detectable (1) by means of human senses, or (2) by accepted tests known in the industry which can reveal its presence to the employer. An example of a "recognized hazard" in the second category is excessive concentration of a toxic substance in the work-area atmosphere, even though such concentration could only be detected through use of measuring devices.

During the course of an inspection, a compliance safety and health officer (CSHO) is concerned primarily with determining whether the employer is complying with the promulgated safety and health standards. However, the officer will also try to discover if the employer is complying with the general duty clause.

The law provides for sanctions against the employer in the form of citations and civil and criminal penalties if the employer fails to comply with the general and specific duties. However, there is no provision for government sanctions against an *employee* for failure to comply with the employee's duty. Although some may view this policy as unjust, it was not one of the controversial issues in the formative stages of the Act.

Both management and organized labor have long agreed that safety and health on the job is a management responsibility. The business community generally did not want the law structured to provide for government sanctions against an erring employee. This is because management can invoke its own measures against an employee who obstructs the employer's efforts to provide a safe workplace.

While the law expressly places upon each employee the obligation to comply with the Act's standards, final responsibility for compliance rests with the employer. As a result, employers should take all necessary actions to ensure that employees follow the promulgated standards. Employers should also establish within their safety system a means to detect when employees are not complying with applicable standards.

The duty of an employer to protect employees against health hazards in addition to safety hazards continues to gain emphasis. The growing awareness of the hazard of chemical exposure has caused OSHA to focus on the employer's obligation to monitor the work environment, to provide periodic medical examinations, and to make a range of protective measures available to guard against hazards of the work environment.

An emerging issue is the question of how far an employer must go to protect employees who are at increased risk of occupational injury or illness. The increased risk may come from lack of experience, language problems, special sensitivity to chemicals, or being a woman of child-bearing age. Medical screening is one method employers have begun to use to identify present and prospective employees who are at increased risk. However, the method became controversial when employees were denied employment, or assigned or reassigned, based on the results of the screening. Can such actions intended to protect the employee become discriminatory or invade privacy? In the early 1980s employees and other groups used the job-discrimination provisions of the Civil Rights Act to protect women workers in general, and pregnant women in particular, against termination and detrimental reassignment because of alleged hazardous working conditions. While not all these issues have been settled, a 1991 U.S. Supreme Court decision struck down the use of fetal protection policies on the basis that they discriminate against women workers.

Employer Rights

An employer has the right to:

1. seek advice and offsite consultation as needed by writing, calling, or visiting the nearest OSHA office
2. request and receive proper identification of the OSHA CSHO prior to inspection
3. be advised by the CSHO of the reason for an inspection
4. have an opening and closing conference with the CSHO
5. file a Notice of Contest with the OSHA area director

within 15 working days after receiving a citation notice and proposed penalty

6. apply to OSHA for a temporary variance from a standard if unable to comply because needed materials, equipment, or personnel are not available to make necessary changes within the required time

7. take an active role in developing safety and health standards through participation in OSHA Standards Advisory Committees, through nationally recognized standards-setting organizations, and through evidence and views presented in writing or at hearings

8. apply for, if a small business employer, long-term loans through the Small Business Administration (SBA) to help bring the establishment into compliance, either before or after an OSHA inspection

9. be assured of the confidentiality of any trade secrets observed by an OSHA compliance officer.

Onsite Consultation

Congress has authorized, and OSHA now provides through a state agency or private contractors, free onsite consultation services for employers in every state. These consultants help employers to identify hazardous conditions and to determine corrective measures.

The service is available upon employer request. Priority is given to businesses with fewer than 150 employees. These firms are generally less able to afford private-sector consultation. OSHA emphasizes companies that use employees in highly hazardous jobs.

The consultative visit consists of an opening conference, a walk-through of the company's facility, a closing conference, and a written summary of findings. During the walk-through, the consultant tells the employer which OSHA standards apply to company operations and what they mean. The employer is informed of any apparent violations of those standards and, where possible, is given suggestions on how to reduce or eliminate the hazard.

Because employers, not employees, are subject to legal sanctions for violating OSHA standards, the employer determines to what extent employees or their representatives will participate in the visit. However, the consultant must be allowed to confer with individual employees during the walk-through in order to identify and judge the nature and extent of hazards.

Voluntary Protection Program

The purpose of OSHA's Voluntary Protection Program (VPP) is to emphasize the importance of, encourage the improvement of, and recognize excellence in employer-provided, site-specific occupational safety and health programs. These programs must not only meet, but exceed, the standards. When employers apply and are accepted, they are removed from the regular inspection list (except for valid formal employee complaints, fatalities, catastrophic events, or spills).

Two voluntary programs have been established: Star and Merit. OSHA's Star Initiative, the agency's most demanding and prestigious voluntary protection program, is for work sites with outstanding workplace safety and health systems.

Safety and health programs can be based primarily on management initiatives or employee participation, although because of the special hazards in construction, firms in that industry must meet employee participation requirements. Star participants are evaluated every three years.

What does it take to qualify for Star? First, a firm's accident record must meet stringent criteria. Star participants must have a three-year average injury incidence and lost-workday-case rate at or below the national average for their industry.

In addition, the work site must have an effective, comprehensive safety and/or health program. Among the key elements OSHA considers are:

- Management commitment and accountability—the agency has found that top-flight safety and health programs have strong corporate commitment and accountability, including written, clearly defined assignment of responsibility for worker protection at every level of management.

- Hazard assessment—this means a comprehensive inventory of potential safety and health hazards and includes periodic review and updating, especially when processes are changed or new substances are introduced. Effective hazard assessment also includes an effective mechanism for inviting and responding to worker notices of possible hazards.

- Safety rules and enforcement—Star participants have their own regulations and procedures that go beyond the protection specified by OSHAct standards.

- Employee training—In addition to formal training, many participants have regular toolbox meetings where safety procedures are reviewed.

- Self-evaluation—Comprehensive safety and health program audits ensure that the program continues to be an effective system for communicating to employees the company's commitment to safety and health and efficient procedures for responding to employee concerns and suggestions.

The Merit Program is a program for companies that have solid safety and health programs but need to make certain specific improvements to qualify for Star. As in Star, companies may use management initiative or employee participation. Merit participants are evaluated every year by OSHA.

When a firm applies for Star or Merit, an OSHA staffer is assigned to facilitate the application, answer any questions, and generally be the liaison between the applicant and the agency. In addition to analyzing written documentation supporting the application, OSHA will visit the workplace, usually for about a day and a half. The onsite review team will examine safety and health records, review the inspection history, conduct a walk-around check, and interview management and employees, if appropriate.

While this application procedure is and must be rigorous, most applicants report that it is not unduly burdensome, and that participation in the program makes the effort worthwhile.

Employee Rights

Although the employee has the legal duty to comply with all the standards and regulations issued under the OSHAct, many employee rights are also incorporated into the Act. Because these rights may affect labor relations as well as labor negotiations, employers should also be aware of the employee rights contained in the Act. These rights fall into three main areas related to (1) standards, (2) access to information, and (3) enforcement.

With respect to standards:

1. Employees may request OSHA to begin proceedings for adoption of a new standard or to amend or revoke an existing one.
2. Employees may submit written data or comments on proposed standards and may appear as an interested party at any hearing held by OSHA.
3. Employees may file written objections to a proposed federal standard and/or appeal the final decision of OSHA.
4. Employees must be informed when an employer applies for a variance of a promulgated standard.
5. Employees must be given the opportunity to participate in a variance hearing as an interested party and have the right to appeal OSHA's final decision.

With respect to access to information:

1. Employees have the right to information from the employer regarding employee protection and obligations under the Act and to review appropriate OSHA standards, rules, regulations, and requirements, which the employer should have available at the workplace.
2. Affected employees have a right to information from the employer regarding the toxic effects, conditions of exposure, and precautions for safe use of all hazardous materials in the establishment. The information can be provided through labeling or other means where such warnings are prescribed by a standard.
3. If employees are exposed to harmful materials in excess of levels set by the standards, the affected employees must be so informed by the employer, who must also tell them what corrective action is being taken.
4. If a CSHO determines that an alleged imminent danger exists, the officer must inform the affected employees of the danger and recommend that relief be sought by court action if the employer does not eliminate the danger.
5. Upon request, employees must be given access to records of their history of exposure to toxic materials or harmful physical agents that must be monitored or measured and recorded. Material Safety Data Sheets must be available to employees onsite or within reasonable time upon request.
6. If a standard requires monitoring or measuring hazardous materials or harmful physical agents, employees must be given the opportunity to observe such monitoring or measuring.
7. Employees have the right of access to (1) the list of toxic materials published by NIOSH, (2) criteria developed by NIOSH describing the effects of toxic materials or harmful physical agents, and (3) industry-wide studies conducted by NIOSH regarding the effects of chronic, low-level exposure to hazardous materials.
8. On written request to NIOSH, employees have the right to obtain the determination of whether a substance found or used in the establishment is harmful.
9. Upon request, the employees should be allowed to review the Log and Summary of Occupational Injuries (OSHA No. 200) at a reasonable time and in a reasonable manner.

With respect to enforcement:

1. Employees have the right to confer in private with the CSHO and to respond to questions from the CSHO during an inspection of an establishment.
2. An authorized employee representative must be given an opportunity to accompany the compliance officer during an inspection to aid in such inspection. (This is commonly known as the "walk-around" provision.) Also, an authorized employee has the right to participate in the opening and closing conferences during the inspection.
3. An employee has the right to make a written request to OSHA for a special inspection if the employee believes a violation of a standard threatens physical harm; the employee has the right to request that OSHA keep his or her identity confidential.
4. An employee who believes any violation of the Act exists has the right to notify OSHA or a compliance officer in writing of the alleged violation, either before or during an inspection of the establishment.
5. If OSHA denies an employee's request for a special inspection, the agency must notify the employee in writing that the complaint was not valid and explain the reasons for this decision. The employee has the right to object to such a decision and may request a hearing by OSHA.
6. If a written complaint concerning an alleged violation is submitted to OSHA and the compliance officer responding to the complaint fails to cite the employer for the alleged violation, OSHA must furnish the employee or an authorized employee representative with a written statement explaining the reasons for its final disposition.
7. If OSHA cites an employer for a violation, employees have the right to review a copy of the citation, which must be posted by the employer at or near the place where the violation occurred. If "system-wide" agreements are made, all locations will be notified.
8. An employee has the right to appear as an interested party or to be called as a witness in a contested enforcement matter before OSHRC.
9. If OSHA arbitrarily or capriciously fails to seek relief to counteract an imminent danger and an employee is injured as a result, that employee has the right to bring action against OSHA for relief as may be appropriate.
10. An employee has the right to file a complaint to OSHA within 30 days if the employee believes he or she has been discriminated against as a result of asserting employee rights under the Act.
11. An employee has the right to contest the abatement period fixed in the citation issued to the employer. This

Figure 2–1. OSHA poster (OSHA 2203), "Safety and Health Protection on the Job," must be posted conspicuously at every plant, job site, or other establishment. At left is the annual Log and Summary of Occupational Injuries and Illnesses, OSHA Form 200, that must be posted by February 1 of the following year and remain in place until March 1.

can be done by notifying the OSHA Area Director who issued the citation within 15 working days of the issuance of the citation.

The OSHA Poster

The OSHA poster (OSHA 2203, Figure 2–1) must be prominently displayed in a conspicuous place in the work environment where notices to employees are customarily posted. The poster informs employees of their rights and responsibilities under the Act.

Occupational Safety and Health Standards

The Act authorizes OSHA to promulgate, modify, or revoke occupational safety and health standards. The rules of procedure for promulgating, modifying, or revoking standards are spelled out in Title 29 *CFR*, Chapter XVII, Part 1911. The current requirements are available at all OSHA area and regional offices.

OSHA is responsible for promulgating legally enforceable standards that may require conditions, or the adoption or use of practices, means, methods, or processes that are reasonably necessary and appropriate to protect employees on the job. It is the employers' responsibility to become familiar with the standards applicable to their firms and to

make sure that employees have and use personal protective equipment required for safety. In addition, employers are responsible for complying with the Act's general duty clause.

In order to get the initial set of standards in place without undue delay, the Act authorized OSHA to promulgate any existing federal standard or any national consensus standard without regard to the usual rule-making procedures prior to April 28, 1973. The initial set of standards, Part 1910, appeared in the *Federal Register* of May 29, 1971. (Subscriptions to the *Federal Register* are obtained through the Government Printing Office, Washington, DC 20402.)

Standards contained in Part 1910 apply to general industry, while those contained in Part 1926 apply to construction. Standards that apply to ship repairing, shipbuilding, shipbreaking, and longshoring are contained in Parts 1915 through 1918, respectively. As new equipment, methods, and materials are developed, these standards are updated via modification.

OSHA standards incorporate by reference other standards adopted by industry organizations. Standards incorporated by reference in whole or in part include those adopted by the following organizations:

American Conference of Governmental Industrial Hygienists

American National Standards Institute
American Petroleum Institute
American Society of Agricultural Engineers
American Society of Mechanical Engineers
American Society for Testing and Materials
American Welding Society
Compressed Gas Association
Crane Manufacturers Association of America, Inc.
The Fertilizer Institute
Institute of Makers of Explosives
National Electrical Manufacturers Association
National Fire Protection Association
National Plant Food Institute
National Institute for Occupational Safety and Health
Society of Automotive Engineers
Underwriters Laboratories Inc.
U.S. Department of Commerce
U.S. Public Health Service

OSHA has the authority to promulgate emergency temporary standards in situations where employees are exposed to grave danger. Emergency temporary standards can take effect immediately upon publication in the *Federal Register.* They will remain in effect until superseded by a standard promulgated under procedures described in the Act. The law requires OSHA to develop a permanent standard no later than six months after publication of the emergency temporary standard. Any person adversely affected by any standard issued by OSHA has the right to challenge its validity by petitioning the United States Court of Appeals within 60 days after its promulgation.

Input from the private sector. Occupational safety and health standards promulgated by OSHA will never cover every conceivable hazardous condition that could exist in any workplace. Nevertheless, new standards and modification of existing standards are of significant interest to employers and employees alike. Industry organizations along with individuals and employee organizations need to express their views in two ways: (1) by responding to OSHA's advance notice of proposed rule making, which usually calls for information upon which to base proposed standards, and (2) by responding to OSHA's proposed standards, since it is within the private sector that most of the expertise and the technical competence lies. To do less means that industry and employees are willing to let the standards development process rest in the hands of OSHA.

Additional sources used by OSHA for the revision of existing occupational safety and health standards or the development of new ones are NIOSH criteria documents and standards advisory committees. These committees are appointed by the Secretary of Labor.

In order to promulgate, revise, or modify a standard, OSHA must first publish in the *Federal Register* a notice of any proposed rule that will adopt, modify, or revoke any standard and invite interested persons to submit their views on the proposed rule. The notice must include the terms of the new standard and provide an interval of at least 30 days (usually 60 days or more) from the date of publication

for interested persons to respond. These persons may file objections to the rule and are entitled to a hearing on their objections if they request one to be held. However, they must specify the parts of the proposed rule to which they object and the grounds for such objection. If a hearing is requested, OSHA must hold one. Based on (1) the need for control of an exposure to an occupational injury or illness, and (2) the reasonableness, effectiveness, and feasibility of the control measures required, OSHA may either issue a rule promulgating an additional standard or modify or revoke an existing standard.

Recordkeeping Requirements

Most employers covered by the Act are required to maintain company records of all occupational injuries and illnesses. Regulations describing how to properly record and report injuries and illnesses are codified in Title 29 *CFR*, Chapter XVII, Part 1904. Such records consist of:

- log and summary of occupational injuries and illnesses, OSHA Form 200
- supplementary record of each occupational injury or illness, OSHA Form 101
- annual summary of the firm's total number of occupational injuries and illnesses—OSHA Form 200 to be used in preparing the summary. The annual summary must be posted by February 1 of the following year and remain posted until March 1 (Figure 2–1).

For details concerning recording and reporting occupational injuries and illnesses, see Chapter 13, Recordkeeping and Incidence Rates. OSHA Forms 200 and 101 are available from all BLS regional offices and from OSHA area and regional offices. In states with an OSHA-approved plan, employers should check for any additional recordkeeping requirements.

OSHA has made an effort to relieve small businesses from many burdensome recordkeeping requirements. As a result, most employers who had no more than 10 employees at any time during the calendar year immediately preceding the current calendar year need not comply with the recordkeeping requirements. However, any employer, regardless of size, can be notified in writing by OSHA that the organization has been selected to participate in a statistical survey of occupational injuries and illnesses. The employer will then be required to maintain a log and summary and to make reports for the time specified in the notice. Further, no employer is relieved of the obligation to report any fatalities or multiple-hospitalization accidents to the nearest OSHA area office.

In 1991, most employers with 10 or fewer employees who engaged in retail trade, finance, insurance, real estate, and services were also exempted from most of the record-keeping requirements. (Standard Industrial Classification 52-89, except 52–50, 70, 75, 76, 79, and 80.)

Reporting Requirements

Within 48 hours after an accident occurs that is fatal to one or more employees or that results in the hospitalization of five or more employees, the employer must report the acci-

dent either orally or in writing to the nearest area director of OSHA. In states with approved state plans, the report must be made to the state agency that has enforcement responsibilities for occupational safety and health. If an oral report is made, it shall always be followed with a confirming letter written the same day. The report must relate the circumstances of the accident, the number of fatalities, and the extent of any injuries.

Variances from Standards

There will be some occasions when, for various reasons, standards cannot be met. In other cases, the protection already afforded by an employer to employees is equal to or superior to the protection that would be provided if the standard were strictly followed. The Act provides an avenue of relief from these situations by empowering OSHA to grant variances from the standards, providing that doing so would not degrade the purpose of the Act. The detailed "Rules of Practice for Variances, Limitations, Variations, Tolerances, and Exemptions" are codified in Title 29 *CFR*, Chapter XVII, Part 1905, 11(b).

OSHA can grant two types of variances—temporary and permanent. Employers may apply for an order granting a temporary variance if they can establish that (1) they cannot comply with the applicable standard because they do not have the personnel, equipment, or time to construct or alter facilities; (2) they are taking all available steps to protect employees against exposure covered by the standard; and (3) their own programs will effect compliance with the standard as soon as possible.

Employer applications for an order for a temporary variance must contain at least the following:

1. name and address of the applicant
2. address(es) of the place(s) of employment involved
3. identification of the standard from which the applicant seeks a variance
4. representation by the applicant that he or she is unable to comply with the standard and a detailed statement of the reasons
5. statement of the steps the applicant has taken and will take, with dates, to protect employees against the hazard covered by the standard
6. statement of when the applicant expects to be able to comply with the standard and what steps have been taken, with dates, to come into compliance with the standard
7. certification that the employer has informed employees of the application and of their right to petition OSHA for a hearing. A description of how employees have been informed is to be included in the certification.

Employers may also apply for a permanent variance from a standard. Variance orders can be granted if OSHA finds that employers have demonstrated, by a preponderance of evidence, that they will provide a place of employment as safe and healthful as the one that would exist if they complied with the standards.

Employer applications for a permanent variance order must contain at least the following:

1. name and address of the applicant
2. address(es) of the place(s) of employment involved
3. description of the countermeasures used or proposed to be used by the applicant
4. statement showing how such countermeasures would provide a place of employment that is as safe and healthful as that required by the standard for which the variance is sought
5. certification that the employer has informed employees of the application
6. any request for a hearing
7. description of how employees were informed of the application and of their right to petition for a hearing.

An employer may request an interim order permitting either kind of variance until the formal application can be acted upon. Again, the request for an interim order must contain statements of fact or arguments why such interim order should be granted. If the request is denied, the applicant will be notified promptly and informed of the reasons for the decision. If the order is granted, all concerned parties will be informed and the terms of the order will be published in the *Federal Register*. In such cases, the employer must inform the affected employees about the interim order in the same manner used to inform them of the variance application.

Upon filing an employer's application for a variance, OSHA will publish a notice of such filing in the *Federal Register* and invite written data, views, and arguments regarding the application. Those affected by the petition may request a hearing. After review of all the facts, including those presented during the hearing, OSHA will publish its decision regarding the application in the *Federal Register*.

Beginning in the early 1980s, OSHA authorized its Regional Administrators to make "interpretations" of standards in a way that essentially became variances for individual employers. Such interpretations had no effect on other employers and reflected specific conditions at a particular workplace. OSHA has also been issuing "clarifications" of standards for employers asking for deviation from standards. While granting less than 10% of employers' requests for variances since enforcement began, OSHA has issued about eight times as many clarifications.

Workplace Inspection

Prior to the United States Supreme Court's decision on the controversial Barlow case—*Marshall v Barlow's Inc.*, 436 U.S. 307 (1978)—the DOL's compliance safety and health officers could enter, at any reasonable time and without delay, any establishment covered by the OSHAct to inspect the premises and all its facilities. (See Title 29 *CFR*, Chapter XVII, Part 1903.) However, since the Barlow decision, the OSHA compliance officer must obtain a search warrant and present it if the employer demands to see the document. OSHA's entitlement to a warrant does not depend on demonstrating probable cause to believe that conditions on the premises violate the OSHA regulations. Rather, the agency merely has to show that reasonable legislative or administrative standards for conducting an inspection have been satis-

fied. An organization needs to determine, well in advance, if indeed a search warrant will be requested. As a general rule, it is not advisable for the employer to refuse entry to a CSHO who has no search warrant. Such action only delays an inspection and increases the officer's suspicion about working conditions. In most of these cases, OSHA will obtain a search warrant within 48 hours.

The OSHAct authorizes an employer representative as well as an authorized employee representative to accompany the CSHO during the official inspection of the premises and all its facilities. Employee representatives also have the right to participate in both the opening and closing conferences.

Usually the authorized employee representative is the union steward or the chairman of the employee safety committee. Occasionally there may be no authorized employee representative, especially in nonunion establishments. In this instance, the CSHO will select employees at random and confer with them.

An employer should not refuse to compensate employees for the time spent participating in an inspection tour and for related activities such as attending the opening and closing conferences.

Inspection priorities. OSHA has established priorities for assignment of staff and resources. The priorities are as follows:

1. *Investigation of imminent dangers.* Allegations of an imminent danger situation will ordinarily trigger an inspection within 24 hours of notification.
2. *Catastrophic and fatal.* Accidents will be investigated if they include any one of the following:
   - one or more fatality
   - five or more employees hospitalized for more than 24 hours
   - significant publicity
   - issuance of specific instructions for investigations in connection with a national office special program.
3. *Investigations of employee complaints.* Highest priority is given those complaints that allege an imminent danger situation. Complaints reporting a "serious" situation are given high priority. If time and resources allow, the CSHO will normally inspect the entire workplace and not just the condition reported in the complaint.
4. *Programmed high-hazard inspections.* Industries are selected for inspection based on their death, injury, and illness incidence rates; employee exposure to toxic substances; and so on.
5. *Reinspections.* Establishments cited for alleged serious violations normally are reinspected to determine whether the hazards have been abated.

General inspection procedures. The primary responsibility of the CSHO, who is under the supervision of the OSHA Area Director, is to conduct an effective inspection to determine if employers and employees are in compliance with the requirements of the standards, rules, and regulations promulgated under the OSHAct. OSHA inspections are almost always conducted without prior notice.

To enter an establishment, the CSHO presents proper

credentials to a guard, receptionist, or other person acting in such a capacity. Employers should always insist on seeing and checking the CSHO's credentials carefully before allowing the individual to enter their establishment for the purpose of an inspection (Figure 2–2). Anyone who tries to collect a penalty or promotes the sale of a product or service is not a CSHO.

The CSHO will usually ask to meet with an appropriate employer representative. It is recommended that employers furnish written instructions to security, the receptionist, and other affected personnel regarding the CSHO's right of entry and initial treatment, whom should be notified, and to whom and where the CSHO should be directed to avoid undue delay.

Opening conference. The CSHO will conduct a joint opening conference with employer and employee representatives. Where it is not practical to hold a joint conference, separate conferences are to be held for employer representatives. If there is no employee representative, then a joint conference is not necessary. When separate conferences are held, a written summary of each conference should be made and the summary provided on request to employer and employee representatives.

Because the CSHO will want to talk with the firm's safety personnel, these employees should participate in the opening conference. The employer representative who accompanies the CSHO during the inspection should also participate in the opening conference.

At the opening conference, the CSHO will:

1. inform the employer that the purpose of the officer's visit is to investigate whether the establishment, procedures, operations, and equipment are in compliance with OSHAct requirements
2. give the employer copies of the Act, standards, regulations, and promotional materials, as necessary
3. outline in general terms:
   - scope of the inspection
   - records the officer wants to review
   - the officer's obligation to confer with employees
   - physical inspection of the workplace
   - closing conference.
4. if applicable, furnish the employer with a listing of the complaint(s)
5. answer questions from those attending the conference.

In the opening conference, the employer representative should find out which areas of the establishment the CSHO wishes to inspect. In some cases the inspection may include areas of the plant in which trade secrets are maintained. If this is the case, the employer representative should orally request confidential treatment of all information obtained from such areas. The employer should follow up with a trade-secret letter to the CSHO requesting the officer to keep information identified in the letter strictly confidential. The CSHO should not discuss any part of this information or provide copies to any person not authorized by law to receive the data without prior written consent of the employer.

During the course of the opening conference, the CSHO

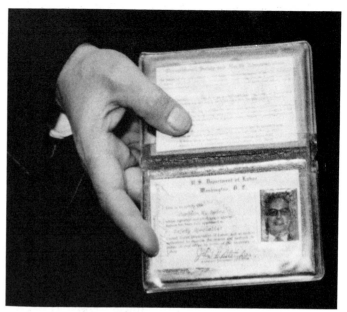

Figure 2–2. Bona-fide OSHA compliance officers are equipped with official identification as shown here. The credentials are signed by the current or former Assistant Secretary of Labor for Occupational Safety and Health. If in doubt about the validity of the credentials, the employer should contact the nearest OSHA area office, determine whether or not the area office has scheduled an inspection at the establishment in question, and verify the serial number on the credentials.

may ask to review company records. The CSHO is authorized to review only the records required to be maintained by the OSHAct, regulations, and standards. In general, these records include the "Log and Summary of Occupational Injuries and Illnesses" (OSHA Form 200) and the "Supplemental Record of Occupational Injuries and Illnesses" (OSHA Form 101). Such records should be made readily available to the officer. Prompt, thorough, and complete cooperation in the inspection will always make a good impression.

The CSHO may request information about the employer's current safety and health program in order to evaluate it. Naturally, a comprehensive safety and health program that shows evidence of effective accident prevention will be impressive to all concerned.

The CSHO will also ask the employer whether workers of another employer (for example, a maintenance or remodeling contractor) are working in or on the establishment. If so, the CSHO will give the authorized representative of those employees a reasonable opportunity to participate in the inspection of their work areas.

During the conference, the CSHO will meet with the employee's authorized representative and explain the person's rights. Generally, the representative will be an employee of the establishment inspected. However, the CSHO may judge that good cause has been shown to require a third party (such as an industrial hygienist or safety consultant) who is not an employee but is still an authorized employee representative to accompany the CSHO on the inspection to ensure an effective and thorough job. The final decision on this matter will rest with the CSHO.

The employer is not permitted to designate the employee representative. Employee representatives may change as the inspection process moves from department to department. The CSHO may refuse to allow any person to accompany him or her whose conduct interferes with a full, orderly inspection. If there is no authorized employee representative, the CSHO will consult with a reasonable number of employees concerning matters of safety and health in the workplace during the course of the inspection.

Inspection of facilities. The CSHO will normally take adequate time to inspect all operations in the establishment. The primary objective of the inspections is to enforce occupational safety and health standards as well as to enforce other promulgated regulations such as placement of the OSHA poster (Figure 2–1).

The CSHO will have the necessary instruments for check-

ing items such as noise levels, certain air contaminants and toxic substances, electrical grounding, and the like. During the course of inspection, the CSHO will note and usually record any apparent violation of the standards, including its location, and any comments regarding the violation. The officer will do the same for any apparent violation of the general-duty clause. These notes will serve as a basis for the Area Director's citations or proposed penalties. For these reasons, the employer representative should find out what apparent violations the CSHO has detected during the actual inspection of the facilities. The employer representative also should take the same notes as the CSHO during the inspection so that the employer will have exactly the same information as the CSHO.

It should be noted that the CSHO is only required to record apparent violations and is not required to present a solution or method of correcting, minimizing, or eliminating the violation. OSHA, however, will respond to requests for technical information regarding compliance with given standards. In such cases, the employer is urged to contact the regional or area office.

During the course of an inspection, the CSHO may receive a complaint from an employee regarding a condition alleged to be in violation of an applicable standard. The CSHO, even though the complaint is brought via an informal process, will normally inspect for the alleged violation.

In the course of a normal inspection, the CSHO may make some preliminary judgments regarding environmental conditions affecting occupational health. In such cases, the officer will generally use direct-reading instruments. Should this occur, and if proper instrumentation is available, it would be prudent for the employer to have qualified personnel at the establishment make duplicate tests in the same area at the same time under the same conditions. In addition, the employer representative should again take careful notes on the CSHO's methods as well as the results. If the inspection indicates a need for further investigation by an industrial hygienist, the CSHO will notify the Area Director, who may assign a qualified industrial hygienist to investigate further. If a laboratory analysis is required, samples will be sent to OSHA's laboratory in Salt Lake City and the results will be reported back to the Area Director.

Closing conference. Upon completion of the inspection, the CSHO will hold a joint closing conference with employee representatives and representatives of the employer. If a joint conference is not possible, separate conferences will be held. Again, the employer's safety personnel should be present at the closing conference. At this time the CSHO will advise the employer and employee representatives of all conditions and practices that may constitute a safety or health violation. The officer should also indicate the applicable section or sections of the standards that may have been violated.

The CSHO will normally advise that citations may be issued for alleged violations and that penalties may be proposed for each violation. The authority for issuing citations and proposed penalties, however, rests with the Area Director or the director's representative.

The employer will also be informed that the citations will fix a reasonable time for abatement of the violations alleged. The CSHO will attempt to obtain from the employer a reasonable estimate of the time required to control or eliminate the alleged violation. The officer will take such estimates into consideration when recommending a time for abatement. Although the employer is not required to do so, it might be advantageous to give the officer copies of any correspondence or orders concerning equipment to achieve compliance. This act of good faith may help to establish a reasonable abatement period and may reduce the proposed penalty. The CSHO should also explain the appeal procedures with respect to any citation or any notice of a proposed penalty.

Informal postinspection conferences. Issues raised by inspections, citations, proposed penalties, or notice of intent to contest may be discussed at the request of an affected employer, employee, or employee representative at an informal conference held by the Assistant Regional Director. Whenever the employer or employee representatives request an informal conference, both parties shall be afforded the opportunity to participate fully.

Follow-up inspections. Follow-up inspections will always be conducted for those situations involving imminent danger or where citations have been issued for serious, repeated, or willful violations. Follow-up inspections for all other cases will be ordered at the discretion of the Area Director.

The follow-up inspection should be limited to verifying compliance of the conditions alleged to be in violation. The follow-up inspection is conducted with all of the usual formality of the original inspection, including the opening and closing conferences and the walk-around rights of the employer and employee representative.

Violations

In addition to the general-duty clause, OSHAct occupational safety and health standards are used to determine alleged violations. There are four types of violations: imminent danger, serious, nonserious, and de minimis (very minor).

Imminent danger. The OSHAct defines imminent danger as "any condition or practice in any place of employment which is such that a danger exists which could reasonably be expected to cause death or serious physical harm immediately or before the imminence of such danger can be eliminated through the enforcement procedures otherwise provided by this Act." Therefore, for conditions or practices to constitute an imminent danger, there must be a reasonable certainty that within a short time such conditions or practices could result in death or serious physical harm. According to the Act, "serious physical harm" includes the permanent loss or reduction in efficiency of a part of the body, or inhibition of a part of the body's internal system such that life is shortened or physical or mental efficiency is reduced.

Normally a health hazard would not constitute an imminent danger except in extreme situations. These might include the presence of potentially lethal concentrations of

airborne toxic substances that pose an immediate threat to the lives or health of employees. If, during the course of inspection, the CSHO deems that a particular set of conditions appears to constitute an imminent danger, the officer will immediately inform the employer or employer representative and will attempt to have the danger corrected at once through voluntary compliance. Further, if any employees appear to be in imminent danger, they will be informed of the danger, and the employer will be requested to remove them from the area.

An employer will be deemed to have controlled or eliminated the imminent danger by (1) removing employees from the danger area and assuring the CSHO that employees will not return until the hazardous condition has been eliminated or (2) eliminating the conditions or practices creating the imminent danger. When the employer voluntarily eliminates the danger, no imminent danger procedure is instituted and no Notice of Imminent Danger is issued. However, citations and proposed penalties are, nonetheless, issued.

An employer may refuse to voluntarily address the alleged imminent danger. In such cases, the CSHO will inform the affected employees of the hazard and will tell both the employer and affected employees that he or she will recommend the Area Director take civil action (in the form of a court order) for appropriate relief (e.g., to shut down the operation). The CSHO will personally post the imminent danger notice at or near the area in which the exposed employees are working. However, the federal CSHO alone has no authority to order an operation closed down or to direct employees to leave the area of imminent danger or the workplace. Nevertheless, a prudent employer will take these actions voluntarily.

In such cases, the Area Director normally will request the Regional Solicitor to obtain an injunction permanently restraining the employer's practices. To protect employees until the hearing on the injunction, a temporary restraining order, issued without notice to the employer and effective for up to five days, may be issued. The Act vests jurisdiction in the U.S. district courts to restrain any condition or work practice in imminent danger situations.

Serious violation. A serious violation involves hazardous conditions that could cause death or serious physical harm to employees, conditions that the employer knew, or should have known, existed.

OSHA's *Field Operations Manual* (Chapter IV, Violations) sets forth four steps for the CSHO to follow to determine whether a violation is serious or other-than-serious. Section 17(k) of the Act provides

a serious violation shall be deemed to exist in a place of employment if there is a substantial probability that death or serious physical harm could result from a condition which exists, or from one or more practices, means, methods, operations, or processes which have been adopted or are in use, in such place of employment unless the employer did not, and could not with the exercise of reasonable diligence, know of the presence of the violation.

The CSHO shall take four steps to make the determination that a violation is serious. The first three steps determine whether there is a substantial probability that death or serious physical harm could result from an accident or exposure relating to the violative condition. (The probability that an accident or illness will occur is not to be considered in determining whether a violation is serious.) The fourth step determines whether the employer knew or could have known of the violation.

Apparent violations of the general duty clause shall also be evaluated on the basis of these steps to ensure that they represent serious violations. The four elements the CSHO shall consider are as follows:

Step 1. The type of accident or health hazard exposure which the violated standard or the general duty clause is designed to prevent.

Step 2. The type of injury or illness which could reasonably be expected to result from the type of accident or health hazard exposure identified in Step 1.

- In making this determination, the CSHO shall consider all factors which would affect the severity of the injury or illness which could reasonably be predicted to result from an accident or health hazard exposure. (The CSHO shall not give consideration at this point to factors which relate to the **probability** that an injury or illness will occur.) The following are examples of a determination of the types of injuries which could reasonably be predicted to result from an accident:
- If an employee falls from the edge of an open-sided floor 30 ft to the ground below, that employee could break bones, suffer a concussion, or experience other more serious injuries.
- If an employee trips on debris, that employee could experience abrasions or bruises, but it is only marginally predictable that the employee could suffer a substantial impairment of a bodily function.
- If an employee is exposed regularly and continually to beryllium at .004 mg/m^3, it is reasonable to predict that berylliosis or cancer could result.
- If an employee is exposed regularly and continually to acetic acid at 20 ppm, it is reasonable to predict that the illness which could result (irritation to nose, eyes, throat) would not involve serious physical harm.

Step 3. Whether the types of injury or illness identified in Step 2 could include death or a form of **serious physical harm.**

- Impairment of the body in which part of the body is made **functionally useless** or is **substantially reduced in efficiency** on or off the job. Such impairment may be permanent or temporary, chronic or acute. Injuries involving such impairment would usually require treatment by a medical doctor. Examples include:
  - Amputation (loss of all or part of a bodily appendage which includes the loss of bone).
  - Concussion.
  - Crushing (internal, even though skin surface may be intact).

Examples of illnesses which constitute serious physical

harm include:

- Cancer
- Poisoning (resulting from the inhalation, ingestion, or skin absorption of a toxic substance which adversely affects a bodily system)
- Lung diseases, such as asbestosis, silicosis, anthracosis
- Hearing loss.

Step 4. Whether the employer knew, or with the exercise of reasonable diligence, could have known of the presence of the hazardous condition.

The knowledge requirement is met if it is determined that the employer actually knew of the hazardous condition which constituted the apparent violation.

As a general rule, if the CSHO was able to discover a hazardous condition, it can be presumed that the employer could have discovered the same condition through the exercise of reasonable diligence.

Other-than-serious violations. This type of violation shall be cited in situations where the accident or illness that would be most likely to result from a hazardous condition would probably not cause death or serious physical harm but would have a direct and immediate relationship to the safety and health of employees.

Willful violations. The following definitions and procedures apply whenever the CSHO suspects that a willful violation may exist:

A willful violation exists under the Act where the evidence shows either an intentional violation of the Act or plain indifference to its requirements.

It is not necessary that the violation be committed with a bad purpose or an evil intent to be deemed "willful." It is sufficient that the violation was deliberate, voluntary, or intentional as distinguished from inadvertent, accidental, or ordinarily negligent.

The determination of whether to issue a citation for a willful or repeated violation will frequently raise difficult issues of law and policy and will require the evaluation of complex factual situations. Accordingly, a citation for a willful violation shall not be issued without consultation with the Regional Administrator, who shall, as appropriate, discuss the matter with the Regional Solicitor.

Criminal/willful violations. Section 17(e) of the Act provides that: "Any employer who willfully violates any standard, rule or order promulgated pursuant to Section 6 of this Act, or of any regulations prescribed pursuant to this Act, and that violation caused death to any employee, shall, upon conviction, be punished by a fine of not more than $10,000 or by imprisonment for not more than six months, or by both; except that if the conviction is for a violation committed after a first conviction of such person, punishment shall be a fine of not more than $20,000 or by imprisonment for not more than one year, or by both."

Repeated violations. An employer may be cited for a repeated violation if that employer has been cited previously for a *substantially similar condition* and the citation has become a final order.

Identical standard. Generally, similar conditions can be demonstrated by showing that in both situations the identical standard was violated.

Different standards. In some circumstances, similar conditions can be demonstrated when different standards are violated.

De minimis violations. De minimis violations refer to conditions that represent no immediate or direct threat to safety or health. "De minimis" is short for the legal maxim, *De minimis non curat lex,* "The law does not concern itself with trifles."

Citations

An investigation or inspection may reveal a condition that is alleged to be in violation of the standards or general-duty clause. In such instances, the employer may be issued a written citation that describes the specific nature of the alleged violation, cites the standard allegedly violated, and fixes a time for abatement. The employer must prominently post each citation, or copy thereof, at or near the place where the alleged violation occurred. All citations are issued by the Area Director or a designee and will be sent to the employer by certified mail.

A "Citation for Serious Violation" will be prepared to cover those violations which fall into the "serious" category. This type of violation must be assessed a monetary penalty.

A citation used for other-than-serious violations may or may not carry a monetary penalty. A citation may be issued to the employer for employee actions that violate the safety and health standards.

A notice, in lieu of a citation, is issued for *de minimis* violations that have no direct relationship to safety and health. The employer is not required to post this notice, unlike the citation.

If an inspection has been initiated in response to an employee complaint, the employee or authorized employee representative may request an informal review of any decision not to issue a citation. However, employees may not contest citations, amendments to citations, penalties, or lack of penalties. Employers may contest the time for abatement of a hazardous condition specified in a citation. They also may contest an employer's Petition for Modification of Abatement (PMA), which requests an extension of the abatement period. Employees must contest the PMA within 10 working days of its posting or within 10 working days after an authorized employee representative has received a copy.

Within 15 working days after the employer receives the citation, an employee may submit a written objection to the citation to OSHA. The OSHA Area Director then forwards the objection to OSHRC. Employees may request an informal conference with OSHA to discuss any issues raised by an inspection, citation, notice of proposed penalty, or employer's notice of intention to contest.

Petition for modification of abatement. Upon receiving a citation, the employer must correct the cited hazard by the prescribed date. However, factors beyond the employer's reasonable control may prevent the work from being com-

pleted on time. In such a situation, the employer who has made a good-faith effort to comply may file for a PMA date.

The written petition should specify (1) all steps the employer took to achieve compliance, (2) the additional time needed to complete the work, (3) reasons why additional time is needed, (4) all temporary steps being taken to safeguard employees against the cited hazard during the intervening period, (5) the fact that a copy of the PMA was posted prominently or at least near each place where a violation occurred, and (6) the employee representative (if there is one) who received a copy of the petition.

Penalties

General policy. The penalty structure provided under Section 17 of the Act is designed primarily to provide an incentive toward correcting violations voluntarily, not only to the offending employer but, more especially, to other employers who may be guilty of the same infractions of the standards or regulations.

1. OSHA has always taken the position that the penalties are not designed as punishment for violations nor as a source of income for the Agency. The Congress has made clear its intent, however, that penalty amounts should be sufficient to serve as an effective deterrent to violations.
2. Large proposed penalties, as Congress has clearly recognized, serve the public purpose intended under the Act; criteria guiding approval of such penalties by the Assistant Secretary are based on meeting this public purpose.

Civil Penalties

Type of violation as a factor. In proposing civil penalties for violations, a distinction is made between serious violations and other-than-serious violations. There is no statutory requirement that a penalty be proposed when the violation is not serious; but a penalty must be proposed when the violation is serious.

- The maximum penalty that may be proposed for a serious or an other-than-serious violation is $7,000.
- In the case of willful or repeated violations, a civil penalty of up to $70,000 may be proposed; but the penalty may not be less than $5,000 for a willful violation.
- For other specific violations of the Act, civil penalties of up to $7,000 may be proposed.
- Penalties for failure to correct a violation may be up to $7,000 for each calendar day that the violation continues beyond the final abatement date.

Statutory authority. Section 17 provides the Secretary with the statutory authority to assess civil penalties for violations of the Act.

Minimum penalties. The following guidelines apply:

- The proposed penalty for any willful violation shall not be less than $5,000. This is a statutory minimum and not subject to administrative discretion.
- When the adjusted proposed penalty for an other-than-serious violation (citation item) would amount to less than $100, no penalty shall be proposed for that violation.
- When, however, there is a citation item for a posting

violation, this minimum penalty amount does not apply with respect to that item since penalties for such items are mandatory under the Act.

Penalty factors. The Act provides that penalties shall be assessed on the basis of four factors:

- gravity of the violation
- size of the business
- good faith of the employer
- employer's history of previous violations.

Gravity of violation. The gravity of the violation is the primary consideration in determining penalty amounts. It is the basis for calculating the basic penalty for both serious and other-than-serious violations.

To determine the gravity of a violation the following two assessments shall be made:

(1) The **severity** of the injury or illness which could result from the alleged violation. The most serious injury or illness which is reasonably predictable as a result of an employee's exposure to the safety or health hazard cited shall be assigned a severity assessment in accordance with the following factors:

- *High severity:* Death from injury or illness; injuries involving permanent disability; or chronic, irreversible illnesses.
- *Medium severity:* Injuries or temporary, reversible illnesses resulting in hospitalization or a variable but limited period of disability.
- *Low severity:* Injuries or temporary, reversible illnesses not resulting in hospitalization and requiring only minor supportive treatment.
- *Minimal severity:* Other-than-serious violations. Although such violations reflect conditions which have a direct and immediate relationship to the safety and health of employees, the injury or illness most likely to result would probably not cause death or serious physical harm.

(2) The **probability** that an injury or illness could occur as a result of the alleged violation. Probability shall be categorized either as greater or as lesser probability.

- **Greater probability** results when the likelihood that an injury or illness will occur is judged to be relatively high.
- **Lesser probability** results when the likelihood that an injury or illness will occur is judged to be relatively low.

For serious violations, the gravity-based penalty (GBP) shall be assigned on the basis of the following scale:

Severity	Probability	GBP
High	Greater	$ 5,000
Medium	Greater	$ 3,500
Low	Greater	$ 2,500
High	Lesser	$ 2,500
Medium	Lesser	$ 2,000
Low	Lesser	$ 1,500

- For other-than-serious safety and health violations, there is no severity assessment.

Penalty adjustment factors. The GBP may be reduced by as must as 95% depending upon the employer's "good faith," "size of business," and "history of previous violations." Up

to 60% reduction is permitted for size; up to 25% reduction for good faith, and 10% for history.

- Since these rates are based on the general character of a business and its safety and health performance, the rates shall generally be calculated only once for each employer—after the classification and probability ratings have been determined for each violation and the general character of the employer's performance is apparent.

- Penalties assessed for violations that are classified as high severity and greater probability shall be adjusted only for size and history.

- Penalties assessed for violations that are classified as repeated shall be adjusted only for size.

- Penalties assessed for violations classified as willful shall have been adjusted only for size and history.

- The rate of penalty reduction for size of business, employer's good faith, and employer's history of previous violations shall be calculated on the basis of the criteria described in the following paragraphs:

Size. A maximum penalty reduction of 60% is permitted for small businesses. The rates of reduction to be applied are as follows:

Employees	Percent reduction
1–25	60
26–100	40
101–250	20
251 or more	None

When a small business has one or more serious violations of high gravity or a number of serious violations of moderate gravity, indicating a lack of concern for employee safety and health, the Area Director may determine that only a partial reduction in penalty shall be permitted for size of business.

Good faith. A penalty reduction of up to 25% is permitted in recognition of an employer's "good faith."

A reduction of 25% shall normally be given if the employer has a written safety and health program (as documented during the inspection) that has been effectively implemented in the workplace.

A reduction of 15% shall normally be given if the employer has a documented safety and health program, but with more than only incidental deficiencies.

No reduction shall be given to an employer who has no safety and health program.

History. A reduction of 10% shall be given to employers who have not been cited by OSHA for any serious, willful, or repeated violations in the past three years.

Total. The total reduction will normally be the sum of the reductions for each adjustment factor.

Good faith effort to abate. When the CSHO believes and so documents in the case file that the employer has made good faith efforts to correct the violation and had good reason to believe that it was fully abated, the Area Director may reduce or eliminate the daily proposed penalty that would otherwise be justified.

Repeated violations. Section 17(a) of the Act provides that an employer who repeatedly violates the Act may be assessed a civil penalty of not more than $70,000 for each violation.

Penalty increase factors. The amount of the increased penalty to be assessed for a repeated violation shall be determined by the size of the employer.

- *Smaller employers.* For employers with 250 or fewer employees, the GBP shall be **doubled** for the first repeated violation and **quintupled** if the violation has been cited twice before.

- *Larger employers.* For employers with more than 250 employees, the GBP shall be multiplied by **5** for the first repeated violation and multiplied by **10** if the violation has been cited twice before.

Other-than-serious, no initial penalty. For a repeated other-than-serious violation that otherwise would have no initial penalty, a penalty of $200 shall be assessed for the first repeated violation, $500 if the violation has been cited twice before, and $1,000 for a third repetition.

Regulatory violations. For repeated regulatory violations, the initial penalty shall be **doubled** for the first repeated violation and **quintupled** if the violation has been cited twice before. If the Regional Administrator determines that it is appropriate to achieve the necessary deterrent effect, the initial penalty may be multiplied by 10.

Willful violations. Section 17(a) of the Act provides that an employer who willfully violates the Act may be assessed a civil penalty of not more than $70,000 but not less than $5,000 for each violation.

- *Serious violations.* For willful serious violations, the **adjusted** GBP shall be multiplied by **seven.**

- In no case shall the proposed penalty be less than $5,000.

- *Other-than-serious violations.* For willful other-than-serious violations, the minimum willful penalty of $5,000 shall be assessed.

Regulatory violations. In the case of regulatory violations that are determined to be willful, the unadjusted initial penalty shall be multiplied by **seven.** In no event shall the penalty, after adjustment for size and history, be less than $5,000.

Annual summary. If an employer fails to post the summary portion of the OSHA-200 Form during the month of February, as required by 29 *CFR* 1904.5(d)(1), an other-than-serious citation shall be issued with an unadjusted penalty of $1,000.

Citation. If an employer received a citation that has not been posted as prescribed in 29 *CFR* 1903.16, an other-than-serious citation shall normally be issued. The unadjusted penalty shall be $3,000.

Reporting and recordkeeping requirements. Section 17(c) of the Act provides that violations of the recordkeeping and reporting requirements may be assessed civil penalties of up to $7,000 for each violation.

(1) *OSHA-200 and OSHA-101 Forms.* If the employer does not maintain the Log and Summary of Occupational Injuries and Illnesses, OSHA-200 Form, and the Supplementary Record, OSHA-101 Form (or equivalent), as prescribed in 29 *CFR* 1904, an other-than-serious citation shall be issued. There shall be an unadjusted penalty of $1,000 for each OSHA form not maintained.

(2) *Reporting.* Employers are required to report either orally or in writing to the nearest Area Office within 48 hours, any occurrence of an employment accident which is fatal to one or more employees or which results in the hospitalization of five or more employees.

(a) An other-than-serious citation shall be issued for failure to report such an occurrence. The unadjusted penalty shall be $5,000.

(b) If the Regional Administrator determines that it is appropriate to achieve the necessary effect, an unadjusted penalty of $7,000 may be assessed.

Access to records.

(1) *29 CFR 1904.* If the employer fails upon request to provide records required in Section 1904.2 for inspection and copying by any employee, former employee, or authorized representative of employees, a citation for violation of 29 *CFR* 1904.7(b)(1) shall normally be issued. The unadjusted penalty shall be $1,000 for each form not made available.

(a) Thus, if the OSHA-200 for the three preceding years is not made available, the unadjusted penalty would be $3,000.

(b) If the employer is to be cited for failure to maintain these records, no citation of 1904.7 shall be issued.

(2) *29 CFR 1910.20.* If the employer is cited for failing to provide records as required under 29 *CFR* 1910.20 for inspection and copying by any employee, former employee, or authorized representative of employees, an unadjusted penalty of $1,000 shall be proposed.

Proposed penalties. Once the proposed penalties have been calculated, each is listed in the "Citation and Notification of Penalty" (OSHA-2 Form). This document is used to officially inform the employer of violations found during the inspection and penalties proposed for those violations. This is sent to the employer by certified mail. An information copy is sent to an employee representative or to the employee organization.

Egregious Policy

When OSHA considers the apparent violations flagrant, the agency instead of grouping similar violations levies a separate penalty for each.

Some of the factors used to determine the situation include:

- the number of worker fatalities, a work-site catastrophe, or a large number of similar injuries or illnesses
- a violation which results in high rates of injuries or illnesses
- the organization has an extreme history of workplace violations
- the employer seriously disregarded workplace safety and health responsibilities
- a large number of violations is found at the work site.

States can adopt their own penalty structure, which may even exceed the one set up under OSHA.

CONTESTED CASES

An employer has the right to contest any OSHA action. The employer may contest one or more of the following: a cita-

tion, a proposed penalty, a notice of failure to correct a violation, or the time allotted for abatement of an alleged violation. OSHAct regulations that cover procedures for contesting cases are codified in Title 29 *CFR*, Chapter XX, Part 2200. On the other hand, an employee or authorized employee representative may contest only the time allotted for an abatement of an alleged violation.

Prior to formally initiating a contest, employers should request an informal hearing with the Area Director or the Assistant Regional Director. Many times such informal sessions will resolve questions and issues, thus avoiding the formal contested case proceedings.

However, the informal conference may fail to resolve the dispute between OSHA and the employer. If the employer elects to contest the case, affected employees or the authorized employee representative are automatically deemed to be parties to the proceeding. In contesting an OSHA action, the employer must comply with the following rules that apply to a specific case:

1. The employer must notify the Area Office which initiated the action that the employer is to contest the case. This must be done within 15 working days after receiving OSHA's notice of proposed penalty; it must be sent by certified mail. If the employer does not contest within the required 15 working days, the citation and proposed assessment of penalties are deemed to be a final order of OSHRC and are not subject to review by any court or agency. As a result, the alleged violation must be corrected within the abatement period specified in the citation.

2. If any of the employees working at the site where the alleged violation exists are union members, a copy of the notice of contest must be served upon their union.

3. If employees who work on the site are not represented by a union, a copy of the notice of contest must either be posted at a place where the employees will see it or be served upon them personally.

4. The notice of contest must also list the names and addresses of those parties who have been personally served a notice, or, if such notice is posted, it must contain the address of the posted location.

5. In some cases the employees at the site of the alleged violation are not represented by a union and have not been personally served with a copy of the notice to contest. If so, posted copies must specifically advise the unrepresented employees that they may not be able to assert their status as parties to the case if they fail to properly identify themselves to the Commission or to the Hearing Examiner before the hearing begins or when it first opens.

6. There is no specific form for the notice of contest. However, such notice should clearly identify what is being contested—the citation, the proposed penalty, the notice of failure to correct a violation, or the time allowed for abatement—for each alleged violation or combination of alleged violations.

If the employer contests an alleged violation in good faith, and not solely for delay or variance of penalties, the

abatement period does not begin until the OSHRC enters the final order. When a notice of contest is received by an Area Director from an employer, an employee, or an authorized employee representative, the Director will file with OSHRC the notice of contest and all contested citations, notice of proposed penalties, or notice of failure to abate.

Upon receiving the notice of contest from the Area Director, the Commission will assign the case a docket number. Ultimately, an ALJ will be assigned to the case and will conduct a hearing at a location reasonably convenient to those concerned. At the hearing, OSHA presents its case and is subject to a cross-examination by other parties. The party contesting then presents its case and is also subject to a cross-examination by other parties. Affected employees or an authorized employee representative may participate in the hearings. The decision by the ALJ will be based only on what is in the record. Therefore, if statements go unchallenged, they will be assumed to be fact.

After the hearings are completed, the ALJ will submit the record and a report with recommendations to OSHRC. If no Commissioner orders a review of an ALJ's recommendations, they will stand as OSHRC's decision. If any Commissioner orders a review of the case, the Commission itself must render a decision to affirm, modify, or vacate the judge's recommendations. The Commission's orders become final 15 days after issuance, unless stayed by a court order.

Any person adversely affected or aggrieved by an order of the Commission may obtain a review of the order in the United States Court of Appeals. However, the person must seek a review within 60 days of the order's issuance.

SMALL BUSINESS LOANS

The Act enables small businesses to obtain economic assistance for health- and safety-related issues. It amends the Small Business Act to provide for financial assistance to small firms that must make changes to comply with the standards promulgated under the OSHAct or by a state under a state plan. Before approving any assistance, the SBA must first determine that the small firm is likely to suffer substantial economic injury without financial help.

An employer can apply for a loan under one of two procedures: (1) before federal or state inspection in order to come into compliance, or (2) after federal or state inspection to correct alleged violations.

When an employer has not been inspected and requests a loan to bring the establishment into compliance prior to an inspection, the employer must submit to the SBA:
- a statement of the conditions to be corrected
- a reference to the OSHA standards that require the employer to make corrections
- a statement of the firm's financial condition showing that a loan is needed.

The employer should submit this information to the nearest SBA field office along with any background material. The SBA will then refer the application to the appropriate OSHA Regional Office, Office of Technical Support. The OSHA Regional Office will review the application and advise SBA whether the employer is required to correct the described conditions in order to come into compliance and whether the proposed use of funds will accomplish the needed corrections. Direct contact with the applicant will be initiated by OSHA only after clearance with the SBA.

If the employer is making an application after an inspection to correct alleged violations, the procedure is the same as before inspection, except that the applicant also must furnish SBA a copy of the OSHA citation(s). SBA then refers the application to the OSHA Area Office that conducted the inspection. That office will notify SBA whether the proposed use of loan funds will adequately correct cited violations.

Forms for loan applications may be obtained from any SBA field office. In some instances, private lending institutions will be able to provide the form for SBA/bank participation loans.

FEDERAL-STATE RELATIONSHIPS

The OSHAct encourages the states to assume the fullest responsibility for administering and enforcing their own occupational safety and health laws. However, in order to assume this responsibility, such states must submit a state plan to OSHA for approval. If such a plan satisfies designated conditions and criteria, OSHA must approve the plan. The regulations pertaining to state plans for the development and enforcement of state standards are codified in Title 29 *CFR*, Chapter XVII, Part 1902.

The states and possessions listed in Figure 2–3 have approved plans.

The basic criterion for approval of state plans is that the plan must be "at least as effective as" the federal program. It was not Congress's intent to require that state programs be a "mirror image" of the federal program. Congress believed rules for developing state plans should be flexible to allow consideration of local problems, conditions, and resources. The Act provides for funding up to half the costs of the implementation of the state program.

A state plan must include any occupational safety and health "issue" (industrial, occupational, or hazard group) for which a corresponding federal standard has been promulgated. A state plan cannot be less stringent, but it may include subjects not covered in the federal standards. However, state plans that do not include those issues covered by the federal program, in effect, surrender such issues to OSHA. For example, a state plan may cover all industry except construction. If such is the case, the state surrenders its jurisdiction for safety and health programs in construction operations to OSHA. It is then OSHA's obligation to enforce the federal standards for those operations not covered by the state plan.

Following approval of a state plan, OSHA will continue to exercise its enforcement authority until it determines on the basis of actual operations that the state plan is indeed being satisfactorily carried out. If the implementation of the

Alaska Department of Labor
P. O. Box 21149
Juneau, ALASKA 99802-1149
(907) 465-2700

Industrial Commission of Arizona
800 W. Washington
Phoenix, ARIZONA 85007
(602) 255-5795

California Department of Industrial Relations
525 Golden Gate Avenue
San Francisco, CALIFORNIA 94102
(415) 557-3356

Connecticut Department of Labor
200 Folly Brook Boulevard
Wethersfield, CONNECTICUT 06109
(203) 566-5123

Hawaii Department of Labor and Industrial Relations
830 Punchbowl Street
Honolulu, HAWAII 96813
(808) 548-3150

Indiana Department of Labor
1013 State Office Building
100 North Senate Avenue
Indianapolis, INDIANA 46204
(317) 232-2663

Iowa Division of Labor Services
1000 E. Grand Avenue
Des Moines, IOWA 50319
(515) 281-3447

Kentucky Labor Cabinet
U.S. Highway 127 South
Frankfort, KENTUCKY 40601
(502) 564-3070

Maryland Division of Labor and Industry
Department of Licensing and Regulation
501 St. Paul Place
Baltimore, MARYLAND 21202-2272
(301) 333-4176

Michigan Department of Labor
309 N. Washington
P. O. Box 30015
Lansing, MICHIGAN 48909
(517) 373-9600

Michigan Department of Public Health
3423 N. Logan Street
Box 30195
Lansing, MICHIGAN 48909
(517) 373-1220

Minnesota Department of Labor and Industry
443 Lafayette Road
St. Paul, MINNESOTA 55101
(612) 296-2342

Nevada Department of Industrial Relations
Division of Occupational Safety and Health
Capitol Complex
1370 S. Curry Street
Carson City, NEVADA 89710
(702) 885-5240

New Mexico Environmental Improvement Division
Health and Environmental Department
1190 - St. Francis Drive - N2200
Santa Fe, NEW MEXICO 87503-0968
(505) 827-2850

New York Department of Labor
One Main Street
Brooklyn, NEW YORK 11201
(718) 797-7668

North Carolina Department of Labor
4 W. Edenton Street
Raleigh, NORTH CAROLINA 27603
(919) 733-7166

Accident Prevention Divison
Oregon Department of Insurance and Finance
Labor and Industries Building
Salem, ORGEON 97310
(503) 378-3304

Puerto Rico Department of Labor and Human Resources
Prudencio Rivera Martinez Building
505 Munoz Rivera Avenue
Hato Rey, PUERTO RICO 00918
(809) 754-2119-22

South Carolina Department of Labor
3600 Forest Drive
P. O. Box 11329
Columbia, SOUTH CAROLINA 29211-1329
(803) 734-9594

Tennessee Department of Labor
501 Union Building
Suite "A" - 2nd Floor
Nashville, TENNESSEE 37219
(615) 741-2582

Utah Occupational Safety and Health
160 E. 300 South
P. O. Box 5800
Salt Lake City, UTAH 84110-5800
(801) 530-6900

Vermont Department of Labor and Industry
120 State Street
Montpelier, VERMONT 05602
(802) 828-2765

Figure 2–3. States with approved plans.

<table>
<tr><td>

Virgin Islands Department of Labor
Box 890
Christiansted
St. Croix, VIRGIN ISLANDS 00820
(809) 773-1994

Virginia Department of Labor and Industry
P. O. Box 12064
Richmond, VIRGINIA 23241-0064
(804) 786-2376

</td><td>

Washington Department of Labor and Industries
General Administration Building
Room 334 - AX-31
Olympia, WASHINGTON 98504-0631
(206) 753-6307

Wyoming Department of Occupational Health and Safety
604 E. 25th Street
Cheyenne, WYOMING 82002

</td></tr>
</table>

Figure 2–3. *(Concluded.)*

state plan is satisfactory during the first three years after the plan's approval, then the relevant federal standards and OSHAct enforcement of such standards no longer apply to issues covered under the state plan. This means that for the interim period when dual jurisdiction exists, employers must comply with both state and federal standards.

While the state agencies administering the state plan are vitally concerned with its success, members of the state legislature do not always share their enthusiasm. The legislature must appropriate not only an adequate budget, but in many cases must pass legislation enabling the state agency to carry out all the functions incorporated in the state plan. At times the state agency responsible may fail to fully implement the state plan, and the state's performance falls short of being "at least as effective as" the federal program. In such instances, OSHA has the right and obligation to withdraw its approval of the state plan and once again assume full jurisdiction in that state.

MEDICAL ACCESS AND RIGHT-TO-KNOW

In the 1980s, two important standards came into effect: (1) final rules for access to exposure and medical records and (2) hazard communication. Both were designed to provide employees with information about the hazardous conditions to which they are, or have been, exposed, and to give them access to their own medical and exposure records and documented information about chemicals used in the workplace.

Medical Access Standard

Employers must maintain records on the exposure employees have had to dangerous substances during their working time. These records, together with an employee's medical records, must be available to the employee, or a designated representative, upon request. The standard embodies certain restrictions on the extent of the information that must be provided and on the procedure for providing it to designated representatives of employees.

Hazard Communication Standard

Over 20 states and local communities enacted statutes and ordinances, which in some instances do not go as far as the federal standard but in other cases go much further. The indication is that employers will face more stringent re-

quirements to comply with *both* federal and state/local laws, where the two do not conflict.

Approximately one-fourth of the U.S. workforce is exposed to one or more chemical hazards. Exposure to these hazards may cause or contribute to many serious health effects. Due to the seriousness of safety and health problems associated with these chemicals and the lack of information available to many employees, OSHA issued the Hazard Communication Standard (HCS).

The Standard applies to all employers covered by OSHA in both manufacturing and nonmanufacturing sectors including construction. It also covers workers who may be exposed to hazardous materials under normal conditions or in a foreseeable emergency.

The basic purpose of the HCS is to establish uniform requirements to make sure that the hazards of all chemicals produced, imported, or used within the United States are evaluated. This hazard information must be transmitted to affected employers and employees.

This is accomplished through:
- Hazard evaluation
- Employee training
- Container labeling
- Material Safety Data Sheets (MSDSs)
- A written hazard communication program.

Hazard Evaluation

The quality of a hazard communication program depends on how accurate the initial hazard assessment is. The primary responsibility for hazard evaluation lies with the chemical manufacturer or importer. If your company uses a process that produces a chemical to which employees are exposed, you are considered a "chemical manufacturer" and must evaluate the chemical's hazards. This is true for chemical intermediates or for decomposition products such as welding fumes.

Hazard evaluation involves developing an inventory of chemicals used or produced. You can accomplish this inventory by reviewing purchase orders and performing a physical inventory of all containers of chemicals. After developing an inventory, your next step is to determine which substances are hazardous. The products in your inventory and their components should be compared to the following lists:
- 29 *CFR* 1910.1000–1047, Toxic and Hazardous Substances, OSHA

- "Threshold Limit Values for Chemical Substances in the Work Environment," American Conference of Governmental Industrial Hygienists (ACGIH), (last edition)
- National Toxicology Program (NTP), "Annual Report on Carcinogens" (latest edition).
- International Agency for Research on Cancer (IARC), "Monograph" (latest edition).

Next you must determine whether any of the remaining chemicals on the inventories possess physical or health hazards. If you have a limited knowledge of chemistry, consult the MSDS provided by the manufacturer, importer, or distributor, or discuss your questions with knowledgeable industrial hygienist or safety personnel (call the National Safety Council).

Employee Training

An employer must provide training for employees exposed to hazardous chemicals. Training must be done when employees are first assigned to an operation and whenever a new hazard is introduced into the work area. You must train employees on:

- Existence and requirements of the Standard
- Operations in the work area where hazardous chemicals are present and the hazards of these chemicals
- How the hazard communication program is implemented in the workplace, how to read and interpret information on labels and MSDSs, and how employees can obtain and use available hazard information
- Measures employees can take to protect themselves from hazards
- Specific procedures you have adopted to provide protection, such as work practices and the use of engineering controls or personal protective equipment.

Labeling

Chemical manufacturers, importers, and distributors must be sure that containers of hazardous chemicals leaving the workplace are labeled with:

- Identity of the product
- Written hazard warnings (in English)
- Name and address of manufacturer or other responsible party.

In the workplace, each container of hazardous chemicals must be labeled, tagged, or marked with:

- Identity of the product (able to reference this name to a MSDS)
- Hazard warnings (written in English)
- Graphic symbols. However, OSHA cites studies indicating that graphic symbols are not as quickly recognized as word statements. The warnings may be printed in other languages in addition to English.

There are several exemptions for in-plant labeling of containers. You may use a sign or placard for a number of stationary containers within a work area with similar contents. Operating procedures, process sheets, batch tickets, blend tickets, and similar written materials can be substituted for container labels if they contain the same information and are readily available in the work area to the employees.

Portable containers which are intended for immediate use by the employee who makes the transfer are also exempted. You are not required to label pipes or piping systems. However, the means you will use to train employees on contents of piping systems must be described in the written hazard communication program.

Material Safety Data Sheets

Chemical manufacturers and importers must develop a MSDS for each hazardous chemical they produce or import. You must obtain MSDSs for every hazardous chemical in your workplace. Copies must be readily available to employees. The HCS requires that specific information be on the MSDS:

- Section I—manufacturer's name, address, phone number, and date sheet prepared
- Section II—Hazardous Ingredients/Identity Information—chemical identity of components, exposure limits (OSHA, PEL, ACGIH, TLV, and other recommended limits)
- Section III—Physical/Chemical Characteristics—boiling point, vapor pressure and density, specific gravity, melting point, etc., and the physical and chemical data which indicate the potential for vaporization
- Section IV—Fire and Explosion Hazard Data—flash point, flammable limits, extinguishable, media, unusual fire and explosion hazards, special firefighting procedures
- Section V—Reactivity Data—stability of product, potential for polymerization and decomposition, materials and conditions to avoid
- Section VI—Health Hazards—acute and chronic hazards, carcinogenicity, signs and symptoms of exposure, emergency and first-aid procedures
- Section VII—Precautions for Safe Handling and Use—procedures to be used for spills, waste disposal, handling, and storage
- Section VIII—Control Measures—personal protective equipment, ventilation, special worker or hygienic practices.

The producer of the MSDS is responsible for the information on it and must ensure that all sheets are up to date.

Written Hazard Communication Program

You must establish a written, comprehensive hazard communication program which includes provisions for container labeling, MSDS availability, and employee training. The program must also include a list of hazardous chemicals in each work area, how you will inform employees of the hazards of nonroutine tasks and unlabeled pipes, and how you will inform contractors in manufacturing facilities of the hazards to which their employees may be exposed. The program need not be lengthy or complicated, and must be available to employees.

WHAT DOES IT ALL MEAN?

Congressional action in creating the OSHAct is only one step toward achieving the full purpose underlying the Act.

Getting it to work with reasonable efficiency is the second and more difficult task. Achieving this purpose of providing a safe, healthy environment on and off the job will depend on the willingness and cooperation of all concerned—employees and organized labor as well as business and industry.

There is no doubt that the Act has given new visibility to the whole realm of occupational safety and health. Because many employee rights are incorporated into the OSHAct, it has given employees a significant role to play in occupational safety and health matters. It has moved the laggards from "little or no safety" to "some safety," but not to "optimum safety." It has raised occupational safety and health issues to a higher priority in business management. It has given new status and responsibility to professionals working in the occupational safety and health field. Management is now relying more heavily on these safety professionals for advice. And, the Act has bestowed a new status to nationally recognized organizations that develop industry standards.

The OSHAct has also given new impetus to the field of occupational health, a much more difficult discipline when compared with occupational safety. Much more needs to be done to determine what kinds of exposures are indeed hazardous to humans and under what conditions. Further, a great deal more needs to be done to determine what countermeasures are not only adequate but also reasonable and feasible to eliminate or minimize exposures to occupational health hazards. Far more research and data about occupational health will be required to achieve the best occupational safety and health programming.

The OSHAct has encouraged greater training for professionals in occupational safety and health. Several universities have developed new curricula and programs leading to various degrees in this field—and more are yet to come.

The OSHAct also gave new emphasis to the product safety discipline. Until the passage of the U.S. Consumer Product Safety Act, the OSHAct was the most significant piece of legislation affecting product safety ever passed by the Congress. Designers and manufacturers of equipment now used by industry have a moral (but not legal) obligation to design, deliver, and install such equipment in accordance with the applicable standards.

The OSHAct is not without limitations. Mere compliance with the requirements of the Act will not achieve optimum safety and health in terms of cost, benefits, and human values. All those concerned must recognize that occupational safety and health cannot be handed to the employer or to the employee by legislative enactment or administrative decree. At best, state or federal occupational safety and health standards can cover only those areas that are enforceable—namely, control over physical conditions and environment.

As a matter of hard reality, enforcement standards simply do not adequately relate to the human in the human-machine-environment system. Important elements of a complete safety program, such as (1) establishment of work procedures to limit risk, (2) supervisory training, (3) job instruction training for employees, (4) job safety analysis, and (5) human factors engineering, by and large have not been included in the standards promulgated under the OSHAct. Neither do the standards address such issues as employee attitudes, morale, or teamwork.

For the most part, the occupational safety and health standards developed under the OSHAct are minimal criteria and represent a floor rather than a goal to achieve. Thus, to rely on mere compliance with these standards is to invite disaster since the residual risk after compliance often remains unacceptable. Effective accident prevention and control of occupational health hazards must go beyond the OSHAct.

Generally, a violation of a standard is only symptomatic of something wrong with the management safety system as a whole. Only complete occupational safety and health programming as described elsewhere in this *Manual* will achieve a level of risk acceptable to employers *and* employees. The real objective and the purpose of the OSHAct is improved occupational safety and health performance and not merely compliance with a set of standards.

ENVIRONMENTAL IMPACT

Managers responsible for occupational safety and health are increasingly affected by developments in environmental law. In some cases, responsibility for compliance with environmental laws and regulations rests on the occupational safety and health manager or on a member of the same department. It is becoming more difficult to draw a clear line between safety and health in the plant and the safety and health of the surrounding community. Further, environmental laws and regulations contain provisions addressing issues of protection for anyone—workers or the public—who comes into contact with dangerous waste products and other hazardous substances.

The two most important pieces of legislation of concern to those responsible for occupational safety and health are the Resource Conservation and Recovery Act (RCRA) and the Comprehensive Environmental Response, Compensation and Liability Act (CERCLA), commonly known as Superfund. The 1986 amendments to Superfund, which set up procedures for reporting and controlling those hazardous substances with an environmental impact, are particularly important. These environmental laws impose requirements for the use of protective equipment and procedures, and for extensive reporting to federal, state, and local government agencies. They also affect the design and operation of many manufacturing processes and waste disposal systems.

As a result, occupational safety and health and environmental law compliance have become inextricably interrelated—and will continue to be so. (See Chapter 5, Environmental Management.) A memorandum of understanding to coordinate enforcement activities was signed by OSHA and the EPA, an accord OSHA described as a "precedent setting" agreement.

1. The agreement provides, among other measures, for EPA to assist OSHA with its special emphasis program for the petrochemical industry, and for OSHA to assist EPA in enforcement actions directed at lead pollution.

2. The memorandum is the second signed by the two agencies in recent months. In September, 1990, OSHA and EPA signed an agreement to set out procedures for joint inspection at some 30 hazardous waste incinerators.
3. The two agencies will work jointly on other programs affecting employers.

PART II
The Mine Safety and Health Act

On November 9, 1977, President Jimmy Carter signed into law the U.S. Mine Safety and Health Act of 1977, Public Law 95–164. The Act became effective March 9, 1978.

The U.S. Mine Safety and Health Act of 1977 (subsequently referred to as the Mine Act) is intended to ensure, so far as possible, safe and healthful working conditions for miners. It applies to operators of all types of mines, both coal and metal/nonmetal and both surface and underground. The Mine Act states that mine operators are responsible for preventing unsafe, unhealthful conditions or practices in mines that could endanger the lives and health of miners.

Mine operators are required to comply with the safety and health standards promulgated and enforced by the Mine Safety and Health Administration (MSHA), an agency within the DOL. Like OSHA, MSHA may issue citations and propose penalties for violations. Unlike the employees under OSHAct, miners (employees) are subject to government sanctions for violating safety standards relating to smoking in or near mines and mining machinery. Similarly, employers and other supervisory personnel may be held personally liable for civil penalties or may be prosecuted criminally for violations of Mine Act standards.

LEGISLATIVE HISTORY

Historically, the Bureau of Mines within the Department of the Interior administered the mine safety and health laws. Before Congress passed the Mine Act, mine operators were governed by two separate laws, the Federal Coal Mine Safety and Health Act of 1969 and the Federal Metal and Nonmetallic Mine Safety Act of 1966.

Because the Bureau of Mines was also charged with promoting mine production, critics charged that this responsibility produced an inherent conflict of interest with respect to enforcement of safety and health laws. The establishment of the Mine Enforcement Safety Administration (MESA) in 1973 within the Interior Department failed to answer the criticism. Congress looked for alternative solutions, including the transfer of mine safety and health to the OSHAct. Finally, Congress resolved the issue by adopting the Mine Act, which repealed the Federal Coal Mine Safety and Health Act of 1969 and the Federal Metal and Nonmetallic Mine Safety Act of 1966.

ADMINISTRATION

The administration and enforcement of the U.S. Mine Safety and Health Act are vested primarily with the Secretary of Labor and the Mine Safety Health Review Commission. The agency that administers the investigation and prosecution aspects of the enforcement process is MSHA. The Mine Safety and Health Review Commission, an independent agency created by the Mine Act, reviews contested MSHA enforcement actions.

The Mine Act distinguishes between health research and safety research. Miner health research and standards development is the responsibility of NIOSH, in cooperation with MSHA. The Department of the Interior is responsible for mine safety research in cooperation with MSHA, and the DOL is responsible for mine inspector training.

Mine Safety and Health Administration

The Mine Safety and Health Administration, located within the DOL, administers and enforces the Mine Act. MSHA is headed by an Assistant Secretary of Labor for Mine Safety and Health, who is appointed by the president with the advice and consent of the Senate. The Assistant Secretary acts on behalf of the Secretary of Labor. For the purposes of this chapter, MSHA is also synonymous with the term Secretary or Assistant Secretary of Labor.

MSHA is authorized to adopt procedural rules and regulations to carry out the provisions of the Mine Act. The agency also has the responsibility and authority to perform the following:

- promulgate, revoke, or modify safety and health standards
- conduct mine safety and health inspections
- issue citations and propose penalties for violations
- issue orders for miners to be withdrawn from all or part of the mine
- grant variances
- seek judicial enforcement of its orders.

Aiding the Assistant Secretary in carrying out the provisions of the Mine Act are, among others, (1) an Administrator for Coal Mine Safety and Health and (2) an Administrator for Metal and Nonmetal Mine Safety and Health. Each administrator is responsible for a Division of Safety and a Division of Health.

Mine Safety and Health Review Commission

The five-member Mine Safety and Health Review Commission serves as the administrative adjudication body. The Review Commission is completely independent from the DOL. The commission has the authority to assess all civil penalties provided in the Mine Act. It reviews contested citations, notices of proposed penalties, withdrawal orders, and employee discrimination complaints. Commission members are appointed by the president for six-year terms with the advice and consent of the Senate. The first commissioners took office for staggered terms of two, four, and six years.

The Commission appoints ALJs to conduct hearings on behalf of the Commission. The decision of an ALJ becomes

a final decision of the Commission 40 days after its issuance unless the Commission directs a review.

National Institute for Occupational Safety and Health

The functions carried out under the Mine Act by NIOSH include the following:

- supporting miner health research
- recommending standards to MSHA for adoption
- conducting health hazard evaluations at a mine upon request.

To carry out its responsibilities, NIOSH is given authority to enter workplaces to gather information for research and to conduct health hazard evaluations. NIOSH also has the authority to provide medical examinations for miners at government expense for research purposes, to develop recordkeeping regulations relating to toxic exposure, and to require mine operators to make additional reports from time to time.

NIOSH is also responsible for reviewing toxic materials or harmful physical agents that are used or found in mines and for determining whether such substances are potentially toxic at the concentrations found. Further, NIOSH must review the toxicity of new substances brought to its attention, and submit criteria documents on toxic substances to assist MSHA in setting its standards.

Department of the Interior Duties

The Mine Act assigns responsibility for mine safety research to the Department of the Interior and responsibility for training mine inspectors, operators, and miners to the DOL. Mine safety and health inspectors and technical support personnel of MSHA are trained by the DOL's National Mine Safety and Health Academy located in Beckley, West Virginia. The DOL is also authorized to conduct education and training programs for operators and miners in safety and health matters.

MAJOR PROVISIONS OF THE MINE ACT

Coverage

The Mine Act covers all mines that affect commerce. The Act defines "mines" as all underground or surface areas from which minerals are extracted and all surface facilities used in preparing or processing the minerals. Structures, equipment, and facilities including roads, dams, impoundments, and tailing ponds used in connection with mining and milling activities are also included.

OSHA and MSHA established an interagency agreement which, among other things, delineates certain areas of authority and provides for coordination between OSHA and MSHA in all areas of mutual interest. (This was published in 44 *FR* 22827, April 17, 1979.) In case of jurisdictional disputes between OSHA and MSHA, the Secretary of Labor is authorized to assign enforcement responsibilities to one of the agencies.

Advisory Committees

The Act requires the Secretary of the Interior to appoint an Advisory Committee on Mine Safety Research. The Secretary of Health and Human Services is required to appoint an Advisory Committee on Mine Health Research. The Secretary of Labor or the Secretary of Health and Human Services may appoint other advisory committees as needed to aid in carrying out the provisions of the Act.

Miners' Rights

The Act affords miners (employees) a number of rights, including the following:

- Miners may request an inspection in writing if they believe a violation of a standard or an imminent danger situation exists in the mine. Similarly, written notification of alleged violations or imminent danger situations may be given to an inspector before or during an inspection.
- An authorized representative of miners must be given the opportunity to accompany the inspector during the inspection process. Also, miners have the right to participate in postinspection conferences held by the mine inspector on the premises.
- At least one representative of the miners who accompanies the inspector during the inspection must be paid at the regular rate of pay for time spent with the inspector.
- Miners are entitled to observe monitoring and to examine monitoring records when the standards require monitoring exposure to toxic materials or harmful physical agents.
- Miners, including former miners, must be given access to medical and other records documenting their own exposures.
- Operators must notify miners if they are exposed to toxic substances in concentrations that exceed prescribed limits of exposure. Further, those miners must be informed of the corrective action being taken.
- Miners given new work assignments for medical reasons because of their exposure to hazardous substances must be paid at their regular rate if the related standard so provides.
- Miners who are not working because of a withdrawal order are entitled to be compensated subject to certain limits.
- Miners or their authorized representatives may contest the issuance, modification, or termination of any MSHA order or the time period set for abatement.
- Miners adversely affected or aggrieved by an order of the Review Commission may obtain judicial review.
- Miners may file a complaint with the Review Commission concerning their compensation for not working as the result of a withdrawal order, or for acts of employee discrimination.
- Miners, through their authorized representative, may petition for a variance from mine safety standards.
- MSHA is required to send to the miners' authorized representative copies of proposed safety or health standards. In addition, the mine operator must post a copy of such standards on its office bulletin board.

- To keep miners informed, mine operators are required to post copies of orders, citations, notices, and decisions issued by MSHA or the Review Commission.
- Miners are entitled to receive training for their specific jobs and must be given refresher training annually. They are entitled to normal compensation while being trained. Miners who leave the operator's employ are entitled to copies of their training certificates.
- Operators may not discriminate against miners or representatives of miners.
- Miners suffering from black lung disease are entitled to extensive black lung benefits.

Duties

Mine operators are required to comply with the safety and health standards and other rules promulgated under the Act and are subject to sanctions for failing to comply. Similarly, every miner is required to comply with the safety and health standards promulgated under the Act. However, no sanctions are imposed against miners except for willful violation of safety standards relating to smoking or to carrying smoking materials, matches, or lighters.

Miner Training

Mine operators are required to have a safety and health training program approved by MSHA which provides the following:

- At least 40 hours of instruction for new underground miners. The training must include the statutory rights of miners and their representatives under the Act, use of the self-rescue device and respiratory devices, hazard recognition, escapeways, walk-around training, emergency procedures, basic ventilation, basic roof control, electrical hazards, first aid, and the safety and health aspects of the task assignment.
- Twenty-four hours of instruction for new surface miners. The training must include all of the items for underground miners just listed, except escapeways, basic ventilation, and basic roof control, none of which are essential to surface mining.
- At least eight hours of annual refresher training for all miners.

The Mine Act requires that the training must be conducted during normal working hours and that the miners must be paid at their normal rate during the training period. Regulations concerning training and retraining of miners are codified at Title 30 *CFR*, Part 48. (See Chapter 16, Safety Training.)

MINE SAFETY AND HEALTH STANDARDS

The Mine Act authorizes MSHA to promulgate, modify, or revoke mine safety and health standards. To implement the initial set of standards without delay, the safety and health standards under the Coal Mine Safety and Health Act of 1969 were adopted under the Mine Act. These standards are codified in Title 30 *CFR*, Parts 70, 71, 74, 75, 77, and 90.

Similarly, the Mine Act adopted the mandatory standards

that prevailed under the Metal and Nonmetallic Mine Safety Act of 1966. Later, many of the advisory standards were adopted as mandatory standards under the Mine Act. All of the metal/nonmetal standards are codified in Title 30 *CFR*, Parts 55, 56, and 57. (A list of all of the standards promulgated under authority of the Mine Act is provided in the References, at the end of this chapter.)

If MSHA should determine that a standard is needed, it may propose a standard or seek assistance from an advisory committee. MSHA must publish the proposed standard in the *Federal Register*, and establish a time period of at least 30 days for public comment. MSHA may hold public hearings if objections are raised about a proposed standard. Upon adoption by MSHA, the standard must be published in the *Federal Register*. The new standard becomes effective upon publication or at a date specified.

Judicial Review

Any person adversely affected by any standard issued by MSHA has the right to challenge its validity by petitioning in the United States Court of Appeals within 60 days after promulgation of the standard. Although filing such a petition does not stay enforcement of the standard, the court may order a stay before conducting a hearing on the petition. Objections that were not raised during rule making will not be considered by the court, unless good cause is shown why an objection was not raised.

Input from the Private Sector

Mine safety and health standards promulgated by MSHA can never cover every conceivable hazardous condition that might exist in mines. Nevertheless, new standards and modification or revocation of existing standards are important to mine operators and miners alike. Mining operator organizations, miner organizations, and individuals should express their views during the rule-making process by responding to MSHA's proposed standards. Their participation is important because most expertise and technical competence lies in the private sector. To do less means that mining operators and miners are willing to let the standards-development process be controlled solely by MSHA.

Emergency Temporary Standards

MSHA has the authority to publish emergency temporary standards if it deems that immediate action must be taken to protect miners "exposed to grave danger" from toxic substances or physically harmful agents. The emergency temporary standard is effective immediately upon publication in the *Federal Register* and remains in effect until superseded by a permanent standard developed under normal rule-making procedures. MSHA is required to establish a permanent standard within nine months after publication of an emergency temporary standard.

Variances

Upon petition by an operator or a representative of miners, MSHA may modify the application of any mandatory

safety standard. The Act does not allow for variances of health standards.

A petition may be granted under two conditions. First, MSHA must find that an alternative method of compliance will achieve the same measure of protection for miners as the standard would provide. Second, MSHA must determine that the standard in question will provide less safety to miners.

A variance petition should be filed with the Assistant Secretary of Labor for Mine Safety and Health. If the mining operator submits a petition, a copy must be served on the miners' representative. Similarly, if the miners' representative petitions for a variance, a copy must be served on the mine operator. The petition must include the name and address of the petitioner and the mailing address, identification, and name or number of the affected mine. It must also identify the standard, describe the desired modification, and state the basis for the request.

MSHA will publish a notice of the petition in the *Federal Register*. The notice will summarize information contained in the petition. Interested parties have 30 days to comment. MSHA then will conduct an investigation on the merits of the petition, and the appropriate Administrator will issue a proposed decision. The proposed decision becomes final 30 days after service, unless a hearing request is filed within that time.

Accident, Injury, and Illness Reporting

For the purpose of reporting accidents, injuries, and illnesses under the Mine Act, the term "accident" includes:

- a fatality at a mine
- an injury that may result in death
- entrapment for more than 30 minutes
- an unplanned ignition or explosion of gas or dust
- an unplanned fire not extinguished within 30 minutes of its discovery
- an unplanned ignition or explosion of a blasting agent or an explosive
- roof fall in active work areas where roof bolts are in use or a roof fall that impairs ventilation or impedes passage
- coal or rock outbursts that disrupt mining activities for more than one hour
- conditions requiring emergency action or evacuation
- damage to hoisting equipment in a shaft or slope that endangers an individual or interferes with use of equipment for more than 30 minutes
- an event at the mine that causes a fatality or bodily injury to an individual who is not at the mine at the time of occurrence.

"Occupational injury" means an injury that results in death, loss of consciousness, medical treatment, temporary assignment to other duties, transfer to another job, or inability to perform all duties on any day after the injury. "Occupational illness" is an illness or disease that may have resulted from work at a mine or for which a compensation award is made.

All mine operators are required to immediately report accidents (as defined earlier) to the nearest MSHA district or subdistrict office. Similarly, operators must investigate and submit to MSHA, upon request, an investigation report on accidents and occupational injuries. The investigation report must include:

- date and hour of occurrence
- date the investigation began
- names of the individuals participating in the investigation
- description of the site
- explanation of the accident or injury
- name, occupation, and experience of any miner involved
- if appropriate, a sketch of the accident site, including dimensions
- description of actions taken to prevent a similar occurrence
- identification of the accident report submitted.

All mine operators must submit to MSHA within 10 days of the incident a report of each accident, occupational injury, or illness on Form No. 7000–1. A separate form is to be prepared for each miner affected.

Accident investigation reports and the injury/illness reports filed by means of Form 7000–1 must be kept for five years at the mine office closest to the mine in which the accident, injury, or illness occurred.

Inspection and Investigation Procedures

Inspections of a mine are conducted by MSHA to determine if an imminent danger exists in the mine and if the mine operator is complying with the safety and health standards and with any citations, orders, or decisions issued. Mine inspectors from MSHA or representatives of NIOSH have the right to enter any mine to inspect the site or to conduct an investigation. However, NIOSH representatives have no enforcement authority. The Mine Act's provision for conducting inspections without securing a search warrant has been held valid.

As in OSHA, NIOSH and MSHA inspectors cannot give employers advance warning of an inspection conducted to determine compliance. However, NIOSH may give advance notice of inspections carried out for research or other purposes.

Frequency. MSHA must inspect underground mines in their entirety at least four times a year. Surface mines are to be inspected at least two times a year. Normally, these inspections require multiple visits. MSHA must also conduct spot inspections based on the number of cubic feet of methane or other explosive gases liberated in mining operations during a 24-hour period. The Act authorizes MSHA to develop guidelines for additional inspections based on other criteria.

Miner complaints. A miner's authorized representative, or any individual miner if there is no authorized representative, may request in writing an immediate inspection by MSHA if he or she has reasonable grounds to believe that a violation of a standard or an imminent danger situation exists. MSHA will normally conduct a special inspection soon after receiving the complaint. If MSHA determines that a violation does not exist, it must notify the complainant in writing. Similarly, before or during an inspection, the miners' authorized representative, or an individual miner if no representative exists,

may notify the inspector in writing of any alleged violation or imminent danger situation believed to exist in the mine.

Health hazard evaluations. Upon written request of an operator or authorized representative of miners, NIOSH is authorized to enter a mine to determine whether any toxic substance, physical agent, or equipment found or used in the mine is potentially hazardous. A copy of the evaluation will be submitted to both the operator and the miners' representative.

The inspection procedure. A MSHA inspector will normally begin the inspection at the mine office. The officer will inform the mine operator of the reason for the inspection and request all needed records. The inspector's review of the records will likely focus on the preshift or on-shift examination record. Such records help the inspector determine where to concentrate attention during the inspection.

An operator's representative and a representative authorized by the miners must be given the opportunity to accompany the MSHA inspector during the inspection. Similarly, each must be given the opportunity to participate in the postinspection conference. One miner representative (who is an employee of the operator) must be paid the regular wage for the time spent accompanying the inspector.

Whenever the inspector observes a condition that appears to be a violation of the standards, he or she must issue a citation. If, in the opinion of the mine inspector, an imminent danger condition exists, then the inspector must issue a withdrawal order.

After completing the inspection, the inspector will hold a closing conference with the representatives of the mine operator and the miners to discuss all findings. Occasionally, in the interests of those concerned, a separate closing conference may be held with the mine operator and another with the miners' representative.

Withdrawal Orders

MSHA has the authority, under specified conditions, to order an operator to withdraw the miners from all or part of a mine. Miners idled by such an order are entitled to receive compensation at their regular rate of pay for specified periods of time. All miners in the affected area must be withdrawn except those necessary to eliminate the hazard, public officials whose duty requires their presence in the area, representatives of the miners qualified to make mine examinations, and consultants.

If an imminent danger is found to exist during an inspection, MSHA is required to order the withdrawal of all persons from the affected area, except those referred to in Section 104 of the Act, until the danger no longer exists. The order must describe the conditions or practices both causing and constituting the imminent danger and the area affected. The withdrawal order does not preclude issuance of a citation and proposed penalty.

Other situations for which MSHA may issue a withdrawal order include the following:
- If, during a follow-up inspection, MSHA finds that a mine operator has failed to abate a cited violation and there is

no valid reason to extend the abatement period, MSHA must issue a withdrawal order until the violation is abated.
- If a mine operator fails to abate a respirable dust violation for which a citation has been issued and the abatement period has expired, MSHA must either extend the abatement period or issue a withdrawal order.
- If two violations constituting "unwarrantable failures" to comply with the standards are found during the same inspection, or if the second unwarrantable violation is found within 90 days of the first, a withdrawal order must be issued. An unwarrantable failure violation (second within 90 days) refers to a situation in which the operator knew or should have known that a violation existed and yet failed to take corrective action.
- Miners may be ordered withdrawn from a mine if they have not received the safety training required by the Act. Miners withdrawn for this reason are protected by the Mine Act from discharge or loss of pay.

Except for withdrawal orders issued for respirable dust violations and imminent danger situations, an operator or a miner may file a written request for a temporary stay of the order with the Review Commission. Also, they may request temporary relief from any modification or termination of a withdrawal order. The Review Commission may grant a stay of a withdrawal order provided granting such relief would not endanger the safety and health of miners.

Both operators and miners, or their representatives, may contest an imminent danger withdrawal order or any modification or termination of such an order. They must file with the Review Commission an application for review of the order within 30 days after receiving it or after receiving any modification or termination of such an order.

Citations

If a MSHA inspector or the inspector's supervisors believe that the mine operator is in violation of any standard, rule, order, or regulation promulgated under the Mine Act, they must issue a citation to the operator with "reasonable promptness." A citation may be issued immediately at the site of the alleged violation. In any case, the inspector must provide a citation for each alleged violation before leaving the mine property, unless mitigating circumstances exist.

Citations must be in writing, describe the nature of the violation, and include a reference to the provision of the Mine Act, standard, rule, regulation, or order allegedly violated. The citation, based on the inspector's opinion, will establish a reasonable time for the employer to correct the violation.

The Act requires the operator to post all citations on the mine's bulletin board. Copies are sent to the miners' representative, to the state agency charged with administering mine safety and health laws, and to those designated by the operator as having responsibility for safety and health in the mine.

Within 10 days after an operator receives a citation or an order for an alleged violation, the operator and/or miners' representative has the right to request a safety and health

conference with MSHA management. The conferencing process is designed to allow parties an opportunity to present evidence surrounding an alleged incident that may have influenced the action taken by the inspector. The conferencing officer considers information from these sources and has the authority to modify, vacate, or affirm the citation.

This procedure enables mine operators and/or representatives of miners to resolve some issues prior to the penalty stage.

Penalties

MSHA must assess a civil penalty of not more than $50,000 for each violation of the Act. The agency may assess penalties up to $5,000 per day for each day the operator fails to correct a cited violation. If an operator is convicted of willfully violating a standard, a federal judge may assess a fine of $25,000 and/or one year imprisonment. A penalty of up to $250 per occurrence may be assessed miners who willfully violate a standard that prohibits smoking or carrying smoking materials, matches, or lighters into or near a mine or mining equipment.

After the alleged violation has been corrected and after any safety and health conference has been conducted, MSHA will issue the proposed penalty. Nonsignificant and substantial (commonly termed "Non-S&S") violations that are abated in a timely fashion usually result in a minimal penalty. In determining the amount of the penalty for all violations, MSHA considers six criteria:

1. operator's history of previous violations
2. size of the operator's business
3. evidence of operator negligence
4. impact on the operator's ability to remain in business
5. gravity of the violation
6. demonstrated good faith to achieve rapid compliance after notification of the alleged violation.

Most of the proposed penalty assessments are assigned a range of penalty points based on each of the above criteria. The total points are then converted into a dollar penalty. In addition, serious violations or those involving negligence are usually given a special penalty assessment.

CONTESTED CASES

Operators and miners (or miners' representatives) have 30 calendar days after receiving notice to contest a citation, a withdrawal order, or a proposed penalty. The notice of contest must be sent by registered or certified mail to the Assistant Secretary of Labor for Mine Safety and Health at the MSHA headquarters located at 4015 Wilson Boulevard, Arlington, VA. 22203. The miners' representative must also receive a copy of the notice.

If a mine operator fails to notify MSHA within the 30-day period and no notice is filed by any miner or miners' representative, the citation and/or the proposed penalty is deemed a final order of the Mine Safety and Health Review Commission and is not subject to review by any court or agency. However, it should be understood that the citation

and the penalty have separate 30-day periods within which they may be contested. For instance, if a mine operator fails to contest the citation, he is not without options. When the proposed penalty is received at some later date, the mining operator has another 30 days to contest that penalty. In addition, if the penalty is contested, the citation may be reopened for negotiation at the same time.

A citation may be contested before the operator receives a notice of proposed penalty, even if the alleged violation has been abated. A notice of contest states what is being contested and the relief sought. A copy of the order or citation being contested must accompany the notice of contest.

Upon receiving the notice of contest, MSHA immediately notifies the Review Commission. Then a docket number and an ALJ are assigned to the case. The Review Commission will provide an opportunity for a hearing via the ALJ. The ALJ may hold an informal conference with all parties involved to clarify and settle the issues. If the issues are not resolved, then a formal hearing conducted by the ALJ will take place.

Mine operators, miners or representatives of miners, and applicants for employment may be parties to the Review Commission proceedings. Miners or their representatives may become parties by filing a written notice with the Executive Director of the Review Commission prior to the hearing.

The Review Commission's ALJs are authorized, among other things, to administer oaths, issue subpoenas, receive evidence, take depositions, conduct hearings, hold settlement conferences, and render decisions. The decision will include findings of facts, conclusions of law, and an order. A copy of the decision will be issued to each of the parties involved and to each of the Commissioners. Any person aggrieved by the decision of the ALJ may, within 30 days after an order or decision is issued, file a petition for a discretionary review by the Review Commission.

The Review Commission on its own motion and with the affirmative vote of two members may direct review of an ALJ's decision within 30 days of issuance only under two conditions: (1) when the decision may be contrary to law or to Commission policy or (2) when the petition raises a novel question of policy.

Any person adversely affected by a decision of the Review Commission, including MSHA, may appeal to the United States Court of Appeals within 30 days of issuance. The court may affirm, modify, or set aside the Commission's decision in whole or in part.

SUMMARY

- The purpose of the Occupational Safety and Health Act of 1970 and the Mine Safety and Health Act of 1977 is to ensure safe, healthful working conditions and to preserve the nation's human resources. For the first time, these two acts established nationwide standards and guidelines.
- Administration and enforcement of OSHAct rests primarily with the Secretary of Labor and the Occupational

Safety and Health Review Commission. Research and education activities are vested in the Secretary of Health and Human Services and are carried out by the National Institute for Occupational Safety and Health (NIOSH). The Assistant Secretary of Labor for Occupational Safety and Health assists the Secretary and acts as the chief of the Occupational Safety and Health Administration (OSHA).

- OSHA's primary responsibilities are (1) to promulgate, modify, and revoke safety and health standards; (2) to conduct inspections and investigations and to issue citations, including proposed penalties; (3) to require employers to keep records of safety and health data; (4) to petition the courts to restrain imminent danger situations; and (5) to approve or reject state plans for programs under the Act.

- OSHA also provides educational programs, consulting services, funding for state plans, and statistical records on occupational accidents, injuries, and illnesses. Ten regional offices help OSHA carry out its duties, each headed by a Regional Administrator and Assistant Regional Administrator.

- The Occupational Safety and Health Review Commission (OSHRC) is a quasi-judicial, three-member board that hears cases when OSHA actions are contested by employers or employees. OSHRC has two levels of adjudication: administrative law judges and the three-member Commission. Although the Commission is the final authority on OSHA cases, its decisions can be reviewed by state and federal courts.

- NIOSH is the principal federal agency engaged in research, education, and training. Its primary functions are (1) to develop and establish recommended occupational safety and health standards, (2) to conduct research experiments and demonstrations, and (3) to develop educational programs to provide qualified safety and health personnel. NIOSH representatives are not authorized to enforce OSHA regulations, only to inspect establishments and ensure compliance.

- Responsibility for conducting statistical surveys and establishing methods used to acquire injury and illness data is placed in the Bureau of Labor Statistics.

- With some exceptions, OSHAct applies to every employer in all 50 states and U.S. possessions who has one or more employees and who is engaged in a business affecting commerce. All federal, state, and local government employees and mine operators and their employees are exempted from coverage.

- Under OSHA, the employer has a general and specific duty to provide safe, healthy work environments and comply with all applicable standards. Employees, in turn, must comply with all standards that apply to their situation and conduct on the job. Although the employer is liable for state and federal sanctions for violating standards, employees are subject only to their employer's sanctions.

- Sanctions are in the form of citations for violating standards and civil and criminal penalties if the employer failed to comply with the general and specific duties under the Act.

- Employers are granted certain rights to seek advice and consultation with OSHA staff, participate in inspections, take an active role in developing safety and health standards with the OSHA Standards Advisory Committees, apply for financial assistance to bring an establishment into compliance, and appeal or contest OSHA findings and decisions.

- Employees have the right to request new standards or amend or revoke old ones, participate in OSHA hearings, contest federal standards or decisions, request an inspection of a potential hazard, and be informed of variance-from-standard petitions and take part in variance hearings.

- Employees also have the right to be informed about OSHA regulations and standards, all hazardous and toxic materials in the workplace, any imminent danger in their work area, any monitoring for hazardous and toxic materials, and access to all of their medical and employment records to determine the amount and length of exposure to any substance.

- All standards promulgated by OSHA must be published in the *Federal Register*. The agency has the authority to develop emergency temporary standards and to order the withdrawal of employees from work areas when OSHA officers believe they are in imminent danger of injury, illness, or death. These emergency temporary standards must also be published in the *Federal Register*.

- OSHA requires employers to keep records of all occupational injuries and illnesses. Within 48 hours after an accident occurs, employers must make a report either to OSHA or to their state plan.

- OSHA can grant two types of variances from standards—temporary and permanent. Employers can apply for either of these variances, but they must show just cause for the variance and must also inform employees of the application. Regional administrators have authority to "interpret" standards in a way that can become variances.

- Workplace inspections are carried out by compliance safety and health officers and generally conducted without notifying the employer beforehand. Investigations of imminent dangers receive top priority, then catastrophic and fatal accidents, followed by employee complaints of hazards, programmed high-hazard inspections, and reinspection.

- General inspection procedures include an opening conference, walk-through of the establishment, documenting alleged violations, interviewing workers, and a closing conference. The employer, or employer's representative, and designated employees or employees' representatives usually accompany the inspector on the tour. The officer can request company records for review, but must keep trade secrets confidential. All citations and proposed penalties are issued by the Area Director or the director's representative.

- Employers or employees can request informal post-inspection conferences to discuss problems and solutions.

Follow-up inspections are conducted to ensure that alleged violations have been abated.

- Citations and penalties vary according to whether the violation is an imminent danger, serious violation, other than serious violation, or *de minimis* violation. Willful violation and repeated violation also fall under OSHA jurisdiction and carry their own penalties. Penalties are calculated according to a ranking system.

- All citations except those for *de minimis* violations must be posted near the place where the violation occurred. The employer can object to or contest a citation and petition for a modification of the conditions for correcting the violation. If a citation or penalty is not contested, it takes immediate effect.

- If an employer wishes to contest a case, he or she must notify the Area Office that initiated the action. The employer should also inform employees that the decision is being contested. The Review Commission hears all contested cases, and will examine both sides of the issue. An ALJ will render a decision, which may be reviewed by the three-member Commission or the courts.

- The OSHAct encourages states to assume the fullest responsibility for administering and enforcing their own occupational safety and health laws. The basic criterion for approval of a state plan is that it be "at least as effective" as the federal program. When a plan falls short of this goal, OSHA may withdraw its approval and reassert jurisdiction in the state.

- In the 1980s, two important standards came into effect: (1) final rules for access to exposure and medical records and (2) hazard communication. Employers are responsible for maintaining records on employees' exposure to hazardous and toxic materials and making them available to workers. Employers must also inform employees of any hazardous materials used in the workplace and train them in methods of handling these materials and self-protection.

- Safety and health issues also extend to the environment. The Resource Conservation and Recovery Act and the Comprehensive Environmental Response, Compensation, and Liability Act (Superfund) set up procedures for reporting and controlling hazardous substances that affect surrounding environments. They also mandate the use of protective equipment and affect the design and operation of manufacturing processes and waste disposal.

- The Federal Mine Safety and Health Act of 1977 is intended to ensure safe, healthful working conditions for miners in all types of mines. Administration and enforcement of the Mine Act are vested with the Secretary of Labor and the Federal Mine Safety and Health Review Commission, which reviews contested cases. Standards are promulgated and enforced by the Mine Safety and Health Administration (MSHA). Health research is carried out in conjunction with NIOSH, while safety research and training is handled by the Department of the Interior.

- MSHA is headed by the Assistant Secretary of Labor for Mine Safety and Health. Its primary duties are (1) to develop, revoke, or modify safety and health standards, (2) conduct inspections, (3) issue citations and propose penalties for violations, (4) order miners withdrawn from parts or all of mines, (5) grant standard variances, and (6) seek judicial enforcement of its orders.

- The Mine Act covers all mines that affect commerce and the various structures, equipment, and facilities used in mining and milling operations. Several advisory committees are appointed to help carry out provisions of the Mine Act.

- Miners (employees) have the right to request inspections, accompany inspectors during their plant tour, observe monitoring procedures, gain access to their own medical and employment records, be informed of their exposure to toxic substances, contest any alteration in MSHA orders, and receive training for their jobs.

- Mine operators must comply with safety and health standards and are subject to sanctions if they fail to do so. They must also provide safety and health training for their employees and provide proper protective equipment and clothing.

- MSHA has the authority to propose new standards, emergency temporary standards, or variances, which are published in the *Federal Register*. Employers or employees can contest these standards by petitioning the United States Court of Appeals or apply for a variance of a safety standard.

- Within 10 days of an accident, mine operators must file a report on the incident with the nearest MSHA district or subdistrict office. They must follow up with a report on their investigation of the accident's causes.

- Mine inspections are usually carried out without prior warning to the operator. All underground mines must be completely inspected at least four times a year, while surface mines are inspected at least twice a year. The inspector will conduct an opening conference and a walk-through of the premises, document all alleged violations, and hold a closing conference. Operator and miner representatives usually accompany the inspector and may have separate closing conferences.

- Inspectors or their supervisors can issue citations for violations. Penalties are assessed for each violation and for each day the violation remains uncorrected. Operators, miners, or others affected have 30 days in which to contest the citations either at an informal conference or through formal procedures. The Review Commission hears all cases, which are decided by an ALJ. All decisions may be reviewed by the Commission or the courts.

Directory of Federal Agencies

THE OCCUPATIONAL SAFETY AND HEALTH ADMINISTRATION

National Headquarters:

OSHA
Occupational Safety and Health Administration, U.S. Department of Labor, Department of Labor Building, 200 Constitution Avenue NW, Washington, DC 20210; (202)523–8017.

NIOSH
National Institute for Occupational Safety and Health, U.S. Department of Health and Human Services, 1600 Clifton Road NE, Atlanta, GA 30333; (404)639–3061.

BLS
Bureau of Labor Statistics, U.S. Department of Labor, 200 Constitution Avenue NW, Washington, DC 20210; (202)523–7943.

OSHRC
Occupational Safety and Health Review Commission, 1825 K Street NW, Washington, DC 20006; (202)634–7943.

OSHA Regional Offices

Region I (Connecticut, Maine, Massachusetts, New Hampshire, Rhode Island, Vermont)
1st Floor, 133 Portland Street,
Boston, MA 02114; (617)565–7159.

Region II (New York, New Jersey, Puerto Rico, Virgin Islands, Canal Zone)
(Room 670) 201 Varick Street,
New York, NY 10014; (212)944–3432.

Region III (Delaware, District of Columbia, Maryland, Pennsylvania, Virginia, West Virginia)
Gateway Building, 3535 Market Street,
Philadelphia, PA 19104; (215)596–1201.

Region IV (Alabama, Florida, Georgia, Kentucky, Mississippi, North Carolina, South Carolina, Tennessee)
1375 Peachtree Street NE,
Atlanta, GA 30367; (404)881–3573.

Region V (Illinois, Indiana, Michigan, Minnesota, Ohio, Wisconsin)
J.C. Kluczynski Federal Building,
230 South Dearborn Street,
Chicago, IL 60604; (312)353–2220.

Region VI (Arkansas, Louisiana, New Mexico, Oklahoma, Texas)
555 Griffin Square Building, Griffin & Young Streets,
Dallas, TX 75202; (214)767–4731.

Region VII (Iowa, Kansas, Missouri, Nebraska)
Old Federal Office Building,
911 Walnut Street, Kansas City, MO 64106; (816)374–5861.

Region VIII (Colorado, Montana, North Dakota, South Dakota, Utah, Wyoming)
Federal Building, 1961 Stout Street,
Denver, CO 80294; (303)844–3061.

Region IX (Arizona, California, Hawaii, Nevada, Guam, American Samoa, Trust Territory of the Pacific Islands)
Federal Building, 450 Golden Gate Avenue,
San Francisco, CA 94102; (415)995–5672.

Region X (Alaska, Idaho, Oregon, Washington)
Federal Office Building, 909 First Avenue,
Seattle, WA 98174; (206)442–5930.

THE MINE SAFETY AND HEALTH ADMINISTRATION

Mine Safety and Health Administration, U.S. Department of Labor, Room 601, 4015 Wilson Boulevard, Arlington, VA 22203; (202)235–1452

National Institute for Occupational Safety and Health, U.S. Department of Health and Human Services, 1600 Clifton Boulevard NE, Atlanta, GA 30333; (404)329–3061

National Mine Safety and Health Academy, P.O. Box 1166, Beckley, WV 25801; (304)255–0451

Mine Safety and Health Review Commission, 1730 K Street NW, Washington, DC 20006; (202)653–5625

Compilations of Regulations and Laws

The safety and health professional and industrial hygienist should be familiar with three U.S. government publications:

- The *Federal Register (FR)*
- The *Code of Federal Regulations (CFR)*
- The *United States Code (USC)*.

The first two are published by the Office of the Federal Register, National Archives and Records Service, General Services Administration. All three publications are available from the Superintendent of Documents, U.S. Government Printing Office, Washington, DC 20402. Every safety office should obtain them.

THE *FEDERAL REGISTER*

The *Federal Register*, published daily Monday through Friday, provides a system for making publicly available regulations and legal notices issued by all federal agencies. In general, an agency will issue a regulation as a proposal in *FR*, followed by a comment period, and then will finally promulgate or adopt the regulation in *FR*. Reference to material published in *FR* is usually in the format A *FR* B, whereby A is the volume number, *FR* indicates *Federal Register*, and B is the page number. For example 43 *FR* 58946, indicates volume 43, page 58946.

THE *CODE OF FEDERAL REGULATIONS*

The *Code of Federal Regulations*, published annually in paperback volumes, is a compilation of the general and permanent rules and regulations that have been previously released in *FR*.

The *CFR* is divided into 50 different titles, representing broad subject areas of federal regulations, for example: Title 29—"Labor"; Title 40—"Protection of Environment"; Title 49—"Transportation"; etc. Each title is divided into chapters (usually bearing the name of the issuing agency), and then further divided into parts and subparts covering specific regulatory areas. Reference is usually in the format 40 *CFR* 250.XX, meaning Title 40 *CFR* Part 250 (Hazardous Waste Guidelines and Regulations), or 49 *CFR* 172.XX, (Hazardous Materials Table and Hazardous Materials Communications Regulations). The "XX" refers to the number of the specific regulatory paragraph.

The *Code of Federal Regulations* is kept up to date by the individual issues of the *Federal Register*. These two publications must be used together to determine the latest version of any given rule or regulation.

THE *UNITED STATES CODE*

Whereas the two previously described publications contain rules and regulations authorized by a law, the *United States Code (U.S. Code)* describes the actual law. The *U.S. Code* is the current official compilation by subject of the "public, general and permanent laws of the United States in force. No new law is enacted and no law is repealed. It is *prima facie* the law. It is presumed to be the law. The presumption is rebuttable by production of prior unrepealed Acts of Congress at variance with the *Code*." (Preface to the *United States Code*.) In other words, the *Code* is an arrangement by subject of the federal legislation of a public and permanent nature, from 1789 to date in force today; it does not include repealed and expired acts.

Editorial selection of statutes included. Inclusion of legislation in the *Code* is under the supervision of a committee of the House of Representatives. *Code* sections included in one edition may be omitted from the next or may be changed from one title of the *Code* to another by editorial fiat. To avoid possible confusion, the date or supplement number of the *Code* edition cited should be given.

Editorial notes in the *Code* are printed with it but not technically a part of it. This material, set out in fine print beneath the respective *Code* sections, is meant to help the reader understand those sections.

Text of Code sections. Sections are copied from the original enactment but often with changes of form, not of substance. Introductory words of the act, such as "Provided," or "That," at the beginning of the original statute section may be omitted; and the *Code* title and section numbers are substituted in the body of the text for the official title and sections of the act from which they derived, when these are mentioned in the act. Other changes are

also permitted. Statutory authority is cited in parentheses at the end of each *Code* section or group of sections. Such authority may be Congressional acts or joint resolutions, presidential executive orders, or reorganization plans. The *Code* is well indexed.

Tables of contents to Code titles. Each of the 50 titles into which the *Code* is divided is preceded by a table of contents, consisting of a table of chapters in that title, by number and caption. At the beginning of each chapter, there is a similar expanded table of contents.

REFERENCES

Bureau of National Affairs, Inc., 1231 25th Street NW, Washington, DC 20037. *Occupational Safety and Health Reporter*.

Commerce Clearing House, Inc., 4025 West Peterson Avenue, Chicago, IL 60646. *Employment Safety and Health Guide*.

La Dou J, ed. *Occupational Safety and Health*. Chicago, IL: National Safety Council, 1992.

National Institute for Occupational Safety and Health, 5600 Fisher Lane, Rockville, MD 20857.
  "The Advisor" (newsletter).
  "Occupational Safety and Health Directory."

National Safety Council, 444 North Michigan Avenue, Chicago, IL 60611.
  Fundamentals of Industrial Hygiene
  Safety and Health (magazine)
  "OSHA Up-to-Date" (newsletter).

Price MO and Bitner H. *Effective Legal Research*, 4th ed. Boston, MA: Little, Brown, and Co., 1979.

Rothstein M. *Occupational Safety and Health Law*, 2nd ed. Minneapolis, MN: West Publishing Co., 1983 (with yearly updates).

Superintendent of Documents, U.S. Government Printing Office, Washington, DC 20402.
  Annual List of Toxic Substances.
  Directory of Federal Agencies.
  Federal Register.
  Field Operations Manual.
  Industrial Hygiene Technical Manual.
  Occupational Safety and Health Act of 1970 (P.L. 91–596).
  Occupational Safety and Health Regulations, Title 29 *Code of Federal Regulations* (CFR).

Part 11—Department of Labor, National Environmental Policy Act (NEPA) Compliance Procedures.

Part 1901—Procedures for State Agreements.

Part 1902—State Plans for the Development and Enforcement of State Standards.

Part 1903—Inspections, Citations and Proposed Penalties.

Part 1904—Recording and Reporting Occupational Injuries and Illnesses.

Part 1905—Rules of Practice for Variances, Limitations, Variations, Tolerances and Exemptions.

Part 1906—Administration Witnesses and Documents in Private Litigation.

Part 1907—Accreditation of Testing Laboratories.

Part 1908—On-Site Consultation Agreements.

Part 1910—Occupational Safety and Health Standards.

Part 1911—Rules of Procedure for Promulgating, Modifying, or Revoking Occupational Safety or Health Standards.

Part 1912—Advisory Committees on Standards.

Part 1912a—National Advisory Committee on Occupational Safety and Health.

Part 1913—OSHA Access to Employee Medical Records.

Part 1915—Occupational Safety and Health Standards for Shipyard Employment.

Part 1917—Marine Terminals.

Part 1918—Safety and Health Regulations for Longshoring.

Part 1919—Gear Certification.

Part 1920—Procedure for Variations under the Longshoremen's and Harbor Workers' Compensation Act.

Part 1921—Rules of Practice in Enforcement Proceedings under Section 41 of the Longshoremen's and Harbor Workers' Compensation Act.

Part 1922—Investigational Hearings under Section 41 of the Longshoremen's and Harbor Workers' Compensation Act.

Part 1924—Safety Standards Applicable to Workshops and Rehabilitation Facilities Assisted by Grants.

Part 1925—Safety and Health Standards for Federal Service Contracts.

Part 1926—Safety and Health Regulations for Construction.

Part 1928—Occupational Safety and Health Standards for Agriculture.

Part 1949—Office of Training and Education, Occupational Safety and Health.

Part 1950—Development and Planning Grants for Occupational Safety and Health.

Part 1951—Procedures for 23(g) Grants to State Agencies.

Part 1952—Approved State Plans for Enforcement of State Standards.

Part 1953—Changes to State Plans for the Development and Enforcement of State Standards.

Part 1954—Procedures for the Evaluation and Monitoring of Approved State Plans.

Part 1955—Procedures for Withdrawal of State Plan Approval.

Part 1956—Safety and Health Provisions for Public Employees in Non-approved Plan States.

Part 1960—Safety and Health Provisions for Federal Employees.

Part 1975—Coverage of Employees under the Williams-Steiger Occupational Safety and Health Act of 1970.

Part 1977—Discrimination against Employees Exercising Rights under the Williams-Steiger Occupational Safety and Health Act of 1970.

Part 1990—Identification, Classification and Regulation of Potential Occupational Carcinogens.

Part 2200—Review Commission Rules of Procedure.

Part 2201—Regulations Implementing the Freedom of Information Act.

Part 2202—Standards of Ethics and Conduct of Occupational Safety and Health Review Commission Employees.

Part 2203—Regulations Implementing the Government in the Sunshine Act.

Part 2204—Implementation of the Equal Access to Justice Act.

Part 2205—Enforcement of Nondiscrimination on the Basis of Handicap.

Mine Safety and Health Act of 1977 (PL 95–164).

Mine Safety and Health Regulations and Standards.

29 *CFR* Part 2700—Mine Safety and Health Review Commission, Rules of Procedure.

30 *CFR* Part 11—Certification of Vinyl Chloride Respiratory Protective Devices.

Part 40—Representatives of Miners at Mines.

Part 41—Notification of Legal Identity of Mine Operators.

Part 43—Procedures for Processing Hazardous Condition Complaints.

Part 44—Procedures for Processing Petitions for Modification of Safety Standards.

Part 45—Independent Contractors.

Part 46—State Grants for Advancement of Safety and Health in Coal and Other Mines.

Part 47—National Mine Safety and Health Academy.

Part 48—Training and Retraining of Miners.

Part 49—Mine Rescue Teams.

Part 50—Notification, Investigation, Reports and Records of Accidents, Injuries, Illnesses, Employment, and Coal Production in Mines.

Part 56—Safety and Health Standards: Metal and Nonmetallic Mines.

Part 57—Safety and Health Standards: Metal and Nonmetallic Underground Mines.

Part 70—Mandatory Health Standards: Underground Coal Mines.

Part 71—Mandatory Health Standards: Surface Work Areas of Underground Coal Mines and Surface Coal Mines.

Part 74—Coal Mine Dust Personal Sampler Units.

Part 75—Mandatory Safety Standards: Underground Coal Mines.

Part 77—Mandatory Safety Standards: Surface Coal Mines and Surface Work Areas of Underground Coal Mines.

Part 90—Procedure for Transfer of Miners with Evidence of Pneumoconiosis.

Part 100—Civil Penalties for Violation of the Mine Safety and Health Act of 1977.

42 *CFR* Part 37—Specifications for Medical Examinations of Underground Coal Miners.

Part 85—Requests for Health Hazard Evaluations.

Part 85a—NIOSH Policy on Workplace Investigations.

PART 2
PROGRAM
ORGANIZATION

Hazardous waste cleanup.

3

Hazard Control Program

To be effective, a loss, or hazard, control program must be monitored, planned, directed, and controlled; it cannot simply develop on its own. Management needs to establish program objectives and safety policies and to assign line management responsibility for the hazard control program. Safety personnel and others must be able to perform specific steps to identify and control hazards. These steps are discussed later in the chapter. First, however, those charged with designing and participating in the hazard control program need to understand the nature of hazards, their effects on the work process, and the basic causes of accidents and ways they can be controlled.

ACCIDENTS AND LOSS CONTROL

Definition of Hazards

A workable definition of hazard is any existing or potential condition in the workplace that, by itself or by interacting with other variables, can result in deaths, injuries, property damage, and other losses (Firenze, 1978). This definition carries with it two significant points.

- First, a condition does not have to exist at the moment to be classified as a hazard. When the total hazard situation is being evaluated, potentially hazardous conditions must be considered.
- Second, hazards may result not only from independent failure of workplace components but also from one workplace component acting upon or influencing another. For instance, if gasoline or another highly flammable substance comes in contact with sulfuric acid, the reaction created by the two substances produces both toxic vapors and sufficient heat for combustion.

Hazards are generally grouped into two broad categories: those dealing with safety and injuries and those dealing with health and illnesses. However, hazards that involve property and environmental damage must also be considered.

Effects of Hazards on the Work Process

In a well-balanced operation, workers, equipment, and materials are brought together in the work environment to produce a product or to perform a service. When operations go smoothly and time is used efficiently and effectively, production is at its highest.

When an accident interrupts an operation, it sets in motion a different chain of events and carries its own price tag. An accident increases the time needed to complete the job, reduces the efficiency and effectiveness of the operation, and raises production costs. If the accident results in injury, materials waste, equipment damage, or other property loss, there is a further increase in operational and hidden costs and a decrease in effectiveness. These cost factors are discussed in Chapter 12, Accident Investigation, Analysis, and Costs.

Safety Management and Productivity Improvement

The process of identifying and eliminating or controlling

hazards in the workplace is one way of making the best use of human, financial, technological, and physical resources. Optimizing these resources results in higher productivity. For purposes here, productivity is defined as producing more output with a given level of input resources.

Loss control, like productivity, quality, costs, and personal relations, is a strategic process. To be effective, it must be integrated into the day-to-day activities and management systems of the organization and must become institutionalized—an operating norm and a strategic part of the organization's culture.

There are other similarities between efforts aimed at hazard control and productivity improvement. To achieve both objectives, an organization must intelligently manage its financial and human resources and use the most appropriate technology. It must illustrate innovative, enlightened, and efficient use of its facilities, equipment, materials, and workforce, and have a trained, educated, and skilled workforce.

An accident interrupts the production process. It not only increases the time needed to complete a production task, it may also reduce the efficiency and effectiveness of the overall operation and increase production costs. Sometimes a succession of interruptions, or one long one, will prevent the production schedule or desired product quality from being met. Such conditions make it difficult to attract new business.

Production accomplishment and control. Control of an operation, by definition, means keeping the system on course and preventing problems from occurring. However, it also implies some allowance for variations within the system, provided they remain within controlled limits. Any production system has built-in control limits, both upper and lower. These limits provide direction and also any acceptable leeway for the system's operation.

There are many aspects to control, including control over the quality of products and services, personnel, capital, energy, materials, and the plant environment. Each of these factors interacts with the other factors to produce the desired effect.

Determining accident factors. In order to set realistic goals for its process, the organization should first determine the major factors likely to cause loss of control. It should then identify their location, importance, and potential effects. Control measures can then be instituted to help reduce risk and potential losses. Factors responsible for accident losses may be identified by either inspection or detailed hazard analyses. The control measures may be some type of process innovation or machine safeguarding, personal protective equipment, training, or administrative change. In addition to the control measures, monitoring systems should be used to continuously assess the effectiveness of these hazard-reducing controls.

Controlling Hazards: A Team Effort

Traditionally, most managers have relied solely on their safety and operations people to locate, evaluate, and control hazardous situations. However, the more that is learned about

hazard and loss control, the more evident it becomes that the job is too large for any individual or small group to do alone. Accident and hazard reduction requires a team effort by employees and management.

Here is how several departments and employee teams can work together.

- The engineering departments can design facilities to be free of uncontrolled hazards and provide technical hazard identification and analysis services to other departments. Their designs must comply with federal, state or provincial, and local laws and standards.
- Manufacturing departments can reduce hazards through efforts such as effective tool design, changes in processes, job hazard analysis and control, and coordinating and scheduling production.
- Quality control can test and inspect all materials and finished products. It can conduct studies to determine whether alternate design, materials, and methods of manufacture could improve the quality and safety of the product and the safety of the employees making the product.
- Purchasing departments can ensure that materials and equipment entering the workplace meet established safety and health standards, and that adequate protective devices are an integral part of equipment. They should disseminate information received from suppliers to line management and workers about safety and health hazards associated with workplace substances and materials.
- Maintenance can perform construction and installation work in conformance with good engineering practices and with acceptable safety and health criteria. This department also can provide planned preventive maintenance on electrical systems, machinery, and other equipment to prevent abnormal deterioration, loss of service, or safety and health hazards.
- Industrial relations often administers programs directly related to health and safety.

Input also can come from the joint safety and health committee (discussed later in this chapter) and from quality circles and safety circles (see Chapter 18, Promotions and Campaigns).

Loss Control and Management

To coordinate the organizational and departmental efforts, a program of loss control is necessary as part of the management process (Windsor, 1979). Such a program provides hazard control with management tools such as programs, procedures, audits, and evaluations. Sometimes hazard control program teams neglect the basics in their rush to be competitive and innovative, to deal with complex employee relations issues and government involvement, and to address the technical aspects of the programs. A program of hazard control ensures that safety fundamentals also will be addressed. These basics include sound operating and design procedures, operator training, inspection and test programs, and communicating essential information about hazards and their control.

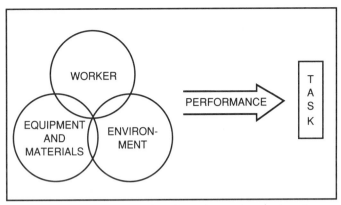

Figure 3–1. A system approach to hazard control recognizes the interaction between worker, equipment, materials and environment in the performance of work.

A loss control program establishes facility-wide safety and health standards and coordinates responsibilities among departments. For example, if one department makes a product and another distributes it, they share responsibility for hazard control. The producer knows the nature of the process, its apparent and suspected hazards, and how to control the hazards. The producer and the distributor are responsible for making sure this information does not end with the production department but is available to the purchaser or the next unit in the manufacturing process.

Coordination is also important when manufacturing responsibility is transferred from one department to another (as when a pilot program becomes a complete manufacturing unit). In addition, when a process is phased out, departments need to coordinate efforts to ensure that personnel who know the hazards are retained throughout the phase-out and that appropriate hazard control activities continue until the end.

Worker-Equipment-Environment System

Those involved in establishing effective loss control programs must understand the interrelationships in the worker-equipment-environment system. (Chapter 5, Ergonomics in the Workplace, in the *Engineering and Technology* volume, examines the system in greater detail.) The present chapter explains the elements of the system (Figure 3–1). As shown in Figure 3–2, an accident can intervene between the system and the task to be accomplished.

Worker. In any worker-equipment-environment system, the worker performs three basic functions: (1) sensing, (2) information processing, and (3) controlling.

- As a *sensor*, the worker serves to monitor or gather information.
- As an *information processor*, the worker uses the information collected to make a decision about the relevance or appropriateness of various courses of action.
- The third function, *control*, flows from the first two. Once information is collected and processed, the worker keeps the situation within acceptable limits or takes the necessary action to bring the system back into an acceptable or safe range.

Evaluating an accident in the light of these three functions can pinpoint the causes. Did the error occur while the worker was gathering information as a sensor? Was the worker able to gather information accurately, for example, in adequate illumination without glare? Did the error occur as a result of faulty information processing and decision making? Did the error occur because an appropriate control option was not available or because the worker took inappropriate action?

In order for the system to move toward its production objectives, the employee must perform work effectively and avoid taking unnecessary risks. To do this, workers must be educated about the following (Firenze, 1978):

1. necessary requirements of the task and the steps needed to accomplish it
2. personal knowledge, skill, and limitations and how they relate to the task
3. what will be gained if the worker attempts the task and succeeds
4. what will result if the worker attempts the task and fails
5. what will be lost if the worker makes no attempt to accomplish the task.

Equipment. Equipment (materials), the second component in the system, must be properly designed, maintained, and used. Hazard control can be affected by the shape, size, and thickness of tools; the weight of equipment; operator comfort; and the strength required to use or operate tools, equipment, and machinery. These variables influence the interaction between worker and equipment. Other equipment variables important in hazard recognition include speed of operation and mechanical hazards (Plog, 1988).

Environment. Special consideration must be given to environmental factors that might detract from the comfort, health, and safety of the worker. Emphasis should be placed on factors such as:

1. layout: the worker should have sufficient room while performing the assigned task
2. maintenance and housekeeping
3. adequate illumination: poorly lit areas increase eyestrain and also the chance of making an accident-causing mistake
4. temperature, humidity, noise, vibration, and control of

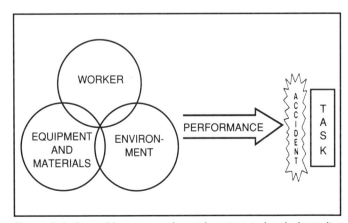

Figure 3–2. An accident causes the work system to break down. It intervenes between the worker, equipment, and environment and the task to be performed.

emission of toxic materials.

Interpersonal relationships are another system factor that plays an important role in operational effectiveness. The task performed by one worker is related to tasks performed by others. Special consideration must be given to coordinating information, materials, and human effort (Hannaford, 1976).

ACCIDENT CAUSES AND THEIR CONTROL

Close examination of each accident shows that it can be attributed, directly or indirectly, to an oversight, omission, or malfunction of the management system regarding one or more of the following three items (Firenze, 1978; refer also to discussion later in this chapter):

1. human factors; either the worker or another person
2. situational work factors, for example, facilities, tools, equipment, and materials
3. environmental factors, such as noise, vibration, temperature extremes, illumination.

If an adequate line management hazard control system is properly designed for the organization's workers, equipment, and environment, then the likelihood of accidents occurring in the workplace is greatly reduced.

Human Factors

The human factor refers to any person who by commission (action) or by omission (failure to act) causes an accident. Both workers and management can cause accidents by commission; for example, a worker may sharpen a wood gouge on a grinder without placing the tool on the grinder's rest. In this case, ask questions such as was the worker trained properly or pressured to rush a job, or were procedures enforced. Conversely, a supervisor contributes to the cause of an accident by omission when failing to have an oil spot on the floor wiped up. An unsafe practice is generally described as either a human action departing from prescribed hazard controls or job procedures or practices, or as an action causing a person unnecessary exposure to a hazard.

An unsafe practice often is a deviation from the standard job procedures. Examples of such actions include:

1. using equipment without authority
2. operating equipment at an unsafe speed or other improper way
3. removing safety devices, such as guards, or rendering them inoperative
4. using defective tools.

Unsafe practices also can be a deviation from safety rules or regulations, instructions, or job safety analyses. Why the deviation occurred is the real issue. Some causes and their countermeasures are given in the following examples. When implementing a hazard control program, emphasis should be placed on the countermeasure. The following eight examples show how to do this.

1. No known standard for safe job procedure exists. Countermeasure: Perform a job safety analysis (JSA) and develop a good procedure through job instruction training (JIT).
2. The employee did not know the standard job procedure. Countermeasure: Train in the correct procedure.
3. The employee knew, but did not follow, the standard job procedure. Countermeasure: Consider an employee performance evaluation. Test the validity of the procedure and motivation.
4. The employee knew and followed the procedure. Countermeasure: Develop a safer job procedure.
5. The procedure encouraged risk-taking, such as incentive pay for piecework. Countermeasure: Change the unsafe job design, procedure, or incentive program.
6. The employee changed the approved job procedure or bypassed safety equipment. Countermeasure: Change the method or safeguards so safety measures cannot be bypassed.
7. The employee did not follow the correct procedure because of work pressure or the supervisor's influence. Countermeasure: Counsel employee and supervisor; consider change in work procedures or job requirements.
8. Individual characteristics, which may involve a disability, made the employee unable or unwilling to follow the correct procedure. Countermeasure: Counsel employee; consider a change in work procedures, workstation design, or job requirements; also consider in-depth training.

Many accidents are the result of someone deviating from the standard job procedures, doing something prohibited, or failing to do something that should be done. In other situations, however, the worker unfairly becomes the target for criticism when other factors actually caused the mishap (Firenze, 1981). The following example illustrates this point. Suppose a newly hired worker, after receiving what was thought to be sufficient instruction on the use of a table saw guard, is required to make a particular cut that cannot be made with the guard in proper position. In this case, the required task causes the worker to remove the guard temporarily so the cut can be made. While removing the guard, the worker's hand slips off the wrench and is cut on the saw blade. Obviously the worker was instrumental in the accident situation, and consequently many people would view the procedure as unsafe. A closer analysis, however, reveals that the primary cause was the failure of training to identify equipment limitations.

In this instance, a failure in the management system contributed to the accident. First, a better guard should have been purchased. Second, the new worker should have been instructed more carefully in the use of the guard, including how to remove it when necessary. Most important, a contingency plan should have provided protection for times when the saw has to be used without adequate safeguarding.

An important first step in loss control is distinguishing between worker error and supervisory error, then addressing what caused the system to break down. Don't just look for a "scape goat"; look for the cause(s). Human error is reduced when

1. supervisors and workers know the correct methods and procedures to accomplish given tasks;
2. workers demonstrate a skill proficiency before using the

particular piece of equipment;

3. higher management and supervisors consider the relationship between worker performance and physical characteristics and fitness;
4. the entire organization gives top priority and continuous regard to potentially dangerous situations and the corrective action necessary to avoid accidents; and
5. supervisors provide proper direction, training, and surveillance. The supervisor must be aware of the worker's skill level with each piece of equipment and process, and adjust the supervision of each worker accordingly. The supervisor shapes worker attitudes and actions by letting employees know that nothing less than safe work practices and the safest possible workplace will be accepted.

Situational Factors

Situational factors are another major cause of accidents. These factors are materials that make accidents likely, and unsafe operations, tools, equipment, and facilities. Examples are unguarded, poorly maintained, and defective equipment; ungrounded equipment that can cause shock; equipment without adequate warning signals; poorly arranged equipment, buildings, and layouts that create congestion hazards; and equipment located in positions that expose more people to a potential hazard.

Some causes of situational problems are:

1. Defects in design—for example, a lightweight, unvented metal container for use with flammable materials or no guard on a power press
2. Poor, substandard construction—for example, a ladder built with defective lumber or with a variation in the space between its rungs
3. Improper storage of hazardous materials—for example, oxygen and acetylene cylinders stored in an unstable manner and ready to topple over with the slightest impact
4. Inadequate planning, layout, and design—for example, a welding station located near combustible materials or placed where many workers without eye protection are exposed to the intense light of the welding arc.

An example of a situational problem occurred in a light industrial manufacturing plant where maintenance workers too often found themselves replacing a bearing on an expensive machine. Something had to be done to save downtime, labor, and the cost of the bearing. The industrial engineering and maintenance departments jointly devised the solution: a system that fed oil to the bearing at set intervals, keeping it well lubricated. It was no longer necessary to replace the bearing so often.

But the solution created new hazards. When oil was fed to the bearing, it dripped onto the floor in the aisle adjacent to the machine. Workers could slip on the oil spot and sustain serious injuries. Forklifts drove over the oil. With oil on the rubber wheels, the driver might not be able to stop the vehicle.

Had the maintenance and industrial engineering organizations been thinking of accident prevention, they could have avoided situational hazards by correcting their design. As an interim step, they might have collected the oil by placing a pan under the motor where the bearing was housed. They would then have time to install a tube that would return the oil to the system, thus saving oil while eliminating the hazard.

Environmental Factors

The third factor in accident causation is environmental, that is, the way in which the workplace directly or indirectly causes or contributes to accident situations. Environmental factors fall into four broad categories: physical, chemical, biological, and ergonomic.

Physical factors. Noise, vibration, radiation, illumination, and temperature extremes are examples of factors having the capacity to influence or cause accidents and illnesses. Operations on a machine lathe, for example, may produce high noise levels that prevent the worker from hearing other sounds and impair communication with others or may damage the workers' hearing over time. Thus, workers may be unable to warn one another of a hazard in time to avoid an accident.

Chemical factors. Classified under this category are toxic gases, vapors, fumes, mists, smokes, and dusts. In addition to causing illnesses, these often impair a worker's skill, reactions, judgment, or concentration. A worker exposed to the narcotic effect of some solvent vapors, for example, may experience a loss of judgment and fail to follow safe procedures.

Biological factors. Biological factors refer to those items capable of making a person ill through contact with bacteria, viruses, fungi, or parasites. For example, workers may suffer boils and inflammations caused by staphylococci and streptococci or experience grain itch caused by parasites.

Ergonomic factors. See Chapter 5, Ergonomics in the Workplace, in the *Engineering and Technology* volume.

Sources of Situational and Environmental Hazards

Actions by purchasing agents; those responsible for tool, equipment, and machinery placement and for providing adequate machine guards; and those responsible for maintaining shop equipment, machinery, and tools may result in situational and environmental hazards.

Employee contributions to situational and environmental hazards include disregarding safety rules and regulations by (1) making safety devices inoperative, (2) using equipment and tools incorrectly, (3) using defective tools rather than obtaining serviceable ones, (4) failing to use engineering controls such as exhaust fans when required, and (5) using toxic substances in unventilated areas or without proper protection.

Purchasing agents can be instrumental in creating situational and environmental hazards if they disregard safety engineering recommendations. These agents may acquire tools, equipment, and machinery without adequate guards and other safety devices, especially if such items are selected with only cost in mind. Sometimes highly toxic and hazardous materials are purchased when less toxic and hazardous

materials could be substituted. Other times purchasing agents fail to acquire from the vendor the necessary warning and control information that must be given to those in charge of the particular process. In many companies, however, the purchasing agent's choices are controlled by engineers, safety professionals, and government or consensus standards and other regulations. The safety and health professional, in conjunction with engineers, should provide the necessary criteria, specifications, etc. to assist the purchasing agents.

Those involved in layout, design, and placement of equipment and machinery also must consider adequate safeguarding and safety devices or equipment. Otherwise, they contribute to hazardous situations in the workplace. Examples are:

1. placing equipment and machinery with reciprocating parts where workers can be crushed between the equipment and substantial objects
2. installing electrical control switches on machinery where the operator will be exposed to the hazards of cutting tools or blades in order to start and stop the equipment
3. installing equipment without providing for adequate lock-out/tagout
4. installing equipment and machinery guards that interfere with work operations
5. locating high hazard workstations where they expose workers unnecessarily. For example, placing a welding station in the middle of a floor area instead of locating it in a corner or along a wall where better control over the welding arc is possible.

Those responsible for maintenance, both management and employees among others, sometimes cause hazards in the workplace. Examples are:

1. improperly identifying high and low pressure steamlines, compressed air, and sanitary lines
2. failing to detect or replace worn or damaged machine and equipment parts, such as abrasive wheels on power grinders
3. failing to adjust and lubricate equipment and machinery on a scheduled basis
4. failing to inspect and replace worn hoisting and lifting equipment
5. failing to replace worn and frayed belts on equipment
6. over-oiling motor bearings, resulting in oil being thrown onto the insulation of electrical wiring and onto the floor, and possible damage to the bearings
7. failing to replace guards
8. failing to lock out and tag unsafe equipment.

More details are included in this chapter under the heading Responsibility for the Hazard Control Program.

Need for a Balanced Approach

Before the concept of loss control was developed, accidents were regarded either as chance occurrences and acts of God—a view still held by some—or as an inherent consequence of production. Such approaches accept accidents as inevitable and, therefore, yield no information about causation and prevention. Control strategies are limited to mitigating the consequences of the occurrence.

In the early days of loss control, accident prevention activities focused on the human element. Findings indicated that a small proportion of workers accounted for a significant percentage of accidents. Control strategies were devised to reduce human error through training, education, motivation, communication, and other forms of behavior modification. During World War II, industrial psychology was aimed at matching employees to particular jobs. Personnel screening and selection were seen as the primary ways to prevent accidents. However, accident-proneness and other behavior models have a glaring weakness: while useful for understanding human behavior, they do not consider the interaction between the worker and the other parts of the system. (See the discussion in Chapters 15 and 16.)

The 1950s and 1960s saw the emphasis change to engineering and control programs aimed at machines and equipment. Further, the implementation of the Occupational Safety and Health Act of 1970 (OSHAct) placed emphasis on preventing accidents through control of the work environment and the elements of the workplace. This act, along with other legislation, specified standards and compliance rules and regulations.

In the 1980s and early 1990s, there is the realization that even with all of the above, accidents still occur. In the 1980s concern with the commitment and culture of the organization were important issues. Clearly emphasis on any one area does not bring about meaningful changes. The balanced, personal, and human-interest approach stands to benefit the organization.

Management Oversight and Omission

Over the past few decades, many organizations seeking to reduce hazards have focused on system defects, which result from management oversight or omission, or on malfunction of the management system. A balanced approach to loss control looks at each component of the system and includes such weaknesses as inadequate training and education, improper assignment of responsibility, unsuitable equipment, or failure to fund hazard control programs. Because managers are responsible for the design, implementation, and maintenance of systems, management errors can result in system defects.

Examining Accident Causation

There are two basic approaches to examining how accidents are caused, after-the-fact and before-the-fact.

After-the-fact. This approach relies on examining accidents after they have occurred to determine the cause and to develop corrective measures. Evaluation of past performance uses information derived from accident and inspection reports and insurance audits. Too often, this approach is used only after a serious accident has resulted in injury or damage, or system ineffectiveness. Furthermore, accident frequency and severity rates do not answer the crucial questions, how, what, why, and when incidents occur.

Before-the-fact. This method relies on inspecting and sys-

LIST OF TYPICAL INCIDENTS

An incident is any observable human activity sufficiently complete in itself to permit references and predictions to be made about the persons performing the act.

1. Adjusting and gaging (calipering) work while the machine is in operation
2. Cleaning a machine or removing a part while the machine is in motion
3. Using an air hose to remove metal chips from table or work (a brush or other tools should be used for this purpose, except on recessed jigs)
4. Using compressed air to blow dust or dirt off clothing or out of hair
5. Using excessive pressure on air hose
6. Operating machine tools (turning machines, knurling and grinding machines, drill presses, milling machines, boring machines) without proper eye protection (including side shields)
7. Not wearing safety glasses in a designated eye-hazard area.
8. Failing to use protective clothing or equipment (face shield, face mask, ear plugs, safety hat, cup goggles)
9. Failing to wear proper gloves or other hand protection when handling rough or sharp-edged material
10. Wearing gloves, ties, rings, long sleeves, or loose clothing around machine tools
11. Wearing gloves while grinding, polishing, or buffing
12. Handling hot objects with unprotected hands
13. No work rest or poorly adjusted work rest on grinder ($1/8$ in. maximum clearance)
14. Grinding without the glass eye shield in place
15. Making safety devices inoperative (removing guards, tampering with adjustment of guard, beating or cheating the guard, failing to report defects)
16. Using an ungrounded or uninsulated portable electric hand tool
17. Improperly designed safety guard, for example, a wide opening on a barrier guard which will allow the fingers to reach the cutting edge

Figure 3–3.

tematically identifying and evaluating the nature of undesired events in a system. One such method is the critical incident technique.

Critical incident technique. This technique can identify the cause of an accident before the loss occurs. To obtain a representative sample of workers exposed to hazards, management selects persons from various departments of the plant. An interviewer questions a number of workers who have performed particular jobs within certain environments. They are asked to describe only those existing hazards and unsafe conditions they are aware of. These are called "incidents." Figure 3–3 lists some incidents that might be typically described. Management then classifies incidents into hazard categories, and identifies problem areas.

The critical incident technique measures safety performance and identifies practices or conditions that need to be corrected.
The investigative team can also analyze the management systems that should have prevented the occurrence of unsafe practices or the existence of unsafe conditions. The technique can lead to improvements in loss control program management.

The procedure needs to be repeated because the worker-equipment-environment system is not static. Repeating the technique with a new sample of workers can reveal new problem areas and measure the effectiveness of the accident prevention program.

Safety sampling. Also called behavior or activity sampling, safety sampling is another technique that uses the expertise of those within the organization to inspect, identify, and evaluate hazards. This method relies on personnel—usually management or safety staff members—who are familiar with operations and well trained in recognizing unsafe practices. While making rounds of the plant or establishment, they record on a safety sampling sheet both the number and type of safety defects they observe. A code number can be used to designate specific unsafe conditions, such as hands in dies, failure to wear eye protection and protective clothing, failure to lock out source of power while working on machinery, crossing over belt conveyors, working under suspended loads, improper use of tools, or transporting unbanded steel.

Safety personnel or managers should make observations at different times of the day, on a planned or random basis in the actual work setting, and throughout the various parts of the plant. In a short time, they can easily convert observations to a simple report showing what specific unsafe conditions exist in which areas and what supervisors and foremen need help in enforcing good work practices. The information is unbiased and therefore irrefutable. What has been recorded is what has been observed.

PRINCIPLES OF LOSS CONTROL

Loss control is the function directed toward recognizing, evaluating, and eliminating, or at least controlling, the destructive effects of occupational hazards. These hazards generally result from human errors and from the situational

and environmental aspects of the workplace (Firenze, 1978). The primary function of a loss control system is to locate, assess, and set effective preventive and corrective measures for those elements detrimental to operational efficiency and effectiveness.

The process exists on three levels:
1. National—laws, regulations, exposure limits, codes, and standards of governmental, industrial, and trade bodies
2. Organizational—management of hazard control program, safety and health committees, task groups, teams, etc.
3. Component—worker-equipment-environment.

Loss control can be thought of as "looking for defects." In the first place, there are fewer defects, or failures, than successes. Second, *it is* easier to agree on what constitutes failure than on what constitutes success. Failure is the inability of a system or a part of a system to perform as required under specified conditions for a specific length of time. The causes of failures often can be determined by answering a series of questions. What can fail? How can it fail? How frequently can it fail? What are the effects of failure? What is the importance of the effects? The manner in which a system, or portion of a system, can exhibit failure is commonly known as the *mode of failure*.

The opposite of failure is not necessarily total success. After all, totally error-free performance is an ideal state, not a reality. Rather, the opposite of failure is the *minimum acceptable success*. This is the condition in which operations are run with a minimum number of losses and interruptions, keeping efficiency and effectiveness of the operation within acceptable limits of control.

Management builds into each of its systems lower and upper limits of control. Each of these interfacing subsystems—maintenance, quality control, production control, personnel, purchasing, to name a few—is designed to move the system within acceptable limits toward its objective. This concept of keeping operations within acceptable limits gives substance and credibility to the process of loss control. In addition to familiarizing management with the full consequences of system defects, loss control can pinpoint hazards before failures occur. The anticipatory character of loss control increases productivity.

PROCESSES OF LOSS CONTROL

An effective hazard control program has six steps or processes (Firenze, 1978):
1. hazard identification and evaluation
2. ranking hazards by risk
3. management decision making
4. establishing preventive and corrective measures
5. monitoring
6. evaluating program effectiveness.

Hazard Identification and Evaluation

The first step in a comprehensive hazard control program is to identify and evaluate workplace hazards. These hazards are associated with machinery, equipment, tools, operations, materials, and the physical plant.

There are many ways to acquire information about workplace hazards. A good place to begin is with those who are familiar with plant operations and the hazards associated with them. (See the Appendix, Sources of Help, for a description of many organizations that can be of help.) The critical incident technique (described earlier) is useful for obtaining information from workers and supervisors. Insurance company loss control representatives know those hazards most likely to cause damage, injuries, and fatalities. In addition to the National Safety Council (NSC), professional societies such as the American Society of Safety Engineers (ASSE), American Industrial Hygiene Association (AIHA), and the American Conference of Governmental Industrial Hygienists (ACGIH) have information about safety and health experience. Manufacturers of industrial equipment, tools, and machinery offer information about the hazards associated with their products, as do suppliers of materials and substances. Labor representatives and business agents can offer a perspective on hazards overlooked by others. Safety and health personnel in organizations doing similar work can be of inestimable value.

A second place to look would be old inspection reports, either internal (by a safety and health committee or company management and specialists) or external (by local, state or provincial, or federal enforcement agencies). The Occupational Safety and Health Administration (OSHA) can supply information describing violations uncovered in similar operations and outlining compliance regulations. (See Chapter 2, Regulatory History and Compliance, and the Appendix, Sources of Help, for descriptions of state agencies and private concerns that give onsite inspection and consultation services under OSHA and the National Institute for Occupational Safety and Health (NIOSH).)

Hazard information also can be obtained from accident reports. Information explaining how a particular injury, illness, or fatality occurred often will reveal hazards requiring control. Close review of accident reports filed in the past three to five years will identify the individuals and specific operations involved, the department or section where the accident occurred, the extent of supervision, and possibly the injured person's deficiencies in knowledge and skill.

OSHA incident rates also are useful. Although they are historical and reflect what has happened, not the current status of safety performance, they provide, from a large sample, data that reflect what actually has occurred in the workplace. Other valuable sources can be found in other chapters of this volume, in the Industrial Safety Data Sheets of the NSC, in the specifications for particular equipment and machines published by the American National Standards Institute (ANSI), Underwriters Laboratories Inc. (UL), American Society for Testing and Materials (ASTM), and the National Fire Protection Association (NFPA). Information about work activities, facilities, and equipment is distributed by the NIOSH.

Hazard analysis is another way to acquire meaningful hazard information and a thorough knowledge of the de-

Figure 3–4. Relative consequences of various hazard categories.

Hazard Consequence Category	Explanation
I.	**Catastrophic**—may cause death or loss of a facility.
II.	**Critical**—may cause severe injury, severe occupational illness, or major property damage.
III.	**Marginal**—may cause minor injury or minor occupational illness resulting in lost workday(s), or minor property damage.
IV.	**Negligible**—probably would not affect personnel safety or health and thus, less than a lost workday, but nevertheless is in violation of specific criteria.

mands of a particular task. Analysis probes operational and management systems to uncover hazards that (1) may have been overlooked in the layout of the plant or the building and in the design of machinery, equipment, and processes; (2) may have developed after production started; or (3) may exist because original procedures and tasks were modified.

The greatest benefit of hazard analysis is that it forces those conducting the analysis to view each operation as part of a system. In doing so, they assess each step in the operation while keeping in mind the relationship between steps and the interaction between workers and equipment, materials, the environment, and other workers. Other benefits of hazard analysis include (1) identifying hazardous conditions and potential accidents; (2) providing information with which effective control measures can be established; (3) determining the level of knowledge and skill as well as the physical requirements workers need to execute specific shop tasks; and (4) discovering and eliminating unsafe procedures, techniques, motions, positions, and actions.

The topic of hazard analysis—its underlying philosophy, the basic steps to be taken, and its ultimate use as a safety, health, and decision-making tool—will be treated in Chapter 11, Acquiring Hazard Information.

Ranking Hazards by Risk (Consequence and Probability)

The second step in the process of loss control is to rank hazards by risk. Such ranking takes into account both the consequence (the severity) and the probability (the frequency). The purpose of this second process is to address hazards according to the principle of "worst first." Ranking provides a consistent guide for corrective action, specifying which hazardous conditions warrant immediate action, which have secondary priority, and which can be addressed in the future. The classification scheme outlined in Figure 3–4 is suggested for rating hazards by consequence.

Once safety personnel or others have ranked hazards according to their potential destructive consequences, the next step is to estimate the probability of the hazard resulting in an accident situation. Quantitative data for ranking hazard probability are desirable, but almost certainly they will not be available for each potential hazard being assessed. Whatever quantitative data exist should be part of the risk-

Figure 3–5. Qualitative probability estimate for use in decision making.

Hazard probability category (qualitative estimate)

A. Likely to occur immediately or within a short period of time when exposed to the hazard
B. Probably will occur in time
C. Possible to occur in time
D. Unlikely to occur

rating formula used to estimate probability. Qualitative data—estimates based on experience—are a necessary supplement to quantitative data. Figure 3–5 shows how probability estimates should be made.

After estimating both consequence and probability, the next and final step is to estimate worker exposure to the hazard. The exposure classification scheme in Figure 3–6a is suggested for rating exposure.

Figure 3–6a. Exposure category

Exposure—the number of persons regularly exposed to the hazard. Here the team must evaluate how many people would ordinarily be exposed to the hazard. The following scheme is used to estimate exposure:
1. Greater than 50 different persons regularly exposed to the hazard
2. From 10 to 49 different persons regularly exposed to the hazard
3. From 5 to 9 different persons regularly exposed to the hazard
4. Less than 5 different persons regularly exposed to the hazard.

Risk Assessment

When the hazards have been ranked according to all three criteria—consequence, probability, exposure—the next step is to assign a single risk number or risk assessment code (RAC). Figure 3–6b illustrates the RAC numbers and their designations.

Figure 3–6b. Risk assessment code (RAC)

RAC No.	Title
1	Critical
2	Serious
3	Moderate
4	Minor
5	Negligible

Management Decision Making

The third step involves providing management with full and accurate information, including all possible alternatives, so managers can make intelligent, informed decisions concerning loss control. Such alternatives will include recommendations for training and education, better methods and procedures, equipment repair or replacement, environmental controls, and—in rare cases where modification is not enough—recommendations for redesign. Information must be presented to management in a way that clearly states the actions required to improve conditions. The person who reports hazard information must do so in a manner that promotes, rather than hinders, action.

RECORD OF OCCUPATIONAL SAFETY AND HEALTH DEFICIENCIES

Location _____ Shipping _____

Pete Varga
Inspector

Deficiency No.	Date Recorded	Description of Hazardous Condition	Specific Location	Identification of Acceptable Standard	Hazard Rating		Corrective Action	Estimated Cost of Correction	Date Deficiency Corrected	Resources Used for Correction
					Conse-quence	Proba-bility				
S - 1	12/11/9-	Ungrounded Tools and Equipment	Throughout Shop	OSHA; Subpart S National Electrical Code, Article 250; 4S	I	A	Provide receptacles with the 3-prong outlet. Test each to make certain it is grounded. Make sure that all tools (other than double-insulated) have a grounding plug.	$5,000	1/9/9-	$4,900

Figure 3–7. One approach to recording and displaying hazard information for decision making. (Reprinted with permission from RJF Associates, Inc.)

After management's decision-makers receive hazard reports, they normally have three alternatives:

1. take no action
2. modify the workplace or its components
3. redesign the workplace or its components.

When management chooses to take no positive steps to correct hazards uncovered in the workplace, it usually is for one of three reasons:

1. It feels that it cannot take the required action. Immediate constraints—be they financial, crucial production schedules, or limitations of personnel—loom larger than the risks involved in taking no action.
2. It is presented with limited alternatives. For example, it may receive only the best and most costly solutions with no less-than-totally-successful alternatives to choose from.
3. It does not agree that a hazard exists. However, the situation can require additional consultation and study to resolve any problem.

When management chooses to modify the system, it does so with the idea its operation is generally acceptable but, with the reported deficiencies corrected, performance will be improved. Examples of modification alternatives are the acquisition of machine guards, personal protective equipment, or ground-fault circuit interrupters to prevent electrical shock; a change in training or education; a change in preventive maintenance; isolating hazardous materials and processes; replacing hazardous materials and processes with nonhazardous or at least less-hazardous ones; and purchasing new tools.

Although redesign is not a popular alternative, it sometimes is necessary. When redesign is selected, management must be aware of certain problems. Redesign usually involves substantial cash outlay and inconvenience. For example, assume that the air quality in a plant is found to be below acceptable standards. The only way to correct this situation is to completely redesign and install the plant's

general ventilation system. The cost and inconvenience can be formidable.

Another problem is the fact that the new designs usually contain hazards of their own. For this reason, whenever redesign is offered as an alternative, those making the recommendation must establish and execute a plan to detect problems in design and the early stages of construction so hazards can be eliminated, reduced, or controlled.

One way to expedite decision making regarding actions for loss control is to present findings clearly so that management understands the nature of the hazards, their location, their importance, the necessary corrective actions, and the estimated cost. Figure 3–7 shows a record of occupational and safety health deficiencies, and illustrates one approach for recording and displaying hazard information for decision making. It indicates the hazard ranking, the specific location and nature of the hazard, and what costs are likely to be incurred. It also clearly states the recommended corrective action. At a glance, it shows if the corrective action has been taken and the final cost.

Establishing Preventive and Corrective Measures

After the safety team or others have identified and evaluated hazards and provided data for informed decisions, the next step involves implementing control measures.

Controls are of three kinds:

- administrative (through personnel, management, monitoring, limiting worker exposure, measuring performance, training and education, housekeeping and maintenance, purchasing)
- engineering (isolation of source, lockout procedures, design, process or procedural changes, monitoring and warning equipment, chemical or material substitution)
- personal protective equipment (body protection, fall protection, etc.). See Chapter 14 in the *Engineering and Technology* volume.

Before control installation takes place, it is essential that those involved in safety and health activities understand how hazards are controlled. Figure 3–8 illustrates the three major areas where hazardous conditions can be either eliminated or controlled.

- The first and perhaps best control alternative is to attack a hazard at its source. One method is to substitute a less harmful agent for the one causing the problem. For example, if a certain solvent is highly toxic and flammable, the first step is to determine whether the hazardous substance can be exchanged for one that is nontoxic and nonflammable, yet still capable of doing the job. If a nonhazardous substance meeting these criteria is not available, then a less toxic, less flammable substance can be substituted and additional safeguards employed.

- The second alternative is to control the hazard along its path. This is done by erecting a barricade between the hazard and the worker. Examples of engineering controls are (1) machine guards, which prevent a worker's hands from making contact with the table saw blade; (2) protective curtains, which prevent eye contact with welding arc flashes; and (3) a local exhaust system, which removes toxic vapors from the breathing zone of the workers.

- The third alternative is to direct control efforts at the receiver, the worker. Removing the worker from exposure to the hazard can be accomplished by (1) employing automated or remote control options (for example, automatic feeding devices on planers, shapers); (2) providing a system of worker rotation or rescheduling some operations to times when there are few workers in the plant; or (3) providing personal protective equipment when all options have been exhausted, and the hazard cannot be corrected through substitution or engineering redesign.

Protective equipment may be selected for use in two instances: when there is no immediate way to control the hazard by more effective means, and when it is employed as a temporary measure while more effective solutions are being installed. There are, however, major shortcomings associated with the use of personal protective equipment:

1. Nothing has been done to eliminate or reduce the hazard.
2. If the protective equipment (such as gloves or an eye shield) fails for any reason, the worker is exposed to the full destructive effects of the hazard.
3. The protective equipment may be cumbersome and interfere with the worker's ability to perform tasks, thus compounding the problem.

Chapter 14, Personal Protective Equipment, in the *Engineering and Technology* volume, discusses these subjects more fully.

Monitoring

The fifth step in the process of hazard control deals with monitoring activities to locate new hazards and assess the effectiveness of existing controls. Monitoring includes inspection, industrial hygiene testing, and medical surveillance. These subjects are covered in Chapter 11, Acquiring Hazard Information.

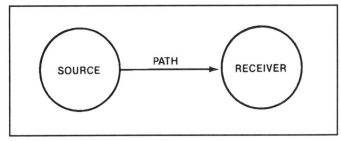

Figure 3–8. Three major areas where hazards can be controlled—the contaminant source, the path it travels, and the employee's work pattern and use of personal protective equipment.

Monitoring is necessary (1) to provide assurance that hazard controls are working properly, (2) to ensure that modifications have not so altered the workplace that current hazard controls can no longer function adequately, and (3) to discover new or previously undetected hazards.

Evaluating Program Effectiveness

The final process in hazard control is to evaluate the effectiveness of the safety and health program. Evaluation involves answering the following questions. What is being done to locate and control hazards in the plant? What benefits are being received, for example, reduction of injuries, workers' compensation cases, and damage losses? What impact are the benefits having on improving operational efficiency and effectiveness?

The evaluation team examines the program to see if it has accomplished its objectives (effectiveness evaluation) and whether they have been achieved in accordance with the program plan (administrative evaluation, including such factors as schedule and budget). Evaluation must be adapted to (1) the time, money, and kinds of equipment and personnel available for the evaluation; (2) the number and quality of data sources; (3) the particular operation; and (4) the needs of the evaluators.

Among the criteria management can use to determine the effectiveness of its safety and health program effort are the number and severity of injuries to workers compared with work hours; the cost of medical care; material damage costs; facility damage costs; equipment and tool damage or replacement costs; and the number of days lost from accidents.

An indicator of the effectiveness of a hazard control program is the experience rating given a company by the insurance carrier responsible for paying workers' compensation. Experience rating is a comparison of the actual losses of an individual (company) risk with the losses that would be expected from a risk of such size and classification. Experience rating determines whether the individual risk is better or worse than the average and to what extent the premium should be modified to reflect this variation. Experience modification is determined in accordance with the Experience Rating Plan (ERP) formula, which has been approved by the insurance commissioners in most states. Loss frequency is penalized more heavily than loss severity because it is assumed that the insured can control the small loss more easily than the less frequent, severe loss.

ORGANIZING AN OCCUPATIONAL SAFETY AND HEALTH PROGRAM

The purposes of a loss control program organization are to assist management in developing and operating a program designed to protect workers, to prevent and control accidents, and to increase effectiveness of operations. Figure 3–9 illustrates the major organizational components of a safety and loss control program.

Establishing Program Objectives

Critical to the design and organization of a safety and health program is the establishment of objectives and policy to guide the program's development. If the organization has a joint safety and health committee, it could be the body to set objectives. It is assumed that those making recommendations to management would be employee representatives, supervisors, middle management, and safety and health professionals (safety directors, managers, supervisors, and administrators; industrial hygiene technicians and professionals; and fire protection engineers).

Among the program objectives should be the following:
1. gaining and maintaining support for the program at all levels of the organization
2. motivating, educating, and training the program team to recognize and correct or report hazards located in the workplace
3. engineering hazard control into the design of machines, tools, and facilities
4. providing a program of inspection and maintenance for machinery, equipment, tools, and facilities
5. incorporating hazard control into training and educational techniques and methods
6. complying with established safety and health standards.

Establishing Organizational Policy

Once the objectives have been formulated, the second step is for management to adopt a formal policy. A written policy statement, signed by the chief executive officer/president of the organization, should be made available to all personnel. It should state the purpose of the hazard control program and require the active participation of all those involved in the program's operation. The policy statement also should reflect:
1. importance that management places on the health and well-being of employees
2. management's commitment to occupational safety and health
3. emphasis the company places on efficient operations, with a minimum of accidents and losses
4. intention to integrate loss control into all operations, including compliance with applicable standards
5. necessity for active leadership, direct participation, and enthusiastic support of the entire organization.

The policy of the American Telephone and Telegraph Company, for example, is probably the shortest statement, but it drives home the point that:

No job is so important and no service is so urgent— that we cannot take time to perform our work safely.

The National Safety Council publishes a 12-page Industrial Data Sheet, *Management Safety Policies*, 12304-0585, that covers this subject more fully.

After management (in cooperation with labor) has established a safety policy, it should be publicized so that each employee becomes familiar with its content, particularly how it applies directly to him or her. Ways to publicize the statement include meetings, letters, pamphlets, and bulletin boards. The policy should also be posted in management offices to serve as a constant reminder of management's commitment and responsibility.

Responsibility for the Hazard Control Program

Responsibility for the safety program can be established at the following levels: board of directors, chief executive officers, managers, and administrators; department heads, supervisors, foremen, and employee representatives; purchasing agents; housekeeping and maintenance personnel; employees; safety personnel; staff medical personnel; and safety and health committees.

Management and administration. Before any safety program gets underway, it must receive full support and commitment from top management and administration. The president, board members, directors, and other management personnel have primary responsibility for the safety program, which involves the continuing obligation to monitor the program's effectiveness. Management provides the motivation to get the program started and to oversee its operations. Management must initiate discussions with personnel during preplanning meetings and periodically review the performance of its safety program. Discussions should cover program progress, specific needs, and a review of company procedures and alternatives for handling emergencies in the event an accident occurs.

Specifically, responsibility at this level consists of setting objectives and policy and supporting safety personnel in their requests for necessary information, facilities, tools, and equipment to conduct an effective safety program and to establish a safe, healthy work environment. Management must realize that it is not fulfilling its organization's potential efficiency and effectiveness until it brings its operations at least into compliance with mandatory and voluntary regulatory safety and health standards.

Management and administration must delegate the necessary tasks at all organizational levels to ensure the safety program's success. Although management cannot delegate to others its responsibility for employee safety and health, it can assign responsibility for certain parts of the safety program. However, the authority to act must be delegated along with responsibility. While authority always starts with those in the highest administrative levels, it eventually must be passed down to middle and line management to achieve the desired results. If safety professionals, safety and health committees, department heads, supervisors, foremen, and employee representatives are to conduct a vigorous and thorough safety program, and if they are to accept and assert the authority delegated to them when circumstances warrant it,

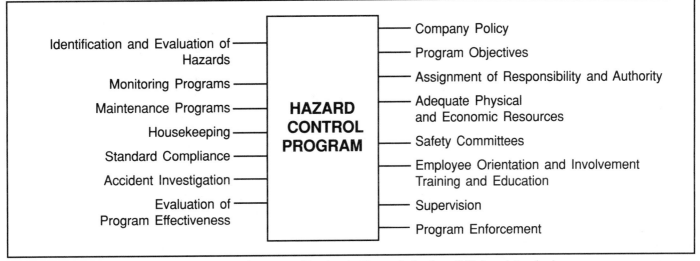

Figure 3–9. Major components of a hazard control program. (Reprinted with permission from RJF Associates, Inc.)

they must be fully confident that they have administrative support.

Management must understand that, although it can assert authority, it may encounter resistance unless it has enlisted employee support from the earliest stages of the program. If supervisors, employees, and their representatives are not aware of the reasons for and the benefits of an effective safety program, they may resist any changes in their methods of operation and instruction and may do as little as possible to assist the overall program effort.

Management must insist that safety and health information is an integral part of training, methods, materials, and operations. It must guarantee a system where loss control is considered an important part of equipment purchase and process design, operation, preventive maintenance, and layout and design. It must make sure that effective fire prevention and protection controls exist. Management also is responsible for informing subcontractors, at the time of negotiating the contract, of all applicable safety and health standards. Management must see that subcontractors comply fully with company and other safety regulations.

Management is required to safeguard employees' health by seeing to it that the work environment is adequately controlled. They must be aware that those operations producing airborne fumes, mists, smoke, vapors, gases, dust, noise, and vibration have the capacity to cause impaired health or discomfort among their workers. Management must be aware that occupational illnesses beginning in the workplace can take their toll later, even after a worker retires. To protect the future of its employees, management must maintain an effective industrial hygiene monitoring system.

Management should also provide meaningful criteria to measure the success of the safety program and to provide information upon which to base future decisions. It must decide the safety program goals for reduced accidents, injuries, and illnesses, and their associated losses.

There are many concrete ways management can show evidence of its commitment to safety: managing safety and health programs, attending safety and health meetings, peri-odic walk-through inspections, reviewing and acting upon accident reports, investigating accidents, reviewing safety records through conferences with department heads and joint employee-management committees, providing awards, and by setting a good example. Management must also audit the program to be certain that it was not "lost" in the management chain.

Department heads, supervisors, foremen, and employee representatives are in strategic positions within the organization to implement safety policies. Their leadership and influence should ensure that safety and health standards are enforced and upheld in each work area and that standards and enforcement are uniform throughout the workplace.

What are some responsibilities of department heads? They make certain that materials, equipment, and machines slated for distribution to their areas are hazard free or that adequate control measures have been provided. They make certain that equipment, tools, and machinery are being used as designed and are properly maintained. They keep abreast of accident and injury trends occurring in their areas and take proper corrective action to reverse these trends. They investigate all accidents occurring within their jurisdiction. They should see to it that all safety and health rules, regulations, and procedures are enforced in their departments. They require that a hazard analysis be conducted for certain operations, particularly those that they regard as dangerous, either from past accident history or from their perception of accident potential. They require that hazard recognition and control information be included in instruction, training, and demonstration sessions for both supervisors and employees. They actively participate in and support the safety and health committee and follow up on its recommendations.

Supervisors, foremen, and employee representatives carry great influence. With their support, top management can be assured of an effective safety and health program. Supervisors have a moral and professional responsibility to safeguard, educate, and train those who have been placed under their direction. Thus, they are generally responsible for creating a safe, healthy work setting and

for integrating hazard recognition and control into all aspects of work activities. By their careful monitoring, they can prevent accidents.

For all practical purposes supervisors, foremen, and employee representatives are the eyes and ears of the workplace control system. On a day-to-day basis, they must be aware of what is happening in their respective areas, who is doing it, how various tasks are being performed, and under what conditions. As they monitor their areas, they must prevent accidents from occurring. If, despite their controls, they see danger, they must be prepared to intervene in the operation and take immediate corrective action. What are the chief safety and health responsibilities of supervisors? They train and educate workers in safe working methods and techniques. They should be certain that employees understand the properties and hazards of the materials they store, handle, or use. They make sure that employees observe necessary precautions, including proper guards and safe work practices. They furnish employees with the proper personal protective equipment, instruct them in its use, and make certain it is worn.

Supervisors demonstrate an active interest in and comply with loss control policy and safety and health regulations. They actively participate in and support the safety and health committees. They supervise and evaluate worker performance, with consideration given to safe behavior and work methods. They should first try to convince employees of the need for safe performance, and then, unpleasant though it may be, they should administer appropriate corrective action when health and safety rules are violated. Correcting employees requires tact and good judgment. Enforcement should be viewed as education rather than discipline. However, if a supervisor feels that a worker is deliberately disobeying rules or endangering his or her life and the lives of others, then prompt and firm action is called for. Laxity in the enforcement of safety rules undercuts the entire safety and health program and allows accidents to happen. Supervisors monitor their area on a daily basis for human, situational, and environmental factors capable of causing accidents. They should make sure that meticulous housekeeping practices are developed and used at all times. They should correct hazards detected in their monitoring or report such hazards to the persons who can take corrective action. They should investigate all accidents occurring within their areas to determine causes.

Foremen and employee representatives share much of the responsibility for safety and health with upper management. Their job is to inspect, detect, and correct. What are the specific responsibilities of foremen and representatives? They should encourage other workers to comply with the organization's safety and health regulations. They detect safety violations and hazardous machinery, tools, equipment, and other implements. They take corrective action when possible and report the hazard to the supervisor, along with the corrective action taken or still required. They may participate in accident investigations. They represent the interests of the workers on the safety and health committee.

Practical training aids for employee representatives, foremen, and supervisors do exist. The Council's *Supervisors' Safety Manual* and accompanying "Supervisors' Development Program" have been widely accepted by the industry. The Council's Joint Safety and Health Committee Training Course is also a good training tool and review for safety committees.

Housekeeping and preventive maintenance. This can be regarded as two sides of the same coin. No safety program can succeed if housekeeping and maintenance are not seen as important to the effort.

Good housekeeping reduces accidents, improves morale, and increases efficiency and effectiveness. Most people appreciate a clean and orderly workplace where they can accomplish their tasks without interference and interruption.

An industrial organization, by its very nature, contains tools that must be kept clean. In operations, workers use flammable substances and materials requiring special storage and removal. The processes generate dust, scrap metal filings and chips, waste liquids, scrap lumber, and countless other by-products that must be disposed of.

Housekeeping is a continuous process involving both workers and custodial personnel. A good housekeeping program incorporates the housekeeping function into all processes, operations, and tasks performed in the workplace. The ultimate goal is for each worker to see housekeeping as part of their job performance and not as an extra task that someone else should do.

When the workplace is clean and orderly and housekeeping becomes a standard part of operations, less time and effort will be spent keeping it clean, making repairs, and replacing equipment, fixtures, and the like. When the worker can concentrate on required tasks without excess scrap material, tools, and equipment interfering with work, operations will be more efficient and the product of higher quality. Time will be used for work, not searching for tools, materials, or parts. When a plant is clean and orderly as well as safe, employee morale is heightened.

When everything has an assigned place, there is less chance that materials and tools will be taken from the plant or misplaced. In a few moments before quitting time, the supervisor can determine what is missing. Different colors of paint can be applied to tools to identify the department to which they belong. Tool racks or holders should be painted a contrasting color as a reminder to workers to return the tools to their proper places. When stored on a rack, the space directly behind each tool should be painted or outlined in color to call attention to a missing tool.

Savings and efficiency are increased when workers treat materials with the care they deserve by minimizing spillage and scrap, saving pieces of material for use in future projects, and returning even small parts to their storage area. When aisle and floor space is uncluttered, movement within the plant is safer, and workers can easily clean and maintain their machinery and equipment. When the plant has adequate work space and when oil, grease, water, and dust are removed from floors and machinery, workers are less likely

---

COMPANY SAFETY RULES

All Employees Will Abide By The Following Rules:

1. Report unsafe conditions to your immediate supervisor.
2. Promptly report all injuries to your immediate supervisor.
3. Wear hard hats on the job site at all times.
4. Use eye and face protection where there is danger from flying objects or particles, such as when grinding, chipping, burning and welding, etc.
5. Dress properly. Wear appropriate work clothes, gloves, and shoes or boots. Loose clothing and jewelry should not be worn.
6. Never operate any machine unless all guards and safety devices are in place and in proper operating condition.
7. Keep all tools in safe working condition. Never use defective tools or equipment. Report any defective tools or equipment to immediate supervisor promptly.
8. Properly care for and be responsible for all personal protective equipment.
9. Be alert and keep out from under overhead loads.
10. Do not operate machinery if you are not authorized to do so.
11. Do not leave materials in aisles, walkways, stairways, roads or other points of egress.
12. Practice good housekeeping at all times.
13. Do not stand or sit on sides of moving equipment.
14. The use of, or being under the influence of, intoxicating beverages or illegal drugs while on the job is prohibited.
15. All posted safety rules must be obeyed and must not be removed except by management's authorization.
16. Comply at all times with all known federal, state and local safety laws as well as employer regulations and policies.
17. Horseplay causes accidents and will not be tolerated.

Violations of any of these rules may be cause for immediate disciplinary action.

Figure 3–10. Reprinted with permission from the Construction Advancement Foundation SAFE Committee.

---

to slip, trip, or fall, or inadvertently come into contact with dangerous parts of machinery.

Chances of fires also are minimized when a workplace is kept free from accumulations of combustible materials that can burn upon ignition or, in the case of certain material relationships, spontaneously ignite. Furthermore, an orderly plant permits easy exit by keeping exits and aisles leading to exits free from obstructions. A neat, well-organized workplace also makes it easier to locate and obtain fire emergency and extinguishing equipment.

Preventive maintenance is the orderly, uniform, continuous, and scheduled action to prevent breakdown and prolong the useful life of equipment and buildings. Preventive maintenance is a shared responsibility. Workers caring for tools and equipment accomplish specific maintenance tasks. Other maintenance duties, such as oiling, tightening guards, adjusting tool rests, and replacing wheels, are routinely performed by workers. Advantages to be gained from preventive maintenance include safer working conditions, decreased downtime of equipment because of breakdown, and increased life of the equipment.

Satisfactory production depends on having buildings, equipment, machinery, portable tools, safety devices, and the like in operating condition and maintained so that production activities will not be interrupted while repairs are being made or equipment replaced. Preventive maintenance prolongs the life of the equipment by ensuring its proper use. When tools are kept dressed or sharpened and in satisfactory condition, the right tool will be used for the job. When safe and properly maintained tools are issued, workers have an

added incentive to give the tools better care. When repairs are made quickly, workers do not need to improvise—for example, using a strip of metal as a screwdriver or crowbar. Sound, efficient maintenance management anticipates machine and equipment deterioration and sets up overhaul procedures designed to correct defects as soon as they develop. Such a repair and overhaul system obviously requires close integration of maintenance with inspection.

Preventive maintenance has four main components:
1. scheduling and performing periodic maintenance functions
2. keeping records of service and repairs
3. repairing and replacing equipment and equipment parts
4. providing spare parts control.

Maintenance schedules can be set up on several different bases. Factors usually considered include:
- manufacturer's recommendations
- age of the machine
- number of hours per day the machine is used
- past experience
- machine changes with use.

The manufacturer's specifications contain standards that need to be maintained for safe and economical use of the machine. These specifications give maintenance personnel definite guidelines to follow. Examples of scheduled activities include lubricating each piece of equipment; replacing belts, pulleys, fans, and other parts; and checking and adjusting brakes.

Management should keep two types of maintenance records: The first is a maintenance service schedule for each piece of equipment. This schedule indicates the date the

equipment was purchased or placed in operation, its cost (if known), where it is used, each part to be serviced, the kind of service required, the frequency of service, and the person assigned to the servicing. Each piece of equipment also requires a repair record, which includes an itemized list of parts replaced or repaired and the name of the person who did the work.

In addition to scheduled adjustments and replacements, maintenance personnel must repair malfunctioning or broken equipment in accordance with the manufacturer's specifications. Sometimes equipment must be sent back to the manufacturer or its service representative for repair. Maintenance personnel should be aware of their limitations and recognize that their experience and expertise are not sufficient to do all repairs. Those assigned repair responsibilities require special safety training. Many of the jobs to be performed include testing or working on equipment with guards and safety devices removed. Therefore, a statement of necessary precautions should accompany the repair directive. Maintenance personnel (along with others) have a responsibility to lock out and tag defective equipment, or simply tag if the equipment, such as a ladder, cannot be locked out. (See Chapter 13, Safeguarding, in the *Engineering and Technology* volume for the proper lockout/tagout procedure.)

The second maintenance record itemizes spare parts stock. Management should conduct routine surveys of spare parts requirements. To keep needed repair parts on hand, management should schedule review and reordering of stocked spare parts. When maintenance personnel keep purchasing agents informed of their anticipated stock needs, they help to reduce or eliminate downtime that occurs while workers wait for parts.

The difference between a mediocre maintenance program and a superior one is that while the former is aimed at maintaining facilities, the latter is designed to improve them. If conditions are good, a mediocre program will keep them that way but will not make them better; if conditions are not good, a mediocre program will not improve them. Preventive maintenance, on the other hand, is a program of mutual support that creates safe conditions, eliminates costly delays and breakdowns, and prolongs equipment life.

Employees. Employees make the safety and health program succeed. Well-trained and educated employees are the greatest deterrent to damage, injuries, and health problems in the plant or establishment.

What are the specific ways a safety program can be rooted in employee involvement and concern? Employees can observe safety and health rules and regulations and work according to standard procedures and practices. They can recognize and report to the foreman or supervisor hazardous conditions or unsafe work practices in the plant. They can develop and practice good habits of hygiene and housekeeping. They can use protective and safety equipment, tools, and machinery properly. They can report all injuries or hazardous exposure as soon as possible. Employees can help develop safe work procedures and make suggestions for improving work procedures. Management should encourage employees to participate on safety committees.

What shapes an employee's attitude toward safety? From the first day an employee goes to work, the employee starts to form attitudes about the job and organization. Substantial if subtle influence is exerted by the attitudes they observe in management, supervisors, and fellow workers. If these individuals regard safety as vital to an effective operation, a mark of skill and good sense, if they all participate actively and cooperatively in the safety program, if employees are recognized for having good safety records, then the employee will regard safety as something important, not as window dressing or an empty policy to which others pay lip service.

Introduction to the safety program should come on the employee's first day on the job. A three-pronged approach is suggested (Kane, 1979):

1. Coverage of general company policy and rules; discussion of various benefit programs, for example, hospitalization, pension plan, holidays, sick leave. The personnel department usually is responsible for giving this information immediately after the new employee is added to the payroll.

2. Discussion of general safety rules and the safety program. This part of the program should be the responsibility of a safety professional. The company's safety handbook should be given to the new employee, and the company's safety policy statement should be explained. The reasons behind the general safety rules should be explored with the employee, who is more likely to follow rules when they are understood.

3. Explanation of specific safety rules that apply to the new employee's department. At this point the supervisor's role overlaps with that of the safety professional. The supervisor can show how some hazards have been eliminated while others, which could not be designed out of the operation, are guarded against. This discussion provides an opportunity to talk about safe work practices and emergency procedures as well as to show how engineering controls and personal protective equipment can further reduce the effects of the hazard. The employee's responsibilities for safety and health also must be stressed. These responsibilities include reporting all accidents and incidents, checking equipment and tools before use, operating equipment only with proper authorization and prior instruction, and asking questions when the procedure is unclear.

Figure 3–10 lists the company safety rules developed by the Construction Advancement Foundation SAFE Committee and distributed to construction employees in Indiana.

Purchasing agents. Those responsible for purchasing items for organizations are in a key position to reduce hazards associated with operations. The purchasing department has much latitude in selecting machinery, tools, equipment, and materials used in the organization. In maintaining standards of quality, efficiency, and price, the purchasing department must make certain that safety has received adequate attention in the design, manufacture, and shipping of items.

Depending upon the company organization, other departments—such as safety, engineering, quality control, maintenance, industrial hygiene, and medical—should indicate to the purchasing department what equipment and materials meet with their approval. The purchasing department is responsible for soliciting such guidance and direction. (Purchasing responsibilities are covered in depth later in this chapter under Purchasing.)

First, the purchasing agent must make certain all items comply with regulatory standards. A statement to this effect must be part of the purchase order. Purchasing agents also will be guided by (1) the standards of ANSI, Canadian Standards Association (CSA), and other standards and specifications groups; (2) products approved or listed by such agencies as UL and the NFPA; and (3) recommendations by such agencies as the NSC, insurance carriers or associations, the Factory Mutual System in its *Factory Mutual Handbook of Industrial Loss Prevention and Loss Prevention Data*, and trade or industrial organizations. (See the Appendix, Sources of Help.)

Second, the purchasing agent must make certain tools, equipment, materials, machinery, and chemicals are purchased with adequate regard for safety. This requirement applies even to ordinary items such as boxes, cleaning rags, paint, and common hand tools. It is essential to compare safety features among the various brands when purchasing personal protective equipment and larger items, especially machines. Sometimes the cost of an adequately guarded machine seems out of proportion to that of an unguarded machine to which makeshift guards can be added. But experience has repeatedly proven that the best time to eliminate or minimize a hazard is in the design stage. Safeguards that are integral parts of a machine are the most efficient and durable.

Third, the purchasing agent must be cost conscious, realizing that every accident has both direct and indirect costs. The agent should understand that the organization cannot afford bargains that later result in accident losses and occupational disease.

Professionals in Loss Control

The roles of various professionals in the loss control program are described in National Safety Council's *Fundamentals of Industrial Hygiene*, 3rd ed. The following discussion, therefore, describes briefly the responsibilities of the safety and health professional, the industrial hygienist, and staff medical personnel.

Safety and health professionals (loss control specialists). To ensure the effectiveness of the safety program, management usually places program administration in the hands of a safety director or manager of safety and health. The number of full-time safety professionals is increasing as the nature of their duties is better understood.

To administer a safety program effectively requires considerable training and many years of experience. A safety program has many facets: occupational health, product safety, machine design, plant layout, security, damage control, fire

prevention. Safety as a profession combines engineering, management, preventive medicine, industrial hygiene, and organizational psychology. It requires knowledge of system safety analysis, job safety analysis, job instruction training, human factors engineering, biomechanics, and product safety. The professional must have thorough knowledge of the organization's equipment, facilities, manufacturing process, and workers' compensation, and must be able to communicate and work with all types of people. Other desirable traits are the ability to see both management and employee viewpoints, and the skills of a good trainer (Figure 3–11).

The passage of the 1970 OSHAct requires that certain safety standards be met and maintained. Generally, organizations with moderate or high hazards and/or employing 500 or more persons need a full-time safety professional. However, the nature of the operation may indicate the need for a full-time professional, regardless of the number of people employed.

The safety professional—whether called a safety engineer, safety director, hazard control specialist, loss control manager, a safety and health professional, or some other title—normally functions as a specialist on the management level. The loss control program should enjoy the same position as other established activities of the organization, such as sales, production, engineering, or research. Its budget reflects top management's commitment to the safety and health of its employees and includes the safety professional's salary, the salary of staff to help him or her, travel allowance, cost of safety equipment, cost of training and continuing education, and other related items. Safety professionals must define the needs of their program and, according to priorities, make short-range and long-range budget projections. With such projections in hand, they can present their needs to those with fiscal responsibility and stand a better chance of acquiring the resources to make their programs work. Figure 3–11 is the ASSE's description of the basic duties of the safety and health professional.

In general, the safety professional advises and guides management, supervisors, foremen, employees, and such departments as purchasing, engineering, and personnel on all matters pertaining to safety. Formulating, administering, monitoring, evaluating, and improving the accident prevention program are other responsibilities of the safety and health professional.

The safety professional usually investigates serious accidents personally or through the staff, reviews supervisors' accident reports, checks corrective actions taken to eliminate accident causes, makes necessary reports to management, and maintains the accident report system, including files. The safety and health professional recommends safety provisions in (1) plans and specifications for new building construction; (2) repair or remodeling of existing structures; (3) safety equipment; (4) designs of new equipment; and (5) processes, operations, and materials. The safety professional provides (or cooperates with the training supervisor to provide) safety training and education for employees. The safety staff also take continuing education courses and attend

THE SCOPE OF THE PROFESSIONAL SAFETY POSITION

A safety and health professional brings together those elements of the various disciplines necessary to identify and evaluate the magnitude of the safety problem. He or she is concerned with all facets of the problem, personal and environmental, transient and permanent, to determine the causes of accidents or the existence of loss producing conditions, practices or materials.

Based upon the information collected and analyzed, he or she proposes alternate solutions, together with recommendations based upon his or her specialized knowledge and experience, to those who have ultimate decision-making responsibilities.

The functions of the position are described as they may be applied in principle to the safety and health professional in any activity.

In performing these functions the safety and health professional will draw upon specialized knowledge in both the physical and social sciences. He or she will apply the principles of measurement and analysis to evaluate safety performance. He or she will be required to have fundamental knowledge of statistics, mathematics, physics, chemistry, as well as the fundamentals of the engineering disciplines.

He or she will utilize knowledge in the fields of behavior, motivation, and communications. Knowledge of management principles as well as the theory of business and government organization will also be required. His or her specialized knowledge must include a thorough understanding of the causative factors contributing to accident occurrence as well as methods and procedures designed to control such events.

The safety and health professional of the future will need a unique and diversified type of education and training if he or she is to meet the challenges of the future. The population explosion, the problems of urban areas, future transportation systems, as well as the increasing complexities of man's every day life will create many problems and extend the safety and health professional's creativity to its maximum if he or she is to successfully provide the knowledge and leadership to conserve life, health, and property.

Functions of the Professional Safety Position

The major functions of the safety and health professional are contained within four basic areas. However, application of all or some of the functions listed below will depend upon the nature and scope of the existing accident problems, and the type of activity with which the safety and health professional is concerned.

The major areas are:
A. Identification and appraisal of accident and loss producing conditions and practices and evaluation of the severity of the accident problem.
B. Development of accident prevention and loss control methods, procedures, and programs.

C. Communication of accident and loss control information to those directly involved.
D. Measurement and evaluation of the effectiveness of the accident and loss control system and the modifications needed to achieve optimum results.

A. Identification and Appraisal of Accident and Loss Producing Conditions and Practices and Evaluation of the Severity of the Accident Problem
These functions involve:
1. The development of methods of identifying hazards and evaluating the loss producing potential of a given system, operation or process by:
   a. Advanced detailed studies of hazards of planned and proposed facilities, operations and products.
   b. Hazard analysis of existing facilities, operations and products.
2. The preparation and interpretation of analyses of the total economic loss resulting from the accident and losses under consideration.
3. The review of the entire system in detail to define likely modes of failure, including human error and their effects on the safety of the system.
   a. The identification of errors involving incomplete decision making, faulty judgment, administrative miscalculation and poor practices.
   b. The designation of potential weaknesses found in existing policies, directives, objectives, or practices.
4. The review of reports of injuries, property damage, occupational diseases or public liability accidents and the compilation, analysis, and interpretation of relevant causative factor information.
   a. The establishment of a classification system that will make it possible to identify significant causative factors and determine needs.
   b. The establishment of a system to ensure the completeness and validity of the reported information.
   c. The conduct of thorough investigation of those accidents where specialized knowledge and skill are required.
5. The provision of advice and counsel concerning compliance with applicable laws, codes, regulations, and standards.
6. The conduct of research studies of technical safety problems.
7. The determination of the need of surveys and appraisals by related specialists such as medical, health physicists, industrial hygienists, fire protection engineers, and psychologists to identify conditions affecting the health and safety of individuals.
8. The systematic study of the various elements of the environment to assure that tasks and exposures of the individual are within his psychological and physiological limitations and capacities.

Figure 3–11. *(Continued on next page.)*

B. Development of Accident Prevention and Loss Control Methods, Procedures, and Programs

In carrying out this function, the safety and health professional:

1. Uses specialized knowledge of accident causation and control to prescribe an integrated accident and loss control system designed to:
   a. Eliminate causative factors associated with the accident problem, preferably before an accident occurs.
   b. Where it is not possible to eliminate the hazard, devise mechanisms to reduce the degree of hazard.
   c. Reduce the severity of the results of an accident by prescribing specialized equipment designed to reduce the severity of an injury should an accident occur.
2. Establishes methods to demonstrate the relationship of safety performance to the primary function of the entire operation or any of its components.
3. Develops policies, codes, safety standards, and procedures that become part of the operational policies of the organization.
4. Incorporates essential safety and health requirements in all purchasing and contracting specifications.
5. As a professional safety consultant for personnel engaged in planning, design, development, and installation of various parts of the system, advises and consults on the necessary modification to ensure consideration of all potential hazards.
6. Coordinates the results of job analysis to assist in proper selection and placement of personnel, whose capabilities and/or limitations are suited to the operation involved.
7. Consults concerning product safety, including the intended and potential uses of the product as well as its material and construction, through the establishment of general requirements for the application of safety principles throughout planning, design, development, fabrication and test of various products, to achieve maximum product safety.
8. Systematically reviews technological developments and equipment to keep up to date on the devices and techniques designed to eliminate or minimize hazards, and determine whether these developments and techniques have any applications to the activities with which the safety and health professional is concerned.

C. Communication of Accident and Loss Control Information to Those Directly Involved

In carrying out this function the safety and health professional:

1. Compiles, analyzes, and interprets accident statistical data and prepares reports designed to communicate this information to the appropriate personnel.
2. Communicates recommended controls, procedures, or programs designed to eliminate or minimize hazard potential, to the appropriate person or persons.
3. Through appropriate communication media, persuades those who have ultimate decision-making responsibilities to adopt and utilize those controls which the preponderance of evidence indicates are best suited to achieve the desired results.
4. Directs or assists in the development of specialized education and training materials and in the conduct of specialized training programs for those who have operational responsibility.
5. Provides advice and counsel on the type and channels of communications to insure the timely and efficient transmission of useable accident prevention information to those concerned.

D. Measurement and Evaluation of the Effectiveness of the Accident and Loss Control System and the Needed Modifications to Achieve Optimum Results

1. Establishes measurement techniques such as cost statistics, work sampling or other appropriate means, for obtaining periodic and systematic evaluation of the effectiveness of the control system.
2. Develops methods that will evaluate the costs of the control system in terms of the effectiveness of each part of the system and its contribution to accident and loss reduction.
3. Provides feedback information concerning the effectiveness of the control measures to those with ultimate responsibility, with the recommended adjustments or changes as indicated by the analyses.

Figure 3–11. *(Concluded.)*

professional meetings like the National Safety Congress and Exposition (Figure 3–12).

Further, the safety professional makes certain that line managers know and understand their responsiblitity to comply with regulatory standards relating to safety and health and that standards, whether mandatory or recommended, are met; accepts responsibility for preparing the necessary reports (for example, OSHA Forms 100, 102, and 200) for management. On jobs involving subcontractors, the safety professional informs line managment of their responsibility to inform each subcontractor of relevant hazard control responsibilities; this is often spelled out in the contract. (See the discussion in the *Engineering and Technology* volume, Chapter 2, Construction of Plant Facilities.) The safety professional also supervises disaster control, fire prevention, and firefighting activities when they are not responsibilities of other departments and stimulates and maintains employee interest in and commitment to safety.

Progressive management will give the safety professional the authority to order immediate changes on fast-moving and rapidly changing operations or in cases where delayed action could endanger lives (as in construction, demolition, or emergency work, fumigation, and some phases of manufacturing explosives, chemicals, or dangerous substances). Where the safety professional exercises this authority, it is done with discretion by accepting accountability to management for errors in judgment. However, the professional realizes that erring on the side of caution is more easily justified than taking unnecessary risks with worker health and safety.

Industrial hygienists. The industrial hygienist is trained to anticipate, recognize, evaluate, and control health hazards—particularly chemical, physical, biological, and ergonomic agents—that exist in the workplace and have injurious effects on workers. Specialists work in fields such as toxicology, epidemiology, chemistry, ergonomics, acoustics, ventilation engineering.

What are the specific responsibilities of the industrial hygienist in the hazard control program? The industrial hygienist (1) recognizes and identifies those chemical, physical, and biological agents that can adversely affect the physical and mental health and well-being of the worker; (2) measures and documents levels of environmental exposure to specific hazardous agents; (3) evaluates the significance of exposures and their relationship to occupationally and environmentally induced diseases; (4) establishes appropriate controls and monitors their effectiveness; and (5) recommends to management how to correct unhealthy (or potentially unhealthy) conditions. The roles of the industrial hygienists are detailed further in the Council's *Fundamentals of Industrial Hygiene.*

(Staff) medical personnel. Occupational health services vary greatly from one organization to another, depending on number of employees, nature of the operation, and commitment of the employer. One location may include a full-time physician, nurses, and technicians, with treatment rooms and a dispensary. Another may have only the required first

aid kit and a person trained to render first aid and cardiopulmonary resuscitation (CPR).

The small operation has special need for persons thoroughly trained in first aid and CPR to take care of the employees on all shifts. Sometimes small facilities in the same locality share the services of a qualified physician on either a part- or full-time basis.

How much responsibility a part- or full-time nurse assumes generally depends on the availability of a licensed physician. It is not unusual for an occupational nurse to be solely and completely responsible for the occupational safety and health program with medical direction from a physician who rarely visits the workplace.

The American Association of Occupational Health Nurses in 1977 defined occupational health nursing as "the application of nursing principles in conserving the health of workers in all occupations. It involves prevention, recognition, and treatment of illness and injury, and requires special skills and knowledge in the fields of health, education and counseling, environmental health, rehabilitation, and human relations." In addition to being responsible for health care, the occupational nurse may on occasion monitor the workplace, perform industrial hygiene sampling, act as a consultant on sanitary standards, and be responsible for health education.

Chapter 4, Occupational Health Programs, deals specifically with the duties of the industrial physician, nurse, and first aid attendants. Regardless of size, health programs share common goals: to maintain the health of the workforce, to prevent or control diseases and accidents, and to prevent or at least reduce disability and resulting lost time.

Staff medical personnel are a vital part of the total health and safety program. Their most obvious contribution is to provide emergency medical care for employees who are injured or become ill on the job. An outgrowth of this responsibility is to provide follow-up treatment of employees suffering from occupational disease or injuries. For many, periodic examinations are required by OSHA regulations. A further outgrowth is medical personnel's promotion of health education programs for employees and their families. Some ways that medical personnel can encourage the health and well-being of employees include assisting with problems of alcohol and drug abuse; keeping immunizations up to date; and promoting mental health, weight control, and regular exercise.

Medical personnel foster a healthful environment by occasional tours of the workplace. Such tours also familiarize medical personnel with the materials and processes used and procedures performed by the employees under their care.

Another way that medical personnel are important to the health and safety organization is in the placement of employees. Proper placement matches each worker to the right job. Often a preplacement examination gives the medical staff an opportunity to recommend where a prospective worker is to be assigned. This placement should take into consideration the physical and mental capacities of the employee so that no one is subjected to unnecessary safety and health risks.

Figure 3–12. Each year, safety and health professionals continue their professional development at specialty sessions offered at the National Safety Council Congress and Exposition.

PURCHASING

The safety department should have excellent liaison not only with the engineering department but also with the purchasing department, as discussed in the previous section. It should be the duty of the safety department to coordinate or develop written safety standards to guide the purchasing department. These standards should help to eliminate the hazards associated with a particular kind of equipment or material, for example, by substituting a safe material for a dangerous one, or safeguarding equipment for the protection of worker, machine, and product.

The purchasing agent is not closely concerned with educational and enforcement activities but is vitally concerned with many phases of the engineering activities. It is the agent's duty to select and purchase the various items of machinery, tools, equipment, and materials used in the organization. It is also the agent's responsibility—in part or to a considerable degree—to see that in the design, manufacture, and shipment of all these items, safety and health have received adequate attention.

In one facility, a lead hazard occurred when workers unloaded litharge (lead oxide) that had been shipped in 10-gallon paint pails with covers. Some of the litharge had leaked, coating the pails with a film of litharge on the outside. When the pails were moved, litharge was released into the air, creating a lead concentration 30 to 40 times the permissible limit, and contaminating the skin and clothing of workers.

Management and safety workers tried several solutions before finding the best approach. They eliminated the hazard by instructing the purchasing department to specify a rubber gasket under the pail lid as a part of the purchasing requirements. Thus the leakage, which created a serious health hazard, was easily controlled. The company also required that the shipper label the containers with information meeting the requirements of the Hazard Communication Standard.

Specifications

The engineering department, with the help of the safety department, should specify the necessary safeguarding to be built into a machine before it is purchased. Persons responsible for purchasing in an industrial plant are necessarily cost conscious. Consequently, the safety professional must become aware of accident costs associated with specific machines, materials, and processes. For instance, if the individual recommends spending several thousand dollars for a superior grade of tool, he or she must have evidence to justify the investment.

Because of highly competitive marketing, manufacturers

of machine tools and processing equipment often list safety devices as accessories. It is important that the safety professional be familiar with regulatory-required auxiliary equipment and be able to justify its inclusion in the original order.

In some organizations, the safety professional is charged with checking all plans and specifications for machinery and other equipment. In many organizations, particularly where certain items, such as goggles or safety shoes, are to be reordered from time to time, various operating officials cooperate to prepare standard lists, and purchases are selected only from among the types and companies shown on these approved lists. In still other establishments, the responsibility for design, quality, safety, and other features rests with the employees who are to use the articles. In such cases, the purchasing agent is responsible only for price, date of delivery, and similar details.

In many companies where purchases are made in huge quantities and at a great investment of money, important duties are placed in three coordinate departments: (1) the engineering department, whose staff prepare plans and specifications for all machinery and equipment to be purchased; (2) the safety department, where the staff carefully check these plans and specifications for safety and carry out final inspections of articles purchased; and (3) the purchasing department, which still has latitude in making selections as well as in determining standards of quality, efficiency, and price.

Still another variable must be mentioned. Many companies have both a full-time purchasing agent and a full-time safety professional. However, in many other companies, especially smaller ones, these important duties are assumed by executives who devote part of their time to other activities. Nevertheless, the measures that should be taken to prevent accidents in the small plant are substantially the same as those taken in the large plant. The part the purchasing agent can play in the safety program is similar; the interest will be the same and the activities will vary only by degree. Success will lie in adopting as fully as possible all the suggestions that are presented here and in cooperating closely with others in the company to promote safety for all workers.

Specification of shipping methods. When the purchasing agent orders materials, it may be desirable to specify they be shipped in a particular manner. If safe and efficient shipping methods are worked out and then specified in the orders, the suppliers will be better able to deliver materials on time, in good condition, and in a shape or form that can be easily and safely handled by employees. The agent should specify that all hazardous materials must be labeled with Department of Transportation (DOT)-authorized shipping labels. Material Safety Data Sheets should accompany all chemicals or products that contain chemicals, as well as some substances like solid metal that are designed to be remelted or equipment like welding rods.

Codes and Standards

In purchasing, the safety professional will a need to have a thorough knowledge of the facility's accident history, the costs involved in accidents, and the probable benefits of changes suggested. To fulfill this function in cooperation with the purchasing department, it is important that the safety professional be familiar with codes and standards. When a specific item of equipment is recommended, the safety professional should be able to state that it is a type approved by authoritative bodies and that it meets regulatory requirements.

Generally, the safety professional should consult with everyone concerned before setting up company standards to guide the purchasing department. There are many guidelines and standards that can be used as models. Accordingly, the safety professional (and all others concerned with setting company standards) should be familiar with the following:

1. codes and standards approved by the ANSI and other standards and specifications groups—see the Appendix, Sources of Help
2. codes and standards adopted or set by federal, state, and local governmental agencies, such as OSHA, the Bureau of Mines, and the National Bureau of Standards
3. codes, standards, and lists of approved or tested devices published by agencies such as the NIOSH, the Mining Safety and Health Administration, UL, and fire protection organizations. For fire protection, the standards and codes of the NFPA should be followed (see Sources of Help).
4. safe practice recommendations of such agencies as the NSC, insurance carriers or their associations, and trade and industrial organizations.

Purchasing-Safety Liaison

With a background of knowledge gathered from materials listed above, the safety professional should be well prepared to advise the purchasing department.

What purchasing can expect. The purchasing agent can reasonably expect that the safety professional will:

1. give specific information about process and machine hazards that can be eliminated by change in design or by installing manufacturer-designed guarding
2. supply similar information about other equipment, tools, and materials along with facts about injuries caused
3. give specific information about health and fire hazards in the workplace
4. provide information on federal and state safety requirements
5. supply on request additional special information on accident experience with machines, equipment, or materials when such articles are about to be reordered
6. request assistance in the investigation of accidents that may have been caused by faulty equipment or material
7. request a list of equipment and materials requiring safety approval before going out for bids.

What the safety department can expect. Where there is effective liaison between the safety and purchasing departments, the safety professional can expect that the purchas-

ing agent will:

1. become familiar with the departmental and plant process hazards, especially in relation to machinery, equipment, and materials
2. ask the safety department for information on hazards and accident costs, for federal and state safety requirements, for lists of approved devices and appliances before making purchases
3. become acquainted with the specific location and departmental use of machinery or equipment about to be ordered
4. participate in accident investigations where injuries may have been caused through the failure of machinery, equipment, or materials.

Safety Considerations

The purchasing staff should always review applicable safety specifications and guidelines before purchasing supplies and equipment. For many articles, there is no need to consider safety. Some items, however, have a more important bearing upon safety than may be suspected.

The agent must exercise utmost caution when purchasing personal protective equipment, such as eye protection, respirators, and masks; equipment to move suspended loads, such as ropes and chains; equipment to move and store materials; and miscellaneous substances and fluids for cleaning and other purposes that might constitute or aggravate a fire or health hazard. The agent should specify adequate labeling that identifies contents and calls attention to hazards.

Investigation, however, may show that unsuspected hazards arise in the purchase of ordinary items, such as common hand tools, reflectors, tool racks, cleaning rags, paint for shop walls and machinery, and even filing cabinets. Among the factors purchasing agents need to consider include maximum load strength; long life without deterioration; reduction of sharp, rough, or pointed characteristics; less frequent need for adjustment; ease of maintenance; reduction of fatigue-causing characteristics; and minimal hazard to workers' health.

Here are a few examples of hazards created by purchased items that were considered safe. Goggles supplied to one group of workers were found to have imperfections in the lenses that caused eyestrain and headache, which led to fatigue and accidents. The toes of a laborer were crushed because his safety shoe had an inadequate metal cap and collapsed under a weight that would have been easily supported by a shoe meeting the ANSI Z41 standard. In another plant, workers were supplied with wooden carrying boxes, when a proper type of metal box could have eliminated the hazard of splinters and perhaps an infected hand. It is in the purchase of larger items, especially machines, that the more impressive examples of purchasing for safety are found. Today, machines of many types are manufactured and can be bought with adequate safeguards in place as integral parts of the machines. The enclosed motor drive is an outstanding example of engineering machine construc-

tion for safety.

When an order for equipment is about to be placed, the purchasing agent, if possible, should not consider any machine that has been only partly guarded by the manufacturer and needs to be fitted with makeshift safeguards. The agent should be in frequent consultation with the safety department before making any purchase where safety is a factor. The agent also should be particularly careful to see that every purchased machine complies fully with the safety regulations of the state in which it is to be operated.

Price Considerations

When considering plant purchases, the purchasing agent must struggle to reconcile quality, work efficiency, and safety with the price of an item. At times it may seem that the cost of a well-safeguarded machine is out of proportion to the cost of an unguarded machine, including the estimated expense of adding home-made safeguards. But the experience of many industrial organizations has proved again and again that the best time to safeguard a machine or process is in the design stage. Safeguards planned and built as integral parts of a machine are the most efficient and durable.

The purchasing agent, through accident information supplied by the safety department, should be familiar with the costs of specific accidents in the plant, especially those in which a purchased item has been found to contribute to an accident. As a result, the purchasing agent should be able to defend the decision to buy a slightly higher priced component where appropriate. Cooperation between the safety, engineering, manufacturing, and purchasing departments is absolutely necessary if accidents are to be eliminated.

These arguments should appeal to all executives responsible for the success of the industrial organization. The executive who is already sold on safety is agreeable to expenditures reasonably justified in the interest of accident prevention. If an executive does not have this attitude, the purchasing agent should find ways to stimulate the executive's interest in the organizational safety program. In this undertaking, the purchasing agent can undoubtedly count upon the active cooperation of the safety professional.

In some instances, the purchase of machinery or equipment involves important engineering details. For such purchases, the company undoubtedly will have a system whereby engineers will first prepare definite specifications, perhaps including drawings. The safety professional then carefully checks these plans and specifications before the purchasing agent solicits bids and cost estimates. The purchasing agent will have the plans and specifications at hand when asking for prices. He or she will want to keep in close touch with the safety professional throughout the negotiations to use the latter's knowledge and experience in accident prevention.

After the purchase order has been made but before it is signed, there is one other important detail that should not be overlooked. This is a statement, in language that cannot possibly be misinterpreted, that the articles ordered must comply fully with the applicable federal and state safety laws

and regulations of the locality in which they are to be used. This statement must be made a part of the purchase order.

SAFETY AND HEALTH COMMITTEES

Safety and health committees can be invaluable to the loss control program by providing the active participation and cooperation of many key people in the organization. They also can be unproductive and ineffective. The difference between success and failure lies with the original purpose of the committee, its staffing and structure, and the support it receives while carrying out its responsibilities.

A safety and health committee is a group that aids and advises both management and employees on matters of safety and health pertaining to plant or company operations. In addition, it performs essential monitoring, educational, investigative, and evaluative tasks.

Committees may represent various constituencies or levels within the organization or may be management or workplace committees. The joint safety and health committee (discussed here) is responsible for:

- actively participating in safety and health instruction programs and evaluating the effectiveness of these programs
- regularly inspecting the facility to detect unsafe conditions and practices and hazardous materials and environmental factors
- planning improvements to existing safety and health rules, procedures, and regulations
- recommending suitable hazard elimination, reduction, or control measures
- periodically reviewing and updating existing work practices and hazard controls
- assessing the implications of changes in work tasks, operations, and processes
- field-testing personal protective equipment and making recommendations for its use or alteration based on the findings
- monitoring and evaluating the effectiveness of safety and health recommendations and improvements
- compiling and distributing safety and health and hazard communications to the employees
- immediately investigating any workplace accident
- studying and analyzing accident and injury data.

The OSHAct in Section 2(b)(13) clearly contemplates the possibility of joint safety and health initiatives as a supplementary approach to accomplishing OSHA's objectives more effectively. Joint committees have considerable potential for reducing injuries and illnesses, thus leaving OSHA free to target enforcement according to the worst-first principle.

The joint committee concept stresses cooperation and a commitment to safety as a shared responsibility between management and workers. Employees can become actively involved in and make positive contributions to the company's safety and health program. Their ideas can be translated into actions. The committee serves as a forum for discussing changes in regulations, programs, or processes, and potential new hazards. Employees can communicate problems to management openly and face to face, allowing information and suggestions to flow both ways. The knowledge and experience of many persons combine to accomplish the objectives of creating a safe workplace and reducing accidents. The approach can produce effective solutions to safety problems more easily. Because joint committees facilitate communication and cooperation, they usually raise employee morale as well.

Even though a joint committee represents both employees and management, the committee's analyses and recommendations—whether they pertain to policy or practice—should be reviewed and confirmed by experts when they relate to specialized areas (for example, electrical safety, exposure levels).

Labor-management cooperation was discussed in Chapter 1, Historical Perspectives. Committee organization and operation are covered in the Council publication "You Are the Safety and Health Committee."

OFF-THE-JOB SAFETY PROGRAMS

There is a certain amount of confusion as to what off-the-job (OTJ) safety really includes. Essentially, off-the-job safety is a term used by employers to designate the part of their safety program directed to employees when they are not at work.

The principal aim of off-the-job safety is to get an employee to follow the same safe practices used on the job while pursuing outside activities. Experience indicates, however, that many individuals tend to leave their safety training at the workplace when they go home. Therefore, off-the-job safety should not be a separate program but rather an extension of a company's on-the-job safety program. While companies have a legal responsibility to prevent injuries on the job, they have a moral responsibility to try to prevent injuries away from the job. The other reason for an off-the-job safety program is cost. Operating costs and production schedules are affected as much when employees are injured away from work as when they are injured on the job. (These costs are discussed in detail in Chapter 12, Accident Investigation, Analysis, and Costs.)

What Is Off-the-Job Safety?

Off-the-job safety is a logical extension of the occupational safety program. Accident/illness prevention at work makes good business sense (it saves dollars), while fulfilling an organization's moral and legal responsibilities. An effective off-the-job safety program meets these same needs: reduction of costly employee absences due to accidents, injuries, or deaths; and commitment to employee well-being. Preventing off-the-job incidents that could result in injury or illness can be accomplished by using methods proven successful for increasing safety awareness at work.

Promoting Off-the-Job Safety

Techniques for promoting off-the-job safety are essentially the same principles and techniques used on the job. From a

safety standpoint, operating power tools at home involves the same risks as operating the equipment at work; likewise, driving the family car is the same as driving a company vehicle.

The only difference between these two safety programs is that organizations do have to depend more on education and persuasion to get their message across. This is because once employees leave the office, plant, or job site, they tend to believe the risks they assume are a private matter and no longer fall under the organization's policies and guidelines.

As with any other program—whether it be attendance, quality control, or waste reduction—management support and guidance is essential. Once safety personnel show management the seriousness of the problem (through experience and cost records), there should be little difficulty in obtaining support.

Program Benefits

Many benefits are derived by developing or revitalizing an off-the-job safety program. These include:
- fewer off-the-job accidents, injuries, and deaths
- fewer employee absences
- reduced operating costs
- safety awareness at home carries over to the workplace
- improved work efficiency and performance
- enhanced employee and employer relationships
- participation of employees and their families in community safety actions
- positive, viable demonstration of organization's commitment to employee-family well-being and social issues affecting the employee's family. As with any effective occupational safety program, management involvement and participation at all levels in off-the-job safety must be vocal, visible, and continuous.

Off-the-Job Safety Policy

The organization should communicate management commitment to employees and their families. For example, a written policy statement concerning off-the-job safety or reference to off-the-job safety should be in the occupational safety policy statement signed by the top organizational official.

Getting Started

Various methods and sources of information on off-the-job accidents are available to an organization. For example, management can keep records documenting the causes of employee absences due to accidents away from the workplace; health and accident insurance claims may record the accident cause on the form for payment; the NSC's annual publication *Accident Facts* can be used to pinpoint the leading causes of accidental deaths and injuries nationally (these data help to determine the most common accidents occurring in a particular organization).

It is important to tailor an off-the-job safety program to the special needs of an organization. The location of the plant site, its environment, and the special interests of employee groups such as skiing, boating, mountain-climbing, cave exploring, hunting, camping, or flying are factors that can help in selection of topics and activities.

Select Program Details

A good topic breakdown provides a solid structural framework for the development of off-the-job safety programs. For example, content may be based on seasonal hazards; on home, traffic, and public accidents; on health risk assessment, such as exercise and fitness, stress management, alcohol and drugs, community right-to-know; or according to accident types and causes.

Timely, interesting, and practical topics will attract the attention of employees, create and maintain enthusiasm, and encourage active participation to develop patterns of safe behavior.

A seasonal emphasis outline might include:
- spring—good housekeeping, lawn mowers, garden tools, do-it-yourself activities, pruning/planting trees, plowing, bicycles
- summer—sunburn, swimming, camping, boating, hiking, field sports, fishing, poison ivy, insects, vacation hazards
- fall—hunting, home power tools, back-to-school hazards, home-heating equipment, yard cleanup, repair and storage of tools
- winter—winter sports, Christmas holiday safety, severe weather exposure, overexertion, winter driving.

Topics can tie into programs of national scope or interest. These national programs provide radio, TV, newspaper, and other forms of publicity that help promote program content. National Child Passenger Safety Awareness Week in February, National Safe Boating Week in June, National Fire Prevention Week in October, and National Drunk and Drugged Driver Awareness Week in December are examples.

A program on home fire safety might include showing a film on the proper use of fire extinguishers or on general fire prevention or safety practices as well as handout literature on fire safety topics. Combined with this program could be a company discount for employee purchase of smoke alarms, fire extinguishers, and fire escape ladders for home, workshop, and auto use.

Employee, Family, and Community Involvement

The assistance of a special off-the-job safety group may be valuable. Membership in this group can include employees, family members of employees, local civic and school groups, or community people with an interest or role in safety in general. Organizations that use such a committee often find that its members contribute immeasurably to the success of the program by providing special information that represents their background and understanding of off-the-job safety. Another benefit of the group is that members share serious safety convictions with peers, friends, neighbors, and their families.

Every community has special interest groups and organizations already concerned with various phases of off-the-job safety that will lend their resources and personnel to assist

in company activities. Among such groups are the safety council; Chamber of Commerce including the Jaycees; service clubs; Red Cross chapters; local newspapers; radio and TV stations; health, police, and fire departments; parent-teacher associations; rescue squads; emergency service groups.

The local women's clubs can be an especially strong source of support for off-the-job safety. The clubs may adopt home, traffic, or recreation safety as a project for the year. In so doing, they ensure the participation of many homemakers and family members who are not exposed to a formal safety program.

A representative from safety and health disciplines, such as the ASSE, can present an off-the-job safety program to an organization's members or employees. He or she can obtain assistance from state safety organizations; medical, visiting nurse, and other associations; poison control centers; state police; insurance companies; and public health groups.

A few of these groups maintain accident records, some participate in special programs, some publish bulletins on health and safety subjects, and some conduct courses in subjects related to off-the-job safety. Safe-behavior booklets and posters on seasonal activities are available from the NSC.

An organization can cooperate with the municipal recreation department in promoting swimming classes or courses in boating safety. Also, it can request that local police supervise an auto inspection clinic. Often insurance companies will provide the necessary equipment for testing drivers' physical qualifications and skills.

Organization personnel can, in turn, offer leadership and support for community activities. For instance, employees can help form local safety councils, assist schools and churches by making safety inspections upon request, or act as volunteer members of local fire departments.

Several small companies in a community may consider the possibility of pooling their resources and talents in a joint off-the-job safety program, as is sometimes done with disaster and rescue programs. Advance publicity ensures employee support by communicating objectives, plans, and activities. An informal letter from top management sent to employees' homes personalizes the organization's concern for the safety of not only the employees but their families as well.

The initial meeting for employees can be followed by other meetings that include family members. Many organizations hold picnics and other outings featuring various types of entertainment and safety exhibits as a means of reaching the families. An organization's newsletter, magazine, or paper is an effective way to carry the word to employees. Company bulletin boards often reinforce these messages. The community-at-large can be kept informed of plans and progress through spot announcements and stories on home, traffic, and recreational safety carried by the local radio and TV stations and the local newspapers. Preparation of such publicity can be financed by several companies together or by the local safety council.

Meetings of local organizations, such as church groups, women's and service clubs, PTA, and similar groups, are also good resources to educate the community-at-large about off-the-job safety programs and activities.

Programs That Worked

The following sections discuss promotional methods that have proven successful. They can easily be tailored to the needs of your organization.

Contest. Offering some incentive, as minimal as a savings bond or as generous as a college scholarship, can usually elicit good participation from employees' youngsters. The resulting posters, essays, calendars, or slogans can then become vehicles for promoting safety.

The two most widely used contest ideas involve essays (with a limit on the number of words) or posters (with restrictions on size and materials), such as, "What My Dad's/ Mom's Safety Means to Me," "Vacation Safety," "Community Safety," and "How Father/Mother Can Be Safe at Work." Competitors should be divided into age groups—ages running from about 5–7 years old for the youngest, up to 16–18 years for the oldest. Each entrant should receive a token of participation, like a keychain, certificate, embroidered patch, or even a model of the organization's product. To ensure impartiality, judges may be from outside the community or the organization.

Traffic safety. Offering the NSC's Defensive Driving Program, at the organization's expense and possibly with organization facilities and instructors, to all driving members of the employees' families is one of the most positive home, off-the-job, or community safety efforts an organization can make. It can improve the driving habits of those who drive, and can also promote car pooling as a safe answer to energy, traffic, and pollution problems.

Auto safety checks tie in well with vacation and holiday programs and work best on weekends. Have plenty of qualified inspectors available so participants will not be discouraged by long lines.

Company picnic. Safety picnics should involve the whole family and can be as expensive as a completely catered affair or as simple as a family picnic. A safety theme can be included in drawings for door prizes, activities for all ages, and presentations of awards to or recognition of employees for safety achievements.

Family night. Quite different from picnics, family-night gatherings are built around the presentation of some discussion or audiovisual presentation on safety, accompanied by refreshments. Sometimes the theme shifts from general safety to a program to make the family aware of and gain support for the safety efforts made at work. Sometimes an "open house" tour of the plant facilities can be combined with the family-night get-together.

Family first aid programs, including CPR training, can be held at business locations after business hours, or on weekends to involve the family. First aid kits available at a company discount purchase price could be offered for family use in the home, auto, or workshop.

Youth activities. Sponsorship of activities like softball and football teams and bicycle rallies usually is not aimed

directly at promoting safety. However, a poor safety record among young people taking part in the activity may hurt the sponsor's image. Therefore, financial sponsorship of such activities should be only part of an organization's participation. Employee leaders, who are knowledgeable in the activity and trained in safety and first aid (at the organization's expense), should be on hand to teach and help participants. Such actions can produce a strong, positive image of the organization in the community. Involvement of groups, such as the Boy and Girl Scouts, 4–H Clubs, Campfire, Inc., FFA, and FHA, will expand off-the-job safety efforts.

Recreational programs. In any organization, many employees and their families are sports enthusiasts. Their pastimes may include hunting, boating, camping, swimming, fishing, skiing, to name a few. At the season openings of these activities, organizations may sponsor clinics featuring registered/competent instructors to check equipment and to provide instruction on improving skills. Sources of help include the local gun clubs, powerboat squadrons, the Coast Guard Auxiliary, National Recreation and Park Association, the President's Council on Physical Fitness and Sports, the National Red Cross, YMCA, police and fire departments, and health organizations. All of these groups can provide ideas for safety activities.

Vacation-holiday program. Some organizations close down for regular summer vacation; others offer year-around vacation periods and three-day weekends. A good time to offer safe driving tips and safety literature to employees is at vacation times.

Other promotional methods include:

- Publicity distributed in-house that reinforces safety, both on-the-job and off-the-job (pamphlets, press releases, posters, and billboards are ideal vehicles for in-house promotions). The NSC has an excellent selection of this material, aimed at a variety of off-the-job safety topics. Employee-generated, original posters are also effective in personalizing safety efforts.
- Nearly every organization has some sort of in-house journal or newsletter that is either given to employees at work or sent to their homes. No other publication enjoys wider readership within an organization. As a result, it provides an excellent forum for safety education and safety program promotion; both employees and their families see it. Articles can be written on all aspects of a safety program, and the employee and family can be solicited for safety-related story ideas as well.
- NSC's periodical *Family Safety and Health* is an excellent way to promote off-the-job safety.

Benefits of Off-the-Job Safety

There are three benefits a company can realize from expanding its safety program to include off-the-job safety. The first is a reduction in lost production time and operating costs from both on-the-job and off-the-job injuries. Second, companies have found that efforts in off-the-job safety increase employees' interest in their on-the-job safety program. The third benefit, often overlooked, is that of better public relations.

The aim of safety education, namely, changing the employee's behavior, is especially true of off-the-job safety. No asset is more important to a company than its employees. They should not only be protected during working hours but also be given every incentive to practice safety off the job.

SUMMARY

- Hazards are defined as any existing or potential condition in the workplace that, by itself or interacting with other variables, can result in deaths, injuries, property damage, and other losses. Hazards are grouped into two broad categories: those dealing with safety and injuries and those dealing with health and illnesses. Accidents resulting from hazards interrupt production, harm the workforce, and add to operational costs.
- Loss control exists on the national, organizational, and worker-equipment-environment levels. Its primary function is to help management locate, assess, and set effective preventive and corrective measures for hazards that may interfere with operational efficiency and success.
- Loss control is a strategic process that by eliminating, reducing, or controlling hazards helps an organization make the best use of human, financial, technological, and physical resources. Controlling hazards is a team effort between management and employees.
- Loss control programs set facility-wide safety and health standards and coordinate responsibility among departments. To set realistic goals for loss control programs, the organization must determine the major factors likely to cause accidents and then institute control measures and monitoring systems to reduce, eliminate, or control these hazards.
- Effective loss control programs must take into consideration the worker-equipment-environment system. Workers must be educated to perform work effectively and safely, and equipment must be properly designed, maintained, and used. Finally, the environment should be as safe and healthful as possible.
- Accidents can be attributed to the oversight, omission, or malfunction of the management system regarding situational work factors, human factors, and environment factors. Human error is reduced when workers are properly trained, matched to the right jobs, and possess skills necessary to use equipment and when management gives top priority to safety.
- Environmental factors fall into four broad categories: physical, chemical, biological, and ergonomic. These hazards can be reduced, eliminated, or controlled by attention to layout, design, and placement of equipment, machinery, and furnishings; by properly using and storing chemicals; and by maintaining work area cleanliness and personal hygiene.
- Accidents, once regarded as an inevitable part of doing

business, are now viewed as controllable. Two basic approaches to examining accident causation are after-the-fact and before-the-fact investigations. The after-the-fact approach simply investigates the causes of accidents after they have occurred. In before-the-fact investigations, critical incident and safety sampling techniques are used to measure safety performance and to identify practices or conditions that need to be corrected before loss occurs.

- An effective loss control program has six processes: hazard identification and evaluation, hazard ranking, management decision making, establishment of preventive and corrective measures, monitoring, and evaluating program effectiveness.

- In organizing an effective safety and health program, management must establish program objectives, develop and publish an organizational policy, and delegate responsibility and authority for promoting the program. Employee support for the program must be enlisted from the earliest stages.

- Management and administration, along with housekeeping and preventive maintenance, are a few of an organization's most potent weapons against hazards. All levels of management must educate and train workers in safety and health regulations and practices. Management from top levels to department heads to supervisors must monitor their work areas for human, situational, and environmental factors that may cause accidents. Housekeeping and preventive maintenance programs can help to prevent accidents by maintaining an orderly, neat environment and by keeping machines and equipment in good repair.

- Employees are a key component in the success of any health and safety program. Employees take their cue from management attitudes about health and safety issues. A three-pronged approach to employee safety programs involves presenting general company policy and rules, discussing overall safety rules and the safety program, and explaining those safety rules that apply to each employee's department.

- Safety and health professionals, industrial hygienists, and staff medical personnel advise and guide organizations on safety matters, investigate accidents, maintain accident records, ensure line management's knowledge of their responsibility to comply with all safety and health regulations, and perform other duties as required to assist line management in safeguarding employees.

- Purchasing agents play a key role in safety by (1) ensuring that all items comply with federal and state or provincial regulations and ordinances, (2) specifying that machines and equipment be designed, built, and shipped with adequate safeguards, and (3) balancing cost-effectiveness with safety measures in purchases.

- Organizational safety and health committees can be invaluable to the loss control program by providing the active participation and cooperation of key people and employees. The committee's purpose is to aid and advise management and employees on safety and health, and

perform monitoring, educational, investigative, and evaluative tasks in the safety program.

- Off-the-job safety programs educate employees to apply outside the job many of the same safety practices they use at work. Organizations can promote off-the-job safety through a variety of company, community, and sports activities; educational programs; and contests. Benefits from such programs include reduction in lost production time and operating costs, greater employee interest in on-the-job safety, and better public relations with the community.

REFERENCES

Blankenship LM. *Nonoccupational Disabling Injury Cost Study*, K/DSA-457. Martin Marietta Energy Systems, Inc., October 1981.

Board of Certified Safety Professionals of the Americas, 208 Burwash, Savoy, IL 61874. "Curricula Development and Examination Study Guidelines," Technical Report No. 1.

Boylston RB. "Managing Safety and Health Programs." Speech given before the Textile Section, National Safety Congress, October 1989.

Construction Advancement Foundation, Hammond, IN. *Safety Manual*.

Factory Mutual Engineering Corp. *Loss Prevention Data*. Norwood, MA 02062.

Dennis LE and ML Onion. *Out in Front: Effective Supervision in the Workplace*. Chicago: National Safety Council, 1990.

Firenze RJ. *Guide to Occupational Safety and Health Management*. Dubuque, IA: Kendall/Hunt Publishing Co., 1973.

—. *The Process of Hazard Control*. Dubuque, IA: Kendall/Hunt Publishing Co., 1978.

—. *Safety and Health in Industrial/Vocational Education*. Cincinnati: National Institute for Occupational Safety and Health, 1981.

Johnson WG. *MORT Safety Assurance Systems*. New York: Marcel Dekker, Inc., 1980.

Kane A. Safety begins the first day on the job, *National Safety News* 53, January 1979.

Manuele FA. How effective is your hazard control program? *National Safety News* 53–58, February 1980.

National Association of Suggestion Systems, 230 North Michigan Avenue, Chicago, IL 60611.
   Performance Magazine (6 times a year).
   "Suggestion Newsletter" (6 times a year).

National Safety Council, 444 North Michigan Avenue, Chicago, IL 60611.

Accident Facts (published annually).

Management Safety Policies, Industrial Safety Data Sheets, 12304-0585.

Supervisors' Safety Manual, 7th ed., 1990.

"You Are the Safety and Health Committee."

Peters GA. Systematic safety, *National Safety News* 83–90, September 1975.

Plog B., ed. *Fundamentals of Industrial Hygiene,* 3rd ed. Chicago: National Safety Council, 1988.

Tainter SA and KM Monro. *The Secretary's Handbook.* 10th ed. New York: The Macmillan Co., 1988.

U.S. Department of Human Resources, National Institute for Occupational Safety and Health, Division of Technical Services, Cincinnati, OH 45226. *Self-Evaluation of Occupational Safety and Health Programs,* Publication 78-187, 1978.

U.S. Department of Labor, Occupational Safety and Health Administration. *Organizing a Safety Committee,* OSHA 2231, June 1975.

U.S. Department of Transportation, Office of Hazardous Materials, Washington, DC 20590. "Newly Authorized Hazardous Materials Warning Labels." (Latest edition.) (Based on Title 49, *Code of Federal Regulations,* sections 173.402, —403, and —404; import or export shipments are covered in Title 14, *CFR,* section 103.13.)

Windsor DG. "Process Hazards Management," a speech given before the Chemical Section, National Safety Congress, October 17, 1979.

4

Occupational Health Programs

An integral part of every safety and health program is the medical and health services necessary to cover the employees in the organization. This chapter will discuss the necessary functions and structure of occupational health programs. Each organization will need to decide how to provide these services.

Occupational health programs range from the truly elaborate to the bare minimum. One establishment may have a full-time staff of physicians, nurses, and technicians housed in a model dispensary; another may have only the basic first aid kit with a trained person to provide first aid.

Ideally, occupational health programs, regardless of size, are composed of elements and services designed to maintain the health of the workforce, to prevent or control occupational and nonoccupational diseases and accidents, and to prevent and reduce disability and the resulting lost time. A good program should provide for the following:

1. maintenance of a healthful environment through setting up a comprehensive safety program
2. health examinations, including preplacement testing, testing when employee transfers are internally processed, periodic surveillance for employees performing particular jobs as required by regulatory standards, fit-for-duty evaluations, and testing related to an employee returning to work. Drug testing may be mandated by regulations.
3. diagnosis and treatment
4. immunization programs
5. medical records, which should be confidential and kept separate from personnel records
6. health education and counseling
7. open communication between the plant or company physician and an employee's personal physician.

Any treatment of ill or injured persons falls under the practice of medicine. All states, therefore, regulate medical and nursing practices and provide guidelines and standards for patient care. Ideally, all services provided by company health care personnel related to employee health, injuries, first aid, and medication of any kind should be approved by a licensed physician.

OCCUPATIONAL HEALTH

Occupational health programs are concerned with all aspects of the employee's health and the employee's relationship with the environment.

The basic objectives of a good occupational health program should be:

- to protect employees against health hazards in their work environment
- to facilitate placement and ensure the suitability of individuals according to their physical capacities, mental abilities, and emotional makeup in work that they can perform with an acceptable degree of efficiency and without endangering their own health and safety or that of their fellow employees
- to assure adequate health care and rehabilitation of the occupationally injured

- to encourage personal health maintenance.

The achievement of these objectives benefits both employees and employers by improving health, morale, and productivity.

By applying occupational health principles to the workplace, all employees, including the severely disabled, are placed in jobs according to their physical capacities, mental abilities, and emotional makeup (see Chapter 20, Workers with Disabilities). This approach also ensures continuing medical care and rehabilitation of occupationally ill and injured workers.

Experience has shown a strong relationship between accident prevention and occupational health. For example, some industrial chemicals, when improperly handled, represent serious hazards to health, property, and the environment. Depending on various environmental and workplace conditions, the vapor from a chemical can ignite, explode, or, if inhaled, cause dizziness or death. Dermatitis can be caused by the contact of a chemical with a worker's skin. (For details on the effects of specific chemicals, see the National Safety Council's *Fundamentals of Industrial Hygiene,* 3rd ed.)

Safety and health professionals have ably demonstrated their ability to reduce the rates of accidental injuries by controlling many phases of the industrial environment through worker education and through improved supervisory techniques. These two elements must allow for the different physical and emotional characteristics of individual workers. Such characteristics account for variations in workers' job attitudes, productivity, safety practices, and absence for personal health reasons.

Management often needs the services and skills of several additional professions to attain the best results in its health and safety programs. These professions include:

- Medicine plays an important role, particularly when physicians are trained in industrial and preventive medicine. They are assisted by specialists in orthopedic surgery, ophthalmology, radiology, surgery, dermatology, psychiatry, and other areas.
- Occupational health nursing also requires specialized knowledge, not only of good basic nursing procedures and health maintenance, but also of the legal, economic, social mores, and labor laws within which the nurse must provide health care for workers. Often, nurses are the only full-time medically trained employees in a company or plant.
- The industrial hygienist applies specialized knowledge to the recognition, evaluation, and control of health hazards in the work environment. For additional details, see the National Safety Council's *Fundamentals of Industrial Hygiene,* 3rd ed. (see References).
- Other aspects of occupational health may require the expertise of the community and other professional disciplines.

Working both individually and collectively, these specialists have helped to improve the occupational health and safety record of many industries. In some companies, these professionals are so well organized and effective in anticipating and correcting hazards that employees are actually safer and healthier at work than they are at home. In fact, the work environment is so well controlled, in many cases, that the problem many companies now face is educating workers to recognize and avoid the hazards of home life, recreation, and travel so that they will be able to return safely to work each day.

As a result, a company's occupational health services should be involved in off-the-job safety programs without being obviously intrusive. Health care personnel can be influential in extending this program beyond the facility to include outside activities of employees and their families. If employees are injured off the job, they are as much a loss to the operation as if they were injured on the job.

Some insurance carriers offer a consulting service to help organizations set up an occupational health program suitable for their needs. The consultants usually know which physicians and clinics are available for this kind of service. The basic work, however, has to be done by the organization establishing such a program.

Any justification for these services lies in their record of accomplishments. Prevention is not only better than cure—it is easier and less expensive. Off-the-job safety programs and activities can benefit a company, plant, or shop of any size.

Employee Health Services

Good industrial medicine can be practiced in any clean, private location in the plant. However, experience indicates that the medical unit commands greater respect if management pays careful attention to suitable and efficient housing, appearance, and equipment. Adequate space must be provided for multiple casualties during emergencies. The entire unit should be painted in light colors and kept spotlessly clean. The dispensary should have hot and cold running water and be adequately heated, ventilated, and illuminated; and be located near toilet facilities. Management must make suitable provisions for both men and women.

Health service office (dispensary). The unit should contain a minimum of three rooms: a waiting room, a treatment room, and a room for physical examinations or consultation (Figure 4–1). Rooms for special purposes can be added according to the needs and size of the company. The surgical treatment room should be large enough to treat more than one person at a time. Small dressing booths can give workers some degree of privacy.

First Aid

Good administration of first aid is an important part of every safety program. These services may range from a deluxe first aid kit to a well-staffed first aid and health facility.

First aid kits and supplies approved by a physician should be stored where they are readily accessible. Medical supplies should not be spread about the plant for self-administration by employees. When no nurse is in charge, a supervisory employee for each shift should be given responsibility for all first aid supplies. Health care personnel should keep a care-

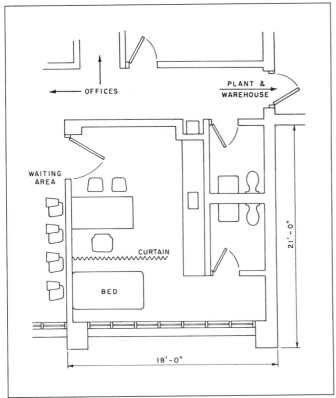

Figure 4–1. The floor plan for a health service office. The office has a waiting room, a treatment room, and a consultation room.

ful record of each administration of first aid and send an incident investigation report to the injured worker's supervisor at the time first aid is administered.

Establishments that do not employ a full-time medical doctor should maintain good liaison with a local physician, or physicians, designated to handle plant injuries. The physician should be invited to the plant on a regular basis to evaluate the quality of first aid procedures, to make recommendations for improvements, and occasionally to tour the establishment. The company-designated physician(s) should be familiar with the type of work done so he or she can evaluate injuries, illnesses, and other complaints from patients. The physician, nurse, or designated supervisor should routinely inspect first aid supplies, stretchers, and stretcher locations.

Definition and limitations. In many small organizations and in field operations, it is neither practical nor justifiable to have qualified professional health care personnel available full time. In such cases, the best arrangement is to use suitable first aid attendants who follow procedures and treatments outlined by a physician. However, a physician should be available on an on-call or referral basis to take care of serious injuries. In some jurisdictions, injured employees have their choice of a physician. The employer should comply with all requests, if possible.

There are two kinds of first aid treatment: emergency and prompt attention.

Emergency treatment. According to the National Safety Council's First Aid Institute first aid textbook, "First aid is the immediate, temporary care given to the injured or suddenly ill before the proper medical treatment can be given." Proper first aid measures reduce suffering and make sure the injured person is in better condition to receive subsequent treatment.

Prompt attention. This type of first aid is used to treat minor injuries such as cuts, scratches, bruises, and burns. Ordinarily, the injured person would not seek medical attention for these injuries.

By requiring that all employees immediately report for treatment when injured, regardless of the extent of the injury, companies can reduce infection and disability and avoid false claims of injury and disability.

A first aid program should include:
1. properly trained and designated first aid personnel on every shift
2. first aid unit and supplies or first-aid kit approved by a physician
3. first aid manual
4. list of reactions to chemicals via routes of exposure
5. posted instructions for calling a physician and notifying the hospital that a patient is en route
6. posted method for transporting ill or injured employees
7. instructions for calling an ambulance or rescue squad
8. adequate first aid record system and follow-up.

NOTE: organizations need to have recordkeeping systems that meet regulatory standards and provide information so that accidents can be prevented. An understanding among safety and health staffs may be needed so proper recordkeeping can be maintained.

Health care staff may often have different opinions regarding proper treatment. As a result, first aid procedures, approved by the consulting physician, should specify the type of medication, if any, to be used on minor injuries such as cuts and burns. These procedures frequently also state that in areas where chemicals are stored, handled, or used, emergency flood showers and eyewash fountains (Figure 4–2) should be available and clearly identified.

Management should choose equipment and supplies in accordance with the physician's recommendations. Health care staff should render service only as covered by written standard procedures, signed and dated by the physician. The physician should specify the procedures to be followed when medication for temporary relief of minor nonoccupational ailments, such as colds and headache, is administered. The limitations of first aid treatment for illness or injury must also be thoroughly understood.

Most regulatory agencies have medical practice acts limiting health care providers to certain definite procedures when attending anyone who is sick or injured—except, of course, under the direct supervision of a physician. It is important, therefore, that anyone responsible for first aid treatment fully understand the limits of this work. Because improper treatment might involve the company in serious legal problems, the first aid attendant should be duly qualified and certified by the Mine Safety and Health Administration (MSHA), the National Safety Council, or the American Red Cross. These

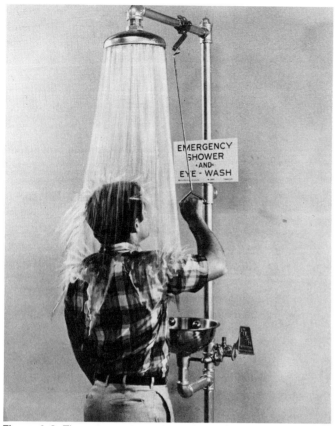
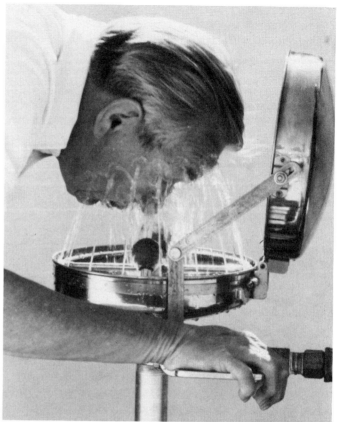

Figure 4–2. The emergency shower and eye wash should be well identified. Harmful chemicals are quickly washed away and clothing fires doused immediately by the drenching action of the shower (left). First–aid treatment to the eye must be prompt and consists of prolonged irrigation of the exposed eye with low–pressure water. (Left photo reprinted with permission from Western Drinking Fountains. Right photo reprinted with permission from Haws Drinking Faucet Co.)

certificates must be renewed at specific intervals.

First aid training. The National Safety Council First Aid Institute's first aid textbooks and the MSHA manual of first aid instruction are recommended for training employees. Many feel that accidents occur less frequently and are usually less severe among workers trained in first aid procedures. It is, therefore, advisable that as many industrial workers as possible be given this training (Figure 4–3).

The National Safety Council publishes posters and booklets that can be used for training employees. Another valuable reference is *Emergency Care of the Sick and Injured*, by the Committee on Trauma, American College of Surgeons (see References).

First aid room. It is always advisable to set aside a conveniently located room for the sole purpose of administering first aid treatment. The person administering first aid should have a proper place to work.

A first aid room is similar to the dispensary. It should be equipped with the following items:

1. examining table
2. cot for emergency cases, enclosed by movable curtain
3. dustproof cabinet for supplies
4. waste receptacle and biohazards-disposable containers
5. small table
6. chair with arms and one without arms
7. magnifying light on a stand

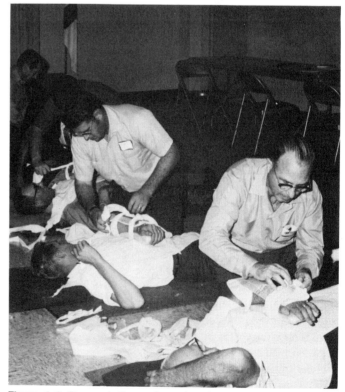

Figure 4–3. Demonstration and practice play an important role in first–aid training.

8. dispensers for soap, towels, cleansing tissues, and paper cups
9. wheelchair
10. stretcher
11. blankets
12. bulletin board to post all important telephone numbers for emergencies
13. bed(s).

Health care personnel often use oxygen in the treatment of many first aid cases. Because of the danger of fire or explosion, smoking should be prohibited when oxygen is administered. Any type of resuscitating device should be used only by trained persons.

First aid kits. Many types of first aid kits are available to fill every need, depending on the type of accident that might occur (Figure 4–4). Commercial or cabinet first aid kits, as well as unit kits, must meet regulatory requirements. MSHA outlines specific first aid items as a requirement at certain locations in mines and processing facilities.

Kits vary in size from the pocket model to almost a portable first aid room. The size and contents depend on the intended use and the types of injuries to be treated. For example, personal kits contain only essential articles for the immediate treatment of injuries while departmental kits meet the needs of a group of workers. As a result, the quantity of material each kit contains depends on the size of the working force. Trunk kits are the most complete. They can be carried easily to an accident site or can be stored near working areas far from well-equipped emergency first aid rooms. Although a trunk kit can include such bulky items as

a wash basin, blankets, splints, and stretchers, it can still be carried by only two persons.

Keeping first aid kits strategically located throughout the plant seems to work best when such kits are supervised properly (Figure 4–5). A single, trained individual should be responsible for maintaining each kit as well as for treating serious injuries and giving first aid measures for minor cuts and scratches. This employee never provides more than immediate, temporary care. By combining well maintained first aid kits with a trained first aid provider, companies can ensure that many minor injuries that might otherwise have been ignored receive prompt, proper treatment.

In some organizations whose activities are widely scattered, employees may need to use first aid kits and administer treatment themselves. Under these circumstances, the organization can control first aid service by having the attendant in charge properly instructed in first aid and by seeing that the service as a whole has medical supervision. All miners, for example, receive first aid training and annual refresher training.

In general, medical personnel should supervise the maintenance and use of all first-aid kits and a consultant physician should approve all materials provided. A member of the health services group should regularly inspect all first aid materials and report on their content level and serviceability.

Maintaining quantities of materials in the first aid kit is easier if each kit contains a list showing the original contents and the minimum quantities needed. Health care workers should clearly label and date all bottles or other containers in the kit.

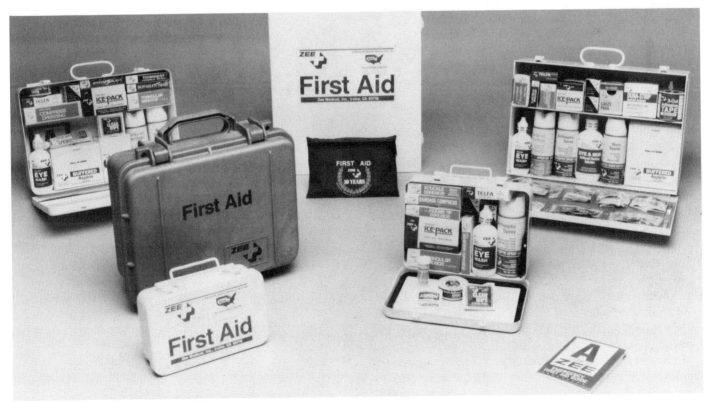

Figure 4–4. First–aid kits can be obtained for special purposes.

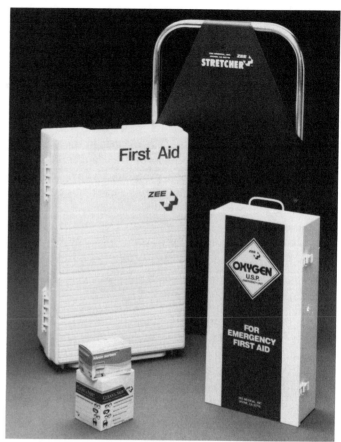

Figure 4–5. First-aid kits should be well-distributed throughout a plant. *Left:* They should be supervised by a trained person so employees do not try to "doctor" themselves. *Right:* Stretchers and oxygen kits should also be available in areas distant from the first aid room. (Courtesy Zee Medical.)

Recommended materials for first aid kits are listed in the Council's First Aid Institute, the U.S. MSHA, and the American Red Cross first aid textbooks. Suggestions are also available from the American Medical Association, the American Petroleum Institute, and from manufacturers of first aid materials.

Stretchers. Company health services must be able to quickly transport a seriously injured person from the scene of an accident to a first aid room or a hospital. Such promptness can help physicians determine the gravity of the injury and perhaps even increase the victim's chances for survival.

The stretcher provides the most acceptable method of hand transportation for accident victims. It can be used as a temporary cot at the scene of the accident, during transit, and in the first aid room or dispensary.

Although there are several types of stretchers, the commonly used army stretcher is satisfactory in most cases. However, when workers must hoist or lower the injured person out of an awkward place, it is better to use a specially shaped stretcher to which the patient may be strapped and kept immovable. A lightweight stretcher can serve this purpose.

Stretchers should be conveniently stored near places where employees are exposed to serious hazards. It is customary to keep stretchers, blankets, and splints in clearly marked, prominent cabinets. Stretchers should be clean and ready to use at all times. They should be protected against destructive vapors, gases, dust, or other substances, and against mechanical damage. If the stretcher is made of materials that will deteriorate, health care staff should test it periodically for durability and strength.

Any victim of a severe injury, such as a fall, head injury, spinal injury, or any injury resulting in unconsciousness, should be secured to and transported on a rigid spine board. Regardless of the transportation device, the patient should be secured to the device with straps or cravats.

HEALTH SERVICES

The health department should be easy for workers to locate. If management places it near the greatest number of employees, more workers are likely to report all minor injuries and have them treated. If possible, the department should also be connected with the human resources and safety departments. This arrangement will facilitate prompt physical examinations of job applicants, the sharing of clerical services, and the interchange of ideas and plans regarding employment, accident, and health problems.

Another potential location is near the plant entrance. This will make it possible to bring an ambulance to the door, if necessary. Also, injured workers who are off duty but under treatment may come and go through a separate entrance.

Management should try to put this department in a protected area so that a major company or plant disaster does not destroy the first aid or dispensary facilities.

The Occupational Physician

Physicians in industry may be employed at several different facilities (part time at each). The number depends on such considerations as the hazards in the operations, the number of employees, and the type and extent of the occupational health program. The company can make arrangements for full-time, part-time, on-call, or consulting services.

Some large organizations have a full-time medical director on their headquarters staff, with part-time physicians serving their decentralized operations.

Some physicians devote a scheduled number of hours, either daily or weekly, to the medical service needs of a company and are available at other times for emergencies. Others arrange their service by telephone on the basis of current company needs, for example, when job applicants require examinations, injured employees need medical care, or other medical problems arise.

The on-call physician arrangement is most often used by companies (or establishments) with fewer than 500 employees, a low incidence of accidental injuries, or a minimum health services program for employees. The on-call physician is usually located nearby and is available in emergencies.

This physician cares for most injuries not requiring hospitalization or for those cases requiring hospitalization that fall within his or her competence. On-call physicians often use specialists as consultants. Frequently, the on-call physician has similar arrangements with a number of companies in the area—a particularly convenient system for small-plant or small-company clusters, such as in industrial parks. Also, it is not unusual to find physicians specializing in the care of industrial injuries and diseases and providing occupational medical services in some manufacturing centers or in associated free-standing or hospital-based occupational medicine clinics.

A physician who has a consulting service is not usually called in except to diagnose and treat serious injuries, illnesses, toxicological problems, and special kinds of injuries or disorders (such as eye injuries) that require the services of a specialist.

State and local health departments often supply, without charge, medical, nursing, and engineering consultation. They also make industrial hygiene and radiological surveys for industries within their areas. Many insurance and private consulting companies offer this service to their clients.

Duties of the occupational physician. An effective medical service program should be planned by the health department, as developed with management cooperation. The program must have the full support of top management if it is to be successful. Only then is it possible to establish and maintain an adequate professional staff and facilities for examinations, emergency cases, and record storage.

Occupational physicians must be given enough authority so that workers will respect their judgment and follow their instructions on personal health and safety matters. These physicians should be familiar with all jobs, materials, and processes that are used within the company. An occasional inspection visit will help them keep abreast of what is going on. During these visits, the physician can develop an inventory of hazards and point out to the safety professional those potentially harmful conditions from which the employees should be protected.

Physicians can also be involved in other company services that relate to the health of workers, such as food service, welfare service, safety programming, sanitation, and mental health. They can initiate and supervise company-wide immunization programs (against tetanus, polio, or flu), blood donor drives, blood pressure screening, and wellness programs.

Maintaining a true physician-patient relationship (with fairness to both employee and employer) is essential to the success of any occupational health program. For example, employee patients deserve the same courtesy and professional honesty as do private patients. The first meeting of physician and employee usually occurs at the introductory physical examination. This meeting may be followed by subsequent examinations. The examining physician—within the framework of professional discretion—should tell the worker the results of all examinations and, if necessary, refer the worker to a personal physician for further treatment.

The occupational physician should provide emergency medical care for employees who are injured or become ill on the job (Figure 4–6). Necessary follow-up treatment for employees suffering from occupational disease or injuries also should be arranged. The treatment of nonwork-related employee injuries or diseases is the function of private medical practice. Therefore, the occupational physician should not render such services except under the following conditions: (1) when independent facilities are not readily available, (2) when the ailment or discomfort is so minor that the employee ordinarily would not seek medical attention, or (3) when rendering such service would enable an employee to complete a shift.

In most settings the occupational physician should not devote time or facilities to diagnose or treat dependents of employees. Instead, health education programs should be promoted for employees and their families so that they will be encouraged to seek proper private consultation.

Medical and surgical management in every case of industrial injury or disease should aim to restore disabled workers to their former earning power and occupation as completely and rapidly as possible. To help achieve these goals, the physician should promptly submit medical reports to those agencies that need to receive them. Furthermore, equitable administration of Workers' Compensation often depends on a physician's medical testimony regarding the injury and its possible consequences.

The industrial physician is also responsible for maintaining necessary records and reports. These records act as a guide to management and keep both management and employees informed regarding the success of the health program. Records and reports are necessary to direct and evaluate

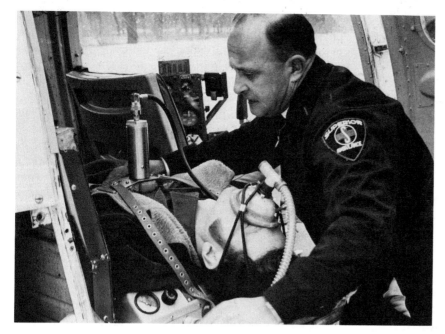

Figure 4–6. When speed is essential, a helicopter can be used to transfer a patient.

preventive medical and safety engineering techniques, to chart progress in the reduction of accidents, and to meet the regulatory recordkeeping requirements (see Chapter 2, Regulatory History and Compliance, and Chapter 13, Recordkeeping and Incidence Rates).

The physician is responsible as well for properly instructing nurses and paramedical personnel and directing their activities. Their duties, therefore, should be described in clear, concisely written directives, a copy of which should be posted in the health department.

The Occupational Health Nurse

Occupational health nursing is a specialized branch of the nursing profession. The position requires a registered professional nurse who is licensed to practice where employed. In addition, the occupational health nurse should have some knowledge of workers' compensation laws, insurance, health and safety regulations, occupational diseases, sanitation, first aid, and recordkeeping.

The occupational health nurse works with the company physician, who provides written directives that the two have discussed, mutually understood, and agreed upon. When the physician is employed only part time, the nurse works with the safety and health professional in planning and conducting accident and illness prevention programs.

Working with a full- or part-time physician, the occupational health nurse can provide a variety of nursing services, such as initial care for injuries or illnesses, counseling, health education, consultation about sanitary standards, and referral to community health agencies. The occupational health nurse also can participate in programs that evaluate employee health (such as physical examinations) or prevent disease (such as immunizations).

The occupational health nurse can perform excellent employee health education services by distributing literature

on healthy heart measures, weight control, cancer, tuberculosis prevention, and the prevention and treatment of AIDS, venereal disease and hepatitis. The nurse can also provide information about smoking cessation programs and may even facilitate such programs. The occupational health nurse may play a large role in implementing many OSHA-mandated programs, such as hazard communication and hearing conservation. For example, the nurse should have ready access to Material Safety Data Sheets (MSDS) for chemicals at the facility, and may participate in providing employee training.

The occupational health nurse must maintain a confidential, professional relationship with the employees in conformity with legal and ethical codes. The nurse must not divulge information contained in individual employee health records unless the employee gives written permission. The medical files should be accessible only to medical personnel.

It must be clear that establishing medical diagnoses and defining treatment are the physician's functions. The nurse is not a substitute for a physician. Each has a legally defined area of practice and responsibility. Ideally, a company will have the services of both a physician and nurse.

Physical Examination Program

Surveillance of workers' health status by qualified personnel is essential if an occupational health program is to provide maximum benefits for both employer and employee (Figure 4–7). Therefore, the health department should establish a regular physical examination program for all employees. The examining physician should discuss all significant findings with each worker. With the employee's consent, a transcript of the data may be supplied to the worker's personal physician or to an insurance company. Courts, workers' compensation commissions, or health authorities may request this information by legal means, but employee consent is a more

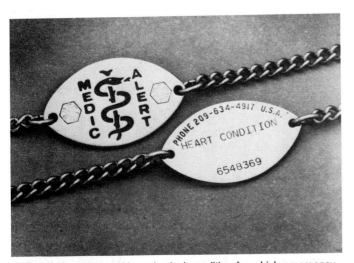

Figure 4–7. Persons with a physical condition for which emergency care may be needed should wear an identification tag. The tag provides a general indication of the problem and a phone number to call for more details if the tag wearer is unable to supply them. (Reprinted with permission from Medic Alert Foundation, Turlock, CA.)

agreeable method. Health services personnel must strictly preserve the confidential character of health examination records.

The physician or nurse should inform the employer of potentially harmful exposures or hazards detected through their examination of workers.

Scope of the examination. It is impossible to state categorically what constitutes a complete examination or even a suitable examination. Physicians have different opinions regarding the relative values of various test procedures, based on their own training, experience, and individual needs under consideration.

The scope of a physical examination should be determined by the company physician. This individual is familiar with the operations involved—the nature of the industry; its inherent hazards; and the variations in jobs, physical demands, and health exposures. The values of different test procedures and their cost in time and dollars must be carefully weighed. Examinations will probably be different for different jobs; for example, the physical requirements for an ironworker who will be engaged in construction work are different from those of an office worker who will sit at a desk. However, certain basic examination considerations apply to each.

The various kinds of examinations may be classified as follows—preplacement, periodic, and special examinations such as job transfer, return-to-work, termination, X-ray, vision, or hearing examinations.

Preplacement examinations. The preplacement examination is done to determine and record the physical condition of the prospective worker in order to match the worker's mental and physical capabilities to the right job. The applicant (or the personal physician with the applicant's approval) should be told of any conditions that need attention. Medical department follow-up may be necessary. The primary purpose of the preplacement examination program is selec-

tion and placement of workers—not merely selection of the physically perfect and rejection of all others.

From the public and occupational health standpoint, one of the few bars to immediate employment in nonhazardous occupations should be communicable diseases. It is obvious that communicable diseases must be controlled, and this may involve the assistance of public health officials. One of the values of an examination program is the detection of disease in its early stages when it is most easily treated. Applicants with incipient but still nondisabling disorders often can be employed while being treated by their personal physicians.

The Americans with Disabilities Act requires employers to make reasonable accommodations for people with other disabilities. See Chapter 20, Workers with Disabilities.

It is not the physician's function to inform applicants whether they are to be employed. This is the prerogative and duty of management. Other factors besides physical qualifications determine whether a candidate is suitable for employment in the company.

Periodic examinations. Periodic examinations of all employees may be on a required or voluntary basis. Management should institute a program for workers who are exposed to hazardous processes or materials or whose work involves responsibility for the safety of others, such as vehicle operators. Substances such as lead, carbon tetrachloride, or many others that can cause occupational diseases are usually subjected to process controls that will keep the workers safe from poisoning. However, caution dictates that the industrial physician periodically examine such workers to be certain that the engineering and hygiene controls are effectively safeguarding employees. Periodic examination also permits early detection of hyper-susceptible individuals and practices or procedures that circumvent safety devices and policies.

How often workers are examined varies according to the quality of engineering controls, the nature of the exposure (influenced by how rapidly the hazardous substance acts on the human body), and the findings produced on each examination, as well as the regulatory standards in some cases. Thus, worker exposure to substances might require that employees receive examinations or laboratory tests on a weekly, monthly, quarterly, annual, or biannual basis.

In many cases, laboratory tests of blood or urine can serve as the major portion of a periodic examination program, with complete examinations being made less frequently. The type of special examination (laboratory, X ray, etc.) necessary for any exposure and the interpretation of the results are decisions requiring expert medical personnel. Guidance for tests will come from regulatory standards, medical practice, and other sources.

Special examinations. Employees who experience on-the-job health problems often benefit from special examinations. Job transfer also may require a complete medical evaluation. Some organizations find it worthwhile to require "return to work" examinations of employees who have been absent more than a specified number of days as a result of a nonoccupational illness or injury. This type of examination is done to control communicable disease, as well as to deter-

mine whether an employee is fit to return to work. The same disease can affect different people in widely different ways, depending in part on their personality.

When an employee returns to work following serious injury, either occupational or nonoccupational, it may be necessary to reevaluate the person's physical capacities. On the other hand, rehabilitation procedures may have not only reduced the worker's disability but actually improved the range of jobs or tasks the person can perform.

Exit examinations. Upon termination of employment, some organizations examine employees and document the findings. This procedure is particularly appropriate where operations exposed workers to health hazards such as lead, benzene (benzol), silica, and asbestos dust, to excess noise levels, or where required by standards.

Laboratory tests. Urine and blood tests are a good investment in detecting liver and kidney disease, diabetes, and anemia, to name the more common diseases that are detected. Where workers will be exposed to toxic substances, appropriate laboratory tests may be indispensable.

X-ray tests. While not to be used routinely, X-ray tests may be appropriate in alerting a physician to a particular condition. X rays are used much less frequently than previously in surveillance of workers for pulmonary disease. Less potentially hazardous examinations, such as pulmonary function tests are being used to detect decrements in function before the disease reaches the stage where it can be seen on an X ray.

X-ray screening tests, as part of the preplacement medical examination, are of no significant value in predicting the onset of future physical conditions, such as back pain or injury, and should be avoided to reduce unnecessary expense as well as unnecessary exposure to X radiation.

Vision tests. Special devices have been developed for routine testing of several aspects of vision. Near-sighted and far-sighted vision should be documented. If the company fails to match the visual requirements of a job with the visual abilities of employees, workers may become easily fatigued, inefficient, and accident-prone. If the job demands working with colors or color-coded materials, color vision should also be tested.

If the organization does not have testing facilities available on-site, management can make arrangements with outside doctors for annual eye examinations and fitting prescription safety glasses, if needed.

Hearing tests. The industrial physician or an audiologist or hearing conservationist working with the physician, should examine workers who will be exposed to high noise levels (generally 80 dBa or greater for an 8-hour day) for hearing acuity before placement to determine prior hearing loss, if any. They should examine employees periodically thereafter to detect early hearing loss due to noise. Protective hearing devices can reduce noise reaching the auditory nerve, but hearing tests and records are also valuable in preventing premature hearing loss. The audiometric examination is the accepted method of testing. A special booth will usually be required to administer these tests.

Health history. A carefully taken personal health and occupational history may give as much information about the worker's health status as the physical examination. The history also will indicate the need for special tests and perhaps placement restriction. A nurse or other specially trained person can secure basic data from the examinee, such as height, weight, age, and the history. Nurses may also assist in other portions of the examination when specially trained and authorized to do so.

Emergency Medical Planning

The industrial physician, nurse, and safety and health professional should confer with management to plan emergency procedures for handling large numbers of seriously injured employees in the event of disaster, such as explosion, fire, or other catastrophe. Management should coordinate these plans with community plans for such events (see Chapter 6, Emergency Preparedness).

Procedures should include the following:
1. selection, training, and supervision of auxiliary nursing and other personnel
2. transportation and caring for the injured (Figure 4–6)
3. transfer of seriously injured to hospitals
4. coordination of these plans with the safety department, security, police, road patrols, fire departments, and other interested community groups.

Medical Records

Regulatory standards require employers to maintain accurate records of work-related deaths, injuries, and illnesses (see Chapter 13, Recordkeeping and Incidence Rates).

The Access to Medical Records standard permits the worker, the worker's representative, and regulatory authorities access to employer-maintained medical and toxic exposure records. These specific conditions under which access is allowed apply to all employers in general industry, maritime, and construction whose employees are exposed to toxic substances or harmful physical agents. Exposure records include records of the employee's past or present exposure to toxic substances or harmful physical agents, exposure records of other employees with past or present job duties or working conditions related to those of the employee, records containing exposure information concerning the employee's working conditions, and MSDSs.

Medical records contain an employee's medical history, examination and test results, medical opinions and diagnoses, descriptions of treatments and prescriptions, and employee medical complaints. Some exposure records may have to be maintained for 30 years and medical records for the duration of employment plus 30 years.

Industrial medical records also provide data for use in job placement, in establishing health standards, in health maintenance, in treatment and rehabilitation, in workers' compensation cases, in epidemiologic studies, and in helping management with program evaluation and improvement. Such data are collected in the history interview, from the preplacement examination and any subsequent examinations,

and from all visits the worker makes to the dispensary or first aid room. They serve to establish a medical profile of each worker.

The key to accurate diagnosis and treatment often lies in the adequacy and completeness of this medical profile; therefore, record maintenance is a professional responsibility. Further, to compile a complete history, medical records should include absences caused by illness or off-the-job injury. Thus, recordkeeping of nonoccupational and occupational incidents often reveals chronic or recurrent conditions. In such cases, early treatment (referral to family doctors) and preventive measures can reduce absenteeism and decrease accident rates.

Maintenance of health records, however, should not be so burdensome that the occupational health nurse (or first aid attendant) becomes a file clerk. Recording forms and filing systems must be simple enough that they can be used and interpreted by a physician, nurse, or first aid attendant.

Although the employer needs to know a worker's limitations for job placement purposes, only people who have a legitimate or legal right to know should have access to the records. All medical records are confidential. If a company does not have a resident medical director or nurse, its medical records are usually filed in the personnel department.

Neck or Wrist Tags for Medic Alert

A universal symbol for emergency medical identification has been developed by the American Medical Association (Figure 4–7). The object of the symbol is to identify its wearer immediately as a person with a physical condition requiring special attention. If the wearer is unconscious or otherwise unable to communicate, the symbol will indicate that vital medical facts are recorded on a health information card in the bearer's purse, wallet, or elsewhere. A telephone number is given on the identification tag for obtaining more detailed information.

These details should be known by anyone before attempting to help an individual struck down by an accident or sudden illness (see Medic Alert Foundation and *Emergency Medical Identification* in References).

SPECIAL HEALTH CONCERNS

Health Care of Women

There is no reliable evidence to support the view that women are more susceptible than men to occupational health hazards, except those that might affect the fetus during pregnancy.

In the past, many companies used fetal protection policies to prevent women from working in areas where toxic chemicals that might harm a developing fetus were present. A 1991 Supreme Court decision struck down these policies as discriminatory to women, since they may also bar them from higher paying jobs. Critics of such policies had also argued that they did not deal with reproductive hazards posed to men by chemicals such as lead. The Court decision puts the responsibility on the employer to provide information to the worker on the hazards she faces and let her make a determination of the risks she will accept. It said that as long as there is no negligence in this regard, it is unlikely that the employer would be found liable for any injury to a fetus. Critics of the Supreme Court decision disputed this assertion.

The thrust of these decisions and other laws regarding physical impairment is to treat pregnancy as any other temporary disability. While it is a complex subject, informing women of the hazards they face is essential. Chemicals such as lead, mercury, some solvents, and therapeutic agents containing synthetic hormones may produce harmful effects on the fetus or lead to spontaneous abortion. Biological agents such as rubella (German measles) or physical hazards such as ionizing radiation may also produce such damage. Pregnancy may limit a woman's ability to do some physical work, since a pregnant woman may tear more readily, have poorer balance, and respond more slowly to the physiological demands of strenuous physical work. For these reasons a woman should consult her doctor about any restrictions that should be placed on her work during pregnancy.

In line with the Pregnancy Discrimination Act and the Americans with Disabilities Act, an employer needs to make reasonable accommodations to a woman during pregnancy. Employers need to consult the latest regulatory interpretations of laws and court decisions in developing programs to deal with issues posed by workers who are pregnant or planning to conceive a child. For example, the U.S. OSHA lead standard requires that employers offer medical examinations and consultations to both men and women on request, if the employee desires medical advice regarding lead exposure and the ability to procreate a healthy child. New regulations on the use of video display terminals that are being enacted in various localities also may require that accommodations be made for workers who do not want to work on VDTs during pregnancy. (There has been concern about possible reproductive effects of VDT usage, but the question has not been resolved.)

Alcohol and Drug Control

Alcoholism and drug abuse continue to be among the workers' leading illnesses. Therefore, it is essential that the industrial physician and the employer understand what they can do to help employees who have these illnesses in the work setting (see Chapter 7, Employee Assistance Programs).

Managers and supervisors should be alert to employees whose work and performance are deteriorating because of an alcohol problem. These employees should be referred for medical care as quickly as possible.

When alcohol and drugs are combined, they produce a variety of effects that severely impair workers. Concentrations of alcohol and drugs remain in the bloodstream much longer than most users realize, and the effects of this combination may appear unexpectedly.

Dependence on alcohol or other drugs is a major cause of family strife, impaired job performance, morale and disciplinary problems, increased insurance rates, occupational accidents, increased absenteeism, and the rising crime rate.

These illnesses know no boundaries, nor is there any "generation gap" among abusers—all ages, races, and socioeconomic groups are susceptible to their destructive effects.

Every time an applicant or employee visits a physician or nurse, some health education and/or counseling should be given. Health departments in industry are an ideal way to provide these services. Employers are paying an increasingly large part of the health care costs for employees and their dependents. The earlier medical problems are detected, the less they are likely to cost. As a result, it makes economic sense for employers to maintain a sound health department to provide proper care for employees and to offer timely referrals for further care, if needed. (See also Chapter 7, Employee Assistance Programs.)

SUMMARY

- Occupational health programs are concerned with all aspects of employees' health and workers' relationships with their environment. These programs, regardless of size, should provide for the following: a healthful environment, health examinations, diagnosis and treatment, medical records, health education and counseling, and communication between company and personal physicians. All health services provided by organizations must meet federal, state, and local codes.

- The principal goals of an occupational health program are to protect employee health, to match the right worker to the right job, to treat and rehabilitate injured employees, and to encourage personal health maintenance. Companies should also be involved in off-the-job safety and health programs.

- Management may work with other health professionals such as industrial hygienists and medical specialists in delivering health care to employees. Experience has shown a strong relationship between occupational health and accident prevention.

- First aid services are generally mandated by regulations. A first aid program includes properly trained personnel; a first aid unit and supplies; company manual; list of reactions to chemicals; instructions for calling physicians, hospitals, and ambulances or rescue squads; methods for transporting ill or injured employees; and an adequate record system of injuries and treatment.

- First aid treatment can be either emergency (before a physician can be summoned) or prompt attention (for minor injuries and illnesses). Those administering first aid measures should understand the limitations of such treatment. One person should be responsible in each work area to make sure that supplies are adequate and in good condition and that all worker injuries are treated promptly.

- Many companies establish health departments to oversee employee occupational health programs. The department should be located near work areas and, if possible, be connected with the employment and safety departments.

- In some organizations, a full-time, part-time, or on-call physician is hired to supervise employee health activities,

coordinate with safety programs, provide medical diagnoses and treatment, and maintain accurate and timely medical records. Often this person will head up the health department.

- Occupational health nurses often work with physicians in providing medical services and health education to workers. Like physicians, nurses must maintain a confidential relationship with employee patients and maintain accurate, timely records.

- One of the most important functions of employee health programs is setting up a regular schedule of physical examinations of all workers. The basic purpose is to establish a baseline for worker health at the outset of employment, to detect or monitor any work-related illnesses or injuries, and to match the right worker with the right job.

- Physicians and nurses conduct preplacement, periodic, transfer, promotion, special, and termination examinations. The findings of these examinations are confidential and may be released only with a worker's signed consent. Workers have a right to review these records to determine how they may have been affected by chemical, radiological, or biological or other hazards while on the job.

- Medical records must be maintained for up to 30 years in industries where workers have been exposed to hazardous materials. Such records contain the results of laboratory, X-ray, vision, and hearing tests along with a complete health and occupational history, medical opinions and diagnoses, treatments, and employee complaints.

- Emergency medical planning is a responsibility of industrial physicians, nurses, safety professionals, and management. These plans include procedures for training personnel in emergency measures, transporting and treating injured people, transferring the seriously injured to hospitals, and coordinating plans with other plant and community departments and agencies.

- Most companies now have some type of drug and alcohol screening and referral or treatment programs in place. Abuse of alcohol and drugs greatly affects a worker's job performance, places other workers at greater risk, and cost industry several billion dollars a year in lost time, accidents, and lower productivity.

REFERENCES

American College of Surgeons, Committee on Trauma. *Emergency Care of the Sick and Injured*. Philadelphia, PA: W.B. Saunders Company, 1982.

Brown ML. "Occupational Health Nurse: A New Perspective." In: *Occupational Medicine—Principles and Practical Applications*, 2nd ed., Chicago: Year Book Medical Publishers, Inc., 1987.

Occupational Health Nursing. New York: Springer Publishing Company, Inc., 1981.

Code of Federal Regulations, 29. Washington DC, Government Printing Office, 1985.

A Comprehensive Guide for Establishing an Occupational Health Service. Atlanta: American Association of Occupational Health Nurses, Inc., 1987.

Emergency Medical Identification, American Medical Association, 535 North Dearborn Street, Chicago, IL 60610.

Guidotti TL et al. *Occupational Health Services: A Practical Approach*. Chicago: American Medical Association, 1989.

Hannigan L. Nurse, health assessment—The new 'physical' in American industry. *Occupational Health Nursing*, August 1982.

LaDou J, ed. *Occupational Health and Safety*. Chicago: National Safety Council, 1992.

Medic Alert Foundation, 2323 Colorado, Turlock, CA 95380.

Plog BA, ed. *Fundamentals of Industrial Hygiene*, 3rd ed. Chicago: National Safety Council, 1988.

5

Environmental Management

Names such as "Bhopal," "Chernobyl," and "Valdez" have come to symbolize large-scale environmental disasters and to remind those whose job it is to control emissions and hazardous waste disposal that their task is vital to the well-being of the planet. All governments, as well as industry, are charged with protecting public health and welfare from a growing range of toxic materials (Figure 5–1).

U.S. GOVERNMENT REGULATIONS

Clean Air Act

The original Clean Air Act (CAA) was passed by Congress in 1967. This act provided the U.S. Environmental Protection Agency (EPA) with the authority to establish air quality standards. Following this, Congress passed the Clean Air Act of 1970 and amended it in 1974, 1977, and 1990.

The Clean Air Act refers to the discharge of pollutants from motor vehicles and industry into the atmosphere, into outdoor "ambient" air. There is no discussion in the original act or its amendments of indoor air pollution. Nevertheless, in the late 1980's EPA set a recommended action level for indoor levels of radon gas of four picocuries/per liter of air.

The Clean Air Act of 1970 laid the foundation for regulatory efforts to improve ambient air quality nationwide. It empowers the EPA to set and periodically review National Ambient Air Quality Standards (NAAQS), which define maximum concentrations of certain air pollutants allowable in ambient air to protect public health and welfare. The EPA identified primary air quality standards for ozone, carbon monoxide, particulate matter, sulfur dioxide, nitrogen dioxide, and lead. The national strategy was to establish threshold levels of these air pollutants, below which there would be no adverse effect on human health or the environment. The Act requires EPA to establish New Source Performance Standards (NSPS), which are allowable emissions limits for various stationary sources. The Clean Air Act also charges EPA with setting National Emissions Standards for Hazardous Air Pollutants (NESHAPs), pollutants for which no ambient air quality standards exist. Some NESHAPs have been published for arsenic, asbestos, benzene, beryllium, mercury, radionuclides, and vinyl chlorides.

The Energy Supply and Environmental Coordination Act of 1974 modified some of the NAAQS to compensate for shifts in energy sources caused by the Middle Eastern oil embargo. In 1977, Congress enacted additional amendments to the Clean Air Act. These include restrictions on new pollution sources installed in areas where the ambient air quality standards have not been attained and restrictions on industrial expansion to prevent significant deterioration of the ambient air quality.

Individual state governments are responsible for meeting and maintaining NAAQS that have been set by EPA. State governments measure local air quality and define ways to meet national standards via the State Implementation Plan (SIP). The 1970 Clean Air Act required that the original

SIPs be promulgated by 1972 to attain federal air quality standards by 1977. The Clean Air Act amendments of 1977 required that the original state implementation plans be upgraded to achieve compliance by 1982, or 1987 at the latest.

When states develop an implementation plan, it must be approved by municipal and state governments, and then approved by the EPA. Therefore, it becomes part of both the federal and state laws and may be enforced by either the EPA or the state regulatory agency.

Congress intended to create a partnership between federal and state governments in the CAA and in NSPS. EPA must develop regulations that include criteria for preconstruction permit programs for new and modified stationary sources of air emissions. Congress recognized that state governments best understand local air quality and local economic circumstances, and directed them to issue and enforce permits.

The Clean Air Act also gives EPA responsibility to prevent and control air emissions from mobile sources. EPA must prescribe and revise emission standards for new motor vehicle engines for certain vehicle categories and develop programs to test and certify these engines for compliance with national standards. EPA offices and staff also are charged with enforcing provisions related to in-use emissions from vehicles.

In 1990, Congress passed the Clean Air Amendments of 1990, which contained possibly the largest and most complex amendment to air pollution regulations since 1970. These amendments affected nonattainment areas, mobile sources, air toxic chemicals, acid rain, chlorofluorocarbons, permits, and enforcement. The provisions of these amendments change the approach the EPA and state regulatory agencies must take to control air pollution. The act identifies 189 "air toxic" chemicals that will require control under Maximum Achievable Control Technology (MACT), and regulates based on the "maximum exposed individual" (Table 5–A). It removes the "grandfather" exemption, which limited the need for air emission permitting.

Clean Water Act

The Clean Water Act was passed in 1972 and amended in 1977, 1980, 1981, and 1987. Its genesis was in the Federal Water Pollution Act of 1956 which, with its amendments, is the basis for the federal water pollution control program. The objective of the Clean Water Act is "to restore and maintain the chemical, physical, and biological integrity of the Nation's waters."

To meet that objective, Congress charged EPA with developing criteria for water quality standards, technology-based effluent limitation guidelines, pretreatment standards, new source performance standards, and a national permit program to regulate pollutant discharge. Congress gave individual states responsibility for developing water quality management programs and for setting water quality standards. In addition to regulations, the Clean Water Act established a large grant program to assist municipalities in meeting the act's requirements.

INTERNATIONAL AGREEMENTS

ANTARCTICA
1959 Antarctic Treaty and related instruments have sparked a substantial body of law. The 1988 agreement, not yet in effect, would regulate mineral exploitation. More recent proposals would turn Antarctica into a wilderness park and ban mineral activities.

ATMOSPHERE
Vienna Convention on the Protection of the Ozone Layer and its **Montreal Protocol** is a global effort on reducing ozone depletion. Parties to the protocol met in June 1990 to strengthen the controls on depleting substances and establish financial assistance for developing countries.

Convention on Long-Range Transboundary Air Pollution and its protocols controlling nitrogen-oxide and sufur-dioxide emissions cover the United States and Europe.

HAZARDOUS-WASTE EXPORT
A global treaty on exports of hazardous waste was adopted in 1989 in Basel, Switzerland, but is not yet in force.

INTERNATIONAL INSTITUTIONS
United Nations Environment Programme (UNEP).

International Maritime Organization (IMO) is set up to control marine pollution, primarily from ships, and serves MARPOL and the London Ocean Dumping Convention agreements.

Food and Agriculture Organization (FAO) deals with forestry, fisheries, and pesticide issues.

Organization for Economic Cooperation and Development (OECD) consists of the United States, Canada, Western Europe, Australia, New Zealand, and Japan. It is empowered to make binding decisions. Its primary successes have been in developing international environmental law.

United Nations Economic Commission for Europe (ECE) develops East-West treaties on environmental impact assessment and protection of watercourses.

Multilateral Development Banks include the **World Bank** and are crucial in protecting the environment through lending policies.

World Conservation Union is a combined governmental/nongovernmental institution that plays a significant role servicing agreements, including those on Earth's biological diversity.

MARINE ENVIRONMENT
London Ocean Dumping Convention, a direct outcome of the Stockholm Conference, regulates the disposal of wastes in the world's oceans. One unresolved issue is whether the oceans can be used for disposal of radioactive wastes, including decommissioned U.S. nuclear submarines.

International Convention for the Prevention of Marine Pollution from Ships (MARPOL) regulates the discharge of oil, chemicals, and garbage, including plastics, from ships. The convention will play an increasing role in the problem of marine debris, including debris in the Gulf of Mexico.

United Nations Environment Programme's Regional Seas Program is a broad marine-conservation treaty with supporting legal arrangements. It has sparked a series of regional agreements for the United States concerning such areas as the Gulf of Mexico, Caribbean, and South Pacific. These agreements are the only ones currently addressing the increasing problem of land-based pollution of the seas.

1982 United Nations Convention on the Law of the Sea is not in force. The United States objects to the conventions's seabed-mining provision.

WILDLIFE AND HABITAT
1973 Convention on International Trade in Endangered Species (CITES) regulates all international trade in endangered species. The protection of the African elephant through bans on ivory trade is a recent initiative.

1971 Ramsar Convention on Wetlands is an increasingly recognized vehicle for conserving wetlands worldwide.

Figure 5–1. Recently several other nations have passed agreements to manage and prevent environmental pollution.

The act classifies direct dischargers of pollutants as either point or nonpoint sources. To control point sources, those dischargers of pollutants who release into the atmosphere through a single point of conveyance, the act outlines the National Pollutant Discharge Elimination System (NPDES). The system incorporates and applies effluent limitations in individual permits for municipal and industrial dischargers. These permits subject dischargers to both technology-based treatment requirements and, where necessary to protect an area, to controls based on water quality standards.

The Clean Water Act's technology-based guidelines prescribe minimum standards of performance for all municipal and technology-based effluent limitations. The guidelines do not direct EPA to prescribe specific control technologies. EPA reviews treatment technology presently used or available in each industrial sector and determines which pollution limitations are achievable. Once EPA sets specific effluent limitations for a particular industrial category, each company can choose any method to reach them that does not violate other environmental laws.

By contrast, state water quality standards identify intended use of various bodies of water. Based on water-quality criteria guidance developed by EPA, the standards set forth biological and chemical conditions necessary to sustain those uses. Sometimes technology-based limitations are not adequate to achieve a particular state-designated use. Then state water quality standards are used to develop water-quality-based effluent limitations to meet the designated use.

The 1987 Amendment to the Clean Water Act provides up to $9.6 billion in grants and $8.4 billion in revolving loan funds to assist municipalities in constructing water treatment works projects through 1994. Programs up to $2 billion were authorized to clean up surface water bodies of chemical contamination and to curb nonpoint source pollution. It also gave EPA and the Secretary of the Army new authority to levy administrative penalties against offenders.

Amendments to the Clean Water Act emphasize control of toxic pollutants by requiring each state to list water bodies whose existing controls cannot meet water quality standards. The amendments require these states to develop individual control strategies to reduce toxic loading from point sources into listed water bodies. The law created new requirements for municipal sewage sludge management, to be implemented through the NPDES permit.

On November 16, 1990, the EPA published the new Stormwater Discharge Regulations under the authority of the Clean Water Act. In their most basic form, these laws regulate the discharge of stormwater "associated with industrial activities" into the waters of the United States. The rules were published in the *Federal Register,* and all affected facilities must file a Form 1 and Form 2F no later than November 18, 1991.

RCRA and HSWA

The Resource Conservation and Recovery Act (RCRA) was passed in 1976. In 1984, Congress reauthorized it as the Hazardous and Solid Waste Amendments (HSWA).

RCRA was the first statutory framework designed to provide comprehensive federal and state hazardous-waste management. It requires industry to identify and list all hazardous wastes and to note the toxicity, persistence, degradability in nature, potential for accumulation in tissue, and other characteristics of these waste materials. RCRA directs standards establishment for hazardous waste generators to protect human health and the environment. These standards must include requirements for recordkeeping, container labeling, component disclosure, use of a manifest system to track hazardous waste movement from "cradle to grave," and reporting to EPA. Similar standards are described for hazardous waste transporters in cooperation with the Department of Transportation.

In addition to standards for hazardous waste generators, RCRA prescribes performance standards development and establishment of a permit system for owners and operators of hazardous waste treatment, storage, and disposal facilities (TSDs). This provision was designed to ensure that all facilities that handle hazardous wastes will operate under the conditions specified in an RCRA permit.

HSWA added a major provision to RCRA in 1984 when it required corrective action at sites with continuing hazardous waste releases. Every TSD facility seeking a RCRA permit must now take corrective action for releases of any hazardous wastes or constituents from any solid waste management unit (SWMU), regardless of when the waste was placed in the SWMU. HSWA enables EPA and authorized states to issue orders for requiring a firm to take corrective action when hazardous waste release occurs either at a permitted facility or at an interim status facility (a TSD that is operating pending approval of its permit application).

Subtitle D of the Act provides for developing and encouraging environmentally sound solid-waste disposal methods that also conserve valuable resources. This objective is to be supported by federal technical and financial assistance to state and regional authorities pursuant to federal guidelines. By using these guidelines and federal grants, each state is to develop its own solid-waste management plan.

A major 1984 HSWA amendment to the act is Subtitle I, Regulation of Underground Storage Tanks (USTs). This law outlines how to develop and implement a program to regulate USTs that store petroleum (including gasoline and crude oil) and other substances defined as hazardous under the Comprehensive Environmental Response, Compensation, and Liability Act of 1980 (CERCLA). The UST program defines the types of tanks permitted and initiates a tank notification system. It requires EPA to issue technical standards for release detection, prevention, and corrective action for all tanks. The UST program is intended to be state run, with EPA issuing regulations, guidance, and providing financial assistance.

The 1986 reauthorization of CERCLA amended Subtitle I to address leaking USTs. EPA, or states in a cooperative agreement with EPA, can (1) require an owner or operator of a UST to take corrective action if it is determined that the action will be taken promptly and properly and (2) under-

take corrective action if it is necessary to protect human health and the environment. A $500 million Leaking Underground Storage Tanks Trust Fund was established to cover the cost of EPA or state cleanup responses. The amendment resulted in EPA-issued regulations for maintaining evidence of financial responsibility, mitigating releases, and compensating third parties for bodily injury and property damage.

Protection of groundwater was a major theme of HSWA, which mandated:

1. new technological standards for land disposal facilities, including double liners, leachate collection systems, and groundwater monitoring
2. new requirements for managing and treating smaller quantities of hazardous waste, such as those generated by auto repair shops and dry cleaners
3. new release detection, prevention, and correction regulations for USTs that contain liquid petroleum or chemical products
4. upgraded performance and design criteria for disposing of municipal solid waste in municipal or industrial landfills
5. restrictions on the land disposal (i.e., "land ban rule") of many untreated hazardous wastes.

CERCLA and SARA

Congress enacted the CERCLA in 1980 after several incidents of uncontrolled, dangerous disposal of toxic chemicals made it apparent that RCRA's prospective regulatory framework could not meet the remedial needs of some sites. By 1980, thousands of uncontrolled sites had been identified, but the lack of funds and legal authority impaired any real progress in cleaning up these areas. Among other things, CERCLA, which soon became known as the "Superfund Law," established a $1.6 billion fund to clean up abandoned hazardous chemical sites.

CERCLA was envisioned as a five-year program to spearhead federal and state response to hazardous substance releases into the environment. Legislative goals were to eliminate the most serious threats to public health and the environment posed by hazardous substance spills and uncontrolled chemical waste sites in a cost-effective manner.

Title I of CERCLA deals with the release of hazardous substances, liabilities imposed for those releases, and compensation to be paid for costs and damages resulting from the releases. Title II of the original CERCLA imposed certain "environmental taxes" on the petroleum and chemical industries and set up the Hazardous Substance Response Trust Fund. Taxes were also imposed on the owners and operators of qualified hazardous waste disposal facilities to establish a second fund, known as the Post-Closure Liability Trust Fund.

CERCLA defines "hazardous substance" by incorporating the substances listed in other environmental statutes, including CAA, the Clean Water Act, RCRA, and the Toxic Substances Control Act. In addition, CERCLA directs EPA to issue and revise regulations designating as hazardous other substances that pose a substantial danger to public health. Under CERCLA, regulations were also issued to establish the threshold quantity of a hazardous substance spill. Envi-

ronmental releases or spills in excess of the reportable quantity (RQ) trigger notification and response requirements under the Act.

The Act requires any person in charge of a vessel or facility who has knowledge of a hazardous-substance release greater than the RQ from that vessel or facility to immediately notify the National Response Center. CERCLA also requires owners or operators of hazardous substance storage, treatment, and disposal sites to notify EPA of the facilities' existence, the amount and type of hazardous substances on site, and whether known or suspected releases have occurred. If a hazardous substance is released, the National Contingency Plan (NCP) sets forth the procedures to be followed.

The NCP procedures provide for the discovery, investigation, evaluation, and removal (where necessary) of hazardous substances. Three possible CERCLA actions following hazardous-substance release are provided by the NCP:

- immediate removal actions—a prompt response, within hours or days, to prevent immediate and significant harm to human life, health, or the environment
- planned removal actions—those that allow time to plan the cleanup activities
- remedial actions—intended to achieve a permanent remedy or cleanup of hazardous waste sites.

Under CERCLA, the EPA has authority to remediate any facility ranked and placed on the National Priority List as a remedial action site. It may also step in on an emergency basis at any site where, in its judgment, there is an imminent, significant threat to human health or the environment. Once these actions are complete, the EPA looks at the cost recovery section to identify potentially responsible parties. The agency then initiates action against these parties to recover EPA costs on a joint and several liability basis. This means offenders may be liable either individually or together for all costs.

MANAGING FOR A HEALTHY ENVIRONMENT

Industry experts Christopher Hunt and Ellen Auster describe five distinct stages of environmental management program development (Table 5–B and References). Their survey of corporations revealed departments that range from the "beginner" (stage one) through the "proactivist" (stage five). Small companies or ones that face minimal environmental risks do not require a "stage five, proactivist" environmental management program. Companies that use hazardous materials in their manufacturing processes or those with a variety of processes will, however, require a more active management approach. A program can be proactive for one company, but inadequate for another. A bakery may be able to operate successfully with a few guidelines and periodic reviews by a consultant. However, large manufacturers or chemical companies often require a couple of hundred staff members who are environmental professionals.

List of 189 Chemicals Selected for Air Toxics

75070	Acetaldehyde
60355	Acetamide
75058	Acetonitrile
98862	Acetophenone
53963	2-Acetylaminofluorene
107028	Acrolein
79061	Acrylamide
79107	Acrylic acid
107131	Acrylonitrile
107051	Allyl chloride
92671	4-Aminodiphenyl
62533	Aniline
90040	o-Anisidine
1332214	Asbestos
71432	Benzene (including benzene from gasoline)
92875	Benzidine
98077	Benzotrichloride
100447	Benzyl chloride
92524	Biphenyl
117817	Bis(2-ethylhexyl)phthalate (DEHP)
542881	Bis(chloromethyl)ether
75252	Bromoform
106990	1,3-Butadiene
156627	Calcium cyanamide
105602	Caprolactam
133062	Captan
63252	Carbaryl
75150	Carbon disulfide
56235	Carbon tetrachloride
463581	Carbonyl sulfide
120809	Catechol
133904	Chloramben
57749	Chlordane
7782505	Chlorine
79118	Chloroacetic acid
532274	2-Chloroacetophenone
108907	Chlorobenzene
510156	Chlorobenzilate
67663	Chloroform
107302	Chloromethyl methyl ether
126998	Cloroprene
1319773	Cresols/Cresylic acid (isomers and mixture)
95487	o-Cresol
108394	m-Cresol
106445	p-Cresol
98828	Cumene
94757	2,4-D, salts and esters
3547044	DDE
334883	Diazomethane
132649	Dibenzofurans
96128	1,2-Dibromo-3-chloropropane
84742	Dibutylphthalate
106467	1,4-Dichlorobenzene(p)
91941	3,3-Dichlorobenzidene
111444	Dichloroethyl ether (Bis(2-chloroethyl)ether)
542756	1,3-Dichloropropene
62737	Dichlorvos
111422	Diethanolamine
121697	N,N-Diethyl aniline (N,N-Dimethylaniline)
64675	Diethyl sulfate
119904	3,3-Dimethoxybenzidine
60117	Dimethyl aminoazobenzene
119937	3,3'-Dimethyl benzidine
79447	Dimethyl carbamoyl chloride
68122	Dimethyl formamide
57147	1,l-Dimethyl hydrazine
131113	Dimethyl phthalate
77781	Dimethyl sulfate
534521	4,6-Dinitro-o-cresol, and salts
51285	2,4-Dinitrophenol
121142	2,4-Dinitrotoluene
123911	1,4-Dioxane(1,4-Diethyleneoxide)
122667	1,2-Diphenylhydrazine
106898	Epichlorohydrin (1-Chloro-2,3-epoxypropane)
106887	1,2-Epoxybutane
140885	Ethyl acrylate
100414	Ethyl benzene
51796	Ethyl carbamate (Urethane)
75003	Ethyl chloride (Chloroethane)
106934	Ethylene dibromide (Dibromoethane)
107062	Ethylene dichloride (1,2-Dichloroethane)
107211	Ethylene glycol
151564	Ethylene imine (Aziridine)
75218	Ethylene oxide
96457	Ethylene thiourea
75343	Ethylidene chloride (1,1-Dichloroethane)
50000	Formaldehyde
76448	Heptachlor
118741	Hexachlorobenzene
87683	Hexachlorobutadiene
77474	Hexachlorocyclopentadiene
67721	Hexachloroethane
822060	Hexamethylene-1,6-diisocyanate
680319	Hexamethylphosphoramide
100543	Hexane
302012	Hydrazine
7647010	Hydrochloric acid
7664393	Hydrogen fluoride (Hydrofluoric acid)
123319	Hydroquinone
78591	Isophorone
58899	Lindane (all isomers)
108316	Maleic anhydride
67561	Methanol
72435	Methoxychlor
74839	Methyl bromide (Bromomethane)
74873	Methyl chloride (Chloromethane)
71556	Methyl chloroform (1,1,1-Trichloroethane)
78933	Methyl ethyl ketone (2-Butanone)
60344	Methyl hydrazine
74884	Methyl iodide (Iodomethane)
108101	Methyl isobutyl ketone (Hexone)
624839	Methyl isocyanate
80626	Methyl methacrylate
1634044	Methyl tert butyl ether
101144	4,4-Methylenebis(2-chloroaniline)
75092	Methylene chloride (Dichloromethane)
101688	Methylene diphenyl diisocyanate (MDI)
107779	4,4-Methylenedianiline
91203	Naphthalene
98953	Nitrobenzene
92933	4-Nitrobiphenyl
100027	4-Nitrophenol
79469	2-Nitropropane
684935	N-Nitroso-N-methylurea
62759	N-Nitrosodimethylamine

Table 5–A. List of 189 chemicals selected for air toxics, 1990.

List of 189 Chemicals Selected for Air Toxics

59892	N-Nitrosomorpholine
56382	Parathion
82688	Pentachloronitrobenzene
	(Quintobenzene)
87865	Pentachlorophenol
108952	Phenol
106503	p-Phenylenediamine
75445	Phosgene
7803512	Phosphine
7723140	Phosphorus
85449	Phthalic anhydride
1336363	Polychlorinated biphenyls
	(Aroclors)
1120714	1,3-Propane sultone
57578	beta-Propiolactone
123386	Propionaldehyde
114261	Propoxur (Baygon)
78875	Propylene dichloride
	(1,2-Dichloropropane)
75569	Propylene oxide
75558	1,2-Propylenimine
	(2-Methyl aziridine)
91225	Quinoline
106514	Quinone
100425	Styrene
96093	Styrene oxide
1746016	2,3,7,8-Tetrachlorodibenzo-p-dioxin
79345	1,1,2,2-Tetrachloroethane
127184	Tetrachloroethylene
	(Perchloroethylene)
7550450	Titanium tetrachloride
108883	Toluene
95807	2,4-Toluene diamine
584849	2,4-Toluene diisocyanate
95534	o-Toluidine
8001352	Toxaphene (chlorinated camphene)
120821	1,2,4-Trichlorobenzene
79005	1,1,2-Trichloroethane
79016	Trichloroethylene
95954	2,4,5-Trichlorophenol
88062	2,4,6-Trichlorophenol
121448	Triethylamine
1582098	Trifluralin
540841	2,2,4-Trimethylpentane

108054	Vinyl acetate
593602	Vinyl bromide
75014	Vinyl chloride
75354	Vinylidene chlonde
	(1,1-Dichloroethylene)
1330207	Xylenes (isomers and mixture)
95476	o-Xylenes
108383	m-Xylenes
106423	p-Xylenes

Antimony Compounds
Arsenic Compounds
(inorganic including arsine)
Beryllium Compounds
Cadmium Compounds
Chromium Compounds
Cobalt Compounds
Coke Oven Emissions
Cyanide Compounds [1]
Glycol ethers [2]
Lead Compounds
Manganese Compounds
Mercury Compounds
Fine mineral fibers [3]
Nickel Compounds
PolycylicOrganic Matter [4]
Radionuclides (including radon) [5]
Selenium Compounds

[1] X'CN where X = H' or any other group where a formal dissociation may occur, for example, KCN or Ca (CN) 2.
[2] Includes mono- and di- ethers or ethylene glycol, diethylene glycol, and triethylene glycol R-(OCH2CH2) n-OR' where
 n = 1, 2, or 3
 R = alkyl or aryl groups
  R' = R, H, or groups which, when removed, yield glycol ethers with the structure R- (OCH2CH) n-OH. Polymers are excluded from the glycol category.
[3] Includes mineral fiber emissions from facilities manufacturing or processing glass, rock, or slag fibers (or other mineral derived fibers) of average diameter 1 micrometer or less.
[4] Includes organic compounds with more than one benzene ring, and which have a boiling point greater than or equal to 100°C.
[5] A type of atom which spontaneously undergoes radioactive decay.

Table 5–A. *(Continued.)*

Hunt and Auster describe "stage one, beginner" environmental programs as those of either older companies established before the environmental acts were passed or of firms, like banks or real estate developers, that do not deal directly with toxic materials, but that may encounter related risks from them. "Beginners" often ignore environmental concerns or deal with them by adding responsibility to an existing job description, such as plant manager or senior engineer. "Beginners" do not define corporate environmental policy, nor do they consider the potential effects of failing to have such a policy.

"Stage two: the fire fighter" companies are described by Hunt and Auster as having a few people who spend time on environmental concerns or a small group of professionals who help individual installations respond to problems. Given this structure, environmental professionals have no option but to work at a crisis-intervention level. They have no opportunity to consider risks from serious problems that might occur in the future. Small or medium-sized firms that are at this level engage in environmentally hazardous activities and have not considered the potential benefits of active environmental management.

Hunt and Auster describe "stage three: the concerned citizen" companies as those whose established environmental departments are either too understaffed or not high enough in the corporate structure to make significant changes. They are often staffed by technically competent professionals, like chemists, biologists, or geologists, whose backgrounds do not prepare them to offer business, legal, and public relations expert advice for environmental management. These companies, concerned about their environmental responsibilities, have funded programs that operate without serious

Table 5–B. Developmental Stages of Corporate Environmental Management Programs.

Criteria	Stage One "Beginner"	Stage Two "Fire Fighter"	Stage Three "Concerned Citizen"	Stage Four "Pragmatist"	Stage Five "Proactivist"
Degree to which Program Reduces Environmental Risk	No protection	Minimal protection	Moderate protection	Comprehensive protection	Maximum protection
Commitment of Organization					
–General Mindset of Corporate Managers	Environmental management is unnecessary	Environmental issues should be addressed only as necessary	Environmental management is a worthwhile function	Environmental management is an important business function	Environmental management is a priority item
–Resource Commitment	Minimal resource commitment	Budgets for problems as they occur	Consistent, yet minimal budget	Generally sufficient funding	Open-ended funding
–Support and Involvement of Top Management	No involvement	Piecemeal involvement	Commitment in theory	Aware and moderately involved	Actively involved
Program Design					
–Performance Objectives	None	Resolve problems as they occur	Satisfy corporate responsibility	Minimize negative environmental impacts	Actively manage environmental matters
–Integration with Company	Not integrated	Involved with other departments on piecemeal basis	Minimal interaction with other departments	Moderate integration with other departments	Actively involved with other departments
–Reporting to Top Management	No reporting	Exceptions reporting only	Generates voluminous reports that are rarely read	Consistent and targeted reporting	Personal meetings with managers and board of directors
–Reporting Structures	None	Exceptions reporting only	Internal reporting only	Mostly internal with some external reporting	Formalized internal and external reporting mechanisms
–Involvement with:					
•Legal Counsel	None	Moderate	Moderate	High	Daily
•Public Relations	None	None	Moderate	High	Daily
•Manufacturing/ Production	None	None	None	Moderate	Daily
•Product Design	None	None	None	Minimal	Daily

senior management commitment to integrate them with the operating units.

At "stage four, the pragmatist" companies, environmental department staffs have sufficient expertise, funding, and authority. Staff members review all facilities and design better ways to limit toxic releases. They look to the future, evaluating potential risks and designing solutions whenever appropriate. They train key workers in environmental protection and write policy manuals for operating personnel. Formal reporting relationships are established. Although these departments are aggressive, their authority and program funding may be limited. At this stage of development, companies have not made environmental management a top-level corporate concern.

Environmental management is a top priority for companies that have reached "stage five: the proactivist." The environmental department is staffed with assertive leaders whose commitment to the environment goes beyond meeting regulations and planning for their firm's future environmental needs. These professionals are active in industry roundtables, sharing information and solutions with competitors to ensure that their performance does not result in polluted resources, adverse publicity, and stricter regulations. They participate in local, state, and federal policymaking meetings, aggressively pursuing solutions as they help policymakers determine how best to regulate their industry.

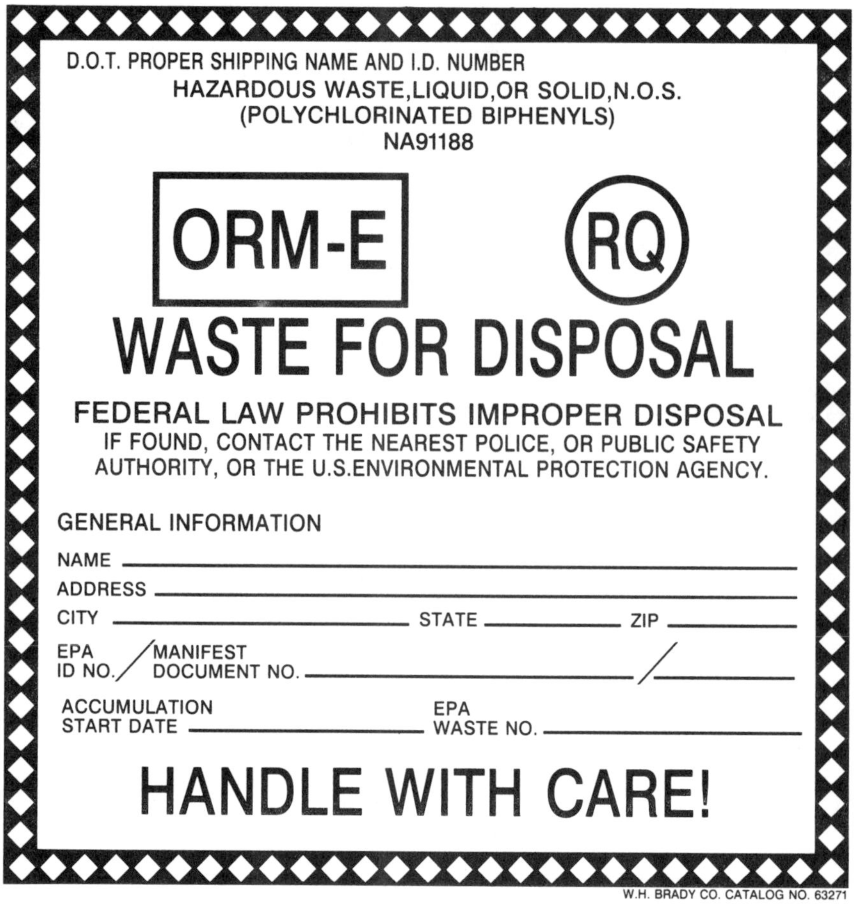

Figure 5–2. Sample hazardous waste label. (Courtesy W.H. Brady Co.)

Not every organization needs to budget for a sophisticated, proactive environmental management department. The level of commitment appropriate for each company should be determined by its inherent environmental risks and its size.

Organizing an Environmental Management Program

Establishing an environmental compliance program can be an overwhelming proposition. However, any company whose business is regulated by U.S. EPA, the Occupational Safety and Health Administration (OSHA), Department of Transportation (DOT), other government agencies, or state or local authorities cannot afford to be without an environmental management program.

Corporate culture—the size of the company, diversity of its business units, and how it interacts with headquarters—will determine if the environmental management program will be small and basic, large and sophisticated, centralized or decentralized. Whether the company is large or small, however, senior management must communicate a strong commitment to responsible environmental practices through the ranks, down to the line worker and platform loader. Unfortunately, a sincere environmental management commitment that alters old practices often occurs only after a major disaster or stringent penalty. This does not have to be the case. Top management teams can learn from other companies' misfortunes.

Deciding where to begin can be the biggest obstacle to organizing an environmental management program. The ways to comply with governmental regulations can seem as endless as the thick tomes that transmit those regulations. William Friedman outlines five basic activities essential to any successful environmental compliance program (see References).

Prevent common violations. Mislabeling of chemicals and hazardous waste is probably the most commonly cited violation, and one that can most easily cause an environmental accident. It is also the easiest violation to avoid. Companies should attach proper labels to all containers. In the case of hazardous waste, use a label that meets the DOT's requirements (Figure 5–2).

Regulations require storage containers to be in good condition. Leaking containers are a common source of fines. Inspect tanks daily, and check drums and similar containers weekly. Make sure aisle space is adequate to give inspectors easy access to the facilities. When waste is stored in tanks, regulations require secondary containment to prevent leaks or spills (Figure 5–3).

Figure 5–3. Tank storage in this chemical operation is being inspected. Note the railings and toeboard on the stairway.

Recordkeeping. Environmental regulations have mandated that plants alter the way they handle and dispose of chemicals and hazardous wastes. They also mandate that company personnel create and maintain extensive records that document how the firm has met handling and disposal regulations. During an inspection by a regulatory agency, the mere existence of the records, their easy accessibility, and identifiability may be all an inspector needs.

It is also important to document all environmental management decisions. For example, whenever a facility has analyzed its waste streams and found them to be nonhazardous, records of the information supporting that determination should be maintained and made readily available to an inspector. Relying on a very sketchy explanation from a compliance manager (who may or may not be employed or present at the time of the inspection) can lead to an inspection report listing possible violations. Staff must then reconstruct the information and transmit it to the regulatory agency, which may still issue a violation. Accurate, available, and complete records save companies time, money, and legal headaches.

Create a spill reporting plan. Under numerous statutes

and regulations (notably the U.S. federal Superfund law), companies must notify the U.S. National Response Center immediately after release of a reportable quantity of any hazardous substances or pollutants. As regulators decrease the time allowed between an incident and the reporting time, environmental authorities are strictly enforcing existing laws. In some cases, like air pollution discharges, the permitted time lag is only minutes.

An unintentional spill or release of a toxic material usually occurs during a plant emergency when personnel are literally putting out fires or responding to explosions or other emergencies. For that reason, every facility must have written instructions for employees to follow whenever a release or spill occurs. The reporting plan should detail the plant's procedures, telephone numbers of all agencies that should be notified, and who has the authority to report spills. Employees should be aware of these instructions and where they can be found at various locations. Spill reporting plans not only permit companies to respond immediately to emergencies, but their existence is also evidence of a firm's environmental responsibility to regulatory agencies that might question the company's ability to respond to an emergency.

Set realistic limits and schedules. Environmental managers should suggest realistic compliance schedules and discharge limitations to environmental regulators. Often in a spirit of cooperation, a manager will offer overly ambitious schedules and limits that ultimately cannot be met. Do not overcommit. Once limits are set in discharge permits and orders, it is difficult to have them changed. Before such limits are written into a company's legal documents, management should consult with government regulators to set realistic dates and limits.

When reporting a release that exceeds set limits, always include the reason and make an attempt to solve the problem. Companies have reported pollutant releases above limits, received no response from the regulatory agency and, over time, have assumed that the increased rate is acceptable. The company's own reports then establish its lack of compliance, which can prompt an enforcement action by the agency or serve as evidence in a citizen suit against the company. Whenever it appears that legal requirements cannot be met in the future, the compliance manager should request a change in limits or schedules, and detail the reasons why compliance is not possible.

Motivate employee action. Spills and releases are almost never caused by the person who writes a company's environmental compliance program. Mistakes that become environmental problems are usually made by the platform loader, line, or storage worker. If these employees are not educated or trained in environmental management practices, they often focus on doing their jobs quickly at the expense of environmentally important details and filling out "bothersome" reports.

Although training for all employees who handle environmentally sensitive substances is essential, it is all too often insufficient. Including, adherence to environmental protection policies in employees' annual reviews and holding them

responsible for cited violations brings home the importance of "little details" and "bothersome" reports. It is crucial, as well, to make employees aware that in this age of strict environmental enforcement, it is less costly for a company to adhere to regulations than to defend itself against a lawsuit.

Staff Skills and Backgrounds
The key player in any successful environmental management department is the manager. Whether his or her background is in engineering, science, or law is immaterial. The complex and diverse issues that the manager must deal with call for a person with top managerial skills, good internal and external networks, rapport with senior management, and the respect of others within the organization. Because of the diversity and sensitivity of day-to-day business, the manager should have a background in environmental issues, whether it be legal or scientific.

Most of the environmental management department staff will be scientists and engineers. Depending upon the industry, a company's environmental staff will include chemists, geologists, biologists, and so on. Because many corporate environmental issues are legal, companies also need a close association with an appropriate attorney. Without such a relationship, the company may overlook key environmental data or legal facts that can plague it far into the future. An attorney can take preventive legal measures that will avoid problems which could prove costly both financially and in terms of corporate image and public relations.

Some environmental management departments have an attorney on staff. In others, the corporate general counsel's office includes an attorney who is an environmental specialist. Smaller companies may rely entirely on outside counsel. Whichever approach is taken, it is essential to establish a firm relationship with an attorney skilled in environmental issues who will review environmental data and regulations and oversee all company transactions. Counsel must have a close relationship with the environmental manager so as to respond to concerns and questions in a timely fashion.

A strong relationship must also be established between the environmental management department and the company's public relations department or outside public relations firm. A three-way partnership among the environmental manager, public relations executive, and attorney is the best assurance that the company's environmental efforts meet corporate guidelines, various regulations and statutes, and are reported to the public in the best light possible.

SOURCES OF INFORMATION
The U.S. federal government publishes all of the regulations concerning environmental protection in Volume 40 of the *Code of Federal Regulations*. These regulations are amended and updated periodically by notice in the *Federal Register*. It is advisable to keep a current set of 40 *CFR* with all sections applicable to a firm's operations and to keep informed of proposed rule changes and notices of rule makings. This can be done by subscribing to the *Federal Register,* or by en-

listing the aid of private services and newsletters.

All U.S. states have environmental regulations, and companies should maintain a current set of these regulations available for access. As with the federal regulations, these will remain current only through the date they are published. There are few services and newsletters that do an in-depth job of keeping abreast of state regulations. However, some are available, including the *State Register*, similar to the *Federal Register*, which publishes rule changes and notices of rule makings.

Air Emissions

In evaluating a facility's air emissions, management must obtain some preliminary information. First, acquire a current copy of the state air regulations and review them to determine if any restrictions in excess of the federal minimums apply to the area in which the company exists or is to be constructed. Find out if the facility is in an area that has attained compliance with the NAAQS. If the facility is in a nonattainment zone, restrictions may apply to the existing emissions and may restrict future expansion. Determine if the facility has adequate baseline data to estimate total emissions accurately. Review all existing air emission permits to determine if all emissions are accounted for and whether they are consistent with the baseline data.

The Clean Air Act places most of the responsibility on the states to achieve compliance with the NAAQS. In order to accomplish this, the states developed SIPS. Each state divided their geographical areas into Air Quality Control Regions (AQCR) and attempted to calculate how much loading of each pollutant each area could withstand before exceeding the NAAQS. Then, states imposed control requirements in areas where pollution loading exceeded the calculated threshold level. Because of this, regulations vary from state to state and even from area to area within a state. Therefore, to assess a facility's position, management must know the regulations applicable to the area in which the plant is located.

Air emissions are categorized as either "point source" or "fugitive." Point source emissions are those that are released into the atmosphere through a single point of conveyance. Examples of this would be a stack or ventilation fan exhaust. Fugitive emissions enter the atmosphere through uncontrolled leakage at valves, flanges, pumps, and fittings. Baseline data for point sources can be developed through application of stack testing, mass balance calculations, or engineering knowledge. Fugitive emissions are much more difficult to estimate. There are published emission factors that managers can use to calculate fugitive emissions.

After management has assembled the baseline data to determine the total plant emissions and reviewed the permits to determine what the plant is allowed to emit, they should conduct a review to ensure that the plant is in compliance with all permit requirements and that the existing baseline data are accurate as necessary. In addition, the facility management should be consulted to determine if there is any planned expansion or process alterations. With this information, it is possible to make an assessment of the present facility position and the adequacy of its air emission control equipment to comply with future requirements.

The Clean Air Act Amendments of 1990 will have a profound impact on most industries, particularly in the area of planned industrial expansion. Prior to the 1990 amendments, the principal concerns of industry in planned expansion focused on "prevention of significant deterioration" requirements and "new source performance" standards.

Prevention of significant deterioration (PSD) is an extremely complex area of the federal air emission regulations. The basic purpose is to prevent unlimited industrial expansion from degrading the existing air quality in areas where ambient air quality standards already are being met. This is accomplished by requiring all new plants and major modifications to install the Best Available Control Technology (BACT). BACT is determined on a case-by-case basis. Any plant that falls within eight specified industrial categories and has potential emissions of any regulated pollutant exceeding 100 tons per year, or any other facility with potential emissions exceeding 250 tons per year, is classified as a major source. A "major modification" generally is defined as any significant net increase emission of a regulated pollutant. The definition of significant net emission increase varies, but for sulfur dioxide, nitrogen oxide, and volatile organic compounds (VOCs), an increase of 40 tons per year is deemed significant.

The EPA has published NSPS for specific industrial categories, limiting emissions by production-based standards or specifying BACT. As an example, emissions from polyethylene, polystyrene and polypropylene, crystal, and high-impact grade continuous process production lines, were limited to "0.0036 lb/1,000 lb of production, or the final temperature on the final condenser in the final material recovery section of -25 C." Any new construction or major modification of an existing facility within an industry for which a NSPS has been established must demonstrate conformity with those standards.

United States areas that have not attained the primary NAAQS are subject to additional restrictions. These restrictions apply to the construction of new sources or modification of existing sources. These regulations require that affected sources install state-of-the-art control technology and also provide supplemental reductions in emissions from like-kind sources to offset whatever emissions will result from the new source, even after BACTs are used.

These sources are subjected to the use of the Lowest Available Emission Rate (LAER). This requirement imposes standards requiring that either the most stringent emissions limitations contained in any SIP or the most stringent emission limitation achieved in practice within the industrial category, whichever is more restrictive, be used. In no event can LAER be less stringent than any applicable NSPS. It may be equal to or more stringent than the published BACT standard. Facilities are required to offset the emissions increase caused by the new or modified source. These offsets can be derived from the installation of advanced controls, producing extra reductions in emissions from existing

sources, or from the shutdown of sources. These offsets must be measured against prior actual emission data and must be equal to, or greater than, the potential emission increase from the new source.

The CAA Amendments of 1990 will have profound effects on all forms of air pollution. They have been projected to cost approximately $25 billion per year. These amendments modify the requirements for nonattainment areas, tightening the controls on industrial facilities by using the LAER requirements on sources not previously covered. The 1977 amendments required that major sources install Reasonably Available Control Technology (RACT). Major sources were defined as those with the potential to emit 100 tons per year or more.

Under the 1990 amendments, nonattainment areas would be divided into four categories: extreme, severe, serious, and moderate/marginal. In an extreme nonattainment area, RACT would be imposed on sources emitting 10 tons per year or more. Severe nonattainment areas would have RACT imposed on sources of 25 tons per year or more. Serious nonattainment areas would have RACT requirements imposed on sources of 50 tons per year. Moderate/marginal nonattainment areas would continue to have RACT imposed on sources of 100 tons per year or more. The EPA will issue control technique guidelines to provide a generic definition of RACT for specific industrial categories. In addition, states must revise their SIPs for all but the marginal nonattainment areas by November 15, 1993, and provide controls to achieve 15% reduction in VOC emissions in the first six years.

The air toxics provisions of the 1990 amendments are likely to have the most profound effect on industry. These requirements set standards governing the emissions of 189 identified air toxic chemicals. Table 5–A lists those chemicals and their CAS numbers. Prior CAA efforts addressed the six NAAQS.

Air toxics provisions are intended to address small concentrations of air toxic chemicals generally present in localized areas due to industrial concentrations. Generally, any industry that emits 10 tons per year of any single air toxic chemical, or 25 tons per year of all air toxic chemicals will be required to comply with the MACT standards. These standards will require major investments in new control technology to reduce the releases of these substances wherever possible.

Within the first two years, the EPA is mandated to promulgate MACT standards for 40 industrial categories. MACT standards will require the maximum reduction in emissions achievable through measures that will not be less stringent than those achieved in practice by the best similar sources for new sources, and for existing sources shall not be less stringent than the best performing 12% of existing sources. In addition, the air toxics provisions require that the EPA promulgate a list of at least 100 air pollutants that pose the greatest risk from accidental release, considering the severity of acute health effects, the likelihood of accidental releases, and the potential magnitude of human exposure.

The CAA Amendments of 1990 also addressed additional restrictions on mobile sources, acid rain, chlorofluorocarbons, permits, and enforcement. Although some of these issues will not have a direct effect on industry, the permit provisions will require that most sources acquire a permit to operate their facilities. The permits will be issued by state agencies and will require compliance with federal regulations. These permits will include a fee of not less than $25 per ton of all regulated pollutants except carbon monoxide. These permits will probably be due by 1995. Enforcement is strengthened to include administrative penalties of up to $25,000 per day, field citations imposing penalties of up to $5,000 per day for each violation, and various other enforcement provisions.

Effluent Discharges
Effluent discharges can be divided into two broad categories. The first category includes discharges associated with the NPDES permitting program. The second category is discharges associated with an industrial pretreatment program. Wastewater discharges associated with the NPDES permitting program cover discharges of wastewaters into the waters of the United States at any point source. Wastewater discharges associated with industrial pretreatment include wastewater that is discharged through any point source into a publicly owned treatment works (POTW). Some industrial pretreatment programs are administered by the state and others by the receiving POTW.

Industrial wastewater is divided into two categories: process contact wastewater or noncontact wastewater. Industrial wastewater is identified as a process contact wastewater if it comes into direct contact with, or results from, the production or use of any raw material, intermediate finished product, byproduct, or waste product. Noncontact wastewater includes cooling water blowdown, (when the cooling water does not come into direct contact with any raw material, intermediate product, waste product, or finished product), blowdown from industrial boilers or furnaces, and similar discharges.

Effluent discharges are regulated under the U.S. Clean Water Act, which is detailed in 40 *CFR*. If these discharges are directly into a navigable waterway through a point source, they require a NPDES discharge permit. In addition, state regulatory agencies may require state water discharge permits. An NPDES discharge permit takes into consideration the pollution loading on the receiving water body. The EPA has established pollution standards for various types of water systems based mainly on the use of the aquatic habitat. For example, a water system used for high-quality aquatic life and contact recreation would have a lower pollution loading limit than would a noncontact aquatic habitat located outside of any known fishing area. All NPDES discharge permits contain some federal pollution standards. These include such things as a biological oxygen demand, chemical oxygen demand, total organic carbon, total suspended solids, pH, ammonia, and chlorine. The EPA has published categorical effluent guidelines covering specific industrial categories. These effluent standards restrict

Figure 5–4. This analyzer assists compliance with U.S. EPA residual discharge requirements by accurately and continuously monitoring both sulfite and chlorine residuals in wastewater.

specific pollutants in the discharge from those industries.

Industrial pretreatment programs require that industry discharging into a POTW comply with specific discharge restrictions (Figure 5–4). In instances where the POTW administers its own industrial pretreatment program, compliance is normally governed through an interconnect permit. This permit establishes the effluent guidelines that must be achieved prior to discharge into the sewer system. In areas where the POTW does not have an industrial interconnect program, the state regulates industrial interconnects through the state pretreatment program. As with NPDES discharge permits, the EPA has established industrial categorical pretreatment standards, requiring that specific industrial categories pretreat their effluent water prior to discharge to a POTW.

The stormwater discharge regulations require that any "stormwater associated with wastewater activities" discharged through a permit source be permitted under the NPDES permit program. "Stormwater associated with industrial activities" includes, generally, everything that falls inside the fence line of the facility. If it rains on any area used to load, unload, store, handle, process, produce, package, transport, or ship any raw material, intermediate, finished product, or chemical listed in 40 *CFR*, Part 302.4 or Part 335, a facility must apply for a stormwater discharge permit regardless of how the stormwater is conveyed to the waterway and whether or not it passes through a treatment works or sewer system. Although there are exceptions, they are few, and it is doubtful that most will apply to a signifi-

cant number of facilities. Owners and operators of facilities should assume that they need a permit until they establish that they are exempt.

The most important provisions of the permit application forms are the following three requirements:

1. Site plan, detailing all drainage on and off the facility
2. Sampling of each outfall or drainage path during a major storm event
3. Certification under penalty of law that each outfall has been evaluated for presence of nonstormwater contamination.

Although these provisions sound simple enough, they may not prove that simple in practice. Many plants and facilities have cross-contamination problems that may not have been detected in the past. Many plants know where some of their stormwater goes but cannot account for it all.

The EPA does offer group permitting. Under this provision, a group of similar plants may apply for a permit covering all of their plants. Testing is limited to 10% of the outfalls. Missouri is the only state to notify the EPA officially that it will not accept group permits to date, but the following states have indicated that they may refuse to accept them:

Alabama	Michigan
Delaware	Mississippi
Hawaii	Nevada
Illinois	New York
Iowa	Ohio
Kansas	South Carolina

Therefore, if a facility is located in one of these states, it may not be able to obtain a group permit. For facilities in other states, management can explore the possibility that similar facilities are preparing a group permit, which will save the firm considerable expense. Group permit applications were due in September 1991. Individual permits were due on November 18, 1991.

Waste Management

All states have regulations governing waste disposal. United States federal regulations are contained in 40 *CFR*. Most states have regulations as restrictive or more so than the federal statutes. The first step in developing a waste management program for any facility is for management to inventory all residues and byproducts created in the production process. Next, they should determine whether each waste stream qualifies as a solid waste under the RCRA. According to 40 *CFR* 261.2, a RCRA solid waste is a solid, liquid, semi-solid, or containerized gas that is discarded, has served its intended purpose, or is a manufacturing or mining byproduct, and is not either domestic sewage, a Clean Water Act point source discharge, irrigation return flow, Atomic Energy Commission source, or in-situ mining waste. If the material is garbage, refuse, or sludge, it is also a RCRA solid waste. It is a solid waste regardless of whether you discard it, use it, reuse it, recycle it, reclaim it, or store or accumulate it for any of these purposes.

Once a firm determines which of the wastes meet the definition of solid waste, it must establish which of the solid wastes meet the definition of hazardous waste. First, management should refer to 40 *CFR* 261.4(b) to determine if the solid wastes identified in the facility review are excluded from the definition of hazardous waste. These would include items such as household wastes, certain agricultural wastes, mining overburdens, fly ash, drilling fluids, and several others. If the solid wastes are not exempt under 40 *CFR* 261.4(b), the company must determine if the solid wastes are listed in 40 *CFR* Part 261, Subpart D, or if they are a mixture that contains a waste listed in Subpart D. Subpart D lists wastes from nonspecific sources, specific sources, acutely hazardous wastes, and toxic wastes. If the solid waste is listed in Subpart D, or is mixed with a waste listed in Subpart D, it is a hazardous waste. If it is not listed in Subpart D, the firm needs to determine if it exhibits any of the characteristics defined in 40 *CFR* Part 261, Subpart C, which includes ignitability, corrosivity, reactivity, and toxic constituent (TC) toxicity. If the solid waste exhibits any of these characteristics, it is a hazardous waste.

If a firm determines that its facility generates any RCRA-regulated hazardous waste, the next step is to establish the proper facility generator classification. There are four potential classifications: conditionally exempt small-quantity generator, small-quantity generator, ninety-day accumulation generator, and permitted storage facility. Generators that produce no more than 100 kg of hazardous waste per month should refer to 40 *CFR*, Part 261.5 to determine the conditions under which they are exempt from most of the require-

ments imposed on other generators. Conditionally exempt small-generator status allows for the generation of no more than 1 kg of acutely hazardous waste.

A generator of greater than 100 kg, but less than 1,000 kg, of hazardous waste in a calendar month is subject only to limited recordkeeping and reporting requirements. Generators within this category fall under 40 *CFR*, Part 261.44. A generator that generates over 1,000 kg per month of hazardous waste must apply for a TSD facility permit, unless it complies with the ninety-day accumulation time exemption, detailed in 40 *CFR*, Part 262.34. This requires that a generator accumulate hazardous waste for no more than ninety days. A small-quantity generator may accumulate waste on site for 180 days, unless it must transport its waste over 200 miles or more for off-site treatment storage or disposal, in which case it may accumulate hazardous waste on site for 270 days or less without a permit.

After determining the generator status of the facility, it will be necessary to select transportation and disposal for hazardous waste. Probably one of the largest areas of liability risk for any corporation is the disposal of by products of manufacturing that are classified as hazardous waste by federal or state regulations. This is because CERCLA makes contributing parties jointly and severally liable for all or a portion of the cleanup costs for any site where waste from their facility has been deposited, regardless of whether their waste constitutes in any way to a threat to human health or to the environment. This includes any party that produced the waste, arranged for its disposal, transported the waste, arranged for its transportation, or handled the waste at any time from source to waste site. This is a broad exposure to millions of dollars of liability for any firm producing any hazardous waste.

An approved hazardous waste disposal facility receives waste from various generators, and each contributes to the potential cost of cleanup (Figure 5–5). Each firm accepts this combined liability, but many do not know that they may be liable for the entire cost of cleanup. Many times small firms with few assets produce highly toxic wastes. When these firms find themselves with environmental liability, they collapse, and their liquidated value does little to compensate for their apportioned share of the resulting liabilities. These costs fall to the larger, more successful firms that have "deep pockets." Therefore, it is extremely important to carefully select disposal and destruction options for hazardous waste. It is usually a good idea to contract for transportation and disposal separately. If there is an indemnity from liability given by the disposal firm, the facility will want to have that indemnity in their name and not in the name of the broker or transporter. In some cases, the disposal site has picked up the transporter liability, while the generator has been left liable because of the way the contractual indemnity was written.

It is a good idea to have a company inspection of each disposal site and transporter before they are used. This is a firm's best defense against using sites that may some day produce liabilities under CERCLA. Just because a company

Figure 5–5. This crew is removing hazardous materials. Note the full personal protective equipment with a self-contained oxygen supply.

uses incineration or fuel programs, it should not feel exempt. What does the incinerator do with its scrubber waste and ash? What does the fuel blender do with the fuel oil and the residues from their process? These are also wastes that can be traced back to a facility. A company could be liable for the cleanup costs of the site that its incinerator firm uses for disposal of scrubber waste. Any transporter hauling hazardous waste should provide the generator with evidence of $5,000,000 in liability insurance (an MCS-90 environmental liability endorsement form provides such evidence).

Nonhazardous solid waste consists of normal plant refuse and scraps. In evaluating the options for solid waste disposal, a company should give preference to reuse and recycling. It should investigate each solid waste stream to determine if, by point source segregation, it can produce a waste stream that can be reused or recycled. If so, it should institute a reuse or recycling program. For those nonhazardous solid waste streams that must be shipped offsite, the firm should be evaluate them for potential offsite reuse or recycling. Finally, the remaining material should be sent offsite to a permitted solid waste disposal site. The company must review the state regulations governing solid waste disposal, because many states impose restrictions on the disposal of specific nonhazardous solid waste.

In selecting a disposal site for nonhazardous solid waste, the facility should review potential sites located in its area. These sites should be evaluated by an inspection process similar to that used for hazardous waste facilities. Some of the sites currently being investigated for remedial activities under CERCLA are solid waste disposal facilities. Therefore, although these wastes are excluded from the requirements of RCRA, the same CERCLA liabilities apply.

CHEMICAL RISK MANAGEMENT

U.S. federal, state, and local authorities generally impose certain regulatory reporting and employee communications requirements when hazardous substances are present in the workplace. Federal agencies include EPA and OSHA. Each state has set up its own environmental regulatory body, as have certain counties and cities. Additionally, certain states are empowered to administer some portions of the federal regulations.

Federal Regulations

Occupational Safety and Health Administration. Federal regulations administered by OSHA require that all employers in Standard Industrial Classification (SIC) Codes 20

through 39 provide information to employees about the hazardous chemicals to which they are exposed by means of a hazard communications program, labels, and other forms of warnings, Material Safety Data Sheets (MSDSs), and information and training. These communications and training requirements are covered in 29 *CFR*, 1910–1200. Minimum requirements include the following:

- Written hazard communication programs that include an inventory of all hazardous chemicals used in the facility
- MSDSs for each hazardous chemical
- Appropriate facility placards and warnings
- Hazard communication training plan
- Training to employees who are potentially exposed to hazardous chemicals in their facilities.

U.S. Environmental Protection Agency. EPA administers the Superfund Amendments and Reauthorization Act of 1986 (SARA), which established new authorities for emergency planning and preparedness, community right-to-know reporting, and toxic chemical release reporting. SARA requires facilities subject to OSHA workplace regulations to submit MSDSs to the appropriate local emergency planning committees, state emergency response commission, and local fire departments. EPA also prescribes exposure limits to hazardous chemicals and mandates specific reporting requirements.

State Regulatory Bodies

Each state has set up a separate environmental regulatory body that generally administers federal regulations in permitting and enforcement. State regulations are patterned after U.S. federal rules and regulations, although certain states enforce statutes that are much tougher than federal regulations.

Emergency Planning

All facilities should prepare and update emergency and contingency plans. These plans should be given to local emergency groups and fire and police departments.

Hazardous Waste Management

All facilities handling hazardous material should develop a formal training program and emergency procedures. Each facility should designate an individual to be responsible for all phases of hazardous chemical regulatory reporting, training, and regulatory compliance.

Hazardous waste training. The training requirements are outlined in 29 *CFR*, 1910–1200, and are administered by OSHA. Each training plan should include the following:

- Communicate in writing a list of the hazardous chemicals known to be present, the methods the employer will use to inform its employees of the hazards of nonroutine tasks, the methods the employer will use to inform any contractor whose employees may be exposed to the hazardous chemicals while performing their work, and any suggestions for appropriate protective measures
- Provide a complete library of the MSDSs for all hazardous chemicals present at the workplace. All employees should know the location of this information.
- Each employee who could come into contact with hazard-

ous chemicals should attend training courses and be informed of the properties of the chemicals, the procedures to be followed in an emergency, and the location of all MSDS. A log signed by each employee should be kept of all employees who receive the training. The company should provide training manuals to each employee.

Emergency plans. All facilities should prepare an emergency plan and distribute it to the appropriate local authorities, including the emergency planning groups, fire departments, police departments, and medical authorities. Copies of the MSDS should be given to the local fire department. In this way, should an emergency occur, the local authorities will be able to assist the firm with the maximum efficiency.

WASTE REDUCTION AND MINIMIZATION

Waste minimization can produce several benefits. It can save money by reducing waste treatment and disposal costs, raw materials purchases, and other operating costs. It can reduce potential environmental liabilities. It can protect public health and worker health and safety and protect the environment. Waste minimization can be accomplished by source reduction and recycling. Of the two approaches, source reduction is usually preferable to recycling because it has a lower total effect on the environment. The RCRA requires that generators of hazardous waste have a program in place to reduce the volume and toxicity of waste generated, to the extent that such a program is economically practical.

A waste minimization program must begin with an assessment, which requires a company to evaluate all solid waste streams to determine if they can be reduced or recycled. Next, the company must calculate the economic impact of reuse or recycling.

Technical evaluation of a waste stream should begin in the production process where the waste is generated. Factors such as changes in raw materials, production process, equipment, and operating conditions could change the characteristics of a solid waste to make it a better candidate for reuse or recycling.

Economic evaluation should consider the capital and operating costs associated with any onsite collection or treatment of the waste. In addition, it should consider disposal fees, transportation costs, raw material costs, and operating and maintenance costs. Economic projections should be as realistic as possible.

In choosing whether to reuse or recycle solid wastes, a company should consider the stability of the minimization technology. If reuse is the selected alternative, evaluate the effect minor changes in waste characterization will have on the reuse process. Waste is, by nature, inconsistent in composition. Therefore, if a company finds that minor fluctuations in waste composition will adversely affect the finished product or preclude the material from reuse, it should consider other alternatives.

In evaluating a solid waste for recycling, remember that recycling is a commodity-driven business, possible only if a

market exists for the recycled product. An excellent example of a solid waste that is particularly suitable for recycling is the aluminum can. It takes less energy to recycle an aluminum can than it does to produce aluminum by conversion of bauxite ore. In this industry, manufacturers have done an excellent job of establishing facilities capable of recycling aluminum cans. Therefore, there is an excellent market for aluminum cans collected for recycling.

Glass is another material suitable for recycling. Once again, it is less expensive to produce glass through recycling than to manufacture new glass. However, the market for glass recycling is not as well developed. For example, a facility may find that at times a recycler is unable to accept its material because the recycled product cannot be sold. This does not preclude the use of recycling as a disposal alternative, however. If a facility selects recycling as a minimization technique, it must be prepared to store waste when the material cannot be recycled, or to develop disposal alternatives.

Groundwater Contamination

One of the most serious and costly forms of pollution is groundwater contamination. When industrial pollution seeps into the groundwater and drinking water supplies, it can, at best, be costly to correct and, at worst, present long-term chronic health risks for people who live downstream from the facility.

The strata underneath the surface of the earth contain "horizons," some of which are water bearing. These water-bearing horizons vary in flow, characteristics, and water quality. The largest horizons are referred to as aquifers. These aquifers can range anywhere from continental to regional to local in size. Many are high-yield, drinking water-quality aquifers that serve as the primary drinking water source for a large portion of the population. In evaluating groundwater contamination, a firm should identify an area's geology and hydrology at least to a depth that penetrates the uppermost aquifer.

A company should classify site geology in accordance with the Unified Soil Classification System. This is accomplished by drilling bore holes and collecting soil samples at predetermined depth intervals. In good, cohesive soil conditions, a company can sometimes collect samples from the auger flites. However, when cohesive soil conditions do not exist and there is sidewall sloughing into the bore hole, samples from the auger flites will not allow proper soil classification. To properly classify the soils in noncohesive conditions, it is necessary to use a split spoon or sample tube device to collect samples at predetermined depths. Site hydrology includes the identification of groundwater flow paths, determination of groundwater flow directions (including horizontal and vertical flows), and determination of multihorizonal interconnects.

Horizontal groundwater flow direction is usually found by installing piezometers to determine the potentiometric surface of the groundwater. With these data, a team can identify the horizontal direction of flow by constructing a piezometric surface map.

The vertical flow component is determined by placing vertically nested piezometers in closely spaced, separate bore holes. The piezometers are screened at different depths to measure vertical variations in the hydraulic head. The data are used to construct flow nets, which are graphical representations of the vertical flow components.

After characterizing the site hydrology, an investigative team must determine temporal influences that might alter the piezometric surface. These influences might include location of offsite pumping wells, tidal variations, offsite and onsite land pattern changes, and seasonal variations in groundwater recharge.

Hydraulic interconnects between horizons can affect the horizontal and vertical flow path of groundwater contamination. Hydraulic interconnects are usually detected by installing a pumping well into the lower horizon, pumping at a predetermined rate, and monitoring the potentiometric surface of the upper horizon. Obviously, if there is a drawdown of the potentiometric surface at any of the piezometers during pumping of the lower horizon, hydraulic interconnect might exist.

Groundwater assessments require a great deal of experience in data collection and interpretation to produce meaningful results. If a facility is considering conducting a groundwater assessment, it is recommended management seek the assistance of an experienced groundwater assessment firm.

U.S. REGULATORY INSPECTIONS

Handling regulatory inspections properly can be one of the most critical procedures necessary to avoid enforcement actions, violations, consent orders, and fines. If a regulatory representative is competent and motivated to do so, almost any plant, on any given day, could be the subject of an enforcement action. This is true in any industry where byproducts of production are released, emitted, or discharged into the environment. Therefore, management should view any visit by a regulatory agency representative as a potential enforcement action. By developing a sound policy for handling regulatory representatives when they arrive, a company can avoid many of the problems associated with such events.

There are several considerations in preparing for a regulatory visit. Remember that the agency has a responsibility to enforce their rules. They cannot concern themselves with production, costs, profits, or jobs. If these concerns entered into their considerations, it would be necessary to make special rules for each plant. Therefore, these concerns should not be used to defer the visit or excuse deficiencies noted.

Regulatory inspections can be done under many names, ranging from an "informal visit" to a full "CERCLA 104 audit." Many visits are unannounced and unscheduled. Regardless of the circumstances of the visit, the agency has the right of access and the right to collect samples and review documents. If the company denies access for any reason, it will cause the agency to obtain an administrative warrant to conduct the inspection. This may create suspicion and mo-

tivate the representative to find a violation during the inspection.

Train several people to accompany government inspectors, and designate a minimum of two people to be present during any inspection. One person should be knowledgeable in environmental issues and the other in process knowledge. They should keep notes of all questions and comments. With two people present taking notes, it will be easier to corroborate statements if required at a later date. If samples are collected, ask for split samples. This way the company has sample material if questions arise later. Remember, any samples must be collected, preserved, and analyzed in accordance with the requirements of U.S. EPA S.W. 846, as revised. Review the sample protocol and be prepared to follow chain of custody and preservation requirements for evidence. (Chain of custody is a written document showing any individual who has handled the substance in question.)

Inspectors are entitled to inspect all documents related to their visit. Keep a record of any documents they review or copy. The documents copied by or supplied to the inspector become public information unless the company requests confidentiality. This is done by marking each page with the words, "TRADE SECRET" or "COMPANY CONFIDENTIAL." Although these documents may be found later to be unprotected, a company is always better off making the claim when supplying the data rather than trying to protect an improperly disclosed document at a later time.

Laboratory reports and inspection reports should be available if requested. Find out before the inspector leaves how to obtain a copy of all documents associated with the inspection. It is always a good idea to get copies of anything the agency is placing in the company's record. Additionally, this request places the inspector on notice that the firm will be reviewing what he or she writes about the inspection.

If everything goes wrong and an inspection results in a citation for a violation, it is not the end of the world. The normal procedure will be for the agency to issue a "Notice of Violation." This notice will give a company the opportunity to request a hearing. Always notify the legal department or the company attorney as soon as this notice arrives. It is usually best to request the hearing. Otherwise, guilt is assumed, and the order that is attached to the notice is regarded as a final citation. If the company requests a hearing, there are usually several opportunities to meet and negotiate with the agency before the hearing.

If the hearing goes against a company and management wants to contest the findings, the orders of the agency can usually be appealed in the court system of the county in which the main office of the agency is located. From that point, the normal appeal process of the judicial system is available through the state or federal appellate courts (depending on the agency that issues the notice). Usually, the issues are settled by agreed orders with the agency before any hearing takes place. Remember to act in good faith, but keep in mind that regulatory visits are serious matters and proper planning can help avoid unnecessary problems.

ENVIRONMENTAL AUDITS

The EPA defines an environmental audit as "an investigation into the history and current status of a particular piece of property (site)." The purpose of an environmental audit is to:

1. Identify the presence and extent of environmental contamination or hazardous materials from current or previous site activities
2. Determine the level of compliance with current standards or regulations
3. Provide a general review of environmental risks associated with the site and its operations.

Companies that handle environmentally sensitive substances should conduct internal audits on a consistent, company-wide basis and requests audits by outside consultants as needed. An audit is an important review of a site's environmental risks. Its minimum goal is to determine a company's level of compliance with regulations and standards.

Internal Audits

The main purpose of an internal audit is to ensure that a facility or department is safe and that the company is in compliance with environmental standards set by various regulations and statutes. Environmental management programs should institute consistent, comprehensive internal audits. They can be either or "top-down" or self-audits. Each has its advantages and disadvantages.

In a "top-down" audit, staff from a centralized environmental management department review procedures, recordkeeping, and so on in operating departments or facilities throughout the organization. This type of review ensures professional auditing expertise. It also enables corporate staff to make sure of company-wide compliance with corporate environmental policy. Its disadvantage is a feeling of a "police action" that can create hostile relationships with operating staff who often feel that the environmental professionals are not familiar enough with their department or facility to conduct a comprehensive study.

Self-audits enable employees who work in the department or facility to assess problems in the operations with which they are familiar. However, they may be too close to the operations to be able to conduct an objective audit or to spot operational processes that could be improved to go beyond minimal environmental release standards. Operations staff may also lack incentive to identify problem areas, believing such problems will reflect badly upon themselves. Self-audits also do not help corporate staff become more aware of individual operations and areas that could be improved.

A company policy that combines self-audits with "top-down" audits (periodic department or facility self-assessments with follow-up audits by environmental professionals) can combine the advantages of the two approaches while minimizing their disadvantages. Environmental audits conducted on an internal basis, whether "top-down" or self-audit, should follow some predefined protocol. Records of this protocol should be retained in the company's records for review by environmental regulatory agencies and management.

As part of the procedure to develop the internal audit protocol, the company should develop a procedure for addressing deficiencies or violations detected. There is no point in developing self-examination information if the firm has no program to address remediation of the deficiencies. In fact, an internal audit can present a problem if the regulatory agency detects a deficiency or violation and determines, through a company's internal audit document, that it had prior knowledge of the fact but took no further action.

The primary purpose of any audit program is to translate the information gathered into new environmental programs or change existing programs to help the company resolve problems swiftly.

Consultant Audits

Environmental audits by outside consulting firms are conducted for a variety of reasons. Some companies use these audits as the first step in determining what needs to be done to comply with regulations. The main advantage to a consultant audit is that the company obtains an expert, unbiased opinion of the environmental condition of a facility. These audits are normally viewed by regulatory agencies to be more complete, with a higher degree of accuracy than "top-down" or self-audit protocols. They are also used as a periodic review of a company's present overall performance on controlling air emissions, discharges to storm and sanitary sewers, groundwater, storage and handling of hazardous materials, and so on. Such a periodic audit can be integrated with an internal audit program.

An environmental consulting firm can be specifically contracted when there are serious concerns about compliance with environmental regulations and statutes. This can be a complete review of a company's compliance with all pertinent regulations, or it can focus on a specific section of the Clean Water Act, SARA, and so on. These audits often pay for themselves by reducing future compliance costs and avoiding costly penalties. An environmental consultant also can be engaged to do a site assessment before a real estate transaction.

CERCLA holds current and former property owners liable for environmental contamination of the property, no matter how long the hazard has existed or who caused it. Without an environmental audit before a real estate closing, an innocent buyer can own property that cannot be sold or that must be cleaned up at enormous cost. Many lending institutions have acquired environmentally contaminated property through foreclosure. Because of their potential liability, some of them are requiring an environmental audit before approving a loan.

United States federal regulations do not require an environmental audit to be part of every real estate transaction. Environmental audits of commercial, industrial, and some residential property may be required by state law. Whether mandated or not, an environmental audit will enable both the buyer and seller to be aware of the condition of the property.

When SARA was passed in 1986, it gave an "innocent purchaser" of contaminated property a narrow way of escaping liability. Liability may be avoided if the buyer can prove that an inquiry into previous ownership and uses of the property (an environmental audit) was made before its purchase. The completed audit form can be submitted as an exhibit in a court case, proving that the buyer did not know that any hazardous substances were released or disposed of at the site before the property was purchased.

Uses of Audits

Environmental audits serve many purposes. Site assessments before a real estate purchase help companies to avoid costly cleanup charges. When conducted to audit a company's compliance with regulations, they can prevent fines. If they include an assessment of environmental hazards to employees, audits promote good labor relations. When their results are reported to the community, they enhance a company's public image with citizens who may have been suspicious of the facility's activities.

In all cases, other than prepurchase site assessments, environmental audits serve as a "report card" showing where a facility has scored "A+" as well as areas that "need improvement." During one company's audit, a drum of trichloromethane was found set apart from other hazardous waste because it was going to be recycled. Only during the audit process was the fact that the drum had no secondary containment system discovered.

Whenever management plans an audit, the first step is a concise statement of company goals. Once management has determined what information it needs to gain from the audit, it can decide whether to conduct an internal audit or hire a consulting firm. Time and expertise are the deciding factors. Complex environmental regulations can require the knowledge of someone other than the environmental manager or department personnel. An extensive environmental audit also can take more time than busy internal staff have to give.

The Audit Form

Figure 5–6 is a sample environmental audit protocol, which provides an example of a thorough audit procedure. Review it before each use to ensure that it is current and covers all applicable regulations. Questions should be answered "yes," "no," or "not applicable" ("N/A"). Avoid entries such as, "This facility is totally out of compliance," or "Major violations are present." These comments can be interpreted broadly by a regulatory agency or court, when the company's intention may have been more limited and the remark made for effect and not accuracy.

GLOBAL SOLUTIONS FOR GLOBAL PROBLEMS

No part of the globe is immune from chemicals that nations release into the ground and waterways or emit into the air. Water currents like conveyor belts deliver toxic chemicals to the Arctic Ocean. Carbon, sulfur, and other pollutants float on air currents from Eurasia to hover over the Arctic. The "imported" pollutants combine with "home-grown" nitro-

SAMPLE ENVIRONMENTAL AUDIT PROTOCOL

DATE: _____

　　　　　　　Month　　　　　　　　　　　　　　Day　　　　　　　　　Year

FACILITY NAME / ADDRESS:

_____
Name

_____
Address

_____
City / State / Zip

_____
Telephone Number

AUDIT TEAM MEMBERS:

_____

_____

_____
(Include Affiliation and Position)

AUDIT PARTICIPANTS:

_____

_____

_____
(Include Position)

FACILITY EPA IDENTIFICATION NUMBER: _____

FINAL AUDIT AND APPROVAL: _____

FACILITY MANAGER:

　　　　　　　　　　　　　　　　　　　　　　　　　　DATE:
_____
Signature

SENIOR AUDIT MEMBER:

　　　　　　　　　　　　　　　　　　　　　　　　　　DATE:
_____
Signature

Figure 5–6. Sample environmental audit protocol. (Courtesy Huntsman Chemical Corp.)

SAMPLE ENVIRONMENTAL AUDIT PROTOCOL

TABLE OF CONTENTS

Figure 5–6. *(Continued.)*

SAMPLE ENVIRONMENTAL AUDIT PROTOCOL

1.0 GENERAL

1.1   Have the reasons for this audit been discussed with the Facility Manager?

1.2   Has the Facility Manager been made aware of how the audit will be conducted?

1.3   In the opinion of the Facility Manager and the Environmental Manager what are the:

      1.3.1   Major environmental issues at this site?

      1.3.2   Major concerns of the Audit Procedure?

1.4   Does the site's Environmental Manager devote all of his/her time to environmental affairs?

1.5   Training

      1.5.1   Are there any formal training programs or materials available to employees that either:

      1.5.1.1 Inform them of their responsibilities under environmental statutes?

      1.5.1.2 Provide training in the conduct of environmentally acceptable work practices?

      1.5.2   Is there a formally led environmental orientation program for new employees?

      1.5.3   Has any statement of environmental policy been issued by the facility to the staff or to the public?

1.6   Is there a central location for all environmental records including permits, compliance reports, etc.?

Figure 5–6. *(Continued.)*

SAMPLE ENVIRONMENTAL AUDIT PROTOCOL

2.0 WASTE WATER DISCHARGES

    2.1   Discharges to Municipal Waste Water Treatment Facilities (POTW)

        2.1.1   Is the permit on file current?

        2.1.2   Has the facility received any notice concerning the permit?

        2.1.3   Is water analysis documentation available for the last two years?

        2.1.4   Are records available showing the average and maximum daily discharges?

        2.1.5   Are the pollutant characteristics defined and documented?

        2.1.6   Is the facility in compliance with the U.S. EPA Categorical Industrial Pretreatment Regulations (if required)?

        2.1.7   Has the facility received any lawsuits or fines within the last year?

        2.1.8   Has the facility received any complaints from neighbors, regulatory agencies or the POTW within the last year?

        2.1.9   Has the facility exceeded any permit limitations?

        2.1.10   Has the facility had any other permit non-compliance incidences (late reporting, bypass, etc.)?

        2.1.11   Have any local, state, or federal agencies conducted any compliance inspections within the last year?

        2.1.12   Have any local, state, or federal agencies required or participated in any wastewater testing at this site?

    2.2   Discharges to Waterways

        2.2.1   Is the current NPDES permit available?

        2.2.1.1 Has the facility applied for a stormwater discharge NPDES?

        2.2.1.2 Is the facility eligible for participation in a group stormwater discharge permit?

        2.2.2   Are there notices of violation within the year?

        2.2.3   Is the current SPDES permit available?

        2.2.4   Are the Discharge Monitoring Reports (DMRs) for the past two years available?

        2.2.5   Is the test data available for:

Figure 5–6. *(Continued.)*

SAMPLE ENVIRONMENTAL AUDIT PROTOCOL

    2.2.5.1   Biochemical Oxygen Demand (BOD)?

    2.2.5.2   Chemical Oxygen Demand (COD)?

    2.2.5.3   Total Organic Carbon (TOC)?

    2.2.5.4   Total Suspended Solids (TSS)?

    2.2.5.5   Ammonia?

    2.2.5.6   Temperature?

    2.2.5.7   pH?

    2.2.5 8   Total Metals?

    2.2.5.9   Cyanide?

    2.2.5.10 Phenols?

    2.2.5.11 Organic pollutant fractions?

2.2.6   Are analytical and calibration methods available?

2.2.7   Is backup power available for control or testing?

2.2.8   Are required alarms installed and functioning?

2.2.9   Is training record documentation in order?

2.2.10 Is the inventory for environmental control equipment and spare parts available?

2.2.11 Are control equipment files current?

2.2.12 Are the waterway discharge points inspected regularly for:

    2.2.12.1 Oil?

    2.2.12.2 Sheen?

    2.2.12.3 Grease?

    2.2.12.4 Turbidity?

    2.2.12.5 Foam?

    2.2.12.6 Color?

Figure 5–6. *(Continued.)*

SAMPLE ENVIRONMENTAL AUDIT PROTOCOL

2.2.13   Is the facility in compliance with the U.S. EPA Categorical Industrial Discharge Pretreatment Regulations (if required)?

2.2.14   Has the facility received any lawsuits or fines within the last year?

2.2.15   Has the facility received any complaints from neighbors or regulatory agencies within the last year?

2.2.16   Has the facility exceeded any permit limitations?

2.2.17   Has the facility had any other permit non-compliance incidences (late reporting, bypass, etc.)?

2.2.18   Have any local, state, or federal agencies conducted any compliance inspections within the last year?

2.2.19   Have any local, state, or federal agencies required or participated in any discharge water testing at this site?

2.3   Public Drinking Water

2.3.1   Are cross connection inspections conducted?

2.3.2   If drinking water is treated by the facility, is it tested by the facility? Is it tested by a quality program as required?

2.3.3   If a well (on or off site) is used, are the following documents available?:

   2.3.3.1   Distance to potential contamination

   2.3.3.2   If access controlled and secure as required

   2.3.3.3   If area subject to flooding

   2.3.3.4   If testing and analysis documented

2.3.4   Have there been any non-compliance problems with Safe Drinking Water Standards?

2.3.5   Have any local, state or federal agencies conducted any compliance inspections within the last year?

2.3.6   Have any local, state, or federal agencies required or participated in any water testing at the facility?

2.4   U.S. Coast Guard (As Applicable)

Figure 5–6. *(Continued.)*

SAMPLE ENVIRONMENTAL AUDIT PROTOCOL

2.4.1 Has the facility received a warning or citation?

2.4.2 Is the welding permit current?

2.4.3 Are notices to the Coast Guard regarding welding or burning operations on file?

2.4.4 Are the Certificates of Adequacy for oil reception current and signed?

2.4.5 Is there a current Certificate of Adequacy (COA) for noxious liquid substances on file?

2.4.6 Is there a current Certificate of Adequacy for a garbage reception facility on file?

3.0 AIR EMISSIONS

3.1 Does each process emission point have a current operating permit from the state or federal E.P.A.?

3.1.1 Has this facility been audited for the emission of any chemical listed as "Air Toxic?"

3.1.2 Does the facility emit more than 10 tons per year of any single air toxic chemical?

3.1.3 Does the facility emit more than 25 tons per year of all air toxic chemicals?

3.1.4 Does the facility comply with the requirements of the Clean Air Act Amendments of 1990?

3.2 Are all air emissions monitoring devices currently operating properly?

3.3 Is the inspection documentation of the air emission monitoring devices available?

3.4 Are there any visible or odorous sources of air emissions?

3.5 Asbestos insulation:

3.5.1 Present in any building?

3.5.2 Air Analysis Reports available?

3.5.3 Proper procedures being utilized for removal and disposal?

Figure 5–6. *(Continued.)*

SAMPLE ENVIRONMENTAL AUDIT PROTOCOL

3.6   Emission Analysis Documentation:

    3.6.1   Sulfur Dioxides

    3.6.2   Particulates

    3.6.3   Carbon Monoxide

    3.6.4   Ozone

    3.6.5   Hydrocarbons

    3.6.6   Nitrogen Oxides

    3.6.7   Lead

    3.6.8   Benzene

    3.6.9   Beryllium

    3.6.10   Mercury

    3.6.11   VOC

    3.6.12   Fluorides

    3.6.13   Hydrogen Sulfide

    3.6.14   Radionuclides

    3.6.15   Asbestos

    3.6.16   Vinyl Chloride

3.7   Is the following available for E.P.A. inspection?:

    3.7.1   Design of air emission control system

    3.7.2   Operating manuals for control systems

    3.7.3   Emission testing data per required reference methods

3.8   Has the facility received any lawsuits or fines within the last year?

3.9   Has the facility received any complaints from neighbors or regulatory agencies within the last year?

Figure 5–6. *(Continued.)*

3.10   Has the facility exceeded any permit limitations?

3.11   Has the facility had any other permit non-compliance incidences (late reporting, bypass, etc.)?

3.12   Have any local, state, or federal agencies conducted any compliance inspections within the last year?

3.13   Have any local, state or federal agencies required or participated in any air emissions testing at this site?

3.14   Is the facility located in a non-attainment or a transition zone into a non-attainment area?

      3.14.1   Are there any special restrictions imposed on the facility as a result of non-attainment?

      3.14.2   Are there any pending changes that will affect the facility because of non-attainment concerns?

4.0 HAZARDOUS MATERIAL AND OIL STORAGE MANAGEMENT

4.1   Are the purchases of all chlorinated solvents backed up by disposal documentation?

      4.1.1   Are Material Safety Data Sheets available on these materials?

4.2   Is there a training program available for appropriate personnel?

4.3   Are written procedures available for the safe handling of hazardous materials?

4.4   Are hazardous materials safely segregated as required?

4.5   Are hazardous materials properly identified?

4.6   Spill Prevention Control and Counter Measure Plan (SPCC):

      4.6.1   Is it current?

      4.6.2   Is it certified by a professional engineer?

4.7   Are adequate spill clean-up materials available as required?

Figure 5–6. *(Continued.)*

SAMPLE ENVIRONMENTAL AUDIT PROTOCOL

4.8 Has the facility reported any spills, releases, or discharges of oil or hazardous substances within the past year?

4.9 Has the facility received any notices of violation for any spills, releases, or discharges?

4.10 Has the facility received any fines for spills, releases, or discharges?

4.11 Has the facility received any complaints for spills, releases, or discharges?

4.12 Are there any visible signs of past spills around the grounds of the facility?

4.13 Is there a file of previous hazardous substance spills?

5.0 UNDERGROUND STORAGE TANKS (WHERE APPLICABLE)

5.1 Are tanks properly registered:

5.1.1 With state?

5.1.2 U.S. E.P.A.?

5.1.3 Local jurisdiction?

5.2 Were tanks tested for leaks within the past year?

5.3 Are adequate files available for tank location, age, size, type, and use?

5.4 Is documentation available for:

5.4.1 Implementing ground testing?

5.4.2 Leak detection procedures?

5.4.3 Showing tank level alarms?

5.4.4 Initiating removal?

6.0 TRANSPORTATION OF HAZARDOUS MATERIALS (D.O.T.)

6.1 Are all regulated substances properly labeled with names, labels, and placards?

6.2 Have there been any citations within the past year relating to the transportation of hazardous materials?

6.3 Is there one person on site knowledgeable and responsible for D.O.T. regulations?

Figure 5–6. *(Continued.)*

SAMPLE ENVIRONMENTAL AUDIT PROTOCOL

6.4 Are personnel training records available?

6.5 Are transfer checklists used?

6.6 Is manifest documentation available?

6.7 Have transporters provided evidence of environmental impairment insurance in accordance with MCS 90?

7.0 SOLID WASTE

7.1 Is there a quantitative inventory of all hazardous waste generated?

7.2 Is there a quantitative inventory of all non-hazardous waste generated?

7.3 Is laboratory analysis available to verify non-hazardous waste classification?

7.4 Is there an annual waste survey performed (R/C)?

7.5 Is there a recordkeeping and tracking system available?

7.6 Has a baseline been established?

7.7 Has a waste minimization program been implemented (R/C)?

8.0 HAZARDOUS WASTE (RCRA)

8.1 Is documentation available of official Acknowledgement of Notification of Hazardous Waste Activity?

8.2 E.P.A. Identification Number assigned?

8.3 Are all hazardous wastes:

    8.3.1 Stored properly?

    8.3.2 Labeled properly?

8.4 Is data available to track hazardous waste generation and disposal?

8.5 Is the hazardous waste shipped to licensed disposal facilities (TSD)?

    8.5.1 Is there a copy of the TSD permit on file?

    8.5.2 Have qualified personnel visited and inspected the TSD?

Figure 5–6. *(Continued.)*

SAMPLE ENVIRONMENTAL AUDIT PROTOCOL

8.5.3 Are licensed (documented) transporters used?

8.5.4 Is a written signature from the transporter obtained and filed?

8.5.5 Are required uniform hazardous waste manifests used?

8.6 Are any hazardous wastes on the National Priority List?

8.7 Were there any violations on the latest RCRA site inspection?

8.8 Were the last two bi-annual RCRA reports submitted as required?

8.9 Are "exception reports" available? (For use if site has not received a signed copy of the manifest from the TSD within forty-five days.)

8.10 Is there an Emergency Action Plan available?

8.10.1 Has training been implemented?

8.11 Are waste characteristics documented by:

8.11.1 Test data?

8.11.2 Process knowledge?

8.12 Does each manifest contain a waste minimization statement?

8.13 Is the waste area secured and provided a sign indicating:
DANGER — UNAUTHORIZED PERSONNEL KEEP OUT?

8.14 Are weekly logs of inspection of the following items maintained for three (3) years on:

8.14.1 Monitoring equipment?

8.14.2 Physical equipment?

8.14.3 Security?

8.14.4 Operating equipment?

8.14.5 Recommendation follow-up?

8.15 Training program includes the following:

8.15.1 Trained instructor

8.15.2 Annual audit

8.15.3 Job descriptions and titles available

Figure 5–6. *(Continued.)*

SAMPLE ENVIRONMENTAL AUDIT PROTOCOL

8.16   Do the following visit the site and have Hazard Information?:

    8.16.1   Fire Department
    8.16.2   Hospital or Emergency Medical Services

8.17   Are smoking and open flames controlled?

8.18   Does the waste tank storage program include the following?:

    8.18.1   Operating procedures

    8.18.2   Audit program

    8.18.3   Closure cost procedure

    8.18.4   Waste compatible

    8.18.5   Tank diked

    8.18.6   Drainage control system

    8.18.7   Emergency stop or cutoff or bypass provided on continuous feed tank

    8.18.8   Daily inspection program for control equipment, monitoring equipment and corrosion, cracks and leaks

8.19   Does incineration include the following?:

    8.19.1   Written operating procedures

    8.19.2   Monitoring and control equipment

    8.19.3   Control procedures

    8.19.4   Audit program

    8.19.5   Closure cost procedure

    8.19.6   Steady state before adding waste

    8.19.7   Waste analysis, including heat value, halogen, sulfur, lead, mercury content

    8.19.8   Monitor waste feed, auxiliary fuel feed, airflow, temperature, scrubber flow and pH, oxygen and carbon monoxide

    8.19.9   Stack plume monitored every hour for color and opacity

Figure 5–6. *(Continued.)*

SAMPLE ENVIRONMENTAL AUDIT PROTOCOL

9.0 TREATMENT, STORAGE, AND DISPOSAL FACILITIES

9.1 Are Part A and/or Part B permits for TSD current?

9.2 Is there a written waste analysis plan?

9.3 Does the operation include the following?:

9.3.1 Analytical sampling procedures

9.3.2 Recordkeeping

9.3.3 Security

9.3.4 Warning signs

9.3.5 Audit program with follow-up

9.3.6 Written training plan

9.3.7 Emergency action plan

9.3.8 Manifest system

9.3.9 Compliance with OSHA Hazardous Waste and Emergency Response Regulations

9.4 Does operation provide for ground water protection?:

9.4.1 Well monitoring

9.4.2 Sample and analysis

9.4.3 Recordkeeping of data

9.4.4 Site hydrology

9.4.5 Response plan

9.4.6 Closure cost procedure

9.5 Has the facility received any lawsuits or fines within the last year?

9.6 Has the facility received any complaints from neighbors or regulatory agencies within the last year?

9.7 Has the facility exceeded any permit limitations?

9.8 Has the facility had any other permit non-compliance incidences (late reporting, bypass, etc.)?

Figure 5–6. *(Continued.)*

SAMPLE ENVIRONMENTAL AUDIT PROTOCOL

9.9   Have any local, state, or federal agencies conducted any compliance inspections within the last year?

9.10   Have any local, state or federal agencies required or participated in any hazardous waste testing at this site?

10.0 SUPERFUND

10.1   Has the facility received a Section 104 information request from the U.S. EPA or other requests from a state EPA about any disposal site?

10.2   Are any state or federal Superfund lawsuits pending against the facility?

10.3   Has the facility received notice of any kind from the state EPA or U.S. EPA of that agency's intent to look to the facility for a contribution for the cleanup of a site?

11.0 SARA TITLE III

11.1   Did the facility notify the state as required of any Extremely Hazardous Substances on site in greater amounts than the threshold planning quantity?

11.2   If required, did the facility notify the local planning committee of its designated facility representative?

11.3   Were the MSDSs, or a list of the MSDS chemicals, submitted to the State Emergency Response Commission (SERC), local planning committee, and the local fire department, as required?

11.4   For any applicable spills during the past year, was the Emergency Notification made as required under this law?

11.5   Was the TIER I Inventory performed for the past year?

    11.5.1   Did it include maximum amount each category anytime during the past year?

    11.5.2   Did it include average daily amount each category?

    11.5.3   Did it include general location of each chemical?

    11.5.4   Has it been filed annually?

11.6   Has TIER II information been requested by the LEPC, SERC or Fire Department? Did it include:

    11.6.1   Chemical name?

Figure 5–6. *(Continued.)*

SAMPLE ENVIRONMENTAL AUDIT PROTOCOL

11.6.2   Maximum amount?

11.6.3   Manner of storage?

11.6.4   Location?

11.7   Form 313 to EPA and State by July 1st of each year for emissions.

11.8   Are training records available on employee training for spill emission incidents?

11.9   Has a formal audit of process, storage and transfer operations been made? Did it include hazard recognition/evaluation, control systems, monitoring devices and physical controls?

11.10   Is dispersion modeling available for potential releases?

11.11   Have drills been conducted to assess on-site and community exposure?

11.12   Have local agencies been given: tours, training and drills within the last year?

11.13   Have information presentations been given to community groups and schools, etc. on the company emergency response organization and notification system? Has company literature been distributed?

11.14   Is there a written description on how public notification and communication are implemented?

12.0 TOXIC SUBSTANCES CONTROL ACT (TSCA)

12.1   New Chemicals

12.1.1   Was the EPA notified 90 days before any new chemical was manufactured or imported? This is called a Pre-manufacture Notification (PMN).

12.1.2   Are records maintained of adverse effects of products and kept for at least five (5) years?

12.1.3   Is documentation available on the substantial risk of our products to health or the environment? (lab tests & toxicological tests, flammability, etc.)

12.2   PCBs

12.2.1   Are PCB transformers properly labeled?

12.2.2   Are the quarterly inspection forms current?

12.2.3   Is the latest annual inventory current?

Figure 5–6. *(Continued.)*

SAMPLE ENVIRONMENTAL AUDIT PROTOCOL

12.2.4   Has the fire department been properly notified about the location, etc., of the PCB transformers?

12.2.5   Are there plans to eliminate PCB-containing equipment?

12.2.6   Is there secondary spill containment for each of the transformers?

12.2.7   Is the storage location above the 100-year flood plain?

12.2.8   Have any PCBs been disposed of in the past several years?

12.2.9   Are any PCBs on site illegally?

12.2.10   Is documentation available that describes how PCB content is determined?

12.2.11   Have there been any leaks, spills and waterways contaminated?

12.2.12   Is storage area protected from rainfall?

12.2.13   Are diked areas void of drains, valves, expansion joints or openings?

12.2.14   Is the storage area marked with signs?

12.2.15   Are all storage items dated?

12.2.16   Is sampling of solvent used for PCB removal maintained and controlled?

12.2.17   Are records maintained of removal to storage and transportation?

13.0 NOISE

13.1   Has any regulatory agency cited the facility for noise pollution within the past year?

13.2   Has environmental sampling been conducted and documented?

13.3   Has the facility received any lawsuits or fines within the last year?

13.4   Has the facility received any complaints from neighbors or regulatory agencies within the last year?

13.5   Has the facility exceeded any permit limitations?

Figure 5–6. (Continued.)

SAMPLE ENVIRONMENTAL AUDIT PROTOCOL

13.6   Has the facility had any other permit non-compliance incidents (late reporting, bypass, etc.)?

13.7   Have any local, state, or federal agencies conducted any compliance inspections within the last year?

13.8   Have any local, state or federal agencies required or participated in any noise testing at this site?

14.0 COMMENTS

Figure 5–6. *(Concluded.)*

gen oxides from Alaskan oil fields to produce Arctic haze. The Artic's annual mean level of photochemical smog rivals that of Los Angeles.

In the once-pristine Arctic, high levels of toxic substances have been discovered in seals and polar bears in recent years. These pollutants have made their way to the top of the food chain, where mercury is now found in mammalian milk in many parts of the Arctic. Sulfur dioxide, produced mainly by coal-fueled electrical utilities, and nitrogen oxide, the product of transportation sources and utilities, are chemically transformed in the atmosphere and transported as acid rain over national borders by prevailing winds.

Carbon dioxide, chlorofluorocarbons, methane, and nitrous oxide, produced in quantity by industrialized nations over the last 200 years, gather in the atmosphere and absorb the infrared waves emitted from the earth. This "greenhouse effect" may be causing global warming—the rise of the mean surface temperature of the earth by 0.6 C in the last 100 years. Increases of 2 to 5 C over the next 50 to 100 years are predicted. Those few degrees could turn arable land in the higher latitudes into deserts and could melt polar icecaps, causing a one-meter rise in sea levels. That increase would flood large population centers, leaving 50 million people homeless worldwide.

The last two decades have proven that we are all part of the world community. The effects of one country's pollution are felt around the globe. With this understanding, international policymakers have joined together to protect our earth and our atmosphere from environmentally hazardous practices (Figure 5–1).

International Agreements

The international community joined forces to meet the environmental challenge at the 1972 Stockholm Conference and has continued to make giant strides. The "Stockholm Declaration" articulated several principles of international environmental law. Principle 21, for example, holds that while countries have the right to develop, they have a responsibility not to damage the environment outside their borders. This includes the oceans and Antarctica, as well as other countries. Principle 21 now represents customary international law, law instituted by the agreements and practices of states. Its application to some worldwide environmental problems, however, is difficult. It is not easy to assess an individual country's responsibility regarding global warming.

Serious environmental accidents have spurred nations to cooperate more rapidly in developing international environmental protection laws. Following several major oil spills in the 1960s and 1970s, the International Maritime Organization reached rapid accords on oil-spill liability and regulations for oil discharges from ships. When scientific data about ozone layer damage from synthetic chemicals were released, the Montreal Protocol was negotiated and amended in record time. The Basel Convention's agreement was accelerated as soon as the hazardous waste and incinerator ash shipment to a Nigerian dump site came to light.

Immediately after the Chernobyl nuclear accident, the International Atomic Energy Agency swiftly finished new treaties on responsibility for notification and assistance. A key feature codified a country's obligation to notify other nations if the risk of transboundary damage existed.

SUMMARY

- Toxic chemicals and pollutants from industry in nations around the world are posing serious threats to the environment and to public health. Today's companies must not only comply with government environmental regulations and statutes but often must create an environmental management function within their organizations.

- The federal government has passed numerous laws and regulations over the past three decades to control pollutants and to regulate industrial wastes. These laws include the Clean Air Act and its amendments, Clean Water Act, Resource Conservation and Recovery Act, Hazardous and Solid Waste Amendments, and Comprehensive Environmental Response, Compensation and Liability Act.

- These acts created various environmental offices such as the Environmental Protection Agency (EPA) and set government restrictions for hazardous chemical emissions and toxic waste removal. The laws also give the government the power to inspect facilities, cite companies for violations, and levy penalties and fines for noncompliance.

- Industry experts have identified five stages of environmental management program development that can be found in many companies. These stages range from beginner to firefighter, concerned citizen, pragmatist, to proactivist. A company's stage often depends on its size, amount of hazardous or toxic material, and financial resources.

- For companies that have reached stage five, environmental management is a top priority. Top management must demonstrate a strong commitment to responsible environmental practices and communicate this priority throughout the company for the program to succeed. Program staff must include skilled professionals with backgrounds in the sciences, regulatory law, and environmental issues.

- Steps in the environmental management program include preventing common violations, maintaining accurate and thorough records, creating a spill reporting plan, setting realistic limits and schedules for meeting environmental regulations, and training employees in program objectives and procedures. In addition, the company must establish a strong relationship with an attorney skilled in environmental law.

- Companies must obtain reliable, up-to-date information on current regulations, guidelines, and environmental data. The *Federal Register* is a primary source of such information, but companies can also use government agency publications and private services, newsletters, and consultants. Some states publish the *State Register*, which carries the latest information on state environmental regu-

- lations and guidelines.
- Accurate, reliable data will help companies avoid citations for violating clean air and water regulations. Amendments and changes in these regulations are generally becoming more restrictive and can have a major impact on a company's method of handling emissions and discharge of wastewater. The regulations also outline current procedures for acquiring permits to operate within a state or municipality.
- Disposal of toxic or hazardous wastes is a growing problem in many nations, one that various governments are seeking to regulate more closely. Current regulations define how companies should handle their hazardous and nonhazardous wastes and the methods they must use to transport and dispose of such byproducts.
- Companies must also comply with state and federal regulations regarding chemical risk management in their facilities. They must make sure workers know their exposure risk to toxic chemicals and the safety procedures for handling such materials. Information such as material safety data sheets must be distributed to all workers and displayed throughout the workplace. EPA and OSHA have strict guidelines regarding workers' exposure to and protection from hazardous chemicals and the reporting obligations of management.
- Companies must establish emergency and contingency plans for handling accidents involving toxic or hazardous materials. They must also develop formal training programs to teach employees safe practices and emergency measures in the workplace.
- Efforts to reduce and minimize waste can save companies money, reduce potential environmental liabilities, protect public health and worker health and safety, and protect the environment. Waste minimization can be accomplished through source reduction and recycling. Source reduction is generally more cost effective and less of an impact on the environment than is reuse/recyling.
- Groundwater contamination is one of the more serious environmental issues of the past decade. Seepage of toxic and hazardous materials into aquifers has prompted local and federal agencies to force companies to take stronger measures in assessing and controlling this contamination.
- All regulatory agencies conduct formal, often unannounced inspections of facilities to determine if companies are in compliance with state and federal regulations. Companies should train and designate staff members to accompany inspectors on their rounds, provide inspectors with free access to company records and reports, and be familiar with the company's legal rights under state and federal regulations.
- Government agencies or companies themselves can conduct environmental audits of facility operations to (1) identify comtamination or hazardous materials from current or previous activities, (2) determine compliance with current standards or regulations, and (3) provide a general review of environmental risks associated with a site and its operations.

- Environmental audits can be internal (conducted by the company) or external, conducted by an outside firm. Internal "top-down" and self-audits do not always result in objective treatment of a company's environmental problems and risks. Audits performed by outside consultants provide objectivity and may pay for themselves in reduced fines and penalties.
- Environmental audits serve many purposes: avoiding costly cleanup charges, preventing fines, promoting good labor relations, enhancing a company's public relations with the community, and serving as a "report card" on a firm's environmental management functions.
- The problem of environmental contamination and destruction is a global one calling for a global solution. In the past three decades, nations have been cooperating more fully to create international regulations and standards to protect the environment. Such cooperation must increase in the near future if ecological disasters, such as global warming, are to be avoided.

REFERENCES

Friedman W J. "Avoiding environmental liability in five simplified steps." *Chemical Processing*, (April 1988).

Hajost S. "The challenge to international law and institutions." *EPA Journal*, Vol. 16, No. 4, (July/August 1990).

Hunt CB and Auster ER. "Proactive environmental management: avoiding the toxic trip." *Sloan Management Review*, Vol. 31, (Winter 1990).

Keenan T. "Why is everyone talking about environmental audits?" *Industrial Safety & Hygiene News*, Vol. 24, (June 1990).

Leaf A. "Potential health effects of global climatic and environmental changes." *New England Journal of Medicine*, Vol. 321, (December 1989).

Main J. "Here comes the big new cleanup." *Fortune*, (November 21, 1988).

Quarles J. and Lewis WH, Jr. *The New Clean Air Act: A Guide to the Clean Air Programs as Amended in 1990*. Morgan, Lewis & Bockius, 1990.

Rhodes D. "Safety, environmental crimes and the tough new laws." *Professional Safety*, Vol. 35, (January 1990).

U.S. Department of Health and Human Services, Public Health Service, Centers for Disease Control and National Institute for Occupational Safety and Health. *Occupational Safety and Health Guidance Manual for Hazardous Waste Site Activities*. Washington, DC: U.S. Government Printing Office, 1985.

U.S. Environmental Protection Agency, Region 5, 230 S. Dearborn, Chicago, IL 60604.
An Introductory Guide to the Statuary Authorities of The United States Environmental Protection Agency; Our Air,

Our Land, Our Water (April 1988).
EPA Property Searches for Buyers of Real Estate (Fact Sheet) February, 1991.

U.S. Occupational Safety and Health Administration, *CFR,* Title 29, Labor (for OSHA Hazardous Waste Training and Communications Regulations). Washington, DC: U.S. Government Printing Office, 1990.

——CFR, Title 40, Protection of Environment, Washington, DC: U.S. Government Printing Office, 1990.

Young OR. "Saving the Arctic: challenge to eight nations." *EPA Journal,* Vol. 16, No. 4, July/August 1990.

6

Emergency Preparedness

No industrial, commercial, mercantile, or public sector organization is immune from disaster. Emergencies can arise at any time and from many causes, but the potential loss is the same—injury and damage to people, the environment, and property. Advance planning for emergencies is the best way to minimize this potential loss.

A written comprehensive emergency management plan is intended to provide preplanned response to those unexpected or disastrous events that can strike any organization, such as a fire or natural disaster. Emergency planning is often assigned to the safety and health professional. In some instances this may be acceptable; however, ultimately the responsibility lies with the highest levels of management who best know a facility's resources, operations, and capabilities. The safety and health professional should act as the consultant, guiding line management through the process of identifying potential emergency events and developing primary and contingency plans to respond to them.

The safety of employees and the public must be the first concern in planning for an emergency. Preplanning should take into account the immediate short-term needs people are likely to have (i.e., injuries) as well as the long-term needs that may result from an emergency event.

Next, management should consider ways of protecting the property, operations, and the environment. In new plants, this may mean locating certain buildings or operations well away from others or locating operational response teams in a more central area to allow for a quicker response. In general, all emergency plans must include salvage, overhaul, and possible decontamination operations.

The final steps in the plan should involve restoring business operations to normal. In emergencies, a plant's operations are likely to be damaged or shut down altogether. Management will need to decide how and when to resume operations in the face of such obstacles as temporary wiring, lack of heating, or the need for significant rebuilding. Moreover, the probability of environmental damage to the immediate and surrounding community will demand considerable attention from the organization's top management. Operations that could adversely affect the health of employees or community neighbors must meet the emergency planning requirements of regulatory agencies.

Regardless of the size or type of organization, management is responsible for developing and operating an emergency planning program designed to meet the eventualities. An effective plan requires the same good organization and administration as any business undertaking. There is no one emergency plan that will do all things for all organizations. Each company must therefore decide on a plan that fits its needs and budget.

Emergency plans involve organizing and training small groups of people to perform specialized services, such as evacuation, fire fighting, rescue, spill response, or first aid. These small, well-trained groups can serve as a nucleus that can be expanded to meet any kind of emergency. Even with outside help available, a self-help plan is the best assurance that losses will be kept to a minimum. Although

planning should occur within each unit of the organization, the plant's emergency management staff should develop and implement a large-scale response scheme that encompasses all departments.

This comprehensive emergency management plan should meet all the different types of contingencies. Certain basic elements such as command functions, communications, and emergency staff personnel will be common to all emergency operations. Of course, the comprehensive plan should contain many procedures to deal with natural, technological, and nuclear hazards.

Organizations should also plan for an emergency where they may have to stand alone with no outside help. This is not a remote possibility in the event of a major disaster such as a tornado, hurricane, flood, or brushfire, that may strike an entire area. At times like these, communities will either have to stand alone or join with others in a mutual-aid pact.

Self-help plans should include provisions for recall of off-duty personnel. Many of these call-back lists look good on paper, but one should try a practice run to see if they really work. Long holiday weekends can present real problems should an emergency occur during these times. Further, consider how you will telephone the required personnel and how long it may take to secure 20 or more employees.

Before an organization initiates an emergency plan, it should identify and evaluate the potential disasters that might occur. The section Types of Emergencies later in this chapter discusses these in detail.

The next step is to assess and prioritize the potential harm to people, the environment, and property. The time of day and workshift patterns are other factors that should be considered in assessing the potential damage. Planning should take into consideration the impact of a catastrophe that might occur during weekends or holidays when no one or only a skeleton staff may be on hand.

To estimate potential damage to property, one should look not only at the general structures but their surroundings as well. For example, a building may be strong enough to resist an earthquake but a sudden 7-in. (18-cm) rainstorm might cause dangerous flooding that could short out all electricity. On the other hand, an exploding boiler located in an adjacent building would probably not harm the main plant.

Next, probable warning time should be considered. For example, a flood may build up over a period of several days while a bomb scare may afford only a few minutes warning from a telephone call. Warning time should give management a chance to alert personnel and to mobilize the plan. It is desirable to have a number of different plans, depending upon the nature of the emergency and the actual (or estimated) time available.

Another factor is how much company operations must be changed to meet the emergency. For example, in anticipation of a heavy snowstorm, it may be necessary to send employees home early. Some equipment may be left on or idling instead of being shut down completely. Finally, consider what power supplies and utilities may be needed, particularly those used for fire protection, lighting, ventilation, and communications.

A basic emergency preparedness plan will usually include a chain of command, an alarm system, medical treatment plans, a communications system, shutdown and evacuation procedures, and auxiliary power systems. This chapter points out the various elements involved in developing emergency and personnel-protective plans. Not every element discussed will apply to every organization. Also, several of the functions may well be combined and handled by one person, particularly in a smaller company. Generally, the text is directed to the more elaborate and expanded type of organization and planning.

WHAT KIND OF PLAN DO I NEED?

Once you have an idea of the company's risk to various hazards, have assessed the response capabilities, and have reviewed existing or previously written plans, then it is time to decide what type of plan is best for your facility.

The type of facility and its associated hazards is the ultimate determinant of the complexity of the plan, but there are other factors too—such as the availability of qualified personnel to write and maintain the plan, and the availability of funds to produce it. Nuclear power plants, for example, are required to develop extremely sophisticated plans that cost utilities millions of dollars to develop, implement, and maintain. Other facilities, such as small manufacturing plants where few chemicals are located in an area not prone to natural hazards, often require less complicated plans. Choosing the correct type of plan for your facility is very important, and there are a variety to choose from. The most common types used in industry include:

- action guides/checklists
- response plans
- emergency management plans
- mutual aid plans.

(In general, the most comprehensive emergency plan would state who does what, when, where, and how before, during, and after a disaster. The plans described here may address only a few or even just one of these questions.)

Action guides or checklists are generally a short and simple means to describe basic procedures that must be followed, such as whom to call, necessary information, and basic response functions. They are intended to be more of a reminder and should, therefore, not be used by an untrained individual. These guides or checklists are sometimes used in conjunction with other types of plans. This type of plan states what to do during a disaster.

Response plans are usually very detailed and tell all responsible individuals actions that must be taken to mitigate the problem at hand. Sometimes a response plan is written for each type of possible hazard at the site. For example, there might be a chemical spill response plan as well as a hurricane response plan, etc. These plans usually deal with only the actions necessary to respond to an emergency and do not cover requirements for actions before a disaster (training, drills, and exercises) or after a disaster (i.e., recovery plans, business interruption plans,

etc.). This type of plan states who does what and how during a disaster.

An emergency management plan is the most comprehensive plan used in business. It usually states who does what, and when—before, during, and after a disaster. Often, this type of plan will incorporate "implementing procedures" that state how something is to be done.

Finally, a mutual aid plan can be developed with other nearby firms. Such a plan calls for firms to share resources and help one another during an emergency. This type of plan demands a lot more coordination, but can be very useful to small firms with limited resources or larger firms with high hazard potential. This type of plan usually states who does what during an emergency.

Selecting the proper type of plan is important, but do not lock yourself into one of the mentioned plans. Choose a type that answers the questions you want it to answer. Be sure to consider whether you want it to address only response activities during an emergency or if you want it to consider preparedness activities before and recovery activities after an emergency. Be sensitive to the cost and time associated with the development and maintenance of each type of plan, as well.

TYPES OF EMERGENCIES

The first step in the emergency planning process begins with determining what types of hazards may affect the organization. Targeting specific hazards allows the company to create a comprehensive planning, organizing, and implementation program. Hazards posing a threat to the organization will vary according to its location, production/engineering processes, and work practices.

Many sources of hazard information can help management determine the likelihood of specific emergency events in their locality. These sources include historical knowledge and records of accidents, fire statistics, National Weather Service logs, U.S. Geological Survey studies, location of rail lines and airports, and local emergency management U.S. Hazard Vulnerability Community Analysis documents. The organization's management should agree upon a set of hazards that appear to have some chance of occurring. After this assessment, emergency response operations planning can begin, based on a realistic study of the potentially disastrous threats facing personnel, property, and environment. A discussion of types of emergencies follows.

Fire and Explosion

Except where fires result from large-scale explosions, warfare, civil strife, or hazardous chemicals, the fire emergency usually allows some time for marshaling firefighters and organizing an evacuation, if necessary. Many large fires originate as small blazes that, if caught early, could be controlled by trained personnel. Therefore, prompt action by a small, properly trained and equipped group can usually handle most situations. However, plans should include the marshaling of extensive firefighting forces upon the first indication of any fire growing beyond the "small fire" stage. Plans should also include procedures for controlling chemical contamination that may mix with the runoff from water used to put out the fire. Also, firefighting and evacuation procedures should take into consideration toxic gases, smoke, and fumes that may be produced by the burning materials.

The main point is this: small fires must be extinguished as soon as they start. The first five minutes are considered the most important. Good housekeeping, prompt action by trained people, proper equipment, and common sense precautions will prevent a small fire from becoming a disastrous blaze.

Specific information on fire extinguishing and control is in Chapter 6, Fire Protection, of the *Engineering and Technology* volume of this *Manual*.

Both small and large fires have the potential for producing significant environmental problems. These include:

- production of toxic gas and dust from combustion and decomposition
- thermal plumes that can carry the materials substantial distances
- disposal problem involving large quantities of contaminated water used in extinguishing the fire.

Commonly used construction materials, furniture, carpets, and industrial materials can release significant quantities of toxic combustion products.

The combustion of industrial and agricultural chemicals and materials may also be a source of hazardous gases, vapors, fumes, or dusts. The dispersion of these materials may be enhanced by the buoyancy effects of the hot gases caused by the fire. Depending on weather conditions, these plumes may travel substantial distances and remain concentrated. Ash and dusts generated from some forest fires in the western part of the United States have been detected, for example, in the Midwest. To minimize these effects, those responsible for responding to fires must be aware of the potential hazardous airborne materials, be able to track their movement, and be prepared to measure their ground-level concentration. These activities may require access to or on-site collection of meteorological data and air sampling.

An additional environmental hazard in responding to fires is the disposal of the potentially contaminated wastewater produced in extinguishing a fire. For example, 100,000 gal of water can easily be used in fighting a simple residential fire. If a thousand gallons of a 10% concentrated pesticide is released in the course of the fire, the resultant concentration could be a 100 ppm solution. Introduction of such a liquid waste flow into a sewer system or surface waterway could have significant environmental ramifications. Therefore, diking, chemical neutralization, use of chemical absorbants, or a controlled burn may be considered necessary containment methods.

Floods

When a company or plant is located in a floodplain, it should have the protection of dikes of earth, concrete, or brick construction. The probable high-water mark can be obtained

from the U.S. National Weather Service or the Army Corps of Engineers. The latter group also provides valuable assistance in planning floodwater control.

Floods—except flash floods caused by torrential cloudbursts or the bursting of a storage tank, dam, or water main—do not strike suddenly. Ordinarily, there is enough time to take protective measures when a flood seems imminent.

Another source of information on flooding in a particular area is the local emergency planning commission or the U.S. Geological Group. Many facilities are not in floodplains. However, the U.S. Geological Group has classified different types of floods that may hit a region at specified times, such as the flood that strikes every 100 years. These infrequent events should be included in a company's planning process, even if the chances of their occurring seem remote.

Precautions that should be considered in anticipating flood conditions include:
1. electrocution hazards and the need for proper grounding and electrical fault protection
2. bracing or storing important equipment, materials, and chemicals off the ground
3. availability of pumps and emergency power sources
4. protection against soil erosion, which may cause serious structural damage
5. provision and protection of potable water supplies.

Hurricanes and Tornados

Areas most frequently exposed to winds of destructive hurricane force are the Atlantic and Gulf coasts. However, some inland locations are not immune to this type of disaster.

The National Weather Service and other agencies have developed improved methods of detecting and tracking hurricanes; thus ample warning can be given for maximum protection of property and evacuation of personnel from threatened areas.

Organizations regularly exposed to this hazard have developed a system of tracking hurricanes on a map. At predetermined locations, a specified alert condition becomes effective, and each supervisor completes a checklist for that alert. As the hurricane progresses toward the facility through the 100–mile (160–km) circle, 50–mile (80–km) circle, etc., the plant is shut down in an orderly manner.

Buildings constructed in areas where hurricanes occur should be built to withstand these destructive winds and tides. Basic preventive measures include equipping the facilities with storm shutters or battens that can be promptly attached, at least on the side from which the storm is expected to approach. If this is not done, the facilities could have windows shattered, the roof torn away, or other parts of the building destroyed. If the roof is lost or damaged, building contents can be drenched by the heavy rain that accompanies the storm and by water from broken sprinkler pipes. Therefore, roofs should be securely anchored and tall structures (such as chimneys, water towers, and flagpoles) designed to withstand high winds.

Although the central Mississippi Valley is considered the top tornado area of the U.S., almost every state has experienced these destructive storms. They can inflict enormous damage quickly, although it is usually restricted to a small area.

It is not possible for the U.S. National Weather Service to give as much advance warning or to pinpoint the strike area as accurately as they can with hurricanes. Therefore, a company must be prepared to protect its personnel on short notice and to take quick action to protect and restore undamaged equipment and materials.

The National Weather Service radio bands can be monitored during storm conditions. In one midwestern city, several large companies have set up a cooperative warning network. A lookout is stationed atop the city's tallest building. Through a central network, not only the member companies but also the city's radio stations and civil defense are alerted in the event of an approaching tornado. Private weather consultants are also available.

Tornado and hurricane experience indicates that emergency plans should include:
1. Establishing procedures for alerting and getting personnel to a safe place. If the building is not constructed to withstand these natural forces, emergency shelters should be located close to the work area. All personnel should be instructed in the procedure to follow, with and without advance warning.
2. Assigning trained personnel to take care of power lines—dangling wires are a serious hazard.
3. Assigning trained people to remove wreckage to prevent injury to salvage and repair workers.
4. Scheduling regular meals and rest for the repair crews.

Earthquakes

Several areas of the United States have the potential to experience damaging earthquakes. In particular, the Pacific Coast contains the San Andreas fault network and the Midwest contains the New Madrid fault system. Other major fault areas are located in North and South Carolina, Utah, and New England. No reliable earthquake warning system exists. This factor, coupled with the widespread damage that a quake can bring, makes this geological hazard a difficult phenomenon for planners to address.

"Earthquake-resistant construction" consists of building a structure so that it "floats above the bedrock and ballasting it as a ship is ballasted, by making lower stories heavy and upper stories light." Utility lines and water mains should be flexible and laid in trenches that are free of the building, rising in open shafts, and connected to fixtures by flexible joints. Lockers, cabinets, shelves, etc., should be securely installed with seismic bracing and safety restraining strips on shelves containing bottles of chemicals.

The principal dangers of earthquakes come from structural failures of buildings, bridges, and other items; fires; flooding; broken utility lines; hazardous materials releases; water shortages and contamination; health problems; and damage to transportation facilities such as interstate highways, airport runways, navigable waterways, and railways. Water reser-

voirs or emergency water sources should be available as backup support for firefighting operations. Also, a system should be established for shutting down gas mains supplying companies. The method of turning off the gas supply depends on the size of the main and the pressure on the line that comes into the plant. Company officials should obtain expert advice when devising these plans.

See Figure 6–1 for recommended steps to take if an earthquake strikes. Be sure your emergency management plan provides for the actions recommended in Figure 6–1 if your facilities are in an earthquake risk area.

Civil Strife and Sabotage

Riot or civil strife is another item on the list of emergencies for which a company should plan in advance.

Civil strife. An emergency involving civil strife raises the questions of the right to protect property versus individuals' legal right to assemble. A company should obtain from an elected legal authority in the community (the district attorney, for example) a statement explaining the company's rights in protecting its property and the company's legal responsibility for the safety of employees and other people—such as customers, supplier salespeople, and visitors—who may be on company property. A company's legal department can be helpful in determining such a position, but its opinion does not have the force of law.

Some of the problems involved include disruption of business when an office or plant area is invaded by outsiders, protection against a mob intent on destroying company property, requests from neighboring companies for assistance during a riot, and the rights and responsibilities of armed company guards. Civil strife emergencies can be just as disastrous as any other type and should receive advance planning by manufacturing, mercantile, or commercial establishments.

Sabotage. Protection against sabotage is also an important consideration. The saboteur may be a highly trained professional or an amateur. He or she may be anyone—usually one of the least-suspected members of the organization. Because physical sabotage is frequently an inside job or requires the assistance—knowingly or unknowingly—of someone inside the plant, the principal measures of defense must be denying entry to suspicious persons. Evidence of sabotage should be reported to the U.S. FBI and, if military related, to the U.S. Department of Defense.

Work Accidents and Rumors

The "chain reaction" from a so-called "routine work accident" can result in an emergency situation. For example, a break in a chemical line or accidental emission of toxic vapors may create an emergency. Panic caused by rumors or lack of knowledge can also create an emergency. Plans for such situations should include establishing auxiliary areas in the building for medical treatment, a method of notifying employees of the situation, a method of quickly taking a head count of workers, and sources of oxygen supplies available on short notice. A public relations coordinator should deal with the press, the public, and families.

Work accidents. The potential for work accidents to cause emergencies has increased due to the complexity of processes, the proliferation of industrial and agricultural chemicals, and the often close proximity of residential areas to industrial activities. For example, a small break in a chemical line could allow hazardous air contaminants to enter a plant's ventilation system creating a work-site emergency. A larger release of a chemical or toxic vapor may endanger a neighborhood. The toxic release may cause direct injuries or result in panic and create an emergency situation.

The potential for such events should be investigated and evaluated to protect individuals on site and in the surrounding community. The methods used in this evaluation may include the application of risk-assessment techniques. These techniques require management to identify hazards, calculate their probability of occurrence, and develop means to respond to these events. Various existing U.S. governmental regulations that require these types of analysis include:

- 40 *Code of Federal Regulations (CFR)*, Parts 330 and 335 "Extremely Hazardous Substance List and Threshold Planning Quantities; Emergency Planning and Release Notification Requirements; Final Rule." Vol. 52, No. 77, Washington, DC, pp. 13378–13410.
- 29 *CFR*, Part 1910.120 "Hazardous Waste Operations and Emergency Response."
- *Hazard Communication Guidelines for Compliance.* U.S. Department of Labor, Occupational Safety and Health Administration (OSHA), 29 *CFR*, 1910.1200.
- *Emergency Planning and Community Right-To-Know Act*, Title III, Superfund Amendments and Reauthorization Act, Public Law No. 99–499.

Shutdowns

An emergency situation may occur following an unscheduled action, such as a disaster or strike; hence a fast shut-down procedure should be covered under an emergency plan. This plan should be based on a priority checklist, that is, all the tasks to be assigned and functions to be performed should be arranged in order of importance ahead of time so that if only a short notice is given, at least the most vital precautions are completed. This "crash procedure" is usually an adaptation of the routine procedure used for scheduled shutdowns, such as for vacation or renovation. Naturally, the amount of warning time controls the speed of shutdown.

Whenever a plant or other unit or building must be shut down, safeguards against fire, explosion, or chemical release take on added importance. The extent of these measures will vary with the size and purpose of the plant. It is important to organize a formal program for instructing personnel in emergency shutdown procedures. In particular, workers should remove lint, dirt, and rubbish from the area; drain and clean dip and mixing tanks and other equipment where flammables have been used; clean spray booths, ducts, and flammable liquid storage; close gas, chemical, and fuel line

ACTIONS TO TAKE IF AN EARTHQUAKE STRIKES

The greatest threat during an earthquake is from falling debris. Earthquakes are unpredictable and strike without warning. Therefore, it is important to know the appropriate steps to take when one occurs, and to be thoroughly familiar with these steps to be able to react quickly and safely.

IN THE OFFICE

During the earthquake	
Step	**Action**
1	**Remain inside** the building.
2	**Seek immediate shelter** under a heavy desk/table, or brace yourself inside a doorframe or against an inside wall. ▪ Get at least 15 ft away from windows.
3	**Stay there**. If shaking causes the desk or table to move, be sure to move with it.
4	**Resist the urge to panic**. Organize your thoughts; mentally review the established psychological considerations for earthquake safety. ▪ Don't be surprised if the electricity goes out, fire or elevator alarms begin ringing, or the sprinkler system is activated. ▪ Expect to hear noise from broken glass, creaking walls, and falling objects.

Immediately after the earthquake	
Step	**Action**
1	**Remain in the same position** for several minutes after the earthquake in case of aftershocks.
2	**Do not attempt to evacuate** or leave your immediate area unless absolutely necessary or when instructed to do so by a proper authority.
3	**Check for injuries** and administer first aid. Recognize and assist co-workers who are suffering from shock or emotional distress.
4	**Implement your survival plan**. Establish a temporary shelter if rescue teams are expected to be delayed.
5	**Use stairway** when instructed to exit building.

AT HOME

During the earthquake	
Step	**Action**
1	**Remain inside** your house.
2	**Seek protection** from flying debris or fixtures. Brace yourself inside a doorframe or against inside walls. Seek cover beneath a table, desk, or bed.
3	**Stay in position** until the shaking stops.

After the earthquake	
Step	**Action**
1	**Remain calm**. Organize your thoughts by reviewing your home earthquake survival plan.
2	**Check for injuries** and administer first aid. Be prepared to respond to the psychological aftereffects generated by a major earthquake.
3	**Check water, gas and electric lines**. If you suspect damage, turn off the main valves and leave them off until advised by a utility company representative or other competent source.

Figure 6–1. In earthquake-prone areas, emergency preparedness training should include what to do if an earthquake occurs. (Courtesy Pacific Bell Safety Staff.)

AT HOME (CONTINUED)

After the earthquake

Step	Action
4	**Do not use** candles, matches, or other open flames, or turn lights on/off, either during or after the tremor because of possible gas leaks.
5	**Turn on the radio** to receive emergency instructions. Reserve telephone usage for emergency calls only.
6	**Check your house for structural and internal damage**. Wear boots, if possible, to protect against shattered glass. ■ Chimneys are earthquake prone if not well enforced and should be approached with caution. Look for separation down the sides or for loose bricks. Unnoticed damage could result in a fire. ■ Check the interior of the house for dislodged items. The continued swaying motion from an earthquake will cause loose doors of medicine and kitchen cabinets/drawers to open and their contents to spill out. ■ Check closets and bulk storage areas (garage/basements) for items which may have toppled or collapsed during the earthquake spilling substances which could produce fumes or become potential fire hazards.

IN PUBLIC

During the earthquake

Step	Action
	■ **On the street**
1	**Enter the closest structure immediately**—do not look up. Enter a store, terminal, office building, etc., just get inside. NOTE: The greatest danger during an earthquake is falling debris.
2	**Remove yourself from windows** that may shatter.
3	**Brace against** an inside doorframe or against inside walls.
	■ **In a stadium, amphitheater or church**
1	**Remain in your current location. Do not** rush to exits. The chaotic fleeing of large crowds diminishes the effectiveness of an evacuation procedure and frequently results in unnecessary injuries or deaths.
2	**Seek cover** under a bench or chair. If unavailable, crouch down, and cover your head with your arms to protect against falling debris.
3	**Keep away from overhead electric wires** or anything that might fall.
	■ **In a vehicle**
1	**Stop the vehicle** if it is currently in motion. **Avoid** stopping either **on** or **under** a bridge or overpass.
2	**Remain inside vehicle** until the shaking stops.

After the earthquake

Step	Action
1	**Remain calm**.
2	**Check for injuries** and administer first aid. Recognize and assist individuals who are suffering from shock or emotional distress.
3	**Await emergency evacuation instructions**.
4	**Watch for hazards** created by the earthquake when traveling to another location such as downed electrical wires, broken or undermined roadways, collapsed freeways, overpasses, or bridge structures.
5	**Stay away** from waterfronts or beach areas. Tsunamis may result as an aftereffect of the earthquake.
6	**Avoid sightseeing**. Emergency vehicles will need ready access to respond to emergency situations.

Figure 6–1. *Concluded.*

valves; open switches on power circuits that may be out of service; check serviceable condition of sprinkler systems, fire extinguishers, hydrants, alarms, and other protective apparatus; and anchor cranes. Prior to the closing, employees are alerted by special instructions to keep their workstations clean and fire safe.

During the shutdown, continuous inspection of any maintenance or special operations, such as remodeling, must be maintained. Gas cutting and welding should be carefully supervised. Employees who remain on duty—the plant protection force, security staff, maintenance workers, supervisors, or executives—should be briefed in effective countermeasures in case a fire breaks out.

If there has not been sufficient notice to effect a normal shutdown, it may become necessary to allow personnel into the area to perform emergency functions. Emergency shutdown and spill containment procedures should be clearly established for high- and low-staffing situations and all personnel involved should be appropriately trained.

Company management should designate someone who can admit those personnel necessary to handle emergencies arising within the area. The security chief or the fire chief should arrange with local police and fire department officials for assistance if an emergency gets beyond local control. It is especially important that arrangements be completed to speed up admitting firefighters, security, or cleanup personnel and their equipment to the disaster site.

Some companies use plant protection service agencies to prevent loss from theft, fire, and accident hazards during shutdowns. Similar plans should be worked out with these people so that police and fire assistance can be obtained quickly when it is needed.

Wartime Emergency Management Planning

One of the main differences between planning for peacetime emergencies and planning for wartime emergencies is that war may cripple an entire community. This difference makes it more important that emergency plans take into account the fact that the plant may have to stand alone. In war, outside sources of help—fire and police departments, hospitals and doctors, regular sources of supply for material and equipment—are not so readily available.

Emergency management planning (EMP) consists of the plans and preparations of business and industry managements to achieve a state of readiness during times of disaster. This would enable their plants, facilities, and employees to cope with the effects of a nuclear or conventional incident.

Even if a particular area is not attacked, plants in that area may be requested to furnish transportation to evacuate the injured from damaged areas and to house and feed the evacuees. Plant emergency squads may also be required to assist other stricken facilities. In a major catastrophe, there would probably not be enough hospital space available, making it necessary to keep the injured in temporary shelters for some time. In such cases, employees with the proper training might be required to administer sedatives and plasma and to treat injuries.

Hazardous Materials

As most organizations use a variety of chemical substances, management must be concerned about potential usage, handling, and disposal problems. Although there are many rules and procedures in place, the questions to ask are: What if a safeguard fails? What if the container cracks and substances leak out? In addition to normal hazards, can chemicals react with other substances to create further hazards to people and property? (See OSHA requirements in 29 *CFR*, 1910.1200 Hazard Communication.)

Chemical hazards are discussed in the National Safety Council's *Fundamentals of Industrial Hygiene*, 3rd ed. and *CFR*, Title III, 1910.1200(g).

Radioactive Materials

Fires and other emergencies involving radioactive materials are becoming more common with the widespread, peaceful use of isotopes.

Radioactive elements and fire. Giraud (1973), (see References) makes the following observations. Radioactivity cannot by itself cause fires, nor can it be destroyed or modified by fire. However, a fire may change the state of a radioactive substance and render it more dangerous by converting it to a gas, aerosol, smoke, or ash and allowing it to contaminate a wide area.

Furthermore, fires can cause structural disruptions in stocks of fissionable materials and in the special equipment for their treatment or use. Such disruptions may, at worst, result in a nuclear chain reaction and initiate a critical nuclear accident.

Radioactive elements are found in various forms, depending on their uses. The human eye can detect no difference between an inactive element and the same element when rendered radioactive. Both appear equally harmless. A fundamental distinction must, however, be made between so-called "sealed" and "unsealed" sources.

- In the case of sealed sources, the radioactive substance is not accessible. The container has sufficient mechanical strength to prevent the substance from spreading during normal conditions of use. The capsule is made of stainless steel. The sources are of small dimension—approximately one centimeter.

- In unsealed sources, however, the radioactive substance is accessible. In normal conditions of use, there is no means of preventing it from spreading. Solid substances are kept in aluminum tubes, liquids are kept in flasks, and gases in glass ampules.

The fact that a substance is radioactive does not affect its general physical properties nor its behavior when heated to an abnormally high temperature—as, for instance, during a fire. The substance will, on contact with fire, undergo the normal transformations, depending on its initial form—i.e., solid, liquid, or gas. Melting, boiling, and sublimation can be expected, with the formation of combustion products corresponding to the chemical properties of the substance: slag, ash, powder, dust, mists, aerosols, fumes, or gases.

These combustion products are generally finer and less

dense than the original substance, so they disperse more easily. Although the change in the physical state of the substance will not affect its radioactivity, the radiation hazard will be more difficult to control.

The protective containers currently in use have a widely varying resistance to fire. Therefore, the protection afforded to the contents will depend on the type of container used. In general, sealed sources are strongly fire resistant, and radioactive elements thus contained are well protected (Figure 6–2).

Unsealed sources, however, and solutions or gases in fragile containers easily fall victim to fire. What type of immediate action needs to be taken in the event of an accident with radioactive materials can be determined by the firefighting staff once the type of container is known. Actions to deal with the hazard will depend on the properties of the radioactive substance. (See OSHA regulations in 29 *CFR*, 1910.1200 (f), Hazard Communication, 'Labels and Other Forms of Warning' and 29 *CFR* 1910.38, Employee Emergency Plans and Fire Preventing Plans.)

When, as the direct or indirect result of a fire, the protective container has been broken, the radiation hazards for rescue workers at the fire, or for personnel in the vicinity, are likely to be more serious than the danger of a conventional fire. Accordingly, the person in charge of the rescue work will sometimes be obliged to override the normal firefighting procedures to ensure proper confinement of any radioactive elements released. If they are already affected by the fire, further hazards may arise.

The release of radioactive elements may result in contamination of surface areas. This may be caused by the spilling or splashing of radioactive substances or by the spreading of solid radioactive substances in paste, powder, or dust form. All possible precautions must be taken to prevent any further spread of the contamination. The means to be used, however, will differ with each case. In the first (spilling or splashing), absorbent materials should be used—such as powder, earth, sand, etc. In the case of spreading, the substances should be slightly dampened with a spray of water—unless it is otherwise specified on the container. (See OSHA regulation 29 *CFR*, 1910.1200 (g), Material Safety Data Sheets.)

Radioactive liquids can be contained by the methods normally used by the firefighting brigade. The contaminated area must be clearly marked and roped off to prevent the entry of unauthorized personnel (Figure 6–3).

Contamination of the atmosphere is caused by radioactive elements in the form of dust, aerosols, fumes, and gases. The spreading of such contamination is determined mainly by the prevailing weather conditions, and it is difficult to control. Such atmospheric contamination may lead to other toxic or corrosive hazards associated with the particular chemical. The most serious danger is that of inhaling the substance when it is suspended in the air. Firefighters, accordingly, should wear self-contained breathing apparatus.

The danger of internal irradiation is always present whenever there is contamination by a source of penetrating ra-

Figure 6–2. "Radiation Yellow—III" label, which is affixed to each package of highly radioactive material. Different labels are required for different intensities and quantities of radioactive material. For details see Title 49—Transportation, *Code of Federal Regulations*, Part 172, Hazardous Materials Table and Hazardous Materials Communications Regulations.

diation. It may also occur by accidental release of an alpha or beta emitter from its protective container, or by the destruction (even partial) of the protective container.

Weather-Related Emergencies

Throughout any year, unusually severe and unexpected weather events can require some changes in normal operations. For example, in North Dakota, the temperature occasionally may drop to –35 F (–30 C), yet most activities and travel are not normally affected. But, if the wind increases in strength or the temperature drops suddenly, people may need help as they travel or participate in other outside activities (Figure 6–4 provides a windchill chart). Employees could be alerted to the danger prior to leaving work and told when or how they are to be notified about whether the company will be open in the morning.

Management must plan for a variety of weather-related emergencies appropriate for the locale. In the event of extremely heavy snowfall, for example, what changes might be needed in operations? What should employees be told prior to leaving for home? An unusually heavy rainstorm may strand hundreds of customers in a store just a few minutes before closing. Are supervisors and clerks prepared to handle the situation? May they allow telephone calls in and out? How do they control the crowd? Hail or wind may start breaking glass windows while customers are shopping. What is the immediate action? Suppose that adverse weather caused a power failure in a company or someone suddenly shut off all power and lights while crowds were shopping? The emer-

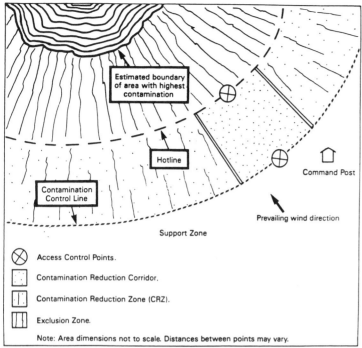

Estimated boundary
of area with highest
contamination

Hotline

Command Post

Contamination
Control Line

Prevailing wind direction

Support Zone

⊗ Access Control Points.

▫ Contamination Reduction Corridor.

▯ Contamination Reduction Zone (CRZ).

▯▯ Exclusion Zone.

Note: Area dimensions not to scale. Distances between points may vary.

Figure 6–3. This diagram shows site work zones for an emergency situation involving a radioactive hazard. Note that decontamination facilities are located in the Contamination Reduction Zone. (Courtesy NIOSH/OSHA/USCG/EPA, *Occupational Safety and Health Guidance Manual for Hazardous Waste Site Activities.*)

gency lighting system may operate as intended, but employees, particularly key supervisors, must understand emergency plans and be prepared to act responsibly.

Other weather emergencies include droughts and extreme heat. These harsh conditions require attention to outdoor work schedules, water resources, and health effects. In addition, lightning is a major cause of weather-related deaths in the United States. Planning should include emergency procedures for halting work in areas vulnerable to lightning strikes or the loss of utilities when this particular danger is imminent.

PLAN-OF-ACTION CONSIDERATIONS

Once the risk assessment of potential emergencies has been completed, the next stage of the planning process should be the preparation of a plan of action. This plan should be supported by management and include input from both the public and private sectors. It is desirable to include union or labor representatives.

Generally, someone should be appointed emergency planning director or coordinator, perhaps with help from an advisory committee. Usually, because of their experience and training, the health and safety, medical, fire, and security departments will be involved. Of course, because production and maintenance will be affected, these departments must be consulted. Also, the legal staff needs to be aware of the plan. Finally, management should contact local law enforcement agencies and fire departments.

Some managers may object to the cost and effort involved in giving immediate attention to emergency planning. However, this activity can be justified by weighing the cost of preparedness against the possibility of yearly losses from accidents, fires, floods, and other catastrophes.

Program Considerations

Once management has completed its advance planning and has evaluated the type of emergencies and their potential harm to people and property, the next step is to develop a working emergency management plan within the organization. In some cases, this step requires working with local agencies to protect a company's operations.

Advance planning is the key. Management must establish a written set of plans for action. The plans should be developed locally within the company (and corporate structure) and be in cooperation with other neighboring or similar organizations and with governmental agencies. It may not always be possible for every organization or agency to cooperate or participate fully, but through planned action each one can be aware of the resources and help available. In some instances when outside assistance is limited, a company may need to depend largely upon its own resources to deal with emergencies.

Often an emergency manual or handbook will be developed for the plant or organization. The following outline covers many of the items that might be included, but other items may be needed as dictated by the expected emergencies and the available resources.

1. company policy, purposes, authority, principal control measures, and emergency organization chart showing positions and functions
2. some description of the expected disasters with a risk statement
3. map of the plant, office, or store showing equipment, medical and first aid, fire control apparatus, shelters, command center, evacuation routes, and assembly areas
4. list (which may also be posted) of cooperating agencies and how to reach them
5. plant warning system
6. central communications center, including home contacts

Estimated wind speed (mph)	Actual thermometer reading (°F)											
	50	40	30	20	10	0	−10	−20	−30	−40	−50	−60
	Equivalent temperature (°F)											
Calm	50	40	30	20	10	0	−10	−20	−30	−40	−50	−60
5	48	37	27	16	6	−5	−15	−26	−36	−47	−57	−68
10	40	28	16	4	−9	−24	−33	−46	−58	−70	−83	−95
15	36	22	9	−5	−18	−32	−45	−58	−72	−85	−99	−112
20	32	18	4	−10	−25	−39	−53	−67	−82	−96	−110	−124
25	30	16	0	−15	−29	−44	−59	−74	−88	−104	−118	−133
30	28	13	−2	−18	−33	−48	−63	−79	−94	−109	−125	−140
35	27	11	−4	−20	−35	−51	−67	−82	−98	−113	−129	−145
40	26	10	−6	−21	−37	−53	−69	−85	−100	−116	−132	−148
Wind speeds greater than 40 mph have little added effect.	Little danger for properly clothed person. Maximum danger of false sense of security.			Increasing danger Danger from freezing of exposed flesh.			Great danger					
Trenchfoot and immersion foot may occur at any point on this chart.												

Figure 6–4. Windchill factors. The human body senses "cold" as a result of both temperature and wind velocity. The numerical factor that combines the effect of these two is called "the windchill factor." Because of the extra clothing that people wear in cold weather, their physical size is greater than it is in warm weather. Be sure that equipment and controls are of adequate size and simplicity so that they can be run effectively and safely by persons wearing heavy clothing. (Courtesy Holmes Safety Association.)

of employees

7. shutdown procedure, including security guard
8. how to handle visitors and customers
9. locally related and necessary items
10. list of equipment and resources that would be available and where they can be reached.

Some of these items will be discussed in more detail in the following pages.

Management should rehearse the plan under realistic conditions to test the plan's effectiveness. For example, emergency lights may fail when needed, or the telephone service break down. These are real conditions that might occur in an actual disaster. Therefore, planning should include all possible as well as probable contingencies.

Chain of Command

Once management has decided to establish a disaster plan, they should appoint a director or coordinator and create an advisory committee representing the various departments. A basic guideline to follow in establishing the chain of command is (1) keep the chain as small as practical and (2) appoint personnel to crisis management positions based not on their title but rather on their ability to respond to a situation under extreme stress.

Experience has shown that the smaller the chain of command, the more efficient and effective its decisions and actions will be during a crisis situation. Normally, the chain of communication will pattern itself after the chain of command. Communications must be conducted in as efficient a manner as possible. Management should quickly provide accurate information or request the proper support. Long delays could affect the outcome of a given emergency situation.

It is normal to think that the ranking manager should assume the crisis manager's position. However, to do so may not be the proper action to take. Many people are excellent managers under normal conditions. They are able to give good direction if allowed time to think through the situation and weigh the pro's and con's. But given a situation where extreme stress is introduced and quick, timely decisions are required, the same individual may not be able to perform. It is, therefore, imperative that the individuals selected to become a part of the crisis team be tested, through training exercises, to see if they can perform under emergency conditions. If the ranking manager is found not to be the proper choice for the crisis manager, he or she should be a consultant to whoever is placed in the crisis manager's position. This

same philosophy applies to any of the positions established within the chain of command.

The emergency director should be a member of top management, whether it be a one-building or one-plant or a national organization. This individual will need to delegate authority and to speak for the organization. The head of the disaster-control organization must be cool and quick thinking and sufficiently healthy to withstand the arduous duties involved in an emergency. The emergency director's regular duties should be such that the greater part of the time will normally be spent at his or her own unit. However, the plan should always name an alternate director. The alternate should be a person who has authority and qualifications similar to the director and should be trained with the director.

The director (and alternate) should be the first to be trained. Management should maintain liaison with local emergency management authorities, if possible, to make sure that the plans are coordinated with those of the community and to keep the company informed on new developments. (See OSHA regulation 29 *CFR*, 1910.156, Fire Brigades; 29 *CFR*, 1910.1200, Hazard Communication; and EPA Regulations 40 *CFR*, Part 311, OSHA 29 *CFR*, 1910.120.)

The director of emergency management is responsible for:

- emergency operation center management
- communications
- firefighting
- security and law enforcement
- rescue operations
- emergency medical services
- transportation
- damage assessment
- mitigation and investigation
- public information and media briefings
- rumor control
- on-scene safety functions at the emergency site
- warning and evacuation of plant and community personnel
- utilities and engineering functions
- sheltering, feeding, and counseling functions
- morgue establishment and notification of survivors
- notification to the SARA Title III authorities in the event of a hazardous materials release.

All of these functions are likely to be essential although some may be combined. The person (and alternates) responsible for each function should be selected with great care and trained by the director. These chiefs should be familiar with all parts of the plan and should have experience in the fields in which they are to serve.

Management should train assigned personnel to carry out their duties in accordance with the overall emergency plan. In small operations without regular security staff or firefighters, operating personnel will be trained to take care of these duties. Of course, the number of members on each of the teams depends on the circumstances of each facility. Each team captain should select personnel from the available volunteers, supervise their training, and procure their equipment. The strongest workers can be assigned to rescue squads because the work usually demands strenuous physical effort; those who are less strong could be useful in light salvage operations.

Because workers' whole-hearted cooperation is necessary to the successful operation of an emergency plan, shop stewards or other employee representatives should take part in the planning. They should understand that measures taken are to protect the lives and jobs of the workers as well as to protect property.

The organization should establish emergency reporting centers so that employees will know where to report should a disaster strike while the employees are away from work. Reporting centers give employees a feeling of security and continuity, and aid in taking a "roll call." To facilitate these arrangements, each employee should carry an identification card containing specific instructions on where to report, list of other reporting centers, basic employment record, and designation of the employee's next of kin in case they must be contacted or receive money due. The reporting center will keep a duplicate record for each employee assigned to the center. A one-plant company or a small company can consider using the home of a member of management, a supervisor, or an employee.

Training

One of the most important functions of the director and staff, on both the corporate and plant levels, is training. Training for each type of disaster is essential in developing a disaster-control plan and keeping it functioning. Employees must be taught to realize that an emergency plan is vital and real—it has no value if it remains simply an idea (Figures 6–5a, 5b and 6–6). Training and rehearsals are time consuming, but they keep the program in good working order.

Simulated disaster drills will help key people and employees respond to emergencies with greater confidence and effectiveness. Sometimes called a paper drill, there is no actual response required. Instead, key personnel operate under the direction of a drill coordinator who feeds information to them on a real-time basis and monitors the response.

Another type of drill is a full-scale dress rehearsal involving all personnel with simulated situations and injuries. Management must be careful to prevent injuries to participants and the public while making the full response as realistic as possible. Feedback from either type of drill is essential to improve emergency management plans.

Management should assure employees that the company is doing everything possible to prevent injury to them, that every employee is an essential and necessary part of the team, and that the disaster-control organization is ready for any emergency. Such assurance will go a long way toward preventing panic when a real disaster strikes. Should such an event occur, emergency forces will snap into action, workers will file quietly into their shelters or other designated areas, firefighters will be ready with hoses and equipment, and first aid squads will stand by ready to aid the wounded. Such planning is further evidence of management's concern for employees.

IN CASE OF FIRE OR OTHER EMERGENCY

✔ **KEEP YOUR HEAD**—avoid panic and confusion.

✔ **KNOW THE LOCATION OF EXITS**—be sure you know the safest way out of the building no matter where you are.

✔ **KNOW THE LOCATION OF NEARBY FIRE EXTINGUISHERS**—learn the proper way to use all types of extinguishers.

✔ **KNOW HOW TO REPORT A FIRE OR OTHER EMERGENCY**—send in the alarm without delay; **notify the CHIEF OF EXIT DRILLS.**

✔ **FOLLOW EXIT INSTRUCTIONS**—stay at your work place until signaled or instructed to leave; complete all emergency duties assigned to you and be ready to march out rapidly according to plan.

✔ **WALK TO YOUR ASSIGNED EXIT**—maintain order and quiet; take each drill seriously—It may be "the real thing."

REMEMBER—IT IS PART OF YOUR JOB TO PREVENT FIRES

Figure 6–5a. Sample emergency exit notice for general posting.

Hazardous Waste/Spills Emergencies (HAZMAT)

The U.S. OSHA regulation entitled Hazardous Waste Operations and Emergency Response: Final Rule became effective on March 6, 1990. It addresses the many aspects of health and safety that are now legally required at hazardous waste sites; treatment, storage, and disposal facilities; and other hazardous materials emergency locations. Although the rule mandates certain requirements for monitoring instrumentation, site safety plans, respiratory and personal protective equipment, medical surveillance, engineering controls, work practices, training requirements, and other operational functions.

Generally, the *CFR*, 1910.120 training standard addresses three categories of employees: hazardous waste site workers (paragraph [e]); treatment, storage, and disposal workers (paragraph [p]); and emergency responders to hazardous substance releases regardless of location. (See also OSHA regulation 29 *CFR*, 1910.156, Fire Brigades.)

The HAZMAT team consists of an organized group of employees, designated by the employer, who are expected to perform work to handle and control actual or potential leaks or spills of hazardous substances requiring possible close approach to the substance. The team members perform responses to releases or potential releases of hazardous substances for the purpose of control or stabilization of the incident. A HAZMAT team is not a fire brigade nor is a typical fire brigade a HAZMAT team. A HAZMAT team, however, may be a separate component of a fire brigade or department.

Recommended minimum training requirements for the HAZMAT team and appropriate courses are as follows:

Hazardous waste site workers. This course is for general workers who are on site full time and probably have been exposed to a hazardous chemical or situation. They are required to have 40 hours of classroom "hands-on" instruction away from the site followed by a minimum of three days actual field experience under a trained supervisor. All site workers must complete eight hours of refresher training each year, and the training should be documented.

Occasional hazardous waste site workers. This course accommodates inspectors, engineers, monitoring technicians, or others who are not likely to be exposed over permissible exposure limits. They are required to have 24 hours of off-site instruction followed by one day of actual field experience.

Hazardous waste site supervisors. These site managers must have eight hours of specialized training after completing the 40-hour course. Supervisory training covers site safety plans, personal protective equipment selection, and health monitoring.

Treatment, storage, and disposal (TSD) facility workers. New employees in these work areas must have at least 24 hours of health and safety training. Current employees must

---

EMERGENCY EXIT INSTRUCTIONS
MACHINE SHOP—DAY SHIFT

Read Carefully

The following persons will be in command in any emergency, and their instructions must be followed:

    CHIEF OF EXIT DRILL—H. C. Gordon, General Sup't.
    MACHINE SHOP EXIT DRILL CAPTAIN—R. L. Jones, Foreman
    MACHINE SHOP MONITORS—Dave Thomas and A. L. Smith

In event of FIRE in machine shop

✔ **NOTIFY THE GENERAL SUPERINTENDENT'S OFFICE**

✔ **PUT OUT THE FIRE, IF POSSIBLE**—If the fire cannot quickly be controlled, follow instructions given by Exit Drill Captain R. L. Jones or by the shop monitors. Leave by the exit door at the south end of the shop; if it is blocked by fire, use the door through the toolroom to the outside stairway.

In event of FIRE or EMERGENCY in other sections of building

The general alarm gong will ring for two 10-second periods as an "alert" signal. Continue work, but be on the alert for the "evacuation" signal, which will be a series of three short rings. At the evacuation signal:

   ✔ SHUT OFF ALL POWER TO MACHINES AND FANS

   ✔ TURN OFF GAS UNDER HEAT TREATING OVENS

   ✔ CLOSE WINDOWS AND CLEAR THE AISLES

   ✔ FORM A DOUBLE LINE IN THE CENTER AISLE AND FOLLOW MONITORS AND EXIT DRILL CAPTAIN TO EXIT—Walk rapidly, but do not run or crowd; do not talk, push, or cause confusion!

After leaving the building, do not interfere with the work of the plant fire brigade or the city fire department. Await instructions from the General Superintendent or your foreman.

Returning to the building

Return-to-work instructions will be given over the loudspeaker system or by telephone from the Superintendent's office.

Figure 6–5b. Sample individual instruction notice for general posting. Some facilities will include after-hours phone numbers and alternate communications protocols.

demonstrate the equivalent amount of training from previous experience. All workers must complete eight hours of TSD refresher training each year.

Hazardous materials emergency responders (regardless of location). These workers are distributed into five categories. The training requirements apply to private plant employees as well as to community emergency services personnel.

First responder awareness. This is a four- to eight-hour course for responders who will only attempt to identify the involved hazardous material and then notify more qualified personnel.

First responder operations. An eight-hour training requirement exists, but 24 hours of instruction is highly rec-

ommended. This category covers responders who will identify materials, perform basic diking and confinement operations, and initiate evacuation. This course requires training in personal protective equipment, decontamination, chemistry, and toxicology. However, employees are taught to respond in a defensive fashion; the course does not qualify them to attempt a patching or plugging operation.

Hazardous materials technician. This is a 24-hour course for workers who will perform the actual plugging, patching, or sealing of a container leaking a hazardous substance. The 24-hour First Responder Operations class is a prerequisite to this course.

Hazardous materials specialist. Additional training be-

Figure 6–6. The clear identification of evacuation routes from all locations is an important aspect of emergency preparedness training.

yond the 48 hours required for the responder and technician levels must be taken in the specialty area of this skilled responder. No specific hours requirement exists, but this individual should be competent in the chemical, toxicological, and/or radiological behavior of the particular material involved.

Incident commander. The incident commander must be a graduate of the 24-hour First Responder Operations course and have additional training (24 hours is recommended) in planning, decontamination, protective clothing, and command systems. The commander does not have to receive instruction in the technician or specialist categories.

Command Headquarters

Although the command headquarters of a company will not withstand catastrophic disasters, management can plan for emergencies that may occur.

The disaster control organization should be coordinated from a well-equipped and well-protected control room. The headquarters should be equipped with telephones, sound-powered phones, public address system, maps of the plant, emergency lighting and electric power, sanitary facilities, a second exit, and two-way radios for communication both locally and with emergency management authorities if required.

Good communications are necessary for effective control and flexibility in a disaster situation. Communications include the (cellular) telephone, radio, messengers, and the plant's alarm system (discussed separately later in this chapter). The disaster plan should provide for adequate telephones in emergency headquarters to handle both incoming and outgoing calls. Panic and disintegration of the organization will develop quickly if these calls are not handled with dispatch. Operators should keep an accurate log of all incoming and outgoing messages.

Some means of communication independent of normal telephone service must be available during an emergency, such as the one provided by a battery-operated radio. Mo-

bile cellular phones may or may not be operable during or following a disaster. The disaster plan must anticipate the possibility of losing normal telephone communications and electric power.

Emergency Equipment

An emergency checklist should include equipment and material to be ordered as well as shutdown actions to be performed. For example, where it is not feasible to keep on hand the necessary emergency equipment and materials, one should maintain a resource or supply list and names of suppliers. The information should include type of equipment they can supply, after-hours information and phone numbers, and, if possible, prices. These sources must be outside of the immediate area. For example, a list of typical emergency equipment and material should include sandbags, battens for windows and doorways, boats, tarpaulins, fuel-driven generating equipment (such as gasoline-powered arc welding machines or motor-generator sets), standby pumping equipment, a supply of gasoline in safety containers to fuel this equipment, lubricating oil and grease, rope, life belts, portable battery-operated radio equipment, and audio speakers, etc.

Some of the items on the shutdown part of the checklist would include closing valves; protecting equipment that cannot be moved; closing and battening doors, windows, and ventilators to keep out looters as well as water; and plugging vents and breather pipes. The checklist should include a list of telephone numbers of supervisors and key employees to be notified.

Some provision should be made for moving tank cars to higher ground and anchoring them. Move portable containers above the high-water mark as well, along with buoyant materials and chemicals soluble in water. Storage tanks under the probable high-water mark (including underground tanks) should be specially anchored to prevent floating. Workers can construct auxiliary dikes of sandbags or dirt around key areas.

Other procedures must include shutting off electric and gas utility services at the main line before any water reaches them. Make sure hot equipment is cooled before water reaches it. Operators should coat all machine surfaces liberally with heavy grease, especially around openings to bearings. (This step applies even to machines that may not be under water because dampness can damage equipment.) Shut off all open flames so that any flammable liquid floating on the floodwaters will not be ignited.

If possible, keep a salvage crew at the site to continue preventive operations after the plant or facility has been shut down. This crew can take further steps if the flood exceeds the estimated high-water level.

Alarm Systems

Most industrial operations have a special fire alarm system using existing signaling systems such as a plant whistle. However, to avoid confusing fire alarms with the regularly used signals, some plants have special codes or other signaling devices for fires. This type of signal also may indicate the location of the fire, or separate signaling devices may be used for each building or working area within the company property.

The alarm system activating the emergency plan may or may not go through the communications center, but it should be touched off in the emergency headquarters office. All buildings should contain alarm systems (Figure 6–7).

In hospitals or other locations where both employees and nonemployees can hear an audible page system, a code name can be used to announce a fire and its location, such as "Doctor Red wanted in—". Employees must be trained to be alert for this subtle signal.

Electric alarms are preferred to mechanical ones, except in a large open shop area with only one alarm-summons station and one alarm-sending device, such as a manually operated gong. Manually operated alarms should supplement electric alarms. Closed-circuit systems of the type specified by NFPA standards are recommended (see References).

Companies in areas where municipal fire departments are available usually have a municipal alarm box close to the firm's entrance or in one of the buildings. Others may have auxiliary alarm box areas, connected to the municipal fire alarm system, at various points on the premises. Another system often used is a direct connection to the control dispatcher. This system may be touched off by a water alarm in the sprinkler system or be activated manually. If possible, connect the fire alarm system to the local firefighting alarm and make sure it has an independent power supply.

In some areas, private central station services are available and provide excellent protection. These central stations receive signals from plant fire alarm boxes, security staff, sprinkler head operations, and other hazard control points in the plant. Because these devices monitor plant security, they can relay information to fire or police departments without delay. The signal received at the fire department or assistance agency should pinpoint the site of the fire, or at least the building or area, so the fire can be found quickly (Figure 6–8). Automatic sending stations (thermostatic detectors) may be used but should not interfere with sending a manual alarm.

Regular checks should be made on the alarm system. Checks should include monthly inspections of all stations and a weekly test of the overall system to ensure it is in proper working condition. These weekly tests should be conducted at a prearranged time and under a variety of wind and weather conditions to determine if the signal can be heard in all parts of the plant at all times.

Fire Brigades

Fire prevention and fire protection must receive major attention in any emergency program because a fire can start from so many causes. Advance planning is important. (NOTE: U.S. organizations may elect to have fire brigades, in which case regulatory standards have to be followed.)

The company fire chief must be able to command people

Figure 6–7. Spacing of alarm systems is important for notification of personnel in remote areas of large plants.

in addition to having special training in fire prevention and protection. A person who has had experience in city or volunteer fire department work, or a military service veteran with experience in firefighting may be a good choice. In a smaller company or plant, a master mechanic, maintenance department head, or other employee with mechanical experience can be a good part-time fire chief.

Most U.S. states in the country have a State Fire Service Training Program. As part of this service, there are now schools specifically oriented toward industrial fire protection. At the very least, all the fire brigade personnel or the guards should have advanced training. They are the ones most likely to use auxiliary firefighting equipment.

The fire chief may need one or more assistants who have complete knowledge of the plant and equipment, command the respect and obedience of those under them, and are qualified to perform the duties of the chief officer if necessary.

The size of the facility and the fire potential presented by the occupancy will determine the kind of firefighting organization (brigade) required. The majority of facilities may need only first aid firefighters under the direction of departmental foremen or managers. In larger facilities, the

fire brigade organization, directed by a full-time fire chief, is composed of full-time and emergency members. The full-time members maintain fire brigade equipment and are responsible for the permanent fire protection of the facility. Emergency members report for fire duty when the alarm is sounded.

Fire brigade apparatus should be selected only after a study has been made of the specific conditions to ensure the equipment will be adequate for any emergency. Management can obtain advice and assistance from the local fire department, the NFPA, State Fire Training Organization, or insurance companies.

The large facility fire brigade is usually organized into squads, each with specific duties. One company's organization chart is shown in Figure 6–9.

The evacuation squad evacuates all employees from the emergency area as quickly and orderly as possible, without injury. They search closed areas, such as washrooms, to determine that everyone has been evacuated.

The environmental monitoring squad identifies and monitors environmental conditions to determine hazardous conditions, to make sure sites are not contaminated, and to determine when an area can be re-entered safely.

Figure 6–8. A command console for computer-based supervision and monitoring of large buildings or building complexes for loss-producing hazards including intrusion, fire, and other emergencies. Based on a central processing unit, the console includes computer display screen, high speed printer, and zoned annunciator panels.

Utility control squad members usually are maintenance personnel familiar with plant piping systems and the control of process gases, flammable liquids, and electricity.

Sprinkler control squad members must understand the automatic sprinkler system—the direction of rotation of the valve they are to operate, the use of sprinkler stops, and the replacement of sprinkler heads, if this is not a maintenance department function.

Extinguisher squad. Portable fire extinguishers are frequently operated by designated employees who work in the vicinity. However, as the size of the facility increases, it is advisable that special squads be selected, trained, and equipped for handling fire extinguishers (Figure 6–10).

Hose squad members are trained to operate fire hydrants and hoses. They should drill frequently with wet hose lines so they have the feel of a charged hose. After they become proficient in handling the hose, drills once or twice a year may be sufficient to maintain their preparedness (Figure 6–11).

The salvage squad is trained to protect as much stock and equipment as possible by controlling the directional flow of water and by covering stock with tarpaulins. Their training should include proper methods of throwing tarpaulins and using them to direct the flow of water. Squad members should also be familiar with the location of sawdust or other absorbent material and know how to use it in controlling water on floors.

The brigade-at-large or *rescue team* consists of maintenance personnel or specially trained workers. The main functions of this unit are to extricate casualties and eliminate hazards that may endanger other workers involved in controlling the emergency. Members respond to all alarms with a utility truck containing such rescue equipment as ropes, chains, block and tackle, ladders, cutting torches, saws, axes, and jacks. The amount of equipment, of course, will depend upon the size of the facility and the hazards involved, but every effort should be made to anticipate possible problems.

Under the direction of the brigade officer, personnel in this unit also control utilities, ventilating fans, and blowers. They close fire doors, windows, and other openings in division walls, and open windows and doors leading to fire exits. If escape equipment is the swinging section type, rescue personnel should be the first to operate the escapes and to secure the steps to the ground.

Depending upon the size and inherent hazards of the plant facility, it may be necessary to train squads in handling and erecting ladders, using foam lines, recharging foam generators, specialized rescue techniques, and chemical spill response procedures.

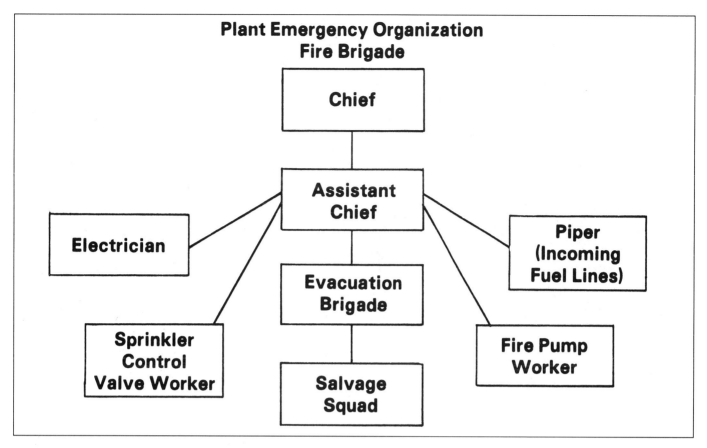

Figure 6–9. Because of the complexity of response needed to cope with an emergency, duties must be divided among plant emergency organization members to avoid confusion. Here is one plant's organizational chart. Current names, phone numbers, and those of alternates should be listed in the boxes. (Reprinted with permission from Textile Section Newsletter.)

Fire pump team. There should be at least two competent people for pump duty in the main pump room. Management must assign one trained person to each pump located elsewhere.

Firefighter training. The brigade should go through complete drills, preferably every week. Later, the leaders can hold less elaborate drills less often. Drills should be unannounced and thorough in every respect, closely approximating fire conditions (Figure 6–11).

No matter how thoroughly the industrial fire brigade is trained, management should make sure they cooperate closely with adjoining or nearby plants and the public fire department. It would be a good idea to do mutual training with the local fire department so that each side knows the expectations of the other. This approach builds a better relationship in the short run and helps to unite the two groups as a single firefighting team for the long term. As stated before, the brigade chief, safety and health professional, or designated employee should immediately call the municipal fire department whenever a fire is reported at the facility.

In fire emergencies, security staff should open the yard gates and be ready to direct fire apparatus to the blaze. If plants have railroad tracks, rail crossings that may be needed in an emergency should be kept free of all cars, hand trucks, and the like.

The brigade chief is in command at a fire until the officer in charge of the public fire department takes over. The brigade chief then serves as an advisor on processes and special hazards.

The fire station itself should be centrally located but not exposed to possible fires. It should be built of fire-resistant material or located in a sprinklered part of the plant and protected with portable extinguishers. A larger facilities may require mobile units, such as light hand-drawn trucks outfitted for special hazards.

Facility Protection and Security

Security is management's responsibility. Government agencies, consultants, and insurance representatives can provide assistance and advice in establishing a policy. Since the basic problems of security are protection of property and control of persons, an organization does not have to establish a new department to handle this function. It could simply build its emergency security force around present security and administrative personnel.

Personnel need training to maintain order, handle crowds, cope with the threat of panic, and prevent on-site looting. They map emergency routes to shelters, both inside and outside the grounds.

Figure 6–10. Fire extinguisher training is very important. Underwriters Laboratories Inc. considers any fire extinguisher in the hand of a trained operator to have 2 times more firefighting capacity than when used by a novice. Here, the fire chief instructs a worker in the use of a pressurized water fire extinguisher. (Courtesy Ansul Fire School.)

Fires have often been caused by security guards either smoking while on duty or overlooking fire hazards. Heavy losses can result from their failure to discover fires promptly, from shutting off sprinklers before determining if fires have been extinguished, or from not knowing the proper sprinkler valves to close after a fire is extinguished.

As a supplement to automatic alarm and signal systems, a firm can employ an intelligent, well-trained, and physically fit security guard to prevent or minimize fire loss. Because fires that start in an idle plant produce more damage, the security guard becomes an important part of a facility's fire prevention and detection organization.

Security functions include protecting the premises against pilferage, burglary, vandalism, and espionage. The time and route of inspection rounds should be scheduled irregularly, to avoid creating any detectable pattern.

The first inspection round immediately after the plant closes is the most important. Evidence shows that most fires are likely to start just after employees have left; they are caused either from machines or processes left running unattended or from carelessly discarded smoking materials.

The security guard should have enough time on the rounds to make a thorough inspection of the premises. The route should require no more than 40 minutes and should include all hazardous areas. The guard should be provided with an approved flashlight or other illuminating device; where practical, lights along the security route should be left on.

The guard can look for violations of smoking rules, improper storage of chemicals or flammable materials; and leaking oil, gasoline, gas, or other flammable materials, and should immediately report unsatisfactory conditions to the

management. Security guards should be physically capable of turning on a fire alarm, dealing immediately with small fires, and with such matters as shutting off gas and closing fire doors. If the guard ever fails to report on schedule at the end of a regular patrol, the incident should be investigated immediately.

Additional protection and security measures call for closing off certain windows and other openings in plants that are not vital to operations and limiting the number of plant entrances and exits. (All measures must be consistent with good fire prevention practices.) Companies might also install protective wire mesh over windows along public thoroughfares and use floodlights in critical areas of plants at night.

Emergency Medical Services (EMS)

First aid and medical services should be headed by the company doctor, if available, as discussed in Chapter 4, Occupational Health Programs. In the organization of the medical phase of the emergency plan, those responsible must select and train personnel; decide what measures, equipment, and supplies are needed; and establish first aid stations and a treatment center.

All employees should be encouraged to enroll in a cardiopulmonary resuscitation (CPR) first aid course. People assigned to first aid and medical units should pass standard and advanced courses in these areas. The National Safety Council First Aid Institute offers training in CPR and first aid. Local chapters of the American National Red Cross also provide training programs. Because more than 140,000 Americans die every year from injuries, and one-in-three suffers a nonfatal injury, the Council believes that every person at some time will encounter an emergency requiring first aid. The Council trains and certifies instructors who in turn teach classes nationwide. The courses meet OSHA requirements for workplace first aid and CPR training. Any time an emergency sprinkler system or some type of shower facility is installed in a chemical work area, management should have a light or alarm installed in case of an emergency. In many locations, especially cold climates, people have been known to suffer severe cases of hypothermia when they had a chemical accident and drenched themselves to wash the chemicals off. They had no way to summon help and suffered from exposure to the cold. In addition, supervisors should keep copies of Material Safety Data Sheets (MSDS) on hand to send with an injured worker when the person is taken to the hospital or medical facility. This information will help emergency and in-house physicians treat the employee. (See also ANSI Z358.1, Emergency Eye Wash and Shower Equipment.)

If a major disaster occurs, there may not be enough trained doctors and nurses available to treat victims. In such an eventuality, care beyond the first aid level will have to be provided by emergency medical technicians (EMTs), paramedics, and first responders who have received additional training. Such a medical team can be developed by recruiting volunteers, preferably with some medical or paramedic experience. In addi-

Figure 6–11. Under direction of an instructor (white helmet), an emergency response drill team approaches a 300-gallon oil tank filled with burning fuel oil. Note the shower spray above the lead firefighters.

tion to first aid training, more advanced instruction by regular medical personnel or local hospitals should be provided. Companies should keep in mind when establishing an EMS program that while most of the medical aid will be given in a central medical station, some may be administered on the job site, possibly under hazardous conditions.

Plans should be made for representatives of the medical team to check all personnel at the disaster scene for trauma. The team can provide a written clearance for victims to leave the plant when they are able to do so. Representatives of the investigation team may interview personnel before they leave to be sure that all eyewitness information is recorded.

Welfare and medical services include devising steps to prevent epidemics and to inspect food and sanitation facilities. In the planning stages for these activities, management should designate certain company trucks as ambulances and make sure they are stocked with the necessary equipment. Two-way radio communication for such ambulance service is essential. Management should also make provisions to have nonperishable food and water rations on hand. There should be close coordination of plant EMS measures with the local civil defense, health, and medical services.

The chemical service responsibility of the EMS unit requires that they provide gas masks and other protective equipment to workers if toxic or corrosive chemical gases are released. The units must also have trained personnel, equipment, and supplies for chemical defense and decontamination. The company should plan a priority sequence for decontamination—that is, water supply first, power plant second, machinery areas third, warehouse areas fourth, and so on. Medical team personnel will also be responsible for any radiological monitoring required after a nuclear incident. The mere knowledge that such monitoring equipment is available can be a morale builder for workers. After a disaster, no one should be permitted to drink water until it has been examined for contaminants.

Warden Service and Evacuation

The warden service is responsible for maintaining employee control during emergencies, including (1) guiding employees to safe areas; (2) directing employees, including physically disabled workers, away from hazardous areas; and (3) averting panic. In smaller organizations, the wardens may take charge of the shelters. In some cases, the warden service

may be responsible for seeing that process and equipment shutdown is carried out smoothly.

This type of service was devised primarily for areas of high population densities, such as commercial structures, factories, and residential areas. For facilities with several hundred or several thousand employees, the use of warden teams can be vital during emergencies. In facilities with only a few personnel, the machine operators will have to be trained in shutdown details.

Management, in checking and providing for safe exits, emergency lighting, and evacuation drills, should refer to the NFPA's standards and to appropriate regulatory standards. Smooth, safe functioning of an evacuation plan requires that those in charge have a thorough knowledge of all operations and employees; the number, type, and location of available exits; alternate exits; and location of hazards, as well as a knowledge of warning and evacuation facilities. The subject of building exits is covered in Chapter 19, Office Safety; in Chapter 1, Buildings and Plant Layout, in the *Engineering and Technology* volume; and in the Life Safety Code, NFPA 101. (See OSHA regulations 29 *CFR*, 1910.157 (a)(2) and 29 *CFR*, 1910.38.)

Most facilities have a rigid rule that only especially appointed people on the fire brigade shall go to the vicinity of the fire. Everyone else should proceed on signal to a safe location or assembly area designated by the evacuation plan.

Transportation

When transportation facilities are disrupted or traffic is temporarily restricted, many employees may find it impossible to get to work. The organization may need to provide transportation with trucks and cars. Advance planning for car pools and pick-up stops will greatly facilitate such a procedure.

The transportation responsibility includes arranging for ambulance service, transporting employees to and from work, and moving emergency service crews and supplies as needed. Any planning for adequate transportation service and traffic control requires cooperation with the public police department, emergency management planning authorities, and possibly the military.

An emergency transportation unit might consist of a group of regularly assigned drivers. Station wagons with the seats removed and company trucks can be used when it becomes necessary to handle stretcher cases in evacuating any injured. The unit will transport auxiliary firefighters, first aid teams, and salvage and rescue workers to the scene of the disaster at the earliest possible instant. The unit can also deliver needed equipment and material from outside suppliers.

When developing emergency transportation plans, the organization must anticipate sources of motor fuel.

PROTECTION FROM PERSONAL ATTACK

Protecting the employee from muggings, rapes, and robberies is a concern of management, especially when such attacks occur on company property. Even if these attacks happen elsewhere, they can affect workers' performance on the job.

The Crime Problem

Safety and health personnel are not professional crime fighters. However, the welfare of the workforce on the job is their responsibility, which may include the security of grounds, buildings, or plant. For example, a woman may be attacked while waiting in her car for her husband to get off work, or a man may be mugged in the parking lot on payday, or some part of the office or facility may be burglarized. In each case the safety of employees or family members is jeopardized.

How should management cope with security and crime problems within the confines of the facility or building? How does a safety and health professional include this area of concern with other duties? Finally, how are any associated costs budgeted?

There are no pat answers to these questions. Each must be answered in terms of a particular industry, location, and workforce. All should be answered, however, with an attitude of common sense and practicality that addresses itself to the welfare of the employee and the organization.

Building and Premises Survey

Management should survey the grounds and buildings of each facility. What does the building or plant look like at night? Is it well lit with floodlights? Does an armed guard control electronic gates to the facility or yard? Or is it set back from the street in an open, landscaped industrial park? Whatever the physical setup, management should study it for exposed areas such as unguarded access to the parking area from a busy street or highway, or viaducts or catwalks that allow easy entrance to a restricted area.

Although not all mugging/burglary crimes take place in dark, secluded areas, many of them happen in those locations simply because discovery is less likely. Thus a security survey of an industrial complex should focus on finding any secluded or remote areas that may place employees at risk. These areas include tunnels between buildings, street underpasses, poorly lighted stairwells, dock areas, and freight elevators. They all offer hiding places for unauthorized personnel who may attack employees. Don't overlook washrooms and locker-room facilities. Burglars have been known to hide in toilet stalls until after hours and then shop through the plant and offices for valuables.

Implementation of the Survey

Once management has drawn up a list of security problems, planners should establish priorities and decide how to implement them. For example, installing lights and a fence around the parking lot may deter a car thief from the area or prevent an assailant from hiding in the back seat of an unlocked car. In addition, the fence and gate may slow up employees leaving the area and make shift-change traffic more organized and safe. Adequate lighting in a stairwell could prevent attacks and reduce the risk of employees slipping or tripping on the stairs.

Often the facility or building survey reveals security gaps that can be closed by the maintenance department. For

example, they can repair or board up broken windows and remove trash barrels or other containers an intruder might use to reach and squeeze through a window or trap door. For small organizations or industrial operations that may be housed in one building, lights around the periphery of the structure are a relatively low-cost method of deterring intruders. Decals can be displayed prominently in a window announcing on-premises security alarms. (Many establishments display such a decal whether or not an alarm system is actually installed.)

Professional Assistance and Community Relations

The experience of local law enforcement agencies can be a major source of help in resolving security needs. In addition, they can offer advice regarding any local laws or codes that require the company's compliance.

Some police departments, particularly in suburban or rural areas, patrol industrial sites throughout the evening hours. Find out what the local police force does. At the same time management should investigate types of crime in the area and recommend special procedures to prevent such crimes from happening on site.

For instance, an adjacent race car track may bring large crowds near the company's warehouse. Burglars may use a race as a diversion to break into the complex without being noticed.

Even the unpredictability of weather should be considered in a security program. Do heavy rains, snow, or fog render some security procedures inoperable? If so, the procedures may have to be changed. The police and fire department may tell the organization of special procedures they are required to take in the event of unusual weather conditions that might affect some of the company's security procedures. For example, the safety and health professional should find out if bus routes or traffic must be rerouted behind the facility or building if the viaduct floods or the surrounding area catches fire.

Based on discussions with the law enforcement agencies, management may decide to work with additional security professionals. These can include surveillance companies, insurance firms, burglar alarm and detector manufacturers, and lighting companies. Naturally, any security equipment management selects should be based on the firm's actual needs.

Just as multiplant industrial complexes share core medical facilities, it may be possible for an individual company or public agency to participate in cooperative security plans. Several organizations can share expenses for additional lighting, fences, and security patrols. However, this step should be taken only with the approval of the organization's insurance carriers.

Personnel Protection

Once the plant facilities are secure under the supervision of security staff and/or detector systems, much of the personal security of employees is also secured. Not all unauthorized persons questioned by a guard or detected by a surveillance

system intend to steal property. Some may wish to harm an employee, perhaps even selecting someone at random.

A well-organized employee identification system is basic to an organization's security. All too often people assume that a person sauntering through an area, be it restricted or not, has a legitimate reason for being there. By challenging the stranger's presence and asking for a company identification card or verification from another employee, workers and managers can help to maintain security. At first, employees may object to "police" overtones of "identify or else," but selling security to them on the basis of, "It's for your own protection," can win them over. Even the strongest resisters can be persuaded if management supports this approach with examples of problems resulting from little or no security measures.

Management can require workers to wear photo-identification cards clipped to a shirt pocket or collar as a condition of employment. In addition, the card can be color-coded to identify first aid personnel, fire brigade personnel, maintenance workers, and others selected for specific tasks during an emergency. This system works especially well when the organization cooperates with other firms or groups. Different colors help people easily identify members of each group. Finally, asking the employees themselves about improving security can often elicit ideas that fit in well with specific operations. The approach "we want you on the job safely" is a more positive way of selling security to employees than the "No-admittance-beyond-this-point" or "Don't-do-this" approach.

Female Employees

Rape is a hideous crime. It can occur at any place, at any time. It is a crime in which the victim is left with serious emotional scars and/or physical injuries.

Security measures to prevent rape should include adequate lighting, visible presence of security staff, and a guard escort service, if necessary. Female employees should be informed that rape victims are usually victims by chance, not necessarily because they know or recognize the assaulter.

Tell female employees about self-protection during a training session set up just for the subject of security. Contact local police or sheriff's departments to come in and do programs on rape prevention. Set the tone of the program with, again, "we want you on the job safely." Encourage comments from the women, particularly suggestions on ways they will feel more secure. Don't let them leave the meeting empty handed. Distribute National Safety Council's *Just Another Statistic* (see References) or similar literature that may be available from your local law enforcement agency. Some programs distribute whistles for the women to take with them. A number of cities report a reduction in rapes after establishing whistle campaigns. While stopping rape is one of the reasons for using the whistle, it can be used by men and women in other emergency situations. For example, the whistle is a good signal to use when a person sees something suspicious happening or is personally threatened and wants to summon help.

Sometimes to enhance the safety of a workforce, management can institute additional security measures for the in-transit employee. In-transit security is a sensible procedure to prevent muggings and robberies as well as rapes. Cabs and minibus rides from the plant gate to public transportation are standard procedures in many work situations. Guard escort from the plant gate to the employee's car is provided in some high-crime areas. For the most part, these efforts have been established primarily to protect female employees who work evening shifts. Similar protection is afforded women office workers by having them work in groups or by requiring the presence of a supervisor or other management staff during overtime hours.

Mutual Assistance

Organizations can cooperate with surrounding organizations and plants, local law enforcement authorities, and transportation systems to develop in-transit security. The cost of operating a minibus service from several establishments to the local bus or train can be shared by the participating parties. Or, a similar agreement can be made with a local taxi company. Some companies located in high-crime areas have borne the expense of taxi rides to employees' homes in an effort to ensure the safety of their workforce.

OUTSIDE HELP

The chance of survival and recovery from a disaster or major accident is greater when organizations pool knowledge, equipment, and personnel with their neighbors. Therefore, emergency plans should include a provision for exchanging aid with other organizations in the community.

These plans should be drawn up ahead of time for both groups. Everyone needs to know beforehand what equipment and services will be available and how it can be called in. Plans should include provisions for calling another organization after hours should a problem develop outside normal working times.

Mutual Aid Plans

A number of industrial communities are organized to assist their members in the event of an emergency or disaster. These organizations include manufacturing plants, large offices, stores, hotels, utility companies, chemical plants, law enforcement organizations, hospitals, newspapers, and radio and television stations. They operate independently of or as supplements to any individual emergency management teams.

It is usually not possible to have adequate supplies available for a really large disaster such as a major earthquake. The best defense is to have adequate supplies in other areas committed for standby use, with communication channels and a plan for rapid transportation of supplies to the stricken area. Adequate emergency medical supplies and firefighting equipment are especially important. Companies should develop a plan for rapid and accurate communication, as discussed earlier in this chapter.

Mutual aid plans with neighboring companies and community agencies should include establishing an organizational structure and communication system; standardizing an identification system, procedures, and equipment (such as fire hose couplings); formulating a list of available equipment; stockpiling medical supplies; sharing facilities in an emergency; and cooperating in test exercises and training.

One item often overlooked in most mutual-aid contracts is who will pay any costs, such as repairs to equipment or workers' compensation for injured employees. Most often, workers' compensation is covered by the person who is considered the full-time employer.

Frequently these "cooperatives" establish a task force composed of personnel from each member company. Their training is supplemented by detailed written instructions. Each plant marks bulldozers, floodlights, and tools for emergency use. Training on a community basis might include instruction by members of the public fire department to plant fire brigade members. In addition, members of a construction or wrecking company may provide actual training to show the salvage and rescue teams how to handle heavy weights and to work safely among debris.

Contracting for Disaster Services

Some companies contract for disaster service. The service is paid for by a fixed annual retainer, plus additional pay for the actual hours worked. For example, a wrecking company can be engaged to supply the men and equipment necessary to clear debris created by a disaster. Contracting for such a service removes the burden from companies of providing trained personnel and maintaining idle emergency equipment that could easily be damaged in the very disaster for which it was designed.

One must remember that sometimes in a major disaster, some of the contractors a company has on standby may go instead to the highest bidding firm. In such cases, management must have contingency plans.

Municipal Fire and Police Departments

Firefighters from the station most likely to respond to an alarm should be fully acquainted with all fire hazards in the facility. The organization may encourage such cooperation by inviting local fire officials to inspect the area prior to an emergency. The officials can become familiar with the location, construction, and arrangement of all buildings, as well as all special hazards, such as toxic, corrosive, or flammable gases, liquids, and materials. Because public fire department rescue equipment can supplement facility rescue units, they can make sure that equipment is compatible with the municipal equipment.

Therefore, the local fire department can formulate an efficient plan of attack before an emergency occurs. Such a procedure is far better than waiting for an incident to happen, then running the risk that local firefighting units will misunderstand the situation and initiate an improper response.

The public police force can help to quell large-scale disturbances and to assist in evacuating the plant should a major disaster occur. Planning for this outside help should include

making arrangements for traffic control, particularly where a parking lot empties immediately onto a public highway.

Industry and Medical Agencies

Details of the following services are given in the Appendix, Sources of Help.

In 1970, the Chemical Manufacturers Association (CMA) created the Chemical Transportation Emergency Center (CHEMTREC). (See Sources of Help.)

The Toxicology Information On-Line Network (TOXLINE) has been designed to provide current and prompt information on the toxicity of substances. It is intended to be used by health professionals and other scientists involved in antipollution, safety, drug, health, and other disciplines. (See Sources of Help.)

A number of other emergency and specialized information sources are listed in the Appendix, Sources of Help.

Governmental and Community Agencies

During a community-wide disaster, a large number of governmental and private agencies are available to assist industries; these include Federal Emergency Management Agency (FEMA), the U.S. Army Corps of Engineers, the Salvation Army, the American Red Cross, the U.S. Public Health Service, the National Weather Service, and National Oceanic and Atmospheric Administration (NOAA). To be effective in coping with a disaster, the efforts of all of these groups must be coordinated and directed toward a common end. Therefore, each organization should have an updated listing of all cooperating agencies; the administrator's name, address, and telephone number; and the task assignment of the agency. If possible, safety and health professionals and emergency planning directors should meet periodically with these administrators to discuss mutual problems and disaster control techniques.

The emergency planning director should become thoroughly familiar with the authority, organization, and emergency procedures that are established by law and which will become effective upon declaration of a civil emergency.

It is in the organization's best interests to be active in the county Local Emergency Planning Committees (LEPCs). In this way, they can cooperate with other organizations from other industries and groups in the community. The main purpose of an LEPC is handling hazardous materials problems along with natural disasters and other crises.

In wartime, the federal, state, and local governments are responsible for relief measures. The American Red Cross has offered to assist the government in providing food, clothing, and temporary shelter on a mass-scale basis during the emergency period immediately following enemy attack. In many communities, local emergency management officials have requested American Red Cross chapters to assume all or part of this responsibility, acting under emergency management authorities.

In natural disasters, the American Red Cross is responsible for assisting families and individuals to meet disaster-caused needs that cannot be met through their own resources. The relief operations of local American Red Cross chapters are coordinated with the activities of the local, state, and federal governments. The resources of the national organization are available to supplement chapter assistance.

SUMMARY

- Advanced emergency management planning is the best way to minimize potential loss from natural or human-caused disasters and accidents. Emergency planning is often assigned to the safety and health professional, although ultimately responsibility always resides with top management.

- Emergency planning must provide for the safety of employees and the public, protect property and the environment, and establish methods to restore operations to normal as soon as possible. Detailed and comprehensive planning must encompass natural, technological, and nuclear hazards that may strike a facility or the immediate area.

- In developing an emergency plan, organizations must (1) identify and evaluate potential disasters; (2) assess the potential harm to people, property, and environment; (3) estimate warning time needed to mobilize the plan; (4) determine what changes must be made in company operations; and (5) consider what power supplies and utilities may be required to handle the emergency.

- Basic emergency management planning usually includes establishing a chain of command, an alarm system, medical treatment plans, communication system, shutdown and evacuation procedures, and auxiliary power systems.

- Organizations can use a variety of government and private sources to find out what types of hazards are most likely to occur in their area. These hazards include fires and explosions; floods, hurricanes, and tornados; earthquakes; civil strife and sabotage; work accidents and rumors; wartime emergencies; hazardous materials; radioactive materials; and weather-related emergencies.

- Once the initial background work has been done, the next step is to develop a working emergency management plan. The plan should be created not only within the facility but in cooperation with other facilities in the area. To administer the emergency plan, someone within the company should be appointed emergency planning director or coordinator. The organization can develop a manual or handbook to cover many of the emergency steps that will need to be implemented.

- The chain of command established for emergency plans should be kept as short as practical and should be staffed with employees selected for their ability to respond well in high-pressure situations. The director of emergency planning should help coordinate facility plans with those of the community.

- One of the most important functions of the director and staff is training employees in disaster control and response. Training includes rehearsals under realistic conditions and simulated disaster drills.

- Emergency plans generally call for establishing a command headquarters or center to coordinate communications and disaster response. The center should be equipped with telephone, public address systems, maps, emergency power and lighting, and two-way radios.
- An emergency equipment and materials checklist will help management organize a rapid, effective response to a disaster or serious accident. Equipment and materials can either be kept on hand or be on standby from other sources in the community. The checklist would also include a shutdown procedure to protect lives and property. A salvage crew can be on hand to make sure these procedures are completed and to save as much equipment and materials as possible.
- Most operations have special alarm systems for fire and security. Each organization should work out a special emergency alarm signal that all workers can recognize. At times, the alarms may also trigger an alert in local fire and police departments. Alarm systems must be regularly inspected and maintained.
- Many organizations establish local fire and hazardous materials brigades known as hazardous materials response teams (HAZMAT), as defined by OSHA and various statutes. If a larger organization elects to have a fire brigade, it is directed by a full-time fire chief. Brigade members must be well trained and equipped. The brigade may be composed of various squads for evacuation, utility control, sprinkler control, extinguisher work, salvage, rescue efforts, and fire fighting.
- Security and protection is a high-priority concern for management. As a result, top management must ensure that security staff are properly trained not only in their duties but in such hazards as fires, chemical spills, personal injuries, and so on that are likely to threaten the safety of the workforce and facilities.
- Emergency medical services are a vital part of any emergency management plan. They should be headed by a doctor, if available, or by the health and safety professional. These individuals select and train personnel, gather equipment and supplies, and establish first aid and treatment centers when disasters or major accidents occur. Some organizations also have emergency medical technicians, paramedics, and other trained personnel on hand to treat victims of emergencies.
- The warden service is responsible for maintaining employee control during emergencies including (1) guiding employees to safety, (2) directing them away from hazards, and (3) preventing panic among workers. Wardens should have a thorough knowledge of all operations and employees, the number and location of all exits, and location of all hazards.
- An emergency transportation unit includes ambulance service, transporting emergency service crew and supplies, and moving employees to and from work as necessary. Emergency transportation plans should be coordinated with local police, emergency planning authorities, and in some cases the military.

- One major responsibility is protecting employees from personal attack—muggings, rapes, and robberies. To help ensure worker safety, the organization can make a security survey of buildings and grounds, implement steps to correct a hazard or breaches of security, enlist the advice and aid of local law enforcement agencies, and train workers in the basics of personal protection.
- In some cases, an organization may need to request outside help to survive and recover from a disaster or major accident. Mutual aid plans with other firms and organizations in the community can pool the knowledge, resources, equipment, and personnel of many organizations. These cooperatives may establish a task force composed of members from each organization to organize and administer emergency responses.
- Other outside agencies include municipal fire and police departments, industry and medical agencies such as CHEMTREC or TOXLINE, and governmental and community agencies such as the American Red Cross, the Federal Emergency Management Agency, U.S. Army Corps of Engineers, the Salvation Army, U.S. Public Health Service, and the National Weather Service. These groups can be contacted for aid by any company or organization hit by a natural disaster or major accident.

REFERENCES

American Insurance Services Group, Engineering and Safety Service, 85 John Street, New York, NY 10038.
Fire Hazards and Safeguards for Metalworking Industries, Technical Survey No. 2.
Fire Safeguarding Warehouses, Technical Survey No. 1.

American National Standards Institute, 11 West 42nd Street, New York, NY 10036.
Emergency Eye Wash and Shower Equipment, Z358.1-1990.

The Conference Board, 845 Third Avenue, New York, NY 10022.
Studies in Business Policy, No. 55, "Protecting Personnel in Wartime."

Pacific Telesis Group. *Earthquake Survival Guide*. San Francisco: 1987.

Environmental Protection Agency, OS–120. "Chemicals in Your Community."

Factory Mutual Research Organization, 1151 Boston Providence Turnpike, Norwood, MA 02062.
Handbook of Industrial Loss Prevention.
Loss Prevention Data.

Federal Emergency Management Agency, 500 C Street SW, Washington, DC 20472. (Available through Superintendent of Documents, U.S. Government Printing Office, Washington, DC 20402.)
Attack Environment Manual, June 1987.
Emergency Planning and Community Right-to-Know Act,

Title III, Superfund Amendments and Reauthorization Act, Public Law No. 99–499.
In Time of Emergency, H–14, Oct. 1985.
Mass Casualty Planning: A Model for In-Hospital Disaster Response, Aug. 1986.

Giraud R. Radioactive elements. *National Safety News*, June 1973. Adapted from author's article in *Revue Technique du Feu, Entreprise moderne d'édition*, 4 rue Cambon, 75 Paris 1, France.

International Association of Fire Chiefs, 1329 18th Street NW, Washington, DC 20036–6516. *Nuclear Hazard Management for the Fire Service*, 1975; "Disaster Planning Guidelines for Fire Chiefs."

Kerr JW. Preplanning for a nuclear incident. *Fire Command!* April, 1977.

National Fire Protection Association, 1 Batterymarch Park, Quincy, MA 02269–9101.
Central Station Service, NFPA 71.
Explosion Prevention Systems, NFPA 69.
Facilities Handling Radioactive Materials, NFPA 801.
Fire Protection Handbook, latest ed.
Guard Operations in Fire Loss Prevention, NFPA 602.
Guard Service in Fire Loss Prevention, NFPA 601.
Handling Radiation Emergencies, 1985.
Health Care Facilities, NFPA 99.
Industrial Fire Hazards Handbook, SPP 57A.
Installation, Maintenance, and Use of Auxiliary Protective Signaling Systems, NFPA 72B.
Installation, Maintenance, and Use of Local Protective Signaling Systems, NFPA 72A.
Installation, Maintenance, and Use of Remote Station Protective Signaling Systems, NFPA 72C.
Installation, Maintenance, and Use of Signalling Systems for Installation of Sprinkler Systems, NFPA 13.
Inspection, Testing and Maintenance of Sprinkler Systems, NFPA 13A.
Life Safety Code, NFPA 101.
Professional Competence of Responders to Hazardous Materials, NFPA 472.

National Petroleum Council, 1625 K Street NW, Washington, DC 20006.
Disaster Planning for the Oil and Gas Industries.
Security Principles for the Petroleum and Gas Industries.

National Safety Council, 444 North Michigan Avenue, Chicago, IL 60611.
Fire Prevention and Control at Construction Sites, Industrial Data Sheet 12304–0491, 1988.
Fire Prevention in Stores, Industrial Data Sheet 12304–0549, 1990.
First Aid Institute.
Fundamentals of Industrial Hygiene, 3rd ed.
Hazardous Materials: Handle with Care, Booklet.
Just Another Statistic, Booklet.
Security from personal attack. *National Safety News*, July 1973.
(See other appropriate topics treated in both volumes of this *Manual*, especially those pertaining to organization, training, medical and nursing services, fire extinguishment and control.

Noll GG, Hildebrand MS, Yvorra JG. *Hazardous Materials: Managing the Incident*. Fire Protection Publications, Oklahoma State University, Stillwater, OK 74078.

OSHA. 29 *CFR*, 1910.38, "Employee Emergency Plans and Fire Preventing Plans."

———. 29 *CFR*, 1910.120, "Final Rule for Hazardous Waste Operations and Emergency Response."

———. 29 *CFR*, 1910.156, "Fire Brigades."

———. 29 *CFR*, 1910. Section 3088, "How to Prepare for Workplace Emergencies."

———. 29 *CFR*, 1910.1200, "Hazard Communication Guidelines for Compliance."

———. 29 *CFR*, 1910.1200 (f), "Labels and Other Forms of Warning."

———. 40 *CFR*, 311, 330, 335, "Extremely Hazardous Substance List and Threshold Planning Quantities; Emergency Planning and Release Notification Requirements; Final Rule."

———. 49 *CFR*, Part 172, "Hazardous Materials Table and Hazardous Materials Communications Regulations."

Underwriters Laboratories Inc., 333 Pfingsten Road, Northbrook, IL 60062.
Classification of Fire-Resistance Record-Protection Equipment.
Gas Shutoff Valves—Earthquake.

U.S. Department of Health & Human Services. *Occupational Safety and Health Guidance Manual for Hazardous Waste Site Activities*.

7

Employee Assistance Programs

A major goal for any organization is to have a highly productive, efficient workforce with the lowest possible accident and lost time rates. Sometimes, however, employees may have problems severe enough to affect job and personal performance, such as substance abuse, marital problems, or financial troubles. These problems can cause family hardship for the employees and lost productivity or safety problems for the organization and its other employees. When employees have problems, it is both more responsible and more cost-efficient to offer assistance rather than to discipline or, as a last resort, release them.

The cost of allowing employee personal problems to go untreated is very high. According to the National Institute on Drug Abuse, 70% of all illegal drug users are employed either full- or part-time. That's over 10 million workers. The U.S. Chamber of Commerce says that typical drug users in the workforce are 3.6 times more likely to injure themselves or another person in a workplace accident. The Employee Assistance Society of North America (EASNA) estimates that up to 40% of industrial fatalities and 47% of industrial injuries can be linked to alcohol abuse and alcoholism. And the National Institute on Alcohol Abuse and Alcoholism estimates that drug and alcohol abuse on the job costs society $102 billion a year. The Chamber of Commerce says drug abusing employees incur 300% higher medical costs and benefits.

The costs are both direct and indirect. Substance abusing workers miss more workdays, are more likely to injure themselves or others, file more Workers' Compensation claims, and are less productive (Figure 7–1). The real, measurable costs are obvious:

- absenteeism
- sick leave
- overtime pay
- tardiness
- insurance claims
- Workers' Compensation.

The hidden costs may not be as obvious:

- diverted supervisory and managerial time
- friction among workers
- damage to equipment
- poor decisions
- damage to the organization's image or reputation
- personnel turnover.

The concern of many employers about these problems has led them to work with the labor force to develop a program complementary to the traditional disease and accident health care benefits. The program may be called an employee assistance program (EAP), a troubled employee program, or a health promotion or wellness program. In this chapter, we will call it an EAP.

Employee assistance programs are cost-effective, humanitarian, job-based strategies to help employees identify problems and resolve them through confidential, short-term counseling, referrals for more specialized services, and follow-up services. EAPs may also offer supervisory training, education and prevention programs, and health promotion activities.

Figure 7–1. Off-the-job substance abuse affects performance on the job in terms of safety, quality, and quantity of work.

More than 85% of the Fortune 500 corporations in the United States and more than half of all those companies with 1,000 employees or more have EAPs. The service is usually provided in addition to employee insurance programs and a full-size human resources department. The EAP should be part of a larger plan that includes written policies, supervisory training, employee training, and, where appropriate, a drug-testing program.

Employee assistance program services may be provided in-house or on a contract basis by many hospitals, other healthcare organizations, or a wide variety of independent EAP vendors. Services provided by EAPs can include counseling for mental problems, financial affairs, substance abuse, legal assistance, family and children concerns, and marital relationships. Some EAPs also provide retirement, promotion and outplacement services, and health promotion and fitness programs dealing with weight control, nutrition, exercise and smoking. Services are often offered both to the employee and members of the employee's immediate family.

What sets EAPs apart from other programs designed to help people understand or overcome personal problems is that EAPs don't intervene in the private lives of those they seek to assist. Rather, they offer aid with the resolution of problems that are affecting job performance. EAPs also are not comprehensive care-givers, but instead resources for limited assistance and referral to outside professionals or agencies whose specific role it is to provide extended care.

EAPs also keep confidential records for the benefit of the employee and employer, and follow up after initial assistance has been sought to ensure that problems are being resolved which are affecting job performance.

BENEFITS OF AN EAP

One EAP consultant reports studies consistently show assistance programs average a return of $3 for every $1 invested in them. The programs can pay off as much as $13 to $1 when employees involved were recommended to the EAP by a supervisor due to poor job performance. EAPs are beneficial because:

- Problem solving can aid productivity and improve employee morale, thus reducing tardiness, absenteeism, and the number of accidents both on and off the job. It can reduce the number of employee grievances and Workers' Compensation claims, improve performance, lower the rate of employee turnover, and decrease the cost of health benefits.

- The existence of an EAP gives managers an alternative to disciplinary action when dealing with problem employees. They can suggest that the employee seek help, or suggest to a union representative or to friends of the employee that he or she get help in dealing with life's difficulties. Such decisive, helpful action can result in improved employee morale as personal problems are met head-on and handled constructively.

- Trusted employees who encounter personal difficulties can get help, improve their performance, and retain a valued job. Employees who do not have a problem also can be assured that confidential help is available if needed.

- An EAP also can help prevent legal issues. A person with a personal difficulty who is discharged because of that difficulty without first being offered help in overcoming it might have grounds for a wrongful discharge lawsuit against his or her employer. A well-run EAP documents the personal assistance offered.

TYPES OF PROGRAMS

Employee assistant programs usually fall into one of five categories. The most appropriate one for a particular organization depends on the organization's size, financial resources, employee characteristics, work-site locations, and design preference.

- *Full-time internal programs.* These are usually centrally located in a large organization (3,000 employees or more). Typically there's a coordinator whose sole responsibility is oversight of the EAP and a staff. The coordinator assesses the personal problems of employees, refers them to appropriate outside professionals, monitors treatment, follows up on individual cases, counsels employees, and promotes the EAP among company workers and management.

- *Part-time internal programs.* These are often provided by smaller organizations with 750 employees or less. The coordinator has both the EAP and other responsibilities. Often this person is in the medical or personnel department or has had personal experience with a problem, plus additional training. The EAP coordinator may have other duties, in addition to the main EAP duties: (1) assessment, (2) referral, and (3) management of the program.

- *Consortiums.* These are usually nonaffiliated service providers (individuals or a firm) serving either a single organization with a number of work sites or several separate organizations in one geographical area. The provider performs assessments and referrals, treatment, follow-up,

aftercare, and program management.

- *Independent service providers.* Smaller organizations often use an independent service provider. Some contract for assessment and referral only. Some purchase the complete range of services from assessment, referral, to treatment, aftercare, and management of the program. Independent service providers are usually offsite.

- *Regional.* This type of program is seen in an organization with work sites spread over a region. The coordinator is usually full-time and holds scheduled visits to the sites for assessment and referral, counseling, and coordination of local contracted services.

HOW THEY WORK

All EAPs operate on pretty much the same basic principles. A person can seek assistance on a voluntary basis or be referred to an EAP by supervisory personnel on the basis of poor job performance. A supervisor also can suggest that an employee seek assistance voluntarily.

A guiding principle of successful EAP operation is that confidentiality be assured when an employee seeks assistance on a voluntary basis. EAP guidelines specify that confidentiality also should be assured when an employee voluntarily seeks assistance at a supervisor's suggestion. Regardless of an employee's job performance, the fact that he or she sought employee assistance is not revealed to supervisory personnel or union representatives and is not made a part of the worker's employment record.

When a person is referred to an EAP by supervisory personnel, however, the fact is recorded in the worker's employment record. The extent of assistance provided, and the degree of employee cooperation with those providing assistance, may or may not be passed along to supervisory personnel depending on company policy.

When an employee seeks assistance, EAP personnel assess the employee's problem, open a confidential EAP assistance record, and decide on a course of action. Depending on the severity of the problem and the capabilities of EAP personnel, it might be decided that the problem can be resolved through short-term in-house counseling, or that referral is needed to an outside resource (a substance abuse program, for example) for longer-term care. If long-term institutional care on an out-patient or in-patient basis is needed, company health care benefits usually cover the cost.

Personnel Responsibilities—Internal Programs

EAP staff responsibilities usually include assessment and referral of clients, confidential records management, benefits coordination, evaluation of outside service providers, contracting for service, monitoring the quality of service, monitoring EAP case follow-up, training supervisors and employee representatives on how the EAP works and how to recognize signs of employee problems, promoting the EAP among employees and others eligible for services, interacting with management, employee representatives, and outside service providers, and writing reports on program utilization and finance.

In addition, some EAPs produce an employee newsletter that contains tips on healthful living and notices of EAP services. In EAPs where staff members assess employee problems and refer employees to outside service providers, staffers should either hold a Bachelors degree (B.A. or B.S.) in social work, psychology, community counseling, educational counseling, nursing, human resources, safety and health or rehabilitation counseling, or have more than three years of full-time experience in an EAP or a documented supervised setting.

Staffers who provide counseling should hold a Masters degree (M.A.) or the equivalent in psychology, social work, pastoral counseling, educational counseling, psychiatric nursing, rehabilitation counseling, or a B.A. or B.S. degree in a relevant field with a valid alcoholism or drug counseling certification.

ASSESSING THE NEED

To determine whether an organization needs an EAP, a committee representing management, labor, and employees, and the human resources, health care, safety and health, financial, legal, and public relations departments would typically be assembled (Figure 7–2). The responsibilities and objectives of the EAP committee would include:

- explore the needs of the workforce, perhaps through confidential interviews or questionnaires
- evaluate existing benefits in light of the needs of employees
- assess available resources, both in the organization and in the community, that might be used to meet workers' needs not already met by existing organization benefits
- analyze data to determine whether current policies and activities are adequate to meet the needs of the workforce.

The first duty of the committee is to assess the need for an EAP. The committee will look at several types of records for evidence of problems. These include:

- insurance benefit claims relating to lifestyle (drug abuse treatment, psychological counseling, car accidents where drugs or alcohol were involved, and heart and gastro-intestinal problems)
- high rates of illness and absenteeism
- accidents and safety violations involving drugs, alcohol, stress or personal problems
- employee turnover due to poor job performance
- employee grievances and suggestion program submissions that an EAP might be able to address
- error rates that suggest substance abuse is a problem.

When the committee considers its second task, assessing existing benefits and resources, it should analyze programs that are offered and those that could be established within the organization as well as services in the community. For example, counseling on family financial problems might be provided by a qualified person already on staff. Services that might be available in the community include:

- alcohol and substance abuse treatment centers and programs (see References)
- mental health professionals, psychologists, psychiatrists,

Management—Management must participate in all stages of assessment and definition to demonstrate support of the program. Open communication between employees and management is essential.

Labor representative—Participation by top level labor representatives is essential during the entire process. In order to achieve eventual acceptance of the program by employees, labor must be involved from initial information gathering to development planning to program implementation.

Employees—Because they are the users of EAP services, employees can give information on what is needed and how it can be accessed most easily.

Human resources/benefits coordinator—By monitoring use of current benefits, personnel staff can provide realistic information on the potential use of services and on the costs involved.

Health care—Advice from health care professionals is necessary for assessment and for selection of services.

Safety and health professional—The safety and health professional should be involved in the assessment of health and safety problems that may affect employee performance. If an EAP is adopted, the safety and health professional should perform follow-up evaluations.

Financial officer—Input from the organization's financial officer is necessary to define the feasibility and scope of the program.

Legal counsel—Legal advice is essential to assure that the EAP structure developed or contracted is fair, accurate, and legally defensible.

Public relations—The public relations department can contribute information on public opinion, from both inside and outside the organizational structure, and on methods of information distribution.

Figure 7–2. Depending on the size and structure of your organization, the EAP task force may include the members discussed here. If these positions do not exist, include persons who do handle the pertinent information or seek outside advice.

social workers, psychiatric nurse practitioners
- financial counselors
- family therapists and family service agencies
- visiting nurse services
- legal aid and public health programs
- veterans' affairs offices and hospitals
- union service committees.

The committee should find out whether existing health and disability benefits cover services such as psychiatric care and counseling. Employees may be more likely to use such services if the cost is covered by insurance.

The committee should also examine the existing corporate philosophy about helping employees with lifestyle issues. Has the company been involved with workers in the past? Is it interested in increasing involvement? Can employees get time off or flexible work hours so they can take advantage of counseling or treatment? How are job performance problems dealt with? Has company policy in that regard been successful?

The input of the financial department will help the EAP committee determine what services can be provided. For example, funding will affect the order in which services are provided. Those that are covered by health insurance and those that can be offered by existing staff can be provided initially. Services that require cash outlays might need to be offered in the next fiscal year.

After the committee has analyzed all data gathered, it will recommend one of four options: improving the existing benefit package to better meet the needs of the workers,

setting up an in-house EAP, contracting for an outside EAP, or making no changes.

If the committee decides an EAP is needed, it must then research and make recommendations on:
- goals of the EAP
- projected number of users
- methods of implementation in terms of staffing, facilities, and management
- fees and benefits
- marketing strategies
- EAP insurance requirements
- legal issues
- a proposed timetable for implementation
- budget and justification.

The committee should issue a report on its work. The report should cover the current situation regarding employee benefits and needs, benefits that will accrue from change, the advantages and disadvantages of the options considered, the rationale for the option selected, and the recommended actions.

SETTING UP AN INTERNAL EAP

The services to be provided, the size, and location of an in-house EAP will depend on the demographics of the workforce. Discover this by charting the number and distribution of employees in the organization. Also record:
- age groups, gender groups, and gender groups within age groups

- ethnic groups and educational levels
- number of dependents who might be eligible for EAP services
- where employees live (pertinent to the location of the EAP office)
- employee interests and work shifts
- work location factors such as travel and weather.

These demographics can indicate, first, employee areas of need. For example, a workforce made up mostly of males ages 35 to 50 will have different needs, obviously, than will a workforce made up mostly of women in their 20s and 30s. (See Figure 7–3 for a discussion of a sample EAP program.)

Location

An EAP office should be conveniently located in a place where workers can come and go without compromising the confidential nature of employee assistance. Hours of operation should be such that employees can visit during their own time, and arrangements have to be made for emergency service when it is needed.

Program Operation

The elements of the day-to-day operation of an EAP depend on the scope of the program. In general, all EAPs will have at least the following operations:

- assessment and referral of clients
- confidential records management
- coordination of benefits
- evaluating and contracting outside service providers and monitoring the quality of services
- monitoring follow-up
- training supervisors and labor representatives about the EAP and how to recognize troubled employees
- promoting the EAP to employees and others eligible for services
- interacting with related departments, employee representatives, management, and entities inside and outside the organization
- writing reports on program use and finances.

In extensive programs, in-house counseling or teaching health classes would be added to that list, along with other administrative duties. Reports would probably be more elaborate on cost/benefit analysis and include research. Many EAPs publish a newsletter with tips on healthful living and leisure activities and notices of EAP services.

Organizational Placement

The placement of an EAP within the organization is very important. To be successful, it must be endorsed by the senior levels of management and employee representatives. Placement is an indicator of their support. Also, services are seldom used by employees who are a level higher than that of the responsible officer, so to reach people who are in the upper levels of management, the administrator of the EAP must report to a fairly high-ranking official.

Often the EAP is housed in an established department to indicate support and lend credibility to the program. Creating a new, separate department can subject the EAP to cutbacks during times of limited resources.

Characteristics of a Successful EAP

When the program is initiated, all employees should receive a general orientation on the purposes of the program and how to use the services. Thereafter, the program should be promoted through frequent communications with posters, paycheck inserts, notices in newsletters, and letters to employees' homes. The program should be promoted at all levels. New employee orientation should include information about the EAP. In addition, the program should meet the following criteria for success.

Written Policies and Procedures

Policy statements should cover all the following issues (Figure 7–4):

- purpose of the policy
- program philosophy
- management and worker commitment to the program and to the well-being of employees (Figure 7–5)
- affirmation that both employees and management were involved in program planning, implementation, and management
- assurance of confidentiality within specific limits
- definition of program objectives
- definition of the types of problems covered by the EAP
- eligibility for participation in the program
- inherently encouraging process for voluntarily seeking help
- definition of process for referrals based on performance problems
- guarantee that participation in the EAP will not affect future employment or advancement
- statement that participation in the program will not exempt employees from remedial action for continued substandard job performance
- definition of responsibilities of the organization and the employee
- statement that alcoholism and drug abuse are diseases responsive to treatment and rehabilitation
- step-by-step process, agreed on in advance by all parties, for handling substance abuse and other job performance problems.

Confidentiality

Confidentiality is a key to the success of an EAP. It should be guaranteed by:
1. written rules regarding handling of records, including:
   a. storage in a secure place separate from all other worker records
   b. designation of who has access to records
   c. release of information, to whom it will be released, and when it will be released
   d. system that protects identity of employee through codes, but allows use of statistics for research, evaluation, and reports

---

General Motors–United Auto Workers
Detroit, Michigan

Union and management representatives at General Motors have been working together on behalf of alcohol- and other drug-impaired employees for more than 16 years. The program began as a joint effort because neither management nor the union working alone could always provide the level of motivation to get employees the necessary help for their addictions. The GM–UAW EAP was initially developed as a joint union-management alcoholism recovery program in 1972. Six years later it became a substance abuse program which evolved into a broad brush EAP by the early 1980s.

Employing more than half a million workers at 168 sites, over 90 percent of the hourly employees are represented by UAW, with the International Union of Electrical Workers (IUE) and the United Union of Rubber Workers (URW) representing the rest. Each worksite has an EAP team composed of local union representatives and local management coordinators. Day-to-day operations are generally under these two officials. Yet overseeing their EAP activities is the local plant medical director.

GM–UAW uses a case management model which includes assessment, referral, aftercare, follow-up, and administration. Intensive treatment is provided by local community facilities. In 1985 the EAP began a Predetermination Program handled through a national contract with Family Services America. This family-oriented organization conducts clinical assessment and psychosocial evaluation of employees who are in need of substance abuse treatment. They determine what level of care is required.

In the last few years, problems with chemical dependency were clearly the major concern, accounting for better than two-thirds of the EAP clientele.

An important measure of program success is the number of clients who agree to accept the specific kind of help suggested by the EAP. Considering the entire range of problems for which employees came to the EAP, eight out of ten clients accepted the program's recommendations. In approximately one-third of the new cases, EAP counseling was the recommendation, with another one-third recommended for outpatient care.

Since 1986, the collaborative efforts between management and union EAP representatives have increased. Not only has the jointly funded program continued to be so, as in the past, but organizationally the relationship has become co-equal. At present, the EAP is housed at the UAW–GM Human Resources Center, with co-directors—one union and one management.

Contact: Daniel Lanier, Jr.; DSW and Ronald L. Murray; UAW–GM Human Resources Center; Auburn Hills, MI 48057; 313/377–2432.

Figure 7–3. Sample EAP Program

e. use of statistics for research, evaluation, and reports
2. separation of EAP files and exclusion from the employee's personnel file
3. requirements that any program that receives federal funds in any way must adhere to federal regulations on the confidentiality of alcohol and substance abuse records (42 *CFR*, Part 2; see also the section on Substance Abuse later in this chapter)
4. program personnel who are discreet and hold confidentiality as a personal and professional duty
5. location and hours of operation that support confidentiality.

Employees should be assured that EAP information will not be revealed to company or outside sources without the written consent of the persons concerned or under certain specific circumstances. Those circumstances might be medical emergencies, a case in which the person involved might harm himself or others, child abuse and criminal activity.

Qualified EAP Staff

The EAP administrator or coordinator must be qualified for the position, and hold the proper licenses and certifications. Staff should have managerial and administrative experience and skills in problem identification and interviewing, motivating, and referring clients. In addition, the staff should have experience and expertise in working with problems related to substance abuse.

Supervisors, employee representatives, and staff must be trained to recognize behaviors that may indicate troubled

Employee Assistance Program Policy

Bowater Carolina Company and Locals 925 and 1924 of the United Paperworkers International Union recognize that a wide range of problems not directly associated with one's job function can have an effect on an employee's job performance. In most instances, the employee will overcome such personal problems independently and the effect on job performance will be negligible. In other instances, normal supervisory assistance will serve either as motivation or guidance by which such problems can be resolved so the employees's job performance will return to an acceptable level. In some cases, however, neither the efforts of the employee or the supervisor have the desired effect of resolving the employee's problems and unsatisfactory performance persists over a period of time, either constantly or intermittently.

Bowater Carolina Company and Locals 925 and 1924 believe it is in the interest of the employee and the employee's family to provide an employee service which deals with such persistent problems. Implementation of the program will be conducted on the basis of urging employees displaying patterns of poor job performance to participate in the program. The existing discipline—grievance and arbitration procedures remain in effect but will be dovetailed as much as possible with progressively stronger urgings to the employee to become involved in an Employee Assistance Program. Therefore, a joint union-management committee has been formed to handle such problems within the following framework:

1. Bowater Carolina Company recognizes that almost any human problem can be successfully treated provided it is identified in its early stages and referral is made to an appropriate modality of care. This applies whether the problem is one of physical illness, mental or emotional illness, finances, marital or family distress, alcoholism, drug abuse, legal problems, or other concerns.
2. When an employee's job performance or attendance is unsatisfactory and the employee is unable or unwilling to correct the situation either alone or with normal supervisory assistance, it is an indication that there may be some cause outside the realm of his/her job responsibilities which is the basis of the problem.
3. The purpose of this policy is to assure employees that, if such personal problems are the cause of unsatisfactory job performance, they will have assistance available to them to help resolve such problems in an effective and confidential manner.
4. Problems causing unsatisfactory job performance will be handled in a forthright manner within the established employer's health and personnel administrative procedures and all records will be preserved in the highest degree of confidence.
5. In instances where it is necessary, a leave of absence may be granted for treatment or rehabilitation for alcoholism and/or drug abuse on the same basis as it is granted for other ordinary health problems.
6. Employees who have a problem they feel may affect work performance are encouraged to voluntarily seek counseling and information on a confidential basis.
7. Employees referred through the program by their supervisor will be encouraged to secure adequate medical, rehabilitative, counseling, or other services as may be necessary to resolve their problem.
8. It will be the responsibility of the employee to comply with the referrals for assessment and his/her problem and to cooperate and follow the recommendation of the diagnostician or counseling agent.
9. Since employee work performance can be affected by the problems of an employee's spouse or other members of the immediate family, the referral source is available to the families of our employees as well.

Figure 7–4. Sample EAP policy statement.

SAMPLE COLLECTIVE BARGAINING LANGUAGE

SUBSTANCE ABUSE POLICY

WHEREAS, the Employer and the Union acknowledge that substance abuse is a serious and complex, but treatable condition/disease that negatively affects the productive, personal, and family lives of employees and the stability of companies; and,

WHEREAS, the Employer and the Union are committed to addressing the problems of substance abuse in order to ensure the safety of the working environment, employees, and the public, and to providing employees with access to necessary treatment and rehabilitation assistance; and

WHEREAS, the Employer and Union have defined a program of employee assistance and have provided coverage to assure that employees requiring treatment and rehabilitation resulting from their substance abuse can receive such services without undue financial hardship;

NOW THEREFORE, the Employer and the Union agree that,

1. Appropriate efforts will be undertaken by the Employer and the Union to establish employee understanding that the experience of alcohol or drug problems is not, of itself, grounds for adverse action. Employees will be strongly encouraged to seek and receive the services of the employee assistance program prior to such problems affecting job performance or resulting in on the job incidents.

   When the Employer has a reasonable suspicion based on objective criteria that an employee is under the influence of alcohol or drugs, hereinafter referred to as 'substances', the Employer may require that the employee immediately go to a medical facility to provide both urine and blood specimens for the purpose of testing and to receive a fitness for work examination by a licensed physician.

   Reasonable suspicion based on objective criteria means suspicion based on specific personal observations that the Employer representative can describe concerning the appearance, behavior, speech or breath odor of the employee. Suspicion is not reasonable, and thus not a basis for testing, if it is based solely on third party observations and reports.

2. The requirement for this testing shall be implemented where practicable, in accordance with the following procedures:

   (a) When the Supervisor has established a reasonable suspicion that an employee may be under the influence of substance(s), based upon specific, individualized observations, the Supervisor shall contact another Supervisor or management employee, for purposes of confirming the reasonable suspicion. The Supervisor shall contact the Business Agent, Union Steward, or other bargaining unit employee for the purpose of informing and involving the appropriate and available Union representative in the immediate situation.

   In the presence of the employee and Union representative, the Supervisor shall present the observations establishing the reasonable suspicion. The employee shall, upon hearing the Supervisor's confirmed observations, receive a written description of his/her rights, obligations, and options and shall be presented with the opportunity to immediately self-refer to the employee assistance program.

   (b) While the observations of the Business Agent, Union Steward, or other bargaining unit employee, may be solicited and are relevant in the context of the joint Employer/Union commitment to addressing the problem of substance abuse, Union representatives will not be expected to give their assent to the Supervisor's decision to require testing or to take other management action.

Figure 7–5. This sample substance abuse policy illustrates collective bargaining language.

(c) An employee who does not self-refer into the employee assistance program and refuses to go to a medical facility, after being informed of the observations establishing reasonable suspicion and of the requirement for immediate fitness for work examination and provision of blood and urine samples, will be discharged.

If requested, the employee shall sign consent forms authorizing: (1) the medical facility to withdraw a specimen of blood and urine; (2) authorizing the testing laboratory to release the results of the testing to the medical facility for physician review and to the Employer; and (3) at the employee's discretion, he/she may authorize the same release as defined in (2) to the Union. By signing these consent forms, the employee does not waive any claim or cause of action under the law. An employee's refusal to sign the release shall constitute a refusal to be examined and tested subject, however, to Section 2. (d) below.

(d) An employee who refuses to be examined and tested shall be encouraged to go to the medical facility for this purpose with the understanding that blood and urine samples drawn will not be tested unless that employee, within twenty-four (24) hours, authorizes that these be tested.

If, at the end of this period, the employee still refuses to have the samples tested, the employee will be discharged unless the employee agrees, within the same twenty-four (24) hour period to self-refer into the employee assistance program.

(e) The employee to be tested shall be taken to the medical facility by an Employer representative and, at the request of the employee, the Business Agent, Union Steward or other bargaining unit employee.

(f) In an effort to protect individual privacy, employees will not be subject to direct observation while rendering urine samples. If the employee provides blood and/or urine samples that contain confirmed evidence of any form of tampering or substitution, the act shall constitute a refusal to be tested and the employee shall be discharged.

(g) Blood and urine samples shall be drawn, subject to the provisions in Section 3 below. Upon receipt of the specimens by the laboratory, one of the two urine specimens will be placed immediately, unopened, in a locked freezer for storage for a period of six months. Employees may, within twenty-four (24) hours of receipt of test results, request the presence of an approved, consulting toxicologist during the full conduct of a second, independent test to be conducted at the laboratory site. Employees requesting independent tests are liable for the costs of the second test and the consulting toxicologist unless the employee's second test results are negative.

In cases of second tests, the urine specimen alone will be used as this fluid better retains the integrity of its chemical contents. Because some drugs/drug metabolites deteriorate or are lost during freezing and/or storage, the retesting of specimens is not subject to the same testing level criteria as were used in the original analysis.

(h) Employees subject to the requirement for testing shall be suspended, effective immediately after receipt of the fitness for work examination and rendering of samples, for the period of time required to process, screen, and confirm test results.

(i) Employees whose test results are negative, and who pass the fitness for work examination, shall be reinstated with backpay for the period of suspension, except as provided in Section 4 (a) below. Employees whose test results are positive shall not be eligible for reinstatement with backpay but shall be given the opportunity to

Figure 7–5. *(Continued.)*

immediately self-refer into the employee assistance program. In the absence of immediate self-referral, such employees will be discharged.

3. The examination and testing procedures and standards to be carried out by the medical facility personnel and testing laboratory shall be those adopted by the Employer and the Union, shall use the blood alcohol level established by State law for intoxication, shall rely in the testing for drugs other than alcohol, on the urine specimen to test for the presence of drugs and/or their metabolites, shall consider 'presence' only and not degree of intoxication or impairment, and shall include the following general components:

(a) Rigorous review, selection and performance monitoring of medical facilities performing the examination and specimen collection and of the laboratory facilities performing the tests.

(a.1) Medical Facilities

Medical facilities performing the examination and specimen collection must be under the direction of a licensed physician. The facility must employ at least one charge nurse who is a registered nurse.

A licensed physician must perform the fitness for work examination and review the laboratory reports of drug tests. The physician must have knowledge of substance abuse disorders and must possess the appropriate medical training to interpret and evaluate all positive test results together with the employee's medical history, including medications use, and any other relevant biomedical information.

The medical facility must possess all necessary personnel, materials, equipment, facilities, and supervision to provide for the collection, security, temporary storage, and transportation (shipping) of blood and urine specimens to the drug laboratory. The medical facility must provide written assurances that the specimen collection space is secure; that chain of custody forms will be properly executed by authorized collection personnel upon receipt of specimens; that the handling and transportation of specimens from one authorized individual or place to another will be accomplished through the use of chain of custody procedures; and that no unauthorized personnel are permitted in any part of the specimen collection or storage spaces.

(a.2) Laboratory Facilities

Laboratory facilities must comply with applicable provisions of any State licensure requirements and must be approved by the BITLC and/or the parties to the agreement. BITLC approval of a laboratory shall be contingent upon successful demonstration and on-site review establishing that the laboratory meets the standards for accreditation promulgated by the National Institute on Drug Abuse and upon the laboratory's ongoing participation in a program of external quality assurance. These standards may be revised as recommended by the National Institute on Drug Abuse.

(b) Specific specimen collection procedures that include safeguards to ensure the employee's right to privacy.

Authorized specimen collection personnel shall request that the employee show positive identification by providing a pictured identification card such as a driver's license and shall assure that the employee signs the waiver agreement that explains the procedures for testing and reporting results. These personnel shall remove all articles and items from the collection space or bathroom, shall assure that toilet water is colored or blued, shall turn off the hot water valve under the sink, shall assume that the tamper-proof specimen

Figure 7–5. *(Continued.)*

collection kit is intact, and shall instruct the employee to wash and dry hands prior to entry. Employees shall remove all excess clothing and leave belongings outside the bathroom and shall provide urine samples in two containers. Employees will not be subject to direct observation while rendering samples. Authorized specimen collection personnel shall, however, be present outside the bathroom and shall receive containers, assure that the quantity is sufficient for testing, check color and measure the temperature of each container and record same. These personnel shall fill in specimen labels in the presence of the employee, shall cap and seal containers with evidence tape and shall secure the employee's initials on the tape.

(c) Flawless chain of custody procedures governing specimen handling throughout the testing process. Chain of custody procedures shall assure that blood and urine samples shall not leave the sight of the employee until each vial has been sealed and initialed and, that at least the following measures are taken by medical facility and laboratory staff:

(c.1) Medical Facilities

Authorized medical facility personnel shall seal specimen tubes with evidence tape in the presence of the employee and the employee shall initial the evidence tape. These personnel shall complete a chain of custody form and shall place the sealed and initialed specimen tubes in the drug collection kit or box provided by the laboratory along with the chain of custody form and signed waiver. The collection kit or box shall be sealed by authorized medical facility personnel and this seal or tape shall be initialed by these personnel and by the employee.

The medical facility shall make prior arrangements for courier pickup of the specimens and shall assure that all specimens are couriered or shipped to the testing laboratory as immediately as possible. The medical facility shall assure that no specimens will be shipped on a Friday or the day before a holiday and that any specimens held at the facility overnight shall be placed in a secured refrigerator until courier pickup.

(c.2) Laboratory

The testing laboratory shall assure that personnel authorized to receive specimens immediately open the package, inspect the sealing tape for initials, and open the kit or box. These personnel shall examine and inspect the chain of custody form, the specimen tubes, and kit or box to assure that it conforms to the requirements of Subsection c.1 (above). If these requirements are not met, the laboratory personnel shall immediately notify the laboratory's scientific director and shall document any and all inadequacies in the chain of custody requirements. The laboratory's scientific director shall immediately notify the medical facility, the Employer and the Union of the inadequacies and shall retain the specimens in a locked freezer pending disposition direction.

If the requirements are met, authorized laboratory personnel shall sign on the appropriate line of the chain of custody form and deliver the specimen kit or box to authorized laboratory technologists for testing. Each technologist shall sign on the appropriate line of the chain of custody form.

All positive samples shall be resecured with evidence tape, signed, and dated by an authorized technologist. Upon completion of testing procedures, testing reports shall be prepared and signed by at least two authorized technologists for the review, approval, and signature of the scientific director.

Figure 7–5. *(Continued.)*

(d) Established levels below which specimens are deemed negative:

Drug Assay	Screening Cut off level	
Blood Alcohol*	100	mg/dl
Cocaine Metabolite	300	ng/ml
Phencyclidine	25	ng/ml
Opiates	300	ng/ml
Amphetamine	1000	ng/ml
Cannabinoids	100	ng/ml

*Subject to Section 3

(e) Laboratory use of appropriate screening and confirmation procedures and technology.

The laboratory shall assure that each specimen will be screened by an immunoassay method, i.e., EMIT, RIA, FPI, for each drug/drug group. Each specimen shall also be analyzed for acid, neutral and basic drugs by thin layer chromatography (TLC).

If either or both of these assays are positive, an intermediate screening procedure shall be performed by a second, authorized laboratory technologist using a more specific TLC procedure, an alternate second immunoassay method, and/or a high pressure liquid chromatography.

Gas chromatography/mass spectrometry (GC/MS) must be used as the final confirmation method. All three tests must be positive before a specimen is reported as positive.

Blood and urine ethanol testing shall be performed by gas chromatography (GC) and, if positive, a second GC column shall be used. If results are positive on both columns, fluorescent polarization immunoassay (FPI) or an enzymatic assay shall be used as the third and confirming test.

Final confirmation by gas chromatography/mass spectrometry (GC/MS) and/or fluorescent polarization immunoassay (FPI) shall be subject to the following levels below which specimens are deemed negative:

Drug Assay	Confirmatory Cut off level	
Blood Alcohol*	100	mg/dl
Cocaine Metabolite	150	ng/ml
Phencyclidine	25	ng/ml
Opiates	300	ng/ml
Amphetamine	300	ng/ml
Cannabinoids	20	ng/ml

*Subject to Section 3

Screening methods measure a group of drugs and/or their metabolites simultaneously. Confirmatory methods, on the other hand, measure single and specific drugs and/or their metabolites. Cut off levels for confirmatory methods, therefore, may be lower than those for initial screening.

(f) Procedures to assure the confidentiality of test results and the treatment of these records as confidential health information or data.

The laboratory shall ensure that testing reports, including the original chain of custody form, are mailed to those personnel authorized by the medical facility, the Employer, and if the employee so chooses, by the Union immediately and shall ensure that, in the

Figure 7–5. (Continued.)

event that telephone reports of testing results are required by the medical facility, the Employer and the Union, a security code system be used to establish that results are being verbally reported only to those individuals authorized by the medical facility, the Employer and by the Union.

4. After examination and specimen testing results, the following shall apply:

    (a) If an employee is subject to discipline or termination under existing practices, such employees shall not utilize the substance abuse policy to circumvent the labor agreement or existing practices or to avoid discipline or termination.

    (b) In the cases not covered in Section 4 (a) above, the employee will have the opportunity for appropriate assistance, assessment, referral, treatment, and aftercare as provided in the employee assistance program and as agreed in the employee assistance program's individual treatment plan with the employee. Failure to seek and receive these services or failure to abide by the terms of the treatment plan shall be grounds for discharge.

    (c) An employee who seeks and receives assistance and who completes the defined employee assistance program shall, upon return to work, be subject to random and mandatory tests for a period of nine (9) months.

    (d) An employee who, on the basis of such random and mandatory tests defined in 4 (c) above, provides samples that contain positive and confirmed evidence of substances at or above the stipulated levels, shall not be given a second opportunity to access the employee assistance program as an alternative to discharge.

    (e) Employees who successfully complete the employee assistance program and their individual treatment plan agreements and who return to work will be encouraged to contact and avail themselves of the employee assistance program's services on a self-referral basis whenever they desire ongoing assistance and support.

    Employees who relapse and for whom reasonable suspicion of substance use is established a second time, and whose test results are positive, will be subject to the disciplinary procedures up to and including discharge. The Union and Employer may agree, however, to consider such mitigating factors as the employee's length of sobriety, job performance, length of service, etc. in such situations.

5. The employee assistance program shall include the following components:

    (a) Full clinical evaluation and appropriate assessment followed by a specific individual treatment plan and regimen for the receipt of counseling, treatment, aftercare, and related services subject to the ongoing monitoring of the employee assistance program staff.

    (b) Active encouragement and procedures for the voluntary and self referral of troubled employees to the employee assistance program in cases in which reasonable suspicion has not been established and in which examination and testing procedures are not invoked.

    (c) Assurances and procedures to protect the confidentiality of employees who voluntarily seek employee assistance program services; procedures governing the management of such employee records as medical information.

6. Any disputes arising under this addendum shall be subject to the grievance procedure established in the labor agreement, up to and including arbitration.

Figure 7–5. *(Concluded.)*

employees. The supervisors, who work daily with their employees, are an important aspect of the problem recognition training program. In addition, practice in human relations skills is important in directing employees into the program. Knowledge of and adherence to program policies are essential. EAP staff must also obtain necessary licensure and certification and must seek continuing education.

Qualified Service Providers

Professionals or facilities, receiving referrals or providing services must be highly qualified clinically (Figure 7–6). They must be able to work within the guidelines of the program to provide cost-effective services.

Adequate Insurance

Employees are more likely to use a program if services are covered by insurance. Because of this, a program's success may largely depend on insurance benefits and coverage. These benefits must adequately cover the problems the EAP addresses. Minimally, health care benefits should cover:

- substance abuse treatment, both in-patient and out-patient
- psychiatric care, in-patient and out-patient, by psychiatrists, psychologists, social workers, and psychiatric nurse practitioners.

Adequate Staff and Space

Staffing must be sufficient to handle the volume of employees seeking the services. A long wait for an appointment will defeat a program that is intended to help people in crisis.

Space should be commensurate with services. EAPs offering counseling require office space, but classes and group meetings need classrooms offsite or in a location convenient to participants.

Protection from Stigma

Employees may feel uncomfortable using the program because of the stigma attached to problems it addresses. Widely publicizing the program may not remove the stigma of substance abuse and mental health, family, marital, and financial problems, but it will show that the organization recognizes these problems and wants to help.

Another way to relieve employees' embarrassment is to locate the EAP office in a place employees can easily reach without being seen. Also, the program should be placed in a department that does not deal with disciplinary problems, so voluntary use is not discouraged. For example, if the EAP where housed in the medical or personnel department, employees could say where they were going without causing raised eyebrows.

A written policy guaranteeing that program participation will not hurt employment or advancement opportunities is also important in dealing with stigma.

Staffing

Professional counselors should hold a Masters degree (MA) ·n counseling from an approved program at an accredited college or university. They should also be licensed in states where counselors are licensed.

EASNA specifies that EAP administrators have two years of EAP experience and be professionally licensed. The group also says there should be at least one counselor for each 6,000 employees; more counselors if services beyond assessment and referral are provided.

Records

EAP case records should be kept for at least five years and should be stored securely and apart from company, personnel, and medical records. If the company makes use of an automated information system, provisions should be made for guarding against unauthorized access to EAP records.

Insurance

EASNA stipulates that an EAP be provided with sufficient liability insurance. Company self-insurance may be an option to purchasing a separate EAP policy.

CONTRACTING FOR EAP SERVICES

Firms with fewer than 2,000 employees generally find it more cost-effective to contract with an independent vendor for EAP services than to develop and maintain a program in-house. Outside EAP suppliers usually can be found who will provide services ranging from consultation to full program management. So a firm first needs to know its EAP requirements to specify clearly a contracting vendor's services.

Services an EAP consultant can provide include:

- aid in program, policy, and procedure development
- training for supervisors
- organization of an EAP advisory committee
- coordination of EAP activities between management and employees
- analysis of employee benefit packages as they relate to EAP
- help with legal questions
- formulation of strategies to increase supervisor referrals
- identification of the needs of special employee groups
- help in resolving organizational difficulties.

Some consultants advertise themselves as specialists in clinical issues and offer to provide extra aid in managing difficult EAP cases. These specialists also offer extra assistance in the selection of referral sources and when providing suggestions concerning recordkeeping, improvement of case management and follow-up techniques, and development of evaluation plans.

While consultants serve primarily as expert independent advisors, program management contractors are vendors who get involved in the day-to-day operation of an EAP. They can:

- conduct supervisory training and employee orientation sessions
- write publicity materials
- assist managers with employee problems
- provide assessment and referral services
- manage individual EAP cases
- maintain records

- implement program evaluation systems
- ensure that confidentiality is maintained.

Assessing an EAP Contractor

One way to help assure quality among contractors under consideration is to look for program accreditation by EASNA. Accreditation at least is evidence that a contractor can meet standards set by the profession. EASNA accredits programs in four areas:

- screening, referral for assessment, support, and follow-up
- assessment, referral, and follow-up
- assessment, referral, short-term counseling, and follow-up
- managed mental, nervous, and addictive care.

If an EAP service is affiliated with a healthcare institution or parent organization, inquire as to the parent organization's reasons for offering EAP services. A hospital's EAP unit may exist to generate in-patient and out-patient referrals, for example. A decision has to be made then whether contracting with that particular EAP provider is in the company's best interests.

Organizations listed in the References at the end of this chapter can provide names of professional consultants. Colleagues with EAPs in place can refer you to their EAP administrator, or you can call local hospitals and ask for information.

The consultant you hire to help with program design should have the following characteristics and expertise:

- Direct experience with EAPs, both in their management and clinical service provision. Especially important is experience in program development, including writing policies, procedures, and training programs.
- Knowledge of alcoholism and other drug abuse and government regulations relating to these problems in the workplace.
- Skills in communication and organizational and group dynamics. Assessment and proposal development depend heavily on the consultant's ability to recognize key issues. The consultant will be working closely with the EAP committee members, management, and employee representatives.
- Knowledge of personnel and legal issues. Development of an EAP requires analysis of benefits packages and integration of the program into personnel policies and overall procedures.
- Marketing expertise. An EAP's success depends largely on employees being aware of the program and its benefits. The consultant must be able to promote the EAP to employees and their families through a variety of methods, such as brochures, posters, and meetings (Figure 7–6).

Contractor Pricing

Most EAP contractors charge for their services either on the basis of a fixed fee or fee for services. A fixed fee frequently is expressed as a per capita charge for each employee covered by an EAP. Some per capita arrangements include one fee for employees and a separate or additional percentage charge for family members. Fees for service differ according to geographic region. A retainer may be sought by a contractor in addition to an agreed-upon schedule of fees.

SUBSTANCE ABUSE

Certain U.S. regulations may require an organization to guarantee a drug-free workplace or set up a drug-testing program. Such a program can be part of an EAP and should include five elements:

- a written substance abuse policy (Figure 7–4)
- a supervisory training program
- an employee education and awareness program
- access to an employee assistance program
- a drug-testing program, where appropriate.

U.S. Drug-Free Workplace Act

The U.S. Drug-Free Workplace Act of 1988 requires that certain federal contractors and grantees maintain a drug-free workplace. Those affected include:

- organizations having a single contract with the U.S. government of $25,000 or more
- organizations receiving a grant from the U.S. government.

The Act does not cover subcontractors and subgrantees.

This Act does not require an organization to establish an EAP or implement a drug-testing program. However, it does require that the company:

- provide a drug-free workplace
- publish a statement notifying employees that the unlawful manufacture, distribution, dispensing, possession, or use of a controlled substance is prohibited in the workplace and what actions will be taken against employees for violations
- establish an ongoing, drug-free awareness program to inform employees of the dangers of drug abuse, the company policy, the availability of any drug-counseling programs, and the possible penalties for drug-abuse violations occurring in the workplace
- require each employee directly involved in the work of the contract or grant to notify appropriate personnel of any criminal drug statute conviction for a violation occurring in the workplace
- notify the U.S. government of such a violation
- require the imposition of sanctions or remedial measures for an employee convicted of a drug abuse violation in the workplace
- show continuing good faith to comply with the above requirements.

U.S. Department of Transportation Regulations

The U.S. Department of Transportation (DOT) has special requirements that affect many U.S. businesses. Since November 1988, a DOT regulation requires employers engaged in commercial transportation industries regulated by DOT to implement workplace drug-free programs. The regulations require drug testing of employees in safety-sensitive positions and drug-free awareness education for supervisors and employees. All drug testing must be conducted in accor-

190

Professionals and facilities that provide services must meet at least the following criteria, but you may want to add others:

- Physicians should:
  a. Hold a degree in medicine (MD)
  b. Hold a license to practice in your state
  c. Have expertise in dealing with emotional and addiction problems, especially if they will be working on substance abuse cases
  d. Be recommended, if possible, by a physician or other reliable source who is familiar with the physician's services
  e. Be on staff at a hospital, with an appointment in his or her specialty.

- Psychologists should:
  a. Hold a state license or certification
  b. Hold a Doctor of Philosophy (PhD) in psychology
  c. Have two years supervised clinical practice.

- Alternate requirements for a psychologist are:
  a. Hold a Masters degree (MA) in psychology
  b. Have three years supervised clinical experience
  c. Have additional training in the treatment they will provide, such as chemical dependency or marital and family therapy.

- Social workers should:
  a. Hold a Masters degree (MA) in social work or a Masters of Social Work degree (MSW) from an accredited school of social work
  b. Have two years supervised clinical experience
  c. Be a Licensed Clinical Social Worker (LCSW) in states where social workers are licensed or an Accredited Clinical Social Worker (ACSW) in other states.

- Nurse practitioners should:
  a. Be licensed by the state as a Registered Nurse (RN)
  b. Comply with state credentials requirements for nurse practitioners
  c. Hold a Masters degree (MS), as recommended by the American Nurses' Association
  d. Have experience and expertise in dealing with the relevant area.

- Professional counselors should:
  a. Hold a Masters degree (MA) in counseling from an approved program at an accredited college or university
  b. Be a Licensed Professional Counselor (LPC) in states where counselors are licensed
  c. Have two years supervised clinical experience
  d. Have special training in the treatment they will provide, such as marital and family therapy.

- Facilities. Visit facilities under consideration and discuss with managers the following criteria:
  a. Proof of Joint Commission for the Accreditation of Hospitals accreditation (for in-patient treatment providers)
  b. Proof of state licensure
  c. Licenses or certifications of personnel
  d. Variety of resources
  e. Written objectives for program, services, and treatment, and for procedures followed to meet these objectives
  f. Individualized program, services, and treatment plans
  g. Assurance of confidentiality and compliance with state and federal confidentiality regulations

Figure 7–6. Criteria for Service Providers

```
   h. Length of time in business (more than one year)
   i. Marketing methods
   j. Client references
   k. Usage rate of other EAP clients
   l. Methods and services consistent with organization's employees
   m. Follow-up provided or available (with behavior change programs)
   n. Location convenient to employees
   o. Service times, length and number of sessions, and frequency convenient to employees
   p. Minimum and maximum enrollment for classes
   q. Flexiblity of service provision logistics
   r. Safety precautions consistent with those of the organization
   s. Waiver form for participants in the program
   t. Evaluation of participants and method of sharing results with the organization
   u. Evaluation of program, services, and treatment
   v. Willingness to negotiate costs
   w. Contract for program, services, and treatment.
```

Figure 7–6. *(Concluded.)*

dance with procedures outlined in 49 *CFR*, Part 40.

These regulations affect employers with employees in the following positions:

- flight crew member, flight attendant, flight instruction or ground instruction, flight testing aircraft dispatch, aircraft maintenance or preventive maintenance, aviation security or screening, or air traffic control duties in commercial aviation
- operators of commercial motor vehicles in interstate commerce when (1) the vehicle has a GVWR of 26,000 or more pounds, or (2) the vehicle is designed to transport more than 15 passengers including the driver, or (3) the vehicle is used in the transportation of hazardous materials in a quantity requiring placarding
- railroad services subject to the Hours of Service Act (45 USC 61–64b)
- operating, maintenance, or emergency-response functions on a pipe line or liquid natural gas facility
- crew members on a commercial vessel licensed, certificated, or documented by the U.S. Coast Guard.

If a drug-testing program is established, it must meet several requirements, including:

- statutory or regulatory requirements
- disability discrimination provisions
- collective bargaining agreements (Figure 7–5).

Legal advice may be required to assure the program is fair, accurate, and legally defensible.

Training Supervisors and Employees

Once an EAP is established or an outside service is contracted, supervisors and other management personnel must be trained in using it. They need to know how to recognize signs of employee need for referral to the EAP, how to detect and document drug or alcohol use, how to talk to employees about work problems and what needs to be done about them, and when to take action. Supervisors

also need to be trained in what not to do: not to diagnose or treat substance abuse problems.

Employees also need education and awareness training. They need to know the organization's policies on workplace substance abuse and the consequences of violations. They need to know that help is available and how to use it. They need specific education on the effects of substance abuse and other personal problems on the organization's productivity, product quality, absenteeism, health care costs, and accident rates.

SUMMARY

- In-house or contracted employee assistance programs (EAP) are cost-effective, humane, and job-related strategies to identify and resolve employee problems before they become a source of major liability. These strategies improve problem solving, offer an alternative to discipline, help valued employees, and prevent legal issues from going to litigation.
- Employee assistance programs consist of full-time internal programs, part-time internal programs, consortiums, independent service providers, and regional programs. Internal programs are administered by company personnel staff.
- To evaluate whether a firm needs an EAP, it must assess employee problems, survey community resources available, examine company insurance benefits and corporate philosophy, and decide on the best type of program to adopt.
- Companies electing to set up an internal EAP must consider the services to be provided, the size and location of the program, and how it will be administered. To be successful, the programs must have top management support, develop formal policies and procedures, and ensure employee confidentiality.
- Smaller firms may find it more cost-effective to contract with an independent vendor for EAP services. EAP con-

sultants can help the firm and community resources.
- Certain U.S. regulations, such as the U.S. Drug-Free Workplace Act of 1988, may require an organization to guarantee a drug-free workplace or set up a drug-testing program.

REFERENCES

Archambault R, et al. *Reaching Out: A Guide to EAP Casefinding.* Troy, MI: Performance Resource Press, Inc., 1982.

Archibald K, et al. *A Guide to Resources for Health Promotion and Illness Prevention.* Chicago: Commonwealth Edison, 1986.

Association of Labor-Management Administrators and Consultants on Alcoholism, Inc. *EAP Starter Kit.* Arlington, VA: ALMACA, 1986.

Blair BR. *Hospital Employee Assistance Programs.* Chicago: American Hospital Publishing, 1985.

Business and Legal Reports. *Supervisor's Guide to On-the-Job Alcoholism and Drug Problems.* Madison, CN: BLR, 1986.

Durkin WS. *The EAP Manual: A Practical, Step-by-Step Guide to Establishing an Effective Employee Alcoholism/ Assistance Program.* New York: National Council on Alcoholism, 1982.

EASNA Standards for Accreditation of Employee Assistance Programs, Berkley, MI: Employee Assistance Society of North America, 1990.

EAP Digest. Performance Resource Press, Inc. 2145 Crooks Road, Suite 103. Troy, MI 48084.

Lanier D, and Gray M. *Employee Assistance Programs: A Guide for Administrators and Consultants.* Troy, MI: Performance Resource Press, Inc., 1986.

Masi DA. *Designing Employee Assistance Programs.* American Management Associations.

National Safety Council, 444 North Michigan Avenue, Chicago, IL 60611.
  "Aids in the Workplace." 1986.
  "Getting Help: A Guide to Problems and Their Solutions." 1987.
  "Guide to Developing an Employee Assistance Program." 1987.
  "Guiding Workers into Employee Assistance Programs." 1987.
  "The Trouble with Alcohol." 1987.
  "The Trouble with Love." 1987.
  "The Trouble with Money." 1987.
  "The Trouble with Stress." 1987.
  "The Trouble with Substance Abuse." 1987.
  "Workplace Drug Testing: Wise Hunt or Witch Hunt?" 1986.

Nye SG. *Employee Assistance Law Answer Book.* New York: Panel Publishers, 1990.

Spicer J., et al. *Evaluating Employee Assistance Programs; A Guide for Administrators and Consultants.* Troy, MI. Performance Resource Press.

Organizations and Government Agencies
Alcoholics Anonymous World Service, Inc., Box 459, Grand Central Station, New York, NY 10017.

Al-Anon Family Groups Headquarters, Inc., P.O. Box 182, Madison Square Station, New York, NY 10010.

Employee Assistance Society of North America, 2728 Phillips, Berkley, MI 48072.

National Institute on Alcohol Abuse and Alcoholism, P.O. Box 2345, Rockville, MD 20852.

National Institute on Drug Abuse, Public Health Service, Alcohol, Drug and Mental Health Administration, U.S. Department of Health and Human Services, Rockville, MD 20857.

(Additional information can be obtained from articles in employee assistance trade magazines, general interest business and health publications, safety journals and literature supplied upon request from EAP providers.)

8

Workers' Compensation

In a typical year on the job, injuries account for one-fifth of the total accidental injuries from all causes in the United States. Injured workers, their families, their employers, and society as a whole suffer substantial economic losses as a result of these accidents. The statistics do not include the physical and mental suffering injured workers and their families undergo. Thus, when a worker dies, is disabled, or merely requires medical attention because of work-connected injury or disease, the economic consequences affect everyone.

Effective loss control can prevent injuries and accidents and reduce their costs, thus benefitting workers, employers, and the entire economy. Many companies are continuing to show much greater interest in controlling the costs related to such incidents. This interest is sparked by the increasing importance that costs are being given in executive decision making. As a result, safety and health professionals have a great opportunity to influence management toward adopting more effective safety measures.

ECONOMIC LOSSES

Workers and their families may suffer two types of economic losses (1) a loss of earnings and (2) additional expenses.

If a worker dies because of a work-related injury or sickness, the survivors lose the income the worker would have earned—less the amount spent on personal expenses—over the remainder of the individual's working career and retirement years. This loss can be substantial.

Total and permanent disability cause even greater earnings losses than death because the worker must be maintained even though unable to work. Permanent partial disability accounts for part of the economic losses due to disability, depending upon the proportion of annual earnings lost because the worker cannot function fully. An employee who is totally disabled temporarily loses any income for the time he or she is recuperating. Loss of even a month's earnings can be a serious financial problem for most workers. In addition to these earnings losses, the deceased or disabled worker often is unable to provide valuable household services that must now be forgone or be taken over by someone else at additional cost.

Although not all injured workers are disabled, nearly all will require some form of medical attention. In general, medical expenses usually amount to less than the total earnings loss; but for many workers, their medical expenses equal or exceed the income lost.

In addition to direct earnings losses, society also loses the taxes that injured employees would have paid and the products or services they would have produced. Some injured employees and their families become public assistance beneficiaries and must be supported by other members of society.

WORKERS' COMPENSATION IN THE UNITED STATES

In the early decades of industrialization in the United States,

efforts to implement a system of compensation for industrial injuries lagged far behind the developments in Europe. However, toward the end of the 19th century, as work-related injuries and diseases and their consequences grew more severe and costly, the public and others began to demand radical change. The first tangible evidence of government interest in workers' compensation laws appeared in 1893 when legislators seized upon John Graham Brooks' account of the German system as a guide for their own efforts at reform. Their interest was further stimulated by the passage of the British Compensation Act of 1897.

Early Laws

In 1902, Maryland passed an act providing for a cooperative accident insurance fund. Although this represented the first legislation to embody the compensation principle in any degree, the scope of the act was restricted. Benefits, which were quite meager, applied only in cases of fatal accidents. Within three years, the courts declared the act unconstitutional. In 1908, a Massachusetts act authorized the establishment of private plans of compensation upon approval of the state board of conciliation and arbitration. This law had no practical significance, and proved to be a dead letter from the start.

By 1908, the United States still had no workers' compensation act. President Theodore Roosevelt, recognizing the injustice, urged passage of an act for federal employees in his message to Congress in January. He pointed out that the burden of an accident fell upon the helpless man, his wife, and children, and declared this state of affairs "an outrage." In 1908, Congress passed a compensation act covering certain federal employees. Though somewhat inadequate by some standards, it was the first real compensation act passed in the United States.

During the next few years, advocates of compensation continued to press for state laws. A law passed in Montana in 1909, applying to miners and laborers in coal mines, was declared unconstitutional. Nevertheless, many states appointed commissions to investigate the feasibility of compensation acts and to propose specific legislation. A significant number of laws resulted from these commission reports—all of which favored some form of workers' compensation legislation—which were combined with recommendations from various private organizations. However, widespread agreement on the need for compensation legislation did not end all conflict over reform. Special interest groups clashed over specific bills and over questions of coverage, waiting periods, and state versus commercial insurance.

In 1910, New York adopted a workers' compensation act of general application, whose coverage was compulsory for certain especially hazardous jobs and optional for others. None of the early state compensation acts expressly covered occupational diseases. Statutes that provided compensation for "injury" were frequently interpreted to include disability from disease. However, those acts that limited benefits to "injury by accident" expressly excluded occupational disease. Every state act except the one passed by Oregon re-

quired uncompensated waiting periods of one to two weeks before benefits were paid; several states provided retroactive payments after a prescribed period.

The 1911 Wisconsin workers' compensation act was the first law to remain effective, and was quickly followed by laws in Nevada, New Jersey, California, and Washington that same year. In 1916, the United States Supreme Court declared workers' compensation laws to be constitutional. Although 24 jurisdictions had enacted such legislation by 1925, workers' compensation was not provided in every state until Mississippi enacted its first law in 1948. Thus, the United States proceeded with a statewide workers' compensation system when other countries enacted nationwide workers' compensation. As a result, there are 50 different statewide workers' compensation acts in the U.S. today.

Compensation Legislation

All 50 states, the District of Columbia, Guam, and Puerto Rico have compensation acts. In addition, the Federal Employees' Compensation Act (FECA) covers all employees of the U.S. Government, while the Longshoremen's and Harbor Workers' Compensation Act covers maritime workers, other than seamen, and workers in certain other groups. The latter act provided compensation for workers in the "twilight zone" between ship and shore, because the U.S. Supreme Court had ruled they could not be covered under state compensation laws. (Each of the Canadian provinces and territories also has a compensation act or ordinance.)

Although economic changes and public policy have prompted increases in benefits and scope of the laws, the basic concepts have remained relatively unchanged. Employers and labor are both dissatisfied with certain aspects of workers' compensation. Labor attacks the system for inadequate benefits, coverage limitations, and exclusion of many injuries, illnesses, and disabilities that labor considers job related. Employers criticize the system for covering some injuries and diseases they do not consider job related and for its high cost relative to its apparent benefits. Thus, while the early advocates of workers' compensation conceived of it as a simple, efficient, equitable remedy to reduce litigation over industrial injuries, both labor and management have expressed doubt that their hopes can be realized.

OBJECTIVES OF WORKERS' COMPENSATION

A U.S. Chamber of Commerce publication cites these six basic objectives underlying workers' compensation laws:

1. Provide adequate, equitable, prompt, and sure income and medical benefits to work-related accident victims, or income benefits to their dependents, regardless of fault
2. Provide a single remedy and reduce court delays, costs, and workloads arising out of personal-injury litigation
3. Relieve public and private charities of financial drains—incident to uncompensated industrial accidents
4. Eliminate payment of fees to lawyers and witnesses as well as time-consuming trials and appeals
5. Encourage maximum employer interest in safety and re-

habilitation through an appropriate experience-rating mechanism

6. Promote frank study of causes of accidents (rather than concealment of fault), reducing preventable accidents and human suffering.

Income Replacement

The first objective listed for workers' compensation is to replace the wages lost by workers who are disabled due to a job-related injury or illness. According to this objective, the replacement should be adequate, equitable, prompt, and sure.

To be adequate, the program should replace lost earnings (present and projected, including fringe benefits), less those expenses such as taxes and job-related transportation costs that would not continue. The worker, however, should share a proportion of the loss in order to provide incentives for rehabilitation and accident prevention. A two-thirds replacement ratio is found in most state statutes.

To be equitable, the program must treat all workers fairly. According to one concept of fairness, most workers should have the same proportion of their wages replaced. However, workers with a low wage may need to receive a high proportion of their lost wages in order to sustain themselves and their families. High-income workers who can afford to purchase private individual protection may have their weekly benefits limited to some reasonable maximum. However, if workers' compensation insurance is regarded primarily as a wage-replacement program, few people should be affected by this maximum.

The first objective also includes medical and vocational rehabilitation and return to productive employment. To achieve this goal, workers should receive quality medical care at no cost, care that will restore them as much as possible to their former physical condition. If complete restoration is impossible, workers should receive vocational rehabilitation that will enable them to maximize their earning capacity. Finally, the system should provide incentives for disabled workers and prospective employers to help employees return to productive employment as quickly as possible.

One of the objectives of workers' compensation is to allocate the costs of the program among employers and industries according to the degree to which they are responsible for the losses. Such an allocation is considered equitable because each employer and industry pays its fair share of the cost. In the long run, this allocation shifts resources from hazardous industries to safe industries and from unsafe employers within an industry to safe employers. Eventually, employers with the most unsafe operations will be driven out of the marketplace.

Critics argue that workers' compensation costs account for such a small part of overall operating expenses that they have little, if any, effect on a firm's resource allocation. As a result, unsafe employers would not need to resort to higher prices, and they would remain in the marketplace.

Accident Prevention and Reduction

Occupational accident prevention and reduction is the final commonly accepted objective of workers' compensation. Those who consider this objective to be important believe that the system should and can provide significant financial and other incentives for employers to introduce safety measures that will decrease the frequency and severity of accidents. More specifically, the pricing of workers' compensation should reward good safety practices and penalize dangerous operations. Employees also should have some incentive to follow safe work practices by sharing some of the losses. Injured workers should have the opportunity and should be encouraged to return to work as soon as they are physically able.

MAJOR CHARACTERISTICS

Compensation laws are compulsory or elective. Under an elective law, the employer may accept or reject the act. However, if an employer rejects the act, it loses the three common-law defenses—assumption of risk, negligence of fellow employees, and contributory negligence. This means that in practice all the laws can be considered "compulsory." A compulsory law requires each employer to accept its provisions and provide for benefits specified. Coverage is elective in three states: South Carolina, New Jersey, and Texas.

Most jurisdictions require employers to obtain insurance or to prove financial ability to carry their own risk. Six states, two U.S. territories, and most provinces require employers to contribute to a monopolistic fund operated by the state or provincial agency. In some instances, employers may qualify as self-insurers. Thirteen states permit employers to purchase insurance either from a competitive state fund or from a private insurance company.

Covered Employment

Although most of the state workers' compensation laws apply to both private and public employment, none of the laws covers all forms of employment and occupation.

For example, a few states restrict compulsory coverage to so-called hazardous occupations. Many laws exempt employers having fewer than a specified number of employees, usually less than three or four in any one location. Most of the laws also exclude workers in farming, domestic service, and casual employment. Many laws contain other exemptions, such as employment in charitable or religious institutions.

Federal workers are covered by FECA. Employees of the District of Columbia are covered by the District of Columbia Workers' Compensation Act, which went into effect in 1982. Its provisions closely follow those of FECA.

Two other major groups excluded from coverage by compensation laws are interstate railroad workers and maritime employees. Railroad workers whose duties involve any aspect of interstate commerce are covered by the Federal Employers' Liability Act (FELA). Maritime workers are subject to the Jones Act, which applies provisions of the FELA to seamen.

The Federal Employers' Liability Act is not a workers' compensation law. Instead, it gives an employee the right to charge an employer with negligence and prevents the employer from pleading the common law defenses that the worker is a fellow servant or assumes part of the risk; moreover, the act substitutes the principle of comparative negligence for the common-law concept of contributory negligence.

It is not known how many state and local employees are covered by workers' compensation or provided with such protection voluntarily by their employers. All states (as well as Puerto Rico, Guam, and the District of Columbia) provide some coverage of public employees but the extent of the benefits varies widely. Some laws specify no exclusions or exclude only such groups as elected or appointed officials. Others limit coverage to employees of specified political subdivisions or to employees engaged in hazardous occupations. In still others, the extent of coverage is left entirely up to the state or to the city or political subdivision employing government workers.

Certain other groups, such as the self-employed, unpaid family members, volunteers, and trainees, generally are not protected by workers' compensation.

"Exclusive Remedy" for Work-Related Disabilities

Before workers' compensation laws were enacted in the states, an employee, in order to recover damages for a work-related injury, had to prove some degree of fault or negligence on the employer's part. Under what is now known as the "quid pro quo of workers' compensation law," employers accepted, or were required to accept, responsibility for injuries arising out of and in the course of employment without regard to fault. In exchange, employees gave up the right to sue employers for unlimited damages. These agreements are usually referred to in the state acts as "exclusive remedy" provisions, a term that is quite misleading.

In no state are workers' compensation benefits necessarily the only remedy available to an injured worker. Depending upon the wording of the applicable statute, workers may bring a negligence action against their employer, fellow workers, another contractor on the same job, or some other entity or individual who caused the compensable injury. For example, workers may sue the manufacturer of a piece of equipment which caused an injury. From the employer's viewpoint, the doctrine should be the "exclusive liability rule." As the employee sees the rule, it remains an "exclusive remedy" for obtaining "worker's compensation" from the employer. Neither liability nor remedy is perfectly exclusive.

Two concepts that are broadening the exclusive remedy provision are (1) the expansion of the dual capacity doctrine and (2) the intentional tort exception.

Under the first concept, an injured employee can sue an employer for an injury—even if it arose out of and in the course of employment—if the injury was caused by the employer's product or a service available to the public. (Example: The driver of a tire company delivery truck who is injured when a defective tire, made by the employer, causes

the truck to have an accident. Or a hospital employee who, after an accident on the job, is injured as the result of negligent treatment by one of the hospital's medical staff.) In both cases the injury did not occur as a result of the employer-employee relationship but rather through a relationship more akin to that of a supplier or service provider and the public.

If an employer commits an intentional tort, i.e., either deliberately causes harm to an employee, is grossly negligent, or engages in reckless behavior that results in an injury, the employee has the right to sue the employer for damages. The rationale is that the exclusive remedy provision should not protect an employer against being sued for an injury resulting from a deliberate harmful action or from gross negligence and recklessness. In some states, failure of an employer to comply with OSHA standards may be evidence of deliberate harmful action by the employer.

Covered Injuries

Workers' compensation is presently intended to provide coverage only for certain work-related conditions, and not for all of an employee's health problems. Statutory definitions and tests have been adopted to distinguish between conditions that are compensable and those that are not. All jurisdictions, when drafting workers' compensation laws, relied to some extent on the English legal system (or on other statutes based upon the English model). Even though the statutory language of these laws is remarkably similar, there are variations in terminology and differences in interpretation; as a result, a condition considered compensable in one state may be held noncompensable in others.

The statutes usually limit compensation benefits to personal injury caused by accidents arising out of and in the course of employment. Although this restriction presents four distinct tests that must be met, in practice these tests are often considered in pairs: the "personal injury" and "by accident" requirements in one set, and the "arising out of" and "in the course of" requirements in the other.

Personal injury and "by accident." If interpreted narrowly, personal injury would refer solely to bodily harm, such as a broken leg or a cut, while the "by accident" test would refer to the cause, such as a blow to the body or an episode of excessive or improper lifting. In practice, however, the distinctions are blurred.

The "by accident" concept is a carryover from English law. Early judicial interpretations of English law made it quite clear that the "by accident" requirement was intended to deny compensation to those who injured themselves intentionally. A number of U.S. jurisdictions, however, have applied the test in order to narrow the range of unintentional injuries that must be compensated.

One of the early casualties of the "by accident" requirement was occupational disease coverage. As the typical judicial holding was that "occupational disease" and "accidental injury" were mutually exclusive concepts, special legislation was required to provide coverage for workers suffering from work-related diseases.

Occupational Disease

Although workers' compensation laws initially had no specific provisions for occupational diseases, all states now recognize responsibility for them. Coverage extends to all diseases arising out of and in the course of employment. Most states do not provide compensation for a disease that is an "ordinary disease of life" or that is not "peculiar to or characteristic of" the employee's occupation. One state does not cover cumulative trauma disorders such as carpal tunnel syndrome because it considers it an "ordinary disease of life."

Generally, compensation is the same as for traumatic injuries, and medical care coverage is unlimited. A few states do not provide permanent partial disability benefits for certain diseases. Occupational diseases usually become evident during employment or soon after exposure. However, as with radiation disabilities, certain diseases may be latent for a long time. Most states have extended periods in which claims may be filed concerning latent, slowly developing occupational diseases.

Some states impose special restrictions regarding disability resulting from exposure to coal dust, asbestos, silica, cotton dust, or radiation. A number of states have established presumptions for police and firefighters who have heart attacks or respiratory conditions, but no attempt is made to chart them.

Hearing

The difficulty of distinguishing between occupational and nonoccupational hearing loss has led to enactment of special coverage provisions in many state statutes. They attempt to isolate the occupational component in the hearing loss and to compensate workers accordingly.

Black Lung Disease

One category of occupational ailment, black lung disease, is covered by a federal benefits program under the Federal Black Lung Act, part of the Coal Mine Health and Safety Act of 1969, as amended. The tremendous cost of compensation for disability arising from this and other occupational diseases has led to pressure to fund such coverage in whole, or in part, through federal programs.

Work-Related Impairment

The term "arising out of and in the course of employment," applied by almost every jurisdiction, is meant to clearly define the relationship between employment and an injury or disease for an employee to be eligible for workers' compensation. The phrase obviously lacks precision. Often it is quite difficult to determine whether a given set of facts can support an award of compensation.

The "course of the employment" aspect of this test refers primarily to the time frame of the injury. Virtually every jurisdiction holds that employees are considered within the course of employment—barring unusual circumstances or unreasonable conduct—from the moment they step onto the employer's premises at the start of the workday to the mo-

ment they leave at the day's end.

Although this test appears to be relatively simple to apply, it has often proved difficult. For example, what is meant by the term "premises"? Injuries that occur off premises but appear to deserve compensation lead plaintiffs to search for exceptions in the laws and encourage courts to modify the basic rules. In addition, many workers are not attached to particular premises. Even though an injury occurs off premises, as in travel to and from work, the employee may be compensated if a sufficient employment relationship can be established; perhaps the employer paid the worker for the time or expense of travel or provided a company vehicle for transportation. In these circumstances, the travel time to and from a worker's home may be included in the course of employment.

The "arising out of" segment of the test is intended to establish a causal relationship between the employment and the injury. For example, an employee cannot simply suffer a heart attack while at work and expect compensation. The person must show that the heart attack arose out of the employment. This means that at minimum (some states have more stringent rules) it must be shown that the stress and strain or exertion of the employment caused the heart attack and that it was not merely a spontaneous breakdown of the cardiovascular system.

The degree of employment relationship required varies from state to state and has been modified as workers' compensation law has evolved. Generally, it was felt that the hazard-causing injury must be peculiar to the particular employment or be increased by the employment before the injury could be said to "arise out of the employment."

Although it is difficult to determine what each jurisdiction will require to meet the "arising out of" test, two additional theories have been developed and followed. The first and more widespread is the "actual risk doctrine," which requires that the hazard resulting in injury be a risk of the particular employment regardless of whether it is a risk to which the general public is exposed. The second or "positional risk doctrine" could also be called the "but for" test. According to this theory, if the employment places the worker in a position where he or she is injured ("but for" the employment the injury would not have occurred), the injury is considered to "arise out of the employment."

Benefits

The three basic types of workers' compensation benefits are (1) loss of income, (2) medical payments, and (3) rehabilitation.

Income Replacement

Although 70% or more of recent workers' compensation cases are for temporary total disability, such cases account for only about 25% of cash benefits paid. At the same time, income benefits in the last few years to workers for permanent partial disabilities accounted for almost 66% of the total dollar amount.

Basic features. In general, the cash benefits provided for

temporary total disability, permanent total disability, permanent partial disability, and death are payable as a wage-related benefit—the weekly amount is computed as a percentage of the worker's wage. Although the benefit varies by state and by type of disability, it is commonly set at 66% to 100% of current wages. In some states, the statutory percentage varies according to the worker's marital status and the number of dependent children, especially for survivors' benefits.

For many beneficiaries, the benefit rate is limited to less than two-thirds of wages by another statutory provision—the maximum ceiling on the weekly benefits payable. Because of this ceiling, disabled workers whose wages are at or above the statewide average receive benefits below the statutory benefit rate in almost all states. However, for such individuals, benefits may exceed preinjury take-home pay because they are tax-free.

Other restrictions on benefits set maximum time periods for receiving compensation or maximum dollar amounts that can be paid. Such limitations in permanent total disability and death cases may cut off benefits to workers or survivors even though their need for income continues. Only fourteen states limit the duration of total dollar benefits to widows and orphans.

To reduce administrative costs and to discourage workers from malingering, all states stipulate in their laws that benefits are payable only after a waiting period following the report of disability. This delay in payment, which ranges from three to seven days, applies only to cash indemnity payments and not to medical and hospital care. In all states, workers who remain disabled beyond the specified minimum waiting period receive payment retroactively for that time. In more than three-fourths of workers' compensation laws, the minimum period before retroactive payment of disability benefits begins is two weeks.

Benefits by type of disability. Income benefits vary depending on whether an employee's disability is temporary or permanent, partial, or fatal. Most compensation cases concern workers who incur temporary disability but recover completely. In a majority of the states, the percentage of the state's average weekly wage for temporary total disability is $100 or more. However, dollar ceilings on weekly income benefits mean that many disabled workers are not fully compensated for their earnings lost.

Benefits for permanent total disability are for those disabilities that prevent employees from performing any work in any well-known area or industry in the labor market and that are of indefinite duration. These benefits are similar to those paid for temporary total disability benefits. In a few states, the weekly payment for permanent disability benefits is less than for temporary disability. A small number of states limit the period of benefits for permanent disabilities, usually from 6 to 10 years.

Residual limitations on a worker's earning capacity after recovering from an injury (that is, after a permanent partial disability) are calculated on a relatively complex basis. Partial disabilities are divided into two categories: "scheduled" injuries, those listed in the law such as loss of specific bodily members; and "nonscheduled" injuries, those which are of a more general nature, such as back and head injuries.

Weekly benefits for scheduled injuries are calculated as a percentage of average weekly wages, usually the same as the benefit rate for permanent total disability. The maximum weekly benefit is for the most part the same as or lower than that for total disability. Nonscheduled injuries are paid at the same or similar rate but as a percentage of *wage loss*. This represents the difference between wages before injury and wages the worker is able to earn after injury.

Scheduled benefits are paid for fixed periods that vary according to the type and severity of the injury. For example, most state laws call for payments ranging from 200 to 300 weeks for loss of an arm and 20 to 40 weeks for loss of a great toe. The maximum benefits period for nonscheduled injuries for each state is either the same as or, more generally, less than the time limit set for permanent total disability. In the majority of states, compensation for permanent partial disability is paid in addition to benefits received during the healing period or while the worker is temporarily and totally disabled.

Death benefits are intended to furnish income replacement for families who depended on the earnings of an employee killed by a work-related incident or disease. As is true for other types of benefits, the amount of survivor benefits and the length of time they are paid vary considerably from state to state. If the survivor is a widow without dependents, benefits computed as a percentage of the deceased worker's wage often are less than those paid for permanent total disability. If the worker had dependent children, the benefits in many states will be augmented to help support them. In most states, the duration of these benefits is unlimited, although nine states have established limits that range from 7 to 20 years. In a number of states, payments to widows continue usually as long as the women do not remarry and to children until they are no longer dependent, usually to age 18. In many states, benefits to children continue to age 23 or 25 if they are in school. Benefits may be terminated earlier in four states that also limit total dollar benefits.

In addition to benefits for widows, widowers, and children, some states pay survivor benefits to dependent invalid spouses, parents, or siblings of the dead worker. Burial expenses are covered in all states.

Medical Benefits

For many years, medical care disbursements provided by workers' compensation have comprised about one-third of total outlay for benefits. Care includes first aid treatment, services of a physician, surgical and hospital services, nursing and drugs, supplies, and prosthetic devices. Some large employers, in addition to first aid facilities, employ staff physicians for workers. Most employers insure their medical care responsibility as they do the income benefits under workers' compensation.

Every state law requires the employer to provide benefits

for medical care to the injured worker. In most jurisdictions, such treatment is provided without limit either through explicit statutory language or administrative interpretation. In the few states that limit the total medical care by specified maximum dollar amounts or maximum payment periods, management can decide to exceed the initial ceiling if circumstances warrant such an action. Also, if specified types of injuries or disease are denied cash benefits, medical care for these conditions also is denied.

An issue many employers face in providing medical benefits for workers' compensation is how to choose the physician who is to furnish care. Almost half of the states give the employer the right to designate the physician. In practice, the insurance company of the employer ordinarily will select the physician because the insurer is the one who handles the claim for benefits. When the doctor is chosen in this way, the medical care furnished may be more highly skilled and effective because of the selected physician's specialized experience. On the other hand, workers often feel that their own family physician will place more emphasis on their personal health and well-being. They believe other considerations, such as company interests or insurance costs, may influence a physician they do not select.

Two other sources of medical benefits for the disabled worker are Social Security and private disability programs provided as part of fringe benefit packages by larger employers, in particular. Any disabled worker whose disability is documented as lasting at least 12 months or results in death may receive (or survivors will receive) disability benefits if eligible under Social Security rules. The combination of Social Security benefits and workers' compensation payments cannot exceed 80% of the disabled worker's earnings prior to disability. These benefits are financed out of Social Security.

Even though qualifying for workers' compensation, disabled workers may not be able to meet Social Security's requirements for proving they cannot perform any kind of gainful work.

Private disability insurance programs usually are coordinated with workers' compensation plans. Frequently, a private disability program requires that employees file workers' compensation and Social Security disability claims before they can qualify for private carrier benefits.

Rehabilitation

Along with industrial safety, medical care, and cash compensation, rehabilitation of workers is recognized, at least theoretically, as one of the primary goals of the workers' compensation system. The most widespread benefits offered through workers' compensation laws to restore a worker to the fullest economic capacity are the special maintenance benefits authorized in more than half the states. These benefits usually are paid (sometimes in addition to the regular disability compensation) for various training, education, testing, and other services designed to speed the injured person's return to work.

Probably the main source of retraining and rehabilitation

is the federal-state vocational program. The federal government, through the Federal Vocational Rehabilitation Act, is a major resource for funding such programs. Facilities operated by this program accept both individuals with work-related disabilities and others injured off the job. In all states, these institutions are directed by state vocational rehabilitation agencies. They provide medical care, counseling, training, and job placement. Unfortunately, not all workers' compensation cases referred for vocational rehabilitation can be accepted promptly, and many others are never brought to their attention. As a result, workers who may have been able to return to some type of job remain disabled.

One notable drawback preventing full use of available rehabilitation facilities is the complex, adversarial proceedings for determining a worker's right to benefits. Because the decision about whether benefits should be awarded for permanent partial or total disability (and how large the benefits should be) is based primarily on the worker's inability to work, the claim may offer a strong incentive for the person to put off rehabilitation. Further, in the many compromise settlements, the employer's (or insurer's) main goal is to pay an agreed amount of money and prevent any future liability for medical, vocational, or other needs arising from the injury. Such settlements also work against a full-fledged effort to restore the worker to full health and productivity.

COVERED EMPLOYMENT

Employers and workers alike generally agree on the need for universal compensation coverage for employees. For employers, such coverage represents a relatively inexpensive way to protect themselves against potential lawsuits should workers be injured or killed on the job. For the worker, compensation coverage represents an important part of one's protection against income loss and medical expenses.

Over time, coverage has increased, and the public has come to accept the rationale for including all types of employment in the workers' compensation system. Nevertheless, for the past 30 years, the proportion of covered civilian wage and salary workers has remained around four-fifths of all U.S. workers.

Limitations on Coverage

In view of the fact that some of the exemptions or exclusions in many state laws have persisted to this day, it may be helpful to review some of the reasons behind the original limitations. Nearly all state acts were prepared and enacted in the face of constitutional challenges and the outright opposition of certain business or government interests. Thus, each act was the result of political compromises.

Initially, workers' compensation was hailed as an innovation that would introduce greater certainty into the calculation and payment of benefits in contrast to the common law system. Under common law, workers could sue employers and, if successful, might be assured of an adequate payment; however, those who lost would be left with noth-

ing but debts. To reduce this risk, the workers' compensation law specified the benefits that would be paid to all regardless of fault. Although the outcome of workers' compensation cases is far more certain than the ordinary suit where negligence must be shown, the law is not "automatically" applied.

In part, this remaining uncertainty arises from the wide variety of permanent partial disability cases that the schedules do not cover satisfactorily. Two factors usually prompt compensation litigation. One is uncertainty about whether an accident arose out of and in the course of employment; the other is the extent of disability. As workers' compensation comes to encompass more of the ailments to which the general population might be susceptible, it becomes difficult to separate impairments that are work related from those that are not. In addition, it requires an exercise of legal skills and medical judgment to assess the extent of disability in such difficult cases as occupational diseases, injuries to the soft tissue of the back, heart conditions, or situations where the only evidence before the commission may be a subjective complaint.

ADMINISTRATION

The goal of workers' compensation is to provide for quick, simple, and inexpensive determination of all claims for benefits and to provide such medical care and rehabilitation services as are necessary to restore the injured worker to employment. Nearly all of the states have agencies to carry out these administrative responsibilities.

Objectives

An agency's responsibilities include close supervision over the processing of cases. The primary objective is to ensure that all parties comply with the law and to guarantee an injured worker's rights under the statute.

A key goal of the agency is to see that the injured worker gets the full benefit due. To do so, the agency must follow an injury case from the first report to the final closing. Some states not only check the accuracy of total payments but also require signed receipts for every compensation payment. Some require the employer to file a final receipt that itemizes the purpose of each part of the total benefits outlaid to facilitate a complete audit of individual payments.

Frequently, however, the legislation itself states that a workers' compensation agency must operate on the assumption that each injured worker is responsible for securing his or her rights and that the agency's primary function is to adjudicate contested claims. Even where the law does not favor this policy, lack of staff may force the agency into this restricted role.

Although many workers are unfamiliar with the provisions of their state's workers' compensation act, in only a few states does the agency administrator (as soon as possible after the injury is reported) advise the worker of his or her rights to benefits, medical and rehabilitation services, and assistance available at the commission's office. Too many states fail to insist that employers report accidents promptly, pay benefits on time, or submit final reports that list the amounts paid and how these amounts were computed. Although prompt reporting is usually required, some states impose no penalty if employers violate this statute.

Handling Cases

Workers' compensation claims may be either uncontested or contested. In uncontested cases, the two main methods followed are the direct payment system and the agreement system.

Under the *direct payment system*, the employer or insurer takes the initiative and begins paying compensation to the worker or dependents. The injured worker does not need to enter into an agreement and is not required to sign any papers before compensation starts. The laws prescribe the amount of the benefits. If the worker fails to receive this compensation, the administrative agency can investigate and correct any error. Jurisdictions whose laws provide the direct payment system include Arkansas, Michigan, Mississippi, New Hampshire, Wisconsin, and the District of Columbia; this feature is also provided for in the Longshoremen's and Harbor Workers' Compensation Act.

Under the *agreement system,* in effect in a majority of the states, the parties (that is, the employer or its insurer, and the worker) agree upon a settlement before payment is made. In some cases, the agreement must be approved by the administrative agency before payments start.

In contested cases, most workers' compensation laws provide for a hearing by a referee or hearing officer. The statutes also provide for either the worker or employer/insurer to appeal a decision of the referees or hearing officer to the commission or appeals board and from there to the courts. As the administrative agency usually has exclusive jurisdiction over the determination of the facts in a case, appeals to the courts usually are limited to questions of law. In some states, however, the court is permitted to consider issues of both fact and law in an appeal.

REHABILITATION

Most employees injured in work accidents return to their jobs after minor medical attention with little if any worktime lost. If the effects of the injury are temporary, the incident usually fades from memory. Even those who suffer days or weeks of disability and possibly endure substantial medical treatment may find the injury is not permanent. Although the loss of income and the medical expenses are distressing, eventually, when workers resume their jobs, they recover economically as well.

Unfortunately, a minority of those injured—as much as 10% of the total—experience injuries that disrupt their lives. Even when these workers receive effective medical care and eventually return to productive jobs, their lives are permanently changed by the event. Injuries for some are so severe that prolonged medical treatment and convalescence fail to restore them completely to full health and functioning. Re-

sidual disabilities prevent them from performing their former jobs. Only retraining and education, combined with special assistance, offer a prospect for future employment.

Some never return to work. If they do not die from their injuries, they live with such severe disabilities that they barely can manage for themselves. Often, the most that health services can do is to lighten the burden on those who must care for these persons. Treatment for workers whose livelihood is threatened by work-related impairments consists of medical rehabilitation and vocational rehabilitation.

Medical Rehabilitation

Each medical rehabilitation program, whether set up by an insurance company or workers' compensation agency, contains its own requirements for treatment, qualifications for eligibility, and definitions of service.

A disabled worker who requires medical rehabilitation receives whatever medical care is needed to treat the impairment and to restore lost function. The worker may report first to the plant nurse or physician for immediate attention. If the injury is serious, the person may then go to a hospital. Workers' compensation laws often obligate employers and insurers to pay the costs of rehabilitation medical care. Costs can be covered by having health service workers on salary, by contractual arrangement with health personnel, or by payment of hospital and doctor bills. The insurer may or may not have much influence in the selection or course of treatment.

For injuries associated with chronic disabilities, the insurer usually attempts to control the selection of the rehabilitation services, frequently by directing the worker to a particular specialist or facility with a particular expertise. Often the insurer pays for transportation to the specialist or facility as well as for rooms during treatment. Some insurance companies operate their own rehabilitation facilities, under individual or joint ownership, with medical personnel on salary, at least part time. When insurers contract to share rehabilitation programs or facilities, they may pay expenses case by case or through a rental agreement.

When informed of the potential need for rehabilitation, some agencies do little more than notify the worker and insurer that medical rehabilitation is worth considering. Other agencies conduct formal evaluations of the need for further medical care and recommend action. They seek to convince disabled workers of the wisdom of rehabilitation. When the workers agree, the insurers can be required to finance the care.

Vocational Rehabilitation

Vocational rehabilitation prepares the injured worker for a new occupation or for ways of continuing in an old one. Usually, vocational rehabilitation is assigned when medical treatment fails to restore the worker to the job held when the individual was injured. The worker's injury may be so severe or the work requirements such that even residual impairment can prevent the person from performing effectively. These workers need training to overcome or to

compensate for their limitations. Many may even enter new occupations. In general, however, the more effective the medical rehabilitation, the less need for vocational rehabilitation.

The current definition of vocational rehabilitation makes a greater distinction between it and medical rehabilitation than is necessary. Although the difference in kinds of treatment offered seem clear enough—retraining as opposed to medical care—the two categories often overlap to a considerable degree. For example, in the public vocational rehabilitation programs in each state, services include medical diagnosis and evaluation, surgery, psychological support, the fitting of prostheses, and other health-related services along with education, vocational training, on-the-job training, and job placement.

The two programs also blend on an employee's medical records. Recordkeeping by workers' compensation insurers does not separate claimants who receive medical rehabilitation from those who receive vocational rehabilitation, although some distinguish between medical rehabilitation and acute medical care. In contrast, records kept by workers' compensation agencies usually separate vocational rehabilitation from other benefits.

Injured workers who need vocational rehabilitation are served by several means. An employer or insurer may channel the worker to whatever sources they think will provide satisfactory service. Some workers are referred to the public vocational rehabilitation program where services may be financed by taxes, although insurers may reimburse the public agency. Other insurers direct workers to private facilities where vocational training is conducted by technical schools or on the job. The cost of these services are always paid by insurers.

As with medical rehabilitation, some workers' compensation agencies support vocational rehabilitation and often will direct the worker into a program, if the insurer fails to do so. Several jurisdictions select candidates either in conjunction with screening for medical rehabilitation or separately. Workers who have serious injuries or permanent disabilities, or who are receiving extended compensation payments are reviewed by the agency for referral to the state's public vocational rehabilitation agency or to the insurer.

Some workers obtain vocational rehabilitation through their own efforts. If no one refers them, they may go directly to the public vocational rehabilitation office. Since 1920, the federal government and the states have cooperated financially (80% federal and 20% state funding) in supporting a vocational rehabilitation program that can be used by anyone with a vocational disability. Rehabilitation counselors, who usually determine a referral's acceptability, simply verify the disability without regard to its cause and explore ways of overcoming a worker's limitations. If the candidate shows relatively good prospects, the counselor designs a program to help restore the worker as fully as possible. For those who are unable to return to a paying job, the objective of vocational restoration may be to help them care for themselves

and to free other members of the family to earn wages.

Once workers are established in a vocational rehabilitation program, they are assigned whatever resources the counselors think best fit their needs. Generally, the resources are not owned and operated by the vocational rehabilitation agency but are offered by private vendors or other public agencies. For example, a worker may be sent to a private rehabilitation center or school, enter a sheltered workshop such as those run by Goodwill Industries of America, or be enrolled in a public institution.

DEGREE OF DISABILITY

The question of determining the extent of disability is perhaps responsible for more litigation than any other single issue in workers' compensation. It requires not only correct application of legal principles but also evaluation of facts, subjective complaints and opinions, and attempts to predict the future.

As a general proposition (some jurisdictions use different terminology and slightly different classifications), disability can be classified in one of four categories: temporary total disability, temporary partial disability, permanent partial disability, and permanent total disability.

Temporary Partial and Total Disability

Temporary total disability occurs when an injured worker, temporarily incapable of gainful employment, has a good prospect of improving to the degree that he or she will be able to return to work either with no disability or with only a partial disability. Temporary partial disability is similar to temporary total in that it assumes the worker's physical condition has not stabilized and is expected to improve. The difference lies in the worker's current abilities. When temporarily partially disabled, the worker is capable of some employment, such as light duties or part-time work, but is expected to improve and regain much of his or her former capability.

Permanent partial disability means the injured worker has attained maximum improvement without full recovery. The worker has benefited from medical and rehabilitative services as much as possible but still suffers a partial disability. Permanent total disability represents the same situation except that the disability is total.

The determination of temporary disability, either total or partial, is the least difficult. It requires merely evaluating the employee's present physical condition in light of the work opportunities available. In practice, evaluation of temporary disability is concerned only with the ability of the employee to return to work for the current or last employer. It is assumed that at some point an employee will be able to return to work for this employer. Given the difficulties involved in obtaining employment for workers still under medical care—and that any new employment probably will be temporary—most adjudicators have either expressly or in practice adopted the position that unless the worker can return to the last job held, or can be supplied with tempo-

rary light or part-time duties with this employer, he or she remains temporarily totally disabled. This is the case even though the worker might be able to perform another job whose duties are within the individual's temporary physical limitations.

Permanent Partial Disability

Determining the extent of permanent partial disability depends on what the jurisdiction chooses to label "permanent partial disability." Three theories are used to establish guidelines for the payment of workers' compensation benefits for such disability. Their underlying philosophies differ somewhat, as do the factors to be considered in applying each one to a specific situation.

"Whole-person" theory. This theory is concerned solely with functional limitations. Here, the only considerations in assessing a case are whether the worker has in fact sustained a permanent physical impairment and, if so, to what extent it interferes with the person's usual functions and abilities. Age, occupation, educational background, and other factors are not considered.

"Wage loss" theory. The aim in applying this theory to determine what wages the worker would have been able to earn had the permanent impairment not occurred. If the worker's earnings dip below the estimated wage figure because of the impairment, he or she is paid compensation equal to some percentage of the difference between the wages that would have been earned and those actually earned. Here the degree of physical impairment is of little or no importance. The only concern is the actual wage loss incurred and whether it is due to the impairment.

"Loss of wage-earning capacity" theory. This theory requires a peek into the future. After the worker has reached maximum physical improvement, many factors (such as impairment, occupational history, age, sex, educational background) are considered in an effort to estimate, as a percentage, how much of the worker's potential earning capacity has been destroyed by a work-related impairment. The worker is awarded benefits on the basis of this computation. Benefits may be paid at the maximum weekly rate for a limited number of weeks or they may be based upon a percentage of the difference between wage-earning capacity before and after disability, to be paid up to a preestablished dollar or time limit.

Combinations. These three basic theories are capable of being used also in combination. For example, some states expressly or in practice use either the "wage loss" theory or the "loss of wage-earning capacity" theory but also provide a benefit floor determined by the worker's actual medical impairment. Thus, an employee who sustains a permanent impairment but no loss of wages or of wage-earning capacity would still receive some permanent disability benefits.

The use of schedules has relieved the tedious or controversial aspects of rating disabilities for a significant proportion of permanent partial disability cases. The typical schedule covers injuries to the eyes, ears, hands, arms, feet, and legs. It states that for 100% loss (or loss of use) of that

body part, compensation at the claimant's weekly rate will be paid for a specified number of weeks. If loss or loss of use is less than total, the maximum number of weeks is reduced in proportion to the percentage of loss or loss of use. Only physical impairment is considered, and the effect of the injury on wages or wage-earning capacity is ignored. If an injury is confined to a scheduled body part, the benefits provided by the schedule are exclusive, even though disability rating on a wage loss or loss of wage-earning capacity basis might result in greater benefits. Although this statement is true generally, some states provide additional benefits in the following circumstances: (1) if using one of the other theories results in higher benefits being paid, (2) if diminished wage-earning capacity continues after the scheduled amount is paid, (3) if the scheduled injury results in permanent total disability, or (4) if several scheduled injuries are sustained in the same accident.

Use of the schedule may also be avoided by showing that the effect of the scheduled injury, such as radiating pain, extends into other parts of the body. A few jurisdictions limit the use of schedules to amputation or 100% loss of use of a body part, as opposed to partial loss of use. Another group not only makes the schedule exclusive for permanent disability awards but requires that the weeks for which benefits are paid during the healing period be deducted from the number of weeks authorized by the schedule before an award is made for permanent partial disability.

Most U.S. states operate primarily on the "loss of earning capacity" theory. Even where statutory language seems to indicate clearly that only functional impairment is to be considered, the courts have managed to hold that loss of earning capacity is the real consideration. Even the use of schedules has been justified on an earning capacity basis as merely a legislative determination of presumed wage loss resulting from the impairment listed in the schedule.

Permanent Total Disability

Permanent total disability evaluation is, in most respects, merely an extension of the determination of permanent partial disability. In fact, it is a part of the same process, as the fact finder's only additional task is to determine whether the worker's wage-earning capacity is so destroyed that he or she is unable to compete in the job market.

Two aspects of the permanent total disability question warrant special attention. First, most states use certain presumptions that make the fact finder's job much easier. For example, it may be presumed that the loss of sight of both eyes or the loss of any two limbs will constitute permanent total disability. This relieves the fact finder of the difficult task of evaluating all the factors previously mentioned for other forms of disability. In some cases, these presumptions may be refuted by providing evidence that the worker has some wage-earning capacity, or the presumptions may be applied only for a limited period of time.

Second, the concept of permanent total disability must be defined. The injured employee need not be completely helpless nor unable to earn a single dollar at a job. The person's limitations need only preclude competing in the open job market and be such that no stable job market exists for a worker with this disability.

Insurance Incentives

The asserted insurance incentive of workers' compensation is based on the merit-rated pricing policy. "Merit rating" includes both experience rating and retrospective rating systems. All state funds use merit rating of some sort. Most states permit private insurers to rate employers on merit, although they are not required to do so. Under this procedure, the firm is charged a premium based on the dollar amount of claims for which it is liable. Consequently, a merit-rated firm has an incentive to reduce the amount of its claims through loss control measures. The strength of this incentive has been challenged. Only about one-fourth of insured firms, usually large ones, are eligible for merit rating.

The yearly accident record of firms with only a few employees is not a sufficiently reliable indication of their characteristic experience to be considered in establishing premium rates. On the other hand, merit-rated firms account for 85% of the dollar volume of premiums paid. In addition, self-insured firms, which pay approximately 14% of all benefits, are implicitly merit rated. If incentive effects are inherent in experience rating, they are not available to a large number of small firms and their employers.

Firms not eligible for merit rating are class rated. Under this procedure, all employers engaged in similar business operations within a state pay the same rate per $100 of payroll. These employers do not have a strong incentive to reduce the rates paid by their industry. The only accident prevention incentive generated for individual employers within an industry is that, as poor risks, they may be unable to obtain workers' compensation coverage. Other considerations include the relative quality of the safety and claims services provided by insurers, by management service organizations, and by employers themselves; by tax factors; and by the opportunity cost of paying an insurer a premium instead of paying losses and expenses as they occur.

Safety Incentives

The greatest contribution a safety and health professional can make to a firm's success is to work with managers throughout the organization in safeguarding employees from disabling occupational injury and disease. These injuries and diseases arising on the job hamper both the employee and those with whom he or she works in meeting various job-related objectives: workers miss target dates set for projects, their performance evaluations are disappointing, and often they must forgo opportunities for pay increases and promotions.

The safety and health professional who seeks to promote safety and health in the workplace should recognize the reasons that each worker, supervisor, and executive wants to succeed in his or her occupational efforts. No matter what these reasons may be, achieving career goals requires that individuals and their colleagues remain productive and effi-

cient on the job. Thus, the perceptive safety and health professional can demonstrate to all employees that avoiding disabling injuries and diseases both on and off the job is in their own—not just their employer's—best interests. Doing one's job safely promotes personal and career success.

MANAGING A WORKERS' COMPENSATION PROGRAM

In the final analysis, the total costs of workers' compensation, in terms of both money and human suffering, are the organization's responsibility. Like every other item of cost, workers' compensation expenditures need to be managed carefully. Cost control should be a top management priority to ensure that dollars are spent as prescribed by the law. Injured employees are entitled to benefits, but businesses do not have to cover questionable or fraudulent claims.

To be successful within the legal limits requires that companies have a plan of action and treat each individual case fairly and consistently. Firms should focus on three goals for their workers' compensation program: (1) to prevent accidents, (2) to control costs, and (3) to respond to accidents promptly and efficiently.

When an organization's management style and philosophy demonstrate cooperation, commitment, teamwork, and communication between labor and management, the firm has set a tone for fairness when accident cases occur. Employees in such firms tend to feel more fulfilled both as workers and persons. Their confidence and self-discipline will be reflected in their work habits and productivity. As a result, such a working environment tends to prevent accidents and promote greater worker safety and productivity.

The organization's philosophy should also provide the basis for administering the workers' compensation program. The following are some points usually addressed in program management.

Hiring

A thorough interview and screening process can yield evidence of applicants' existing physical condition(s) and past medical conditions that may suggest future compensation liabilities. The value of preemployment physical examinations depends on the job, the employer's past experience, the physician's evaluation, and other factors. Although the objective is to hire the best people—without discrimination—companies must realize that at times they will inadvertently hire problem employees.

During the screening process, the interviewer should be trained to elicit complete answers from applicants and to probe more deeply when the responses are not satisfactory. The interviewer should make sure that all information is considered in the hiring decision.

First Report of an Accident

It is vitally important that workers and management report all accidents promptly. Employees must understand the value of the employer knowing when something has occurred so that immediate, efficient action can be taken.

Management should ensure that accident investigation reports are completed promptly and added to existing files. Employees should be accompanied if injured, ill, or if the person must be referred to outside medical services. This action demonstrates the company's concern for its employees.

There are advantages to having a centralized accident reporting system. It can provide needed information quickly, help select appropriate medical care for the worker, and offer better monitoring and control of medical treatment.

Doctors and Medical Institutions

The company must choose doctors and medical institutions to provide care to injured employees. In most situations, the employee can see his or her personal physician; but in many cases, the organization provides at least the basic medical services. Even when outside or in-house special services may be needed, the organization may select the provider. Before selecting medical services, the employer should consult with other firms regarding their experience with specific physicians or institutions and seek second opinions on specific cases. Even though once in a while an employee may file a false claim, proper supervisory procedures and open communication about employee injuries and illnesses should eliminate most if not all fraudulent claims.

Rehabilitation

Getting a disabled worker back to work as soon as possible is a major goal of medical treatment and workers' compensation coverage. Adequate follow-up of those on leave is often necessary to achieve this objective. The added cost for rehabilitation is usually money well spent. The employer should find out how the prospective physician handles paperwork, bills, reports, patient care and treatment, and facilitates patient activities. Some physicians are likely to be more suitable to a company's needs than others.

Follow-up

The employer should review ongoing bills, claims, and reports to make sure that services are being provided as specified in compensation acts. Management should maintain a file on each case until it is closed. Accident information arising from each incident must be fed back into the safety and health prevention program.

The employer should treat workers' compensation employees or agents fairly and consistently. If the company buys its insurance, it should not hesitate to seek bids from several different companies each year. The organization should keep in close touch with injured employees to see that follow-up and rehabilitation services are being properly administered.

An effective safety and health program has been shown to prevent accidents, lower workers' compensation claims, prevent some claims, and reduce the organization's overall costs. In addition, an analysis of high-risk hazards is a valuable tool in accident prevention, Typically, 20% of the acci-

dents cause 80% of the costs. Hazard analysis and past accident experience can identify which hazards represent the highest risks and help an organization reduce, control, or eliminate these hazards. At the least, the company will be able to predict with some certainty the costs of accidents resulting from these hazards.

SUMMARY

- Early efforts to introduce workers' compensation in the United States failed until 1911 when Wisconsin passed the first compensation law that remained effective. Five years later the U.S. Supreme Court declared workers' compensation to be constitutional. Currently, all 50 states and all U.S. territories and possessions have such laws. Compensation laws are undergoing continual reform as both labor and management seek to make coverage more effective and less costly.

- Most compensation laws seek to replace wages lost by disabled workers, pay for medical and vocational rehabilitation, reduce litigation, encourage companies to ensure safe working conditions, and promote accident prevention and investigation.

- Income replacement should be adequate, equitable, prompt, and sure. Such requirements guarantee that workers will share some portion of the loss, that all workers will be treated fairly, and that benefits will be paid as soon as possible after the injury or illness is brought to the company's attention.

- The final goal of workers' compensation is to prevent and reduce occupational accidents and illnesses. Compensation laws are meant to benefit employers who adopt safe work practices and to penalize those who do not.

- Although compensation laws are compulsory or elective, in practice all laws can be considered compulsory. None of the laws covers all workers, although most laws apply to both private and public employment. Excluded workers include those in farming, domestic service, self-employment, some interstate commerce, and all federal and many state government employees (who are covered under government programs).

- In addition to workers' compensation benefits, employees can sue employers under (1) an expansion of the dual capacity doctrine and (2) the intentional tort exception. Under these concepts, employers are liable for injuries to workers caused by the employer's product or service available to the public or caused by the employer's deliberately harmful, reckless, or grossly negligent behavior.

- Workers' compensation covers only work-related injuries or illnesses, determined by the "personal injury" and "by accident" requirements on one hand and the "arising out of" and "in the course of" requirements on the other. These tests are used to determine which injuries and illnesses are compensable.

- Three basic types of workers' compensation benefits are (1) loss of income, (2) medical payments, and (3) rehabilitation expenses. Income replacement begins after an es-

tablished waiting period and ranges from 66% to 100% of a worker's current wages. Benefits are generally established for a maximum dollar value and time period, except in cases of a worker's death when benefits may extend for the life of the surviving spouse. Income benefits vary according to the type of disability—permanent or temporary, partial or total.

- All employers are required to provide medical benefits for employees to cover immediate and long-term care. In most cases, the employer or insurer chooses the health care provider(s), although workers may also use their personal physicians. Medical benefits can be supplemented by Social Security and private disability programs.

- Since the first compensation laws were passed, the number of covered workers has remained about the same—80% to 88% of the workforce. Employers view compensation coverage as a relatively inexpensive way to protect themselves against potential lawsuits, while employees view the coverage as protection against income loss and sizable medical expenses.

- Nearly all states have agencies to administer workers' compensation programs. The agencies closely supervise all cases and ensure that all parties comply with the law to guarantee an injured worker's rights under the statute. Agencies advise workers of their rights, require employers to maintain proper records, and report all accident cases promptly. However, these agencies are only as effective as they are vigorous in enforcing the program.

- In uncontested workers' compensation claims, benefits are paid out through either the direct payment system or the agreement system. In contested cases, claims are heard before a referee or hearing officer, with provisions for an appeal of any decisions rendered.

- Rehabilitation work may be medical or vocational or a combination of the two. Care is usually coordinated among the company, disabled worker, and various state agencies to select the physician(s), medical facilities, and vocational assistance that will provide the best rehabilitation services. Vocational rehabilitation focuses on restoring the worker to former capacity or training the person for new employment. Medical and vocational rehabilitation services and coverage often overlap.

- Determining the degree of a worker's disability is often difficult. Jurisdictions generally use one of four classifications: temporary total disability, temporary partial disability, permanent partial disability, and permanent total disability. Benefits are paid according to the severity and duration of the disability.

- Various theories, such as "whole person," "wage loss," or "loss of wage-earning capacity" theory, are used to help determine the degree of a worker's impairment and remaining capacity to find employment. Most states operate on the "loss of earning capacity" theory. The use of schedules can reduce the time and effort it takes to evaluate a worker's disability and to award compensation.

- Many companies are given insurance and safety incentives to reduce accident and illness rates. Merit-rating

systems and occupational health and safety programs are effective ways to induce organizations to make safety and health concerns a top priority.

- A company's management style and philosophy can set the tone for a fair, consistent management of a workers' compensation program. The firm's goals should be (1) to prevent accidents, (2) to control costs, and (3) to respond to accidents promptly and efficiently. In particular, the program can address compensation issues in hiring, accident investigation and reporting, selection of doctors and medical institutions, medical and vocational rehabilitation services, and follow-up care. In the final analysis, effective accident prevention programs by employers are the best control since there is no injury and no workers' compensation.

REFERENCES

Chamber of Commerce of the United States, Washington, DC. *Analysis of Workers' Compensation Laws,* 1990.

Clifford JA. *Workmen's Compensation—New York.* Binghamton, NY: Gould Publications, 1979.

English W. *Strategies for Effective Workers' Compensation Cost Control.* Des Plaines, IL: American Society of Safety Engineers, 1980.

Gaunt LD and McDonald ME. *Examining Employers' Financial Capacity to Self-Insure Under Workmen's Compensation.* Atlanta: Georgia State University, College of Business Administration, 1977.

La Dou J, ed. *Occupational Health and Safety.* Chicago: National Safety Council, 1992.

Martin RA. *Occupational Disability—Causes, Prediction, Prevention.* Springfield, IL: Charles C. Thomas, 1975.

National Council on Compensation Insurance, 200 East 42nd Street, New York, NY 10017. Rate and rating plan manuals.

National Safety Council, 444 North Michigan Avenue, Chicago, IL 60611. *Accident Facts* (annually).

9

Product Safety Management

Although no one can completely prevent product misuse by customers, the producer or manufacturer can do a great deal to minimize or defend against product liability claims. The principal way to achieve this goal is to sell only reasonably safe products and, when necessary, to include instructions for their proper use. The key to achieving a reasonably safe and reliable product and, at the same time, reducing a company's product liability exposure is to incorporate product safety into all aspects of the design, manufacturing, and marketing stages. This is done by establishing and auditing an effective product safety management (PSM) program.

This chapter provides an overview of several interrelated issues regarding product safety and liability prevention management and the role of the product safety coordinator/auditor. A more comprehensive treatment of the same subject can be found in the National Safety Council book, *Product Safety: Management Guidelines.*

A PSM program must include a comprehensive process of evaluating user injury risks from all product-related sources. As a result, many people from various parts of the organization need to be involved in this risk-management process. From the start, management must perform ongoing risk evaluations and determine what corrective actions are necessary if existing controls prove inadequate.

To establish and audit a PSM program, management must begin by (1) selecting a program coordinator and (2) selecting a program auditor. The same person typically serves as both the program coordinator and auditor, depending on the size, needs, and structure of the individual company.

ESTABLISHING AND COORDINATING THE PROGRAM

Regardless of a company's size, establishing and coordinating a satisfactory PSM program requires a comprehensive systems analysis of all operations and production stages, from design through manufacturing, quality assurance, and shipping. To keep all these functions organized and coordinated, someone must oversee the program, either alone or with a committee's assistance.

Program Coordinator

Because the success of the PSM program requires the cooperation and coordination of all departments, the program head or committee chairman should be carefully chosen. The individual must be able to exert stringent control over all phases of product development from initial product design through eventual product sale and distribution. The extent of the individual's role will depend, of course, on the size of the company. In a small to medium-sized company, management will select either a PSM manager or auditor on the basis of the person's broad experience in areas of safety technology.

It is not important whether the position is full time or part time, and, if part time, what the coordinator's primary function is within the organization. What is important is the PSM coordinator's level of authority. Can the coordinator

take needed action without having to go through several supervisory levels? Does the coordinator have easy access to top management? Is the coordinator permitted to implement most program plans or suggestions?

If the organization has a PSM committee to coordinate program activities, the head of the committee must have a similar level of authority. Likewise, committee members should have the authority to speak for their respective departments.

Responsibilities. A program coordinator must have sufficient authority to take action and to carry out the following responsibilities:
- function as a staff member for corporate management
- assist in setting general PSM program policy
- recommend special action regarding:
  — product recall
  — field modification
  — product redesign
  — special analyses
- conduct and review complaint, incident, or accident analyses
- coordinate appropriate PSM program documentation
- ensure an adequate flow of verbal and written communications
- develop sources of product safety and liability prevention data for use by operating personnel
- maintain liaison with business, professional, and governmental organizations on matters pertinent to product safety and liability prevention
- conduct PSM program audits, where appropriate.

Ground rules. The following ground rules must also be clearly defined by management if the PSM program coordinator is to be effective:
- The purpose of the PSM program coordinator must be clearly defined.
- The authority and responsibility of the PSM program coordinator must be clearly specified by top management and understood by the PSM coordinator.

As previously noted, a PSM program involves most of the departments in a company and requires the coordination of many disciplines. As a result, the PSM program coordinator must ensure a thorough and systematic approach when implementing the program.

Committee coordination. Whenever a PSM program committee is used, a company should keep the group's size within manageable limits, generally no more than five or six members. In a large corporation, management may find it better to appoint a small corporate committee or individual corporate coordinator and assign a separate program coordinator for each corporate division or department.

The committee is especially important when the company is launching a new product safety effort within the organization. The role of the committee is to recommend policies and guidelines, offer advice to top management, and audit product safety performance.

Committee membership varies according to the organizational structure of a particular company. However, key

members of the committee will almost always represent the following areas:
- product safety
- design or engineering
- manufacturing
- purchasing
- quality assurance
- service/installation
- legal.

In addition, sales, marketing, advertising, insurance, personnel, public relations, plant safety, and purchasing department representatives should be designated. These individuals can serve as consultants to the PSM program committee when their expertise is required.

Program Auditor

Program audit is perhaps the most important factor in a successful product safety management program. Because the role is so important, discussion of the program auditor's responsibilities is the main topic of this chapter.

Although the program auditor can be a private consultant or an insurance company product safety or loss control specialist, a company's PSM program probably will be most effective if the program auditor is a member of company management. Again, the program auditor may also serve as the program coordinator; this depends on the needs of the company.

The program auditor's main duty is to evaluate the organization's PSM program activities in relation to actual and potential exposures. This evaluation determines how the organization can prevent product-related losses by comparing what is currently being done against what should be done.

There is a distinction between a product safety loss control evaluation and a program audit. The loss control evaluation is a review of a company's product safety efforts compared to what the company should be doing. In a typical example, based on this evaluation, the organization establishes a safety and health program and identifies and initiates appropriate safety procedures. This set of procedures defines the scope of the program and indicates who is responsible for which tasks.

The program audit, on the other hand, involves determining if, in fact, the procedures are being followed. Many times this distinction is overlooked or becomes unimportant because the product safety manager evaluates, audits, and recommends necessary changes on a continuing basis. Formal evaluations, of course, can be performed periodically by the product safety manager and/or a safety committee.

Improve program effectiveness. When auditing the organization's product management control system, the program auditor should strive to increase PSM program effectiveness. This can be accomplished by doing the following:
- Determine the organization's potential product liability exposures.
- Determine the organization's PSM program deficiencies (for example, how well can an organization eliminate or control any product liability exposures found?).

- Develop concise, realistic procedures to correct or minimize product liability exposures.
- Motivate management to implement the proposed corrective measures (e.g., to reduce losses, improve good records, improve the level of product control, comply with government regulations).
- Communicate all PSM audit information to management for review.

Product safety management program audits can be successfully conducted even when the organization does not have a formal program, as long as the auditor has sufficient knowledge of product management control practices.

The level and quality of coordination among departments should be measured by the auditor while evaluating each departmental activity through observation and questioning.

System analysis. An organization's PSM program philosophy should be one of continuous overview or systems analysis of the entire product management control system. In performing the systems analysis, the PSM program auditor must:

- Evaluate management's commitment to product safety and liability prevention.
- Determine the effectiveness of the organization's PSM program coordinator as direct evidence of management's interest in designing and manufacturing a safe, reliable product. The criteria should not be whether there is a PSM committee or full-time PSM program coordinator. The organization may not need either one. The criteria should be whether management has assigned someone, regardless of title, responsibility for coordinating all activities associated with the design and manufacture of safe, reliable products.
- Evaluate the PSM program coordinator's effectiveness in organizing these activities. When one person serves as both the auditor and coordinator, the individual should attempt to be as objective as possible in evaluating the program coordination functions.

Duties. The PSM program auditor must determine the organization's ability to manufacture safe, reliable products and must evaluate its capabilities to control existing product exposures, eliminate potential exposures, and identify uncontrollable exposures. In addition to seeking purely quantitative responses, the program auditor must form general impressions of the organization's overall product safety and liability prevention program activities. For example, the program auditor should:

- Be alert for obvious deficiencies that reflect management's neglect.
- Observe whether the organization takes immediate action when a deficiency is identified. At the least, responsible management should determine the cause of the problem and indicate what corrective action will be taken.
- Document the organization's performance in correcting deficiencies or problems by preparing and submitting an appropriate report. Recommendations for action should be written and submitted as required.
- Discuss obvious deficiencies with responsible management personnel to ensure that all the facts of a situation are

known. Often when the auditor discusses deficiencies, management personnel can provide additional information that may modify a program auditor's initial conclusions.

- Be aware of oversolicitous members of management who may be attempting to hide deficiencies.
- Ask for documentation of actions taken, particularly if the initial explanations were not detailed or explicit enough.
- Consider the failure to receive management's undivided attention during the audit as evidence that management is not truly concerned about product safety and liability prevention.
- Consider the possibility, when the auditor cannot get management to answer specific questions, that the organization is deficient in these areas. Note this observation in the PSM program audit report.

The function of the PSM program auditor is to interview all key management personnel; to observe the actual manufacturing operation; and to investigate, question, and verify safety performance.

CONDUCTING A PSM PROGRAM AUDIT

The purpose of a PSM program is to develop a means to perform (1) an evaluation of the product during design and manufacture, distribution, sale, and consumer use and (2) to control any accident and hazard potential through good product safety management techniques. The major goal of the program audit is to reduce or eliminate the causes of product liability exposure. Some of these causes are:

- unsafe product designs (failure to review product design safety)
- inadequate manufacturing and quality assurance procedures
- inadequate preparation and review of consumer warnings and instructions
- misleading representation of product or services.

Preliminary Procedures

The auditor should undertake the following preliminary activities before beginning the formal audit. They will make the job easier to perform and enhance its effectiveness.

- Explain the purpose of the audit to management.
- Arrange mutually agreeable audit dates and review with involved members of management.
- Define how the audit's operational plan (schedule) will involve members of management.
- Review appropriate PSM program-related documentation and procedures.
- Review appropriate product-related procedures.
- Review appropriate product-related technical and standards information.
- Acquire all necessary product-related printed instructional and precautionary materials.

By carrying out these preliminary procedures, the program auditor usually can obtain the complete cooperation of involved management in each major department and of safety personnel.

Management's Commitment

Wholehearted commitment and support of the PSM program by all levels of management provides the spark and impetus to get the program moving and keep it rolling. As is the case for an occupational safety and health program, management must have a written policy outlining its responsibilities to provide support, to set basic objectives, and to establish priorities for a product safety and liability prevention program within the organization. Some evidence of management's commitment to the program might include:

- record of shop conversations on the subject
- posted letter or bulletin
- evidence of specific meetings
- distribution of brochures and safety information
- special mailing to employees
- formal statement of management policy on the subject
- use of a PSM program coordinator and committee
- use of a PSM program auditor.

Top management should tell all key people within the organization they have an important role to play in the program and they must commit enough time and effort to make the program successful. In short, management should clearly communicate to all employees by word and deed that controlling product losses and liability is a major company objective.

Ideally, the chief executive officer issues a written policy stating the company's commitment to product safety and liability prevention. It should be distributed to management, all company departments, and every employee. Management should always provide some tangible evidence that it is truly committed to product safety and liability prevention and that employees have been advised of this fact.

In addition, the auditor must work with and cultivate the respect and cooperation of supervisory personnel. Supervisors and middle-management personnel are essential to the PSM program. They have direct contact with the employees who make the products, and, hopefully, follow the PSM program guidelines for product safety.

Role of the Safety and Health Professional

The role played by the safety professional in a PSM program will vary according to the size of the company. However, in all cases, the safety and health professional's role is vital in implementing a successful program. In a small to medium-sized company, the safety and health professional, because of broad experience in areas of safety and health technology, may be selected by management to be either a PSM program coordinator or auditor. However, as the company product line increases, management may designate the head of the engineering or design department for this assignment, appointing the safety and health professional as an assistant. Nevertheless, some companies do hire a full-time product safety director.

Regardless of the specific role the safety and health professional plays within the company's PSM program, the auditor should evaluate the person's performance to determine if he or she is, in fact, adequately contributing to the company's overall PSM program. Following are some of the contributions the safety and health professionals can make based on their knowledge of plant operations, experience in safety training, and understanding of accident investigation techniques:

- Evaluate and offer comments on the company's PSM program.
- Evaluate and comment on the product safety-related training sessions developed under the PSM program.
- Assist those who will be performing product accident investigations.

In addition, members of the safety and health department can provide product safety surveillance in production areas to help prevent accidents. These individuals should be advised that all product safety-related complaints or problems they uncover must be discussed with the plant engineer and the production department and formally documented. Copies of the reports must be sent to the PSM program coordinator.

Because of its past experience in developing and implementing employee safety programs, the safety and health department may be aware not only of potential product hazards but also of ways in which customers can misuse those products. Therefore, a knowledgeable safety and health department representative should serve as a consultant to the team performing design reviews, hazard analyses, and product safety audits.

DEPARTMENTAL AUDITS

Although virtually all departments are involved in the PSM program, certain departments, primarily engineering, manufacturing, service, legal, purchasing, and human resources, play more significant roles. For the purposes of this chapter, only these departments will be discussed.

Engineering or Design Department Functions

The primary function of the engineering or design department should be to design and manufacture salable, reliable products that can be used with reasonable safety. It is more practical and cost-effective for the manufacturer to build reliability and safety into the product than to suffer the consequences of catastrophic product liability losses.

Products should be reasonably safe during (1) normal use; (2) normal service, maintenance, and adjustment; (3) foreseeable uses for which the product is not intended; and (4) reasonably foreseeable misuses. Courts have stated that a manufacturing company is responsible for ensuring that its products are safe for any reasonable foreseeable use or misuse to which the customer might put them.

Because the engineering or design department is such a critical area within PSM program activity, the program auditor should check the following:

- Does the organization evaluate product hazards prior to production?
- Is anyone within the organization formally assigned this responsibility?
- Is there a formal written or prepared design review

procedure, even if it is not referred to as such?

At certain predetermined points in the design and manufacture of all complex products, tests or inspections of the products are made to determine compliance with manufacturing requirements. This is because problems detected at predetermined points in the process can be corrected more quickly and economically than can those detected after manufacture.

Thus, at predetermined points in the design process, the design should be checked for compliance with requirements in order to achieve the best product design possible. The appropriate check for the product engineering design process is called the formal design review (FDR).

Formal design review. The FDR is a scheduled systematic review and evaluation of the product design. The personnel carrying out the review are not directly associated with the product's development. However, as a group, they are knowledgeable about and responsible for all elements of the product throughout its life cycle, including design, manufacture, packaging and labeling, transportation, installation, use and maintenance, and final disposal.

Each designer and reviewer should ask the question, "Would I feel safe using the product or having family members use it?" Unless the answer is an unqualified "yes," each reservation or problem should be recorded and addressed by the designer and manufacturer.

Codes and standards. The PSM program auditor must determine whether products conform to all applicable safety standards (including state, provincial, or federal codes and regulations; testing or inspection laboratory requirements; industry standards; technical society standards; and machine safeguarding standards). These regulations and standards, in most cases, should be considered as minimum requirements for product safety and reliability. In some cases, where no formal standards apply, reviewers should determine whether in-house design standards are being used. If so, what criteria are applied to judge the adequacy of the standard? Before each survey, the auditor should become familiar with all standards applicable to the company's product design process.

Human factors. The PSM program auditor must determine if human factors have been considered in product designs. (See description in Chapter 5, Ergonomics in the Workplace, in the *Engineering and Technology* volume.) Some of these human factors include the physical, educational, and mental limitations of people who will use the products. The program auditor also should determine whether the company has weighed the possibility that customers might use the product in ways other than it was designed to be used, but that might be considered reasonable.

Critical parts evaluation. Critical parts or components are defined as those "whose failure could cause serious bodily injury, property damage, business interruption, or serious degradation of product performance." Individual organizations, however, may have different criteria for defining a critical part, such as "one that is unusually expensive, difficult to acquire, or requires lengthy order lead time." When analyzing critical parts, the program auditor must be certain that everyone is using the appropriate definition.

The auditor determines whether products are being analyzed for critical parts or components. If so, are the critical items receiving any special attention? Have they been field-tested and designed to outlast the product itself? If this is impractical, is a special effort made to warn customers and users of possible hazards of critical-part failure? Has any effort been made to instruct customers and users in inspection techniques to detect impending failures? Have maintenance procedures been outlined for critical parts or components on the product itself and in operating and maintenance instructions? Has careful consideration been given to the expected life of the product?

All parts or components must have a life expectancy compatible with the life expectancy of other parts and of the total product. A part or component whose life expectancy is shorter or incompatible with that of the rest of the product should be considered "critical." This is because it may become the source of a product liability lawsuit, an expensive product recall, or a field-modification program.

Packaging, handling, and shipping. The PSM program auditor determines if the products' packaging will help to prevent deterioration, corrosion, or damage. Packaging requirements must cover conditions affecting the product at the manufacturing site, during transit, and under normal storage at the customer's location. Packaging methods should be established for each individual product. Product packages should be marked to indicate special requirements—for example, DO NOT STORE IN HOT AREAS, THIS END UP, USE NO HOOKS. Special containers and transportation vehicles should be used where necessary to prevent damage.

The program auditor also must carefully evaluate shipping procedures. A product shipped without proper instructions or with hidden damage may cause an accident when used. A company should select its packaging materials, cartons, and carriers not simply for economic reasons, but also to deliver a product in perfect condition to the distributor or customer.

The product, as shipped, must agree in every respect with the purchase order accompanying it. All appropriate labels, manuals, warnings, and descriptive materials must be included and should be checked at the time of shipping.

Warning labels. It is not enough to design and make a satisfactory product; an organization must also label its products correctly and warn potential consumers and users of any dangers involved. The legal duty of a manufacturer to label products and warn of potential danger is increasingly cited as the basis for product liability lawsuits. This is especially true in chemical, drug, and food cases.

The program auditor must carefully examine the product instructions and labeling to be sure they conform to pertinent regulations and recent court decisions affecting a company's field of operations. The basic rule, as stated in *Restatement of Torts,* Second Series (see References), No. 388, is that

a manufacturer or supplier must exercise reasonable care to inform its consumers of a product's dangerous condition or the facts which make it likely to be

dangerous if he knows or has reason to know that the product is likely to be dangerous for the use for which it is supplied and has no reason to believe that those for whose use the product is supplied will realize its dangerous condition.

Generally, a manufacturer has no duty to warn its consumers of danger if the danger is well known. However, duty to warn and the adequacy of warning is a question for a jury. Their interpretation of "duty" and "adequacy" will vary depending on the circumstances of product use and any injury or damage customers/users sustain.

In general, the courts have tended to equate insufficient warning with no warning at all. Therefore, a company must not only give clear instructions on how to use the product, but also must provide specific warnings about any dangers or possible misuses that could result in injury.

The courts have consistently held that a manufacturer has a duty to warn customers about any reasonably foreseeable use of a product beyond the purposes for which it was designed. For a warning to be considered adequate, it must advise the user of the following:

- hazards involved in the product's use
- how to avoid these hazards
- possible consequences of failing to heed these warnings.

Instruction manuals accompanying the product should repeat hazard warnings and show how the hazards can be reduced or avoided. In addition, the manual should instruct consumers on how to inspect the product upon receipt and how to assemble, install, and inspect it periodically. If the manuals include troubleshooting hints, the dangers involved should be thoroughly explained, for example, "use of unauthorized parts," "do not remove back panel," or "high-voltage hazards." The manual needs to describe the preventive maintenance program required to maintain the product in safe working order.

Finally, the program auditor must evaluate sales brochures, product advertising, and all aspects of marketing and selling. They may need to be reviewed by the engineering or design department to ensure the product's capabilities are accurately depicted and show only safe operating and maintenance procedures. In addition, the auditor should determine whether all warning labels, hazards, and instructions developed by the engineering or design department have been reviewed by the company's legal counsel. This step will ensure that product users are receiving adequate instructions for use, warnings about potential product hazards, and instructions for proper maintenance.

Manufacturing Department

After a reasonably safe and reliable product has been designed, the manufacturing department must turn the design specifications into a finished product. If this is not done under proper controls and supervision, manufacturing errors could result in an unsafe and possibly unreliable product reaching customers.

The manufacturing department can contribute to a company's overall PSM program in many ways. The most important are listed below. By evaluating each of them (as it applies to a particular situation), the program auditor can effectively measure the manufacturing department's contribution.

- Motivate manufacturing employees by letting each person know that he or she is making a vital contribution to a quality product.
- Instill pride in employees in their work and in the company's product by:
  — using up-to-date manufacturing equipment
  — keeping the plant's manufacturing capacity within bounds
  — providing adequate work space for each task
  — keeping the workplace clean and well lit
  — implementing good equipment maintenance practices
- Provide standardized on-the-job training procedures.
- Implement zero defects or error-free performance or other error-elimination programs.
- Design a program or procedure to identify and eliminate all production trouble spots. This should be accomplished in cooperation with the inspection and testing personnel of the quality assurance department.
- Prevent unauthorized deviations from design specifications and work procedures.
- Participate in safety audits on new product designs. This is often accomplished by serving on the company's PSM committee, if one exists.

Sometimes product specifications are not realistic for the machines and equipment available in the shop. When this is the case, the production department must advise the engineering department immediately of the problem. Production workers should not try to maintain cost or production schedules by deviating from the specifications without first consulting the engineering department. If deviations are necessary or unavoidable, they should be made only with the approval of the engineering or design department.

Recordkeeping. Manufacturing records should be kept for the life of all products and particularly for their critical parts or components. These records are vital to have on file in the event of product recalls or field-modification programs and for successful defense in product liability suits. Records on critical components should be complete enough to identify the batch, lot, and supplier of the raw material and the finished products in which they were used.

Discontinued and new products. Two other important areas of manufacturing department risk exposure are the existence of discontinued products and the development of new products or product lines. The PSM program auditor must determine if discontinued products (products on the market but no longer being manufactured) or the manufacture of new products present additional hazards to either workers, management, or consumers.

Service Department

Most companies provide service contracts for their customers either through the dealer or distributor or through service subcontractors. These contracts can significantly increase a company's product liability exposure because through them

the company has extended its exposure beyond the controlled environment of its manufacturing facility.

The PSM program auditor must carefully evaluate the company's service department controls. The auditors will determine the effectiveness not only of these controls, and their contribution to the company's PSM program activities, but also of the company's use of service department feedback.

In many companies (depending upon the type of product), service department personnel are required to maintain close contact with customers. Consequently, more than any other employees, they are familiar with customer needs and characteristics. They are most likely to hear customer complaints, reactions, and compliments regarding the company's products. They also see product misuses and usually know something about incidents and accidents that occur. In fact, service department personnel are often the first to hear of product accidents.

Legal Department

The legal department in any organization should play a significant role in both the prevention and defense of product liability cases. Too often, companies use their legal department staff only when product safety problems or product liability litigation is imminent. Firms often fail to recognize the importance of involving legal personnel in the day-to-day aspects of product safety and liability prevention.

Legal personnel should act as advisors to the PSM coordinator. It is also essential that the personnel department ensure that product safety duties and responsibilities are included in job descriptions and their performance considered in personnel reviews. This policy is important and is often overlooked by many companies.

The legal department should be assigned the following responsibilities to maximize its contribution to the company's PSM program activities.

Review for potential liability. Legal personnel should review all product-related literature for potential liability.

Coordination. Legal personnel should coordinate all product investigations and product claim defenses including working with the insurance carrier, retaining local counsel, and retaining statements of negotiations.

Marketing Department

Unfortunately, after a "reasonably" safe product has been designed and manufactured, product claims still can be incurred because of the way in which the company presented the product to customers and users. Customers often must rely on the company's sales personnel; advertising and sales brochures; and operating, service, and maintenance instructions for their knowledge of the product's capabilities and hazards.

If the customer is led to believe that the product has capabilities it does not possess or if the instructions do not adequately warn of the product's hazards, an injured customer may have legal cause for action where none existed before. Therefore, the PSM program auditor must review the company's advertising and sales materials and

evaluate whether:
- They are clear and accurate.
- They overstate the product's capabilities.
- They encourage the customer to believe the product has uses for which it was not designed or intended.
- The product can safely do what the company's advertising and sales materials claim.
- Only safe operating procedures are depicted.
- The company's product is illustrated only with safety devices in place.
- Advertising and sales materials have been reviewed and approved by the engineering or design and legal departments regarding their accuracy and potential liability risks.

Warranties and disclaimers developed by the company must also be reviewed by the program auditor to determine whether they are:
- reasonable and practical for the uses intended
- included with each of the company's products
- clear and concise
- prominently displayed and easily recognizable by the customer or product user
- thoroughly reviewed by the company's legal department or legal counsel.

The program auditor must verify that the marketing department has retained (for the life of the product) all sales and distribution records that can identify purchasers. These records are essential if product recall or field-modification programs are to succeed. In addition, these records should indicate, whenever possible, how the company's products will be used, particularly when the company is selling to subcontractors or assemblers.

The company also must ensure that its sales personnel and dealers know how to describe the capabilities of the products they are selling or distributing without incurring undesired implied or expressed warranties. For example, sales personnel must never exaggerate the capabilities of the company's products or claim uses for which they were never designed.

Purchasing Department

Some companies are not large enough to have a formal purchasing department. However, every company has someone responsible for acquiring the raw materials and components needed to manufacture products.

The purchasing activity carries significant product liability potential for the company. Consequently, the PSM program auditor must carefully examine the company's activities to determine whether it is performing acceptably in this area.

Some of the primary responsibilities of the purchasing department include:
- Becoming familiar with all material specifications set by the engineering, design, and manufacturing departments.
- Always purchasing quality raw materials, parts, and components that meet the specifications set by the various departments.
- Evaluating, in conjunction with the company's quality assurance department, the capabilities and reliability of

suppliers, using vendor rating systems. Compile a list of approved suppliers.

Human Resources Department

Employee job qualifications and work attitudes have a considerable influence on the quality of a company's products. Employees who lack proper job skills or who are unmotivated increase the company's chances of producing defective products.

The company's human resources department must work to (1) select, train, and place new and transferred employees and (2) continually seek to upgrade the morale and performance goals of present employees and others who are involved in critical product safety matters within the company.

Insurance Department

Often companies believe that the only function of the insurance specialist or department is to buy insurance and to report claims to the insurance company. In reality, because of the insurance department's experience in handling product claims, it usually possesses valuable knowledge of the factors that often give rise to product liability lawsuits. As a result, management should use the insurance department as an information clearinghouse in conjunction with the PSM program coordinator.

The insurance department usually reports liability claims to the insurance carrier and coordinates any accident investigations with the carrier. Occasionally, the department will coordinate the preparation of a legal defense (if one is necessary) between the legal department and the insurance carrier. Depending upon the company, the legal department may perform these functions instead.

Public Relations Department

A public relations department, like the safety and health department, generally plays a smaller role in a company's PSM program activities than many of the other departments. Nevertheless, the program auditor should not overlook it in an overall evaluation of the PSM program.

Product accidents, product recalls, or product field-modification programs may be reported by the news media and require the company to prepare press releases on the issue. If these statements are positively written, they can help to prevent unfavorable publicity. In fact, correct handling of publicity on product accidents, product recall, or product field-modification programs can be used by the public relations department to show the public the company's diligent efforts to design, manufacture, and sell a safe, reliable product. The public relations department should also distribute product information to the media when the company has incorporated safety advances into its products.

QUALITY ASSURANCE AND TESTING DEPARTMENT

The term *quality assurance* refers to the actions taken by management to ensure that manufactured, assembled, and fabricated products conform to design or engineering requirements. At one time, the terms quality assurance and statistical quality assurance were considered synonymous, because statistical techniques were considered to be the major tools of quality assurance. Over the years, the definition of quality assurance has broadened to include any actions that ensure product conformity to design and other requirements and the achievement of customer acceptance and satisfaction.

Before discussing the evaluation of a company's quality assurance activities, it must be pointed out that quality assurance policy, department organization, function, and responsibility vary according to a company's size, management policy and organization, type of product, plant location, number of plants, economic resources, and other variables. When evaluating a company's quality assurance department, the important points for the program auditor to determine are:

- Is the quality assurance program adequate to help attain the company's quality objectives? An incomplete quality assurance function does not necessarily mean that the system is inadequate. Necessary functions can be added later. Others may not even be required because of the type of products, manufacturing processes, or size of a company.
- Does the quality assurance program function as planned? To answer this question, the program auditor must go beyond the quality assurance manual and review the actual implementation of the system.

In evaluating this function, the PSM program auditor must ask the basic questions: How? Why? When? Where? What? Who? The auditor should then follow through with "Explain how it works," "Let's see examples," "Show me the records." Asking these questions may reveal that the company has instituted a quality assurance program, but it does not prove the system has actually been installed throughout the plant, nor that it is functioning as planned.

The PSM program auditor must see tangible evidence that the quality assurance program has been implemented and is functioning. To get a total picture of the program, the auditor should review all paperwork, procedures, instructions, and records.

The program auditor should spend some time with the quality assurance shop personnel, if possible. These individuals can expound on the pros and cons of the program, thus providing information about the program's strengths and weaknesses that might not otherwise be available.

The basic quality assurance program functions that must be evaluated by the PSM program auditor are as follows:

- organization and manuals
- engineering/product design coordination
- evaluation and control of suppliers/vendors
- evaluation of manufacturing (in-process and final assembly) quality
- evaluation of special process control
- evaluation of measuring equipment calibration system
- sample inspection evaluation
- evaluation of nonconforming material procedures
- evaluation of material status/storage system
- evaluation of error analysis and corrective action system

- evaluation of record retention system.

One of the most important requirements for effective operation of a quality assurance program is the company's positive interest and concern. This begins with top management. In most instances, top management must delegate quality assurance responsibilities to a coordinator and clearly define the assignment of responsibility and authority throughout all levels of management, supervision, and operation.

Manuals

Managers should develop a quality assurance manual, whose form and content vary according to a company's requirements. The manual is usually divided into two sections: policy and procedures.

Policy:
- states company quality assurance policy and objectives
- establishes organizational responsibilities
- establishes systems for implementing quality assurance policy

Procedures:
- states operational responsibilities
- gives detailed operating instructions.

The policy section of the manual outlines general goals, while the procedures section contains the daily, detailed operations of the quality assurance department.

Engineering and Product Design Coordination

The quality assurance department, by looking at operations from a different viewpoint, can assist the engineering or design department in its research, development, design, and specifications functions. Quality assurance normally is not involved in engineering or design or research unless it has specific data to share. Quality assurance performs such functions as inspection, testing, recording, maintenance, equipment calibration, and data accumulation.

Evaluation and Control of Suppliers

It is just as important to control the quality of purchased materials and services as it is to establish and enforce such controls for internal functions. The degree and extent of quality assurance established for a supplier will depend on the complexity and quantity of the supplier's products and its quality history. The primary criteria in selecting suppliers are their ability to maintain adequate quality and to establish good lines of communication with the company.

Manufacturing Quality
(In-process and Final Assembly)

Quality assurance of items produced in-plant is accomplished by planning, inspection, process control, and equipment calibration.

Manufacturing planning. Planning in advance of production ensures that product and manufacturing instructions will be uniform throughout the process. Planning normally is a joint effort of the manufacturing department and quality assurance department. Manufacturing personnel initiate the internal paperwork, including all data necessary to produce the product. Quality assurance personnel review the paperwork to confirm that the data adhere to established parameters and include quality requirements.

Manufacturing work instructions. All work affecting product quality should be supported by documented, step-by-step instructions appropriate to each task, the working conditions, and employee skill levels. Instructions also must include quantitative and qualitative means for determining that each operation has been satisfactorily completed. The amount of detail depends on the skill levels of the workers and on the complexity of each task.

Inspection instructions. Instructions to inspection personnel should give them specific guidelines to ensure that products conform to design specifications.

Manufacturing inspection. Product inspection usually includes one or more of the following general methods:
- visual inspection
- dimensional inspection
- hardness testing
- functional testing
- nondestructive testing (NDT)
- chemical-metallurgical testing.

Product inspection can be conducted in-process rather than after completion of all operations. If so, the inspection should not be done until all previous inspections have been completed and the product has been certified as acceptable up to this point.

Records. Records should be kept of the manufacturing operations, including incoming materials, assemblies or processes, and final inspection results.

Special Process Control

Special processes used by manufacturing to change the physical, mechanical, chemical, or dimensional characteristics of product production include, but are not limited to, the following:
- heat treating
- plating
- fusion welding
- stamping and forming
- batch mixing
- chemical mixing
- adhesive bonding.

Because special processes greatly influence the quality of the completed product, they must be carefully, yet economically controlled. A 100% inspection process is often not feasible because it might require destructive testing to determine product conformance to specifications or other similar tests. The processes, therefore, should have self-monitoring controls to ensure that all processed items will be of the same desired quality.

Calibration of Measuring Equipment

Measuring and process control equipment and instruments used to ensure that manufactured products and processes conform to specified requirements must be calibrated peri-

odically. Any equipment employed either directly or indirectly to measure, control, or record manufacturing processes should be calibrated against certified standards so it can be adjusted, replaced, or repaired before the equipment impairs product quality.

Sample Inspection Evaluation

A sample inspection is often used by quality assurance to determine the quality of a lot without inspecting all items. This technique can be very beneficial if used correctly.

Even when a 100% inspection is performed under the most favorable conditions, it is only 85 to 90% effective because of human and other error. A sampling inspection is not necessarily 100% effective, but it can approach that level. The sampling disadvantage, which is far outweighed by its advantages, is that occasionally the sample gathering procedure for a lot does not give a true picture of the lot quality—good lots can be rejected or bad lots can be accepted. The goal, however, is to make the acceptance of good lots far more likely than the acceptance of bad lots.

Nonconforming Material Procedures

The term *nonconforming material* is usually applied to products that are rejected because they do not meet established requirements. Raw materials, parts, components, subassemblies, and assemblies can be classified as nonconforming material whenever they fail to meet specifications at any point in the manufacturing process. The company should have a system to control nonconforming material that includes the following:

- identification
- segregation
- disposition
- reinspection
- customer notification
- supplier reporting
- records.

Material Status and Storage

The company must have a way to identify whether a product (or lot) either has not been inspected, has been inspected and approved, or has been inspected and rejected. All raw materials, components, subassemblies, assemblies, and end products should be labelled to note their process, inspection, and test status.

Material storage refers to the temporary holding of raw materials and recently purchased or in-process parts and assemblies. Stores personnel must have a control system that documents at all times the status and condition of stored raw materials, components, and products. The program auditor should evaluate the effectiveness of controls used by the company to make sure the stock on hand is the stock ordered and that the system used to identify the materials in the storeroom is adequate.

Error Analysis and Corrective Action System

The error analysis and corrective action system is a follow-up to the nonconforming material system. The objective of this system is to use the discrepancies reported in the nonconforming material system to perform the following:

- analyze manufacturing errors
- request corrective actions from those responsible for the errors, evaluate responses, and determine the effectiveness of the actions taken
- analyze manufacturing rejection rates and report to management
- the dollar losses due to scrappage, rework or repair, and reinspection costs.

Recordkeeping System

Good records are the backbone of a quality assurance program because they document the history of a product. These records indicate that the company's quality assurance system is functioning as planned, substantiate that a product was inspected, and often contain the actual inspection findings. Records can be vital in the defense of a product in a lawsuit.

Good quality assurance records also allow a company to trace any product from start to finish. These documents should include:

- lists of raw materials from which products are produced
- inspection results for purchased and manufactured products
- inspection results from each inspection station
- special processes control data
- calibration data
- sample inspection data
- nonconforming material data
- error analysis and corrective action data
- shipping data.

Records should be stored in metal cabinets in a low-hazard area and kept for the life of the product.

FUNCTIONAL ACTIVITY AUDITS

These final sections cover two interrelated activities: recordkeeping and field-information systems. These important activities are referred to as functions to differentiate them from the departmental categories previously used (design and manufacture, quality assurance, insurance, and others). The term *function* implies that these activities, to be properly accomplished (from a PSM program standpoint), must be performed by several or possibly all of the departments within a company, rather than just one department.

Recordkeeping

A company must not only manufacture and market safe, reliable products, it also must be able to *prove* in court the safety of those products. Complete, accurate records can be convincing evidence in a court of law. Consequently, a company should retain records that document all phases of its manufacturing, distributing, and importing activities from the procurement of raw materials and components, through production and testing, to the marketing and distribution of

the finished products.

In addition to the documents' usefulness in court, comprehensive PSM program records also enable a company to identify and locate products that its data collection and analysis system indicates may have reached the customer in a defective condition. Should the company need to implement a product recall or field-modification program, these PSM program records can play a key role in the success of such efforts.

Both federal regulations and internal PSM program requirements dictate the form and content of a company's recordkeeping program, the types and quantity of records retained, and how long the company should retain them.

Regulations. A company should be aware of regulations affecting recordkeeping requirements. Some of the more important governmental regulations to consider when developing a PSM recordkeeping system both in the United States and Canada are:

- *United States*
  Consumer Product Safety Act (Public Law No. 92–573)
  Federal Hazardous Substances Act (15 USC 1261)
  Federal Food, Drug and Cosmetic Act (21 USC 321)
  Poison Prevention Packaging Act (Public Law No. 91–601)
  Occupational Safety and Health Act (Public Law No. 91–596)
  Child Protection and Toy Act (Public Law No. 91–113)
  Magnuson-Moss Warranty-Federal Trade Commission Improvement Act (Public Law No. 93–637).

- *Canada*
  Consumer and Corporate Affairs Canada Act
  Consumer Packaging and Labeling Act
  Hazardous Products Act
  Industrial Design Act
  National Trademark and True Labelling Act
  Patent Act
  Seals Act.

Internal PSM program requirements. A company must maintain all pertinent information related to the design, manufacture, marketing, testing, and sales of its products. Of particular value are records documenting design decisions, since faulty design is an almost universal claim in product liability lawsuits. A company can mount an effective defense if it can introduce evidence proving that a certain design or material was chosen with the safety of the consumer in mind. Also, records documenting why certain production techniques were chosen, such as having a part forged rather than cast, should be retained. Records can describe various product tests and the reasons for any design or manufacturing modifications and improvements. When a product hazard cannot be eliminated, a company's records should show conclusively why it cannot.

An important means of product liability protection is a complete, up-to-date set of records that can establish the care taken by a company to manufacture and market safe, reliable products. The PSM program auditor must bear in mind that a company's size, product lines, and organizational structure will determine the type and number of records that should be kept.

Period of retention. It is a difficult problem for a company to determine what records should be retained and for how long. Product liability claims can be brought years after the manufacture or sale of an item has ceased.

Generally speaking, unless a company can accurately define the life of its product, it should be prepared to retain all PSM records in perpetuity. If this creates record storage or maintenance problems, a company can use a microfilm or a computerized recordkeeping system. The PSM program coordinator must ensure that key personnel fully understand why records are being maintained, what they contain, and for how long they should be kept.

Reasons for keeping records. Several reasons exist for keeping PSM records:

- to comply with regulations covering the design, manufacture, and sale of the company's products
- to demonstrate management's commitment to market a quality product
- to avoid wasting time and money redoing what has already been done
- to establish how much care is needed to produce and sell a safe, reliable product
- to enable the company to trace a product or customer
- to establish a sound data base for items such as insurance costs, sources of supply, and product recall or field-modification expense requirements.

Field-Information System

Because a company must receive information from the field about product performance, the PSM program auditor should evaluate the effectiveness of the company's field-information system for each product line and its distribution system. The program auditor must thoroughly analyze (1) the information system's ability to identify and trace a product from raw material form through final sales and distribution stages; (2) its ability to acquire and use field data (complaints, incidents, and accidents); and (3) its ability to use these two factors to implement product field-modification or recall actions, where appropriate.

Data collection and analysis of complaints, incidents, and accidents. Every company should have a reporting system to acquire and evaluate product information from the ultimate testing laboratory—the customer. This information comes from a variety of sources both within and outside the organization (service personnel, salespeople, repair, distributors, and retailers). A reasonably detailed report used with a data collection and analysis system can help the company accurately evaluate each complaint, incident, and accident and assess any problem as it develops.

For maximum effectiveness, data should be directed to one individual in the company, the PSM program coordinator. The program auditor must be able to verify, through company records, the results of any field data analysis and what corrective action, if any, was taken.

Diversification among manufacturers makes it impossible to create one data collection and analysis system applicable to all companies. However, every data reporting system must

provide the answers to the following questions:

- Who is the customer?
- What type of product is involved?
- What is the problem?
- How is the product being used?

Analysis of causes. The complaint report should provide answers to questions such as:

- Is the product being used in the manner for which it was intended?
- How is it mounted or installed?
- What environmental conditions has the product been subjected to?
- Has the product been altered or modified by the customer?

When it is determined that a potential product hazard exists, a company should promptly instigate field investigations to determine its cause (design, manufacturing, or quality assurance). The causes should then be analyzed to discover if they are part of a developing trend or if various product problems are unrelated. This information should be given to the PSM program coordinator, who will decide whether to recommend further action.

The type of action taken might include changes in design, manufacturing, quality assurance, or advertising procedures, or the implementation of a product field-modification or recall program. If a company takes no action, it is actually saying that the data collection and analysis system has no value in the manufacture of safe, reliable products.

The best way to evaluate a company's data collection and analysis system is for the auditor to request a review of the company's complaint, loss, field data, and similar files. If the company does not have such files, or if they are poorly maintained, it indicates that the firm does not make use of this information to help it manufacture safe products.

Product Recall or Field Modification

Any manufacturer may be faced with the need to institute a product recall or field-modification program. This action also can involve those who supply component parts, materials, and services for products. Therefore, the PSM program auditor should determine whether a company is capable of implementing a successful product recall or field-modification program should the firm's data collection and analysis system indicate the need for one.

Whenever a company finds that substantial performance or safety defects exist in some or all of a its shipped products, it must recall or field-modify these products. In addition, recall or modification should be considered in those cases where the company is aware of or has itself developed an improvement or changed its product (for example, added a guard or a fail-safe). If the improvement is developed after the sale of earlier product models, a company should advise its customers of the product improvement, the hazard it eliminates, and how they can incorporate it into their earlier product models.

Without an adequate tracing system established during product planning and manufacture, it may be difficult to identify and locate specific lots, batches, quantities, or units

to be recalled or modified in the field. Every company's recordkeeping system should be extensive enough to enable management to trace a product from design through production to delivery of the product to the customer.

Should a product recall or field-modification program become necessary, a company can minimize the quantity that must be recalled or modified and the cost to do so if it can pinpoint the exact number of defective products, parts, or components that need to be changed.

Field-modification program plan. A product recall or field-modification program plan will be different for each company. The plan must be defined by the PSM program auditor after evaluating the firm's recordkeeping program and product traceability systems capability.

Based on information acquired from complaints, incidents, and accident reports made by customers, distributors, or dealers, and from regulatory agencies, a company must be able to determine immediately if a substantial product hazard exists. Techniques for this step include on-site investigation of the complaint, incident, or accident; hazard and failure analyses of the unit involved in the complaint, incident, or accident by the company or an independent laboratory; analyses of other units of the same product batch; and careful evaluation of in-house tests or other records. If a company finds that a substantial product hazard does exist, it should implement an appropriate product recall or field-modification plan.

Independent Testing

As an integral part of the PSM program, a company should conduct adequate product testing. This can be done either in-house or through affiliation with an independent testing laboratory. In certain instances, companies maintaining bona fide laboratory or adequate testing facilities should be encouraged to certify on their own authority and reputation that their products comply with existing standards. In those instances where no external standards exist, products should comply with appropriate company standards developed by safety and engineering personnel.

PSM PROGRAM AUDIT REPORT

The PSM program auditor will generate a sizable body of information during evaluation. This information must be assembled, organized, and presented to management in a clear and concise format.

For management's benefit during review, the audit report should begin with a summary of findings. This summary section clearly states the program auditor's opinion of the company's overall PSM program and any recommendations to correct deficiencies detected during the audit. If recommendations are made, they must be supported by data in the body of the report and should be practical and straightforward, using the company's already existing systems and procedures where possible. The report should indicate each recommendation's importance and the possible consequences if the company fails to comply. The program auditor should

then estimate a realistic time frame for compliance with the recommendations.

It must be clear to management how the program auditor developed his or her overall opinion about the company's PSM program. If management doubts the auditor's conclusions, the value and impact of any efforts to assess and improve an existing PSM program or control system will be seriously impaired.

SUMMARY

- Although no one can completely prevent product misuse by customers, companies can minimize such risks by striving to produce and sell only reasonably safe, reliable products. To fulfill this goal, a company must establish and audit an effective product safety management (PSM) program.
- The first step in developing a PSM program is to select a program coordinator and program auditor. In some companies, these two functions will be combined into one job. This person must have access to top management, clearly understand the purpose of the job(s), and have sufficient authority to carry out various responsibilities and to implement program plans and suggestions. In some firms, the coordinator will work with a PSM program committee.
- The program auditor's primary duty is to evaluate the organization's PSM program activities to determine how the organization can prevent product-related losses; how it can manufacture safe, reliable products; and how management can contribute to product safety. This audit should be distinguished from a product safety loss control evaluation.
- The major goal of a PSM program audit is to reduce or eliminate the causes of product liability exposure. Preliminary steps include explaining the purpose of the audit to management, defining the scope of the audit, and reviewing all product-related procedures and technical information. The auditor must have the support and co-operation of all management personnel.
- The safety and health professional can play a significant role in a company's PSM program by evaluating the program's content, training, and accident investigation activities; by maintaining safety surveillance in production areas; and by identifying potential hazards related to the manufacture and use of products.
- Although nearly all departments are included in the PSM program, certain departments play more significant roles: engineering or design, manufacturing, product service, legal affairs, purchasing, and human resources. These departments have a direct bearing on producing safe, reliable products and on minimizing customer lawsuits.
- The engineering and design department is particularly important. The auditor should review product designs, codes and standards, human factors, critical parts, packaging and shipping procedures, and warning labels and instruction manuals. Companies have a legal duty to warn

customers of any potential hazards or risks related to using or misusing the product.
- The manufacturing department can contribute to PSM programs by motivating workers to produce a quality product, developing efficient, high-quality production processes, and participating in safety audits on product design and manufacture. Auditors must make sure that this department keeps accurate records on all products and helps to minimize risks associated with discontinued or new products.
- The PSM auditor must evaluate the company's service department controls to ensure the company maintains close contact with customers and handles complaints in a timely fashion. Legal department personnel can act as advisors to the auditor, review products and product-related literature for potential liability issues, and coordinate all product investigations and claims. All marketing and sales personnel must be knowledgeable not only about company products but about liability issues as well.
- The purchasing department must be carefully audited to make sure that purchased materials meet specifications and that suppliers are reliable. Likewise, the human resources department must ensure that workers are carefully selected, trained, and matched to the right job and that employee skills are continually upgraded.
- Quality assurance refers to actions taken by a company to ensure that products are produced in conformity with design and engineering requirements. The PSM auditor must determine if the quality assurance program is adequate to carry out the company's quality objectives. The auditor evaluates quality assurance manuals, policy and procedures, manufacturing instructions and inspections, records, equipment calibration, nonconforming material procedures, material status and storage, manufacturing errors, and recordkeeping systems.
- Functional activity audits cover two interrelated areas: recordkeeping and field-information systems. Records must be retained to satisfy government regulations; to prove the company's commitment to manufacture safe, reliable products and to protect the consumer; and to trace customers and product lots should a product recall or field-modification program become necessary.
- Field-information systems are essential to respond to consumer complaints or suggestions regarding product performance. The auditor should determine if the system can (1) identify and trace products through all manufacturing and distribution stages, (2) acquire and use field data, and (3) apply this information to implement field-modification or recall actions. The auditor should also evaluate company testing procedures used to ensure that products comply with standards.
- The PSM audit report must be presented to management in a clear, concise format. It should begin with a summary of findings, contain adequate data to support all recommendations, and provide realistic, measurable implementation steps to correct deficiencies or to improve current procedures.

REFERENCES

American National Standards Institute, 11 West 42nd Street, New York, NY 10036.

"Catalog of American National Standards" (issued annually). *Product Safety Signs and Labels,* Z535.4–1991.

Baldwin S, et al. *The Preparation of a Product Liability Case.* Boston, MA: Little, Brown and Company, 1981.

Canavan M. *Product Liability for Supervisors and Managers.* Reston, VA: Reston Publishing Company, 1981.

Commerce Clearing House, Inc., 2700 Lake Cook Road, Riverwoods, IL 60615, *Products Liability Reports.*

Eads G and Reuter P. *Designing Safer Products—Corporate Responses to Product Liability Law and Regulation.* Santa Monica, CA: Rand Institute for Civil Justice, 1983.

Hammer W. *Product Safety Management and Engineering.* Englewood Cliffs, NJ: Prentice-Hall, Inc., 1980.

Kolb J and Ross S. *Product Safety and Liability.* New York: McGraw-Hill Book Company, 1980.

Leach G. *Product Safety Checklist.* Aurora, IL: Stephens-Adamson, Inc., 1979.

National Bureau of Standards, Washington, DC 20234, *An Index of U.S. Voluntary Engineering Standards,* Spec. Pub. No. 329, Standards Information Service.

National Safety Council, 444 North Michigan Avenue, Chicago, IL 60611.

Industrial Safety Data Sheets (listing available).
Product Safety: Management Guidelines, 1989.
Product Safety Management Training Course, Establishing and Auditing an Effective Product Safety Management Program.
Product Safety Reference Notes.
"Product Safety Up-To-Date" (published bimonthly).

Product Liability Prevention Technical Committee. *Product Recall Planning Guide.* Milwaukee, WI: American Society for Quality Control, 1981.

Pyzdek T. *What Every Engineer Should Know About Quality Control.* WEESKA Series No. 26. New York: Marcel Dekker, Inc., 1988.

Restatement of Torts, Second Series. Philadelphia, PA: American Law Institute, 1965.

Schaden R and Heldman V. *Product Design Liability.* New York: Practicing Law Institute, 1982.

Seiden RM. *Product Safety Engineering for Managers: A Practical Handbook and Guide.* Englewood Cliffs, NJ: Prentice Hall, Inc., 1984.

Thorpe JF and Middendorf. *What Every Engineer Should Know About Product Liability.* New York: Marcel Dekker, Inc., 1979.

Underwriters Laboratories Inc., 333 Pfingsten Road, Northbrook, IL 60062, *Standards for Safety Catalog.*

Weinstein AS. *Product Liability and the Reasonably Safe Product.* New York: John Wiley and Sons, 1978.

10

Fleet Safety Programs

The fleet equipment discussed in this chapter includes trucks, passenger cars, buses, and motorcycles. Powered industrial trucks and handtrucks are covered in the *Engineering and Technology* volume in Chapters 8 and 11, Manual Handling and Material Storage, and Powered Industrial Trucks, respectively. Haulage and off-road equipment are discussed in Chapter 12 of that volume.

Safe vehicle operation is the result of training, skill, planning, and action, not chance. Unfortunately, many companies fail to pay enough attention to the safe operation of motor vehicles. The reason for this lapse may be the difficulties of organizing an adequate safety program and providing good driver and fleet supervision.

The majority of all motor vehicle accidents are caused by driver error, or poor operating practices; only a small percentage are due to mechanical failure of vehicles or to improper maintenance of equipment. As a result, an organization's vehicle accident prevention efforts should focus on both these principal accident factors—(1) driver error and (2) vehicle failure—because both can be controlled.

Companies can control driver error by implementing a program of driver selection, training, and supervision, while vehicle failure can be reduced by a systematic preventive maintenance program. As experience has shown, the unsupervised fleet usually has higher accident costs than the supervised one.

COST OF VEHICLE ACCIDENTS

The total cost of a vehicle accident usually exceeds the amount recovered from the insurance company. Accident control in a large motor vehicle fleet is critical because increased insurance premiums reduce profits. Insurance premiums can fall or rise with the accident frequency and costs.

As discussed in Chapter 12, Accident Investigation, Analysis, and Costs, property damage is only one of the costs resulting from an accident. There also are indirect costs which, as with most work-related accidents, may be several times the direct costs. Typical indirect costs include the following:

1. salary paid and loss of service of employees injured in an accident
2. added workers' compensation costs resulting from a disabling injury
3. vehicle's commercial value while it is out of service, the cost of replacement vehicle, or rental costs
4. cost of supervisory time spent in investigating, reporting, and cleaning up after the accident
5. poor customer and public relations resulting from a company vehicle having been involved in an accident
6. cost of replacing or retraining an injured employee
7. time lost by co-workers while discussing the nature of the accident and the extent of the victims' injuries.

Besides appealing to the company's profit motive, the safety and health professional can make the most telling argument for controlling accidents by pointing out the company's moral obligation to act reasonably and responsibly

toward its employees and the public. The employer, through the authority to hire, supervise, discipline, and discharge employees, exercises considerable control over their driving performance. Management can require employees to enroll in driver safety education programs and provide proper supervision on the job.

VEHICLE SAFETY PROGRAM

A vehicle safety program should provide for the following:
1. a written safety policy, developed, supported, and enforced by management
2. a person designated to create and administer the safety program and to advise management
3. a driver safety program, including driver selection procedures, driver training, and safety-motivating activities. Proper supervision and implementation are mandatory for success.
4. an efficient system for accident investigation, reporting, and analysis; determination and application of appropriate corrective action; and follow-up procedures to help prevent future accidents
5. a vehicle preventive maintenance program.

An example of a company safety policy statement is shown in Figure 10–1.

Policy Statement

The efficiency of any operation can be measured by its ability to control loss. Accidents resulting in personal injury and damage to property and equipment represent needless suffering and waste. Management's responsibility is to provide the safest conditions and equipment for all employees. The company policy on safety is:
1. The safety of the employee, the public, and the operation is paramount and every attempt will be made to reduce the possibility of accidental occurrence.
2. Safety will take precedence over expediency and shortcuts.
3. The company intends to comply with all safety laws and ordinances.

Every employee will be expected to demonstrate an attitude that reflects this policy as outlined in the company safety program.

Figure 10–1. This American Trucking Associations' recommended safety policy can be adapted to fit individual company needs. (Printed with permission from *Fleet Safety Programs, Policy and Procedures for the Safety Director*. American Trucking Associations, Department of Safety, Washington, DC.)

Responsibility

The first requirement of an effective driver safety program is that all employees from the president on down to the individual vehicle operator accept responsibility for safe operation of company vehicles. Management must define standards of acceptable driver and vehicle performance and establish criteria and procedures to evaluate and correct job performance to meet these standards. Companies must make sure all employees understand that accident prevention is a requirement of employment.

In larger companies, a fleet professional is usually re-

sponsible for supervising the program for the safe operation of all automotive equipment. However, in a small industrial concern, this function may be assigned part time to a supervisor or manager. Some of the duties of the person include the following:
1. Advise management on accident prevention and safety matters.
2. Develop and promote safety activities and work-injury prevention measures throughout the fleet.
3. Study and recommend fleet safety programs regarding equipment and facilities, personnel selection and training, and other phases of fleet operation.
4. Evaluate driver performance and skills requirements.
5. Conduct or arrange for effective safety training and procure or prepare and disseminate safety education material.
6. Review accidents to determine their causes and recommend corrective action to management.
7. Compile and distribute statistics on accident-cause analysis and experience; identify problem persons, operations, and locations.
8. Maintain individual driver-safety records and administer the safe-driver award incentive program.

Driver Safety Program

A driver safety program should include the following five basic accident prevention procedures.
1. Initiate a driver training program.
2. Develop standards to determine ways accidents can be prevented.
3. Require immediate reporting of every accident.
4. Recommend performance goals to management; compute and publish the fleet accident record.
5. Establish competency and skills levels, set objectives, and maintain an accident record card for each driver.

The planning and administration of a safety program for motor fleets are described in greater detail in the National Safety Council's *Motor Fleet Safety Manual*, 3rd ed. (see References).

A motor vehicle accident can be defined as any incident in which the vehicle comes in contact with another vehicle, person, object, or animal, in a way that results in death, any degree of personal injury, or any extent of property damage, regardless of where the incident took place or who was responsible. (Another definition appears in the Preface to this volume.)

This definition includes even minor accidents such as fender scratches. All vehicle accidents, major and minor, are important to the person responsible for vehicle safety, whose major concern is eliminating poor driving habits or careless attitudes. Even minor accidents, not just spectacular ones, provide clues that can help to prevent recurrences.

Accident Reporting Procedures

Drivers should be required to make out a complete accident report on a standard accident report form for every incident in which they are involved. National Safety Council Form

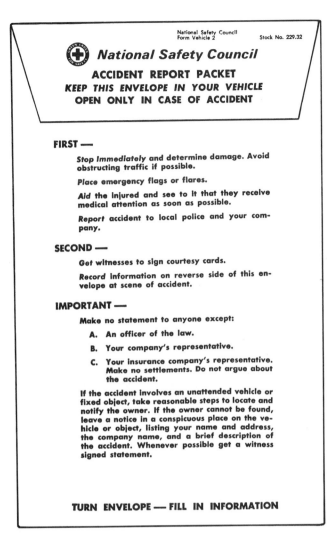

Figure 10–2. This Accident Report Packet is available as NSC Form Vehicle 2. At left is the front side of an envelope that has almost everything a person needs to complete an accurate report and obtain the identification of witnesses.

Vehicle 1 is a good example of an accident report form. (See the National Safety Council's *Motor Fleet Safety Manual*, 3rd ed. for illustrations of forms.) Make sure drivers know how to fill out accident report forms accurately and thoroughly. Drivers should complete the report at the scene of the accident, if possible, and then send the form to their supervisor no later than their next assigned shift.

Companies can supply drivers with an accident report packet to be carried in their glove compartments. This packet (Figure 10–2) should provide for a memorandum report of the accident such as National Safety Council Form Vehicle 2, an accident report checklist, pencils, plain paper, courtesy cards, and telephone numbers of insurance representatives. Drivers will then have what they need to take down all pertinent information at the scene of an accident.

Specially printed accident report cards can help the driver quickly get names and addresses of key witnesses at the accident site (Figure 10–3). Drivers should be told it is important to identify as many witnesses to the accident as possible.

Failure to report an accident, no matter how slight, or falsification of data on an accident report should be cause for disciplinary action against the driver. Management can ensure timely accident reporting by requiring a vehicle inspection before and after each trip.

In the case of serious accidents, especially those resulting in a fatality or severe personal injury, a representative of the company—the supervisor, safety director, or claim agent—should conduct an investigation of the accident at the scene as quickly as possible. The purpose of such an investigation is to verify the driver's information on the accident report form and to obtain other data that might prove valuable for accident prevention work or for defense against liability claims.

Corrective Interviewing

As soon as possible after an accident, the investigator should interview the driver involved to determine whether the accident could have been prevented. If this is determined to be the case, based on the interview and written reports, the accident should be classified as preventable. The fleet professional must then explain to the driver what actions

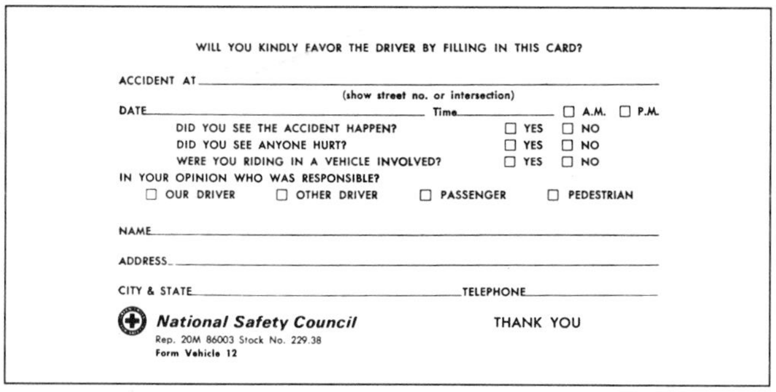

WILL YOU KINDLY FAVOR THE DRIVER BY FILLING IN THIS CARD?

ACCIDENT AT ____________________

(show street no. or intersection)

DATE ____________ Time ________ ☐ A.M. ☐ P.M.

DID YOU SEE THE ACCIDENT HAPPEN? ☐ YES ☐ NO

DID YOU SEE ANYONE HURT? ☐ YES ☐ NO

WERE YOU RIDING IN A VEHICLE INVOLVED? ☐ YES ☐ NO

IN YOUR OPINION WHO WAS RESPONSIBLE?

☐ OUR DRIVER ☐ OTHER DRIVER ☐ PASSENGER ☐ PEDESTRIAN

NAME ____________________

ADDRESS ____________________

CITY & STATE ____________ TELEPHONE ____________

National Safety Council

Rep. 20M 86003 Stock No. 229.38

Form Vehicle 12

THANK YOU

Figure 10–3. When this Courtesy Card is distributed, filled in, and collected, it will help determine who saw the accident in case witnesses are needed later.

contributed to the accident and make sure the driver understands what to do to prevent similar accidents in the future. For example, if the driver's lack of skill contributed to the accident, the safety investigator should recommend remedial driver training.

If, on the other hand, the investigator believes the driver did everything possible to prevent the accident, it should be classified as nonpreventable. Responsibility for preventing accidents involves more than carefully observing traffic rules and regulations. Drivers must practice "defensive driving" to prevent accidents, that is, they must assume other drivers will do the unexpected and will fail to observe normal traffic laws.

Driver Record Cards

Management should maintain a record card or similar ongoing record for each employee who drives a company vehicle. This record not only furnishes a history of all accidents a driver has been involved in but also provides information needed for safe driver awards or other forms of safety recognition (Figure 10–4). The date of each accident and the accident category, preventable or nonpreventable, should be entered on this record. The investigator should review the cards regularly to determine trends and recurring incidents.

When a driver becomes an accident repeater, management should make every effort to rehabilitate the person through counseling, retraining, closer supervision, or reassignment. When all efforts fail to reduce preventable accidents, the driver should be discharged or assigned to nondriving duties in the best interests of both the firm and the employee.

Fleet Accident Frequency

A useful accident control tool is monthly or quarterly computing of the fleet's accident frequency rate per 1,000,000 vehicle miles driven. Vehicle miles should be computed from odometer readings of all vehicles and not left to rough guesses based on route mileages unless the operations of the fleet are stable from day to day. The standard formula for figuring a fleet accident rate is:

$$\text{Fleet accident frequency rate} = \frac{\text{No. of accidents} \times 1{,}000{,}000}{\text{million miles driven}}$$

By keeping a monthly record of frequency rates, the investigator can:

1. analyze changes in group safety performance
2. compare records from several years to find seasonal, or other, trends
3. compare the fleet's performance with similar fleets.

The fleet safety program can be planned based on an analysis of the accident frequency rate, accident causes, and trends. Accident frequency rates also are of interest to management because they can prove whether the safety program is effective.

The National Fleet Safety Contest, conducted annually by the National Safety Council, issues a bulletin giving the accident frequency rates of all participating fleets, thus providing a means of comparison among similar companies.

Selection of Drivers

For some jobs requiring vehicle driving, such as sales representative or technician, other qualifications need to be considered besides a safe driving record. However, if the interviewer does not regard driving as an important part of the job, the prospective employee's competence as a driver may not be investigated completely. As a result, when a person is to be hired for a job in which driving is a regular or even occasional function, the company should make every effort to select an individual who can be expected to drive safely. In making that selection, the following factors should be considered.

Experience. Individuals who have a record of frequent vehicle accidents should not be assigned to drive company vehicles. An individual's safety record should be investigated (1) in a personal interview, (2) by consulting with former employers, and (3) by checking the state motor vehicle department or license bureau for accident reports and the National Driver Register service either for out-of-state or for two or more in-state license violations. A number of private services can provide motor vehicle record information electronically via computer within 24 hours.

Personal traits. Researchers have reported a close relationship between one's ability to drive safely and such personal traits as dependability, good judgment, courtesy, pleasant personality, and the ability to get along with other people. Conversely, those who tend to be antisocial, argumentative, and impulsive often prove to be problem drivers. Likewise, dissatisfied, timid, cocky, troublesome, or otherwise temperamental or unstable individuals often are not consistently good drivers.

Selection Standards

In determining standards of driver selection, management should make a careful analysis of the particular job and of pertinent driver qualifications. The next step is to decide what qualifications the applicant must have to perform the job effectively. Management must be able to justify each qualification imposed. It is helpful to study the employees who are performing the job in an average or better-than-average manner. Their skills and abilities should serve as the

National Safety Council
MOTOR TRANSPORTATION
AWARD AND ACCIDENT RECORD

JAN	FEB	MAR	APR	MAY	JUN	JUL	AUG	SEP	OCT	NOV	DEC
			X								

Name Smith, (Last) Bill (First) E. (Middle)

Company Highway Express

Location Midwest Division

Address 123 (Number) Home St., (Street) Any Town

Badge Number 123 Date Employed Feb. (Month) 1 (Day) 19-- (Year) Age 29 (At Emp.)

DRIVING TESTS

Date	Score	Remarks
2-4-19--	90	Slow reaction time

SAFE DRIVER AWARD RECORD

Earned 2 Year N. S. C. Award during period from 3-2-19-- to 3-2-19-- with Company Overnight Freight, Inc. Certified by F. J. O'Connell, Supt.

Award earned	Certificate Number	Date Award earned	Preventable Accidents File No	Preventable Accidents Date	Non-Driving Time From	Non-Driving Time To
3	468950	3/2/19--	1016	4/10.19--	12/1/19--	1/1/19--
4	501850	3/2/19--				
5	600100	4/2/19--				

REMARKS: 12-1-19-- took on month leave of absence.

Form Vehicle 6 PRINTED IN U.S.A. STOCK No. 229.36

Figure 10–4. An Award and Accident Record should be kept as part of each driver's personnel record. This form is useful in administering a company award plan and in counseling accident repeaters.

basis for what is expected of new employees. In addition, in the U.S. interstate and hazardous materials carriers must comply with driver qualification regulations of the U.S. Department of Transportation (DOT).

Canadian Provinces are encouraged to adopt the National Safety Code. Canadian carriers are required to comply with the provincial regulations adopted from the National Safety Code.

Safe driving skills always should be paramount. Otherwise, accident losses might completely offset any advantages gained through a driver's other special abilities (i.e. sales abilities, etc.).

Information-gathering Techniques

After essential job tasks and employee qualifications have been determined, management's next step is to develop methods for gathering and evaluating data about each applicant. Standard employment procedures include:

- application forms
- personal references
- interviews
- psychological tests
- driving tests
- MVRs
- physical examination.

Illustrations of the forms and more descriptive detail can be found in the National Safety Council's *Motor Fleet Safety Manual*, 3rd ed. (see References).

The application form is used to record details of an individual's past employment and other personal data to help the interviewer determine the best candidate for the job. Because of the Right to Privacy Act, no personal information can be requested on the written application or during the interview that does not have a direct bearing on a position's occupational requirements.

Questions on the form should cover the basic qualifications for the job and should be arranged in logical sequence. Specifically, questions about driving experience should include mileage and years the applicant spent as a driver and types of vehicles operated, seasons of the year and geographical areas in which the candidate operated vehicles, preventable and nonpreventable accidents they experienced, number of convictions received for traffic violations, number and type of driver's licenses held, and safe driver awards they received. The Office of Motor Carriers requires such specific information on the application form, other regulatory agencies may have other requirements.

Personal references. The applicant should furnish the names and addresses of previous employers in the space

provided on the application form; the interviewer should check these references. The interviewer can also look up an applicant's record with the appropriate state motor vehicle department.

Organizations should consult their legal advisor to guard against any liability resulting from personnel investigation activities.

A properly conducted interview is intended to reveal additional facts about the applicant's employment experience, knowledge of traffic regulations, attitude, personality, appearance, family life, and general background in order to help the interviewer make a good decision. The interviewer should conduct the session in private and make sure the applicant is seated and put completely at ease. The interviewer should keep in mind at all times the inventory of basic job qualifications. A written checklist of these can serve as a guide. After the interview, the applicant can be evaluated for each of the qualifications listed.

Driving tests. Each applicant should undergo an actual driving test or a written examination on traffic regulations as part of the selection procedure. Some firms use a written examination to evaluate driving ability. After an individual is hired, the examination can again be used to monitor his or her developing skills when undergoing an extensive driver training course.

There are two types of driving tests: driving range (off the road) and in traffic (on the road). The requirements of a good driving test in traffic can be listed as follows:

1. The test should be designed to sample a number of typical driving situations.
2. The test should include typical maneuvers to test a driver's ability.
3. The test should use a standard scoring procedure and predetermined test route so that all drivers are evaluated equally.
4. The examiner should have an objective checklist concerning the driver's performance (to reduce subjective judgment) and determine if driving errors may be corrected by proper training.

Driver performance measurement test. Michigan State University has developed the old driver performance measurement test. This test can help assess safe driving ability during selection by pointing out the precise unsafe habits likely to result in accidents. Information can be obtained from Michigan State University.

The Michigan State University (MSU) system requires the preparation of a thoroughly calibrated test course on public roads. As many as 50 traffic situations may be charted and correct/incorrect driving procedures determined for each. After the course is validated by MSU, nearby companies can use it to test their drivers, once individual company observers have been trained and certified into the MSU system.

Although the driving range test course obviously requires considerable organization and test time, it does give a standardized measure of driving performance.

Some companies use a truck "roadeo" type of driving course, where maneuvering skill can be measured. This may

be important to measure in a city driver who, for instance, may be backing into tight alleys and loading docks.

In some trucking operations, companies may test job applicants' skills in nondriving tasks, including required paperwork and reports. Experienced tank truck driver applicants may be asked to hook up hoses or load trucks. A driver claiming experience with hauling doubles (trailers) may be required to hook up a set of trailers. The applicant's driving test performance will indicate if driver training is needed and if weaknesses can be corrected during the training period.

Tests for interstate and hazardous materials drivers must meet the requirements of the DOT. The DOT has developed regulations covering driver licensing, drug testing, training, physical examinations, and other qualifications. Depending on the type of motor vehicle operated, organizations must comply with all or some of these requirements. As mentioned earlier, Canadian operations need to check the Provincial Regulations adopted from the National Safety Code.

Medical examination. All applicants should be examined by a physician who is familiar with the various governmental requirements before being hired. For U.S. firms engaged in interstate commerce or hazardous materials, medical examination of all new drivers is mandatory, and drivers must be reexamined at least every two years.

Drivers should also meet certain psychophysical vision, depth-perception, and hearing qualifications. Drivers should be advised of substandard results and how best to compensate for them.

Acceptance interview. Management should contact all applicants who pass the various requirements and tell them whether they have been hired or placed on a waiting list. The interviewer should reinforce applicants' enthusiasm for the job. Applicants should feel that although company employment standards are rigorous, they have attributes the company sincerely wants. The interviewer and others should welcome them into the company as valued new employees. Their new boss or supervisor should give further orientation about the company, what training they will receive, when and where to report for work, and an explanation of wages and working conditions.

Legal and Social Restrictions

The employer must abide by certain hiring requirements. Laws forbid discrimination against any job applicant.

These are not unreasonable restrictions. They certainly should not—in any way—force the employer to hire the unqualified. It is no favor to an applicant to be hired for a position for which he or she is not qualified and runs the risk of accidents, possible injuries, and an unfavorable work record.

Driver Training

Companies should strive to have individual training by a skilled instructor for all employees assigned to drive company motor vehicles. Various types of training can be provided.

- basic—for new employees

- remedial—for drivers who get into trouble
- refresher—for periodic updating of all drivers
- special—for operators of specialized equipment.

The company must plan a course to fit each job, assemble appropriate training materials, and provide training facilities. The driver training course should cover the following points.

State and federal driving rules. Most state and provincial motor vehicle departments publish the rules and regulations for those seeking drivers licenses. The training course should cover the salient points found in such booklets. Employees subject to U.S. licensing requirements and U.S. DOT examinations should attend appropriate training courses. The same is true of employees of Canadian operations as well as those of other countries.

Company driving rules. Each company has rules governing the use of company vehicles: how the vehicles may be obtained, where they may be operated, where parked and under what conditions, speed to be observed, and so on. Instructors should cover these rules thoroughly in the training course and give drivers written guidelines to take with them on the job.

What to do in case of an accident. This topic should include instructions on how to prepare company accident report forms, how individual driver records affect the employee, what to do in case of a vehicle accident away from the plant, mandatory reporting requirements, and so on.

Defensive driving. This concept embraces all the common-sense rules of safe and courteous driving. Defensive driving instruction seeks to impart to prospective drivers a strong sense of responsibility not only for their safety, but for the safety of others who are less skilled and who have had less training and practice (Figure 10–5). The company's safe driver award or incentive plan should be explained, as well as disciplinary action policy, and criteria for determining preventable accidents.

Figure 10–5. Training sessions in the National Safety Council "Defensive Driving Course" uses audiovisual materials to aid retention of good driving tips.

DRIVER'S VEHICLE INSPECTION REPORT
AS REQUIRED BY THE D.O.T. FEDERAL MOTOR CARRIER SAFETY REGULATIONS

CARRIER: _____

DATE _____ TIME: _____ A.M. _____ P.M.

Check any defective item and give details under "Remarks"

TRACTOR NO. _____

- Air Compressor
- Battery
- Body
- Brake Accessories
- Brakes
- Clutch
- Defroster
- Door Handles
- Drive Line
- Engine
- Fifth Wheel
- Front Axle
- Fuel Tanks
- Heater
- Horn
- Lights
- Loses Water
- Mirrors
- Oil Pressure
- Radiator
- Rear End
- Safety Equipment
- Springs
- Steering
- Tachograph
- Tires
- Transmission
- Wheels
- Windows
- Windshield Wiper
- OTHER

TRAILER(s) NO.(s) _____

- Brake Connections
- Brakes
- Coupling Chains
- Coupling (King) Pin
- Doors
- Hitch
- Landing Gear
- Lights
- Roof
- Springs
- Tarpaulin
- Tires
- Wheels
- OTHER

☐ CONDITION OF THE ABOVE VEHICLES IS SATISFACTORY

REMARKS: _____

Driver's Signature _____
☐ Above defects corrected   ☐ Above defects need not be corrected for safe operation of vehicle
Mechanic's Signature _____
Driver reviewing repairs _____ (Signature)

Figure 10–6a. Although maintenance personnel are responsible for giving the driver a vehicle that is in top mechanical condition, the driver must inspect the vehicle at the start of each day. Use of a checklist ensures that not point is forgotten. (Courtesy of JJ Keller & Associates, Inc., Neenah, WI. Used with permission.)

Safe Driving Program

Most fleet professionals regard it as part of their job to provide effective safety motivation so drivers will use more of their driving skills more of the time. To provide this motivation directly or indirectly, the safety program should include:

1. a required detailed report of every accident
2. driver interviews after each accident to determine whether it could have been prevented
3. a record of each driver's safety performance
4. continuous safety instruction and reminders—through the use of all media: company newsletters, bulletins, booklets, posters, bulletin board displays, meetings, and direct conversation
5. safe driving performance recognition.

A cautionary word about safe driver awards and prizes. Although these awards and cash or merchandise prizes for safe driving are strong motivators, company management

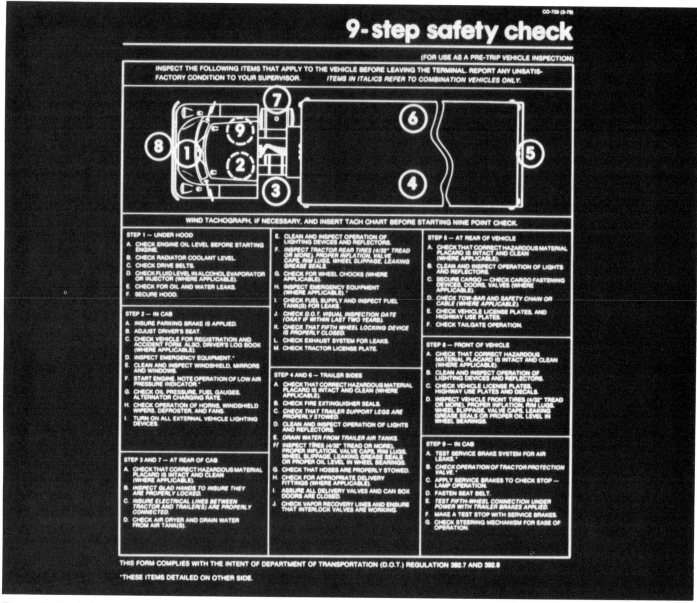

Figure 10–6b. This truck inspection chart will ensure no items are overlooked.

must understand that a recognition program *supports* a safety program, and is not a substitute for it.

Safety Devices

Some motor vehicle accidents are due to lack of safety devices and/or inadequate vehicle maintenance. Therefore, one of the fundamental requirements for safe operation is that all vehicles be equipped with the necessary safety devices. Tires, brakes, and steering mechanisms, headlights, taillights, horn and safety restraints must be maintained in first-class condition and used when required (Figures 10–6 and 10–6b). Additional safety devices include the following:

- directional signals
- windshield wipers
- windshield defroster
- fire extinguisher
- power steering/brakes
- low air-pressure warning system

- rock guards over the drive tires/mud guards on other tires
- adequate outside mirrors
- backup lights/required warning lights
- audible backup signal for trucks/power equipment
- slip-resistant surfacing on fenders, floors, and steps
- safety belts/restraint/protection devices/air bags
- tires/chains (when required)
- automatic sander (buses)
- antijackknife device (where required)
- reflective markings.

In addition, the following devices are recommended for dump trucks:

- light or indicator to show when the body is raised
- "Caution" sign on the rear of packer-loader trucks
- cab protector or canopy
- built-in body prop.

Certain hazardous materials and cargo require placarding and special precautions when they are loaded and unloaded.

Figure 10–7. This truck and loader are properly positioned for safe loading of materials. Loader is approaching the driver's side of the truck, not the blind side, and the driver is in the cab of the truck, not on the ground next to it. (Note the protective canopy over the cab.) The bucket is well positioned to drop the load in the center of the truck bed, reducing spillage. The loader operator has a full view of the truck, the bucket, and the material. Note the excellent housekeeping—loose material is leveled around the entire loading operation and the haul road is well maintained. Courtesy National Safety Council Construction Newsletter.)

The U.S. Department of Transportation (DOT), Environmental Protection Agency (EPA), and/or other regulatory requirements must be followed where applicable.

Accidental injuries incurred in the loading and unloading of materials, such as lumber, pipe, equipment, and supplies, are especially numerous but can be avoided if these precautions are followed:

1. The bulk and weight capacity of the truck should be observed.
2. Loads that may shift should be blocked or lashed. Tiedowns (ropes, chain, boomers) should be tightened on the right side or top of the load.
3. If material extends beyond the end of the tailgate, a red flag (or, at night, a red lamp) should be fastened to the end of the material. No material should extend over the sides.
4. Before loading or unloading a truck, the brakes must be securely set or the wheels blocked to protect the workers both on the truck and on the ground.
5. Loading and unloading of trucks. To reduce the danger to the driver from falling material while the truck is being loaded, the truck should be spotted so the load does not swing over the cab or seat (Figure 10–7).
6. If a truck cannot be so located and does not have a protective canopy over the cab, the driver should dismount and stand clear of the truck.
7. A truck should not be moved until all workers are either off the truck or properly seated on seats provided and are protected from injury if the load should shift during transit.
8. To avoid falling when unloading a flatbed truck, employees should keep away from the sides of the truck, especially when shoes, floors, and loads are wet or muddy.

9. Workers must be alert for pinch points when loads are being pulled, hauled, or lifted.

10. All safe practices for materials handling, such as using mechanical handling equipment, getting sufficient help, and so on, should be observed.

11. Specialized training depending on the classification of the cargo (for example, flammable, corrosive, radioactive) should be included.

12. When loading and unloading detached trailers with a lift truck, be sure the wheels are properly blocked/checked. It is important to place the chock properly when trailers are being loaded and unloaded at the dock. Preferred location of chocks is at the rear set of wheels. If a trailer is not properly blocked, the vehicle may move because of an incline or be set in motion by the loading or unloading operation.

Under a heavy forklift load, landing gears have collapsed from the weight because of dolly metal rust or fatigue, defective struts, or some other cause. If inspection indicates this possibility, the trailer nose can be supported by screw or hydraulic jacks—one on each side of the nose to provide additional support for the landing gear assembly.

Preventive Maintenance

A well-managed motorized equipment program, whether highway or off the road, includes a properly designed and implemented preventive maintenance (PM) program. Such a program, based on either the mileage or the operating hours of the equipment (as recommended by the manufacturer or company procedures), determines when the employer should change the oil, rotate or replace tires, and undertake minor and major routine maintenance.

The objectives of such a program are to:

1. prevent accidents and delays
2. minimize the number of vehicles down for repair
3. stabilize the work load of the maintenance department
4. save money by preventing excessive wear and breakdown of equipment and unscheduled downtime.

The PM program should cover all mechanical factors relating to safe operation of motorized equipment, such as brakes, headlights, rear and stop lights, turn signals, tires, windshield wipers, muffler and exhaust systems, steering mechanisms, glass fixtures, horns, and rearview mirrors. The manufacturer of the equipment and lubricant supplier can help with maintenance specifications.

Drivers and operators can play an important role in a PM program if they are properly instructed and motivated. Because they are most familiar with the vehicle and how it normally operates, they are usually the first to notice when minor or major mechanical defects develop. If at all possible, each driver or operator should be assigned a specific vehicle and be given responsibility for reporting defects. This encourages drivers and operators to take better care of their vehicles.

Drivers regulated by U.S. DOT must perform a pretrip vehicle inspection. However, all drivers should inspect their vehicles thoroughly before beginning the workday. Any major or safety-related defects must be reported and corrected before that vehicle is used. Minor items not affecting the vehicle's operation can be corrected at the next scheduled PM.

The company should provide vehicle condition report forms to all drivers and operators for the inspection, and a copy of the form should be kept in the vehicle cab. The form should cover the following items:

- Brakes should apply evenly to all wheels to prevent a vehicle from swerving when coming to a stop. This even pressure also gives maximum braking effectiveness.
- Headlights should function and be properly aimed to avoid blinding other motorists and to provide maximum lighting of the road ahead and to each side. The dimming switch and upper and lower beams should work properly at all times.
- Connecting cables on a combination vehicle should be fastened securely enough to be unaffected by vehicle vibration. All other cables, such as brake and electrical, should be free of defects.
- Make sure all stop lights, turn lights, rear lights, warning lights, and side-marker lights function properly.
- Tires should be inflated to recommended pressure, and regularly checked for adequate tread and any cuts, breaks, or other defects. Dual tires should be well matched.
- Windshield wipers must wipe clean and not streak or skip any of the glass, which can obscure or blur a driver's view of the road.
- Steering wheels should be free from excessive play. Front wheels should be properly aligned.
- All window glass should be free from cracks, discoloration, dirt, or unauthorized stickers that might obscure vision.
- Horns should respond to a light touch and be easy to locate.
- Rearview mirrors should give the driver a clear view. Portions of outside rearview mirrors can be conventional and convex to provide maximum sight advantage.
- Any stalling or lugging problems of the engine should be investigated and corrected immediately.
- All instruments must work properly.
- Exhaust systems should be checked to protect the driver against carbon monoxide gas leaks. The exhaust manifold, pipe connections, and muffler should be inspected periodically.

Emergency equipment in every vehicle must include a fire extinguisher, essential tools for road repairs, spare bulbs, flares, reflectors, flags, and such other equipment deemed necessary in case of fire, accident, or road breakdown. These items should be checked regularly to ensure they are in good working order and are readily available. Interstate and hazardous materials vehicles must be equipped with emergency items required by the appropriate regulatory agency.

Many governmental agencies require periodic safety tests and inspections for all vehicles. The maintenance facility must know the applicable inspection standards, and the company should see to it that all vehicles meet these requirements.

Figure 10–8. Hazards should be properly identified by signage in battery charging areas.

REPAIR SHOP SAFE PRACTICES

The vehicle maintenance facility must constantly evaluate the work habits of automotive repair employees, determine their safety attitudes, and cooperate with management to create a safe work environment. An effective safety program can help to ensure that all employees engage in safe work practices and that all federal and state or provincial regulations are followed (Figure 10–8).

Servicing and Maintaining Equipment

Workers may suffer a wide range of injuries while servicing and maintaining trucks. The equipment often requires mechanical aids for handling heavy parts.

Serious injuries are likely to occur when equipment undergoing repair shifts or moves unexpectedly. Workers should set equipment brakes and block the wheels. If work must be done under a raised vehicle, the vehicle must be propped up in such a way that it cannot come down should someone or something inadvertently strike the hoist or jack control levers or pedals and release the load.

Workers often use a jack to raise equipment and then leave it in place as an unstable support. Because serious injuries may occur when a vehicle falls off a jack, it is important that workers set the jack on a firm foundation and make sure it is exactly perpendicular to the load. When the vehicle has been raised to the desired height, it should be supported by stanchions, blocking, or other secure support.

Employees should never work near the engine fan or other exposed moving parts until the engine has been stopped. If they must run the engine to inspect or check on moving parts, they should keep a safe distance away and not attempt to make an adjustment. Workers should wear close-fitting unfrayed clothing, safety shoes, and goggles, but remove jewelry, especially rings, while repairing equipment.

Employees risk burn injuries when servicing vehicle engines that have been running. To open a radiator, the employee should use a heavy work glove, bleed off any steam, and then remove the cap. Gasoline or alcohol spilled on hot engines can cause a serious fire. Suitable funnels and safety containers should be used when transferring liquids.

Tire Operations

Workers face a particularly serious hazard when inflating vehicle (i.e., truck) tires should the locking ring blow off at high pressure. Using a tire safety rack will greatly reduce the potential for injury. Workers need to check all tires to make sure the valve has been removed and the tire fully deflated before disassembling it. When removing a tire from a dual wheel, fully deflate both tires before removing the one to be repaired. Workers must inflate tires in steel cages that will restrain flying objects should a blowout occur. A locking ring must be seated properly and must not be yanked free by

being twisted. Always replace a defective locking ring or rim with a sound one and keep ring and rim seats clean. Carefully separate parts, rims, and rings for various types of wheels to eliminate a possible mismatch.

Workers should use inflators they can preset, and that have locking attachments so hands or arms are not in the danger area, even if the tire is in a cage. Blowouts can occur because of overinflation, improper placement of the tire on the rim or wheel (which may pinch or chafe the tire or tube), or improper mounting of lock-rings or rims.

Only employees trained in tire repair and thoroughly familiar with the hazards and safe methods involved in handling tire equipment should inspect, install, repair, and replace tires and rims.

Other potentials for injury include strains or hernias resulting from lifting heavy tire assemblies. Companies should provide mechanical lifting and moving devices so workers are not required to lift heavy tires.

Rubber cement and flammable solvents used for patching inner tubes, and casing compounds used for filling tire cuts, should be kept in safety cans. Routinely inspect electric heating elements used for vulcanizing or branding tires and replace any defective wiring.

Where power-driven rasps or scrapers are used for casings or inner tubes, the operators must wear eye protection and an approved particulate filtering respirator. Attaching a local exhaust or vacuum system to these machines will keep the fine rubber dust out of the workroom air.

Fire Protection

Because of fire hazards, heavy-duty trucks should be equipped with Type B-C fire extinguishers listed by the Underwriters Laboratories for use on burning oil, gasoline, grease, and electrical equipment. Place the extinguisher in a convenient location in the cab or mount it on the vehicle, and train drivers to operate it. Some regulatory agencies require a monthly inspection of fire extinguishers.

The maintenance facility should also have an adequate number of fire extinguishers of the ABC type. Supervisors must show all employees how to use the equipment.

If workers perform cutting, burning, or welding tasks near fuel oil tanks, they should have an extinguisher on hand. A wet tarpaulin may be used to cover fuel, oil tanks, or combustible materials to protect them against sparks and excessive heat. Do not perform such work on a fuel tank or other fuel container until it has been drained and thoroughly purged of vapors. Hydrocarbon vapor meters, properly calibrated and adjusted, will determine vapor levels. Use of the "hot work" permit system is recommended.

Fueling requires certain precautions to avoid fires. Workers must shut off the engine and extinguish all smoking materials. Use safety containers and a grounded fuel hose for fueling. When the tank is being filled, the metal spout of the hose should firmly contact the tank to ensure bonding and to dissipate static charges strong enough to ignite fuel vapors and cause an explosion or fire.

Fires occur in shops each year because gasoline and similar flammable solvents are used for cleaning parts. As a result, use only safe, nontoxic and nonflammable cleaning liquids. Parts cleaning equipment should be provided with a self-closing lid in the event a fire occurs. The risk of a fire in a shop can be reduced further by good housekeeping, especially the proper disposal of oil-soaked combustible waste and similar materials in covered metal containers.

Fire prevention procedures apply as well to all motorized equipment, such as power cranes, shovels, bulldozers, and trucks. Details are covered in the *Engineering and Technology* volume, Chapter 11, Powered Industrial Trucks, and Chapter 12, Haulage and Off-Road Equipment.

Lubrication and Service Operations

In lubrication operations, make sure floors are kept free of grease and oil to prevent slips and falls. Spills should be wiped up immediately or covered with an oil-absorbent compound.

Advise workers to keep their hands away from sharp or rough edges and to obtain immediate first aid treatment for all cuts and scratches. Also, train workers not to put their hands or face in front of the grease gun nozzle when the handle is pulled. Instances have been reported in which quantities of grease have been forced under the skin of workers by high-pressure grease guns. Tops of grease cylinders should be securely fastened into place; otherwise, covers may blow off and seriously injure anyone who is nearby. Inspect all equipment weekly and make repairs when needed.

Workers should be warned of the danger of inhaling sprayed or atomized oils. They should stand clear of the lubricant spray, which settles quickly, and must not direct the spray at other employees. When draining used oil from machinery, many jurisdictions require that it be collected for recycling or properly disposed of by an approved waste handling company. Drained antifreeze and other drained lubricants such as car oils must be considered a hazardous waste and should be treated accordingly.

Wash Rack Operations

When washing vehicles, a person can slip and fall on wet floors, suffer cuts or abrasions from the sharp or rough edges of the vehicle, or experience burns from careless use of hot water or steam. The concrete floor of the wash rack should be rough troweled to produce a slip-resistant surface. While washing vehicles, employees should wear safety toe rubber boots, preferably with slip-resistant soles and heels, and a rubber coat or apron and gloves.

Workers should never point the high velocity streams of hot or cold water at another person as serious injuries can result. Instead, workers should direct the hose, particularly when washing under the vehicle, in such a way as to avoid being struck by a backlashing stream of water and dirt.

Supervisors should provide workers with product safety information covering the washing solutions used. Where a hot water or steam hose is used, employees must wear personal protective equipment to avoid skin contact and to prevent burns. Heavy-duty gloves and face shields should be used when necessary. A portable fan can blow steam away

to enable the operator to see the work. Supervisors should schedule a periodic cleanup of the entire wash rack and associated equipment.

Battery Charging

Although most battery-operated vehicles now are recharged by means of on-board chargers and plug-in cords, some batteries still require out-of-vehicle recharging. Other circumstances occasionally call for battery removal and charging such as when they are being serviced, replaced, or emergency jump-started. The principal hazards of battery charging operations are acid burns, back strains from lifting, electric shocks, slips, falls, and explosions from buildup of hydrogen gas (Figure 10–8). Train workers to observe proper safety procedures whenever they are servicing or charging a battery. They should not charge a frozen battery—most of the charge will go to heat the battery and not to energize it.

Employees should wear safety apparel suitable for battery work. This includes splash-proof eye and face protection and acid-proof gloves, aprons, and boots with slip-resistant soles. Workers should wear rubber boots and aprons when filling batteries. Goggles and face shields should be used when working around batteries to prevent acid burns to the eyes and face.

Flooring should be constructed of wood slats and kept in good condition to prevent slips and falls and to protect against electric shocks from batteries being charged. Install fire doors between battery-charging rooms and adjacent areas where flammable liquids are handled and stored.

Always follow the manufacturer's recommendations about charging rates for various size batteries to prevent rapid generation of hydrogen. Potentially explosive quantities of oxygen and hydrogen accumulate in battery cells. This is particularly true if the battery is defective or if a heavy charge has been or is being applied. The lower the water level in the battery, the larger the cavity in which gas can accumulate. Care should be taken to prevent electrical arcing while batteries are being charged, tested, or handled. Make sure tools and loose metal (and even lifting hoist chains) cannot fall on batteries and cause a short circuit, which in turn can inflict serious burns on workers or cause an explosion.

When manual lifting is necessary, tell employees to request sufficient help to prevent strains, sprains, or hernias. Use hand carts for transporting batteries.

Workers must handle acid carboys with special care to prevent breakage and possible injury caused by splashing acid. Acid carboys should never be moved without their protecting boxes nor stored in excessively warm locations or in the direct rays of the sun. Instead, use carboy tilters. A safety shower and eyewash fountain are required by certain regulatory agencies wherever acids or caustics are used (Chapter 4, Occupational Health Programs).

A summary of recommendations for changing and charging batteries is given in the *Engineering and Technology* volume.

First aid for chemical burns. Many batteries contain an acid electrolyte although some, such as the nickel-iron battery, contain an alkali solution. Whether it be acidic or alkaline, if electrolyte gets on a person's skin or eyes, it must be immediately washed off with large quantities of running water for 15 minutes or more. Supervisors and workers should understand that neutralizing agents are mishandled so often, they can do more harm than good. Use them only if first aid directions are printed clearly on labels or on company or plant directions. Get medical aid at once.

First aid for burns of the eye. Flush the eye thoroughly with large amounts of clean, cool water for 15 minutes or more. Place a sterile dressing over the eye to immobilize the lid and get medical aid at once. Check with the company physician for emergency help.

Gasoline Safety

The handling and storing of gasoline should comply with the provisions of the *National Fire Protection Association Flammable and Combustible Liquids Code*, NFPA 30 and NFPA 30A, which is discussed in Chapter 16 of the *Engineering and Technology* volume. Never use gasoline for any cleaning tasks. Solvents with higher flash points are equally effective and much safer. Even when higher flash-point solvents are used, if carburetors or gas-line parts are cleaned, the solution should be regularly changed. The admixture of even small quantities of gasoline will lower the solvent's flash point and increase the danger of fire and explosion.

Workers can remove grease, oil, and dirt from metal parts with nonflammable solutions or with high-flash point solvents in special degreasing tanks with adequate ventilating facilities and fire protection. Advise workers not to use gasoline to remove oil and grease from garage floors. Nonflammable commercial cleaning compounds are available for such tasks.

Workers should not use gasoline to remove oil and grease from their hands. Strong commercial soaps will effectively remove greasy dirt from the skin without risk of injury. In addition, protective creams and ointments, if applied before starting work, will protect an employee's skin from dirt and grease. Likewise, workers should never use gasoline to clean work clothes.

Gasoline spills should be cleaned up immediately; many absorbent substances can be used to soak up spills. Dirt, sand, or other material should be used to prevent gasoline from entering sewers, drains, and so on. If quantities of gasoline get into the sewage system, notify the fire department at once so the sewers can be flushed. Because gasoline vapor is heavier than air, it collects in low spots, such as basements, elevator pits, and sumps. Ventilate these places until gasoline vapors are dissipated to a safe level. Do not dispose of spilled gasoline by flushing it down a sewer drain, sink, toilet, and the like. Instead, collect spilled gasoline into an approved container and dispose of it through a waste-handling company, or other approved means.

Other Safe Practices

Using jacks and chain hoists. Vehicles jacked up or hung on chain hoists should always be blocked with stan-

chions, pyramid jacks, or wood blocks (make sure these items are carefully inspected before use). The best jack for maintenance garage work is the hydraulic-over-air type—if one system fails, the other will operate. Do not use ordinary pedestal jacks, especially the type supplied for passenger cars, as the vehicle can be tipped or jarred off the jacks and cause injury.

When someone is working under a blocked-up vehicle, other employees should not do any work on the car that might knock it off the blocks. Employees who work under vehicles should be safeguarded from danger when their legs protrude into passageways. Erect barricades around them or make sure the worker's entire body is under the vehicle. Advise employees to wear suitable eye protection such as safety glasses, goggles, or plastic eye shields to avoid injury from dirt and metal chips falling into their eyes. When necessary, fog-resistant eye protection should be used.

Removing exhaust gases. Local exhaust and ventilating equipment can prevent accumulation of vehicle exhaust gases within a repair shop.

Repairing radiators. Where radiators are boiled out or tested for leaks, the operator should have both chemical goggles and a face shield of clear plastic. The entire face needs protection.

Cleaning spark plugs. All mechanics using blast-type spark plug cleaners should wear goggles or face shields.

Replacing brakes. Workers should not use air pressure when cleaning around brake drums and backing plates while replacing brakes. The pressure can send asbestos, metal particles, or dust from the brake area into the air and cause respiratory problems. Instead, workers can clean brakes with vacuums, chemical wash solutions, or steam cleaners. Train workers to wear approved respiratory protective equipment until brake components are free of dust. Often the work area may not be free of dust, and respiratory equipment may still be needed.

Controlling traffic. Supervisors should regulate movement of vehicles inside maintenance facilities. Traffic lanes and parking spaces should be painted on the floors, with arrows and signs indicating the direction of traffic flow. Vehicles with air brakes should not be moved until sufficient air pressure has been built up. Move vehicles using low gear and low speeds inside shop areas, especially up and down ramps.

Drivers must stop their vehicles at entrances or exits and make sure the way is clear or sound their horn before passing through. Signs requiring this procedure should be posted in conspicuous places. Install mirrors at blind corners.

Other potential sources of injury include jumping across open inspection pits, falling off ladders, material handling, and using hand tools improperly. Prevention of all work injuries requires proper selection and training of employees, careful supervision of their work habits, review of all injury causes, and compliance with safety programs.

Training Repair Shop Personnel

Supervisors and managers should train apprentices and new employees to do each job in the most efficient manner. Make sure job instruction includes the safety rules and regulations pertaining to each job and the reasons for such rules. New mechanics must be thoroughly indoctrinated concerning the company's policy toward safety and the safety program.

The new employee, having been indoctrinated and trained to work safely, must be motivated to observe accepted safe practices while performing job tasks. Many motivational methods are available to accomplish this goal, including safety supervision; safety contests; and meetings, posters, safety bulletins, and pamphlets (see Chapter 16, Safety Training).

More details on mechanical and chemical safety can be found in the *Engineering and Technology* volume and in *Fundamentals of Industrial Hygiene*, respectively, the other two books in this series.

SUMMARY

- Most motor vehicle accidents are caused by using improper driving procedures; only a small percentage are the result of mechanical failure. Companies can control driver error by introducing a program of driver selection, training, and supervision, while vehicle failure can be reduced by implementing a preventive maintenance program.

- The total cost of vehicle accidents usually exceeds the amount recovered from the insurance company and includes direct and indirect expenses of accidents. These expenses include loss of service, workers' compensation, time spent investigating and reporting the accident, replacement or retraining of injured workers, and production downtime. Clearly, the cost of a vehicle accident prevention program is more than justified when compared with potential losses related to accidents.

- A vehicle safety program should include a written safety policy, a designated safety program manager, efficient accident investigation and reporting systems, and a preventive vehicle maintenance program. Often the safety and health professional is responsible for supervising the program and reporting on safety issues to top management, although in smaller companies the function may be assigned to the fleet manager.

- A driver safety program should include a training program, accident prevention measures, accident reporting procedures, driver performance goals and incident reports, and a method for establishing competency and skills levels and accident/safety records for each driver. Safe driver awards and recognition can act as strong motivators to help ensure worker compliance.

- A motor vehicle accident can be defined as any incident in which the vehicle comes in contact with another vehicle, person, object, or animal with resulting injury or property damage of any type. All incidents must be reported, regardless of their nature or who was at fault. The reports provide information that may help to prevent similar accidents in the future.

- Accident reporting procedures should enable all employees to follow the same steps and use the same forms to

document details and to record names and addresses of witnesses. Accident reporting packets should be given to each employee. An accident investigator should interview drivers immediately after an accident and compile accurate, timely records to help the company calculate accident frequency rates and identify unsafe drivers.

- When selecting new drivers, companies should look at the applicants' experience and driving record, check their references, and access local, state, and federal agencies to investigate their backgrounds. Applicants can be given driving range (off the road) and in traffic (on the road) tests to evaluate their driving skills. Those who need driver training should receive either basic, remedial, refresher, or special training, depending on their skill levels and job requirements.
- Vehicles should have proper safety devices (directional and other signals, windshield wipers, fire extinguishers, reflectorized triangles, power steering, brakes, etc.) in good working order to help prevent accidents. Workers must take precautions to avoid being injured when loading or unloading trucks or when handling hazardous or toxic materials.
- A preventive maintenance program can be based either on mileage or operating hours of the equipment. Program goals are to prevent accidents and delays, minimize repairs, stabilize maintenance workload, and reduce costs resulting from equipment breakdown and interruption of customer contacts. Drivers can assist in the program by reporting any malfunctions or problems they encounter when driving their vehicles.
- To maintain safe practices in a maintenance facility, management must know employees' work habits and attitudes and cooperate with upper management in enforcing company standards. Workers can avoid injuries and accidents by following proper procedures for working on vehicles, practicing good housekeeping in the shop, donning protective clothing and gear for hazardous tasks, and obtaining a thorough knowledge of repair and cleaning equipment and tools.
- Repair shop workers should receive special safety training for fire protection, lubrication and service operations, wash rack operations, battery charging, gasoline storage safety, and other high-risk work tasks. New employees should not be allowed to work on hazardous jobs until they have been fully trained and have worked for a time under close supervision.

REFERENCES

Alcohol and Drug Problems Association of North America (formerly North American Association of Alcoholism Programs) 444 North Capitol Street NW, Suite 706, Washington, DC 20001.

Alliance of American Insurers, 1501 Woodfield Road, Suite 400 W, Schaumburg, IL 60173. *Code of the Road.*

American Automobile Association, 1000 AAA Drive,

Heathrow, FL 32746. "Driver Training Equipment" (catalog).

American National Red Cross. *Standard First Aid and Personal Safety Manual*, 2nd ed. Washington, DC: American National Red Cross, 1979.

American National Standards Institute, 11 West 42nd Street, New York, NY 10036.
Method of Recording and Measuring the Off-the-Job Disabling Accidental Injury Experience of Employees, ANSI Z16.3–1989.
Lift and High Lift Trucks, ANSI/ASME B56.1–1983 (R1988).
Powered Industrial Trucks, ANSI/NFPA 505–1981 (R1987).

American Society of Safety Engineers, 1800 East Oakton, Des Plaines, IL 60016.
Dictionary of Terms Used in the Safety Profession.
Photographic Techniques for Accident Investigation.
Profitable Risk Control.
Safety Law—A Legal Reference for the Safety Professional.

American Trucking Associations, Inc., 2200 Mill Road, Alexandria, VA 22314.
ATA Hazardous Materials Tariff, 1983.
Bulletin Advisory Service (3 vols.), 1983.
Effective Truck Terminal Planning and Operations, 1980.
Fundamentals of Transporting Hazardous Materials, 1982.
Fundamentals of Transporting Hazardous Waste, 1980.
National Truck Driving Championship, 1986.

Associated General Contractors of America, Inc., 1957 E Street NW, Washington, DC 20006. *Manual of Accident Prevention in Construction.*

Association of American Railroads. *Loading, Blocking and Bracing of Freight in Closed Trailers for Trailer or Flat Bed Service, Suggested Methods for.* Washington, DC: Association of American Railroads.

Association of Casualty and Surety Companies, 110 William Street, New York, NY 10038.
Guide Book, Commercial Vehicle Drivers
Truck and Bus Drivers Rule Book

Baker JS. *Traffic Accident Investigation Manual*, 9th ed. Chicago: Traffic Institute, 1986.

Heavy Construction Contractors Association, P.O. Box 505, Merrifield, VA 22116. (General.)

Kirk-Othmer Encyclopedia of Chemical Technology, 3rd ed. New York: Wiley Interscience, 1984.

The National Committee for Motor Fleet Supervisor Training. *Motor Fleet Safety Supervision, Principles and Practices.* The National Committee, A 364, Engineering Bldg., Michigan State University, East Lansing, MI 48824.

National Fire Protection Association, 1 Batterymarch Park,

Quincy, MA 02269.
Flammable and Combustible Liquids Code, (1987) NFPA 30.
Fire Prevention Code, (1987) NFPA 1.
Fire Protection Handbook, 15th ed., 1981.
Hazardous Chemical Data, (1975) NFPA 49.
Hazardous Materials Transportation Accidents, 1978.
Life Safety Code Handbook, 1981.
National Electrical Code Handbook, 1984.

National Safety Council, 444 North Michigan Avenue, Chicago, IL 60611.
Accident Facts. (Published annually.)
Aviation Ground Operation Safety Handbook, 4th ed.
Chemical Hazard Fact Finder—Slide Rule.
Defensive Driving Program Materials.
Fleet Accident Rates—Manual. (Published annually.)
Industrial Data Sheets
Lead-Acid Storage Batteries, 12304–0635.
Liquefied Petroleum Gases for Industrial Trucks, 12304–0479.
Mounting Heavy-Duty Tires and Rims, 12304–0411.
National Fleet Safety Contest.
Motor Fleet Safety Manual, 3rd ed.
Public Employee Safety Guides
"Street and Highway Maintenance"
"Vehicular Equipment Maintenance"
Safe Driver Award Program.
Standards for School Buses and Operations.
Supervisors' Safety Manual, 7th ed., 1990.

New York University Center for Safety Education, Washington Square, New York, NY 10003. Publications list.

Pilot Judgment Training and Evaluation, Volumes I-III (DOT/FAA/CT–82–56), Bunnell, FL: Embry-Riddle Aeronautical University, June 1982.

National Private Truck Council of America. *Driver Training Manual.* Washington, DC: Private Truck Council of America, 1981.

U.S. Department of Defense, Department of the Army, The Pentagon, Washington, DC 20310.
Driver Selection and Training, TM 21–300.
Drivers' Manual. TM 21–305.
General Safety Requirements. EM 385–1–1, U.S. Army Corps of Engineers.
"Methods of Teaching."
Motor Transportation, Operation, FM 25–10.

U.S. Department of the Interior, Bureau of Mines, 2401 E Street NW, Washington, DC 20241.
Minerals Yearbook.
Also various handbooks, miners' circulars, and other publications.

U.S. Department of Labor, 200 Constitution Avenue NW, Washington, DC 20210.
Occupational Safety and Health Standards, CFR, 1910 and 1926.
OSHA Compliance Guide, Volume 3.
OSHA Compliance Operations Manual.
OSHA Recordkeeping Requirements.
OSHA Job Hazard Analysis.

U.S. Department of Transportation, 400 Seventh Street SW, Washington, DC 20590.
Bureau of Motor Carrier Safety Regulations.
Hazardous Materials Emergency Response Guidebook.
Manual on Uniform Traffic Control Devices for Streets and Highways. (Also identified as American National Standard D6.1).
Model Curriculum for Training Tractor-Trailer Drivers.
Title 49, *Code of Federal Regulations*, Parts 390–397, *Motor Carrier Safety Regulations.*

U.S. Government Printing Office, North Capitol and H Streets NW, Washington, DC 20401.
Code of Federal Regulations:
Title 29—"Labor."
Title 40—"Protection of the Environment."
Title 49—"Transportation."

PART 3
HAZARD
INFORMATION
AND ANALYSIS

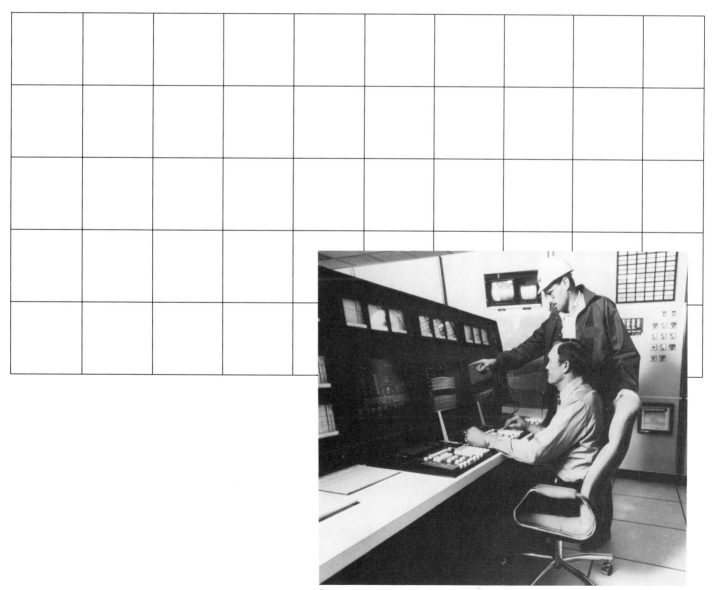

Computer-assisted monitoring of operations. (Courtesy UOP.)

11

Acquiring Hazard Information

Before hazards can be controlled, they must be identified. Monitoring is an effective means of acquiring such hazard information. *Monitoring* can be defined as a set of observation and data collection methods used to detect and measure deviations from plans and procedures in current operations (Johnson, 1980).

Through monitoring it can be ascertained that (1) controls are functioning as intended, (2) workplace modifications have not rendered the original controls ineffective, and (3) new problems have not surfaced since the most recent controls were introduced (Firenze, 1978).

Monitoring can involve four functions: hazard analysis, inspection, measurement and testing, and accident investigation. Including all four functions means that monitoring is performed before the operation begins, during the life cycle of the operation, and after indications that the system has broken down. A systems approach to hazard control uses each of these methods, as described in this chapter.

HAZARD ANALYSIS

Data from hazard analysis can be regarded as a baseline for future monitoring activities. Before the workplace is inspected to ensure environmental and physical factors fall within safe ranges, hazards inherent in the system must be discovered. Hazard analysis has proven to be an excellent tool to identify and evaluate hazards in the workplace.

Analyzing a problem or situation to obtain data for decision making is not new. Good workers and their supervisors constantly make assessments—even if unconsciously—about their work to guide their actions. Written analyses carry the process one step further by providing the means to document hazard information.

Philosophy

Written analyses often serve as the basis for more thorough inspections. They can communicate data about hazards and risk potential to those in command positions. They can educate those in the line and staff organizations who need to know the consequences of hazards within their operations and the purpose and logic behind established control measures. Management can request formal, written analysis for each critical operation. By doing so, it not only gathers information for immediate use, but it also reaps benefits over the long run. Once important hazard data are committed to paper, they become part of the technical information base of the organization. These documents show the employer's concern for locating hazards and establishing corrective measures before an accident happens.

Traditionally, systems have been analyzed during their operational phase to determine failures that impaired system effectiveness. Hazard analytical techniques applied during this phase returned substantial dividends by reducing both accident and overall operational losses.

Hazard control specialists no longer concentrate solely on operations. They look at the conceptual and design stages of the systems for which they are responsible. They use analyti-

239

FMEA Form

COMPONENT	FAILURE OR ERROR MODE	EFFECTS ON		SEVERITY INDEX	FAILURE FREQUENCY INDEX	CRITI-CALITY	DETECTION METHODS	COMPENSATING PROVISIONS AND REMARKS
		OTHER COMPONENTS	WHOLE SYSTEM					

For Reliability:
  Design Characteristic
  Failure Mode
  Failure Probability
  Effect on System
  Essentiality Code
  Control to Minimize:
    Frequency
    Effect

For Priority Problem Lists:
  Energy Sources:
    Kinds
    Amounts
  Potential Targets
  Barriers, Controls
  Residual Risk
  Failure Mode
  Failure Mechanism
  Consequence Potential
  Frequency, Consequence Matrix Class
  Action-Decision Classes
    Authority Level
    Type and Date Action Due

Figure 11–1. Failure mode and effect analysis form used by the Aerojet Nuclear Company (Johnson, 1980).

cal methods and techniques before the process or product is built to identify and judge the nature and effects of hazards associated with their systems.

This wide assessment has significantly altered the direction of hazard control efforts. When potential failures can be located prior to the production or onstream process stage of a system's life cycle, specialists can cut costs and avoid damage, injuries, and death. Systems engineering was initially concerned with increasing *effectiveness*, not profits. Properly applied, however, it can point out profitable solutions to many of management's most perplexing operational problems.

What Is Hazard Analysis?

Hazard analysis is an orderly process used to acquire specific hazard and failure data pertinent to a given system (Firenze, 1978). A popular adage holds that "most things work out right for the wrong reasons." By providing data for informed management decisions, hazard analysis helps things work out right for the right reasons. The method forces those conducting the analysis to ask the right questions and helps to answer them. By locating those hazards that are the most probable and/or have the severest consequences, hazard analyses

provide information needed to establish effective control measures. Analytic techniques assist the investigator in deciding what facts to seek, determining probable causes and contributing factors, and arranging orderly, clear results.

What are some uses for hazard analysis?

- It can uncover hazards that have been overlooked in the original design, mock-up, or setup of a particular process, operation, or task.
- It can locate hazards that developed after a particular process, operation, or task was instituted.
- It can determine the essential factors in and requirements for specific job processes, operations, and tasks. It can indicate what qualifications are prerequisites to safe and productive work performance.
- It can indicate the need for modifying processes, practices, operations, and tasks.
- It can identify situational hazards in facilities, equipment, tools, materials, and operational events (for example, unsafe conditions).
- It can identify ergonomic situations through anthro-prometrics and work design (for example, work table heights, chairs, reaching capabilities).

OPERATIONS HAZARD ANALYSIS									
Process	Operational Step	Task	Source of Potential Hazard	Triggering Event	Potential Effect on Equip., Material Environment	Personal Injury, Property Damage	RAC	Procedural Requirements	Safety & P.P.E.
Turning steel stock between centers on machine lathe.	Rough turning steel stock.	Select cutting tool and place in tool holder.	Improper tool used for rough cutting operation.	Starting lathe.	Tool jams in stock. Stock comes off centers. Uneven cut. Wasted stock.	Operator is hit in face with flying chips of steel.	2	A right-cut tool or roundhouse tool should be held in a straight tool holder. Lathe located to minimize exposure to other work stations.	Select proper tool for job. Operator to wear face protection while operating lathe.
		Place tool holder in the tool post and adjust cutting tool to proper location.	Tool holder extending too far from tool post.	Starting lathe.	Same as above. Breaking cutting tool.	Operator is hit in face with flying chips of steel from stock and broken cutting tool.	2	Tool post should be at end of T-slot. Face of tool must be on center and turned slightly away from headstock.	Operator to wear face protection while operating lathe.

Figure 11–2. This Operations Hazard Analysis (OHA) form is used for industrial operations. (Printed with permission from Indiana Labor & Management Council, Inc.)

- It can identify work practices responsible for accident situations (for example, deviations from standard procedures).
- It can identify exposure factors that contribute to injury and illness (such as contact with hazardous substances, materials, or physical agents).
- It can identify physical factors that contribute to accident situations (noise, vibration, insufficient illumination, to name a few).
- It can determine appropriate monitoring methods and maintenance standards needed for safety.

Formal Methods of Hazard Analysis

Formal hazard analytical methods can be divided into two broad categories: inductive and deductive.

Inductive method. The inductive analytical method uses observable data to predict what can happen within a particular system. It postulates how the component parts of a system will contribute to the success or failure of the system as a whole. Inductive analysis considers a system's operation from the standpoint of its components, their failure in a specific operating condition, and the effect of that failure on the system.

The inductive method forms the basis for such analyses as failure mode and effect analysis (FMEA) and operations hazard analysis (OHA). In FMEA, the failure or malfunction of each component is considered, including the mode of failure. Management can trace throughout the system those effects of the hazard(s) that led to the failure and evaluate the ultimate impact on task performance. However, because only one failure is considered at a time, some possibilities may be overlooked. Figure 11–1 illustrates the FMEA format used at Aerojet Nuclear Co., Idaho Falls, Idaho. Figure 11–2 illustrates the OHA format used for industrial operations.

Once the inductive analysis is completed and the critical failures requiring further investigation are detected, then the fault tree analysis will facilitate an inspection (see Deductive method below). Chapter 16, Safety Training, discusses job safety analysis (JSA), an analysis using the inductive method. That chapter explains the basic steps to be taken and the uses of a job safety analysis.

Deductive method. If inductive analysis tells us **what** can happen, deductive analysis tells us **how**. It postulates failure of the entire system and then identifies how the components could contribute to the failure.

Deductive methods use a combined-events analysis, often in the form of trees. The positive tree calls for stating the requirements for success; see Figure 11–3. Positive trees are less commonly used than fault trees because they can easily become a list of "you shoulds" and sound moralizing.

Fault trees are reverse images of positive trees and show ways troubles can occur. The analyst selects an undesired event, and diagrams, in the form of a tree, all the possible happenings that can contribute to the event. The branches of the tree continue until they reach independent events. The analyst can then determine probabilities for the independent events.

The fault tree requires rigorous, thorough analysis; it involves listing all known sources of failure. The fault tree is a graphic model of the various parallel and sequential combinations of system component faults that can result in a single, selected system fault. Figure 11–3 illustrates three types of analytical trees.

Analytical trees have three advantages:

1. They accomplish rigorous, thorough analysis without wordiness. Using known data, the analyst can identify the single and multiple causes capable of inducing the undesired event.
2. They make the analytical process visible, allowing for the rapid transfer of hazard data from person to person, group

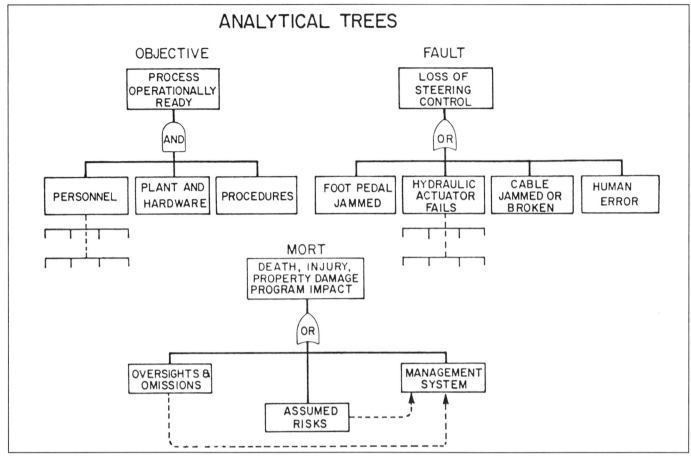

Figure 11–3. Analytical trees are nothing but "structured common sense." Trees are of two major types—the objective or positive trees, which emphasize how a job should properly be done, and the fault trees, which chart those things that can go wrong and produce a specific failure. A fault tree structured for one job can be generalized to cover a wide variety of jobs. The MORT diagram describes the ideal safety program in an orderly, logical manner; it is based on three branches: (1) a branch dealing with specific oversights and omissions at the work site, (2) a branch that deals with the management system that establishes policies and makes the entire system go, and (3) an assumed-risk branch visually recognizing that no activity is completely risk-free, and that risk-management functions must exist in any well-managed organization (Nertney, 1977). (Printed with permission from *Professional Safety*, February 1977.)

to group, with few possibilities for miscommunication during the transfer.

3. They can be used as investigative tools. By reasoning backwards from the accident (the undesired event), the investigator is able to reconstruct the system and pinpoint those elements responsible for the undesired event.

Cost-effectiveness. The cost-effectiveness method can be used as part of either the inductive or deductive approach. The cost of system changes made to increase safety is compared with the decreased costs of fewer serious failures or with the increased efficiency of the system. Cost-effectiveness frequently is used to decide among several systems, each capable of performing the same task.

Choosing which method to use. To decide what hazard analytical approach is best for a given situation, the hazard control specialist will want to answer five questions:

1. What is the quantity and quality of information desired?
2. What information already is available?
3. What is the cost of setting up and conducting analyses?
4. How much time is available before decisions must be made and action taken?

5. How many people are available to assist in the hazard analysis, and what are their qualifications?

Conducting a hazard analysis can be expensive. Before a hazard analysis technique is chosen, it is important to determine what information is needed and how important it is.

It is beyond the scope of this *Manual* to go into detail regarding other applications of system safety (See Johnson, 1980). Figure 11–4 indicates a few other areas in which large-project system safety methods and ideas can, with a little adaptation, be scaled down to apply to industrial jobs (Nertney, 1977).

Who Should Participate in Hazard Analysis?

A hazard analysis, to be fully effective and reliable, should represent as many different viewpoints as possible. Each person familiar with a process or operation has acquired insights concerning problems, faults, and situations that can cause accidents. These insights need to be recorded, along with those of the person initiating the hazard analysis—usually the safety professional. Input from workers and employee representatives can be extremely valuable at this stage.

System Safety Method or Concept	Scaled-Down Version	Comments
Hazard Analysis and Risk Projection	1. Think like an insurance agent—try to reduce hazards to consequences. 2. Use insurance companies. "How much would you charge to insure us?"	Safety people too often try to communicate with line managers in terms of "hazards" rather than reducing the hazards to risks and consequences.
Biomechanics	Looking at the job from the worker's point of view in terms of physical characteristics.	For example, does the vehicle cab "fit" small men and women? Can the largest person, the smallest person, the person with the shortest reach, etc. operate the equipment effectively—or have the designers given us another "average-man" design? Can the job be done on the hottest and coldest days? Are the people going to be completely exhausted half way through the job, etc.?
Human Factors	1. Simple step-by-step walk-through of the work 2. Misuse analysis 3. Select equipment that is easy to operate *and* matches existing equipment in operational logic.	Most of the fancy human factors analyses have shown that a simple walk-through will indicate many very obvious booby traps for the workers in terms of job steps that are hard to do or things that invite the people to do it wrong or to misuse equipment.
Job Information Systems Formal Change Analysis Methods	1. Do job-site surveillance and monitoring in an orderly manner. 2. Pay attention to what the *workers* say. 3. Be sensitive to *any changes* in people, hardware, or plans.	"Following one's nose" can too easily lead to oversights and failure to "pick up" on changes. Studies have shown that in a series of six accidents, workers had reported all major contributing factors in questionnaire-type studies performed prior to the accident. Another analysis indicated two typewritten pages of changes which contributed directly to five major accidents.
Formal Work-Flow Charting for Safety Reasons	1. Formal work-flow charting to get the job done properly. 2. Job Safety Analysis	1. Experience is indicating that the benefits of formal use of work and project flow-charting go far beyond safety benefits. This is particularly true in avoiding foul-ups between different working groups. 2. Use of step-by-step job safety analysis at worker and first-line supervisor level has been adopted by a number of large business-managed tax-paying industrial firms as a cost-effective, efficient method for producing on-the-job safety.

Figure 11–4. Areas in which system safety and ideas can, with a little adaptation, be applied to industrial jobs (Nertney, 1977). (Printed with permission from *Professional Safety,* February 1977.)

What Processes, Operations, and Tasks Need to Be Analyzed?

Many processes, operations, and tasks in any establishment or plant are good candidates for hazard analysis because they have the potential to cause accidents. Eventually, hazard analyses should be completed for all jobs, but the most potentially threatening should have immediate attention. In determining which processes, operations, and tasks receive priority, those making the decisions should take the following factors into consideration:

1. **Frequency of accidents.** Any operation or task with an associated history of repeated accidents is a good candidate for analysis, especially if different employees have the same kind of accident while performing the same operation or task.
2. **Potential for injury.** Some processes and operations can have a low accident frequency but a high potential for major injury (for example, tasks on a grinder conducted without a tool rest or tongue guard).
3. **Severity of injury.** A particular process, operation, or task can have a history of serious injuries and be a worthy candidate for analysis, even if the frequency of such injuries is low.

4. **New or altered equipment, processes, and operations.** As a general rule, whenever a new process, operation, or task is created or an old one altered (because of machinery, equipment, or other changes), the safety professional or supervisor should conduct a hazard analysis. For maximum benefit, the hazard analysis should be done while the process or operation is in the planning stages. No equipment should be put into regular operation until the safety professional has checked it for hazards, studied its operation, installed any necessary additional safeguards, and developed safety instructions or procedures. Adhering to such a procedure ensures that managers can train employees in hazard-controlled, safe operations and help to prevent serious injuries and exposures.

5. **Excessive material waste or damage to equipment.** Processes or operations producing excessive material waste or damage to tools and equipment are candidates for hazard analysis. The same problems causing the waste or damage could also, given the right situation, cause injuries.

INSPECTION

Inspection can be defined as that monitoring function con-

ducted in an organization to locate and report existing and potential hazards having the capacity to cause accidents in the workplace (Firenze, 1978). Inspection works because it is an essential part of hazard control. It is a vital managerial tool, not a gimmick.

Philosophy behind Inspection

Inspection can be viewed negatively or positively:
- fault-finding, with the emphasis on criticism
- fact-finding, with the emphasis on locating potential hazards that can adversely affect safety and health.

The second viewpoint is more effective. This viewpoint depends on three things: (1) yardsticks adequate for measuring a particular situation, (2) comparison of what is with what ought to be, and (3) corrective steps taken to achieve desired performance (Firenze, 1978). Failure to analyze inspection reports for *causes* of defects ultimately means the failure of the monitoring function. Corrective action may fix the specific item but fail to fix the system.

Purpose of Inspection

Its primary purpose is to detect potential hazards so they can be corrected before an accident occurs. Inspection can determine conditions that need to be corrected or improved to bring operations up to acceptable standards, both from safety and operational standpoints. Secondary purposes are to improve operations and thus to increase efficiency, effectiveness, and profitability.

While management ultimately has the responsibility for inspecting the workplace, authority for carrying out the actual inspecting process extends throughout the organization. Obviously supervisors, foremen, and employees fulfill an inspection function, but so do departments as diverse as engineering, purchasing, quality control, human resources, maintenance, and medical.

Types of Inspection

Inspection can be classified as one of two types—continuous or interval.

Continuous, ongoing inspection. This process is conducted by employees, supervisors, and maintenance personnel as part of their job responsibilities. Continuous inspection involves noting an apparently or potentially hazardous condition or unsafe procedure and either correcting it immediately or making a report to initiate corrective action. Continuous inspection of personal protective equipment is especially important.

This type of inspection is sometimes called informal because it does not conform to a set schedule, plan, or checklist. Critics argue that continuous inspection is erratic and superficial, that it does not get into out-of-the-way places, and that it misses too much. The truth is that both continuous and interval inspections are necessary and they complement one another.

As part of their job, supervisors make sure that tools, machines, and equipment are properly maintained and safe to use and that safety precautions are being observed.

Toolroom employees regularly inspect all hand tools to be sure that they are in safe condition. Foremen are often responsible for continuously monitoring the workplace and seeing that equipment is safe and that employees are observing safe practices. When foremen or supervisors inspect machines at the beginning of a shift, a safety inspection must be part of the operation.

Continuous inspection is one ultimate goal of a good safety and health program. It means that each individual is vigilant, alert to any condition having accident potential, and willing to initiate corrective action.

However, in some instances, the supervisor's greatest advantage in continuous inspection—familiarity with the employees, equipment, machines, and environment—can also be a disadvantage. Just as an old newspaper left on a table in time becomes part of the decor, a hazard can become so familiar that no one notices it. The supervisor's blind spot is particularly likely to occur with housekeeping and unsafe practices. Poor housekeeping conditions may be overlooked because the deterioration is gradual and the effect is cumulative. A similar phenomenon may occur with unsafe practices, such as employees not following established production procedures.

In addition, no matter how conscientious the supervisors are, they cannot be completely objective. Inspections of their areas reflect personal and vested interests, knowledge and understanding of the production problems involved in the area, and concern for the employees. A planned periodic inspection of their areas by another supervisor can be used to audit their efforts. Furthermore, the supervisors who inspect another area may return to their own sections with renewed vision. Having looked at the trees day in and day out, they need occasionally to take the long view and see the forest.

Though this section is devoted primarily to discussing planned inspections, continuous inspections should be regarded as a cooperative, not a competitive, activity.

Planned inspection at intervals. This process is what most people regard as "real" safety and health inspection. It is deliberate, thorough, and systematic by design. Planned inspections permit examination of specific items or conditions. They follow an established procedure and use checklists for routine items. These inspections can be any one of three types: periodic, intermittent, and general.

Periodic inspection includes those inspections scheduled at regular intervals. They can target the entire plant, a specific area, a specific operation, or a specific type of equipment. Management can plan these inspections weekly, monthly, semiannually, annually, or at other suitable intervals. Items such as safety guard mountings, scaffolds, elevator wire ropes (cables), two-hand controls, fire extinguishers, and other items relied on for safety require frequent inspection. The greater the accident severity potential, the more often the item should be inspected.

Periodic inspections can be of several different types:
1. Inspections by the safety professional, industrial hygienist, and joint safety and health committees

2. Inspections for preventing accidents and damage or breakdowns (checking mechanical functioning, lubricating, and the like) performed by electricians, mechanics, and maintenance personnel. Sometimes these persons are asked to serve as roving inspectors.

3. Inspections by specially trained certified or licensed inspectors, often from outside the organization (for example, inspection of boilers, elevators, unfired pressure vessels, cranes, power presses, fire extinguishing equipment). These are often mandated inspections required by regulatory agencies, manufacturers, underwriters, or management.

4. Inspections done by outside investigators or government inspectors to determine compliance with government regulations.

The advantage of periodic inspection is that it covers a specific area and allows detection of unsafe conditions in time to provide effective countermeasures. Measurement data collected at regular intervals indicate degenerative trends. The staff or safety committee periodically inspecting a certain area is familiar with operations and procedures and therefore quick to recognize deviations. A disadvantage of periodic inspection is that deviations from accepted practices are rarely discovered because employees are prepared for the inspectors.

Intermittent inspections are those made at irregular intervals. Sometimes the need for an inspection is indicated by accident tabulations and analysis. If a particular department or location shows an unusual number of accidents or if certain types of injuries occur with greater frequency, the supervisor or manager should call for an inspection. When construction or remodeling is going on within or around a facility, an unscheduled inspection may be needed to find and correct unsafe conditions before an accident occurs. The same is true when a department installs new equipment, institutes new processes, or modifies old ones.

Another form of intermittent inspection is that made by the industrial hygienist when a health hazard is suspected or present in the environment. This monitoring of the workplace is covered in detail in Measurement and Testing, later in this chapter. It usually involves:

- sampling the air for the presence of toxic vapors, gases, radiation, and particulates
- sampling physical stresses like noise, heat, and radiation
- testing materials for toxic properties
- testing ventilation and exhaust systems for proper operation.

Intermittent inspections may be initiated because of the following:

- increase in accident or illness rates in an area
- reports from employees in an area
- management directive
- reports of potential hazards from other departments, companies, manufacturers, or regulatory agencies
- random selection
- accident/severity potential
- reaction to an event (such as an accident, threat of sabotage, severe weather warning, and so on).

A *general inspection* is planned and covers places not inspected periodically. This includes those areas no one ever visits and where people rarely get hurt, such as parking lots, sidewalks, fencing, and similar outlying regions.

Many out-of-the-way hazards are located overhead, where they are difficult to spot. Overhead inspections frequently disclose the need for repairs to skylights, windows, cranes, roofs, and other installations affecting the safety of both the employees and the physical plant. Overhead devices can require adjustment, cleaning, oiling, and repairing.

Inspections of overhead areas are necessary to make certain all reasonable safeguards are provided and safe practices observed. Inspectors must verify that workers performing overhead jobs have suitable staging and fall protection equipment. They must also apply this safety directive to themselves during the inspection. They should look for loose tools, bolts, pipelines, shafting, pieces of lumber, windows, electrical fixtures, and other objects that can fall from building structures, cranes, roofs, and similar overhead locations.

Safety conditions change after dark, when the illumination is artificial. Therefore, when an organization has more than one shift, it is important to perform inspections at night to make sure illumination is adequate and the lighting system is well maintained. The safety professional should make this inspection, aided by a photometer and camera, where necessary. Even in organizations with no regular night shifts, some employees—maintenance personnel, firefighters, and night security guards—are required to work after dark. The safety professional occasionally needs to check on their night work conditions.

General inspections are usually required before reopening a plant after a long shutdown.

PLANNING FOR INSPECTION

A safety and health inspection program requires:
1. sound knowledge of the plant
2. knowledge of relevant standards, regulations, and codes
3. systematic inspection steps
4. a method of reporting, evaluating, and using the data.

An effective program begins with analysis and planning. If inspections are casual, shallow, and slipshod, the results will reflect the method. Before instituting an inspection program, these five questions should be answered:
1. What items need to be inspected?
2. What aspects of each item need to be examined?
3. What conditions need to be inspected?
4. How often must items be inspected?
5. Who will conduct the inspection?

The Hazard Control Inspection Inventory

To determine what factors affect the inspection, a hazard control inspection inventory can be conducted. Such an inventory is the foundation upon which a program of planned inspection is based. It resembles a planned preventive maintenance system and yields many of the same benefits.

Management should divide the entire facility—yards,

LIST OF POSSIBLE PROBLEMS TO BE INSPECTED				
Acids	Closets	Forklifts	Piping	Shapers
Aisles	Connectors	Fumes	Pits	Shelves
Alarms	Containers	Gas cylinders	Platforms	Sirens
Atmosphere	Controls	Gas engines	Power tools	Slings
Automobiles	Conveyors	Gases	Presses	Solvents
Barrels	Cranes	Hand tools	Racks	Sprays
Bins	Crossing lights	Hard hats	Railroad cars	Sprinkler systems
Blinker lights	Cutters	Hoists	Ramps	Stairs
Boilers	Docks	Horns and signals	Raw materials	Steam engines
Borers	Doors	Hoses	Respirators	Sumps
Buggies	Dusts	Hydrants	Roads	Switches
Buildings	Electric motors	Ladders	Roofs	Tanks
Cabinets	Elevators	Lathes	Ropes	Trucks
Cables	Explosives	Lights	Safety devices	Vats
Carboys	Extinguishers	Mills	Safety shoes	Walkways
Catwalks	Eye protection	Mists	Scaffolds	Walls
Caustics	Flammables	Motorized carts	Shafts	Warning devices
Chemicals	Floors			

Figure 11–5. This list of possible problems found in the workplace is adapted from *Principles and Practices of Occupational Safety and Health, Student Manual*, Booklet Three, U.S. Department of Labor, OSHA 2215.

buildings, equipment, machinery, vehicles—into areas of responsibility. These areas, once determined, should be listed in an orderly fashion. The analyst might develop a color-coded map or floor plan of the facility. Large areas or departments can be divided into smaller areas and assigned to each first-line supervisor and/or the hazard control department's inspector.

What Items Need to Be Inspected?

Once specific areas of responsibility have been determined, managers should develop an inventory of those items that may become unsafe or cause accidents. These would include:

1. environmental factors (illumination, dust, fumes, gases, mists, vapors, noise, vibration, heat, radiation sources)
2. hazardous supplies and materials (explosives, flammables, acids, caustics, toxic or nuclear materials or by-products)
3. production and related equipment (mills, shapers, presses, borers, lathes)
4. power source equipment (steam and gas engines, electrical motors)
5. electrical equipment (switches, fuses, breakers, outlets, cables, extension and fixture cords, grounds, connectors, connections)
6. hand tools (wrenches, screwdrivers, hammers, power tools)
7. personal protective equipment (hard hats, safety glasses, safety shoes, respirators, hearing protection, gloves, etc.)
8. personal service and first aid facilities (drinking fountains, wash basins, soap dispensers, safety showers, eyewash fountains, first aid supplies, stretchers)
9. fire protection and emergency response equipment (alarms, water tanks, sprinklers, standpipes, extinguishers, hydrants, hoses, self-contained breathing apparatus, toxic cleanup, automatic valves, horns, phones, radios)

10. walkways and roadways (ramps, docks, sidewalks, walkways, aisles, vehicle ways, escape routes)
11. elevators, electric stairways, and manlifts (controls, wire ropes, safety devices)
12. working surfaces (ladders, scaffolds, catwalks, platforms)
13. material handling equipment (cranes, dollies, conveyors, hoists, forklifts, chains, ropes, slings)
14. transportation equipment (automobiles, railroad cars, trucks, front-end loaders, helicopters, motorized carts and buggies)
15. warning and signaling devices (sirens, crossing and blinker lights, klaxons, warning signs, exit signs)
16. containers (scrap bins, disposal receptacles, carboys, barrels, drums, gas cylinders, solvent cans)
17. storage facilities and areas both indoor and outdoor (bins, racks, lockers, cabinets, shelves, tanks, closets)
18. structural openings (windows, doors, stairways, sumps, shafts, pits, floor openings)
19. buildings and structures (floors, roofs, walls, fencing)
20. miscellaneous—any items that do not fit in preceding categories. (List adapted from *Facility Inspection*, see References.)

There are many sources of information about items to be inspected, especially employees in an organization. For instance, maintenance employees know what problems can cause damage or shutdowns. The workers in the area are qualified to point out causes of injury, illness, damage, delays, or bottlenecks. Medical personnel in the organization can list problems causing job-related illnesses and injuries. Manufacturers' manuals often specify maintenance schedules and procedures and safe work methods.

Gathering the information about standards, regulations, and codes is a necessary first step in determining what items need to be inspected. A great deal of work already has been done for the safety inspector.

Building codes, building inspection books, and guides to building and plant maintenance also are useful references.

---

29 CFR 1910
OCCUPATIONAL SAFETY AND HEALTH STANDARDS

Subparts

C — Access to Employee Exposure and Medical Records
D — Walking-Working Surfaces
E — Means of Egress
F — Powered Platforms, Manlifts, and Vehicle-Mounted Work Platforms
G — Occupational Health and Environmental Control
H — Hazardous Materials
I — Personal Protective Equipment
J — General Environmental Controls
K — Medical and First Aid
L — Fire Protection

M — Compressed Gas and Compressed Air Equipment
N — Materials Handling and Storage
O — Machinery and Machine Guarding
P — Hand and Portable Powered Tools and Other Hand-Held Equipment
Q — Welding, Cutting, and Brazing
R — Special Industries
S — Electrical
T — Commercial Diving Operations
Z — Toxic and Hazardous Substances

29 CFR 1926
STANDARDS FOR THE CONSTRUCTION INDUSTRY

Subparts

C — General Safety and Health Provisions
D — Occupational Health and Environmental Controls
E — Personal Protective and Lifesaving Equipment
F — Fire Protection and Prevention
G — Signs, Signals, and Barricades
H — Materials Handling, Storage, Use and Disposal
I — Tools—Hand and Power
J — Welding and Cutting
K — Electrical
L — Ladders and Scaffolding
M — Floors and Wall Openings and Stairways
N — Cranes, Derricks, Hoists, Elevators, and Conveyors

O — Motor Vehicles, Mechanized Equipment, and Marine Operations
P — Excavations, Trenching, and Shoring
Q — Concrete, Concrete Forms, and Shoring
R — Steel Erection
S — Tunnels and Shafts, Caissons, Cofferdams, and Compressed Air
T — Demolition
U — Blasting and Use of Explosives
V — Power Transmission and Distribution
W — Rollover Protection Structures; Overhead Projection

Figure 11–6. Lists of inspection subjects required by 29 *CFR* 1910 and 29 *CFR* 1926.

Publications of the NFPA will help safety personnel ensure that fire hazards are being effectively controlled. In addition, insurance company surveys often contain checklists to help management determine the safety condition of their buildings. Research and reference material is contained in subsequent chapters of this *Manual*. Other publications of the National Safety Council, such as *Accident Facts*, may prove useful as well.

State or provincial and federal governments also publish accident statistics. In one of its publications, the Occupational Safety and Health Administration (OSHA) gave examples of possible problems found in the work area (Figure 11–5). The *Federal Register* and the *Code of Federal Regulations* (*CFR*), Title 29, parts 1900–1950, give OSHA regulations. Figure 11–6 shows some special subjects addressed by the subparts of 29 *CFR* 1910 (General Industry) and the subparts of 29 *CFR* 1926 (Construction).

It is important to remember, however, that federal and state or provincial laws, codes, and regulations usually set up *minimum* requirements only. To comply with company policy and secure maximum safety, management must often exceed these requirements. Some OSHA publications indicate not only what standards are required but also what violations are most frequent. These same sources are helpful in the next step—determining critical factors to be inspected.

What Aspects of Each Item Need to Be Examined?

Particular attention should be paid to the parts of an item most likely to become a serious hazard to health and safety. These parts most often develop problems because of stress, wear, impact, vibration, heat, corrosion, chemical reaction, and misuse. Such items as safety devices, guards, controls, work or wearpoint components, electrical and mechanical components, and fire hazards tend to become unsafe first. For a particular machine, critical parts would include the point of operation, moving parts, and accessories (flywheels, gears, shafts, pulleys, key ways, belts, couplings, sprockets, chains, controls, lighting, brakes, exhaust systems). Also to be checked are items related to feeding, oiling, adjusting, maintenance, grounding, attaching, work space, and location.

The most critical parts of an item are not always the most obvious. When the security of a heavy load depends on a cotter pin being in place, then that pin is a critical part.

What Conditions Need to Be Inspected?

The unsafe conditions for each part to be inspected should be described specifically and clearly. A checklist question that reads "Is ___ safe?" is meaningless because it does not define *what* makes an item unsafe. Inspectors should de-

scribe and not simply list unsafe conditions for each item. Usually, conditions to look for can be indicated by such words as *jagged, exposed, broken, frayed, leaking, rusted, corroded, missing, vibrating, loose,* or *slipping*. Sometimes exact figures are needed; for example, the maximum pressure in a boiler or the percent spread of a sling hook.

Many types of monitoring checklists are available, varying in length from thousands of items to only a few. Each type has its place. These are useful in determining which standards or regulations apply to individual situations. Once the applicable standards are identified, the organization can tailor a checklist to its needs and uses and enter it into the computer system for action and follow-up.

The Centers for Disease Control, U.S. Department of Health and Human Services (DHHS), have devised a suggested checklist for the safety evaluation of shop and laboratory areas. The worksheet is referenced to the OSHA "General Industry Standards."

Checklists serve as reminders of what to look for and as records of what has been covered. They can be used to structure and guide inspections. They also allow on-the-spot recording of all findings and comments before they are forgotten. In case an inspection is interrupted, checklists provide a record of what has and what has not been inspected. Without checklists, inspectors may miss items or conditions they should examine or may be unsure, after inspecting an area, that they have covered everything. Good checklists also help in follow-up work to make sure hazards have been corrected or eliminated.

Of course, merely running through a checklist does little to locate or correct problems. Simply checking off the items on the list is not conducting a safety inspection. The checklist must be used as an aid to the inspection process, not as an end in itself. Any hazard observed during inspection must be recorded, even though it is not part of the checklist.

The following 18 pages comprise a portfolio of various companies' inspection checklists (Figure 11–7). These are used with computer follow-up on inspection results, actions to be taken, and corrections made. Refer to the following portfolio of checklists.

The amount of detail included in the checklist will vary, depending upon the inspector's knowledge of the relevant standards and the nature of the inspection. An experienced inspector with thorough knowledge of the standards will need only a few clues as a reminder of items to be inspected. Checklists for infrequent inspection generally will be more detailed than daily or weekly ones.

Checklists should have columns to indicate either compliance or action date. Space should be provided to cite the specific violation, a way to correct it, and a recommendation that the condition receive more or less frequent attention. Whatever the format of the checklist, space should be provided for the inspector's signature and the inspection date.

Checklists can be prepared by the safety and health committee, by the safety director, or by a subcommittee that includes engineers, supervisors, employees, and maintenance personnel. The safety and health professional and the depart-

ment supervisor should monitor checklist development and make sure all applicable standards are covered. In their final form the checklists should conform to the inspection route.

Choosing the inspection route means inspecting an area completely and thoroughly while avoiding:
- time-consuming backtracking and repetitions
- long walks between items
- unnecessary interruptions of the production process
- distraction of employees.

Often a closed-loop inspection will give good results. Sometimes it is valuable to follow the production path of the material being processed.

How Often Must Items Be Inspected?

The frequency of inspection is determined by four factors.
- *What is the loss severity potential of the problem?* The inspector should ask, "If the item or critical part should fail, what would happen? What injury, damage, or work interruption would result?" The greater the loss severity potential, the more often the item should be inspected. Because a frayed wire rope on an overhead crane block has the potential to cause a much greater loss than a defective wheel on a wheelbarrow, the rope needs to be inspected more frequently than the wheel.
- *What is the potential for injury to employees?* If the item or critical part should fail, how many employees would be endangered and how frequently? The greater the probability for injury to employees, the more often the item should be inspected. For example, a stairway continually used by many people needs to be inspected more frequently than one seldom used.
- *How quickly can the item or part become unsafe?* The answer to this question depends on the nature of the part and the conditions to which it is subjected. Equipment and tools used frequently can become damaged, defective, or worn more quickly than those rarely used. An item located in a particular spot can be exposed to greater damage than an identical item in a different location. The faster an item can become unsafe, the more frequently it should be inspected.
- *What is the past history of failures?* What were the results of these failures? Maintenance and production records and accident investigation reports can provide valuable information about how frequently items have failed and the results in terms of injuries, damage, delays, and shutdowns. The more often an item has failed in the past and the greater the consequences, the more it needs to be inspected.

The Occupational Safety and Health Administration, the Mine Safety and Health Administration (MSHA), the Federal Aviation Administration (FAA), the Nuclear Regulatory Commission (NRC), the Environmental Protection Agency (EPA), and other federal, state, provincial, and local regulatory agencies require periodic inspections. Consult the regulations and the agency responsible for enforcement for current information regarding inspection criteria and inter-

(Text continued on page 267.)

Loading Dock Safety Checklist

Date _____

Company Name _____

Plant Name _____ Plant Location _____

Dock examined/Door numbers _____

Company representative completing checklist:

Name _____ Title _____

A. Vehicles/Traffic Control

1. Do forklifts have the following safety equipment?
   - ☐ Seat belt   ☐ Load backrest
   - ☐ Headlight   ☐ Backup alarm
   - ☐ Horn   ☐ Overhead guard
   - ☐ Tilt indicator
   - ☐ On-board fire extinguisher
   - ☐ Other _____

2. Are the following in use?

	Yes	No
Driver candidate screening	☐	☐
Driver training/licensing	☐	☐
Periodic driver retraining	☐	☐
Vehicle maintenance records	☐	☐
Written vehicle safety rules	☐	☐

	Yes	No
3. Is the dock kept clear of loads of materials?	☐	☐
4. Are there convex mirrors at blind corners?	☐	☐
5. Is forklift cross traffic over dock levelers restricted?	☐	☐
6. Is pedestrian traffic restricted in the dock area?	☐	☐
7. Is there a clearly marked pedestrian walkway?	☐	☐
8. Are guardrails used to define the pedestrian walkway?	☐	☐

9. Comments/recommendations

   _____
   _____
   _____
   _____
   _____

B. Vehicle restraining

	Yes	No
1. If vehicle restraints are used:		
a. Are all dock workers trained in the use of the restraints?	☐	☐
b. Are the restraints used consistently?	☐	☐
c. Are there warning signs and lights inside and out to tell when a trailer is secured and when it is not?	☐	☐
d. Are the outdoor signal lights clearly visible even in fog or bright sunlight?	☐	☐
e. Can the restraints secure trailers regardless of the height of their ICC bars?	☐	☐
f. Can the restraints secure all trailers with I-beam, round, or other common ICC bar shapes?	☐	☐
g. Does the restraint sound an alarm when a trailer cannot be secured because its ICC bar is missing or out of place?	☐	☐

Figure 11–7. The following 18 pages present a sample of inspection forms used in various organizations. Adapt these forms as necessary to your specific needs.

h. Are dock personnel specifically trained to watch for trailers with unusual rear-end assemblies (e.g. sloping steel back plates and hydraulic tailgates) that can cause a restraint to give a signal that the trailer is engaged even when it is not safely engaged? ☐ ☐

i. Are dock personnel specifically trained to observe the safe engagement of all unusual rear-end assemblies? ☐ ☐

j. Do the restraints receive regular planned maintenance? ☐ ☐

k. Do restraints need repairs or replacement? (List door numbers.)

_____

l. Comments/recommendations

_____

_____

2. If wheel chocks are used: Yes No
   a. Are dock workers, rather than truckers, responsible for placing chocks? ☐ ☐
   b. Are all dock workers trained in proper chocking procedures? ☐ ☐
   c. Are chocks of suitable design and construction? ☐ ☐
   d. Are there two chocks for each position? ☐ ☐
   e. Are all trailers chocked on both sides? ☐ ☐
   f. Are chocks chained to the building? ☐ ☐
   g. Are warning signs in use? ☐ ☐
   h. Are driveways kept clear of ice and snow to help keep chocks from slipping? ☐ ☐
   i. Comments/recommendations

_____

_____

C. Dock levelers

   Yes No
1. Are the dock levelers working properly? ☐ ☐

2. Are levelers long enough to provide a gentle grade into trailers of all heights? ☐ ☐

3. Is leveler width adequate when servicing wider trailers? ☐ ☐

4. Do platform width and configuration allow safe handling of end loads? ☐ ☐

5. Is leveler capacity adequate given typical load weights, lift truck speeds, ramp inclines and frequency of use? ☐ ☐

6. Do levelers have the following safety features?
   Working-range toe guards ☐ ☐
   Full-range toe guards ☐ ☐
   Ramp free-fall protection ☐ ☐
   Automatic recycling ☐ ☐

7. Do levelers receive regular planned maintenance? ☐ ☐

8. Do levelers need repair or replacement? (List door numbers.)

_____

_____

9. Comments/recommendations _____

_____

_____

D. Portable dock plates

   Yes No
1. Is plate length adequate? ☐ ☐
2. Is plate capacity adequate? ☐ ☐
3. Are plates of suitable design and materials? ☐ ☐
4. Do plates have curbed sides? ☐ ☐
5. Do plates have suitable anchor stops? ☐ ☐
6. Are plates moved by lift trucks rather than by hand? ☐ ☐
7. Are plates stored away from traffic? ☐ ☐
8. Are plates inspected regularly? ☐ ☐
9. Do plates need repair or replacement? (List door numbers.)

_____

_____

10. Comments/recommendations

_____

_____

E. Dock Doors

	Yes	No
1. Are doors large enough to admit all loads without obstruction?	☐	☐
2. Are door rails protected by bumper posts?	☐	☐
3. Do doors receive regular planned maintenance?	☐	☐

4. Which doors (if any) need repair or replacement?

_____

_____

5. Comments/recommendations

_____

_____

F. Traffic Doors

	Yes	No
1. Are doors wide enough to handle all loads and minimize damage.	☐	☐
2. Does door arrangement allow safe lift truck and pedestrian traffic?	☐	☐
3. Are visibility and lighting adequate on both sides of all doors?	☐	☐

4. Which doors (if any) need repair or replacement?

_____

_____

5. Comments/recommendations

_____

_____

G. Weather sealing

	Yes	No
1. Are the seals or shelters effective in excluding moisture and debris from the dock?	☐	☐
2. Are seals or shelters sized to provide an effective seal against all types of trailers?	☐	☐
3. Are seals or shelters designed so that they will not obstruct loading and unloading?	☐	☐

	Yes	No
4. Are dock levelers weather sealed along the sides and back?	☐	☐
5. In addition to seals or shelters, would an air curtain solve a problem?	☐	☐

6. Do seals or shelters need repair or replacement? (List door numbers.)

_____

_____

7. Comments/recommendations

_____

_____

H. Trailer Lifting

1. How are low-bed trailers elevated for loading/unloading?
   ☐ Wheel risers ☐ Concrete ramps
   ☐ Trailer-mounted jacks
   ☐ Truck levelers

	Yes	No
2. Do lifting devices provide adequate stability?	☐	☐
3. Are trailers secured with vehicle restraints when elevated?	☐	☐

4. Do lifting devices need repair or replacement? (List door numbers.)

_____

5. Comments/recommendations

_____

_____

I. Other Considerations

	Yes	No
1. Dock lights		
a. Is lighting adequate inside trailers?	☐	☐
b. Is the lift mechanism properly shielded?	☐	☐
2. Scissors lifts		
a. Are all appropriate workers trained in safe operating procedures?	☐	☐
b. Is the lift mechanism properly shielded?	☐	☐
c. Are guardrails and chock ramps in place and in good repair?	☐	☐

3. Conveyors
   a. Are all appropriate workers trained in safe operating procedures? ☐ ☐
   b. Are necessary safeguards in place to protect against pinch points, jam-ups and runaway material? ☐ ☐
   c. Are crossovers provided? ☐ ☐
   d. Are emergency stop buttons in place and properly located? ☐ ☐

4. Strapping
   a. Are proper tools available for applying strapping? ☐ ☐
   b. Do workers cut strapping using only cutters equipped with a holddown device? ☐ ☐
   c. Do workers wear hand, foot and face protection when applying and cutting strapping? ☐ ☐
   d. Are all appropriate workers trained in safe strapping techniques? ☐ ☐

5. Manual handling
   a. Is the dock designed so as to minimize manual lifting and carrying? ☐ ☐
   b. Are dock workers trained in safe lifting and manual handling techniques? ☐ ☐

6. Miscellaneous   Yes   No
   a. Are pallets regularly inspected? ☐ ☐
   b. Are dock bumpers in good repair? ☐ ☐
   c. Is the dock kept clean and free of clutter? ☐ ☐
   d. Are housekeeping inspections performed periodically? ☐ ☐
   e. Are anti-skid floor surfaces, mats or runners used where appropriate? ☐ ☐

f. Are stairways or ladders provided for access to ground level from the dock? ☐ ☐
g. Is the trailer landing strip in good condition? ☐ ☐
h. Are dock approaches free of potholes or deteriorated pavement? ☐ ☐
i. Are dock approaches and outdoor stairs kept clear of ice and snow? ☐ ☐
j. Are dock positions marked with lines or lights for accurate trailer spotting? ☐ ☐
k. Do all dock workers wear personal protective equipment as required by company policy? ☐ ☐
l. Is safety training provided for all dock employees? ☐ ☐
m. Are periodic safety refresher courses offered? ☐ ☐

J. General comments/recommendations

_____
_____
_____
_____
_____

For additional copies of this Loading Dock Safety Checklist, write to Rite-Hite Corporation, 9019 North Deerwood Drive, P.O. Box 23043, Milwaukee, WI 53223-0043.
For more information on loading dock safety, write for a complimentary copy of Rite-Hite's Dock Safety Guide.

This loading dock safety checklist is provided as a service by Rite-Hite Corporation, Milwaukee, Wis. It is intended as an aid to safety evaluation of loading dock equipment and operations. However, it is not intended as a complete guide to loading dock hazard identification. Therefore, Rite-Hite Corporation makes no guarantees as to nor assumes any liability for the sufficiency or completeness of this document. It may be necessary under particular circumstances to evaluate other dock equipment and procedures in addition to those included in the checklist. For information on U.S. loading dock safety requirements, consult OSHA Safety and Health Standards (29 CFR 1910). In other countries, consult the applicable national or provincial occupational health and safety codes.

OSHA CHECKLIST FOR HAZARD COMMUNICATION STANDARD

Does Your Company Meet The Requirements of This Standard?

The Occupational Safety & Health Administration (OSHA) requires certain manufacturers, distributors and employers to meet the requirements of the Hazard Communication Standard (29CFR 1910.1200). The following checklist has been developed to help you determine if your company is in compliance with the requirements of this standard.

	OSHA SECTION	YES	NO	ACTION TAKEN
A. HAZARD COMMUNICATION PROGRAM:				
1. The program is in writing.	1910.1200(e)(1)			
Our Written Program Provides The Following:				
2. Describes how hazards will be evaluated and described (employers may rely on the chemical manufacturer or importer).	1910.1200(d)(6)			
3. Tests all hazardous materials in the workplace (employers may rely on the chemical manufacturer or importers).	1910.1200(d)(1)			
4. Describes our labeling system.	1910.1200(f)			
5. Provides a list of hazardous chemicals referenced on MSDS for all hazardous materials used in the workplace (see Section B).	1910.1200(e)(1)			
6. Describes our employee education and training program.	1910.1200(h)			
7. Describes hazards of non-routine tasks.	1910.1200(e)(1)(ii)			
8. Describes how hazards of non-labeled pipes will be handled.	1910.1200(e)(1)(ii)			
9. Includes procedures for informing of on-site contractors of the hazardous substances in the workplace to which their employees may be exposed.	1910.1200(e)(1)(iii)			
10. Is available to employees, their designated representatives, assistant secretary of labor for OSHA, and the director of NIOSH.	1910.1200(e)(3)			
B. LIST OF HAZARDOUS MATERIALS IN THE WORKPLACE:				
Our List contains all hazardous chemicals, including, but not limited to:				
1. Raw materials.	1910.1200(e)(1)(i)			
2. Both isolated and nonisolated intermediates.	1910.1200(e)(1)(i)			
3. Final product.	1910.1200(e)(1)(i)			
4. Cleaning and maintenance chemicals.	1910.1200(e)(1)(i)			
5. Laboratory chemicals for which MSDS information has been received.	1910.1200(b) (ii),(iii)			
6. Waste products not regulated under RCRA, but which are hazardous under this standard.	1910.1200(e)(1)(i)			
7. Impurities and by products.	1910.1200(e)(1)(i)			
8. Waste treatment and products.	1910.1200(e)(1)(i)			
C. HAZARDOUS MATERIALS LABELING SYSTEM:				
1. All products containing hazardous materials leaving the workplace are labeled (applicable to chemical manufacturers, distributors and importers only).	1910.1200(f)(1)			
2. Stationary containers are labeled.	1910.1200(f)(4)			
3. Temporary containers used between workshifts or by different workers are labeled.	1910.1200(f)(6)			
4. A method has been established to insure that our labels are correct and up-to-date.	1910.1200(f)(4)(ii)			
D. CONTENTS OF HAZARDOUS MATERIAL LABEL:				
Our Labels Contain:				
1. A chemical name that coincides with name on MSDS.	1910.1200(f)(1)(i)			
2. The identity of hazards with words (in English), pictures or symbols.	1910.1200(f)(1)(ii)			
3. Hazards of immediate and direct consequences of mishandling are included.	1910.1200(f)(1)(ii)			
4. Information that does not conflict with DOT regulations.	1910.1200(f)(2) 49 CFR 172.101			
5. Other OSHA standards if material is already regulated.	1910.1200(f)(3)			
6. The name and address of a responsible party (or parties).	1910.1200(f)(1)(iii)			
E. IN-PLANT LABELING SYSTEM:				
1. Containers are labeled with the identity of hazardous chemicals and hazard warnings (unless hazard warning materials are used).	1910.1200(f)(4)(i)			
2. Hazard warning materials for hazardous chemicals in stationary process containers are readily accessible to the employee in the workplace.	1910.1200(f)(5)			

(continued on back)

	OSHA SECTIONS	NO	ACTION TAKEN
3. The labels on incoming containers have not been removed or defaced unless immediately replaced with our own labels.	1910.1200(f)(7)		
4. The hazards in pipelines are identified, although they do not have to be labeled under this standard.	1910.1200(e)(1)(ii)		
5. Our labels are legible and in English.	1910.1200(f)(8)		
F. MATERIAL SAFETY DATA SHEETS:			
1. A MSDS is available for every hazardous chemical which an employer uses.	1910.1200(g)(1)		
2. Our MSDS are readily accessible to exposed employees in the work area throughout each work shift.	1910.1200(g)(8)		
G. PROCEDURES HAVE BEEN ESTABLISHED FOR:			
1. Updating our MSDS (or for receiving updated copies from our supplier).	1910.1200(g)(5)		
2. Taking appropriate action if a shipment is received without a MSDS.	1910.1200(g)(1)		
3. Getting new and updated MSDS to employees handling materials.	1910.1200(g)(8)		
4. Advising employees of any changes in MSDS.	1910.1200(h)		
5. Documentation of efforts to obtain MSDS from supplier (recommended practice but not required by this standard).	———		
H. HAZARDS OF NON-ROUTINE TASKS: Procedures have been established assessing the hazards of non-routine tasks as follows:			
1. All non-routine tasks involving the use or exposure to hazardous materials are identified.	1910.1200(e)(1)(ii)		
2. The hazards involved in the performance of non-routine tasks are described in writing.	1910.1200(e)(1)(ii)		
3. A MSDS is prepared or obtained for the hazardous materials involved in these non-routine tasks.	1910.1200(e)(1)(ii)		
4. A labeling system or written operating procedure has been established to identify the hazardous substances and their hazards involved in non-routine tasks.	1910.1200(e)(1)(ii)		
5. Special training has been established for the performance of non-routine tasks, including written operating procedures.	1910.1200(e)(1)(ii)		
I. EMPLOYEE EDUCATION & TRAINING: Procedures have been established to inform employees of:			
1. Covers all manufacturing, quality control, plant service, and R&D employees who may be exposed to hazardous materials.	1910.1200(b)(1)		
2. Requirements of the Hazard Communication Standard.	1910.1200(h)(1)(i)		
3. Operations where hazardous materials are present.	1910.1200(h)(1)(ii)		
4. Location and availability of the written hazard communication program including the hazardous chemical list and material safety data sheets.	1910.1200(h)(1)(iii)		
J. PROCEDURES FOR TRAINING EMPLOYEES INCLUDE:			
1. Information about physical and health hazards of chemicals in work area.	1910.1200(h)(2)(ii)		
2. Detecting the presence of hazardous materials - monitoring procedures, odors, visibility, etc.	1910.1200(h)(2)(i)		
3. Proper use and selection of personal protective equipment.	1910.1200(h)(2)(iii)		
4. Emergency procedures in the event of accidental exposure to hazardous materials, including emergency phone numbers and the location of eye washes and safety showers.	1910.1200(h)(2)(iii)		
5. How to determine hazards by reading a label.	1910.1200(h)(2)(iv)		
6. The location of MSDS and the procedure for reviewing them and/or obtaining a copy.	1910.1200(h)(1)(iii)		
7. How to obtain the correct MSDS for the hazardous substance used by the employee, such as use of the trade name as a key identifier.	1910.1200(h)(2)(iv)		
8. How the MSDS is updated or the procedure for obtaining updated copies from the chemical manufacturer, importer or distributor.	1910.1200(h)(2)(iv)		
9. The significance to the employee of each section of information on the MSDS, how to read it and what it means.	1910.1200(h)(2)(iv)		
10. The measures employees can take to protect themselves from chemical exposure. (Examples include eye washes, face shields, respirators, etc.).	1910.1200(h)(2)(iii)		
11. Training which is done prior to the handling of the hazardous chemical, including employees who may only temporarily do this work.	1910.1200(h)		
12. Updated training is considered when the employee has transferred jobs or departments.	1910.1200(h)		
13. Updated training is considered when significant changes in chemicals or operations have occurred.	1910.1200(h)		

LAB SAFETY SUPPLY CO. • **3430 Palmer Drive** • **Janesville, WI 53546** • **(608) 754-2345**

SRM-115

SELF-INSPECTION CHECKLISTS

General

	OK	ACTION NEEDED

1. Is the required OSHA workplace poster displayed in your place of business as required where all employees are likely to see it? ☐ ☐

2. Are you aware of the requirement to report all workplace fatalities and any serious accidents (where 5 or more are hospitalized) to a federal or state OSHA office within 48 hours? ☐ ☐

3. Are workplace injury and illness records being kept as required by OSHA? ☐ ☐

4. Are you aware that the OSHA annual summary of workplace injuries and illnesses must be posted by February 1 and must remain posted until March 1? ☐ ☐

5. Are you aware that employers with 10 or fewer employees are exempt from the OSHA recordkeeping requirements, unless they are part of an official BLS or state survey and have received specific instructions to keep records? ☐ ☐

6. Have you demonstrated an active interest in safety and health matters by defining a policy for your business and communicating it to all employees? ☐ ☐

7. Do you have a safety committee or group that allows participation of employees in safety and health activities? ☐ ☐

8. Does the safety committee or group meet regularly and report, in writing, its activities? ☐ ☐

9. Do you provide safety and health training for all employees requiring such training, and is it documented? ☐ ☐

10. Is one person clearly in charge of safety and health activities? ☐ ☐

11. Do all employees know what to do in emergencies? ☐ ☐

12. Are emergency telephone numbers posted? ☐ ☐

13. Do you have a procedure for handling employee complaints regarding safety and health? ☐ ☐

Workplace

ELECTRICAL WIRING, FIXTURES AND CONTROLS

	OK	ACTION NEEDED

1. Are your workplace electricians familiar with the requirements of the National Electrical Code (NEC)? ☐ ☐

2. Do you specify compliance with the NEC for all contract electrical work? ☐ ☐

3. If you have electrical installations in hazardous dust or vapor areas, do they meet the NEC for hazardous locations? ☐ ☐

4. Are all electrical cords strung so they do not hang on pipes, nails, hooks, etc? ☐ ☐

5. Is all conduit, BX cable, etc., properly attached to all supports and tightly connected to junction and outlet boxes? ☐ ☐

6. Is there no evidence of fraying on any electrical cords? ☐ ☐

7. Are rubber cords kept free of grease, oil and chemicals? ☐ ☐

8. Are metallic cable and conduit systems properly grounded? ☐ ☐

9. Are portable electric tools and appliances grounded or double insulated? ☐ ☐

10. Are all ground connections clean and tight? ☐ ☐

11. Are fuses and circuit breakers the right type and size for the load on each circuit? ☐ ☐

12. Are all fuses free of "jumping" with pennies or metal strips? ☐ ☐

13. Do switches show evidence of overheating? ☐ ☐

14. Are switches mounted in clean, tightly closed metal boxes? ☐ ☐

	OK	ACTION NEEDED
15. Are all electrical switches marked to show their purpose?	☐	☐
16. Are motors clean and kept free of excessive grease and oil?	☐	☐
17. Are motors properly maintained and provided with adequate overcurrent protection?	☐	☐
18. Are bearings in good condition?	☐	☐
19. Are portable lights equipped with proper guards?	☐	☐
20. Are all lamps kept free of combustible material?	☐	☐
21. Is your electrical system checked periodically by someone competent in the NEC?	☐	☐

EXITS AND ACCESS

	OK	ACTION NEEDED
1. Are all exits visible and unobstructed?	☐	☐
2. Are all exits marked with a readily visible sign that is properly illuminated?	☐	☐
3. Are there sufficient exits to ensure prompt escape in case of emergency?	☐	☐
4. Are areas with limited occupancy posted and is access/egress controlled to persons specifically authorized to be in those areas?	☐	☐
5. Do you take special precautions to protect employees during construction and repair operations?	☐	☐

FIRE PROTECTION

	OK	NEEDED
1. Are portable fire extinguishers provided in adequate number and type?	☐	☐
2. Are fire extinguishers inspected monthly for general condition and operability and noted on the inspection tag?	☐	☐
3. Are fire extinguishers recharged regularly and properly noted on the inspection tag?	☐	☐
4. Are fire extinguishers mounted in readily accessible locations?	☐	☐

	OK	ACTION NEEDED
5. If you have interior standpipes and valves, are these inspected regularly?	☐	☐
6. If you have a fire alarm system, is it tested at least annually?	☐	☐
7. Are plant employees periodically instructed in the use of extinguishers and fire protection procedures?	☐	☐
8. If you have outside private fire hydrants, were they flushed within the last year and placed on a regular maintenance schedule?	☐	☐
9. Are fire doors and shutters in good operating condition?	☐	☐
Are they unobstructed and protected against obstruction?	☐	☐
10. Are fusible links in place?	☐	☐
11. Is your local fire department well acquainted with your plant, location and specific hazards?	☐	☐
12. Automatic Sprinklers:		
Are water control valves, air and water pressures checked weekly?	☐	☐
Are control valves locked open?	☐	☐
Is maintenance of the system assigned to responsible persons or a sprinkler contractor?	☐	☐
Are sprinkler heads protected by metal guards where exposed to mechanical damage?	☐	☐
Is proper minimum clearance maintained around sprinkler heads?	☐	☐

HOUSEKEEPING AND GENERAL WORK ENVIRONMENT

	OK	ACTION NEEDED
1. Is smoking permitted in designated "safe areas" only?	☐	☐
2. Are NO SMOKING signs prominently posted in areas containing combustibles and flammables?	☐	☐
3. Are covered metal waste cans used for oily and paint soaked waste?	☐	☐
Are they emptied at least daily?	☐	☐
4. Are paint spray booths, dip tanks, etc., and their exhaust ducts cleaned regularly?	☐	☐

	OK	ACTION NEEDED
5. Are stand mats, platforms or similar protection provided to protect employees from wet floors in wet processes?	☐	☐
6. Are waste receptacles provided, and are they emptied regularly?	☐	☐
7. Do your toilet facilities meet the requirements of applicable sanitary codes?	☐	☐
8. Are washing facilities provided?	☐	☐
9. Are all areas of your business adequately illuminated?	☐	☐
10. Are floor load capacities posted in second floors, lofts, storage areas, etc.?	☐	☐
11. Are floor openings provided with toe boards and railings or a floor hole cover?	☐	☐
12. Are stairways in good condition with standard railings provided for every flight having four or more risers?	☐	☐
13. Are portable wood ladders and metal ladders adequate for their purpose, in good condition and provided with secure footing?	☐	☐
14. If you have fixed ladders, are they adequate, and are they in good condition and equipped with side rails or cages or special safety climbing devices, if required?	☐	☐
15. For Loading Docks: Are dockplates kept in serviceable condition and secured to prevent slipping?	☐	☐
Do you have means to prevent car or truck movement when dockplates are in place?	☐	☐

MACHINES AND EQUIPMENT

	OK	ACTION NEEDED
1. Are all machines or operations that expose operators or other employees to rotating parts, pinch points, flying chips, particles or sparks adequately guarded?	☐	☐
2. Are mechanical power transmission belts and pinch points guarded?	☐	☐
3. Is exposed power shafting less than 7 feet from the floor guarded?	☐	☐
4. Are hand tools and other equipment regularly inspected for safe condition?	☐	☐

	OK	ACTION NEEDED
5. Is compressed air used for cleaning reduced to less than 30 psi?	☐	☐
6. Are power saws and similar equipment provided with safety guards?	☐	☐
7. Are grinding wheel tool rests set to within 1/8 inch or less of the wheel?	☐	☐
8. Is there any system for inspecting small hand tools for burred ends, cracked handles, etc.?	☐	☐
9. Are compressed gas cylinders examined regularly for obvious signs of defects, deep rusting or leakage?	☐	☐
10. Is care used in handling and storing cylinders and valves to prevent damage?	☐	☐
11. Are all air receivers periodically examined, including the safety valves?	☐	☐
12. Are safety valves tested regularly and frequently?	☐	☐
13. Is there sufficient clearance from stoves, furnaces, etc., for stock, woodwork, or other combustible materials?	☐	☐
14. Is there clearance of at least 4 feet in front of heating equipment involving open flames, such as gas radiant heaters, and fronts of firing doors of stoves, furnaces, etc.?	☐	☐
15. Are all oil and gas fired devices equipped with flame failure controls that will prevent flow of fuel if pilots or main burners are not working?	☐	☐
16. Is there at least a 2-inch clearance between chimney brickwork and all woodwork or other combustible materials?	☐	☐
17. For Welding or Flame Cutting Operations: Are only authorized, trained personnel permitted to use such equipment?	☐	☐
Have operators been given a copy of operating instructions and asked to follow them?	☐	☐
Are welding gas cylinders stored so they are not subjected to damage?	☐	☐
Are valve protection caps in place on all cylinders not connected for use?	☐	☐
Are all combustible materials near the operator covered with protective shields or otherwise protected?	☐	☐
Is a fire extinguisher provided at the welding site?	☐	☐
Do operators have the proper protective clothing and equipment?	☐	☐

Materials

	OK	ACTION NEEDED
1. Are approved safety cans or other acceptable containers used for handling and dispensing flammable liquids?	☐	☐
2. Are all flammable liquids that are kept inside buildings stored in proper storage containers or cabinets?	☐	☐
3. Do you meet OSHA standards for all spray painting or dip tank operations using combustible liquids?	☐	☐
4. Are oxidizing chemicals stored in areas separate from all organic material except shipping bags?	☐	☐
5. Do you have an enforced NO SMOKING rule in areas for storage and use of hazardous materials?	☐	☐
6. Are NO SMOKING signs posted where needed?	☐	☐
7. Is ventilation equipment provided for removal of air contaminants from operations such as production grinding, buffing, spray painting and/or vapor degreasing, and is it operating properly?	☐	☐
8. Are protective measures in effect for operations involved with X-rays or other radiation?	☐	☐
9. For Lift Truck Operations: Are only trained personnel allowed to operate forklift trucks?	☐	☐
Is overhead protection provided on high lift rider trucks?	☐	☐
10. For Toxic Materials: Are all materials used in your plant checked for toxic qualities?	☐	☐
Have appropriate control procedures such as ventilation systems, enclosed operations, safe handling practices, proper personal protective equipment (e.g., respirators, glasses or goggles, gloves, etc.) been instituted for toxic materials.	☐	☐

Employee Protection

	OK	ACTION NEEDED
1. Is there a hospital, clinic or infirmary for medical care near your business?	☐	☐
2. If medical and first-aid facilities are not nearby, do you have one or more employees trained in first aid?	☐	☐
3. Are your first-aid supplies adequate for the type of potential injuries in your workplace?	☐	☐
4. Are there quick water flush facilities available where employees are exposed to corrosive materials?	☐	☐
5. Are hard hats provided and worn where any danger of falling objects exists?	☐	☐
6. Are protective goggles or glasses provided and worn where there is any danger of flying particles or splashing of corrosive materials?	☐	☐
7. Are protective gloves, aprons, shields or other means provided for protection from sharp, hot or corrosive materials?	☐	☐
8. Are approved respirators provided for regular or emergency use where needed?	☐	☐
9. Is all protective equipment maintained in a sanitary condition and readily available for use?	☐	☐
10. Where special equipment is needed for electrical workers, is it available?	☐	☐
11. When lunches are eaten on the premises, are they eaten in areas where there is no exposure to toxic materials, and not in toilet facility areas?	☐	☐
12. Is protection against the effects of occupational noise exposure provided when the sound levels exceed those shown in Table G-16 of the OSHA noise standard?	☐	☐

Mechanical Inspection

Koch Dry Ovens and
Paint Booth Make Up
Air Blowers

To Be Inspected-
Jan-Mar-May-Jul-Sep-Nov

Dept. 862

Accr. No. 26023-61
Ord. No. 759223
Dept. No. 862

Inspec's.
Name: _____
Inspec.
Date: _____

Inspect For: Security, Condition, Operation,
Vibration, Belt Tension, Safety

Mach. No.	Bearings	Belts and Pulleys	Belt Guards	Lube & Lube Lines O.K.	Inspector's Comments
	Sub Assembly Paint System on Mezzanine - North to South				
7227					
7226					
7186					
7187					
7188					
7225					
7189					
	Paint Booth Make Up Air Bldg. - "V" Roof - East Side				
6790					
6791					
7219					
7221					
7220					
7218					
	Work in Process Paint System on Mezzanine - North to South				
6760					
6758					
6795					
6796					

Mach. No.	Bearings	Belts and Pulleys	Belt Guards	Lube & Lube Lines O.K.	Inspector's Comments
	Combine Finish Paint Syst. - Dry Ovens and Paint Booth Make Up Air Units Bldg. - "V1" South to North				
7251					
7252					
7259					
7260					
7261					
7262					
7297					
7298					
7300					
	Touch Up Paint Dry Oven (3-Ovens) Bldg. - "V2" North East Corner				
6390					

MECHANICAL INSPECTION ②

EXHAUST BLOWERS & AIR CONDI. HEAT EXCHANGER & DOOR SEAL BLOWERS

3-Month Inspection
Feb-May-Aug-Nov

Inspector's Name ______________ Inspection Date ________

ACCT. No. 26023 -79
ORD. No. 759223
DEPT. No. -862

Blower Machine Number	Check for Excessive Vibration	Chk. Belts and Pulleys for Ten.& Align.& Condi.	All Safety Shields are in Place and Secure
"V" Bldg. Roof E. Side			
7426-Air Condi.Heat Exchanger			
7429-Air Condi.Heat Exchanger			
Sub.Assemb.Paint & Wash Ovens Mezzanine - North to South			
7228			
7178			
7179			
7181			
7180			
7183			
7182			
7185			
7224			
7223			
7187			
7190			
Flow Coat Booth			
7192			
7193			
7194			
7195			
7196			
7197			
Spray Paint Booth			
7205			
7206			
7208			
7207			
7212			
7213			
7215			
7214			

Blower Machine Number	Check for Excessive Vibration	Chk.Belts & Pulleys for Ten.& Align.& Condi.	All Safety Shields are in Place and Secure
W.I.P. Paint Dip			
5765			
6767			
6766			
6764			
W.I.P. Paint Spray			
6784			
6785			
6786			
6787			
6763			
W.I.P. Wash Dry Off			
6761			
6760			
6759			
6757			
W.I.P. Paint Dry			
6793			
6794			
6798			
6797			
W.I.P. Wash Booth			
6755			
6754			
6753			
Comb. Finish Paint			
7241			
7242			
7243			
7249			
Wash Dry Off			
7250			
7253			

Blower Machine Number	Check for Excessive Vibration	Same as Other	Same
Wash Cool Down			
7255			
7256			
7257			
Combine Paint			
7268			
7269			
7270			
7271			
7272			
7263			
7264			
7265			
7266			
7267			
Combine Flash Off			
7292			
7293			
Combine Paint Dry			
7295			
7296			
7299			
7301			
Combine Cool Down			
7303			
7304			
7305			
7310			
7312			
7306			
7311			
Spray & Dry - N.E. Corner-"V2"			

MECHANICAL INSPECTION ㉝ "HEATER-VENTILATOR BLOWERS" Bldgs. "V" and "V2" Roof Acct. No. 26845-79 Ord. No. 759224 Dept. No. 862	Inspector's Name: **INSPECTIONS TO BE MADE** FEB – APR – JUN AUG – OCT – DEC	Dept. 862

Heater-Ventilator Number / Bearings Check - OK / Lube - OK -Lube Lines O.K. / Belts in Good Condition - / Belt Tension - OK / Pulleys in Good Condition - / Pulleys Secure / Belt Guards "In Place" and Secure / All Filters in Good Condition and Clean

Date of Inspection To Be Inspected - Feb-Apr-Jun-Aug-Oct-Dec

Start Inspection at Southwest Corner of "V" Bldg. Roof. Check Off if Item is "OK". Make Remarks For Needed Repairs.

Inspector's Remarks

Heater-Ventilator Number									
6640									
6641									
6643									
6642									
6644									
6645									
6646									
6650									
6649									
6648									
6647									
6651									
6652									
6653									
6657									
6656									
6655									
6654									
6658									
6659									
6660									
6664									
6663									
6662									
6661									
6665									
6666									
6667									
6378									
6379									
6380									

AIR COMPRESSOR INSPECTION TIRE ROOM DEPT. 945	ACCT. NO. 26021-67 SHOP ORD. 759226 DEPT. 945	DATE: _____ INSPECTOR: _____ SUPERVISOR: _____

MECHANICAL - (862)

IF ITEM IS O.K. USE [✓] MARK. IF WORK IS NEEDED USE [W] AND EXPLAIN. IF INSPECTOR COMPLETES REPAIR USE [R] & EXPLAIN

ITEM NO.	PART TO BE INSPECTED		CHECK OFF					
			7694		7840			
			1stWk	2ndWk	3rdWk	4thWk	5thWk	
1.	CHECK CRANKCASE OIL LEVEL (ANDEROL-500 OIL)	7694						
		7840						
2.	CHANGE CRANKCASE OIL (ANDEROL-500 OIL) (JAN-MAR-MAY-JUL-SEP-NOV)							
3.	CHECK CONDITION OF COMPRESSOR AND AIR RECEIVER							
4.								
5.	CHECK CONDITION OF AIR RECEIVER CONDENSATE TRAP. DRAIN AS NECESSARY	7694	1stWk	2ndWk	3rdWk	4thWk	5thWk	
		7840						
6.	CLEAN OR REPLACE INTAKE AIR FILTER (JAN-JUL)							
7.	DISASSEMBLE COMPRESSOR VALVES. CLEAN OR REPLACE ALL PARTS AS NECESSARY - PER INSTRUCTIONS ON PAGE 25 OF I.R. INSTRUCTION FORM NO. 1050-H (JAN-JUL)							
8.	CHK. CONDITION OF DRIVE BELTS. REPLACE AS NECESSARY. CHECK TENSION.							
9.	CHK. CONDITION, SECURITY, SOUND-ELEC. MOTOR (LUBE-JAN G-1 GREASE)							
10.	CLEAN COMPRESSOR & RECEIVER WITH AIR JET (JAN-JUL) (EXTERNAL SURFACE OF CYLINDERS & INTERCOOLER TUBES)							

TORQUE CHECK ALL CAP SCREWS

ITEM NO.	LOCATION OF CAP SCREWS (JAN-JUL)	QUANT. SCREW	TORQUE	CHECK OFF	
				7694	7840
11.	CONSTANT SPEED UNLOADERS				
12.	AIR HEADS				
13.	CYLINDER BOLTS				
14.	SHAFT END COVER				
15.	CRANKCASE COVER PAN				
16.	DISCHARGE AIR MANIFOLD				
17.					
18.					
19.					
20.					

ITEM NO.	POST ITEM NUMBER AND LIST MAINTENANCE TO BE DONE

PITTSBURGH ROLLS CORPORATION
CRANE INSPECTION REPORT

Crane No.............................Type...................................... Capacity..............................

RUNWAY AND CONDUCTORS
Track Alignment....................Spread........................... Fastenings ....
Line Conductors..Conductor Supports.......................

TRUCKS AND MAIN COLLECTORS
Truck Wheels, Flat Spots?.....................Flanges....................End Play..................
Axle Bearings.................................Lubrication ....
Truck Drive Bearings...................Lubrication.................... Gears ....
Gear Screws..................Pinion.....................Key.............. Collectors ....

GIRDERS AND DRIVE
Drive Shaft..............Couplings.............Bearings...........Lubrication ....
Foot Brake Shaft...........Couplings...........Bearings..........Lubrication ....
Bridge Brake Case.............Adjustment.................. Lubrication ....
Walkway.........................Railing ....Ladder ....
Bridge Motor Support...........Shaft Extension............ Couplings ....
Bridge Drive Gear Case............................Gears.......... Lubrication ....

MOTORS

Location	Armature	Commutator	Brushes	Brush Holders	Bearings	Lubrication
Bridge						
Hoist						
Aux. Hoist						
Trolley						

CONTROLLERS

	Brushes	Brush Holders	Contacts	Wiring	Springs	Resistance
Bridge						
Hoist						
Aux. Hoist						
Trolley						

Trolley Wheels................Trolley Wheel Bearings...................... Lubrication ....
Trolley Gear Case..............Case Support..........Gears........... Lubrication ....
Hoist Gear Case...........Main Gear Train..........Comp. Train.........Lubrication........
Mech. Brake..........Drift.............Elec. Brake............ Adjustment ....Lubrication........
Drum................Cable or Chain.............Cable Pin.........Limit Switch...........
Limit Switch Adjustment................Hook..........Sheaves ....Lubrication.........
Trolley Conductors..............Trolley Collectors..............Cage Roof...........Door.........
Windows.............Foot Brake Treadle..............Cont. Levers.........Load Test..........
Bell or Signal...
Inspected by...Date....................

KEY WORDS—G=Good; F=Fair; W=Worn; A=Need Attention;
C=Need Cleaning; T=Too Tight

BUTLER MANUFACTURING COMPANY
WEEKLY INSPECTION OF FIRE PROTECTIVE EQUIPMENT
Kansas City Plant

Instructions: Fill out this blank while making inspection. Do not report a valve open unless you personally have inspected and tested it. Every valve controlling sprinklers or
water supplies to sprinklers should be listed. When the blank is filled out, it should be
sent to the Safety Department.

Valve No.	AREA CONTROLLED	Location	Open	Shut	Sealed	Pressure
1	Entire West System	Bldg. 57				
2	Bldg. 43-43B	Bldg. 57				
3	Valves 4-5-6-7-8-9-10-11	Bldg. 43				PIV
4	East End Bldg. 2-2B	Bldg. 2				
5	Center 2-2B, Paint Line	Bldg. 2				
6	West End Bldg. 2	Bldg. 2				
7	Valves 8-9-10-11	Bldg. 2				PIV
8	Valves 9-10-11	Bldg. 62				
9	Bldg. 3	Bldg. 62				
10	Bldg. 62-62B-6A-5C-5	Bldg. 62				
11	Bldg. 4-4A-5-5A-5C	Bldg. 62				
12	Paint Booth P34KC	Bldg. 1				
13**	Oven	Bldg. 1				
14	Paint Booth P57KC	Bldg. 1				
15	Paint Booth P58KC	Bldg. 1				
16	Paint Booth P59KC	Bldg. 1				
17	Paint Booth P56KC	Bldg. 1				
18	Locker Room Offices	Bldg. 67				
19	Laboratory Paint Room	Bldg. 58A				
20	Paint Shop Bldg. 23	Bldg. 23				
21	Bldg. 65	Bldg. 65				
22	Paint Booth P49KC	Bldg. 63				
23	Paint Booth P710G	Bldg. 63				
24	Bldg. 17	Bldg. 17				
25	Bldgs. 53-63	Bldg. 17				
26*	Bldg. 57	Bldg. 57				
27	Bldg. 17A Offices	Bldg. 15				
28	Bldg. 17A Balcony	Bldg. 15				
29	Entire East System	12th St.				PIV
30*	Bldg. 52	Driveway				PIV
31*	Bldg. 46	Driveway				PIV
32	Valves 21-22-23-24-25	Driveway				PIV

* Controls Dry System
**Manually Operated, always shut

GENERAL CONDITIONS

HYDRANTS: In good condition? _____
Clear? _____ Remarks _____
_____

AUTOMATIC SPRINKLERS: Any heads missing? _____ Disconnected? _____
Obstructed by high-piled stock? _____
Any rooms not sufficiently heated
to prevent freezing? _____ How
many extra heads available? _____

SPRINKLER ALARMS: Tested? _____ In
good condition? _____ Do not test
hydraulic alarms when temperatures
are below freezing.

EXTINGUISHERS, SMALL HOSE: In good
condition? _____

FIRE DOORS: All inspected? _____
In good order? _____

HOUSEKEEPING: Good throughout? _____
Combustible waste removed before
night? _____

REMARKS on other matters relating to
fire hazard: _____
_____
_____

Date _____ Signed _____

Courtesy Butler Manufacturing Company.

STATIONARY SCAFFOLD SAFETY CHECK LIST

PROJECT: _____

ADDRESS: _____

CONTRACTOR: _____

DATE OF INSPECTION: _____ INSPECTOR: _____

	Yes	No	Action/Comments
1. Are scaffold components and planking in safe condition for use and is plank graded for scaffold use?			
2. Is the frame spacing and sill size capable of carrying intended loadings?			
3. Have competent persons been in charge of erection?			
4. Are sills properly placed and adequate size?			
5. Have screw jacks been used to level and plumb scaffold instead of unstable objects such as concrete blocks, loose bricks, etc.?			
6. Are base plates and/or screw jacks in firm contact with sills and frame?			
7. Is scaffold level and plumb?			
8. Are all scaffold legs braced with braces properly attached?			
9. Is guard railing in place on all open sides and ends above 10' (4' in height if less than 45")?			
10. Has proper access been provided?			
11. Has overhead protection or wire screening been provided where necessary?			
12. Has scaffold been tied to structure at least every 30' in length and 26' in height?			
13. Have free standing towers been guyed or tied every 26' in height?			
14. Have brackets and accessories been properly placed: Brackets?			
Putlogs?			
Tube and Clamp?			
All nuts and bolts tightened?			
15. Is scaffold free of makeshift devices or ladders to increase height?			
16. Are working level platforms fully planked between guard rails?			
17. Does plank have minimum 12" overlap and extend 6" beyond supports?			
18. Are toeboards installed properly?			
19. Have hazardous conditions been provided for: Power lines?			
Wind loading?			
Possible washout of footings?			
Uplift and overturning moments due to placement of brackets, putlogs or other causes?			
20. HAVE PERSONNEL BEEN INSTRUCTED IN THE SAFE USE OF THE EQUIPMENT?			

MECHANICAL INSPECTION

FLOOR CONVEYORS

Acct. No. **90059-67**
Ord. No. *759225*
Dept. No. *862*

Monthly Inspection

Inspecting Dept. - (862)

_____ Inspection Date

Inspector's Name: _____

SUPERVISOR'S NAME: _____

Inspect for: Security – Safety
Operation – Condition

If item is O.K. use [✓] mark. If work is needed use [W] and explain. If inspector completes repair, use [R] and explain.

Inspection items (columns):

- Check 2 Gear Reduction Boxes for Operation, Security, Noise, Lubrication
- Check Drive Couplings and Alignment
- Check Belts and Pulleys for Alignment and Tension - Chain Drives - Replace as Necessary
- Check Chain Drives and Sprockets; "Makeups", Bearings and Sprocket and Pulleys for Alignment
- Check Drive Shafts; Bearings, & Lube
- Check Auto Lube Systems - Mech. Overload Device and Security Lube Systems for proper
- Check Conveyor Chain for Tension, Wear, and Security & Lube
- Adequate Lube; System for Cleanliness, Leaks, Worn
- Check Hyd. Lube; System for Operation, Leaks, Worn Tubing; Hoses and parts
- Check Hyd. Hoist Cylinders and Hoist Hyd. System for Operation
- Return Chain - Main Rollers and Check Chain Guides
- Check Roller or Chain Tracks or Channel Guides and Alignment and Wear - Check Guides and
- Check Chain "Makeup" Mechanism for proper function
- Hydrostatic Drive Unit - Oil Level (w/Hyd. Motor - Oil Level)
- Remarks

Machine Number and Name of Conveyor														Remarks
9160 - East (N. Pit) Assembly Line Conveyor	X	X	X		X	X			X		X	X		
6951 - Hydraulic Hoist System	X	X	X		X	X	X	X			X	X		
9160 South Pit														
9159 - West (N. Pit) Assembly Line Conveyor	X	X	X		X	X			X		X	X		
6949 - Hydraulic Hoist System	X	X	X		X	X	X	X			X	X		
9159 South Pit														
7313 - Combine Paint Conveyor - S. Pit	X		X			X	X					X		
7313 - Combine Paint Conveyor - N. Pit			X				X	X				X		

vals. Frequency of inspections should be described in specific terms: for example, before every use, when serviced daily, monthly, quarterly, yearly.

Who Will Conduct the Inspection?

Answering the four previous questions—the items to be inspected, the aspects of each item to be inspected, the conditions to be inspected, and the frequency of inspections— will help to determine who is qualified to do the inspection. No individual or group should have exclusive responsibility for all inspections. Employees who perform these inspections will benefit from training in hazard recognition. Some items will need to be inspected by more than one person. For example, while an area supervisor may inspect an overhead crane weekly and maintenance personnel inspect it monthly, the operator of the crane will inspect it before each use. When grinding wheels are received, they are inspected by the stockroom attendant, but they must be inspected again by the operator before each use.

As part of the hazard control inspection inventory, management should assign responsibility for each inspection. Figure 11–8 shows how the inventory can designate the proper person by title: area supervisor, operator, foreman, maintenance foreman, and so forth.

A suggested guide for planned inspections is as follows:

- *Daily*—area supervisor and maintenance personnel, who also can request suggestions from employees in their various workstations
- *Weekly*—department heads
- *Monthly*—supervisors, department heads, the safety department, and safety and health committees.

The safety department also may be actively involved in monthly, quarterly, semiannual, and annual inspections.

Five qualifications of a good inspector are:

1. knowledge of the organization's accident experience
2. familiarity with accident potentials and with the standards that apply to his or her area
3. ability to make intelligent decisions for corrective action
4. diplomacy in handling personnel and situations
5. knowledge of the organization's operations—its workflow, systems, and products.

Safety professionals. Clearly, the safety professional should spearhead the inspection activity. During both individual and group inspections, the professional can educate others in inspection techniques and hazard identification by using on-the-spot examples and firsthand contact. Supervisors, foremen, stewards, and safety and health committees can be shown what to look for when making inspections. The organization's fire protection representative or industrial hygienist usually works with the hazard control specialist in conducting inspections.

The number of safety professionals depends on the size of the company and the nature of its operation. Large companies with well-organized accident prevention programs usually employ a full-time staff. Sometimes large companies also have designated employees who spend part of their time on inspections.

In organizations where toxic and corrosive substances are present, the industrial hygienist will be part of the inspection team (see also the following section on Measurement and Testing). When an organization uses chemicals, the chief chemist will need to cooperate closely with the safety professional and fire protection representative in establishing inspection criteria. If the organization has no industrial hygienist, the safety professional needs to obtain training about the hazardous properties of substances, unstable properties of chemicals, and methods of control. An inspection conducted without this knowledge is incomplete and may miss potentially serious problems.

Company or plant management. Safety inspections should

DEPARTMENT *Maintenance*	UNIT *Workshop*	SUPERVISOR RESPONSIBLE *J. P. Smith*	APPROVED BY *Ralph T. Welles*	DATE *4/16/72*	PAGE NO. *1*
1. PROBLEMS	2. CRITICAL FACTORS	3. CONDITIONS TO OBSERVE	4. FREQUENCY	5. RESPONSIBILITY	
1. Overhead hoist	*Cables, chains, hooks, pulleys*	*Frayed or deformed cables, worn or broken hooks and chains, damaged pulleys*	*Daily—before each shift*	*Operators*	
2. Hydraulic pump	*High pressure hose*	*Leaks; broken or loose fittings*	*Daily*	*Shift leader*	
3. Power generator	*High voltage lines*	*Frayed or broken insulation*	*Weekly*	*Foreman*	
4. Fire extinguishers	*Contents, location, charge*	*Correct type, fully charged, properly located, corrosion, leaks*	*Monthly*	*Area safety inspector*	
5. General housekeeping	*Passageways, aisles, floors, grounds*	*Free of obstructions, clearly marked, free of refuse*	*Daily*	*Shift leader foreman*	

Figure 11–8. A hazard control inspection inventory should list the person responsible for each inspection. (Printed from *Principles and Practices of Occupational Safety and Health, Student Manual*, Booklet Three, U.S. Department of Labor, OSHA 2215.)

be considered part of the duties of company or plant management. By participating in inspections, management demonstrates its commitment to maintain a safe working environment. But the psychological effect of inspection by senior executives goes beyond merely showing an interest in safety. When employees know that management is coming to inspect their area, conditions that seemed "good enough" suddenly appear unsatisfactory and are quickly corrected.

First-line supervisor or foreman. Because supervisors and foremen spend practically all their time in the shop or plant, they are continually monitoring the workplace. At least once a day, supervisors need to check their areas to see that (1) employees are complying with safety regulations, (2) guards and warning signs are in place, (3) tools and machinery are in a safe condition, (4) aisles and passageways are clear and proper clearances maintained, and (5) material in process is properly stacked or stored. Although such a spot check does not take the place of more detailed inspections, it emphasizes the supervisor's commitment to maintaining safety in the area. A supervisor also should conduct regular formal inspections to make certain all hazards have been detected and safeguards are in use. Supervisors can perform such inspections weekly on their own and monthly as part of a safety and health committee.

Mechanical engineer and maintenance superintendent. Either as individuals or as members of a committee, the mechanical engineer and the maintenance superintendent also need to conduct regular formal inspections. They can write necessary work orders on the spot for guards or for correcting faulty equipment.

Employees. As mentioned previously, employee participation in continuous inspection is one goal of an effective hazard control program. Before beginning the workday, the employee should inspect the workplace and any tools, equipment, and machinery that will be used. Any defects the employee is not authorized to correct should be reported immediately to the supervisor. Action resulting from this report must be reported to the employee to encourage further participation.

Maintenance personnel. Maintenance employees can be of great help in locating and correcting hazards. As they work, they can conduct informal inspections and report hazards to the supervisor, who in turn should encourage the mechanics to offer suggestions.

Joint safety and health committees. Joint safety and health committees (discussed in Chapter 3) conduct inspections as part of their function. They give equal consideration to accident, fire, and health exposures. By periodically visiting areas, members may notice changed conditions more readily than someone who is there every day. Another advantage provided by the committee is the members' various backgrounds, experience, and knowledge represented.

If the committee is large, the territory should be divided among teams of manageable size. Large groups going through the plant are unwieldy and distracting. (See the Council's *You Are the Safety and Health Committee* for guidelines.)

Other inspection teams. If there is no safety and health committee, a planned, formal inspection is still necessary. Management should assign an inspection team that includes the hazard control specialist, production manager, supervisor, employee representative, fire prevention specialist, and industrial hygienist. The important point is that inspections should be directed by a responsible executive who has the authority to ensure the work is carried out effectively.

Outside inspectors sometimes are needed to perform inspections. For example, insurance company safety engineers and local, state or provincial, and federal inspectors may lend their expertise to specific inspections.

Contractors' inspection services. For some technical systems, notably sprinkler systems, contracting companies furnish inspection services. Companies without either qualified safety professionals or a well-established maintenance program can avail themselves of such services.

For example, a sprinkler contractor may arrange with a customer for periodic inspection and tests of sprinkler equipment. The contractor and the client negotiate how often inspections are to be done. In some cases, the inspection will include other items, such as fire extinguishers, hoses, or fire doors. The contractor furnishes a comprehensive written report. The client can request that the contractor send copies of the report to the insurer.

The basic contract does not include maintenance work or materials required for alterations, repairs, or replacement. However, if the report indicates any maintenance needs, the client can have the contractor perform the work. Contract service does not relieve management of its primary responsibility for inspection and maintenance. Nevertheless, it does provide excellent inspection for small companies, buildings with mixed tenants, and companies with systems too complex for inspection by its own maintenance staff.

CONDUCTING INSPECTIONS

Preparing to Inspect

Inspections should be scheduled for times when inspectors will have the best opportunity to see operations and work practices without much interruption. The inspection route should be planned in advance.

Before making an inspection, the inspector or inspection team should review all accidents that have occurred in the area. At this brief meeting, team members should discuss where they are going and what they will be looking for. During the inspection, it will be necessary to "huddle" before going into noisy areas in order to avoid arm waving, shouting, and other unsatisfactory methods of communication.

In addition to the regular checklist and accident reports, inspectors should have copies of the previous inspection report for that particular area. Reviewing this report makes it possible to check whether earlier recommendations have been followed and reported hazards corrected.

Those making inspections should wear the protective equipment required in the areas they enter: safety glasses and

shoes, hard hats, acid-proof goggles, protective gloves, respirators, and so forth. If inspectors do not have or cannot get special protective equipment, they should not go into the area. They must be careful to "practice what they preach."

Inspectors also should be aware of any special hazards they encounter. For example, because welding crews and other maintenance crews move from place to place, they may be encountered anywhere in the plant. Inspectors should know what precautions are required where these crews are working.

Inspection Tools

Inspectors should have the proper tools ready before the inspection to make the process more efficient and to gather more precise data. Common tools include:

- clipboards
- inspection forms
- pens/pencils
- lockout/tagout supplies
- measuring tape/ruler
- flashlight.

Depending on the inspection area or type, the following equipment may also be useful:

- cameras
- tape recorder
- electrical testing equipment
- sampling devices (air, noise, light, temperature)
- sample containers
- calipers, micrometers, feeler gauges
- special personal protection equipment (see Measurement and Testing section)
- stopwatch.

Relationship of Inspector and Supervisor

Before inspecting a particular department or area, the inspector should contact the department head, supervisor, or other person in charge. This person may have important information for the inspection, particularly if conditions are temporarily altered because of construction, maintenance, equipment downtime, employee absence, and so forth.

If no rules prohibit it, the person in charge may want to accompany the inspector. The inspector can agree but should also emphasize that no tour guide is needed. The inspector must preserve independence and the opportunity to make uninfluenced observations.

If the supervisor of the area does not accompany the inspector, the supervisor should be consulted before the inspector leaves the area. The inspector should discuss each recommendation on particular hazards or unsafe conditions with the supervisor. Usually they can reach an agreement regarding the relative importance of each recommendation. Obviously, an inspector should not focus on numerous trivial items merely to make the report look complete. On the other hand, the inspector does not have the authority to overlook any condition that might cause an accident.

Even minor items that the supervisor can quickly correct should be reported. The inspector can note on the

written report that the supervisor promises to correct the particular condition. This keeps the record clear and serves as a reminder to check the condition during the next inspection.

An inspector should not fail to report hazards merely because a supervisor regards such reporting as criticism. If a supervisor becomes defensive or resentful, the inspector can only repeat what the supervisor knows: the purpose of an inspection is fact-finding, not fault-finding. By retaining objectivity and refusing to let the issue of safety degenerate into a personality conflict, the inspector keeps matters on a proper professional footing and maintains a firm, friendly, and fair attitude.

Sometimes a supervisor will request the inspector's assistance in recommending new equipment, reassignment of space, or transfer of certain jobs from one department to another. When these suggestions deal with safety, the inspector will want to include them in notes and consider whether to make them part of the report. The inspector must be careful, however, not to promise either a supervisor or an employee more than actually can be delivered.

Relationship of Inspector and Employee

Unless company policy or departmental rules prohibit conversation with employees, the inspector can ask questions about operations, taking care, however, not to usurp the responsibility of the supervisor. If, for example, a member of a safety and health committee sees an employee who deviates from established safe work practices, it is better to ask the supervisor rather than the employee about the supposed infraction. The committee member may not fully understand the operation and may be incorrect in the assumption. In another case, the employee may be performing a risky practice that has been sanctioned by those in authority. The employee could become defensive when questioned by the inspector. It is the supervisor's job to require compliance with company safe work procedures; it is the inspector's job to do the inspecting and reporting. If, however, the situation appears to present an immediate danger, the employee and management should be notified immediately.

Regulatory Inspection, Compliance Officer

Each regulatory agency has specific procedures and rules for inspecting an organization's facilities. The safety and health professional needs to know the specific procedures to be observed when an inspector calls. In general, an inspector will:

- call at regular business hours
- present credentials
- identify the purpose for the inspection, its length and scope
- request records, etc., as may be required
- perhaps consult with employees on the inspection
- perhaps request management to remove employees from an area of imminent danger and to take immediate action to correct the danger.

Chapter 3 differentiated between deviations from accepted practices and workplace-induced human error. The inspec-

tion team needs to look for both. The inspector is not concerned with identifying the person who is responsible for the unsafe behavior (fault-finding). The goal is to identify the behavior (fact-finding) and see that it is corrected.

Unsafe behaviors will vary from one area to another. Among common items that might be noted are the following:

- using machinery or tools without authority
- operating motorized vehicles at unsafe speeds or in other violation of safe work practice
- removing guards or other safety devices or rendering them ineffective
- using defective tools or equipment or using tools or equipment incorrectly
- using hands or body instead of tools or push sticks
- overloading, crowding, or failing to balance materials or handling materials incorrectly, including improper lifting
- repairing or adjusting equipment that is in motion, under pressure, or electrically charged
- failing to use or maintain (or using improperly) personal protective equipment or safety devices
- creating unsafe, unsanitary, or unhealthy conditions by improper personal hygiene, using compressed air for cleaning clothes, poor housekeeping, or smoking in unauthorized areas
- standing or working under suspended loads, scaffolds, shafts, or open hatches.

Because the inspector's purpose is to locate unsafe practices, not pinpoint blame, the report should not specify any names. When the report states, "An employee in this area was observed," the supervisor has been advised of the need to enforce safe work practices. The inspector should not be seen as a police officer handing out tickets or, worse, as a snoop from "outside." Nor should information derived from inspections be used for disciplinary measures.

Sometimes it is necessary to closely observe workers on the job to understand their tasks. The inspector should explain to workers why he or she needs to observe the task and should always ask permission to watch. When employees understand that no one is trying to catch them in an error but rather that they have been chosen to demonstrate a task because of their exceptional skills, they probably will agree to being observed.

INSPECTION REPORT

Area Inspected: Building D
Date and Time of Inspection: 11/19/30—11:00 a.m.
Inspector and Title: Ron Baker, Hazard Control Specialist
Date of Report: 11/20/80
Names of Those to Whom Report Is Sent: Bob Firenze (Executive Director); Loren Hall (Department Head); file

No. of Items Carried Over from Previous Report: 3
No. of Items Added to This Report: 4
Total No. of Items on This Report: 7

Item (asterisk indicates old item)	Hazard Classification Consequence	Probability	Hazard Description	Specific Location	Supervisor	Corrective Action Recommended	Corrective Action Taken
*1	II	B	Guard missing on shear blade #2 machine. Work order issued to engineering for new guard 10/16/80. Wooden barrier guard in temporary use 10/23/79. Guard still missing.	S.W. corner, bay #1	Jay Rillo	Contact engineering to replace guard	Engineering says they will have guard by 11/24.
*2	IV	C	Window cracked. Work order issued for replacement 10/30/80.	South wall, bay #3	Joe Whitestone	Have maintenance replace window	Maintenance to replace all broken windows starting next week.
3	II	B	Oil and trash still accumulated under main motor. Was to be cleaned by 10/30/80.	Pump room	Tony Silva	Clean area; have supervisor talk to men	Cleaned out 11/21. Silva told men to keep area clear.
4	III	B	Mirror at pedestrian walk out of line	North end of machine shop	Tom Schroeder	Post temporary warning sign; call maintenance for adjustment	Sign posted 11/21—Butler has scheduled adjustment for 12/1
5	II	A	Three workers at cleaning tank not wearing eye protection	Electric shop	Hank Beine	Have supervisor give more training and education	Discussed with Beine—he held meeting on 11/25
6	I	A	Cable on jib crane badly frayed	Bay #3	Joe Whitestone	IMMEDIATE ACTION REQUIRED	Tagged crane out of Service Cable to be replaced 11/21
7	II	B	Guard rail damaged on stairway to second floor	Bay #1	Jay Rillo	Issue work order to carpenter shop to make replacement	Work issue order 11/21

Figure 11–9. Inspection report form simplifies procedures and emphasizes carryovers, new items, and responsibilities. The column at right is for noting corrective action later taken. (Printed with permission from RJF Associates, Inc. Bloomington, Ind.)

Recording Hazards

Inspectors should locate and describe each hazard found during inspection. A clear description of the hazard should be written down with questions and details recorded for later use. It is important to determine which hazards present the most serious threat and are most likely to occur. The hazard-ranking scheme described in Chapter 3 will simplify the job of classifying hazards.

Properly classifying hazards places them in the right perspective. This approach enables the inspector to briefly describe potential consequences and the probability of such consequences occurring. Inspection reports should enable management to quickly understand and evaluate problems, assign priorities, and make decisions.

On the other hand, the inspector should describe unsafe conditions or deviations from accepted practices in detail and identify machines and operations by their correct names. Also, the inspector must accurately name or number locations and identify in detail the specific hazards within them. Instead of noting "poor housekeeping," for example, the report should give the details: "Empty pallets left in aisles, slippery spots on the floor from oil leaks, a ladder lying across empty boxes, scrap piled on the floor around machines." Instead of noting "guard missing," the report should read, "Guard missing on shear blade of No. 3 machine, SW corner of Bldg. D."

Management must adopt some plan to note intermediate or permanent corrective measures. For example, if intermediate safety measures have been taken, the item could be circled. When permanent measures are taken, the item can be crossed out or marked with an X. Such a system identifies those items requiring further corrective action (Figure 11–9).

If a committee is performing the inspection, one member assumes the task of keeping notes. Without such notes it is almost impossible to write a satisfactory inspection report.

Condemning Equipment

When a piece of equipment presents an imminent danger, the inspector should notify the supervisor immediately and see that the machine or equipment is shut down, tagged, and locked out to prevent its further use. Figure 11–10 shows both sides of a danger tag that can help to prevent unsafe equipment or materials from being used again.

When danger tags and locks are used, those persons authorized to condemn equipment must sign them. Only the inspector who places the tag should be permitted to remove it. This step should occur only when the inspector is satisfied that the hazardous condition has been corrected.

For more on lockouts, see Chapters 13, Safeguarding, and 15, Electrical Equipment, *Engineering and Technology* volume of this *Manual*. No equipment or materials should be placed out of service without notifying the person in authority in the department affected.

Writing the Inspection Report

Every inspection must be documented in a clearly written report furnished by the inspector. Without a complete and accurate report, the inspection would be little more than an interesting sightseeing tour. Inspection reports are usually of three types:

1. Emergency—made without delay when a critical or catastrophic hazard is probable. Using the classification system described in Chapter 3, this category would include any items marked IA or IIA.
2. Periodic—covers those unsatisfactory nonemergency conditions observed during the planned periodic inspection. This report should be made within 24 hours of the inspection. Periodic reports can be initial, follow-up, final, or a combination of all three.
3. Summary—lists all items of previous periodic reports for a given time.

The written report should include the name of the department or area inspected (giving the boundaries or location if needed), date and time of inspection, names and titles of those performing the inspection, date of the report, and the names of those to whom the report was made.

One way to make the report is to begin by copying items carried over from the last report that were not corrected. Each item is numbered consecutively. The item number can be followed by the hazard classification (IB, IIIC, and so on). Carryover items can be marked with an asterisk. The narrative should include the date the hazard was first detected. The inspector should describe each hazard and pinpoint its location. After the hazard is listed, the inspector should recommend corrective action and establish a definite abatement date. There should follow a space for noting corrective action taken later. Figure 11–9 is a sample of an inspection report made by the organization's hazard control specialist after a weekly inspection. In addition, a report should show what is right in a work area as well as what is wrong. When the report is from a committee, it should be reviewed by each member of the inspection team for accuracy, clarity, and thoroughness.

Figure 11–10. Front and back views of a typical tag used when equipment is taken out of service because it has become unsafe.

SAFETY RECOMMENDATION   No. 1053

Date Issued_____19___

Date Ret'd._____19___

To_____

PLEASE HAVE THE FOLLOWING UNSAFE CONDITION OR
DEVIATION FROM STANDARD PROCEDURE CORRECTED:

_____

_____

_____

_____

Please sign and return to Safety Department within ten days, indicating
below what disposition was made of this recommendation.

SAFETY DEPARTMENT

RECOMMENDATION FOLLOWED ( )   WORK COMPLETED_____

                                                    DATE

RECOMMENDATION REJECTED ( )   FOR FOLLOWING REASON :_____

_____

_____

Copy of this recommendation is on
file in the Safety Department. The
Safety Department is instructed to
send a detailed list of all un-
answered recommendations more
than ten days old to the General
Superintendent the first of each   (SIGNED) DEPARTMENT SUPERINTENDENT
period.

Figure 11–11. This special form is padded, numbered in pairs, and carboned in order to save office work and permit follow-up until the recommended work is complete or a procedure correction is made. (Printed with permission from Tennessee Eastman Corporation.)

Generally, inspection reports are sent to the head of the department or area where the inspection was made. Copies are also given to executive management and the manager to whom the department head reports.

Follow-up for Corrective Action

It is after the inspection report has been written and disseminated that the inspection process starts to return benefits. The information acquired and the recommendations made are valueless unless management takes corrective action. Information and recommendations provide the basis for establishing priorities and implementing programs that will reduce accidents, improve conditions, raise morale, and increase the efficiency and effectiveness of the operation.

Inspectors can list recommendations in the order in which the hazards were discovered or group them according to the individuals responsible for their correction. Recommendations are then sent to the proper member of management for approval. Where possible, management should set a definite time limit for correction for each recommendation and follow up to make sure corrective action has been taken.

Often the safety professional is authorized to make recommendations directly to the affected foreman, supervisor, or department if such recommendations do not require major capital outlays. One company has simplified the process of making individual safety recommendations by devising special forms (Figure 11–11). These forms—which have a carbon attached so the safety department can keep its records intact—provide a convenient follow-up file.

Some organizations require that inspection reports be reviewed by the safety and health committee. This is particularly the case when recommendations apply to education and training and directly affect employees.

In making recommendations, inspectors should be guided by four rules (*Facility Inspection*, 1973):

- Correct the cause whenever possible. Do not merely correct the result, leaving the problem intact. In other words, be sure the disease and not just the symptom is cured. If the inspector or supervisor does not have the authority to correct the real cause, they should bring it to the attention of the person who does.

- Immediately correct everything possible. If the inspector has been granted the authority and opportunity to take direct corrective action, he or she should take it. Delays risk accidents.

- Report conditions beyond one's authority and suggest solutions. Inform management of the condition, the potential consequences of hazards found, and solutions for correction. Even when nothing seems to come of a recommendation, it can pay unexpected dividends.

  For example, a company safety and health committee made a detailed proposal about guarding a particularly hazardous location, only to be told that the engineers had planned to move operations to another location. However, instead of feeling that it had wasted its time, the committee pointed out that the organization had serious communication problems, with the right hand not knowing what the left was doing. The committee recommended that effective management techniques be applied to the hazard control program.

- Take intermediate action as needed. When permanent correction takes time, the hazard should not be ignored. Inspectors or supervisors should take any temporary measures they can, such as roping off the area, locking and tagging out equipment or machines, or posting warning signs. These measures may not be ideal, but they are better than doing nothing.

Some of the general categories into which recommendations might fall are setting up a better process, relocating a process, redesigning a tool or fixture, changing the operator's work pattern, providing personal protective equipment, and improving personnel training methods. Recommendations can also call for improvements in the preventive maintenance system and in housekeeping. Cleaning up debris and dirt may be considered the janitor's job, but preventing its accumulation is part of an effective hazard control program.

Management must realize that employees are keenly interested in the attention paid to correcting faulty conditions and hazardous procedures. Recommendations approved and supported by management should become part of the organization's philosophy and program. At regular intervals, supervisors should report progress in complying with the recommendations to the safety department, the company safety and health committee, or the person designated by management to receive such information. Inspectors should periodically check to see what progress toward corrective

action is being made. Unsafe conditions left uncorrected indicate a breakdown in management communications and program application.

Sometimes management will have to decide among several courses of action. Often these decisions will be based on cost effectiveness. For example, it may be cost-effective as well as practical to substitute a less toxic material that works as well as the highly toxic substance presently in use. On the other hand, replacing a costly but hazardous machine may have to wait until funds can be designated. In this case, the immediate alternative may be to install machine guards. In all cases, action taken or proposed must be communicated to all persons involved.

MEASUREMENT AND TESTING

Two special sorts of inspection are conducted by the industrial hygienist and the medical staff. Testing for exposures to health hazards requires special equipment not always available to the hazard control specialist. In such cases, management can often obtain assistance from the state labor department industrial hygiene division or from the provincial department of labor and health. Another source of help can be the industrial hygienists employed by consulting firms and by insurance companies. Conducting physical examinations of employees exposed to occupational health hazards may require medical equipment that the organization's medical staff does not have. The following discussion is a summary of how to recognize, evaluate, and control health hazards in the workplace. Those interested in more details should consult the text *Fundamentals of Industrial Hygiene* (see References). This reference should help safety professionals understand their roles in this area of hazard control.

Kinds of Measurement and Testing

Occupational health surveillance monitors chemical, physical, biological, and ergonomic hazards. Four monitoring systems are used: personal, environmental, biological, and medical.

Personal monitoring. One example of personal monitoring is measuring the airborne concentrations of contaminants. The measurement device is placed as closely as possible to the site at which the contaminant enters the human body. When the contaminant is noise, the device is placed close to the ear. When a toxic substance could be inhaled, the device is placed in the breathing zone.

Environmental monitoring. Environmental monitoring measures contaminant concentrations in the workroom. The measurement device is placed in the general area adjacent to the worker's usual workstation or where it can sample the general room air.

Biological monitoring. Biological monitoring measures changes in composition of body fluid, tissues, or expired air to detect the level of contaminant absorption. For example, blood or urine can be tested to determine excessive lead absorption. The phenol in urine sometimes is measured to determine excessive benzene absorption.

Medical monitoring. When medical personnel examine workers to see their physiological and psychological response to a contaminant, the process is termed medical monitoring. Medical monitoring can include health and work histories, physical examinations, X rays, blood and urine tests, pulmonary function tests, and vision and hearing tests. The aim of such monitoring is to find evidence of exposure early enough to identify especially susceptible workers and to detect any damage before it becomes irreversible.

Biological and medical monitoring provide information after the exposure already has occurred. However, such programs also encompass arrangements to treat an identified health problem and to take corrective action to prevent further damage. To understand how industrial hygienists measure for health hazards, it is necessary to define some basic terms, to distinguish between acute and chronic effects, and to see how safe exposure levels are established.

Measuring for Toxicity

Toxicity. The toxicity of a material is not identical with its potential for being a health hazard. *Toxicity* is the capacity of a material to produce injury or harm. *Hazard* is the possibility that exposure to a material will cause injury or illness when a specific quantity is used under certain conditions. The key elements to be considered when evaluating a health hazard are:
1. amount of material to which the employee is exposed
2. total time of the exposure
3. toxicity of the substance
4. individual susceptibility.

Not all toxic materials are hazardous. The majority of toxic chemicals are safe when packaged in their original shipping containers or contained within a closed system. As long as toxic materials are adequately controlled, they can be safely used. For example, many solvents, if not properly used, will cause irritation to eyes, mouth, and throat. Some also are intoxicating and can cause blistering of the skin and other forms of dermatitis. Prolonged exposure can cause more serious illness. But if workers use the solvents in a well-ventilated area and are given proper protective equipment to prevent the solvents from contacting skin, the substances can be used safely.

The toxic action of a substance can be divided into acute and chronic effects.

Acute effects. These usually involve short-term, high concentrations that can cause irritation, illness, or death. They can be the result of sudden and severe exposure, during which the substance is rapidly absorbed. Acute effects can be related to an accident, which disrupts ordinary processes and controls. For example, sudden exposure to methane gas in a confined space can lead to loss of consciousness, coma, or death by asphyxiation.

Chronic effects. These usually involve continued exposure to a toxic substance over an extended time. When the chemical is absorbed more rapidly than the body can eliminate it, the chemical begins to accumulate in the body. If the level of contaminant is relatively low, the effects, even if they are serious and irreversible, may go unnoticed for long periods. Cancer, for example, may not show up until years after the exposure occurred. Likewise, breathing even low concentrations of carbon monoxide over an extended period can damage the heart muscles and blood vessels.

Inhalation hazards. Inhalation of harmful materials may irritate the upper respiratory tract and lung tissue, or the terminal passages of the lungs and the air sacs, depending upon the solubility of the material. Inhalation of biologically inert gases may dilute oxygen levels below the normal blood saturation value and disturb cellular processes. Other gases and vapors may prevent the blood from carrying oxygen to the tissues or interfere with its transfer from the blood to the tissue, producing chemical asphyxia.

Inhaled contaminants that adversely affect the lungs fall into three general categories: aerosols, toxic gases, and gases that produce systemic effects. Each is discussed later in the chapter in greater detail.
- Aerosols (particulates) are substances that, when deposited in the lungs, may produce either rapid local tissue damage, some slower tissue reactions, eventual disease, or only physical plugging.
- Toxic vapors and gases are hazards that produce adverse reaction in the tissue of the lungs themselves.
- Some toxic aerosols or gases do not affect the lung tissue locally but (1) are passed from the lungs into the bloodstream, where they are carried to other body organs, or (2) have adverse effects on the oxygen-carrying capacity of the blood cells themselves.

An example of the first type (aerosols) is asbestos fiber, which causes fibrotic growth in alveolar tissue, plugging the ducts or limiting the effective area of the alveolar lining. Other harmful aerosols are fungi, such as those found in sugar cane residues, which produce a disorder known as bagassosis.

An example of the second type (toxic gases) is hydrogen fluoride, a gas that directly affects lung tissue. It is a primary irritant of mucous membranes and causes chemical burns. Inhalation of this gas will cause pulmonary edema: after lung tissue is burned, the lungs fill with fluids that directly interfere with the gas-transfer function of the alveolar lining.

An example of the third type is carbon monoxide, a toxic gas passed into the bloodstream without essentially harming the lung. The carbon monoxide passes through the alveolar walls into the blood, where it ties up the hemoglobin so it cannot accept oxygen, thus starving the body of oxygen. Cyanide gas prevents cell enzymes from using molecular oxygen; this state disrupts vital cell processes.

Sometimes several types of lung hazards occur simultaneously. In mining operations, for example, explosives release nitrogen oxides into the air breathed by miners. These compounds impair the bronchial clearance mechanism, so that coal dust (of the particle sizes associated with the explosions) is not efficiently cleansed from the lungs.

Influence of solubility. A compound that is very soluble—such as ammonia, formaldehyde, sulfuric acid, or hydrochloric acid—may pose less of a hazard. Although it is rapidly

absorbed in the upper respiratory tract during the initial phases of exposure, it does not penetrate deeply into the lungs. Consequently, the nose and throat become very irritated, causing workers to leave the exposure area before they suffer serious harm. On the other hand, compounds insoluble in body fluids often cause considerably less throat irritation than do the soluble ones, but may penetrate deeply into the lungs. Thus, a serious hazard can be present without workers being immediately aware of it. With less irritation to the lungs, they have less warning that exposure is building up. Examples of such compounds (gases) are nitrogen dioxide and ozone. The immediate danger from these compounds in high concentrations is acute lung irritation or, possibly later, chemical pneumonia.

Numerous chemical compounds do not follow the general solubility rule. Such compounds are not very soluble in water and yet irritate the eyes and respiratory tract. They also can cause lung damage, even death in many situations. The supervisor must be sure that all hazardous compounds are identified and workers are properly protected.

Threshold limit values (TLVs). Individual susceptibility to respiratory toxins is difficult to assess. Nevertheless, certain recommended limits have been established. A TLV refers to airborne concentrations of substances and represents an exposure level under which most people can work, day after day, without adverse effect. Because of wide variations in individual susceptibility, however, an occasional exposure of an individual at or even below the threshold limit may not prevent discomfort, aggravation of a preexisting condition, or occupational illness. The term TLV refers specifically to limits published by the American Conference of Governmental Industrial Hygienists (ACGIH). The TLV limits are reviewed and updated annually. The National Safety Council's *Fundamentals of Industrial Hygiene* explains this subject in detail. A brief overview follows. There are three categories of TLVs:

1. *Time-weighted average (TLV-TWA)* is the time-weighted average concentration for a normal eight-hour day or 40-hour week. It is believed that nearly all persons can be exposed day after day to airborne concentrations at these limits without adverse effect.

2. *Short-term exposure limit (TLV-STEL)* is the concentration to which persons can be exposed for a period of up to 15 minutes continuously without suffering:
   - irritation
   - chronic or irreversible tissue change
   - narcoses of sufficient degree to reduce reaction time, impair self-rescue, increase the likelihood of accidental injury, or materially reduce work efficiency, provided the daily TLV-TWA is not exceeded.

No more than four 15-minute exposure periods per day are permitted with at least 60 minutes between exposure periods.

A STEL is not a separate, independent exposure limit. Rather, it supplements the time-weighted average (TWA) limit in cases where workers suffer acute reactions to a substance whose toxic effects are primarily chronic. Short-term exposure limits are recommended only where toxic effects have been reported from high short-term exposures in either humans or animals.

3. *Ceiling (TLV-C)* is the concentration that should not be exceeded even for an instant.

4. *"Skin" notation.* Nearly one-fourth of the substances in the TLV list are followed by the designation "skin." This refers to potential exposure through skin absorption. This designation is intended to suggest appropriate measures to prevent absorption of substances through the skin.

Permissible exposure limits (PELs). The first compilation of health and safety standards from the U.S. Department of Labor's OSHA appeared in 1970. Because it was derived from then-existing standards, the compilation adopted many of the TLVs established in 1968 by the American Conference of Governmental Industrial Hygienists. Thus threshold limit values—a registered trademark of the ACGIH—became, by federal standards, permissible exposure limits (PELs). These PELs represent the legal maximum level of contaminants in the workplace air.

The General Industry OSHA Standards currently list about 600 substances for which exposure limits have been established. These are included in subpart Z, "Toxic and Hazardous Substances," Sections 1910.1000 through 1910.1500. The PELs were updated in 1990, again including many of the newer, revised TLVs. This is a very dynamic facet of industrial hygiene. For current information on exposure limits, refer to current regulatory documentation.

U.S. OSHA action level is that point at which employers must initiate certain safety provisions: employee exposure measurement, employee training, and medical surveillance. A U.S. OSHA action level has not been defined for all employees. The action level for some OSHA-regulated chemicals like lead is usually set at about one-half the permissible exposure level (PEL).

Why is an action level set well below the PEL? Setting the action level at one-half the permissible exposure helps to protect employees from overexposure with a minimum burden to the employer. Where employee exposure measurements indicate that no employee is exposed to airborne concentrations of a substance in excess of the action level, employers in effect are exempted from having to initiate certain provisions in the standards.

The action level recognizes that air samples can only estimate the true TLV-TWA. For an extra margin of safety, companies set action levels lower than one-half the PEL. The adequacy of the TLVs to protect workers from illness is controversial, and a lower company action level may be desirable.

When to Measure?

The measurements done by the industrial hygienist can be divided into three phases:
1. problem definition phase
2. problem analysis phase
3. solution phase.

Problem definition phase. In many instances, inspectors take measurements to determine if there is a problem

in the workplace. In addition, some OSHA regulations require measurement at certain specified intervals or any time there is a change in production, process, or control measures. Measurement often establishes that workers are not experiencing excessive exposure to hazardous materials. Such monitoring of the workplace helps to ensure a safe environment.

Monitoring, then, is frequently used to determine that employers are in compliance with OSHAct requirements, state or provincial regulations, commonly accepted standards, TLVs, PELs, and action levels. Newer health standards published by OSHA usually state:

Each employer who has a place of employment in which [toxic substance name] is released into the workplace air shall determine if there is any possibility that any employee may be exposed to airborne concentrations of [toxic substance name] above the permissible level. The initial determination shall be made each time there is a change in production, process, or control measures that may result in an increase in airborne concentrations of [toxic substance name].

When any hazardous substances are released into the workplace air, the employer must take the first step in the employee exposure monitoring program. For OSHA-regulated substances, there must be an actual exposure measurement to see if any employee has been exposed to concentrations in excess of the recommended levels. This step should be taken even if there is only a remote chance that employees have been exposed to a substance above recommended levels.

Where does sampling begin? Should the sample be taken at the worker's breathing zone? Out in the general air? At the machine or process that is emitting the toxic substance? Although OSHA requires sampling only in the worker's breathing zone, sampling at all three sites provides a clearer picture of the situation.

Should the sample be taken for two minutes, two hours, or a whole day? There are two major types of samples:

The *grab sample* is taken over so short a period of time that the atmospheric concentration is assumed to be constant throughout the sample. This usually will cover only part of an industrial cycle. A series of grab samples can be taken in an attempt to define the total exposure. However, doing so requires a sound knowledge of statistical sampling techniques. The *long-term sample* is taken over a sufficiently long period of time that the variations in exposure cycles are averaged. Usually one sample or a series of samples is taken to represent the employee's eight-hour average exposure. OSHA regulations usually require this type of sampling.

An adequate number of tests should be taken to define the TLV-TWA to relate this level to recommended or regulatory exposure levels. But samples also must be taken to characterize the peak emissions during various portions of the process cycle.

If employee measurements indicate exposure at or above the action level, then OSHA requires that all employees so exposed be identified and their exposure measured. This step

clearly determines the population at risk.

When exposure measurements are at or above the action level but not above the PEL or just below the TLV-TWA, the employer needs some statistically reliable means to ensure that exposure levels are not exceeding these values. Management should conduct periodic sampling of the affected area and order medical examinations to determine if any susceptible individuals are exhibiting effects at these exposures.

If employees are exposed above the PEL or TLV, then a more intensive monitoring program is necessary. Medical staff must examine workers exposed to these excessive levels to measure the effects on their health. Noninhalation exposures—such as skin absorption—also may occur. Therefore, accurate exposure evaluation may require breath, blood, and urine sampling.

The problem definition phase is an orderly progression. At each step of the process, employers can decide whether to proceed to the next higher step.

Problem analysis phase. Once the industrial hygienist has defined the problem in the first phase of the measurement process, he or she must determine its causes. Management can identify opportunities for improvements in the workplace, set objectives for solutions, and devise alternative solutions should the initial ones fail to solve the problem.

The following eight methods suggest some ways that exposure hazards can be controlled:

1. substitution of a less harmful material for a hazardous one
2. change or alteration of a process to minimize worker contact
3. isolation or enclosure of a process or work operation to reduce the number of persons exposed
4. wet methods to reduce generation of dust in operations
5. local exhaust at the point of generation
6. personal protective devices (see Chapter 14 in the *Engineering and Technology* volume)
7. good housekeeping, including cleanliness of the workplace, waste disposal, adequate washing, clean toilet and eating facilities, healthful drinking water, and control of insects and rodents
8. training and education—the OSHA hazard communication standard requires training for employees exposed to hazardous chemicals.

Solution phase. Once the problem has been analyzed and a number of solutions proposed, the most effective, timely, and practical solution needs to be selected—one that provides optimum benefits with minimal risks. The details of the solution should be carefully worked out. In effect, management needs to develop a blueprint describing what should be done, how and by whom it should be done, and in what sequence the actions are to take place.

Once controls are installed, they must be periodically checked to be sure they are functioning properly. Follow-up monitoring and inspection will determine if the solution to a given hazardous exposure is controlling it within the specified limits. Thus, managers should regard the monitoring

function as a circular, not horizontal process. If measurements at the solution phase reveal that controls are inadequate, the industrial hygienist must return to the first phase, that of defining the problem.

Who Will Do the Measuring?

Not every organization requires or can afford the services of a full-time industrial hygienist. Independent consultants can be hired to accomplish two major objectives:

1. identify and evaluate potential health risks and accident hazards to workers in the occupational environment
2. design effective controls to protect the safety and health of workers.

Because any person can legally offer services as an industrial hygiene consultant, it is important that the consultant hired is a trained, experienced, and competent professional. A competent industrial hygiene consultant must have detailed knowledge of proper sampling equipment and analytic procedures and will probably hold the designation "certified industrial hygienist" (CIH).

Good sources of information and assistance regarding consultants are the American Industrial Hygiene Association and the American Society of Safety Engineers, the professional associations related to occupational and health safety, respectively (see the descriptive listing in the Appendix, Sources of Help). Regional offices of the National Institute for Occupational Safety and Health (NIOSH) usually have lists of consultants in their area. Many insurance companies have loss prevention programs that employ industrial hygienists. The National Safety Council and its chapters having offices in major cities can offer assistance. The Council offers a full range of consulting services in safety and occupational health management (see the Appendix, Sources of Help). For a state-by-state listing of governmental consulting service offices, see Sources of Help under U.S. Government Agencies.

ACCIDENT INVESTIGATION

A fourth function of monitoring in the total hazard control system is accident investigation, the subject of Chapter 12. The following discussion demonstrates how accident investigation fits into the systems approach to hazard control.

Why Accidents Are Investigated

When viewed as an integral part of the total occupational safety and health program, accident investigation is especially important to determine direct causes, uncover contributing accident causes, prevent similar accidents from occurring, document facts, provide information on costs, and promote safety. Accident investigation concentrates on gathering all information about the factors leading to the accident.

Determine direct causes. Accident investigation determines the direct and contributing causes of incidents. At what points did the hazard control system break down? Were rules and regulations violated? Did defective machinery or factors in the work environment contribute to the accident? Poor machinery layout, for example, or the very design of a job process, operation, or task can contribute to an undesirable situation. Chapter 3 outlined the three primary sources of accidents: human, situational, and environmental factors.

Uncover contributing accident causes. Thorough accident investigation is very likely to uncover problems that indirectly contributed to the accident. Such information benefits accident-reduction efforts. For example, a worker slips on spilled oil and is injured. The oil spill is the direct cause of the accident, but a thorough investigation might reveal other contributing factors: poor housekeeping, failure to follow maintenance schedule, inadequate supervision, or faulty equipment (such as a lathe leaking oil).

Prevent similar accidents. Accident investigation identifies what actions and improvements will prevent similar accidents from occurring in the future.

Document facts. Accident investigation documents the facts involved in an incident for use in any compensation and litigation that may arise. The report produced at the conclusion of an investigation becomes the permanent record of facts about an accident. It may become necessary to reconstruct an accident situation long after the occurrence. To do so, the details of the accident will have to be recorded properly, accurately, and thoroughly.

Provide information on costs. Accident investigation provides information on both direct and indirect costs of accidents. Chapter 12 gives details for estimating accident costs.

Promote safety. Accident investigation yields psychological as well as material benefits. The investigation demonstrates the organization's interest in worker safety and health. It indicates management's sense of accountability for accident prevention and its commitment to a safe work environment. An investigation in which both labor and management participate promotes cooperation between these two groups.

Despite what many people believe, accident investigation is a fact-finding, not a fault-finding, process. When attempting to determine the cause of an accident, the novice investigator may be tempted to conclude that the person involved in the accident was at fault. But if human error is not the real cause, the hazard that produced the accident will go undiscovered and uncontrolled. Furthermore, the person falsely blamed for causing the accident will resent the unjustified accusation as well as any disciplinary action. The worker will be less cooperative in the future and feel less respect for the organization's safety and health program. Investigators should always stress that the intent of accident investigation is to pinpoint causes of error and defects so similar accidents can be prevented.

Conducting an accident investigation is not simple. It can be difficult to look beyond the incident at hand to uncover causal factors, determine the true loss potential of the occurrence, and develop practical recommendations to prevent recurrence. A major weakness of many accident investigations is the failure to establish and consider all factors—human, situational, and environmental—that contributed to the accident. Reasons for this failure include:

- inexperienced or uninformed investigator
- reluctance of the investigator to accept full responsibility for the job
- narrow interpretation of environmental factors
- erroneous emphasis on a single cause
- judging the effect of the accident to be the cause
- arriving at conclusions before all factors are considered
- poor interviewing techniques
- delay in investigating accidents.

The trained investigator must be ready to acknowledge as contributing causes any and all factors that may have led, in any way, to the accident. What at first may appear to be a simple, uninvolved incident can, in fact, have numerous contributing factors that become more complex as analyses are completed. Immediate, on-the-scene accident investigation provides the most accurate and useful information.

When to Investigate Accidents

The longer the delay in examining the accident scene and interviewing the victim(s) and witnesses, the greater the possibility of obtaining erroneous or incomplete information. The accident scene changes, memories fade, and people discuss what happened with each other. Whether consciously or not, witnesses may alter their initial impressions to agree with someone else's observation or interpretation. Further, prompt accident investigation also expresses concern for the safety and well-being of employees.

As a general rule, all accidents, no matter how minor, are candidates for thorough investigation. Many accidents occurring in an organization are considered minor because their consequences are not serious. Such accidents, or incidents, are taken for granted and often do not receive the attention they demand. Management, safety and health committees, supervisors, and employees must be aware that serious accidents arise from the same hazards that produce minor incidents. Usually sheer luck determines whether a hazardous situation results in a minor incident or a serious accident.

In accident investigation, the investigator must give priority to the health and safety of affected personnel (including any victims). When possible, rescue and first aid procedures should be used that disturb the accident scene as little as possible. Measures to protect equipment should also preserve evidence. When the area is secure and victims have received medical attention, appropriate notifications have been made, efforts can be concentrated on investigating the incident.

As with inspections, it is advisable to prepare investigation tools in advance. An investigation kit might include:

- camera and film
- tape recorder
- measuring devices
- sample containers
- interview/investigation forms
- flashlight
- barricade markers/tape
- warning tags and padlocks.

Having the necessary equipment ready will facilitate the investigation and certainly help to eliminate delays and other difficulties.

Who Should Conduct the Investigation?

Chapter 12 discusses the question of who is to make the investigation: the supervisor or foreman, the safety and health professional, a special investigative committee, or a company safety and health committee. As a supplement to that discussion, the following section outlines the roles played by physicians and management in accident investigation. It also covers the responsibility of the safety professional in preventing further accidents from occurring during the investigation itself.

Physician. A physician's assistance is particularly important when human factors have been designated as direct or contributing causes of an accident. The physician can assess the nature and degree of injury and assist in determining the source and nature of the forces that inflicted the injury. The physician also can (1) determine what special biomedical studies, if any, are needed; (2) establish whether the injured person was physically and mentally fit at the time of the accident and whether the screening, selection, and preplacement process is adequate; (3) help judge the adequacy of safety and health protection procedures and equipment; and (4) help evaluate the effectiveness of the plans, procedures, equipment, training, and response of rescue, first aid, and emergency medical care personnel. The physician also can evaluate the effectiveness of measures aimed at early detection of medical conditions, mental changes, or emotional stress.

Management. Management and department heads should help investigate accidents resulting in lost workdays or major property damage. When management actively participates in accident investigation, it can evaluate the hazard control system and determine whether outside assistance is desired or required to upgrade existing structures and procedures. Management also must review accident reports in order to make informed decisions. When accident investigation reveals the need or desirability of specific corrective actions, management must determine whether the recommended action has been implemented.

Safety during the investigation. In many cases the accident scene is a dangerous place. The accident may have damaged electrical equipment, weakened structural supports, and released radioactive or toxic materials.

The safety and health professional must be particularly alert to the hazards encountered by the investigating team and, when necessary, see that proper protective equipment is provided. Investigators need to be alerted to the hazards they may encounter and emergency procedures they should follow (see details in Chapter 6, Emergency Preparedness).

What to Look For

The accident investigation must answer many questions. Because of the infinite number of accident-producing situations, contributing factors, and causes, it is impossible to list all the questions that apply to all investigations. The following questions are generally applicable, however, and will be

considered in most accident investigations (Firenze, 1978).

- What was the injured person doing at the time of the accident? Performing an assigned task? Maintenance? Assisting another worker?
- Was the injured employee working on an unauthorized task? Was the employee qualified to perform the task and familiar with the process, equipment, and machinery?
- What were other workers doing at the time of the accident?
- Was the proper equipment being used for the task at hand (screwdriver instead of can opener to open a paint can, file instead of grinder to remove burr on a bolt after it was cut)?
- Was the injured person following approved procedures?
- Is the process, operation, or task new to the area?
- Was the injured person being supervised? What was the proximity and adequacy of supervision?
- Did the injured employee receive hazard recognition training prior to the accident?
- What was the location of the accident? What was the physical condition of the area when the accident occurred?
- What immediate or temporary actions could have prevented the accident or minimized its effect?
- What long-term or permanent action could have prevented the accident or minimized its effect?
- Had corrective action been recommended in the past but not adopted?

During the course of the investigation, the above questions should be answered to the satisfaction of the investigators. Other questions that come to mind as the investigation continues should be recorded.

Conducting Interviews

Interviewing accident or injury victims and witnesses can be a difficult assignment if not properly handled. The individual being interviewed often is fearful and reluctant to provide the interviewer with accurate facts about the accident. The accident victim may be hesitant to talk for any number of reasons. A witness may not want to provide information that might implicate friends, fellow workers, or the supervisor. To obtain the necessary facts during an interview, the interviewer must first eliminate or reduce an employee's fear and anxiety by establishing good rapport with the individual. The interviewer must create a feeling of trust and establish open communication before beginning the actual interview. Once good rapport has been developed, the interviewer can follow this five-step method.

1. Discuss the purpose of the investigation and the interview (fact-finding, not fault-finding).
2. Have the individual relate his or her version of the accident with minimal interruptions. If the individual being interviewed is the one who was injured, ask what was being done, where and how it was being done, and what happened. If practical, have the injured person or eyewitness explain the sequence of events that occurred at the time of the accident. Being at the scene of the accident makes it easier to relate facts that might otherwise be difficult to explain.
3. Ask questions to clarify or fill in any gaps.
4. The interviewer should then repeat the facts of the accident to the injured person or eyewitness. Through this review process, there will be ample opportunity to correct any misunderstanding that may have occurred and clarify, if necessary, any of the details of the accident.
5. Discuss methods of preventing recurrence. Ask the individual for suggestions aimed at eliminating or reducing the impact of the hazards that caused the accident. By asking the individual for ideas and discussing them, the interviewer will show sincerity and place emphasis on the fact-finding purpose of the investigation, as it was explained at the beginning of the interview.

In some cases, contractual agreements may call for an employee representative to be present during any management interview, if the employee so requests.

Preparing the Accident Investigation Report

Chapters 12 and 13 outline specific ways to record and classify data: how to identify key facts about each injury and the accident that produced it, how to record facts in a form that facilitates analysis and reveals patterns and trends, how to estimate accident costs, and how to comply with regulatory recordkeeping requirements.

An accident in any organization is of significant interest to employees, who will ask questions that reflect their concerns. Is there any potential danger to those in the immediate vicinity? What caused the accident? How many people were injured? How badly?

Those who investigate accidents should answer these questions truthfully and avoid covering up any facts. On the other hand, they must be certain they are authorized to release information, and they must be sure of their data.

Because the accident report is the product of the investigation, it should be prepared carefully and adequately to justify the conclusions reached. It must be issued soon after the accident. When a report is delayed too long, employees may feel left out of the process. If a final report must be postponed pending detailed technical analysis or evaluation, then management should issue an interim report providing basic details of the investigation.

Summaries of vital information on major injury, damage, and loss incidents should be distributed to department managers. Such summaries should include information on accident causes and recommended action for preventing similar incidents. Management should maintain incident and statistical report files as dictated by company policy.

Supervisors need to keep employees informed of significant accidents and preventive measures proposed or executed. Posting accident reports is one way to make information available.

Implementing Corrective Action

The preceding section on inspection emphasized that hazard control benefits accrue only after the inspection report is written and disseminated. Until corrective action is initiated,

any recommendations—no matter how earnest, thorough, and relevant—remain "paper promises."

The same is true of accident investigation when it is used as a monitoring technique. Viewed from the perspective of hazard control, accident investigation serves as a monitoring function only when it provides the impetus for corrective action.

Whenever management and safety professionals review monthly accident reports, they exercise an essential auditing function. Management (including the chief executive officer) can demonstrate interest in safety by requiring prompt reporting of all serious or potentially serious incidents. They use accident reports to make decisions to prevent similar accidents from occurring, and they look for answers to certain key questions. Are all significant accidents being reported? Are all parts of the organization equally committed to the hazard control effort? Are there trends or patterns in accidents or injuries? What system breakdowns predominate? What supervisors require additional training? Are employees advised of the results of accident investigation and of preventive measures being instituted? What management deficiencies are indicated? Accident investigation as a monitoring function occurs after the hazard control system has already broken down. Although no amount of investigation can reverse the accident, accident investigation serves an important monitoring function. Past mistakes can be used to improve future operations. As George Santayana has written, "Those who cannot remember the past are condemned to repeat it."

SUMMARY

- Monitoring is a set of observation and data collection methods used to detect and measure deviations from plans and procedures in current operations. Monitoring can help management ascertain that controls are functioning properly and that changing conditions or new problems have not rendered controls ineffective.
- Monitoring involves four functions: hazard analysis, inspection, measurement and testing, and accident investigation. The four-part system enables monitoring to be performed before and during operations and after an accident or near-miss has occurred.
- Hazard analysis is an orderly process used to acquire specific hazard and failure data pertinent to a given system. Hazard control specialists provide data for informed management decisions about effective control measures to help ensure a safe work environment.
- Formal hazard analytical methods can be divided into inductive and deductive. The inductive method forms the basis for such analyses as failure mode and effect analysis and operations hazard analysis. Deductive methods use a combined-events analysis, often in the form of positive or fault "trees." Hazard analysis, to be effective, should represent as many different viewpoints in an organization as possible.
- In determining which processes, operations, and tasks re-

ceive priority in hazard analysis, the following must be considered: frequency of accidents; potential for injury; severity of injury; new or altered equipment, processes, and operations; excessive material waste or damage to equipment.

- Inspection is a monitoring function conducted in an organization to locate and report existing and potential hazards having the capacity to cause accidents in the workplace. Inspection must be fact-finding and not fault-finding. Its primary purpose is to detect potential hazards so they can be corrected before an accident occurs. Nearly all management levels and many employees are involved in conducting inspections in a firm.
- Inspections must answer four questions: (1) what items are to be inspected, (2) what aspects of each item are to be inspected, (3) what conditions should be inspected, and (4) how frequently should inspections be performed?
- Inspections can be classified as continuous or interval. Continuous inspections of equipment and processes are performed by line managers, supervisors, and their employees as a routine part of their jobs.
- Planned inspections are more formal procedures that may be periodic, intermittent, or general. Periodic inspections are scheduled at regular intervals and can cover the entire plant or only one area. Intermittent inspections are made at irregular intervals in response to an unusual number of accidents, to the installation or modification of equipment or procedures, or to a specific threat to employee health.
- A general inspection is planned and covers places not inspected periodically, such as parking lots, sidewalks, fencing, and similar outlying regions. These inspections are usually required before reopening a plant after a long shutdown.
- Inspections must be carefully thought out and require sound knowledge of the organization and its processes and equipment; knowledge of relevant standards, regulations, and codes; systematic inspection steps; and methods of reporting, evaluating, and using the data. Checklists are effective tools for inspections and can help in follow-up to make sure all hazardous conditions have been corrected.
- Frequency of inspection is determined by five factors: (1) loss severity potential of the problem, (2) injury potential, (3) rate at which the item can become unsafe, (4) past history of failures, and (5) regulatory requirements.
- Inspections should be conducted systematically to avoid overlooking less obvious hazards. The inspector must gain the cooperation of management and employees in the affected areas. The individual reports all hazards—classifying and describing them carefully—and recommends corrective actions. Inspection reports can be emergency for immediate dangers, periodic, or summary.
- Measurement and testing methods are used to monitor chemical, physical, biological, and ergonomic hazards. Four monitoring systems are used: personal, environmental, biological, and medical. These methods are usually employed by an industrial hygienist or hazard control specialist to measure health hazards in the workplace.
- Key hazards monitored by measurement and testing

methods include acute and chronic effects of toxic substances and inhalation hazards (aerosols and toxic gases). Standards used to establish health and safety limits include threshold limit values (TLVs), time-weighted averages (TLV-TWA), short-term exposure limit (TLV-STEL), permissible exposure limits (PELs), and action levels. Organizations generally set their own action level standards well below federal and state requirements.

- Measurements done by industrial hygienists are divided into three phases: problem definition phase, problem analysis phase, and solution phase. To define the problem, the hygienist may take a grab sample or long-term sample to measure concentration and exposure levels. The problem is then analyzed to determine causes, and corrective actions recommended and taken.

- Accident investigations are conducted to determine direct causes, uncover indirect accident causes, prevent similar accidents from occurring, document facts, provide information on costs, and promote safety measures and standards. As is the case in inspections, the purpose of accident investigation is fact-finding, not fault-finding.

- Accidents must be investigated immediately to ensure accurate details and to preserve evidence. All accidents should be investigated, but those involving injuries, fatalities, or serious property damage must be investigated with great care. Accident investigators can be supervisors, safety professionals, special committees, or health and safety committees. Physicians and upper management can advise the investigator(s).

- Investigators must examine all human, situational, and environmental factors in determining accident causes, and apply the five-step method for interviewing accident victims and witnesses. An investigation report should be issued as soon as possible and contain recommendations for corrective action. In this way, past mistakes can be used to improve future operations.

REFERENCES

Boggs RF. Environmental monitoring and control requirements. *National Safety News* 118 (2): August 1978.

Facility Inspection. Philadelphia, PA, Insurance Company of North America, 1973.

Ferry T. *Modern Accident Investigation and Analysis.* New York: John Wiley & Son, 1988.

Firenze RJ. *The Process of Hazard Control*, Dubuque, IA: Kendall/Hunt Publishing Co., 1978.

Industrial hygiene instrumentation. *National Safety News.* 117 (3): March 1978.

Johnson WG. *MORT Safety Assurance Systems.* New York: Marcel Dekker, Inc., 1980. (Also available through National Safety Council.)

National Safety Council, 444 North Michigan Avenue, Chicago, IL 60611.
> *Supervisors' Safety Manual.* 7th ed. Chicago: National Safety Council, 1990.
> *You Are the Safety and Health Committee.*—Booklet.

Nertney RJ. Practical applications of system safety concepts. *Professional Safety* 22 (2): February 1977.

Olshifski J. Air sampling instrumentation. *National Safety News.* 120 (2): August 1979.

———. Selecting and using industrial hygiene consultants. *National Safety News.* 118 (3): September 1978.

Plog B, ed. *Fundamentals of Industrial Hygiene*, 3rd ed. Chicago: National Safety Council, 1988.

Scerbo FA and Pritchard JJ. Fault tree analysis: A technique for product safety evaluation. *Professional Safety.* 22 (5): May 1977.

12

Accident Investigation: Analysis and Costs

This chapter covers the investigation of noninjury accidents as well as injury accidents. Therefore, the term *accident* is used in its broadest sense to include incidents that may lead to property damage, work injuries, or both.

Successful accident prevention requires a minimum of four fundamental activities:

1. study of all working areas to detect and eliminate or control the physical or environmental hazards that contribute to accidents
2. study of all operating methods and practices and administrative controls
3. education, instruction, training, and enforcement of procedures to minimize the human factors that contribute to accidents
4. thorough investigation and causal analysis of every accident resulting in at least a lost-workday injury to determine contributing circumstances; accidents not resulting in personal injury (so-called near-accidents or near-misses) are warnings and should also be investigated thoroughly. This fourth activity, accident investigation and analysis, is a defense against any hazards overlooked in the first three activities, against hazards not immediately obvious, or against hazards resulting from circumstances difficult to foresee.

ACCIDENT INVESTIGATION AND ANALYSIS

The primary purpose of accident investigation and analysis activities is to prevent accidents. As such, the investigation or analysis must produce factual information leading to corrective actions that prevent or reduce the number of accidents. The more complete the information, the easier it will be for management to take effective corrective actions. For example, knowing that 40% of an organization's accidents involve ladders is not as useful as knowing that 80% of the organization's ladder accidents involve broken rungs. A good recordkeeping system, as discussed in Chapter 13, is essential to accident investigation. The system allows the basic facts about an accident to be recorded quickly, efficiently, and uniformly.

All accidents should be investigated, regardless of severity of injury or amount of property damage. The extent of the investigation depends on the outcome or potential outcome of the accident. An accident involving only first aid or minor property damage is not investigated as thoroughly as one resulting in death or extensive property damage, that is, unless the potential outcome could have been disabling injury or death.

For purposes of accident prevention, investigations must be fact-finding, not fault-finding; otherwise, they can do more harm than good. This is not to say responsibility should not be fixed where personal failure has caused injury, nor that such persons should be excused from the consequences of their actions. It does mean the investigation itself should be concerned only with facts. The investigating individual, board, or committee must not be involved with any disciplinary actions resulting from the investigation.

Types of Investigation and Analysis

A variety of accident investigation and analysis techniques are available to the investigator, some of them more complicated than others. The choice of a particular method depends upon the purpose and orientation of the investigation. The Failure Mode and Effect approach discussed in Chapter 11, Acquiring Hazard Information, may be useful when investigating situations involving large, complex, and interrelated machinery and procedures but may be of limited value for investigating accidents involving hand tools. If management procedures and communications and their relationship to accidents are of great interest, the Management Oversight and Risk Tree analysis (MORT, see References at the end of Chapter 11) could prove to be the best choice.

The accident investigation and analysis procedure outlined in this chapter follows the ANSI Z16.2 standard, *Method of Recording Basic Facts Relating to the Nature and Occurrence of Work Injuries,* and focuses primarily on unsafe practices and unsafe conditions. The ANSI Z16.2 standard is discontinued and no longer available. Other similar techniques involve investigation within the framework of defects in man, machine, media, and management (the four Ms) or education, enforcement, and engineering (the three Es). For analysis purposes, these techniques involve classifying the data about a group of accidents into various categories. This approach has been referred to as the statistical method of analysis. Corrective actions are designed on the basis of most frequent patterns of occurrence.

Other techniques discussed in Chapter 11 come under the systems approach to safety. Systems safety stresses a broader viewpoint that takes into account interrelationships between various events that could lead to an accident. As accidents rarely have one cause, the systems approach to safety can point to more than one place in a system where effective corrective actions can be introduced. This process allows the safety professional to choose the corrective actions best meeting the criteria for effectiveness, rapid installation, cost/benefit analysis, and the like. There are additional advantages to using systems safety techniques: management can implement them before accidents occur and can apply them to new procedures and operations.

Persons Conducting the Investigation

Depending on the nature of the accident and other conditions, the investigation should be made by the supervisor. This person can be assisted by a fellow worker familiar with the process involved, the safety and health professional or inspector, the employee health professional, the safety and health committee, the general safety committee, or an engineer from the insurance company. If the accident involves unusual or special features, consultation with a state labor department or federal agency, a union representative, or outside expert may be warranted.

The supervisor or foreman should make an immediate report of every injury requiring medical treatment and other accidents he or she may be directed to investigate. The supervisor is on the scene and probably knows more about the accident than anyone else. It is up to this individual, in most cases, to put into effect whatever measures can be adopted to prevent similar accidents. The accident investigation report in the next chapter illustrates one form that can be used to record the findings of an accident investigation.

The safety and health professional. Ideally, the safety professional should be an adviser and guide to the supervisor on accident investigations and should verify the supervisor's finding and the adequacy of his or her investigation because sometimes a supervisor may attempt to cover up a supervisory error. The safety and health professional should conduct the investigation only in serious cases or where the supervisor is not adequately trained in accident investigation.

Special investigative or review committee. In some companies, a special committee is set up to investigate and report on all serious accidents or to review the quality of accident investigations. To be acceptable to all involved, this committee should be composed of representatives of both management and workers. Thus, a report published by this committee would be more readily accepted not only by workers but also by management than would a report made solely by a safety and health professional. (See the discussion of safety committees in Chapter 3, Hazard Control Program.)

The safety and health committee. In many organizations, especially small or moderate-sized, a number of safety activities, including accident investigation, are handled by a safety and health committee. Ordinarily, such investigation is conducted in a routine manner, but in important cases the head of the committee might call an extra meeting to initiate a special investigation.

Cases to Be Investigated

An accident causing death or serious injury obviously should be thoroughly investigated. Also, the near-accident that might have caused death or serious injury is equally important from the safety standpoint and should be investigated.

Each investigation should be conducted as soon after the accident as possible. A delay of only a few hours may permit important evidence to be destroyed or removed, intentionally or unintentionally. Also, the investigator or committee should present the results of the inquiry as quickly as possible; this greatly increases their value in the safety education of employees and supervisors.

Any epidemic of minor injuries demands study. A particle of emery in the eye or a scratch from handling sheet metal may seem to be a simple case; the immediate cause is obvious, and the loss of time is small. However, if such cases, or any others, occur frequently in the organization or in any one department, they need to be investigated to determine the underlying causes.

The chief value of such an investigation also lies in discovering any contributing causes. The energetic safety and health professional or manager appreciates this type of accident investigation because it can prove more valuable, though less spectacular, to safety efforts than an inquest following a fatal injury.

Fairness and impartiality are absolutely essential. The value of accident investigation can be destroyed if employees suspect its purpose is to place blame or pass the buck. No one should be assigned to investigation work unless he or she has earned a reputation for fairness and is trained and experienced in gathering evidence. They should clearly understand that accident investigations are conducted solely for the purpose of obtaining information to help prevent a recurrence of such incidents.

In the early years of the safety movement, accident prevention usually was a hit-or-miss activity. This approach has been replaced by more scientific techniques—see Chapter 11, Acquiring Hazard Information.

In the earlier years, a reduction in accident rates was prompted primarily by humanitarian appeal to management and workers. The most important methods are aimed at isolating and identifying accident causes in order to permit direct, positive, and corrective action to prevent their recurrence.

Like other phases of modern business management, accident prevention must be based on facts clearly identifying the problem. An approach to the accident prevention problem on this basis not only will result in more effective control over accidents but also will save the organization time, effort, and money. Accident analysis of individual cases identifies the plants, locations, or departments in which injuries most frequently occur, and suggests necessary corrective actions to reduce accidents.

Sometimes an overall high rate is not identified with one or a few departments but instead represents a high frequency of accidents throughout the facility. Under such circumstances, it is even more important that an analysis of the accidents be made. A high accident rate may be hard to spot for several reasons. For example, similar accidents may occur frequently but at widely separated locations, so their high incidence is not apparent. Accidents may be more numerous in some machine operations than in others, or in certain procedures. Some unsafe practices that cause accidents may be committed repeatedly but at different times and in different places, so their importance as accident causes is not immediately recognized.

Analysis of the circumstances of accidents can produce these results:

1. Management can identify and locate the principal sources of accidents by determining, from actual experience, the materials, machines, and tools most frequently involved in accidents, and the jobs most likely to produce injuries.
2. Investigations may disclose the nature and size of the accident problems in departments and among occupations.
3. Results will indicate the need for engineering revision by identifying the principal hazards associated with various types of equipment and materials.
4. The investigation can disclose inefficiencies in operating processes and procedures where, for example, poor layout contributes to accidents, or where outdated, physically overtaxing methods or procedures can be avoided, for instance, by using mechanical handling methods.

5. An accident report will disclose the unsafe practices that need to be corrected by training employees or changing work methods.
6. The report also will enable supervisors to put their safety work efforts to the best use by giving them information about the principal hazards and unsafe practices in their departments.
7. Investigation results permit an objective evaluation of the progress of a safety program by noting in continuing analyses the effect of corrective actions, educational techniques, and other methods adopted to prevent injuries.

The Minimum Data

The purpose of an accident investigation is twofold. First, to identify facts about each injury and the accident that produced it and to record those facts; and second, to determine a course of action to eliminate a recurrence. Accident investigation records, individually and collectively, serve as guides to the areas, conditions, and circumstances to which accident prevention efforts can most profitably be directed.

The following paragraphs describe the eight data elements comprising the minimum amount of information that should be collected about each accident. The Accident Investigation Report Form, Figure 13–2 in the next chapter, shows a minimum data set that was developed to improve the quality of accident investigation and analysis. This minimum data set identifies the why of some of the accident characteristics as well as the who, what, when, where, and how. It acknowledges the existence of multiple causes of accidents by not restricting the investigator or analyst to selecting a single causal act or condition.

In addition, the investigation, instead of focusing solely on the injury and accident type, includes the entire sequence of events leading to the injury, as far back in time as the investigator feels is relevant. This expanded view of the accident sequence allows an employer to identify and implement a wider variety of corrective actions.

- The first of the eight groups of data elements is *employer characteristics*. This includes the type of industry and the size of the company (number of full-time equivalent employees). It is needed when data from one company are compared with those of another.
- Second is *employee characteristics*. The victim's age and sex, the department and occupation in which he or she worked, and whether a full-time, part-time, or seasonal employee. Questions about the victim's experience are also asked. How long has the victim been with the company? How long in current occupation? How often did the employee repeat the activity engaged in when the accident occurred? Employee training records may be examined and assessed during this part of the investigation.
- The third group is about the accident itself. A *narrative description* should be prepared; it should include what the person was doing, what objects or substances were involved, and actions or movements which led to the injury or near-miss. This is elaborated into a detailed accident sequence beginning with the incident and working back-

ward in time through all of the preceding events that directly contributed to the accident. The data also include a description of any product or equipment directly involved with the accident sequence and the task being performed. Any other conditions, such as temperature, light, noise, and weather pertaining to the accident also should be noted here. The description should note contributing acts of coworkers and miscellaneous transient factors as they are identified.

- The fourth group deals with the characteristics of the *equipment* associated with the accident. The description should include the type, brand, size, and any distinguishing features of the equipment, its condition, and the specific part involved.
- Fifth is the characteristics of the *task* being performed when the accident happened: the general task (such as repairing a conveyor) and the specific activity (such as using a wrench). The description should include the posture and location of the employee (for example, squatting under the conveyor) and whether the person was working alone or with others.
- *Time factors* are the sixth group. The investigation should record the time of day and whether it was the victim's first hour of the shift, second hour, or later. Also, what type of shift was it—day, swing, straight, rotating, etc.? Other information includes the phase of the employee's workday: performing work, rest period, meal time, overtime, entering or leaving plant.
- The seventh group, use and nature of *preventive measures,* includes the following questions: What personal protective equipment was being worn, and did the employee's apparel affect the accident sequence? What kind of training did the employee have for the task being performed? Did standards or procedures exist for the task? Were they written? Were they followed? If not followed, how did what happened differ from what should have happened? Were all guards in place and in use? Did any equipment, tools, or materials fail to perform properly? What was the nature of supervision at the time of the accident? What immediate remedial actions were taken to prevent recurrence? By whom? When?
- The last group of questions concerns the *characteristics of the injury.* The nature of the injury or injuries and the parts of the body affected must be recorded as well as the OSHA severity class. If the accident resulted in some permanent impairment, this should be noted.

The answers to the questions in these eight groups constitute the minimum information needed to proceed with an analysis. The nature of a company's operations or the interests of the analyst may suggest other questions to be answered in the investigation.

There are two types of analysis that can be done. First, the investigator can examine the individual accident to determine the corrective action or actions to prevent future occurrences of this specific sequence of events. Second, the investigator can do a statistical analysis to examine a group of similar occurrences for patterns lending themselves to corrective actions. Over time, this statistical analysis can show which corrective actions have been more effective than others.

Identifying Causal Factors and Selecting Corrective Actions

In any accident, many factors are at work that trigger a sequence of events resulting in an injury. The idea behind the corrective action selection procedure is to identify all the factors for which a corrective action is possible. Management then selects the ones likely to be most effective, most cost/beneficial, most acceptable, and so on, and implements those.

Figure 12–1 shows the Guide for Identifying Causal Factors and Corrective Action. This guide is used with the Accident Investigation Report in Figure 13–2. The Guide contains four parts: equipment, environment, people, and management. Although these elements usually combine to produce products and profits, at times they result in accidents. The questions in these four sections help the investigator to analyze the contribution each factor made to the accident.

The structure of the Guide makes it easy to identify the causal factors. Questions are answered by placing an X in a circle or a box. An X in a *circle* means the item is a causal factor, while an X in a *box* indicates the item is *not* a causal factor.

The Comment column provides space to record the specific information about the accident being investigated. The Recommended Corrective Actions column has room to enter specific corrective actions for each causal factor. After listing all of the possible corrective actions identified on the Guide, each action must be evaluated for effectiveness, cost, feasibility, reliability, acceptance, effect on productivity, time required to implement, and any other factor deemed important, before deciding which ones to implement.

This systematic approach to selecting corrective actions ensures three basic steps. First, all major actions are considered; second, the analyst does not stop with familiar and favorite corrective actions; and third, each corrective action chosen for implementation is carefully thought out.

Classifying Accident Data

Although two accidents rarely happen in exactly the same way, accidents do follow general patterns. Because they must be grouped according to pattern for purposes of analysis, finding the patterns and common features of groups of cases is the statistical approach to accident analysis.

Setting up classifications. Before the actual analysis work is begun, classifications must be set up for grouping the various data. For each basic fact, general classifi-

Figure 12-1. The Guide for Identifying Causal Factors and Corrective Actions assists the accident investigator to systematically consider four contributing accident factors: Equipment, Environment, People, and Management.

Answer questions by placing an X
in the "Y" circle or box for YES or
in the "N" circle or box for NO.

GUIDE for IDENTIFYING
CAUSAL FACTORS and CORRECTIVE ACTIONS

CASE NUMBER

PART 1—EQUIPMENT				
1.0 WAS A HAZARDOUS CONDITION(S) A CONTRIBUTING FACTOR? If yes, answer the following. If no, proceed to Part 2.				
CAUSAL FACTORS	**COMMENT**	**POSSIBLE CORRECTIVE ACTIONS**		**RECOMMENDED CORRECTIVE ACTIONS**
Y○ N□ 1.1 Did any defect(s) in equipment/tool(s)/material contribute to hazardous condition(s)?		Review procedure for inspecting, reporting, maintaining, repairing, replacing, or recalling defective equipment/tool(s)/material used.		
Y□ N○ 1.2 Was the hazardous condition(s) recognized? If yes, answer A and B. If no, proceed to 1.3.		Perform job safety analysis. Improve employee ability to recognize existing or potential hazardous conditions. Provide test equipment, as required, to detect hazard. Review any change or modification of equipment/tool(s)/material.		
Y□ N○ A. Was the hazardous conditions(s) reported?		Train employees in reporting procedures. Stress individual acceptance of responsibility.		
Y□ N○ B. Was employee(s) informed of the hazardous condition(s) and the job procedures for dealing with it as an interim measure?		Review job procedures for hazard avoidance. Review supervisory responsibility. Improve supervisor-employee communications. Take action to remove or minimize hazard.		
Y□ N○ 1.3 Was there an equipment inspection procedure(s) to detect the hazardous condition(s)?		Develop and adopt procedures (for example, an inspection system) to detect hazardous conditions. Conduct test.		
Y□ N○ 1.4 Did the existing equipment inspection procedure(s) detect the hazardous condition(s)?		Review procedures. Change frequency or comprehensiveness. Provide test equipment as required. Improve employee ability to detect defects and hazardous conditions. Change job procedures as required.		
Y□ N○ 1.5 Was the correct equipment/tool(s)/material used?		Specify correct equipment/tool(s)/material in job procedures.		
Y□ N○ 1.6 Was the correct equipment/tool(s)/material readily available?		Provide correct equipment/tool(s)/material. Review purchasing specifications and procedures. Anticipate future requirements.		
Y□ N○ 1.7 Did employee(s) know where to obtain equipment/tool(s)/material required for the job?		Review procedures for storage, access, delivery, or distribution. Review job procedures for obtaining equipment/tool(s)/material.		
Y○ N□ 1.8 Was substitute equipment/tool(s)/material used in place of correct one?		Provide correct equipment/tool(s)/material. Warn against use of substitutes in job procedures and in job instruction.		
Y○ N□ 1.9 Did the design of the equipment/tool(s) create operator stress or encourage operator error?		Review human factors engineering principles. Alter equipment/tool(s) to make it more compatible with human capability and limitations. Review purchasing procedures and specifications. Check out new equipment and job procedures involving new equipment before putting into service. Encourage employees to report potential hazardous conditions created by equipment design.		
Y○ N□ 1.10 Did the general design or quality of the equipment/tool(s) contribute to a hazardous condition?		Review criteria in codes, standards, specifications, and regulations. Establish new criteria as required.		
○ 1.11 List other causal factors in "Comment" column.				

Figure 12–1. *(Continued.)*

288

| | | PART 2—ENVIRONMENT | | | |

◯ ☐
Y N
2.0 WAS THE LOCATION/POSITION OF EQUIPMENT/MATERIALS/EMPLOYEE(S) A CONTRIBUTING FACTOR?
If yes, answer the following. If no, proceed to Part 3.

	CAUSAL FACTORS	COMMENT	POSSIBLE CORRECTIVE ACTIONS	RECOMMENDED CORRECTIVE ACTIONS
◯☐ Y N	2.1 Did the location/position of equipment/material/employee(s) contribute to a hazardous condition?		Perform job safety analysis. Review job procedures. Change the location, position, or layout of the equipment. Change position of employee(s). Provide guardrails, barricades, barriers, warning lights, signs, or signals.	
☐◯ Y N	2.2 Was the hazardous condition recognized? If yes, answer A and B. If no, proceed to 2.3.		Perform job safety analysis. Improve employee ability to recognize existing or potential hazardous conditions. Provide test equipment, as required, to detect hazard. Review any change or modification of equipment/tools/materials.	
☐◯ Y N	A. Was the hazardous condition reported?		Train employees in reporting procedures. Stress individual acceptance of responsibility.	
☐◯ Y N	B. Was employee(s) informed of the job procedure for dealing with the hazardous condition as an interim action?		Review job procedures for hazard avoidance. Review supervisory responsibility. Improve employee-supervisor communications. Take action to remove or minimize hazard.	
☐◯ Y N	2.3 Was employee(s) supposed to be in the vicinity of the equipment/material?		Review job procedures and instruction. Provide guardrails, barricades, barriers, warning lights, signs, or signals.	
☐◯ Y N	2.4 Was the hazardous condition created by the location/position of equipment/material visible to employee(s)?		Change lighting or layout to increase visibility of equipment. Provide guardrails, barricades, barriers, warning lights, signs or signals, floor stripes, etc.	
☐◯ Y N	2.5 Was there sufficient workspace?		Review workspace requirements and modify as required.	
◯☐ Y N	2.6 Were environmental conditions a contributing factor (for example, illumination, noise levels, air contaminant, temperature extremes, ventilation, vibration, radiation)?		Monitor, or periodically check, environmental conditions as required. Check results against acceptable levels. Initiate action for those found unacceptable.	
◯	2.7 List other causal factors in "Comment" column.			

| | | PART 3—PEOPLE | | | |

◯ ☐
Y N
3.0 WAS THE JOB PROCEDURE(S) USED A CONTRIBUTING FACTOR?
If yes, answer the following. If no, proceed to Part 3.6.

	CAUSAL FACTORS	COMMENT	POSSIBLE CORRECTIVE ACTIONS	RECOMMENDED CORRECTIVE ACTIONS
☐◯ Y N	3.1 Was there a written or known procedure (rules) for this job? If yes, answer A, B, and C. If no, proceed to 3.2.		Perform job safety analysis and develop safe job procedures.	
☐◯ Y N	A. Did job procedures anticipate the factors that contributed to the accident?		Perform job safety analysis and change job procedures.	
☐◯ Y N	B. Did employee(s) know the job procedure?		Improve job instruction. Train employees in correct job procedures.	
◯☐ Y N	C. Did employee(s) deviate from the known job procedure?		Determine why. Encourage all employees to report problems with an established procedure to supervision. Review job procedure and modify if necessary. Counsel or discipline employee. Provide closer supervision.	
☐◯ Y N	3.2 Was employee(s) mentally and physically capable of performing the job?		Review employee requirements for the job. Improve employee selection. Remove or transfer employees who are temporarily, either mentally or physically, incapable of performing the job.	

Figure 12–1. (Continued.)

	CAUSAL FACTORS	COMMENT	POSSIBLE CORRECTIVE ACTIONS	RECOMMENDED CORRECTIVE ACTIONS
○ □ Y N	3.3 Were any tasks in the job procedure too difficult to perform (for example, excessive concentration or physical demands)?		Change job design and procedures.	
○ □ Y N	3.4 Is the job structured to encourage or require deviation from job procedures (for example, incentive, piecework, work pace)?		Change job design and procedures.	
○	3.5 List other causal factors in "Comment" column.			

○ □ 3.6 WAS LACK OF PERSONAL PROTECTIVE EQUIPMENT OR EMERGENCY EQUIPMENT A CONTRIBUTING FACTOR IN THE INJURY?
Y N If yes, answer the following. If no, proceed to Part 4. NOTE: The following causal factors relate to the *injury*.

	CAUSAL FACTORS	COMMENT	POSSIBLE CORRECTIVE ACTIONS	RECOMMENDED CORRECTIVE ACTIONS
□ ○ Y N	3.7 Was appropriate personal protective equipment (PPE) specified for the task or job? If yes, answer A, B, and C. If no, proceed to 3.8.		Review methods to specify PPE requirements.	
□ ○ Y N	A. Was appropriate PPE available?		Provide appropriate PPE. Review purchasing and distribution procedures.	
□ ○ Y N	B. Did employee(s) know that wearing specified PPE was required?		Review job procedures. Improve job instruction.	
□ ○ Y N	C. Did employee(s) know how to use and maintain the PPE?		Improve job instruction.	
□ ○ Y N	3.8 Was the PPE used properly when the injury occurred?		Determine why and take appropriate action. Implement procedures to monitor and enforce use of PPE.	
□ ○ Y N	3.9 Was the PPE adequate?		Review PPE requirements. Check standards, specifications, and certification of the PPE.	
□ ○ Y N	3.10 Was emergency equipment specified for this job (for example, emergency showers, eyewash fountains)? If yes, answer the following. If no, proceed to Part 4.		Provide emergency equipment as required.	
□ ○ Y N	A. Was emergency equipment readily available?		Install emergency equipment at appropriate locations.	
□ ○ Y N	B. Was emergency equipment properly used?		Incorporate use of emergency equipment in job procedures.	
□ ○ Y N	C. Did emergency equipment function properly?		Establish inspection/monitoring system for emergency equipment. Provide for immediate repair of defects.	
○	3.11 List other causal factors in "Comment" column.			

PART 4—MANAGEMENT

○ □ 4.0 WAS A MANAGEMENT SYSTEM DEFECT A CONTRIBUTING FACTOR?
Y N If yes, answer the following. If no, STOP. Your causal factor indentification exercise is complete.

	CAUSAL FACTORS	COMMENT	POSSIBLE CORRECTIVE ACTIONS	RECOMMENDED CORRECTIVE ACTIONS
○ □ Y N	4.1 Was there a failure by supervision to detect, anticipate, or report a hazardous condition?		Improve supervisor capability in hazard recognition and reporting procedures.	
○ □ Y N	4.2 Was there a failure by supervision to detect or correct deviations from job procedure?		Review job safety analysis and job procedures. Increase supervisor monitoring. Correct deviations.	

Figure 12–1. *(Continued.)*

	CAUSAL FACTORS	COMMENT	POSSIBLE CORRECTIVE ACTIONS	RECOMMENDED CORRECTIVE ACTIONS
☐ ○ Y N	4.3 Was there a supervisor/employee review of hazards and job procedures for tasks performed infrequently? (Not applicable to all accidents.)		Establish a procedure that requires a review of hazards and job procedures (preventive actions) for tasks performed infrequently.	
☐ ○ Y N	4.4 Was supervisor responsibility and accountability adequately defined and understood?		Define and communicate supervisor responsibility and accountability. Test for understandability and acceptance.	
☐ ○ Y N	4.5 Was supervisor adequately trained to fulfill assigned responsibility in accident prevention?		Train supervisors in accident prevention fundamentals.	
○ ☐ Y N	4.6 Was there a failure to initiate corrective action for a known hazardous condition that contributed to this accident?		Review management safety policy and level of risk acceptance. Establish priorities based on potential severity and probability of recurrence. Review procedure and responsibility to initiate and carry out corrective actions. Monitor progress.	
○	4.7 List other causal factors in "Comment" column.			

Developed by the National Safety Council

Figure 12–1. *(Concluded.)*

cations should be established to group similar data. Then, more specific classifications should be set up within each general classification to preserve as many of the details as possible.

For example, among the general classifications for Hazardous Condition are the following:
1. defects of agencies (which means undesired and unintended characteristics)
2. dress or apparel hazards
3. environmental hazards
4. placement hazards
5. inadequate safeguarding
6. public hazards.

Within each one of these general classifications, more specific classifications are set up. Under defects of agencies for example, are:
1. composed of unsuitable materials
2. dull
3. improperly compounded, constructed, or assembled
4. improperly designed
5. rough
6. sharp
7. slippery
8. worn, cracked, frayed, or broken
9. other defects.

It is not always possible to establish classifications before the analysis is begun. In this case, management can develop classifications as they review reports and notice various accident situations.

For example, if an investigator is analyzing ladder accidents and finds that in a number of cases broken rungs caused the accidents, a specific category for "broken rungs" should be set up under the classification "Defects of Agencies."

It is recommended that general and specific classifications be used for most key facts. The point must be emphasized that for an analysis to be of maximum usefulness, classifications must be set up to encompass the situations pertinent to a particular organization.

Use of a numerical code. Regardless of the method eventually used to sort and tabulate the various key facts, the work will be made easier if management assigns code numbers to the different classifications. A numerical code simply refers to numbers assigned in sequence to a list of similar facts. Each basic fact (agency of accident, accident type, hazardous condition, etc.) should have its own number; but for different facts, the numbering series may be repeated.

With this method, the analyst reads each case only once, at which time he or she assigns code numbers to the different facts. Subsequent sorting of the various facts can be completed quickly by referring to the code numbers. After the analyst has reviewed the cases and assigned code numbers to the different key facts, he or she can easily and quickly sort or arrange the reports by any of the facts to reveal the principal data concerning the accidents.

Making the Analysis

Experience has proved that the most effective way to re-

duce accidents is to concentrate on the primary causes of one phase of the accident problem at a time rather than attempt to stop all accidents at once. The problem can be approached in different ways, any one of which should prove effective.

The analyst may group reports by occupation of the injured person. The individual would review each group of reports, then determine the most prevalent accident types, sources of injury, and agencies of accidents among different occupations. Such information is particularly helpful in planning employee training and in developing educational materials and programs.

Injury incidence rates computed by departments may reveal that injuries occur at sharply higher rates in some departments than in others. If this is the case, the analyst can examine accident reports in the high-rate departments to find the sources of the accidents and their causes. This method enables management to concentrate efforts on the locations where the most accidents occur.

If injury incidence rates reveal a high rate of occurrence throughout the organization, the analysis procedure usually starts with information about the injury, goes on to identify the injury-producing event, and then looks at the circumstances and causal factors. The same procedure can be followed to examine injuries occurring within a high-rate occupation or department.

The analyst can crosstabulate the injury data to show the relationship or interaction between the two categories. Table 12–A illustrates an analysis of the Nature of Injury versus the Part of Body. This crosstabulation, in addition to showing what types of personal protective equipment might be useful, also points out common injury patterns needing further investigation.

A crosstabulation can extend in several directions, producing the need for further, separate crosstabulations. For example, the analyst would use the categories in Table 12–A showing the highest frequency of injuries (Cut, laceration, puncture, or abrasion to fingers and Sprain or strain to the back) to construct a second (Table 12–B) and a third crosstabulation.

As shown in Table 12–B, by the second tabulation the number of cases in a category usually is small enough that the analyst can read individual accident reports to find common causal factors and to determine corrective actions. If the number of cases in a category is still too large, then additional crosstabulations, such as Location versus activity or Activity versus occupation, can reduce the number enough to allow study of individual accident reports.

Methods of tabulating. For analyzing a small number of reports (up to about 100), hand sorting and tallying is effective. This method's principal advantage is that the analyst is using original records and has all the information available for reference.

A personal computer or other data-processing equipment is best for large collections of cases. Computers are also useful for smaller data sets as well because they can sort and display cases very quickly. Their efficiency allows the investigator to concentrate on various accidents and to test alternative hypotheses easily.

Using the Analysis

Merely obtaining the information will not prevent recurrence of accidents. Management must correct the underlying and contributing conditions. Thorough analysis of groups of accident investigation reports can point to corrective actions that might not be evident when studying an individual case. In particular, inadequate policies, procedures, or management systems often are apparent after looking at the forest rather than the trees.

The statistical evidence revealed in an analysis can provide the guidance to direct safety efforts along the most effective path. The analysis will provide objective support and justification for budget requests, training programs, or other management safety activities.

ESTIMATING ACCIDENT COSTS

This discussion concerns the elements of cost most likely to result from a work accident and presents a method whereby an organization can obtain an accurate estimate of the total costs of its work accidents. (This procedure for estimating costs was developed by Rollin H. Simmonds, Ph.D., Professor, Michigan State University, under the direction of the Statistics Division, National Safety Council.)

Reliable cost information is one basis for making decisions upon which efficiency and profit depend. Even in so obviously desirable an activity as accident prevention, some proposed measures or alternatives must be evaluated on the basis of their potential effect on profits.

Although most executives want to make their company a safe place to work, they also have a responsibility to run their business profitably. Consequently, they may be reluctant to spend money for accident prevention unless they can see a prospect for saving at least as much as they spend. *Without information on the cost of accidents, it is practically impossible to estimate the savings brought about by expenditures for accident prevention.*

Annual reports stressing dollar savings are more meaningful to management than those using incidence rates. Facts about the costs of accidents also may be effectively used in securing the active cooperation of supervisors. Supervisors usually are cost conscious because they are expected to run their departments profitably. Monthly reports showing the cost of accidents or the savings resulting from good accident records can motivate supervisors to ensure safe operating procedures.

Definition of Work Accidents for Cost Analysis

Work accidents, for the purpose of cost analysis, are unintended occurrences arising in the work environment. These accidents fall into two general categories: (1) incidents resulting in work injuries or illnesses and (2) incidents causing property damage or interfering with production. The inclusion of the no-injury incidents makes "work accidents"

Table 12–A. Nature of Injury versus Part of Body

Nature of Injury	Eyes	Head, Face, Neck	Back	Trunk	Arm	Hand, Wrist	Finger	Leg	Foot, Ankle	Toe	Internal, Other	Total
Amputation	0	0	0	0	0	0	1	0	0	0	0	1
Burn & scald (heat)	0	0	0	0	0	0	0	0	0	0	0	0
Burn (chemical)	2	0	0	0	0	0	0	0	0	0	0	2
Concussion	0	2	0	0	0	0	0	0	0	0	0	2
Crushing	0	0	0	0	0	1	0	0	0	2	0	3
Cut, laceration, puncture, abrasion	0	1	0	0	1	3	18	0	0	0	0	23
Fracture	0	0	0	0	0	2	5	0	1	0	0	8
Hernia	0	0	0	1	0	0	0	0	0	0	0	1
Bruise, contusion	0	2	0	1	0	3	2	3	0	1	0	12
Occupational illness	0	1	0	0	1	2	0	0	0	0	8	12
Sprain, strain	0	0	24	0	0	2	2	3	4	0	0	35
Other	1	0	0	0	0	0	0	0	0	0	5	6
Total	3	6	24	2	2	13	28	6	5	3	13	105

This crosstabulation shows how the nature of injury and the part of body interact. In this example, cuts most often affect the fingers and sprains and strains usually involve the back. Note that bruises and contusions affect several body parts.

roughly synonymous with the type of occurrences a safety department strives to prevent.

Method for Estimating

To be of maximum usefulness, cost figures should represent as accurately as possible the specific experience of the company. It is not useful to have a fixed ratio of uninsured to insured costs representing many different organizations in many different industries. Estimated costs of accidents in general do not take into account differences in hazards from one industry to another or the more important differences in safety performance from one company to another.

Since the distinctions between direct and indirect costs are difficult to maintain, they have been abandoned in favor of the more precise terms "insured" and "uninsured" costs. Using these data, a company can estimate its accident cost with reasonable accuracy.

Insured costs. Every organization paying compensation insurance premiums recognizes such expense as part of the cost of accidents. In some cases, medical expenses, too, may be covered by insurance. These costs are definite and known. They comprise the insured element of the total accident cost.

In addition to these costs, many other costs arise in connection with accidents. Although the expense of damaged equipment is easily identified, others, such as wages paid to the injured employee for downtime on the day of the injury, are hidden. These items comprise the uninsured element of the total accident cost.

Uninsured costs. While insured costs can be determined easily from accounting records, uninsured (frequently called "indirect") costs are more difficult to assess. The method described here is one way to calculate these added expenses associated with many accidents. The first step is to make a pilot study to ascertain approximate averages of uninsured costs for each of the following four classes of accidents:

CLASS 1 — Cases involving lost workdays—days away from work or days of restricted work activity.

CLASS 2 — Medical treatment cases requiring the attention of a physician outside the plant

CLASS 3 — Medical treatment cases requiring only first aid or local dispensary treatment and resulting in property damage of less than $100 or loss of less than eight hours in work time

CLASS 4 — Accidents that either cause no injury or cause minor injury not requiring the attention of a physician, and that result in property damage of $100 or more, or loss of eight or more employee-hours.

Once average costs have been established for each accident class, they can be used as multipliers to obtain total uninsured costs in subsequent periods. These costs are then added to known insurance premium costs to determine the total cost of accidents.

Example of a Cost Estimate

An estimate of costs made by one company is given in the following example. First, a pilot study was made to get the average cost of each class of accident. Included in the study were 20 Class 1 accidents, 30 Class 2 accidents, 50 Class 3 accidents, and 20 Class 4 accidents. Costs were determined and averages developed as in Table 12–C.

During the entire year, the company had 34 Class 1 accidents, 148 Class 2 accidents, and 4,000 Class 3 accidents. No record was kept of the Class 4 accidents after the pilot study was completed. Instead, the ratio of the number of Class 4 to Class 1 accidents found in the pilot study was used. This ratio was shown to be about 1 to 1, and since there were 34 Class 1 accidents during the year, it was assumed there were about 34 Class 4 accidents. (A separate record could be kept of the number of Class 4 accidents.)

The average cost for each accident class was applied to these totals to secure the results shown in Table 12–D.

Since the final total is the sum of many estimates, it should not be implied that the total figure suggests absolute accuracy. The estimate should be rounded to three significant digits—in this case, to the nearest thousand dollars. As a

result, in this instance, the analyst reported to the plant manager, "During the past year, accidents cost this company about $155 thousand in compensation, medical expense, lost time, and property damage."

The average costs determined in this pilot study represent the actual experience of this particular organization. Until important changes take place in this company's safety program, in the kind of machinery used or persons employed, or in other aspects affecting costs, the same average costs can be used.

Adjusting for Inflation

The effects of inflation can quickly render obsolete the cost figures in a pilot study. To account for this effect, the cost factors should be adjusted for inflation each year. The wage-related cost elements can be multiplied by the change in the general level of wages in the company. Other cost elements can be brought up to date by multiplying them by the general inflation rate as measured by the change in the Consumer Price Index. Because these adjustments are only approximate, the pilot study should be repeated at least every five years to establish new benchmarks.

Items of Uninsured Cost

Important to a pilot study is a careful investigation of each accident to determine all the costs arising out of it. The following items of uninsured or indirect costs are clearly the result of work accidents and are subject to reasonably reliable measurement. Less tangible losses, such as the effect of accidents on public relations, employee morale, or on wage rates necessary to secure and retain employees, are not included in this method of estimating costs but can be important factors in some cases.

Information on some of the items is derived from the Department Supervisor's Accident Cost Investigation Report (Figure 12–2). The items are discussed in the order in which they appear on the Investigator's Cost Data Sheet (Figure 12–3).

1. *Cost of wages paid for time lost by workers who were not injured.* These are employees who stopped work to watch or assist after the accident or to talk about it, or who lost time because they needed the equipment damaged in the accident or because they needed the output or the aid of the injured worker.

2. *Cost of damage to material or equipment.* The validity of property damage as a cost can scarcely be questioned. Occasionally, there is no property damage, but a substantial cost is incurred in reorganizing material or equipment. The charge should be confined either to the net cost of repairing or reorganizing material or equipment damaged or displaced or to the current worth of equipment, less salvage value, if damaged beyond repair. An estimate of property damage should have the approval of the cost accountant, particularly if the current worth of the damaged property differs from the depreciated value estab-

lished by the accounting department.

3. *Cost of wages paid for time lost by the injured worker, other than workers' compensation payments.* Payments made under workers' compensation laws for time lost after the waiting period are not included in this element of cost.

4. *Extra cost of overtime work necessitated by the accident.* The charge against an accident for overtime work is the difference between normal wages and overtime wages for the time needed to make up lost production, and the cost of extra supervision, heat, light, cleaning, and other extra services.

5. *Cost of wages paid supervisors for time spent on activities concerning the accident.* The most satisfactory way of estimating this cost is to charge the wages paid to the foreman for the time spent away from normal activities as a result of the accident.

6. *Wage cost caused by decreased output of injured worker after return to work.* If the injured worker's previous wage payments are continued despite a 40% reduction in output, the accident should be charged with 40% of the worker's wages during the period of low output.

7. *Cost of learning period of new worker.* If a replacement worker produces only half as much in the first two weeks as the injured worker would have produced in the same time for the same pay, then half of the new worker's wages for the two weeks' period should be considered part of the cost of the accident. A wage cost for time spent by supervisors or others in training the new worker also should be attributed to the accident.

8. *Uninsured medical cost borne by the company.* This cost is usually that of medical services provided at the plant dispensary. There is no great difficulty in estimating an average cost per visit for this medical attention. The question may be raised, however, whether this expense can be considered a variable cost. That is, would a reduction in accidents result in lower expenses for operating the dispensary?

9. *Cost of time spent by management and clerical workers on investigations or in the processing of compensation application forms.* Time spent by management or supervision (other than the foreman or supervisor covered in Item 5) and by clerical employees in investigating an accident, or settling claims arising from it, is chargeable to the accident.

10. *Miscellaneous usual costs.* This category includes less typical costs, the validity of which must be clearly shown by the investigator on individual accident reports. Among such possible costs are public liability claims, equipment rental, losses due to cancelled contracts or lost orders if the accident causes an overall reduction in total sales, loss of company bonuses, cost of hiring new employees if this expense is significant, cost of above-normal spoilage by new employees, and demurrage. These cost factors and any others not suggested above would need to be well substantiated. Miscellaneous costs were found in less than 2% of several hundred cases reviewed in connection with this study.

Table 12–B. Source of Injury versus Type of Accident

Source of Injury	Type of Accident							
	Fall from elevation	Fall on same level	Struck against	Struck by	Caught in, under, or between	Rubbed or abraded	Bodily reaction	Overexertion
Machine	0	0	0	0	3	0	0	0
Conveyor, elev. hoist	0	0	0	0	0	0	0	0
Vehicle	0	0	0	0	0	0	0	0
Electrical apparatus	0	0	0	0	0	0	0	0
Hand tool	0	0	0	4	0	0	0	0
Chemical	0	0	0	0	0	0	0	0
Working surface, bench, etc.	0	0	0	0	0	0	0	0
Floor, walking surface	0	0	0	0	0	0	0	0
Bricks, rocks, stones	0	0	0	0	0	0	0	0
Box, barrel, container	0	0	0	0	0	0	0	0
Door, window, etc.	0	0	0	0	0	0	0	0
Ladder	0	0	0	0	0	0	0	0
Lumber, woodworking metals	0	0	0	0	0	0	0	0
Metal	0	0	0	9	0	0	0	0
Stairway, steps	0	0	0	0	0	0	0	0
Other	0	0	0	0	0	0	0	0
Unknown	0	0	0	0	0	0	0	0
None	0	0	0	0	0	0	0	0
Total	0	0	0	13	3	0	0	0

This crosstabulation shows Source of Injury versus Type of Accident for one of the most frequent injuries identified in Table 7-A. In this example, being struck by a metal object was the source of most finger cuts. Other causes were being struck by hand tools or becoming caught in machinery.

Table 12–B. (Concluded.)

Source of Injury	Type of Accident							
	Contact with electrical current	Contact with temperature extremes	Radiations, caustics, toxic and noxious substances	Public transportation accident	Motor vehicle accident	Other	Unknown	Total
Machine	0	0	0	0	0	0	0	3
Conveyor, elev. hoist	0	0	0	0	0	0	0	0
Vehicle	0	0	0	0	0	0	0	0
Electrical apparatus	0	0	0	0	0	0	0	0
Hand tool	0	0	0	0	0	0	0	4
Chemical	0	0	0	0	0	0	0	0
Working surface, bench, etc.	0	0	0	0	0	0	0	0
Floor, walking surface	0	0	0	0	0	0	0	0
Bricks, rocks, stones	0	0	0	0	0	0	0	0
Box, barrel, container	0	0	0	0	0	0	0	0
Door, window, etc.	0	0	0	0	0	0	0	0
Ladder	0	0	0	0	0	0	0	0
Lumber, woodworking metals	0	0	0	0	0	0	0	0
Metal	0	0	0	0	0	0	0	9
Stairway, steps	0	0	0	0	0	0	0	0
Other	0	0	0	0	0	0	2	2
Unknown	0	0	0	0	0	0	0	0
None	0	0	0	0	0	0	0	0
Total	0	0	0	0	0	0	2	18

Table 12–C. Average Costs Determined by Pilot Study

Class of Accident	Number of Accidents Reported	Average Uninsured Cost
Class 1	20	$251.10
Class 2	30	80.80
Class 3	50	15.70
Class 4	20	507.10

Making a Pilot Study

The purpose of the pilot study is to develop average uninsured costs for different classes of accidents that can be applied to future accident totals. Therefore, it is desirable not to include the costs of deaths and permanent total disabilities. Such accidents occur so seldom that the costs should be calculated individually and not estimated on the basis of averages.

Management should build in some flexibility in the grouping of classes of cases. If an organization makes no

distinction in the records between medical treatment cases requiring a physician's attention and those not requiring a physician's attention, the pilot study may combine Classes 2 and 3. (See Method for Estimating earlier in this chapter.)

The following discussion assumes the study of costs will be made with the injuries grouped in the recommended classes. The discussion covers Classes 1, 2, and 4. A different method must be applied to Class 3 injuries, and it will be discussed later.

Table 12–D. Estimate of Yearly Accident Costs

Class of Accident	Number of Accidents	Average Cost per Accident (from pilot study)	Total Uninsured Cost
Class 1	34	$251.10	$ 8,537.40
Class 2	148	80.80	11,958.40
Class 3	4,000	15.70	62,800.00
Class 4	34	507.10	17,241.40
Total Uninsured Cost			$100,537.20
Insurance Premiums			54,400.00
Total Accident Cost for the Period			$154,937.20

Classes 1, 2, and 4. To analyze uninsured costs for accidents in Classes 1, 2, and 4, the supervisor in charge of the department where an accident occurs should secure for each accident the information indicated on the Department Supervisor's Accident Cost Investigation Report form (Figure 12–2). These data can be obtained during the supervisor's regular investigation of the accident. As soon as each report form is completed, it should be sent to the safety department.

Safety personnel then transfer information from the department supervisor's report to the Investigator's Cost Data Sheet (Figure 12–3). The safety department then assumes responsibility for securing the supplementary information from the accounting department, industrial relations department, and other departments where records on lost time and other necessary information are kept.

As an alternative, a member of the safety department could secure all information needed on the data sheet. In this case, the supervisor's report form is not used, and the supervisor is required only to report each accident in Classes 1, 2, or 4 to the safety department as soon as it occurs.

The investigator, before computing averages, should be certain that the pilot study has covered a sufficient number of cases of Classes 1, 2, and 4 to be representative. This number will rarely be less than 20 cases. However, more cases should be studied if the costs of the cases in a particular class vary widely. Information should be secured on enough cases of each class so the average cost per case in each class is fully representative of past experience and will, by inference, be applicable to future experience.

Once a sufficient number of cases have been accumulated, the investigation of individual cases can be discontinued. For the data thus collected, separate averages should be calculated for the cases of each class. It is recognized that these costs are averages of the uninsured costs only.

Class 3 injuries. These injuries are the common first aid cases in which no significant property damage results from the accident. They are the most difficult to analyze from the standpoint of cost. This is because the time lost is likely to occur repeatedly and only for short periods. Also, the injuries can occur so frequently as to place an undue burden on the supervisor and safety director if a complete report form and data sheet are required for each case.

The points of essential information needed are the average amount of working time lost per trip to the dispensary, the average dispensary cost per treatment, the average number of visits to the dispensary per case, and the average amount of supervisor's time required per case. The following method of developing averages for each of these items is recommended:

1. *Secure an estimate of average working time lost per trip to the dispensary for first aid.* Departmental time records should be consulted as they may show the amount of time each worker is absent from the job while receiving first aid. If so, a random sample of 50 to 100 records of persons known to have received first aid should be selected from different departments. The average time lost per dispensary visit is calculated by adding the absence time for all visits in the sample and dividing by the total number of visits.

    If departmental records do not contain this information, it will be necessary to assign an investigator to observe a random sample of 50 or more persons visiting the dispensary. As before, to secure the average time, all the estimated time intervals of absence are added and then divided by the total number of persons observed.

2. *Make an estimate of the average cost of providing medical attention for each visit* by dividing the total cost of operating the dispensary for a year by the total number of treatments given during the year.

3. *Calculate the average number of visits to the dispensary per case.* This is done by dividing the number of treatments of Class 3 injuries in a representative period, perhaps a month or six weeks, by the number of Class 3 injuries reported during the same period of time.

4. *Calculate the average amount of supervisor's time required per case.* Where possible, this is accomplished by observing the activities of representative supervisors in connection with first aid cases.

Over time, a sufficient number of cases will have been studied to be representative both of the activities of supervisors in different departments and of different types of first aid cases. The average time spent by a supervisor is then computed by adding all the time intervals recorded and dividing by the number of cases.

In some instances, it may be impossible to make a time study of the supervisor's activities in connection with first aid cases. The only alternative is to secure from each supervisor an estimate of the time spent on the usual first aid case and then to add them and divide by the number of supervisors to find the average time. Determine the

DEPARTMENT SUPERVISOR'S ACCIDENT COST INVESTIGATION REPORT

Injury/Accident_____

Date_____Name of Injured_____Dept._____

TIME LOST

1. How much time did other employees lose by talking, watching, or helping at accident? Number of employees _____ x hours =

2. How much productive time was lost because of damaged equipment or loss of reduced output by injured worker?
Estimate Hours =

3. How much time did injured employee lose for which he was paid on the day of the injury?
Estimate Hours =

4. Will overtime be necessary?   Estimate Hours =

5. How much of the supervisors or other managements' time was lost as a result of this accident?
Estimate Hours =

6. Were additional costs incurred due to hiring and training or replacement?
Training Time Estimate Hours =

7. Describe the damage to material or equipment. _____

_____

8. If machine and/or operations were idle, can loss of production be made up?
Yes_____   No_____

9. Will overtime be necessary? Yes_____   No_____

10. Any demurrage or other cost involved? Yes_____   No_____

ADDITIONAL ACCIDENT COSTS

To compute the total costs of the accident, it is necessary to complete the following costs. Should the supervisor have access to this information it is advised he complete as much as possible. Safety Department will develop those costs not known by supervisor.

Figure 12-2. This cost form (8 x 11 in.) should be prepared by the department supervisor as soon after the accident as information becomes available on the amount of time lost by all persons and the extent of damage to product and equipment. It is sent to the safety department not later than the day after the accident.

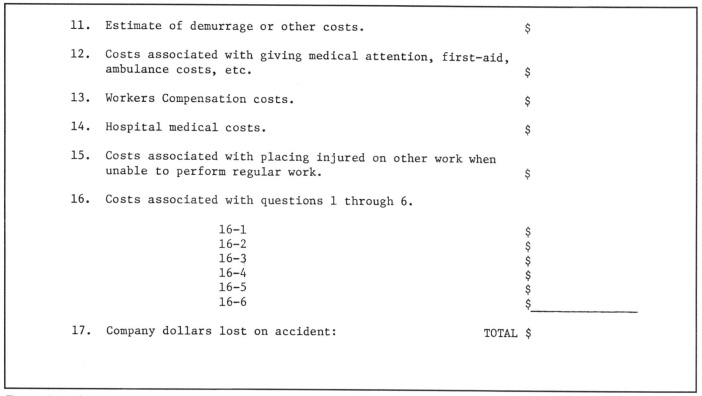

11. Estimate of demurrage or other costs. $

12. Costs associated with giving medical attention, first-aid,
     ambulance costs, etc. $

13. Workers Compensation costs. $

14. Hospital medical costs. $

15. Costs associated with placing injured on other work when
     unable to perform regular work. $

16. Costs associated with questions 1 through 6.

                    16-1 $
                    16-2 $
                    16-3 $
                    16-4 $
                    16-5 $
                    16-6 $_____

17. Company dollars lost on accident: TOTAL $

Figure 12–2. *(Concluded.)*

average value of this time by multiplying it by the average hourly wage of a supervisor.

The average total uninsured cost of a case in Class 3 is estimated from the data accumulated above as follows: the average amount of time lost for a trip to the dispensary (1, above) is multiplied by the plant's average wage rate, secured from the payroll department, to get the average cost per trip for the worker's time lost. To this figure is added the estimated cost of providing medical attention for a single visit (2). This figure is then multiplied by the average number of dispensary visits per medical treatment case (3), and to this result is added the average value of supervisor's time required (4).

This method of recording costs is designed to provide estimates of the average uninsured cost per case for accidents causing localized property damage or, at most, a few injuries.

The method of cost investigation for accidents resulting in deaths, permanent total disabilities, or unusually extensive property damage is essentially the same as for others. However, the main difference is that every one is separately investigated and should be included in the final cost estimate as a separate item. In estimating the cost of a fire, for example, the investigator should bear in mind that the company's fire insurance will probably cover property damage in a major accident while for a less serious incident, fire damage would appear as an uninsured cost.

Development of Final Cost Estimate

Once the average for each class of case has been established, costs for any period in which a sufficiently large number of accidents has occurred to be representative can be estimated

with considerable accuracy. This is calculated by multiplying the average uninsured cost per case for each of the four classes by the number of cases occurring in that class during the period.

If any deaths, permanent total disabilities, or unusually extensive property damage accidents have occurred, the investigator should add the specific uninsured costs of these to the estimated costs of the four classes of accidents.

To these uninsured cost totals the investigator then adds the cost of workers' compensation and insured medical expense. For self-insured companies, this figure will be the total amount paid out in settlement of claims plus all expenses of administering the insurance. For companies not carrying their own insurance, it will be the amount of their insurance premiums plus deductibles paid.

The method will have to be modified in accordance with the recordkeeping systems of different companies. For example, most self-insurers will find it impossible to separate compensated medical expense from dispensary care. In that case, these items should be combined into one, and the dispensary cost omitted from the analysis of noncompensated costs on the data sheets.

For an illustration of the development of a final cost estimate, see the example in Tables 12–C and 12–D.

As stated previously, presentation of accident costs or estimated cost saving is useful in securing management support. Because accident costs represent lost profit opportunities, accident dollars may be treated as lost profit dollars. The general method for achieving additional profits is through additional sales. It is possible then to estimate the

298

INVESTIGATOR'S COST DATA SHEET

Class 1_____
(Permanent partial or temporary
total disability)

Class 2 _____
(Temporary partial disability or
medical treatment case requiring
outside physician's care)

Class 3 _____
(Medical treatment case requiring
local dispensary care)

Class 4 _____
(No injury)

Name _____

Date of injury _____ Its nature _____

Department _____ Operation _____Hourly wage _____

Hourly wage of supervisor $_____

Average hourly wage of workers in department where injury occurred $_____

1. Wage cost of time lost by workers who were not injured, if paid by employer $_____
   a. Number of workers who lost time because they were talking, watching, helping ____
      Average amount of time lost per worker _____ hours _____ minutes
   b. Number of workers who lost time because they lacked equipment damaged in
      accident or because they needed output or aid of injured worker _____.
      Average amount of time lost per worker _____ hours _____minutes.

2. Nature of damage to material or equipment _____

   _____

   Net cost to repair, replace, or put in order the above material or equipment $_____

3. Wage cost of time lost by injured worker while being paid by employer
   (other than workers compensation payments) $_____
   a. Time lost on day of injury for which worker was paid _____ hours _____ min.
   b. Number of subsequent days' absence for which worker was paid _____ days.
      (Other than workers' compensation payments) _____ hours per day.
   c. Number of additional trips for medical attention on employer's time on
      succeeding days after worker's return to work _____
      Average time per trip _____ hours. _____ min. Total trip time _____ hrs. _____ min.
   d. Additional lost time by employee, for which he was paid by company _____ hrs.
      _____ min.

(over)

Figure 12–3. This 8 x 11 in. form can be used to convert time losses into money losses. Initial time losses are obtained from the Department Supervisor's Accident Cost Report (Figure 12–2), and subsequent time losses are obtained from first aid and other departments as necessary. Wage rate information is obtained from the accounting department. The reverse side of the form (shown at the right) contains space for additional costs pertinent to the accident under study. See discussion under the heading, Making a Pilot Study.

4. If lost production was made up by overtime work, how much more did the work cost than if it had been done in regular hours? (cost items: wage rate difference, extra supervision, light, heat, cleaning for overtime.) $_____

5. Cost of supervisor's time required in connection with the accident $_____
   a. Supervisor's time shown on Dept. Supervisor's Report ____ hrs. ____ min.
   b. Additional supervisor's time required later ____ hrs. ____ min.

6. Wage cost due to decreased output of worker after injury if paid old rate $_____
   a. Total time on light work or at reduced output ____ days ____ hours per day
   b. Worker's average percentage of normal output during this period _____

7. If injured worker was replaced by new worker, wage cost of learning period $_____
   a. Time new worker's output was below normal for his own wage ____ days _____ hours per day. His average percentage of normal output during time _____ His hourly wage $_____.
   b. Time of supervisor or others for training _____ hrs. Cost per hour $_____

8. Medical cost to company (not covered by workers' compensation insurance) $_____

9. Cost of time spent by higher supervision on investigation, including local processing of worker's compensation application form. (No safety or prevention activities should be included.) $_____

10. Other costs are not covered above (e.g., public liability claims; cost of renting replacement equipment; loss of profit on contracts cancelled or orders lost if accident causes net reduction in total sales; loss of bonuses by company; cost of hiring new employee if the additional hiring expense is significant; cost of excessive spoilage by new employee; demurrage). $_____

Explain fully:
   Total uninsured cost ....$_____
   Name of Company _____

Figure 12–3. *(Concluded.)*

sales necessary to recover profits lost due to accident expenses. Simply determine the net profit percentage to gross sales and divide accident costs by the net profit percentage. This will provide management with an estimate of the additional gross sales dollar volume necessary to replace profit lost due to accident expenses.

Another method of presenting accident costs to management is by equating lost accident dollars to lost purchasing opportunities for needed supplies or equipment.

OFF-THE-JOB DISABLING INJURY COST

The employer loses the same services whether the employee is injured off the job or on the job and incurs just about the same types of direct and indirect costs. Nevertheless, the costs of off-the-job (OTJ) accidents and illnesses are, at least in part, handled differently. This section explains the difference and presents sample calculations.

When accidents take place off the job, a major portion of the costs are borne by employers. Some of the costs are evident, such as insurance premiums and wages paid to absent employees. Some of the costs are hidden, such as training for new or transferred workers and medical staff time demanded for workers returning to work after an accident. For example, a new worker does not produce at the same level as an experienced worker; thus, the decreased productivity of the new worker indirectly increases the manufacturing overhead.

Other costs are more difficult to assess, although still very real. As accident rates in the community rise, so do insurance rates, taxes, and welfare contributions. Not all organizations are aware of the total costs that can result from off-the-job accidents and the impact they have on operations and profits. Enough experience has been accumulated, however, to develop a simplified plan for estimating such cost.

Categorizing OTJ Disabling Injury Cost

The cost of off-the-job disabling injuries (OTJ DI) to an organization falls into the following two categories: insured and uninsured. These are the same for on-the-job accidents that result in disabling injuries, described in the previous section.

Most uninsured costs are hidden. Aside from wage costs, most organizations do not keep records of uninsured costs. However, these costs are associated with all OTJ DI accidents and, therefore, affect profit margins.

- Insured-worker productivity, cost, product loss, and equipment damage.—Costs directly associated with the employee who sustained the OTJ DI injury are included in this expense subcategory.
- Noninjured-worker productivity cost, product loss, equipment damage, and administrative cost.—Costs incurred by personnel other than the employee who sustained the OTJ DI injury are included in this subcategory.
- Miscellaneous costs.—This subcategory includes loss of profit for cancelled contracts or orders and the costs of demurrage, telephone calls, transportation, or other miscellaneous expenses.

Estimating a Company's OTJ DI Costs

Some experts say the ratio of insured cost to uninsured cost is 3:2, respectively. In order to estimate a company's losses from employee OTJ DIs, management must first determine the insured cost. Next, using the 3:2 factor, it must escalate the insured cost to determine the total (insured and uninsured) employee costs. The insured cost for injuries to dependents of employees is then added to the total employee cost to ascertain total losses. The following examples illustrate calculation procedures.

Example 1. Company A is insured by an outside carrier. Twenty percent or $225,000 of its annual premium charge was required to pay for its previous calendar year OTJ DI accident experience. Of that total, $75,000 was required for 11 employee injuries and the remaining $150,000 was required for 22 employee-dependent injuries. These figures include the administrative fee paid by the company to the carrier.

The total cost for employee-dependent injuries is a conservative figure, since it does not include the administrative cost incurred by the company's insurance office staff to process claims. If this cost is known, it should be added to the employee-dependent injury expense category. When the $75,000 insured cost category is escalated to include the uninsured cost category, the total expense for employee injuries becomes $125,000.

Company A Estimated OTJ DI Costs

Insured cost for employee injuries	$ 75,000
Uninsured cost for employee injuries	50,000
Insured cost for employee-dependent injuries	150,000
Total annual estimated OTJ DI cost	$275,000

Example 2. Company B is insured by an outside carrier. Its carrier stated that $850,000 was paid for 138 employee injuries and $1,800,000 was paid for 279 employee-dependent injuries for its previous calendar year OTJ DI accident experience. Their administrative fee to the carrier was 6% of the total cost, resulting in a cost of approximately $900,000 for employee injuries and $1,900,000 for employee-dependent injuries. If the cost for insurance office staff claim-processing is known, it should be added to the total employee-dependent cost. Escalation of the insured cost category for employee injuries indicated a total of $1,500,000.

Company B Estimated OTJ DI Costs

Insured cost for employee injuries	$900,000
Uninsured cost for employee injuries	600,000
Insured cost for employee-dependent injuries	1,900,000
Total annual estimated OTJ DI cost	$3,400,000

Example 3. Company C is self-insured. Insurance office records indicated that for the previous calendar year, OTJ DI accident experience for the amount of medical and health claims paid for 350 employee injuries was $2,280,000, and $4,700,000 was paid for 690 employee-dependent injuries. Insurance office staff administrative costs for claim-process-

ing should be added to the employee-dependent cost category if that cost is known. Thus, $2,280,000 escalated to include the uninsured cost category for employee injuries resulted in a total cost of approximately $3,800,000.

Company C Estimated OTJ DI Costs

Insured cost for employee injuries	$2,280,000
Uninsured cost for employee injuries	1,520,000
Insured cost for employee-dependent injuries	4,700,000
Total annual estimated OTJ DI cost	$8,500,000

Measuring Effect of OTJ Programs

Calculations of average costs are useful tools to support initiating or accelerating off-the-job safety awareness programs. The calculations also can be used to measure the effects of safety programs.

For example, 350 employees of Company C experienced OTJ injuries during the previous year for a total of approximately $3,800,000 and an average cost per accident of approximately $10,860. Based upon the OTJ DI loss experience, top management allocated $50,000 from the budget to initiate a safety awareness program. At the end of the year, the $10,860 average cost figure will be escalated to adjust for inflation and, in turn, the new figure will be used to calculate losses. For illustration purposes, it is assumed that the new average cost per accident figure is $11,500 and that employee injuries were reduced from 350 to 300. Calculations (300 x $11,500) indicate losses of $3,450,000, a savings of approximately $350,000 from last year's total. The estimated net return is $300,000, which is a 600% return on investment.

To further support justification for operating funds, safety personnel can add savings realized from reduced employee-dependent injuries to the employee savings total. This type of analysis provides a management tool to evaluate the impact of off-the-job disabling injuries on profit margins; thus, it can be used to gain management commitment to support operating budgets for safety awareness programs.

SUMMARY

- Successful accident prevention requires four fundamental activities: (1) study of all working areas to detect and control or eliminate hazards, (2) study of all operating procedures and administrative controls, (3) education, training, and discipline to minimize human factors, and (4) thorough accident investigation and analysis.
- The primary purpose of accident investigation and analysis is to prevent accidents by uncovering facts about each incident and determining how to prevent a recurrence. The investigation should generate complete information to enable management to take effective corrective actions. Emphasis should be on fact-finding, not fault-finding.
- Several accident investigation and analysis techniques

are available to management, including statistical methods. The techniques can be adapted to the needs of each organization.

- Accident investigation generally should be made by the supervisor, who is likely to know the most about the incident. The safety professional advises and guides the supervisor. In some companies, a special investigation or review committee reports on all accidents or reviews the investigation.
- All accidents should be investigated, particularly those involving deaths or serious injuries. Near-accidents and a series of minor injuries also demand study. Such investigations can determine underlying and contributing causes. The investigations must be done fairly and impartially, or the results will not be accepted by workers.
- Analysis of accident investigations can help management identify and locate sources of accidents, pinpoint the nature and size of accident problems, reveal the need for engineering changes or correcting inefficiencies in operating procedures, disclose unsafe practices, enable supervisors to target safety efforts, and evaluate a safety program's effectiveness.
- Eight data elements comprise the minimum amount of information that should be collected about each accident: employer characteristics, employee characteristics, narrative description of the incident, characteristics of the equipment, task being performed, time factors involved, preventive measures required, and characteristics of the injury.
- Two types of accident analysis can be done: (1) to determine corrective actions and (2) to conduct a statistical analysis to uncover accident patterns so corrective actions can be devised. The Guide for Identifying Causal Factors and Corrective Actions is used to analyze how equipment, environment, people, and management contribute to the accident.
- In the statistical analysis approach, classifications are established for various groups of data to identify key factors causing or contributing to accidents. These classifications can be assigned numerical codes.
- Cases and reports can be sorted quickly by any of the classified facts to reveal the principal factors concerning accidents. These data can be used to pinpoint problem areas in an organization and help management devise countermeasures to control or eliminate hazards and unsafe practices.
- Statistical analyses also can be used to compare current accident rates with prior years' experience, with rates of other companies, or with rates for the industry as a whole. This overall picture makes it easier to justify expenditures for safety programs and worker education.
- Most companies have a method for calculating the cost estimates of accidents. Costs can be direct (insured) or indirect (uninsured). Uninsured costs may include effect on public relations and employee morale, certain medical expenses, lost sales, hiring and training new workers, cancelled orders, and the like.

- Organizations generally conduct pilot studies to determine the average uninsured costs for different classes of accidents that can be applied to future accident totals. These figures do not include costs of deaths and permanent total disabilities, because each case is investigated individually.
- To the uninsured cost totals, the investigator then adds the cost of worker compensation and insured medical expense. The method differs slightly with each company, depending on whether they are self-insured or not. These accidents cost figures can also be used to convince management that safety programs are cost-effective in the long run.
- Investigation, analysis, and cost estimates for off-the-job accidents are handled similarly to on-the-job incidents. These activities can show management the impact of off-the-job accidents on organizational operations and profits and help to gain support for worker education and safety training programs.

REFERENCES

Blankenship LM. *Nonoccupational Disabling Injury Cost Study.* K/DSA–457 Oak Ridge, TN: Martin Marietta Energy Systems, Inc., October 1981.

DeReamer R. *Modern Safety and Health Technology.* New York: John Wiley and Sons, Inc., 1981.

Grimaldi JV and Simmonds RH. *Safety Management—Accident Cost and Control,* 5th ed. Homewood, IL: Richard D. Irwin, Inc., 1988.

Kuhlman R. *Professional Accident Investigation: Methods and Techniques.* Loganville, GA: Institute Press, Division of International Loss Control Institute, 1977.

National Safety Council, 444 North Michigan Avenue, Chicago, IL 60611.
   Accident Investigation: A New Approach, 1983.
   Industrial Data Sheet, Off-the-Job Safety, 12304–0601, 1986.
   The Off-the-Job Safety Program Manual, 1990.

13

Record-keeping and Incidence Rates

In this chapter, the terms *accident, incident,* and *injury* are restricted to occupational injuries and illnesses. In other chapters, accident and incident are used in their broad meanings: unplanned events that interrupt the completion of an activity, and that may (or may not) include property damage or injury.

The Williams-Steiger Occupational Safety and Health Act of 1970 (OSHAct) requires most U.S. employers to maintain specific records of work-related employee injuries and illnesses. In addition to these records, many employers also are required to make reports to state compensation authorities. Similarly, insuring agencies may require reports for their records. To administer contest and award programs, management may need to file reports based on Occupational Safety and Health Administration (OSHA) recordkeeping requirements. Occupational injury and illness reports and records are now required of nearly every establishment by its management or the government.

Safety personnel are faced with two tasks—maintaining those records required by law and by their management, and maintaining records useful to an effective safety program. Unfortunately, the two are not always synonymous. A good recordkeeping system requires more data than that contained in most standard federal and state forms.

This chapter deals with both tasks of recordkeeping. However, the discussion does not cover specific legal requirements, which are beyond the scope of this volume because they differ from industry to industry and state to state and are always subject to change. The appropriate federal and state authorities need to be contacted to obtain the most current requirements. An outline of the general recordkeeping regulations under the OSHAct is presented in this chapter. Because the recordkeeping requirements are subject to change, contact an OSHA or Bureau of Labor Statistics (BLS) regional office for the latest information.

Although this chapter covers injuries and illnesses occurring to employees while on the job, a standard for off-the-job injuries to employees (ANSI Z16.3) is briefly summarized in the off-the-job section. The standard itself should be consulted by the safety personnel concerned with that aspect of the overall safety program.

The first section explains general recordkeeping systems and contains sample forms and recommendations for establishing a good system.

ACCIDENT RECORDS

Records of accidents and injuries are essential to maintain efficient and successful safety programs, just as records of production, costs, sales, and profits and losses are essential to efficient and successful business operations. Records supply the information necessary to transform haphazard, costly, ineffective safety work into a planned safety program that controls both the conditions and the acts that contribute to accidents. Good recordkeeping is the foundation of a scientific approach to occupational safety.

Uses of Records

A good recordkeeping system can help the safety professional in the following ways:

1. Provide safety personnel with the means for an objective evaluation of their accident problems and with a measurement of the overall progress and effectiveness of their safety program.
2. Identify high accident rate units, plants, or departments and problem areas so extra effort can be made in those areas.
3. Provide data for an analysis of accidents and illnesses pointing to specific causes or circumstances, which can then be attacked by specific countermeasures.
4. Create interest in safety among supervisors by furnishing them with information about their departments' accident experience.
5. Provide supervisors and safety committees with hard facts about their safety problems so their efforts can be concentrated.
6. Measure the effectiveness of individual countermeasures and determine if specific programs are doing the job they were designed to do.
7. Assist management in performance evaluation.

Recordkeeping Systems

The system presented in this section is a model that can be used to provide the basic items necessary for good recordkeeping. It is designed to dovetail with the present recordkeeping requirements of the OSHAct and attempts to avoid a duplication of effort on the part of personnel responsible for keeping records and filing reports. Some of the forms presented in this section are also constructed with modern data-processing methods in mind. In general, a self-coding check-off form can save time for both the person who fills out the report and the person who is responsible for tabulating and processing the data.

A well-designed form takes into account the person who will fill it out and the way in which the forms will be processed. An accident report should accomplish three things: establish all causes contributing to the accident; reveal questions the investigator should ask to determine all environmental and human causes; and provide a means of accumulating accident data. Such a form is more likely to be filled out accurately and will present fewer problems for those who process and analyze the data. Care in the choice and design of forms will pay dividends in better, more reliable data. Centralized coding of data (e.g., in the safety office) has advantages as well. It increases reliability of the data and makes it possible to change the coding system without retraining as many people. A few coders who work with the system a great deal will have a better understanding of the codes and coding procedures than will supervisors who may code only a few reports each year.

The recordkeeping system in this section is not the only way to keep records, but rather furnishes an example. The accident problems of individual establishments are unique and no one form or set of forms can provide every establishment with all the data for solving all of its individual problems. A system that does a good job of collecting the basic facts, however, makes it easier later on to pinpoint the data relating to a specific problem.

The following sections deal with occupational injuries and illnesses. Property-damage accidents are covered in Chapter 12. Nonemployee accidents are covered in Chapter 21. Specific recordkeeping requirements were explained in Chapter 2, Regulatory History and Compliance.

Accident Reports and Injury Records

To be effective, preventive measures must be based on complete and unbiased knowledge of the causes of accidents. The primary purpose of an accident report is to obtain such information but not to fix blame. The completeness and accuracy of the entire accident record system depend upon information in the individual accident reports. Therefore, management must be sure the forms and their purpose are understood by those who must fill them out. Such employees should be given necessary training or instruction.

The first aid report. The collection of injury data generally begins in the first aid department. The first aid attendant or nurse fills out a report for each new case. Copies are sent to the safety department or safety committee, the worker's supervisor, and other departments as management may wish (Figure 13–1).

The first aid attendant or the nurse should know enough about accident analysis, recordkeeping, and investigation to be able to record the principal facts about each case. The company's occupational physician also should be informed of the basic rules for classifying cases. At times, his or her repeat treatment and opinion of the seriousness of an injury may be necessary to record the case accurately and to help determine if the specific injury could have been caused by this incident.

Accident investigation report. It is recommended that the supervisor make a detailed report about each accident, even when only a minor injury or no injury is the result. For purposes of OSHA summaries, only those reports that meet the minimum severity level need to be separated and tallied. Minor injuries occur in greater numbers than serious injuries, and records of these injuries can help to pinpoint problem areas. By working to alleviate these problems, workers and management can prevent many serious injuries. Minor injuries should not be regarded lightly, however. Complications may arise from these injuries, and their result can be quite serious.

Supervisors should complete the supervisor's accident investigation report form as soon as possible after an accident occurs. They should send copies of these reports to the safety department and to other designated persons. Information concerning activities and conditions that preceded an occurrence is important to prevent future accidents. This information is particularly difficult to get unless it is obtained promptly after the accident occurs.

Generally, analyses of accidents are made only periodically, and often long after the accidents have occurred. Be-

```
Case No. _____                          Date _____

                        First Aid Report

  Name _____  Department _____

  Occupation _____  Supervisor _____

  Date of                    a.m.  Date                      a.m.
  Occurrence _____ Time ____ p.m.  Treated _____ Time ____ p.m.

  Nature of
  Injury or Illness _____

  Treatment Given _____

  Sent:    __ Back to Work    __ To Doctor    __ Home   __ Hospital

  Estimated lost workdays:  ___ away from work  ___ restricted work activity

  Employee's description of Occurrence _____

  _____

  _____

  Report prepared by: _____  Signed:_____
```

Figure 13–1. This First Aid Report (4 x 6 in. or 10 x 16 cm) is prepared by the first aid attendant at the time an injured or ill person comes for treatment. A report serves as a record and permits quick tabulation of such data as department, occupation, and the key facts of the occurrence.

cause at that time workers may find it impossible to recall the precise details of an accident, supervisors and others must record details accurately and completely at once or they may be lost forever.

All information may not be available at the time the accident report is being filled out. As a result, items such as total time lost and dollar amount can be added later. This should not, however, prevent the other items from being answered immediately after the accident occurs.

Three different supervisor's report forms are presented. They fulfill all of the information requirements of the present OSHA 101 form. The forms also include questions in addition to those contained in the present OSHA 101 form (Figure 13–8). These questions ask for additional basic data that should be known about each accident.

The first supervisor's report form (Figure 13–2) is an open-ended, narrative type of form. The next forms (Figures 13–3 and –4) are self-coding to allow keypunching of data items directly from the form without the extra step of recoding this information for data-processing equipment. By using self-coding forms, data-processing equipment can easily be used to enter the information into the system and to produce a variety of summary reports (such as summaries by depart-

ment and by type of accident). Detailed crosstabulations thus can be produced with little effort.

For definitions of the terms regarding severity of injury, what cases are recordable, and the like, please consult the OSHA sections of this chapter. For further clarification of OSHA definitions, see the guidelines section of this chapter. Managers also may want to consult federal or state authorities (if their state has an approved plan in effect).

Employee injury and illness record. After cases are closed, the first aid report and the supervisor's report are filed by agency of injury (type of machine, tool, material, etc.), type of accident, or another factor that will facilitate use of the reports for accident prevention. Another form, therefore, must be used to record the injury experience of individual employees (Figure 13–5). (*Agency* means "undesired and unintended characteristics" in this context.)

This form helps supervisors remember the experience of individual employees. Particularly in large establishments or plants where supervisors have many people working for them, they may not remember the total number of injuries—especially if the injuries are minor—suffered by individual employees. The employee injury card, therefore, fills a real need. It provides space to record such factors as date of

ACCIDENT INVESTIGATION REPORT

CASE NUMBER

COMPANY _____ ADDRESS _____

DEPARTMENT _____ LOCATION (if different from mailing address) _____

| 1. NAME of INJURED | 2. SOCIAL SECURITY NUMBER | 3. SEX ☐ M ☐ F | 4. AGE | 5. DATE of ACCIDENT |

| 6. HOME ADDRESS | 7. EMPLOYEE'S USUAL OCCUPATION | 8. OCCUPATION at TIME of ACCIDENT |

9. LENGTH of EMPLOYMENT
☐ Less than 1 mo.   ☐ 6 mos. to 5 yrs.
☐ 1-5 mos.   ☐ More than 5 yrs.

10. TIME in OCCUP. at TIME of ACCIDENT
☐ Less than 1 mo.   ☐ 6 mos. to 5 yrs.
☐ 1-5 mos.   ☐ More than 5 yrs.

11. EMPLOYMENT CATEGORY
☐ Regular, full-time   ☐ Temporary   ☐ Nonemployee
☐ Regular, part-time   ☐ Seasonal

12. CASE NUMBERS and NAMES of OTHERS INJURED in SAME ACCIDENT

13. NATURE of INJURY and PART of BODY

14. NAME and ADDRESS of PHYSICIAN

15. NAME and ADDRESS of HOSPITAL

16. TIME of INJURY
A. _____ A.M. / P.M.
B. Time within shift
C. Type of shift

17. SEVERITY of INJURY
☐ Fatality
☐ Lost workdays—days away from work
☐ Lost workdays—days of restricted activity
☐ Medical treatment
☐ First aid
☐ Other, specify _____

18. SPECIFIC LOCATION of ACCIDENT

ON EMPLOYER'S PREMISES? ☐ Yes   ☐ No

19. PHASE OF EMPLOYEE's WORKDAY at TIME of INJURY
☐ During rest period   ☐ Entering or leaving plant
☐ During meal period   ☐ Performing work duties
☐ Working overtime.   ☐ Other _____

20. DESCRIBE HOW the ACCIDENT OCCURRED

21. ACCIDENT SEQUENCE. Describe in reverse order of occurrence events preceding the injury and accident. Starting with the injury and moving backward in time, reconstruct the sequence of events that led to the injury.

A. Injury Event _____

B. Accident Event _____

C. Preceding Event #1 _____

D. Preceding Event #2, #3, etc. _____

Figure 13–2. Accident Investigation Report (8 x 11 in. or 22 x 28 cm) provides a record of contributing circumstances to provide a basis for specific remedial action. Users should be trained to properly fill it out.

22. TASK and ACTIVITY at TIME of ACCIDENT

   A. General type of task _____

   B. Specific activity _____

   C. Employee was working:

   ☐ Alone   ☐ With crew or fellow worker   ☐ Other, specify _____

23. POSTURE of EMPLOYEE

_____

24. SUPERVISION at TIME of ACCIDENT

   ☐ Directly supervised   ☐ Not supervised

   ☐ Indirectly supervised   ☐ Supervision not feasible

25. CAUSAL FACTORS. Events and conditions that contributed to the accident. Include those identified by use of the Guide for Identifying Causal Factors and Corrective Actions.

_____

_____

_____

_____

_____

_____

_____

_____

_____

_____

_____

26. CORRECTIVE ACTIONS. Those that have been, or will be, taken to prevent recurrence. Include those indentified by use of the Guide for Identifying Causal Factors and Corrective Actions.

_____

_____

_____

_____

_____

_____

_____

_____

_____

_____

_____

_____

PREPARED BY _____

TITLE _____

DEPARTMENT_____ DATE _____

Developed by the National Safety Council

APPROVED _____

TITLE _____ DATE _____

APPROVED _____

TITLE _____ DATE _____

Figure 13–2. *(Concluded.)*

SERVICE NO. (NSC)
▶ 1-9 _____

CASE OR FILE NO.
▶ 10-15 _____

SUPPLEMENTARY RECORD OF
OCCUPATIONAL INJURIES AND ILLNESSES

OSHA No. 101   NSC revision
(Meets OSHA requirements
when Instruction 1. has
been followed.)

THIS REPORT IS

▶ 16, 1 ☐ First report   2 ☐ Revised report

EMPLOYER

1. NAME _____

2. MAIL ADDRESS _____

3. LOCATION, if different from mail address _____

INJURED OR ILL EMPLOYEE

4. NAME _____

   SOCIAL SECURITY NO. _____

▶ EMPLOYEE NO. 17-26 _____

5. HOME ADDRESS _____

▶ 6. AGE 27-28 _____

▶ 7. SEX 29, 1 ☐ Male   2 ☐ Female

▶ 8. OCCUPATION (specify) _____

   30-31, 01 ☐ Manager, official, proprietor
   02 ☐ Professional, technical
   03 ☐ Foreman, supervisor
   04 ☐ Sales worker
   05 ☐ Clerical worker
   06 ☐ Craftsman—construction
   07 ☐ Craftsman—other
   08 ☐ Machinist
   09 ☐ Mechanic
   10 ☐ Operative (production worker)
   11 ☐ Motor vehicle driver
   12 ☐ Laborer
   13 ☐ Service worker
   14 ☐ Agricultural worker
   15 ☐ Other
   16 ☐ Unknown

9. DEPARTMENT _____
   (Enter the name of department or division in which the injured person is regularly employed.)

CLASSIFICATION OF CASE

A. INJURY OR ILLNESS (see code on Log, OSHA No. 100)
▶ 32, 1 ☐ Injury (10)
   2 ☐ Occupational skin disease or disorder (21)
   3 ☐ Dust disease of the lungs (pneumoconioses) (22)
   4 ☐ Respiratory conditions due to toxic agents (23)
   5 ☐ Poisoning (systemic effects of toxic materials) (24)
   6 ☐ Disorder due to physical agents (other than toxic materials) (25)
   7 ☐ Disorder due to repeated trauma (26)
   8 ☐ All other occupational illnesses (29)

B. EXTENT OF INJURY OR ILLNESS
▶ 33, 1 ☐ Fatality
   2 ☐ Lost workday case
   3 ☐ Nonfatal case without lost workdays

▶ C. Number of workdays lost 34-36 _____

D. Permanently transferred or terminated
▶ 37, 1 ☐ Yes   2 ☐ No

INSTRUCTIONS

1. *Type or print the narrative where requested.*
2. *Check the one box which most clearly describes each narrative statement.*
3. *See also original OSHA No. 101 for more details.*

THE ACCIDENT OR EXPOSURE TO OCCUPATIONAL ILLNESS

10. PLACE OF ACCIDENT OR EXPOSURE (mail address) _____

11. WHERE DID ACCIDENT OR EXPOSURE OCCUR?
   a. On employer premises
▶ 38, 1 ☐ Yes   2 ☐ No   3 ☐ Unknown
   b. Place (specify) _____
▶ 39-40, 01 ☐ Office
   02 ☐ Plant, mill
   03 ☐ Shipping, receiving, warehouse
   04 ☐ Maintenance shop
   05 ☐ General or public area of employer premises (corridor, washroom, lunchroom, parking lot, etc.)
   06 ☐ Retail establishment (store, restaurant, gasoline station, etc.)
   07 ☐ Farm
   08 ☐ Motor vehicle accident
   09 ☐ Other
   10 ☐ Unknown

12. WHAT WAS THE EMPLOYEE DOING WHEN INJURED? (Be specific)
   _____

   a. Task performed at time of accident
▶ 41-42, 01 ☐ Operating machine
   02 ☐ Operating hand tool (power or nonpower)
   03 ☐ Materials handling
   04 ☐ Maintenance & repair—machinery
   05 ☐ Maintenance & repair—building & equipment
   06 ☐ Motor vehicle driver, operator or passenger
   07 ☐ Office and sales tasks, except above
   08 ☐ Service tasks, except above
   09 ☐ Other
   10 ☐ Not performing task
   11 ☐ Unknown

   b. Activity at time of accident
▶ 43-44, 01 ☐ Climbing
   02 ☐ Driving
   03 ☐ Jumping
   04 ☐ Kneeling
   05 ☐ Lying down
   06 ☐ Lifting
   07 ☐ Reaching, stretching
   08 ☐ Riding
   09 ☐ Running
   10 ☐ Sitting
   11 ☐ Standing
   12 ☐ Walking
   13 ☐ Other
   14 ☐ Unknown

Figure 13–3. A self-coding supplementary record of occupational injuries and illnesses.

13. **HOW DID THE ACCIDENT OCCUR?** (Describe fully the events)

_____

_____

_____

_____

a. **AGENCY.** (Object or substance involved)

   ACCIDENT AGENCY (1st column). The first object or substance involved in accident sequence.

   INJURY AGENCY (2nd column). The agency inflicting the injury. See also section 15.

   (Example: Worker fell from ladder and struck head on machine. Check "Ladder" under accident and check "Machine" under injury.)

ACCIDENT	**INJURY**	(Check one box in each column)
45-46, 01 ☐	47-48, 01 ☐	Machine
02 ☐	02 ☐	Conveyor, elevator, hoist
03 ☐	03 ☐	Vehicle
04 ☐	04 ☐	Electrical apparatus
05 ☐	05 ☐	Hand tool
06 ☐	06 ☐	Chemical
07 ☐	07 ☐	Working surface, bench, table, etc.
08 ☐	08 ☐	Floor, walking surface
09 ☐	09 ☐	Bricks, rocks, stones
10 ☐	10 ☐	Box, barrel, container (empty or full)
11 ☐	11 ☐	Door, window, etc.
12 ☐	12 ☐	Ladder
13 ☐	13 ☐	Lumber, woodworking materials
14 ☐	14 ☐	Metal
15 ☐	15 ☐	Stairway, steps
16 ☐	16 ☐	Other
17 ☐	17 ☐	Unknown
18 ☐	18 ☐	None

b. **ACCIDENT TYPE.** (First event in the accident sequence)

49-50, 01 ☐	Fall from elevation
02 ☐	Fall on same level
03 ☐	Struck against
04 ☐	Struck by
05 ☐	Caught in, under or between
06 ☐	Rubbed or abraded
07 ☐	Bodily reaction
08 ☐	Overexertion
09 ☐	Contact with electrical current
10 ☐	Contact with temperature extremes
11 ☐	Contact with radiations, caustics, toxic and noxious substances
12 ☐	Public transportation accident
13 ☐	Motor vehicle accident
14 ☐	Other
15 ☐	Unknown

This space may be used for additional information.

OCCUPATIONAL INJURY OR ILLNESS

14. **DESCRIBE THE INJURY OR ILLNESS** in detail and indicate the part of the body affected.

_____

_____

_____

a. **NATURE OF INJURY OR ILLNESS.** (Check most serious one)

51-52, 01 ☐	Amputation
02 ☐	Burn and scald (heat)
03 ☐	Burn (chemical)
04 ☐	Concussion
05 ☐	Crushing injury
06 ☐	Cut, laceration, puncture, abrasion
07 ☐	Fracture
08 ☐	Hernia
09 ☐	Bruise, contusion
10 ☐	Occupational illness
11 ☐	Sprain, strain
12 ☐	Other

b. **PART OF BODY.** (Check most serious one)

53-54, 01 ☐	Eyes
02 ☐	Head, face, neck
03 ☐	Back
04 ☐	Trunk (except back, internal)
05 ☐	Arm
06 ☐	Hand and wrist
07 ☐	Fingers
08 ☐	Leg
09 ☐	Feet and ankles
10 ☐	Toes
11 ☐	Internal and other

15. **NAME THE OBJECT OR SUBSTANCE WHICH DIRECTLY INJURED THE EMPLOYEE.** Also check one box in injury column under 13a.

_____

_____

_____

16. **DATE OF INJURY OR INITIAL DIAGNOSIS OF OCCUPATIONAL ILLNESS.**

a. **MONTH**

55-56, 01 ☐	Jan.	07 ☐	July
02 ☐	Feb.	08 ☐	Aug.
03 ☐	March	09 ☐	Sept.
04 ☐	April	10 ☐	Oct.
05 ☐	May	11 ☐	Nov.
06 ☐	June	12 ☐	Dec.

b. **DATE OF MONTH** 57-58 _____

17. **DID EMPLOYEE DIE?**

   59, 1 ☐ Yes   Date of Death_____

   2 ☐ No

OTHER

18. **NAME AND ADDRESS OF PHYSICIAN**_____

_____

19. **IF HOSPITALIZED, NAME AND ADDRESS OF HOSPITAL**_____

_____

DATE OF REPORT_____

PREPARED BY_____

OFFICIAL POSITION_____

Figure 13-3. *(Concluded.)*

ZURN / *a step ahead of tomorrow*

DIV. NAME: _____

SITE CODE: _____

OSHA FILE NO.: (Yr.-No.)_____

SUPERVISOR'S ACCIDENT INVESTIGATION REPORT

THIS FORM MUST BE COMPLETED AS SOON AFTER ACCIDENT AS POSSIBLE.

EMPLOYEE(S) _____ CLOCK NO. _____ ACCIDENT DATE ___/___/___

JOB TITLE/CRAFT _____ DEPT./SITE NAME _____

AGE GROUP	TYPE OF ACCIDENT	NATURE OF INJURY	PART OF BODY	UNSAFE CONDITION	UNSAFE ACT	UNSAFE PERSONAL FACTOR
UNDER 20	01 OVEREXERTION	26 BACK STRAIN-SPRAIN	51 HEAD	1 IMPROPERLY GUARDED	1 OPERATING WITHOUT AUTHORITY	1 IMPROPER ATTITUDE
20-29	02 FALL - SAME LEVEL	27 OTHER STRAIN SPRAIN	52 EYE / 53 EAR	2 SAFETY DEVICES INOPERATIVE	2 WORKING ON MOVING OR DANGEROUS EQ	2 LACK OF KNOWLEDGE OR SKILL
30-39	03 FALL - DIFF LEVEL	29 DISLOCATION	54 FACE (OTHER)	3 DEFECTIVE	3 OPERATING AT UNSAFE SPEEDS	3 DEFECTIVE EYESIGHT
40-49	04 STRUCK AGAINST	30 FRACTURE	55 NECK	4 HAZARDOUS ARRANGEMENT	4 MAKING SAFETY DEVICES INOPERATIVE	4 DEFECTIVE HEARING
50-59	05 STRUCK BY	31 CONTUSION	56 SHOULDER	5 IMPROPER ILLUMINATION	5 TAKING UNSAFE POSITION OR POSTURE	5 FATIGUE
60 & OVER	06 CAUGHT BETWEEN	32 AMPUTATION	57 ARM / 58 ELBOW	6 IMPROPER VENTILATION	6 HANDLING MATERIALS INCORRECTLY - MANUAL	6 MUSCULAR WEAKNESS
_____ A M	07 VEHICLE	33 OPEN WOUND	59 WRIST / 60 HAND	7 LACK OF SUITABLE PERSONAL PROT EQ	7 USING DEFECTIVE TOOLS	7 HEART WEAKNESS
_____ P M	10 ELECTRICAL	34 HEART ATTACK	61 FINGER	8 UNSAFE DRESS	8 USING HANDS INSTEAD OF TOOLS	8 EXISTING HERNIA
OVERTIME	11 TEMP EXTREME	35 VISION LOSS	62 BACK / 63 CHEST	9 HAZARDOUS DUST GASES OR FUMES	9 UNSAFE LOADING	9 INTOXICATED
DAY OF WEEK	12 REPETITIVE MOTION	36 BURN	64 LUNGS	10 UNCLASSIFIED INSUFFICIENT DATA	10 FAILURE TO USE PERSONAL PROT EQ	10 UNCLASSIFIED INSUFFICIENT DATA
S M T W Th F Sa	13 RADIATION	40 ASPHYXIA	65 ABDOMEN	11 NO UNSAFE CONDITION	11 DISTRACTING, TEASING OR HORSEPLAY	11 NO UNSAFE PERSONAL FACTOR
TIME ON _____ YRS	14 INHALATION	41 HEARING LOSS	66 GROIN		12 NOT FOLLOWING RULES OR INSTRUCTIONS	
JOB _____ MOS	15 ABSORBTION	42 TENOSYNOVITIS	67 HIPS / 68 LEG		13 UNCLASSIFIED INSUFFICIENT DATA	
MALE	16 INGESTION	43 DERMATITIS	69 KNEE		14 NO UNSAFE ACT	
FEMALE	17 SLIP (NO FALL) TWIST	44 OCCUPATIONAL HEALTH	70 ANKLE			
OSHA RECORDABLE / FATAL	18	45 FOREIGN BODY	71 FOOT / 72 TOES			
YES / LOST TIME	19	46 MULTIPLE INJURY	73 BODY SYSTEM			
/ DR VISIT ONLY	20	47 CUMULATIVE TRAUMA	74 MULTIPLE BODY PARTS			
NO / FIRST AID	24 OTHER	49 OTHER				

DESCRIBE IN DETAIL WHAT HAPPENED (Include name of machine, machine part or section, hand tool or material involved.) _____

_____

_____

_____

REGULAR JOB? YES ☐ NO ☐ WITNESS(ES) _____

REPORT DELAYED? YES ☐ NO ☐ IF YES, WHY? _____

DESCRIBE UNSAFE ACT OR FAILURE TO ACT BY EMPLOYEE(S) OR OTHERS: _____

_____

_____

DESCRIBE ANY MECHANICAL, PHYSICAL OR ENVIRONMENTAL CONDITION THAT CONTRIBUTED TO ACCIDENT _____

_____

_____

DESCRIBE ANY PERSONAL FACTORS THAT CONTRIBUTED TO ACCIDENT _____

WHAT **APPLICABLE** PERSONAL PROTECTIVE EQUIPMENT (PPE) **WAS BEING USED**? _____

_____ DID PPE FAIL? YES ☐ NO ☐

WHAT PERSONAL PROTECTIVE EQUIPMENT **SHOULD HAVE BEEN USED**? _____

_____ WAS PROPER PPE AVAILABLE? YES ☐ NO ☐

DESCRIBE ACTION(S) TAKEN TO PREVENT RECURRENCE _____

INDICATE DATE ACTION WAS OR WILL BE TAKEN: _____ BY WHOM: _____

___/___/___

DATE OF REPORT   NAME AND TITLE OF SUPERVISOR COMPLETING REPORT

SPECIFIC COMMENTS AND RECOMMENDATIONS BY DEPARTMENT HEAD _____

_____

___/___/___

DATE   NAME OF DEPT. HEAD

DATE: _____ **SAFETY SUPERVISOR:** _____

This report should be completed and sent to Safety Dept. **no later than the next regularly scheduled workday** following the accident.

DIV./PLANT SAFETY   Form No. 80-PER, 4/89

CORP. SAFETY (FOR ALL LOST TIME INJURIES)   Form No. 80-PER, 4/89

DIV. SITE FIRST AID DEPT.   Form No. 80-PER, 4/89

EMPLOYEE FILE   Form No. 80-PER, 4/89

Figure 13–4. This Supervisor's Accident Investigation Report must be completed as soon after the accident as possible. The 4–part form has distribution indicated on the bottom. (Courtesy Zurn Industries, Inc. Corp. Safety Dept.)

```
                    INJURY AND ILLNESS RECORD OF EMPLOYEE
    Name _____  ID Number _____
                                                        Date
    Occupation _____  Department _____  Hired _____
```

Case Number	Injury or Illness	Date of Occurrence	Medical, Compensation, and Other Costs	Severity of Case			
				Fatal	Lost Workdays Involved?*		First Aid Only
					Yes	No	

*Enter number of days lost in "yes" column or check mark in "no" column.

Figure 13–5. Injury and Illness Record of Employee is a 4 x 6-in. (10 x 15-cm) card for recording injuries.

injury, classification, costs, and lost workdays.

Because of the importance of the human factor in accidents, much can be learned about accident causes from studying employee injury records. If certain employees or job classifications have frequent injuries, a study of employee working habits, physical and mental abilities, training, job assignments, working environment, and the instructions and supervision given them may reveal as much as a study of accident locations, agencies, or other factors. An increase in the frequency of accidents by an employee bears further investigation.

Filing reports. After injury reports have been used to compile monthly summaries, the incomplete reports can be kept in a temporary file for convenient reference as later information about the injuries becomes available. When injury reports are complete, they should be filed in a convenient system that permits quick access of information for special studies of accident conditions. For example, the reports can be filed by agency of the injury, by occupation of the injured person, by department, or by some similar item.

The employee injury card should be cross-referenced to the file location of the detailed accident report.

Periodic Reports

The forms discussed in the preceding paragraphs are prepared when the accidents occur; they are used to record the

accidents and preserve information about contributing circumstances. Periodically, management should summarize this information and relate it to department or plant exposure in order to evaluate safety work and to identify the principal accident causes.

Monthly summary of injuries and illnesses. Managers and supervisors should prepare a monthly summary of injuries and illnesses to reveal the current status of accident experience. This monthly summary of injury and illness cases (Figure 13–6) allows for tabulating monthly and cumulative totals and the computation of OSHA incidence rates. Space is also provided for yearly totals and rates. This form would be filled out on the basis of the individual report forms that were processed during the month or from OSHA form No. 200, Log and Summary of Occupational Injuries and Illnesses (Figure 13–7).

The monthly summary should be prepared as soon after the end of each month as the information becomes available but not later than the 20th of the following month. Because this report is primarily prepared to reveal the current status of accident experience, it is essential that this information be determined as soon as possible. If an accident report is still incomplete on the 20th of the following month because the employee has not returned to work, or if the classification of an injury is still in doubt at that time, an estimate of the

					OSHA Recordable Cases[1]							
Period	Average Number of Employees	Number of Employee-Hours Worked	Costs (Compensation, other)	Deaths	Lost Workday Cases	Nonfatal Cases Without Lost Workdays	Total Cases	Total Lost Workdays	Total Case Incidence Rate[2]	Lost Workdays Incidence Rate[2]	First Aid Cases
Jan.											
Feb.											
2 mo.											
Mar.											
3 mo.											
Apr.											
4 mo.											
May											
5 mo.											
Jun.											
6 mo.											
Jul.											
7 mo.											
Aug.											
8 mo.											
Sep.											
9 mo.											
Oct.											
10 mo.											
Nov.											
11 mo.											
Dec.											
Year											

MONTHLY SUMMARY OF INJURIES AND ILLNESSES, 19___

Company _____ Plant _____ Department _____

[1]Refer to OSHA Log for numbers of cases: Deaths, columns 1+8; Lost Workday Cases, cols. 2+9; Nonfatal Cases Without Lost Workdays, cols. 6+13; Total Lost Workdays, cols. 4+5+11+12. Total Cases equals Deaths plus Lost Workday Cases plus Nonfatal Cases Without Lost Workdays.
[2]Incidence rate is number of "Total Cases" or "Total Lost Workdays" multiplied by 200,000 and divided by "Number of Employee-Hours Worked."

Figure 13–6. Results of the safety program can be gauged from data on this Monthly Summary of Injuries and Illnesses form. Rates computed for the month, year to date (cumulative), and year permit comparisons between time periods, and departments, plants, and companies during the same period. Changes in the classification of injuries and other adjustments can be easily made.

outcome should be made by the company or consulting physician. The report must then be included with the completed cases in the monthly summary.

When definite information becomes available for estimated cases, any change in classification or OSHA lost workdays should be entered in the appropriate columns in the month of closing of the case. The adjustment is then included in the cumulative figures for the year through that month. This procedure provides reliable monthly data and an easy method of adjusting cumulative data.

Annual report. Every establishment subject to the OSHAct is obliged to post its annual summary by February 1st for 30 days. The cumulative totals on OSHA form No. 200, Log and Summary of Occupational Injuries and Illnesses, serve as the annual report. Those establishments designated as part of the BLS annual survey also must report these figures to the requesting agency.

For management purposes, however, the annual report fulfills a more direct function. Managers and others prepare monthly summaries of injuries and illnesses primarily to show the trend of safety performance during the year. However, annual reports are prepared to compare data for

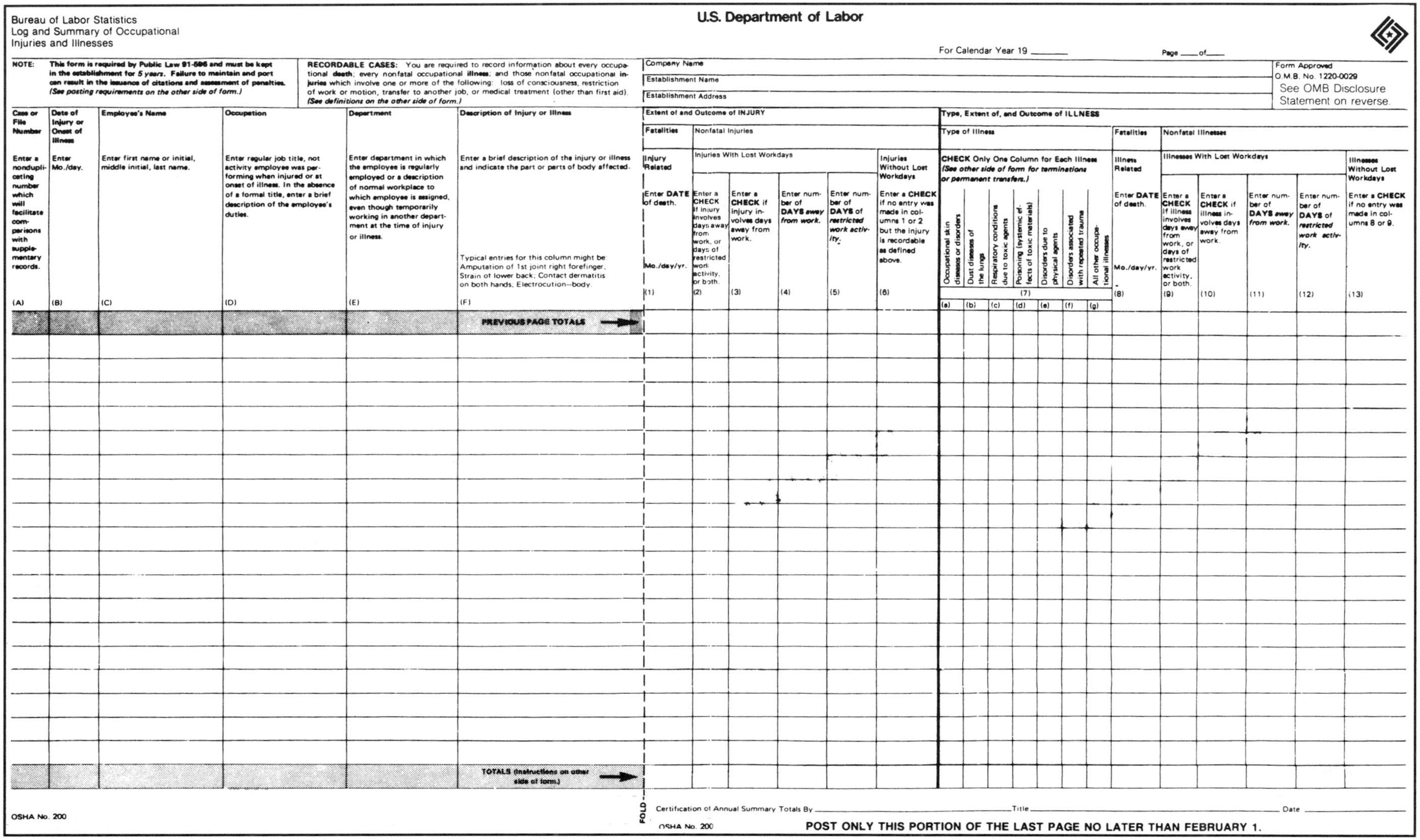

Bureau of Labor Statistics
Log and Summary of Occupational Injuries and Illnesses

U.S. Department of Labor

For Calendar Year 19 ______ Page ___ of ___

NOTE: **This form is required by Public Law 91-596 and must be kept in the establishment for 5 years. Failure to maintain and post can result in the issuance of citations and assessment of penalties.** *(See posting requirements on the other side of form.)*

RECORDABLE CASES: You are required to record information about every occupational **death**; every nonfatal occupational **illness**; and those nonfatal occupational **injuries** which involve one or more of the following: loss of consciousness, restriction of work or motion, transfer to another job, or medical treatment (other than first aid). *(See definitions on the other side of form.)*

Company Name
Establishment Name
Establishment Address

Form Approved
O.M.B. No. 1220-0029
See OMB Disclosure Statement on reverse.

Case or File Number	Date of Injury or Onset of Illness	Employee's Name	Occupation	Department	Description of Injury or Illness
Enter a nonduplicating number which will facilitate comparisons with supplementary records.	Enter Mo./day.	Enter first name or initial, middle initial, last name.	Enter regular job title, not activity employee was performing when injured or at onset of illness. In the absence of a formal title, enter a brief description of the employee's duties.	Enter department in which the employee is regularly employed or a description of normal workplace to which employee is assigned, even though temporarily working in another department at the time of injury or illness.	Enter a brief description of the injury or illness and indicate the part or parts of body affected. Typical entries for this column might be: Amputation of 1st joint right forefinger; Strain of lower back; Contact dermatitis on both hands; Electrocution—body.
(A)	(B)	(C)	(D)	(E)	(F)
					PREVIOUS PAGE TOTALS →
					TOTALS (Instructions on other side of form.) →

Extent of and Outcome of INJURY

Fatalities: Injury Related	Nonfatal Injuries — Injuries With Lost Workdays				Injuries Without Lost Workdays
Enter DATE of death. Mo./day/yr.	Enter a CHECK if injury involves days away from work, or days of restricted work activity, or both.	Enter a CHECK if injury involves days away from work.	Enter number of DAYS away from work.	Enter number of DAYS of restricted work activity.	Enter a CHECK if no entry was made in columns 1 or 2 but the injury is recordable as defined above.
(1)	(2)	(3)	(4)	(5)	(6)

Type, Extent of, and Outcome of ILLNESS

Type of Illness — CHECK Only One Column for Each Illness *(See other side of form for terminations or permanent transfers.)* (7)

Occupational skin diseases or disorders	Dust diseases of the lungs	Respiratory conditions due to toxic agents	Poisoning (systemic effects of toxic materials)	Disorders due to physical agents	Disorders associated with repeated trauma	All other occupational illnesses
(a)	(b)	(c)	(d)	(e)	(f)	(g)

Fatalities: Illness Related	Nonfatal Illnesses — Illnesses With Lost Workdays				Illnesses Without Lost Workdays
Enter DATE of death. Mo./day/yr.	Enter a CHECK if illness involves days away from work, or days of restricted work activity, or both.	Enter a CHECK if illness involves days away from work.	Enter number of DAYS away from work.	Enter number of DAYS of restricted work activity.	Enter a CHECK if no entry was made in columns 8 or 9.
(8)	(9)	(10)	(11)	(12)	(13)

FOLD

Certification of Annual Summary Totals By ______ Title ______ Date ______

OSHA No. 200

POST ONLY THIS PORTION OF THE LAST PAGE NO LATER THAN FEBRUARY 1.

Figure 13–7. Log of Occupational Injuries and Illnesses, OSHA Form No. 200, is used to record injuries or illnesses that result in fatalities, lost workdays, require medical treatment, involve loss of consciousness, or restrict work or motion. Complete instructions for using the Log are on its reverse side (not shown here). The form measures 15 1/4 x 9 1/4 in. (34 x 56 cm). Because forms are subject to change, be sure to ask your OSHA Area Office for the latest forms.

Bureau of Labor Statistics
Supplementary Record of
Occupational Injuries and Illnesses

U.S. Department of Labor

This form is required by Public Law 91-596 and must be kept in the establishment for *5 years*. Failure to maintain can result in the issuance of citations and assessment of penalties.	Case or File No.	Form Approved O.M.B. No. 1220-0029

Employer

1. Name

2. Mail address *(No. and street, city or town, State, and zip code)*

3. Location, if different from mail address

Injured or Ill Employee

4. Name *(First, middle, and last)* — Social Security No.

5. Home address *(No. and street, city or town, State, and zip code)*

6. Age   7. Sex: *(Check one)*   Male ☐   Female ☐

8. Occupation *(Enter regular job title, not the specific activity he was performing at time of injury.)*

9. Department *(Enter name of department or division in which the injured person is regularly employed, even though he may have been temporarily working in another department at the time of injury.)*

The Accident or Exposure to Occupational Illness

If accident or exposure occurred on employer's premises, give address of plant or establishment in which it occurred. Do not indicate department or division within the plant or establishment. If accident occurred outside employer's premises at an identifiable address, give that address. If it occurred on a public highway or at any other place which cannot be identified by number and street, please provide place references locating the place of injury as accurately as possible.

10. Place of accident or exposure *(No. and street, city or town, State, and zip code)*

11. Was place of accident or exposure on employer's premises?   Yes ☐   No ☐

12. What was the employee doing when injured? *(Be specific. If he was using tools or equipment or handling material, name them and tell what he was doing with them.)*

13. How did the accident occur? *(Describe fully the events which resulted in the injury or occupational illness. Tell what happened and how it happened. Name any objects or substances involved and tell how they were involved. Give full details on all factors which led or contributed to the accident. Use separate sheet for additional space.)*

Occupational Injury or Occupational Illness

14. Describe the injury or illness in detail and indicate the part of body affected. *(E.g., amputation of right index finger at second joint; fracture of ribs; lead poisoning; dermatitis of left hand, etc.)*

15. Name the object or substance which directly injured the employee. *(For example, the machine or thing he struck against or which struck him; the vapor or poison he inhaled or swallowed; the chemical or radiation which irriated his skin; or in cases of strains, hernias, etc., the thing he was lifting, pulling, etc.)*

16. Date of injury or initial diagnosis of occupational illness   17. Did employee die? *(Check one)*   Yes ☐   No ☐

Other

18. Name and address of physician

19. If hospitalized, name and address of hospital

Date of report | Prepared by | Official position

OSHA No. 101 (Feb. 1981)

Figure 13–8. Supplementary Record of Occupational Injuries and Illnesses, OSHA Form No. 101, gives details of each recordable occupational injury or illness. Records must be available for examination by representatives of the U.S. Department of Labor and the Department of Health and Human Services and states accorded jurisdiction under the OSHAct. Records must be kept at least five years following the calendar year to which they relate.

the longer periods with the experience of previous years and with the experience of similar organizations and of the industry as a whole.

Monthly injury rates, especially in smaller companies, often show wide variations, making it difficult to evaluate safety performance correctly. For example, a small company may have only two or three injuries in a year. Thus, for the months in which these injuries occur, the rates will jump to extreme highs; in the other months, the rates will be zero. These variations will be smoothed out in annual totals, and the rate for the longer period will become more significant.

Annual reports should be prepared as soon after the close of the year as information becomes available, but again not later than the 20th of the following month. In some cases, an injury report may still be incomplete 20 days after the close of the year because the employee has not returned to work. In others, the classification of an injury may be in doubt at that time. In both these instances, an estimate of the outcome should be made by the company physician and the report included with the completed cases.

When annual records are closed, OSHA requires that the anticipated future lost workdays (both days away from work and days of restricted work activity) be recorded. Then, when a case becomes final, the records can be corrected to show the actual days lost.

If lost workdays that occurred in one year were included in the record for the next year, the latter year's experience would be biased and the record would fail to give a true picture of the severity of injuries.

Use of Reports

Reports to management. Management is increasingly interested in the accident experience of its company. Therefore, monthly and other periodic summary reports showing the results of the safety program should be furnished to the responsible executive. Such reports do not need to contain details or technical language. They can be supplemented by simple charts or graphs to compare current accident rates with those of the preceding period and the rates of other companies in the industry.

In a large company, departmental data help the executive visualize accident experience in various plant operations and provide a yardstick for better evaluation of progress made in the elimination of accidents. It can be particularly valuable to compare cost figures, if they can be obtained, for different periods.

Bulletins to supervisors. Supervisors are primarily interested in their own department and workers. One of the most effective ways to create and maintain the interest of supervisors in accident prevention is to keep them informed about the accident records of their departments. Department injury rates based on sufficient amounts of exposure reflect the effectiveness of the supervisors' safety activities. Because interest increases with knowledge, management can send supervisors bulletins containing analyses of the principal causes of accidents in each department. These bulletins not only can maintain supervisors' interest at a high level but

provide the supervisors with information to help them reduce injuries even further.

The agenda for employee safety meetings should emphasize information about outstanding injury and illness problems, frequent unsafe practices, hazardous types of equipment, and similar data disclosed by analysis of the accidents that have occurred in the department and plant.

Bulletin board publicity. Posting a variety of materials on bulletin boards is one of the best ways to maintain the interest of employees in safety. Accident records furnish many items, such as the following:
- no-injury records
- unusual accidents
- frequent causes of accidents
- charts showing reductions in accidents
- simple tables comparing departmental records
- standings in contests.

Reports to National Safety Council. The Council sponsors two recognition programs for its members requiring periodic reports of their occupational injury and illness experience. The Occupational Safety/Health Award Program, open to all employer members, is a noncompetitive evaluation of a company's annual experience. Awards are given if the records meet predetermined criteria.

The annual reports are tabulated to determine injury and illness incidence rates by industry. An annual "Work Injury and Illness Rates" pamphlet containing these incidence rates by industry is published and distributed to members. Each company can then compare its experience with the average experience of other companies from the same industry that participate in the Occupational Safety/Health Award Program. These rates (along with data from the BLS, state and national vital statistics authorities, state compensation authorities, and other federal, state, and private agencies) are shown in *Accident Facts,* the Council's annual statistical publication.

National Safety Council industrial employer members may also compete in more than 20 employee safety contests administered by the Council. Monthly bulletins are sent to contestants so that they can compare their experience with other competitors. Awards are made to the leaders of the individual contests at the end of each contest year.

The Concept of Bilevel Reporting

As mentioned earlier in this chapter, each company has different accident problems from those of establishments not only in other industries but often within the same industry. No individual form or set of forms can possibly include all of the information necessary to fully investigate the causes of all accidents. With this in mind, and because long forms are rarely welcomed or completed accurately, the concept of bilevel reporting has arisen.

The basic idea behind bilevel reporting is that further details about specific types of accidents are required in addition to the general information contained in the standard report form (such as the Supervisor's Accident Report Form). To obtain this additional data, a supplementary form, con-

taining a few specific questions about the accident type under investigation, is prepared and made available. This supplementary form is then filled out and attached to the regular report, but *only* for those accidents the investigator needs to analyze in detail.

When a sufficient number of the bilevel forms have been collected, the supplementary form is discontinued and the results can be analyzed. This method requires only a minimum of time from the persons who fill out the forms in order to generate useful information. Several bilevel forms, each a different color for easy handling, can be used at any one time. They can be discontinued when their job is done and replaced by other supplementary forms, while the basic report form remains unchanged and in use.

OSHA RECORDKEEPING REQUIREMENTS

The information in this section explains briefly the occupational injury and illness recording and reporting requirements of the OSHAct and Title 29 of the *Code of Federal Regulations*, Part 1904. It is taken from a June 1986 publication of the U.S. Department of Labor, BLS, titled "A Brief Guide to Recordkeeping Requirements for Occupational Injuries and Illnesses."

Note: The requirements and definitions are subject to change, so safety personnel should contact their regional OSHA office for the latest information.

In addition to the requirements of 29 *CFR*, 1904, many specific OSHA standards and regulations require maintenance and retention of records of medical surveillance, exposure monitoring, inspections, accidents, and other activities and incidents, and for the reporting of certain information to employees and to OSHA. These additional requirements are not covered in this section.

Employers Subject to the Recordkeeping Requirements of the OSHAct

The recordkeeping requirements of the OSHAct apply to private sector employers in all states, the District of Columbia, Puerto Rico, the Virgin Islands, American Samoa, Guam, and the Trust Territories of the Pacific Islands.

Employers with 11 or more employees (at any one time in the previous calendar year) in the following industries must keep OSHA records. The industries are identified by name and by the appropriate Standard Industrial Classification (SIC) code:

- agriculture, forestry, and fishing (SICs 01–02 and 07–09)
- oil and gas extraction (SICs 13 and 1477)
- construction (SICs 15–17)
- manufacturing (SICs 20–39)
- transportation and public utilities (SICs 41–42 and 44–49)
- wholesale trade (SICs 50–51)
- building materials and garden supplies (SIC 52)
- general merchandise and food stores (SICs 53 and 54)
- hotels and other lodging places (SIC 70)
- repair services (SICs 75 and 76)
- amusement and recreation services (SIC 79)

- health services (SIC 80).

If employers in any of the industries listed above have more than one establishment with combined employment of 11 or more employees, records must be kept for each establishment.

Employers in other industries are *normally* exempt from OSHA recordkeeping, as are employers with no more than 10 full-time or part-time employees at any one time in the previous calendar year.

Even exempt employers, however, must comply with OSHA standards and display the OSHA poster. They must also report to OSHA within 48 hours any accident that results in one or more fatalities or the hospitalization of five or more employees. Also, some state safety and health laws may require regularly exempt employers to keep injury and illness records, and some states have more stringent catastrophic reporting requirements.

The following employers and individuals *never* have to keep OSHA injury and illness records:

- self-employed individuals
- partners with no employees
- employers of domestics in the employers' private residence for the purpose of housekeeping or child care, or both
- employers engaged in religious activities concerning the conduct of religious services or rites.

Employees engaged in such activities include clergy, choir members, organists and other musicians, ushers, and the like. However, records of injuries and illnesses occurring to employees while performing secular activities must be kept. Recordkeeping is also required for employees of private hospitals and certain commercial establishments owned or operated by religious organizations.

State and local government agencies are usually exempt from OSHA recordkeeping. However, in certain states, agencies of state and local governments are required to keep injury and illness records in accordance with state regulations. Employers subject to injury and illness recordkeeping requirements of other federal safety and health regulations are not exempt from OSHA recordkeeping. However, records used to comply with other federal recordkeeping obligations may also be used to satisfy the OSHA recordkeeping requirements. The forms used must be equivalent to the log and summary (OSHA No. 200) and the supplementary record (OSHA No. 101).

OSHA Recordkeeping Forms

Only two forms are used for OSHA recordkeeping. One form, the OSHA No. 200 (Figure 13–7), serves a dual purpose. It is used as the Log of Occupational Injuries and Illnesses, on which the occurrence and extent of cases are recorded during the year and as the Summary of Occupational Injuries and Illnesses, which is used to summarize the log at the end of the year to satisfy employer posting obligations. The other form, the Supplementary Record of Occupational Injuries and Illnesses, OSHA No. 101 (Figure 13–8), provides additional information on each of the cases that have been recorded on the log.

The Log and Summary of Occupational Injuries and Illnesses, OSHA No. 200. The log is used for recording and classifying occupational injuries and illnesses and for noting the extent of each case. The log consists of three parts: a descriptive section that identifies the employee and briefly describes the injury or illness, a section covering the extent of the injuries recorded, and a section on the type and extent of illnesses.

Usually, the OSHA No. 200 form is used by employers as their record of occupational injuries and illnesses. However, a private form equivalent to the log, such as a computer printout, may be used if it contains the same detail as the OSHA No. 200. It must also be as readable and comprehensible as the OSHA No. 200 to a person not familiar with the equivalent form. It is advisable that employers have private equivalents of the log form reviewed by BLS to ensure compliance with the regulations.

The portion of the OSHA No. 200 to the right of the dotted vertical line is used to summarize injuries and illnesses in an establishment for the calendar year. Every nonexempt employer who is required to keep OSHA records must prepare an annual summary for each establishment based on the information contained in the log for each establishment. The summary is prepared by totaling the column entries on the log (or its equivalent) and signing and dating the certification portion of the form at the bottom of the page.

The Supplementary Record of Occupational Injuries and Illnesses, OSHA No. 101. For every injury or illness entered on the log, it is necessary to record additional information on the supplementary record, OSHA No. 101. However, the OSHA No. 101 is not the only form that can be used to satisfy this requirement. To eliminate duplicate recording, workers' compensation, insurance, or other reports may be used as supplementary records if they contain all of the items on the OSHA No. 101 (Figure 13–9). If they do not, the missing items must be added to the substitute or included on a separate attachment.

Completed supplementary records must be present in the establishment within six workdays after the employer has received information that an injury or illness has occurred.

Location, Retention, and Maintenance of Records

If an employer has more than one establishment, a separate set of records must be maintained for each one. The recordkeeping regulations define an establishment as "a single physical location where business is conducted or where services or industrial operations are performed." Examples include a factory, mill, store, hotel, restaurant, movie theater, farm, ranch, sales office, warehouse, or central administrative office.

Under the regulations, the location of these records depends upon whether the employees are associated with a fixed establishment. The distinction between fixed and nonfixed establishments generally rests on the nature and duration of the operation and not on the type of structure in which the business is located. Generally, any operation at a given site for more than one year is considered a fixed

establishment. Also, fixed establishments are generally places where clerical, administrative, or other business records are kept.

1. *Employees associated with fixed establishments.* Records for these employees should be located as follows:
   a. Records for employees working at fixed locations, such as factories, stores, restaurants, warehouses, etc., should be kept at the work location.
   b. Records for employees who report to a fixed location but work elsewhere should be kept at the place where the employees report each day. These employees are generally engaged in activities such as agriculture, construction, transportation, etc.
   c. Records for employees whose payroll or personnel records are maintained at a fixed location, but who do not report or work at a single establishment, should be maintained at the base from which they are paid or the base of their firm's personnel operations. This category includes generally unsupervised employees such as traveling salespeople, technicians, or engineers.
2. *Employees not associated with fixed establishments.* Some employees are subject to common supervision, but do not report or work at a fixed establishment on a regular basis. These employees are engaged in physically dispersed activities that occur in construction, installation, repair, or service operations. Records for these employees should be located as follows:
   a. Records may be kept at the field office or mobile base of operations.
   b. Records may also be kept at an established central location. If the records are maintained centrally: (1) the address and telephone number of the place where records are kept must be available at the work site, and (2) someone must be available at the central location during normal business hours to provide information from the records.

Although the supplementary record and the annual summary must be located as outlined in the previous section, it is possible to prepare and maintain the log at an alternate location or by means of data-processing equipment, or both. Two requirements must be met: (1) sufficient information must be available at the alternate location to complete the log within six workdays after information is received that a recordable case occurred, and (2) a copy of the log updated to within 45 calendar days must be kept at all times in the establishment. This location exception applies only to the log and not to the other OSHA records. Also, it does not affect the employer's posting obligations.

The log and summary, OSHA No. 200, and the supplementary record, OSHA No. 101, must be retained in each establishment for five calendar years following the end of the year to which they relate. If an establishment changes ownership, the new employer must preserve the records for the remainder of the five-year period. However, the new employer is not responsible for updating the records of the former owner.

In addition to keeping the log on a calendar-year basis, employers are required to update this form to include newly discovered cases and to reflect changes that occur in recorded cases after the end of the calendar year. Maintaining or updating the log differs from simply retaining records, discussed in the previous paragraph. Although all OSHA injury and illness records must be retained, only the log must be updated by the employer. If, during the five-year retention period, there is a change in the extent or outcome of an injury or illness that affects an entry on a previous year's log, then the first entry should be lined out and a corrected entry made on that log. Also, new entries should be made for previously unrecorded cases that are discovered or for cases that initially were not recorded but were found to be recordable after the end of the year in which the case occurred. The entire entry should be lined out for recorded cases that are later found nonrecordable. Log totals should also be modified to reflect these changes.

Deciding Whether a Case Should Be Recorded and How to Classify It

Guidelines presented here for determining whether a case must be recorded under the OSHA recordkeeping requirements should not be confused with recordkeeping requirements of various workers' compensation systems, internal industrial safety and health monitoring systems, the ANSI Z.16 standards for recording and measuring work injury and illness experience, and private insurance company rating systems. Reporting a case on the OSHA records should not affect recordkeeping determinations under these or other systems. Also:

Recording an injury or illness under the OSHA system does not necessarily imply that management was at fault, that the worker was at fault, that a violation of an OSHA standard has occurred, or that the injury or illness is compensable under workers' compensation or other systems.

Employees versus other workers on site. Employers must maintain injury and illness records for their own employees at each of their establishments. However, they are *not* responsible for maintaining records for employees of other firms or for independent contractors, even though these individuals may be working temporarily in their establishment or on one of their job sites at the time an injury or illness exposure occurs. Therefore, before deciding whether a case is recordable an employment relationship needs to be determined.

Employee status generally exists for recordkeeping purposes when the employer supervises not only the output, product, or result to be accomplished by the person's work, but also the details, means, methods, and processes by which the work is accomplished. This means the employer who supervises the worker's day-to-day activities is responsible for recording the worker's injuries and illnesses. Often, part-time employees are supervised by the organization they are working for and thus are considered employees for recordkeeping purposes. Independent contractors are not considered employees; they are primarily subject to supervision by the using firm only for the results to be accomplished or end product to be delivered. Independent contractors keep their own injury and illness records.

Other factors that may be considered in determining employee status are (1) whom the worker considers to be his or her employer, (2) who pays the worker's wages, (3) who withholds the worker's Social Security taxes, (4) who hired the worker, and (5) who has the authority to terminate the worker's employment.

Method used for case analysis. The decision-making process consists of five steps:
1. Determine whether a case occurred; that is, whether there was a death, illness, or injury
2. Establish that the case was work-related; that it resulted from an event or exposure in the work environment
3. Decide whether the case is an injury or an illness
4. If the case is an illness, record it and check the appropriate illness category on the log, or
5. If the case is an injury, decide if it is recordable based on a finding of medical treatment, loss of consciousness, restriction of work or motion, or transfer to another job. Figure 13–10 presents this methodology in graphic form.

Determining whether a case occurred. The first step in the decision-making process is to determine whether an injury or illness has occurred. Employers have nothing to record unless an employee has experienced a work-related injury or illness. In most instances, recognition of these injuries and illnesses is a fairly simple matter. However, some situations have troubled employers over the years. Two of these are:
1. Hospitalization for observation. In some instances, an employee visits or is sent to a hospital for a brief time for observation. However, the case is not recordable, provided no medical treatment was given or no illness was recognized. The determining factor is not that the employee went to the hospital, but whether the incident is recordable as a work-related illness or as an injury requiring medical treatment or involving loss of consciousness, restriction of work or motion, or transfer to another job.
2. Differentiating a new case from the recurrence of a previous injury or illness. Employers are required to make new entries on their OSHA forms for each new recordable injury or illness. However, new entries should not be made for the recurrence of symptoms from previous cases. It is sometimes difficult to decide whether a situation is a new case or a recurrence. The following guidelines address this problem:
   a. Injuries. The aggravation of a previous injury almost always results from some new incident involving the employee (such as a slip, trip, fall, sharp twist, etc.). Consequently, when work related, these new incidents should be recorded as new cases.
   b. Illnesses. Generally, each occupational illness should be recorded with a separate entry on the OSHA No. 200. However, certain illnesses, such as silicosis, may have prolonged effects which recur over time. The recurrence of these symptoms should *not* be recorded as new cases

FOR CARRIER USE ONLY: Insd. Location No.　　　　　　　**Insd. Report No.**

FORM 45: Employers First Report of Injury or Illness　　　　　PLEASE TYPE OR PRINT

Filing of this report does not affect your liability under the Workers' Compensation Act and is not incriminatory in any sense.

A	***45**	ILLINOIS UNEMPLOYMENT COMPENSATION NUMBER		DATE OF REPORT　　MONTH　DAY　YEAR	CASE OR FILE NUMBER

B	EMPLOYER'S NAME　　　　　　　　　　　　　　　　　　Is this a lost ☐Yes / workday case? ☐No

C	DOING BUSINESS UNDER THE NAME OF　　　　　/ CITY, STATE　　　/ ZIP CODE

D	MAIL ADDRESS　　　　　　　/ CITY, STATE　　　/ ZIP CODE

E	EMPLOYER LOCATION IF DIFFERENT FROM MAIL ADDRESS

F	NATURE OF BUSINESS OR SERVICE　　　SIC CODE　　TOTAL NUMBER OF EMPLOYEES AT THE LOCATION WHERE ILLNESS OR INJURY OCCURRED

G	NAME OF WORKERS' COMP. INSURANCE CARRIER　　POLICY NUMBER　　SELF INSURED　　COUNTY WHERE INJURY OCCURRED **LIBERTY MUTUAL INSURANCE CO.**　　　　　　YES ☐　　☐NO

H	EMPLOYEE'S NAME　(LAST, FIRST, MIDDLE)　　　　SOCIAL SECURITY NUMBER

I	HOME ADDRESS　　　　　　　/ CITY, STATE　　　/ ZIP CODE

J	MALE ☐　FEMALE ☐　MARRIED ☐　SINGLE ☐　WIDOW(ER) ☐　DIVORCED ☐　BIRTH DATE　MONTH DAY YEAR　NUMBER OF DEPENDENT CHILDREN UNDER 18 AT TIME OF INJURY OR ILLNESS

K	DATE AND TIME OF THE INJURY OR EXPOSURE　MONTH DAY YEAR　☐a.m. ☐p.m.　EMPLOYEE'S AVERAGE WEEKLY EARNINGS　LAST DAY EMPLOYEE WORKED　MONTH DAY YEAR

L	JOB TITLE OR OCCUPATION　　　　DEPARTMENT NORMALLY ASSIGNED

M	ADDRESS OF LOCATION WHERE INJURY OR EXPOSURE OCCURRED　　/ CITY, STATE　/ ZIP CODE

N	DID EMPLOYEE DIE AS A RESULT OF THE INJURY OR ILLNESS? YES ☐ NO ☐　IF EMPLOYEE DIED AS A RESULT OF THE INJURY OR ILLNESS, GIVE DATE OF DEATH　MONTH DAY YEAR

O	WAS THE INJURY OR EXPOSURE ON THE EMPLOYER'S PREMISES? ☐YES ☐NO　DID THIS INCIDENT RESULT IN: ☐ OCCUPATIONAL INJURY ☐ OCCUPATIONAL DISEASE　Was Employee given Industrial Commission Handbook? YES ☐ NO ☐

P	NATURE OF THE INJURY

Q	PART OF THE BODY AFFECTED (BE SPECIFIC)

R	WHAT TASK WAS EMPLOYEE PERFORMING WHEN ILLNESS OR INJURY OCCURRED?

S	OBJECT OR SUBSTANCE RESPONSIBLE FOR INJURY OR ILLNESS (SOURCE)

T	HOW DID ACCIDENT OR ILLNESS OCCUR (TYPE)?

U	WHAT HAZARDOUS CONDITIONS, METHODS OR LACK OF PROTECTIVE DEVICES CONTRIBUTED?

V	WHAT UNSAFE ACT BY A PERSON CAUSED OR CONTRIBUTED TO THE INJURY OR ILLNESS?

W	HAVE MEDICAL SERVICES BEEN RENDERED TO THE EMPLOYEE? YES ☐ NO ☐　IS OR HAS THE EMPLOYEE BEEN HOSPITALIZED? YES ☐ NO ☐

X	NAME AND ADDRESS OF PHYSICIAN　　/ CITY, STATE　/ ZIP CODE

Y	NAME AND ADDRESS OF HOSPITAL　　/ CITY, STATE　/ ZIP CODE

Z	REPORT PREPARED BY: (NAME—PRINT OR TYPE)　SIGNATURE　TITLE AND TELEPHONE NUMBER

ACCIDENT REPORTING DEPT., ILLINOIS INDUSTRIAL COMMISSION, 160 North LaSalle Street, Chicago, Illinois **60601**

12-CSF-1 R5　　WITHOUT WRITTEN APPROVAL OF COMMISSION, THIS FORM MAY NOT BE REPRODUCED　　(Rev. 1-81)

Figure 13–9. Employers First Report of Injury or Illness is an equivalent to the OSHA No. 101 because it includes all of the same items. (Compare to Figure13–8.)

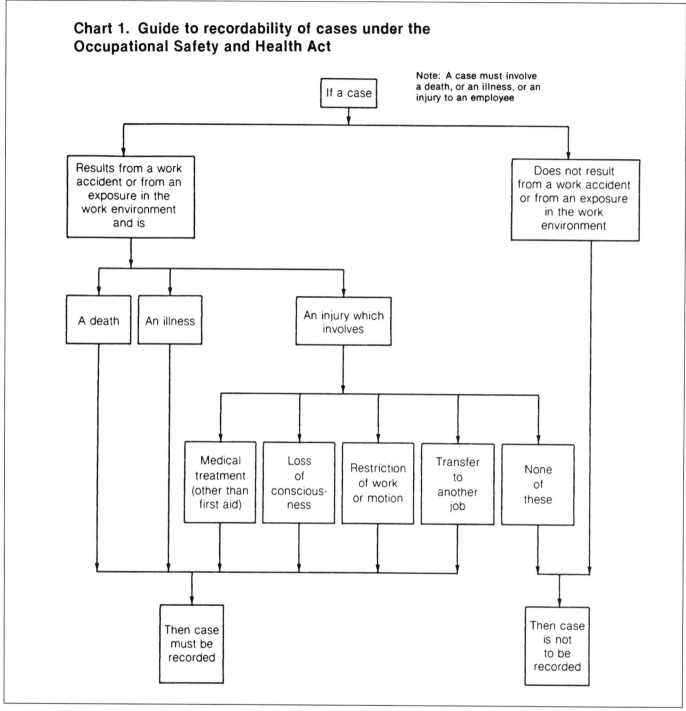

Chart 1. Guide to recordability of cases under the Occupational Safety and Health Act

If a case

Note: A case must involve a death, or an illness, or an injury to an employee

Results from a work accident or from an exposure in the work environment and is

Does not result from a work accident or from an exposure in the work environment

A death

An illness

An injury which involves

Medical treatment (other than first aid)

Loss of consciousness

Restriction of work or motion

Transfer to another job

None of these

Then case must be recorded

Then case is not to be recorded

Figure 13–10. Chart 1 of OSHA Guide.

on the OSHA forms. The recurrence of symptoms of previous illnesses may require adjustment of entries on the log for previously recorded illnesses to reflect possible changes in the extent or outcome of the particular case. Some occupational illnesses, such as certain dermatitis or respiratory conditions, may recur as the result of new exposures to sensitizing agents, and should be recorded as new cases.

Establishing work relationship. The OSHAct requires employers to record only those injuries and illnesses that are work related. Work relationship is established under the OSHA recordkeeping system when the injury or illness results from an event or exposure in the work environment (Figure 13–11). The work environment is primarily composed of (1) the employer's premises and (2) other locations where employees are engaged in work-related activities or are present as a condition of their employment. When an employee is off the employer's premises, work relationship

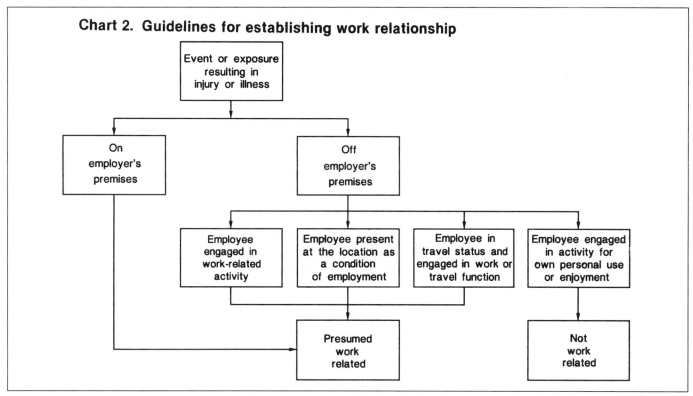

Figure 13–11. Chart 2 of OSHA Guide.

must be established; when on the premises, this relationship is presumed. The employer's premises encompass the total establishment, including not only the primary work facility but also such areas as company storage facilities. In addition to physical locations, equipment or materials used in the course of an employee's work are also considered part of the employee's work environment.

1. Injuries and illnesses resulting from events or exposures on the employer's premises. Injuries and illnesses that result from an event or exposure on the employer's premises are generally considered work related. The employer's premises consist of the total establishment. They include the primary work facilities and other areas considered part of the employer's general work area.

The presumption of work relationship for activities on the employer's premises is rebuttable. Situations where the presumption would not apply include (1) when a worker is on the employer's premises as a member of the general public and not as an employee, and (2) when employees have symptoms that surface on the employer's premises but are actually the result of a nonwork-related event or exposure off the premises.

The following subjects warrant special mention:
  a. Company restrooms, hallways, and cafeterias are all considered to be part of the employer's premises and constitute part of the work environment. Therefore, injuries occurring in these places are generally considered work related.
  b. For OSHA recordkeeping purposes, the definition of

work premises excludes all employer controlled ball fields, tennis courts, golf courses, parks, swimming pools, gyms, and other similar recreational facilities. These areas are often apart from the workplace and used by employees on a voluntary basis for their own benefit, primarily during off-work hours. Therefore, injuries to employees in these recreational facilities are not recordable unless the employee was engaged in some work-related activity or was required by the employer to participate.
  c. Company parking facilities are generally not considered part of the employer's premises for OSHA recordkeeping purposes. Therefore, injuries to employees on these parking lots are not presumed to be work related and are not recordable unless the employee was engaged in some work-related activity.

2. Injuries and illnesses resulting from events or exposures off the employer's premises. When an employee is off the employer's premises and suffers an injury or an illness exposure, work relationship must be established; it is not presumed. Injuries and illness exposures off premises are considered work related if the employee is engaged in a work activity or if they occur in the work environment. The work environment in these instances includes locations where employees are engaged in job tasks or work-related activities, or places where employees are present due to the nature of their job or as a condition of their employment.

Employees who travel on company business shall be con-

sidered to be engaged in work-related activities all the time they spend in the interest of the company. This includes, but is not limited to, travel to and from customer contacts and entertaining or being entertained for the purpose of transacting, discussing, or promoting business, etc. However, an injury/illness would not be recordable if it occurred during normal living activities (eating, sleeping, recreation); or if the employee deviates from a reasonably direct route of travel (side trip for vacation or other personal reasons). However, the employee would be considered in the course of employment once again when returning to the normal route of travel.

A traveling employee who checks into a hotel or motel establishes a "home away from home." Thereafter, the individual's activities are regarded in the same manner as for nontraveling employees. For example, if an employee on travel status is to report each day to a fixed work site, then injuries sustained when traveling to this work site would be considered off the job. The rationale is that an employee's normal commute from home to office would not be considered work-related. However, in some situations, employees in travel status report to or rotate among several different work sites after they establish their "home away from home" (such as a salesperson traveling to and from different customer contacts). In these situations, the injuries sustained when traveling to and from the sales locations would be considered job-related.

Traveling sales personnel may establish only one base of operations (home or company office). A salesperson with the home as an office is considered at work when in that office or when leaving the premises in the interest of the company.

Figure 13–11 provides a guide for establishing the work relationship of cases.

Distinguishing between injuries and illnesses. Under the OSHAct, all work-related illnesses must be recorded, while injuries are recordable only when they require medical treatment (other than first aid) or involve loss of consciousness, restriction of work or motion, or transfer to another job. The distinction between injuries and illnesses, therefore, has significant recordkeeping implications.

Whether a case involves an injury or illness is determined by the nature of the original event or exposure causing the case and not by the resulting condition of the affected employee. Injuries are caused by instantaneous events in the work environment. Cases resulting from anything other than instantaneous events are considered illnesses. This concept of illnesses includes acute illnesses that result even from relatively short exposure times.

Some conditions may be classified as either an injury or an illness (but not both), depending upon the nature of the event that produced the condition. For example, a loss of hearing resulting from an explosion (an instantaneous event) is classified as an injury; the same condition arising from exposure to industrial noise over time would be classified as an occupational illness.

Recording occupational illnesses. Employers are required to record the occurrence of all occupational illnesses, defined in the instructions of the log and summary as:

any abnormal condition or disorder, other than one resulting from an occupational injury, caused by exposure to environmental factors associated with employment. It includes acute and chronic illnesses or diseases that may be caused by inhalation, absorption, ingestion, or direct contact.

The instructions also refer to recording illnesses which were "diagnosed or recognized." Illness exposures ultimately result in conditions of a chemical, physical, biological, or psychological nature.

Occupational illnesses must be diagnosed to be recordable. However, they do not necessarily have to be diagnosed by a physician or other medical personnel. Diagnosis may be by a physician, registered nurse, or a person who by training or experience is capable of making such a determination. Employers, employees, and others may be able to detect some illnesses such as skin diseases or disorders without the benefit of specialized medical training. However, a case more difficult to diagnose, such as silicosis, would require evaluation by properly trained medical personnel.

In addition to recording the occurrence of occupational illnesses, employers are required to record each illness case in one of the seven categories on the front of the log. The back of the log form contains a listing of types of illnesses or disorders and gives examples for each illness category. These are only examples, however, and should not be considered as a complete list of types of illnesses under each category.

Recording and classifying occupational illnesses may be difficult for employers, especially the chronic and long-term latent illnesses. Many illnesses are not easily detected; and once detected, it is often difficult to determine whether an illness is work related. Also, employees may not report illnesses, either because the symptoms may not be readily apparent or because they do not think their illness is serious or work related.

The following material is provided to assist in detecting occupational illnesses and in establishing their work relationship:

1. Detection and diagnosis of occupational illnesses. An occupational illness is defined in the instructions on the log as any work-related abnormal condition or disorder in workers' health (other than an occupational injury). Detection of these abnormal conditions or disorders—the first step in recording illnesses—is often difficult. When an occupational illness is suspected, employers may want to consider the following:
   a. A medical examination of the employee's physiological systems. For example:
      - head and neck
      - eyes, ears, nose, and throat
      - endocrine
      - genitourinary
      - musculoskeletal
      - neurological
      - respiratory

- cardiovascular
- gastrointestinal.

b. Observation and evaluation of behavior related to emotional status, such as deterioration in job performance which cannot be explained

c. Specific examination for health effects of suspected or possible disease agents by competent medical personnel

d. Comparison of date of onset of symptoms with occupational history

e. Evaluation of results of any past biological or medical monitoring (blood, urine, other sample analysis) and previous physical examinations

f. Evaluation of laboratory tests: routine (complete blood count, blood chemistry profile, urinalysis) and specific tests for suspected disease agents (e.g., blood and urine tests for specific agents, chest or other X rays, liver function tests, pulmonary function tests)

g. Reviewing the literature, such as Material Safety Data Sheets and other reference documents, to ascertain whether the levels to which the workers were exposed could have produced the ill effects.

2. Determining whether the illness is occupationally related. The instructions on the back of the log define occupational illnesses as those "caused by environmental factors associated with employment." In some cases, such as contact dermatitis, the relationship between an illness and work-related exposure is easy to recognize. In other cases, where the occupational cause is not direct and apparent, it may be difficult to determine accurately whether an employee's illness is occupational. In these situations, it may help employers to ask the following questions:

a. Has an illness condition clearly been established?

b. Does it appear that the illness resulted from, or was aggravated by, suspected agents or other conditions in the work environment?

c. Are these suspected agents present (or have they been present) in the work environment?

d. Was the ill employee exposed to these agents in the work environment?

e. Was the exposure to a sufficient degree and/or duration to result in the illness condition?

f. Was the illness attributable solely to a nonoccupational exposure?

Deciding if work-related injuries are recordable. Although the OSHAct requires that all work-related deaths and illnesses be recorded, the recording of nonfatal injuries is limited to certain specific types of cases: (1) those requiring medical treatment or involving loss of consciousness, (2) those involving restriction of work or motion, and (3) those requiring a transfer to another job. Minor injuries needing only first aid treatment are not recordable.

1. Medical treatment. It is important to understand the distinction between medical treatment and first aid treatment, because many work-related injuries are recordable only if medical treatment was given.

The regulations and the instructions on the back of the log and summary, OSHA No. 200, define medical treatment as any treatment, other than first aid treatment, administered to injured employees. Essentially, medical treatment involves the provision of medical or surgical care for injuries that are not minor through the application of procedures or systematic therapeutic measures.

The act also specifically states that work-related injuries involving only first aid treatment should not be recorded. First aid is commonly thought to mean emergency treatment of injuries before regular medical care is available. However, first aid treatment has a different meaning for OSHA recordkeeping purposes. The regulations define first aid treatment as:

any one-time treatment, and any follow-up visit for the purpose of observation, of minor scratches, cuts, burns, splinters, and so forth, which do not ordinarily require medical care. Such one-time treatment, and follow-up visit for the purpose of observation, is considered first aid even though provided by a physician or registered professional personnel.

The distinction between medical treatment and first aid depends not only on the treatment provided but also on the severity of the injury being treated. First aid (1) is limited to one-time treatment and subsequent observation and (2) involves treatment of only minor injuries, not emergency treatment of serious injuries. Injuries are not minor if:

a. they must be treated only by a physician or licensed medical personnel

b. they impair bodily function (i.e., normal use of senses, limbs, etc.)

c. they result in damage to the physical structure of a nonsuperficial nature (e.g., fractures)

d. they involve complications requiring follow-up medical treatment.

Physicians or registered medical professionals, working under the standing orders of a physician, routinely treat minor injuries. Such treatment may constitute first aid. Also, some visits to a doctor do not involve treatment at all. For example, a visit to a doctor for an examination or other diagnostic procedure to determine whether the employee has an injury does not constitute medical treatment. Conversely, medical treatment can be provided to employees by lay persons, i.e., someone other than a physician or registered medical personnel.

The following classifications list certain procedures as either medical treatment or first aid treatment.

Medical treatment: The following are generally considered medical treatment. Work-related injuries for which this type of treatment was provided or should have been provided are almost always recordable:

- treatment of INFECTION
- application of ANTISEPTICS during second or subsequent visit to medical personnel
- treatment of SECOND OR THIRD DEGREE BURN(S)
- application of SUTURES (stitches)
- application of BUTTERFLY ADHESIVE DRESSING(S) or STERI STRIP(S) in lieu of sutures
- removal of FOREIGN BODIES EMBEDDED IN EYE

- removal of FOREIGN BODIES FROM WOUND; if procedure is COMPLICATED because of depth of embedment, size, or location
- use of PRESCRIPTION MEDICATIONS (except a single dose administered on first visit for minor injury or discomfort)
- use of hot or cold SOAKING THERAPY during second or subsequent visit to medical personnel
- application of hot or cold COMPRESS(ES) during second or subsequent visit to medical personnel
- CUTTING AWAY DEAD SKIN (surgical debridement)
- application of HEAT THERAPY during second or subsequent visit to medical personnel
- use of WHIRLPOOL BATH THERAPY during second or subsequent visit to medical personnel
- POSITIVE X–RAY DIAGNOSIS (fractures, broken bones, etc.)
- ADMISSION TO A HOSPITAL or equivalent medical facility FOR TREATMENT.

First aid treatment: The following are generally considered first aid treatment (e.g., one-time treatment and subsequent observation of minor injuries). These cases should not be recorded if the work-related injury does not involve loss of consciousness, restriction of work or motion, or transfer to another job:

- application of ANTISEPTICS during first visit to medical personnel
- treatment of FIRST DEGREE BURN(S)
- application of BANDAGE(S) during any visit to medical personnel
- use of ELASTIC BANDAGE(S) during first visit to medical personnel
- removal of FOREIGN BODIES NOT EMBEDDED IN EYE if only irrigation is required
- removal of FOREIGN BODIES FROM WOUND; if procedure is UNCOMPLICATED, and is, for example, by tweezers or other simple technique
- use of NONPRESCRIPTION MEDICATIONS and administration of single dose of PRESCRIPTION MEDICATION on first visit for minor injury or discomfort
- SOAKING THERAPY on initial visit to medical personnel or removal of bandages by SOAKING
- application of hot or cold COMPRESS(ES) during first visit to medical personnel
- application of OINTMENTS to abrasions to prevent drying or cracking
- application of HEAT THERAPY during first visit to medical personnel
- use of WHIRLPOOL BATH THERAPY during first visit to medical personnel
- NEGATIVE X–RAY DIAGNOSIS
- OBSERVATION of injury during visit to medical personnel. The following procedure, by itself, is not considered medical treatment:

1. administration of TETANUS SHOT(S) or BOOSTER(S). However, these shots are often given in conjunction with more serious injuries; consequently, injuries requiring these shots may be recordable for other reasons.
2. Loss of consciousness. If an employee loses consciousness as the result of a work-related injury, the case must be recorded no matter what type of treatment was provided. The rationale behind this recording requirement is that loss of consciousness is generally associated with the more serious injuries.
3. Restriction of work or motion. Restricted work activity occurs when the employee, because of the impact of a job-related injury, is physically or mentally unable to perform all or any part of his or her normal assignment during all or any part of the workday or shift. The emphasis is on the employee's ability to perform normal job duties. Restriction of work or motion may result in either a lost-work-time injury or a nonlost-work-time injury, depending upon whether the restriction extended beyond the day of injury.
4. Transfer to another job. Injuries requiring transfer of the employee to another job are also considered serious enough to be recordable, regardless of the type of treatment provided. Transfers are seldom the sole criterion for recordability. This is because injury cases are almost always recordable on other grounds, primarily medical treatment or restriction of work or motion.

Categories for Evaluating the Extent of Recordable Cases

Once the employer decides that a recordable injury or illness has occurred, the case must be evaluated to determine its extent or outcome. There are three categories of recordable cases: fatalities, lost workday cases, and cases without lost workdays. Every recordable case must be placed in only one of these categories.

Fatalities. All work-related fatalities must be recorded, regardless of the time between the injury and the death, or the length of the illness.

Lost workday cases. Lost workday cases occur when the injured or ill employee experiences either days away from work, days of restricted work activity, or both. In these situations, the injured or ill employee is affected to such an extent that (1) days must be taken off from the job for medical treatment or recuperation, or (2) the employee is unable to perform his or her normal job duties over a normal work shift, even though the employee may be able to continue working.

1. Lost workday cases involving days away from work refer to cases when the employee ordinarily would have been on the job but could not work because of the job-related injury or illness. The focus of these cases is on the employee's inability, because of injury or illness, to be present in the work environment during his or her normal work shift.
2. Lost workday cases involving days of restricted work activity are those cases in which, because of injury or illness, (1) the employee was temporarily assigned to another job, (2) the employee worked at a permanent job less than full time, or (3) the employee worked at his or

her permanently assigned job but could not perform all the duties normally connected with it. Restricted work activity occurs when the employee, because of the job-related injury or illness, is physically or mentally unable to perform all or any part of the normal job duties over all or any part of the normal workday or shift. The emphasis is on the employee's inability to perform normal job duties over a normal work shift.

Injuries and illnesses are not considered lost workday cases unless they affect the employee beyond the day of injury or onset of illness. When counting the number of days away from work or days of restricted work activity, do not include the initial day of injury or onset of illness, or any days on which the employee would not have worked even if able to be on the job (holidays, vacations, etc.).

Cases not resulting in death or lost workdays. These cases consist of the relatively less serious injuries and illnesses that satisfy the criteria for recordability. However, they do not result in death or require the affected employee to have days away from work or days of restricted work activities beyond the date of injury or onset of illness.

Employer Obligations for Reporting Occupational Injuries and Illnesses

This section focuses on the requirements of Section 8(c)(2) of the OSHAct and Title 29, Part 1904, of the *Code of Federal Regulations* for employers to make reports of occupational injuries and illnesses. It does not include the reporting requirements of other standards or regulations of OSHA or of any other state or federal agency.

The *Annual Survey of Occupational Injuries and Illnesses.* The survey is conducted on a sample basis. Firms required to submit reports of their injury and illness experience are contacted by OSHA or a participating state agency. A firm not contacted need not file a report of its injury and illness experience. Employers should note, however, that even if they are not selected to participate in the annual survey for a given year, they must still comply with the recordkeeping requirements listed in the preceding sections as well as with those for reporting fatalities and multiple hospitalization cases provided in the next section.

Reporting fatalities and multiple hospitalizations. All employers are required to report accidents resulting in one or more fatalities or in the hospitalization of five or more employees. (Some states have more stringent catastrophic reporting requirements.) The report is made to the nearest office of the Area Director of OSHA, U.S. Department of Labor. However, a different reporting procedure applies if the state in which the accident occurred is administering an approved state plan under Section 18(b) of the OSHAct. Those 18(b) states designate a state agency to which the report must be made.

The report must contain three pieces of information: (1) circumstances surrounding the accident(s), (2) the number of fatalities, and (3) the extent of any injuries. If necessary, the OSHA Area Director may require additional information on the accident.

The report should be made within 48 hours after the occurrence of the accident or fatality, regardless of the time lapse between the occurrence of the accident and the death of the employee.

Access to OSHA Records and Penalties for Failure to Comply with Recordkeeping Obligations

The preceding sections describe recordkeeping and reporting requirements. This section covers subjects related to ensuring the integrity of the OSHA recordkeeping process: access to OSHA records and penalties for recordkeeping violations.

Access to OSHA records. All OSHA records, which are being kept by employers for the five-year retention period, should be available for inspection and copying by authorized federal and state government officials. Employees, former employees, and their representatives are allowed access only to the log, OSHA No. 200.

Government officials with access to the OSHA records include the following: representatives of the Department of Labor, including OSHA safety and health compliance officers and BLS representatives; representatives of the Department of Health and Human Services while carrying out that department's research responsibilities; and representatives of states accorded jurisdiction for inspections or statistical compilations. "Representatives" may include Department of Labor officials inspecting a workplace or gathering information, officials of the Department of Health and Human Services, or contractors working for the agencies mentioned above, depending on the provisions of the contract under which they work.

Employee access to the log is limited to the records of the establishment in which the employee currently works or formerly worked. All current logs and those being maintained for the five-year retention period must be made available for inspection and copying by employees, former employees, and their representatives. An employee representative can be a member of a union representing the employee, or any person designated by the employee or former employee. Access to the log is to be provided in a reasonable manner and at a reasonable time. Employees can obtain redress through OSHA for any company's failure to comply with the access provisions of the regulations.

Penalties for failure to comply with recordkeeping obligations. Employers committing recordkeeping and/or reporting violations are subject to the same sanctions as employers violating other OSHA requirements such as safety and health standards and regulations.

OSHA INCIDENCE RATES

Safety performance is relative. Only when a company compares its injury experience with that of its entire industry, or with its own previous experience, can it obtain a meaningful evaluation of its safety accomplishments. To make such comparisons, a method of measurement is needed that will adjust for the effects of certain variables contributing to differences in injury experience. Injury totals alone cannot be

used for two reasons.

First, a company with many employees may be expected to have more injuries than a company with few employees. Second, if the records of one company include all the injuries treated in the first aid room, while the records of a similar company include only injuries serious enough to cause lost time, obviously the first company's total will be larger than the second company's figure.

A standard procedure for keeping records, which provides for these variables, is included in the OSHA record-keeping requirements. First, this procedure uses incidence rates that relate injury and illness cases, and the resulting days lost, to the number of employee-hours worked; thus these rates automatically adjust for differences in the hours of exposure to injury. Second, this procedure specifies the kinds of injuries and illnesses that should be included in the rates. These standardized rates, which are easy to compute and to understand, have been generally accepted as procedure in industry, thus permitting the necessary and desired comparisons.

A chronological arrangement of these rates for a company will show whether its level of safety performance is improving or worsening. Within a company, the same sort of arrangement by departments not only will show the trend of safety performance for each department but will reveal to management other information to render safety work more efficient.

If it is found, for example, that the trend of incidence rates in a company is up, a review of the rate trends by department may reveal that this change is accounted for by the rates of just a few departments. With the sources of the highest company rates isolated, management can concentrate safety efforts at these points.

A comparison of current incidence rates with those of similar companies and with those of the industry as a whole serves a critical function. This step provides the safety professional with a more reliable evaluation of the company's safety performance than could be obtained merely by reviewing numbers of cases.

Formulas for Rates

Incidence rates are based on the exposure of 100 full-time workers using 200,000 employee-hours as the equivalent (100 employees working 40 hours per week for 50 weeks per year). An incidence rate can be computed for each category of cases or days lost depending on what number is put in the numerator of the formula. The denominator of the formula should be the total number of hours worked by all employees during the same time period as that covered by the number of cases in the numerator.

$$\text{Incidence Rate} = \frac{\text{No. of injuries \& illnesses} \times 200,000}{\text{Total hours worked by all employees during period covered}}$$

or

$$\frac{\text{No. of lost workdays} \times 200,000}{\text{Total hours worked by all employees during period covered}}$$

There are two other formulas that can be used to measure the average severity of the recorded cases:

$$\text{Average lost workdays per total lost workday cases} = \frac{\text{Total lost workdays}}{\text{Total lost workday cases}}$$

$$\text{Average days away from work} = \frac{\text{Total days away from work}}{\text{Total cases involving days away from work}}$$

If these numbers are small, then it is known that the cases are relatively minor. If, however, the numbers are large, then the cases are of greater average severity and should receive serious attention. For example, to calculate the incidence rate for total recordable cases at the end of the year, one would simply multiply the number of recordable cases by 200,000 and divide that by the number of hours worked by all employees for the whole year.

The incidence rates may also be interpreted as the percentage of employees who will suffer the degree of injury for which the rate was calculated. That is, if the incidence rate of lost workday cases is 5.1 per 100 full-time workers, then about 5% of the establishment's employees incurred a lost-workday injury.

SIGNIFICANCE OF CHANGES IN INJURY AND ILLNESS EXPERIENCE

The incidence rate for lost workday cases is the best measure for comparing the occupational injury and illness experience among companies of various sizes. However, it is often not sufficient for determining the significance of month-to-month changes in the actual number of cases within a single company. For this comparison, all cases must usually be used, not merely the lost workday cases. This provides a more objective measure for determining the significance of month-to-month fluctuations in the number of cases, especially when a seemingly large increase or decrease from the monthly average appears.

Since the average number of cases per month is calculated from numbers of cases that are larger or smaller than the average itself, variation from the average is to be expected. The variation can either be random or caused; caused variation is significant and random variation is nonsignificant. Therefore, the significance of variations can easily be determined by distinguishing between those that are random or caused.

Manufacturing organizations are already faced with the task of determining the significance of variations in such factors as dimensions, weight, or performance of their products. To ease this task, they frequently employ a tool

known as a quality control chart. This chart identifies and distinguishes between a random variation, which is said to be "in control," and caused variation, which is said to be "out of control." Being able to distinguish between the two types of variation permits management to concentrate safety efforts on out-of-control variations.

A similar control chart can be developed to evaluate the significance of changes in injury and illness experience. The first step in developing the chart is to calculate the average number of cases per month. Because this monthly average will fluctuate from year to year, several years' experience should be used to develop a stable average. Preferably, the average should be calculated using 60 months' experience. After the average number of cases per month has been determined, the upper and lower control limits (UCL and LCL, respectively) are calculated using the following equation:

$$\text{UCL and LCL} = n \pm 2\sqrt{n}$$

where n is the average number of cases per month.

Figure 13–12 shows a control chart developed for one company. Using 60 months' experience, it was determined that the company had an average of 25 cases per month. Substitution of 25 for n in the equation yielded the upper and lower control limits shown below:

$$25 \pm 2\sqrt{25} = 25 \pm 10 = 25 + 10 = 35 \text{ for the upper limit}$$
$$25 - 10 = 15 \text{ for the lower limit}$$

The control chart was then constructed and the company recorded the actual number of cases each month. Note that the actual monthly number of cases, with the exception of February, falls within the upper and lower control limits. These variations are random and do not represent significant changes from the monthly average of 25 cases. The variation for February is probably a caused variation, indicating that the experience for February is out of control and that corrective measures should be taken. Suppose, for example, an investigation revealed that an unguarded machine was the cause of the increase in cases for the month of February. It can be assumed that the installation of a mechanical guard corrected this cause and brought the injury and illness experience for March back in control.

The control chart can be a useful tool when it is used properly. As with any tool, it depends on the skill of the user to yield good results. Follow these rules when constructing a control chart:

1. Always use several years' experience when calculating the average number of cases per month to ensure that a stable average will be developed. Sixty months' experience is preferable, except as noted in (3), (4), and (5) following.
2. Count all cases when constructing a control chart. The use of lost workday cases alone may not provide enough data. A control chart cannot be constructed if the average number of cases per month is less than four because the lower limit will be negative. Since a plant cannot experience a negative number of cases per month, a chart

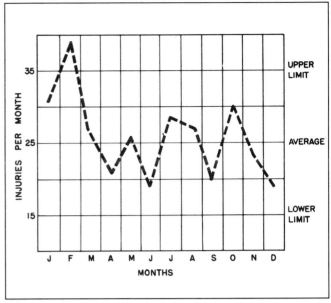

Figure 13–12. Upper and lower limits of this control chart were based on 60 months' experience, and show points at which the number of injuries can statistically be shown to have a contributing cause rather than be by chance. Thus, action can be taken where it will be most effective.

based on an average of less than four would not yield meaningful results.
3. Calculate a new average each year—adding the latest year's cases and subtracting the oldest year's cases—to reflect any change in the monthly average number of cases.
4. Construct more than one chart if there are seasonal changes in employment. For example, if an operation employs 100 people for the first six months of each year and 400 people for the last six months, a separate chart must be constructed for each six-month period.
5. Construct a new control chart if external factors, such as a change in the level of employment, a change in the work environment, or a change in work hazards, bring about a permanent change in injury and illness experience. When calculating a new average in such a situation, do not use experience prior to the change.

OFF-THE-JOB INJURIES

In recent years, off-the-job disabling injuries of employees have far exceeded on-the-job disabling injuries. Any unscheduled absence of employees can cause production slowdowns and delays, costly retraining and replacement, or costly overtime by remaining employees. As a result, many safety specialists are very concerned with off-the-job injuries their employees suffer. Moreover, activity to reduce off-the-job accidents should help to promote interest in this area of safety.

The ANSI Z16.3 standard, as revised in 1989, provides a means for recording and measuring these off-the-job injuries. Definitions and rates used under the ANSI Z16.3 standard are compatible with those of the OSHA recordkeeping re-

quirements. Because the data on off-the-job injuries are not as easy to obtain, however, certain simplifications are introduced in ANSI Z16.3. Exposure (for use in rates per 200,000 employee-hours) is standardized at 312 employee-hours per employee per month (equal to 4 1/3 weeks less 40 hours per week at work and 56 hours per week for sleeping). Provision is also made in ANSI Z16.3 for recording home, public, and transportation injuries separately to allow for concentrated effort in problem areas.

SUMMARY

- Records of work-related employee injuries and illnesses are required by OSHA, state compensation authorities, and insurers. Safety personnel must maintain records (1) required by law and by management and (2) useful to an effective safety program.
- Good recordkeeping systems can help safety professionals by providing data to evaluate accident problems and overall safety program effectiveness, identify high accident-rate areas, create interest in safety among supervisors, enable the company to concentrate efforts on the more serious accident problems, and measure effectiveness of countermeasures against hazards and unsafe practices.
- The primary purpose of an accident report is to obtain accurate, objective information about the causes of accidents in order to prevent the incident from reoccurring, not to fix the blame. Accident records include first aid reports and accident investigation reports. The supervisor's accident investigation report form should be completed as soon after the incident as possible to ensure accurate and detailed information.
- After accident cases are closed, the first aid report and supervisor's report are filed by agency of injury, type of accident, or other factor that will make the report useful for accident prevention. The supervisor keeps an employee injury record for each individual worker.
- Injury reports are used to compile monthly summaries, and the reports themselves are filed for easy access. Monthly summaries show the current status of accident experience in an organization.
- Every establishment subject to the OSHAct is required to provide an annual summary of accident rates by February 1st. For management, annual reports can be used to compare current-year data with the experience of previous years, and with the experience of similar organizations and of the industry as a whole.
- Accident reports to management help executives understand accident experience in various plant operations and provide a yardstick for better evaluation of progress in eliminating or preventing accidents. For supervisors, such reports can help to maintain their interest in safety and accident prevention and to target specific hazards or work practices for countermeasures. Employees should also be informed about accident experience in the company.
- The National Safety Council uses annual reports to determine injury and illness incident rates by industry. Safety

awards are made on the basis of organizations' records.
- Bilevel reporting is used to gain further details about specific accidents in addition to the general information contained in standard report forms. This method saves time and focuses on those accidents an investigator wishes to study in depth.
- OSHA recordkeeping requirements apply to private sector employees in all states, District of Columbia, Puerto Rico, and other U.S. territories and possessions. Employers with 11 or more employees in specified industries must keep OSHA records in addition to any state or local requirements. Employers who are not in these specified industries or who have fewer than 11 employees are exempt from OSHA recordkeeping regulations.
- However, all employers must report to OSHA within 48 hours any accident that results in one or more fatalities or the hospitalization of five or more employees.
- Two of the most common forms in OSHA recordkeeping are the Log of Occupational Injuries and Illnesses (OSHA No. 200) and the Supplementary Record of Occupational Injuries and Illnesses (OSHA No. 101). The log is used for recording and classifying occupational injuries and illnesses and for noting the extent of each case.
- For every injury and illness entered on the log, management must record additional information on the supplementary record. This form must be completed within six workdays after an injury or illness has occurred.
- The location, retention, and maintenance of these records depends on the number of establishments and whether employees are associated with the fixed establishment. Employers are required to retain records for five years and to update them as new information about accidents and illnesses arises.
- To decide whether a case is recordable, employers must (1) establish that an employment relationship exists with the worker(s) and (2) that a work-related injury or illness has actually occurred and is not merely an aggravation of a previous injury and illness.
- All work-related illnesses must be recorded, while injuries are recorded only when they require medical treatment (other than first aid) or involve a loss of consciousness, restriction of work or motion, or transfer to another job. Lost workdays cases are recorded when the employee cannot work because of an injury or illness, is temporarily reassigned, or has restrictions because of the injury or illness.
- All work-related fatalities and multi-hospitalization accidents must be recorded. The report contains the circumstances surrounding the accident, number of fatalities, and extent of injuries. It must be made within 48 hours of the incident.
- All OSHA records should be available for inspection and copying by authorized federal and state governments. Employees, former employees, and their representatives are allowed access only to the log.
- OSHA incident rates help companies compare their safety performance with the performance of previous years or of

the entire industry to evaluate their safety programs. OSHA recordkeeping requirements use incidence rates that relate injury and illness cases, and the resulting days lost, to the number of employee-hours worked. These rates automatically adjust for different accident experiences among companies and specify the kinds of injuries and illnesses that should be included in the rates.

- Quality control charts help distinguish between random variations, which are in control, and caused variations, which are out of control. This distinction allows management to concentrate safety efforts on out-of-control variations. A similar control chart can help management to evaluate whether changes in injury and illness trends are due to random or caused variations.
- The ANSI Z16.3 standard provides a means for recording and measuring off-the-job injuries to help management develop safety programs for employees. This standard uses the same formula for calculating incidence rates as the one used on OSHA forms.

REFERENCES

American National Standards Institute, 11 West 42nd Street, New York, NY 10036.

American National Standard for Injury Statistics—Employee Off-the-Job Injury Experience—Recording and Measuring, ANSI Z16.3–1989.

National Safety Council, 444 North Michigan Avenue, Chicago, IL 60611.
    Accident Facts (published annually).
    Work Injury and Illness Rates (published annually).

Recht JL. Bilevel reporting. *Journal of Safety Research* 2(2): 51–54, 1970.

Tufte ER. *The Visual Display of Quantitative Information.* Cheshire, CT: Graphics Press, 1983.

U.S. Dept. of Labor, Bureau of Labor Statistics, Washington, DC 20212. "A Brief Guide to Recordkeeping Requirements for Occupational Injuries and Illnesses." April 1986.

14

Computers and Information Management

The computer has become an integral part of organizational safety and health departments. Once the exclusive domain of highly trained personnel in data processing, computer systems have become smaller, more affordable, and more accessible to more people than ever before. In many cases, the personal computers (PCs) in use have more power than the early models that once filled an entire room and required water cooling to function!

Computer systems, the combination of hardware (the physical components) and application software (the instructions) are everywhere in the business environment. They have replaced typewriters for correspondence and report writing and can even function as sophisticated publishing machines, replacing typesetting equipment.

Within a safety and health program, computer systems can ease the burden of compliance, maintain better records, analyze more information for decision making, and help prevent work-related deaths, injuries, illnesses, and toxic exposure.

Computers are used by safety and health professionals to:

▪ compile, store, analyze, and report on occupational injuries and illnesses—needed by both the corporation management and the Occupational Safety and Health Administration (OSHA) for compliance purposes

▪ produce follow-up reports pointing to hazards identified during inspections, but not yet abated (Figure 14–1)

▪ assist with accident investigation and analysis to uncover the underlying cause of workplace injuries and identify corrective actions to eliminate them

▪ maintain employee health records as they relate to workplace exposure to hazardous substances

▪ maintain records needed to document regulatory compliance, such as training mandated by the OSHA Hazard Communications standard (Figure 14–2)

▪ store and report inventory information about regulated chemicals, again to comply with provisions of the Superfund and Recovery Act (SARA), administered by the Environmental Protection Agency

▪ develop and maintain an emergency response plan for handling unintentional release of hazardous substances into the environment

▪ train employees in safe procedures and emergency procedures (Figure 14–2)

▪ store and retrieve complex regulatory information that is often difficult to find in federal code books

▪ maintain predictive maintenance, inspection, and repair reports to prevent equipment failure, loss of productivity, or a regulatory violation (Figure 14–3)

▪ communicate with other safety and health professionals and share information, documents, and ideas affecting workplace safety.

Computers are tools to help management obtain the information necessary to make better decisions; to help standardize recordkeeping and reporting functions; assist with regulatory compliance; and protect employees and organizations.

The importance of information management to occupational safety and health professionals became critical with

331

Figure 14–1. This computer-assisted program is used to monitor ongoing operations. (Courtesy UOP.)

the passage of the Occupational Safety and Health Act (OSHAct) in 1970. Environmental laws such as the Toxic Substances Control Act, the Resource Conservation and Recovery Act, and the Superfund Act created a regulatory environment that overwhelmed the resources of existing occupational safety and health programs.

Right-to-know laws further added to the burden of proper compliance. All of these new regulations carried complex recordkeeping and reporting requirements and stiff penalties for noncompliance. Not only must workers be kept informed of hazardous substances being used in the workplace, but the community as well has a right to the information.

As public awareness of workplace hazards grew, litigation involving occupational illnesses and deaths began to have an impact on organizations. Employers were asked to document their compliance in training, monitoring, and reporting procedures. Unfortunately, many

Figure 14–2. An interactive training simulator instructs employees about operations, standard procedures, and emergency procedures. (Courtesy UOP.)

employers found their paper records to be inadequate and incomplete.

Internal pressures within an organization to control costs is another reason for seeking better recordkeeping systems. Many companies have reduced their operations for various reasons—economic forces, increased competition, change of ownership, and mounting personnel costs. The needs have increased, but in many cases budgets have not kept pace with them.

These factors—the need to comply with stiffer regulations, the need for better internal records management, and the need to control costs—have led many organizations to purchase a computer system to manage the workload and to control information systems. As an effective tool for safety management, the computer has proven its worth and is definitely here to stay.

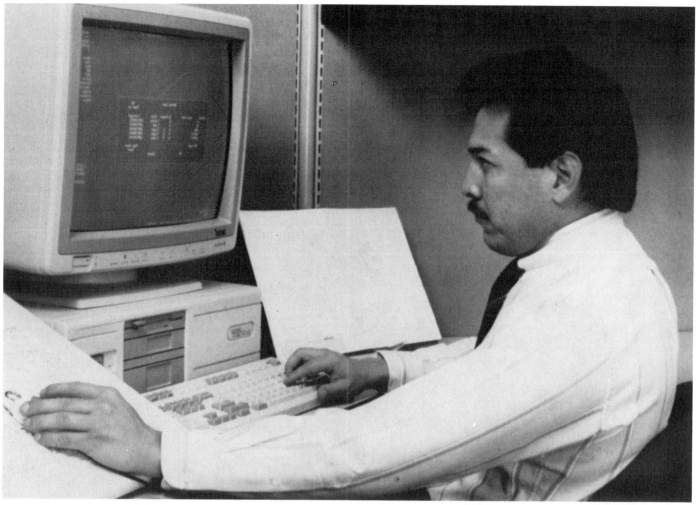

Figure 14–3. Cost- and time-effective safety management will increasingly require the use of computers. Shown here is an application that establishes current corrosion/erosion rates of piping, equipment, and tanks. (Courtesy United States Testing Company, Inc.)

ADVANTAGES OF AUTOMATED INFORMATION SYSTEMS

Computer systems, small and large, consist of physical components called hardware and instructional components called software. One component cannot operate without the other. This chapter addresses issues surrounding the use of microcomputers—also called PCs—as they are the most likely system to be available for the safety and health professional.

The past decade has seen a rapid change in computer technology until, today, the vast power of large computer systems is now available within a desktop system at a fraction of the cost. Highly skilled technicians once dominated the implementation and maintenance of computer systems, but today many office workers use computers for word processing, financial analysis, accounting, records maintenance, and more. Computers are now commonplace in every job function and with minimal capital investment and employee training (Figure 14–4).

Improved Availability of Data

A well-designed computer system (combining both computer software and hardware) can compile and store huge amounts of data and retrieve it easily for reporting and analysis. For example, an employer might want to know how many back injuries affect the workplace population. With a manual system, a time-consuming search through medical files and/or injury files would be necessary. With an automated system, a worker can complete this search in minutes with the help of specialized software developed for the task. Automated systems can quickly provide emergency information, such as first aid and clean-up procedures for chemical spills.

Elimination of Duplication

The time spent maintaining duplicate records within an organization can be reduced with an automated system. For example, with a manual system, it could be necessary to keep records of occupational injuries in four offices: safety, medical, human resources, and area supervision. With a shared information system, one comprehensive file on each injury can be kept in the computer.

Improved Communications

With manual systems, important data can be lost in unfinished piles of paperwork. Often, statistics generated by one

COMPUTER TERMS

The definitions used here are meant to define some of the more basic computer terms as they are commonly used.

Bit A unit of data, a bit is the smallest piece of information that the computer can handle. Technically, a bit represents whether an electronic impulse is turned on or off.

Bug An error or irritation in a software program—usually not so severe that it causes a program to fail.

Byte This unit of measuring/SIZE is equivalent to eight bits, about the amount of information needed to store a single letter, number, or other character.

Central Processing Unit Part of the insides of the computer that controls all of the work the computer does.

Data Single pieces of information stored in a file. For example, in a file of accident records, the date of the accident would be one piece of data.

DOS Disk Operating System is the software that acts as the intermediary between the computer hardware and other software.

File Collection of information stored on a disk. For example: a letter of correspondence might be saved as a single file.

Floppy Disk or Diskette A flexible, round electromagnetic device that stores computer programs and data. When used in a floppy drive (the part of the computer where the diskette is inserted) the information is read into the computer's memory for processing. Often the storage capacity of the disk exceeds the memory of the computer. The information on the diskette is permanent until erased or replaced.

Graphics A display of information in picture form. For example: a pie chart or a bar graph on-screen are graphic capabilities of a computer.

Hardware All of the physical components of the computer system. This includes the monitor, printer, keyboard, central processing unit, and everything else that you can touch; not floppy disks, however.

Hard Disk Resembling stacked floppy diskettes, this component is actual hardware that sits inside the computer. Hard disks, then, can hold much more information than a floppy. Information on the hard disk is read into the computer's memory for processing, and it is permanent until erased or replaced.

Hard Copy The paper copy of information, as printed out by the computer's printer.

K This unit of measuring/SIZE is equal to 1,024 bytes. The term is usually used to express the memory capacity of the microcomputer. For example: 64K, 128K, 256K, 512K. If a software program requires 256K to run, and your computer has only 128K, then your computer cannot run the software program unless you purchase additional memory-expansion hardware.

meg This unit of measuring/SIZE equals 1,048,576 bytes or 1,024 K. It is often used to express the capacity of a hard disk. For example: 10, 20, or 30 meg.

Memory The part of the computer that stores programs and data currently being processed.

Modem The MOdulator-DEModulator or the hardware that allows computer-generated signals to be transmitted and received on telephone lines.

Monitor The hardware part that includes the screen.

Printer The hardware that produces hard copies.

Program A set of computer instructions on a disk.

Software A series of programs that, together, allow the computer to function as a word processor, data base manager, or spreadsheet.

Figure 14—4.

of the company's offices do not match those generated by another. Again take occupational injuries as an example: injury reports from one facility may not reach other facilities, decreasing the chance that all groups within the organization can benefit from the "lesson learned." A good database management system makes data available to all authorized users. In addition, electronic mail systems can facilitate communication within and between facilities.

Data Standardization and Accuracy

Standardizing the data input and the way in which it is organized is particularly important in organizations with many computer users. When the same data are being input in the same way by everyone, the analysis and decision-making capabilities of the database are enlarged. For instance, accident and incident rates can be analyzed for all locations at the same time, rather than analyzing each location separately and then standardizing and entering the collective data for final analysis.

Accuracy of data can be vastly improved by checking entries for the completeness of records. For instance, a computer system can alert the medical department that part of a worker's OSHA-required physical examination was not conducted.

Improved Analytical Capabilities

With most manual systems, compilation and analysis of data from various records consumes professional time that could be better spent on program management activities. A good automated system performs analytical tasks quickly and formats the results for reports to management. For example, the ability to correlate workplace exposure and medical records leads to improved medical surveillance. Trends in employee health can become apparent in time to act to prevent injuries and illnesses. Analyses not practical with manual systems can be done routinely using an automated system.

Cost Savings

While hard to document, automated systems can save organizations money above and beyond the costs of system development and operation. Increased employee productivity, decreased incident rates with attendant savings, and more effective program management can lead to better loss control overall.

SAFETY AND HEALTH FUNCTIONS

Today's occupational safety and health systems support single functions, such as management of training records, or a comprehensive set of functions. This section is an overview of the system functions and tools in use by safety and health professionals today.

Incident Management

Many organizations are using automated systems to store, analyze, and report on incident data. Data for all types of

incidents are entered, such as property damage, occupational injury, near-misses, and transportation. These data usually parallel the organization's first report and incident investigation forms. The system can produce (1) individual incident reports; (2) summary reports that categorize incidents by location, type, rate, severity, loss, and other factors; (3) OSHA Form 200 logs; and (4) analytical reports that pinpoint major causes and types of accidents within organizational subsets. For example, an organization might determine through use of its automated system that a disproportionate number of accidents is occurring in a particular operation, then arrange for employees in that operation to have additional training.

Accident Investigation

With the advent of powerful, low-cost computer systems and artificial intelligence capabilities, it is now possible today to uncover the underlying cause of industrial incidents. Incidents routinely attributed to poor housekeeping, sloppy procedures, or indifference by workers and supervisors can be fully investigated and corrected once the root cause is identified. Many such systems can be made portable with a lightweight laptop computer and taken directly to the scene of an explosion, fire, or spill. In this way, computer applications can reach far beyond the confines of physical facilities and provide benefits never before considered possible.

Workplace Conditions

Many systems manage safety, industrial hygiene, health, physics, environmental, and other sampling, audit, and inspection data capable of describing and quantifying workplace conditions. Industrial hygiene sampling results can be entered and used to produce exposure profiles for each workplace.

Also, if safety inspection findings, recommendations, target dates for completion, and completion dates are automated, a system can produce follow-up reports highlighting identified deficiencies that have not been corrected. In comprehensive systems, workplace data can be correlated with employee health data to support health surveillance programs.

Right-to-Know

Automated information systems greatly help environmental agent monitoring, including the inventories of environmental agents, toxicology data, and material safety information on hazardous substances within the workplace. These agents include chemical, biological, and physical hazards. Inventories of agents can be maintained by department, area, process, job, or any combination of these.

Right-to-know and hazard communications requirements make it mandatory that organizations identify substances in each workplace, train employees in their safe use, maintain material safety data sheets, and promptly respond to requests for information. Automated systems can facilitate compliance by maintaining inventories, employee training records, material safety data sheets, and information on right-to-know requests (Figures 14–1 and 14–2).

Protective Measures

Computer systems can maintain information on measures taken to protect employee safety and health through (1) engineering controls such as hoods for ventilation, (2) administrative controls for health hazards, and (3) personal protective equipment assignment and fit testing. These functions allow an organization to maintain the information needed to facilitate regulatory compliance and document the measures taken to protect employees.

Employee Health

Monitoring employee health is a key function of comprehensive occupational safety and health systems. The protection of employees is, after all, a major goal of safety and health programs. This function is used primarily by occupational medicine specialists to manage and store records related to employee health surveillance.

As with other system functions, individual components of employee health programs can be automated. Hearing conservation program records are a frequent candidate. Noise monitoring, audiogram, and employee history data can be entered and stored. These data can be used to produce monitoring schedules, threshold shift evaluations, notices to employees, and summary reports for program management.

Computer-Assisted Training

More and more, computer systems are being called upon to deliver information to workers, supervisors, and managers. When that information is repetitive, highly complex, or technical, computer-based training (CBT) is an ideal, cost-effective way to reach a large audience. Chemical handling safety, lockout/tagout, safe practices, and emergency action planning are examples of the training being performed with a computer. The computer records student responses, repeats material until mastery is obtained, and tests for comprehension. Students set the pace with CBT and control how material will be presented, an ideal solution to a diverse workplace population (Figure 14–2; see also Chapter 16, Safety Training, and Chapter 17, Audiovisual Media, in this volume).

REGULATORY ITEMS

Another function of computers is to provide information on regulatory requirements, including OSHA permissible exposure limits (PEL) or action levels or internally adopted standards, such as the American Conference of Governmental Industrial Hygienists (ACGIH) threshold limit values (TLVs). This computer function also can be used to track ongoing events or problems requiring follow-up, such as hazard abatement plans and schedules.

Scheduling

The computerized system can be used to schedule physical examinations, workplace monitoring or inspections, fit testing of personal protective equipment, training, or any other event documented in an occupational safety and health system.

The schedules can be generated by system logic, or they can be generated by users and maintained by the system.

CAMEO

The National Safety Council offers CAMEO DOS and Macintosh programs to assist in management of safety from hazardous materials. The new CAMEO chemical database includes an additional 675 chemicals, for a total of over 3,300, as well as petroleum and explosive products.

The CAMEO system can include databases on: facility information and floor plans; Tier II Cards; a facility map out to 10 miles; vulnerability/risk screening and scenarios for each environmental health system (EHS); emergency contacts; fire hazard data sheets; and electrical isolation data sheets. ALOHA, the CAMEO Air Model, was used to develop the scenarios.

In the event of a fire, CAMEO would allow an incident commander quick access to the area's floor plan, where further data on chemicals and other hazards could be obtained. Several special databases or stacks were developed for this system. The "locator" stack consists of multi-layered floor plans of the reactor building and even subdivides these into fire zones. The "Fire Hazards" stack contains key data on special hazards, such flammable solids, firefighting suppressant, radiological concerns, protective clothing, and environmental safeguards. And, an "Isolation" stack provides information to turn off electrical current.

CAMEO permits access to information by emergency response personnel much more rapidly than traditional systems. And with CAMEO on computers located at both the Emergency Operations Center and on an emergency response vehicle, personnel will be able to respond more effectively during an incident.

DEVELOPMENT AND IMPLEMENTATION OF A COMPUTER SYSTEM

There are five steps to the successful development and implementation of an automated system:
1. understand the organization's operations and needs
2. identify and review software and hardware meeting those needs
3. purchase or develop one's own system
4. implement the system
5. evaluate the system.

Understanding Needs

When an organization first considers buying a computer system, management should ask how the system will be used, who will use it, and what features are necessary to satisfy the needs. Those planning the system should make a wish list of all the things they would like to see automated. They need to look at their filing systems and ask, can this be done more efficiently with a computer system?

For instance, in the manual system for gathering data on workplace injuries, data collection may be inconsistent, sometimes incomplete. As a result, the preparation of the

OSHA 200 report takes a considerable amount of time. There are a number of questions to answer before converting to a computer system:

- Where will the incident records come from? If there are branch plants or offices, will each branch forward its records to a central office for compilation or maintain them separately?
- Who will input the records? Will a clerical person be given the task or will the safety manager handle it?
- How many injury cases are recorded each year?
- What information must be maintained on each injury?
- What kinds of analysis does management want from a system? If the records are consistently maintained, what do they reveal about safety in the workplace?

If right-to-know compliance needs to be automated, the following questions should be answered:

- For how many chemicals must material safety data sheets (MSDSs) be maintained?
- Where is the chemical stored? In one plant or many? How will managers know if they have reached the reportable threshold?
- How is the chemical inventory maintained and where? Does the safety manager receive an MSDS with each new shipment of chemicals?
- How does management track which employees have been trained on which chemicals?
- Is there a need to prepare Tier reports or Form R reports for hazardous chemicals at the work site? How will they be done? Who will perform the calculations?

Examining needs will set the direction for choosing the proper system. However, an organization may find that so many functions could benefit from automation, the task may appear overwhelming at first glance. Many safety professionals have prioritized the wish list and tackled one function at a time before proceeding to the next. In this way, automating becomes manageable and later implementations proceed smoothly with the benefit of early experience. Budgetary considerations may also dictate the need for "going slow."

Identifying the System

After developing a list of needs, management should have enough information to identify the computer system capable of meeting those needs. Basically, there are two options: (1) purchase commercially available software products, sometimes known as "off-the-shelf" programs, or (2) develop one's own unique system with the help of skilled programmers.

Since the late 1970s, occupational safety and health software has been commercially available. Many good packages are offered for sale that, in some cases, have been in use for years. Two of the biggest advantages to choosing off-the-shelf software are cost and convenience. As the development cost of creating a software product is quite high, commercial sale allows those costs to be spread over many more users. An organization can benefit from a product that has been well tested, accepted by the profession, and made affordable. As for convenience, most programs can be installed by

nonexperts with little or no training.

The major drawback to off-the-shelf packages is their lack of flexibility. It is rare to find a commercial package that can be customized to fit every situation. Packaged software may force a company to change the way certain records are maintained or change some operating procedures.

While some software vendors do offer their product in what is known as "source code" form to allow for modifications, the cost is generally higher. And once a company has modified a program, it is on its own. The developer may offer little or no assistance if modifications fail to perform as expected.

The alternative is custom-developed programs that fit an organization's needs exactly. For a significantly higher cost, management will have a set of programs that mimic paper systems and that generate all the reports needed to exact specifications. Management must be prepared to wait patiently, however, while the system is designed, developed, tested, and implemented. Management involvement will be greater, too, because the software programmer will need highly detailed information about company needs and reporting requirements.

Developing the System

If an organization decides to develop its own system, managers must participate in the design and development process. It helps at this stage to get some assistance from a systems analyst. The intricacies of system development will depend on the size and complexity of the company's system and the hardware and software environment in which it will be built. Naturally, a small PC-based system for management of one or more unrelated functions will require less effort and systems expertise than will a comprehensive safety and health system.

Many organizations build and implement large systems in phases. This can work well if management prepares an overall plan for total system design that is followed during the phased development. Too often, groups have built system modules without considering total system requirements and necessary connections between modules, thus leading to project failure.

Implementing the System

Implementing the new information system should be a joint effort between system developers and users. While there are many activities associated with implementation, the most important one is user training. This training should establish realistic expectations among users about the system's capabilities and requirements. Any changes required in record-keeping procedures and potential transitional difficulties should be thoroughly explained.

Training should be conducted as part of the everyday work routine. Naturally, managers should hold some general training sessions, but the most effective approach is on-the-job training. Safety and health professionals can learn to use terminals for data input and output in the context of their actual work.

Training should be timed to prevent any gap between training and actual use of the system. When something entirely new is learned, most people need to practice it immediately to retain the knowledge. The longer the gap between training and system use, the less retention there will be.

Follow-up training also is important. Periodically, the system manager should determine if users' expectations for the system are realistic and if they are using the system correctly. Error rates should be studied. If they are too high, the reasons should be identified and the users retrained. System success depends just as much on follow-up training as it does on initial training.

Every system should have a manager or administrator. This person has ultimate responsibility for the day-to-day operation of the system, including (1) management of system security, (2) supervision of the data content, (3) problem solving, (4) coordination of changes to the system, (5) archiving of data, and (6) planning for future needs and applications. The system manager should know the application well and work closely with users to make sure the system meets user expectations.

Evaluating the System

After an information system has been implemented, management should evaluate it annually. Estimating an information system's value to an occupational safety and health program can be difficult for the same reasons that conducting a cost/benefit analysis is difficult. However, evaluation is an important follow-up. The main factors to consider when evaluating a system include:

- completeness
- reliability
- user acceptance
- costs
- improved availability of information
- new capabilities and flexibility.

If problems are identified in any but the last area, corrective steps should be taken. Problems can result from human, hardware, or software elements of the system.

SYSTEM SOFTWARE

Many companies are making PCs available to managers for information management, word processing, spreadsheet manipulation, and electronic mail. They are part of an overall strategy for office automation.

Word processing is one of the most favored software packages because it is easy to learn and has a range of office applications. The user can create and change text on-screen before printing it out. The text is saved, then retrieved at a later time for possible editing before reprinting. The computer does indeed become a typewriter, but with additional features that enable users to complete work in about half the time it would take a typist to accomplish the same task. Typical word processing safety applications include quarterly reports, audit reporting, mail merges for multiple mailings, and custom notices and posters for safety committee activities.

Database management systems are an integral component of most contemporary occupational safety and health systems. All database management systems allow the storage, sorting, and retrieval of information in useful ways. They have tools for building files and reports, and they offer programming languages or commands for building complex programs.

Many different database management systems are available. Some safety and health professionals are using them to develop automated functions on PCs. There are database management systems available for all types of computers—some are easier to use than others. If an organization decides to develop its own applications software, management should look to the organization's information center or data processing department to find out what is available and if development support is offered.

Applications are prewritten programs that perform specific functions, most often unique to a particular departmental task. Safety software is available to create and retrieve MSDSs, track and record injuries and illnesses, evaluate chemical protective clothing, perform in-depth accident investigation, and even plot a potential chemical spill.

Spreadsheets allow users to create analytic tools that often involve multiple columns and rows of data. Financial spreadsheets look at budgets, profit and loss, and other corporate accounting functions. The safety professional might use a spreadsheet to calculate incidence rates, compare lost work-time costs among several departments, or prepare Tier reports.

SYSTEM HARDWARE

The types of computers to be used for an occupational health and safety information system must be determined during the requirements stage. Today, there are many alternatives for computers, ranging from large mainframe computers to PCs, and information processing can be accomplished on one or many of these.

Most safety and health departments are using PCs for information management. Personal computer technologies are evolving at a rapid pace. Every year, vendors offer greater speed, more memory, and better compatibility among hardware and software system components.

The key features to look for in a computer are speed and capacity. Speed is determined through use of fast processing chips that handle large chunks of data.

Capacity of a computer is measured by its memory and storage components. Memory is the work space of the computer, much like a manager's desk is his or her workspace area. A computer may have as little as 640 kilobytes of workspace memory (also known as RAM for random access memory) and perhaps as much as 16 megabytes—a very large desk indeed!

Storage is accomplished with hard disk drives, which are like large filing cabinets. Capacities range from a meager 20 megabytes (about 1,000 pages) to 40, 80, and 100 megabyte sizes to several hundred megabytes. The greater the storage,

the more programs and data can be stored for rapid access.

More speed and more capacity generally mean more money. Because of the rapid advances in PC technologies and declining prices, many companies have developed occupational safety and health systems relying solely on PCs. Some use local area networks (LANs) to link the PCs to one another either directly or through a shared mainframe computer.

Other configurations also are being used in the development of safety and health systems. They include (1) use of a mainframe or minicomputer for all system functions, with terminals distributed to users and communication conducted over telephone lines; and (2) use of mini-frame or maxi-frame computers for regional database management, with terminals or PCs used for data input and communication lines among the larger machines for corporate-wide analyses.

Many factors influence an organization's choice and location of computer hardware for its information system. These include currently available hardware, corporate standards for computer hardware, telecommunications, software, and the information requirements of system users. All system options and constraints should be scrutinized during the requirements study to create an economical, efficient, and integrated system.

Increased internal and external pressures have made efficient data management more important than ever. Regulatory compliance, rising costs, and the desire to reduce worker injuries and deaths have prompted many companies to adopt state-of-the-art computer systems. Today, the safety professional can choose from among many programs and computer systems, or develop a system tailored to specific company needs. Regardless of the choice, a well-thought-out system will provide better information for creating and maintaining a safer workplace.

LOSS CONTROL IN THE COMPUTER ROOM

As the electronic data processing (EDP) system is expanded and tailored to meet the firm's specific needs, it becomes increasingly critical to protect it. Many firms would be in serious trouble if their EDP systems went down for any length of time. Even an organization with a relatively small system could suffer devastating losses if its system became unable to function. Lost data, business interruption, failure to meet customer needs, subsequent loss of customers and replacement costs of EDP equipment—these are only a few of the losses that could occur.

Just as the intelligent businessperson recognizes that EDP equipment must be used to stay competitive in today's marketplace, management needs to recognize the potential for costly loss of that equipment. Managers should take positive actions to reduce this loss potential. The safety professional can be the appropriate person to point out to management not only the possible hazards but also safeguards that can substantially reduce the exposure.

The National Fire Protection Association (NFPA) outlines recommended safe procedures to be used when installing EDP equipment. This chapter contains a summary of the more important requirements outlined by NFPA. More specific information can be obtained by reviewing the NFPA standard, *Protection of Electronic Computer/Data Processing Equipment*, 75. (See also Chapter 6, Fire Protection, in the *Engineering and Technology* volume for more details on fire protection and suppression systems for the computer room.)

The Room

Physical location and construction of a computer room should be considered from both a fire and security standpoint. Management should choose interior finish, furnishings, and layout of the room prior to equipment installation, because any necessary changes will be less costly at this time. Naturally, the location of the room is decided during the design stage and after thoroughly reviewing the NFPA 75 standard.

To minimize exposure to fire, water, corrosive fumes, heat, and smoke from adjoining areas and activities, the computer room should not have other processes above, below, or adjacent to it. It should be located away from street-side windows and exterior walls and above ground level for security reasons and to minimize dampness.

Entries to the computer room should be lockable from the outside and equipped with panic bars so operators can quickly exit from the area in an emergency. The operators will need to use keys to enter and exit the room at other times. The emergency exits should have local alarms.

Utility lines, including electric, gas, and phone lines and air-conditioning cables, should be placed to prevent tampering. The floors should be raised and contain easily accessible cable raceways designed with adequate drainage to handle potential domestic water, sprinkler, or coolant leakage or water and chemical runoff from firefighting operations. Cable openings should be covered or protected so debris and other combustible materials cannot fall into the openings. Smoking in the computer room should be strictly prohibited.

Contents. Only the EDP equipment, approved furnishings, and auxiliary electronic equipment should be kept in the computer room. To minimize the chance of a fire starting, furniture should be metal and the amount of paperwork and paper storage allowed in the room should be limited. The interior room finish should be made of noncombustible materials with no exposed cellular plastics.

The Equipment

After management has decided on the location of the computer room, the next step is to consider equipment requirements. The arrangement and wiring of equipment—such as electric shutoff devices, air-handling equipment, and smoke and heat detectors—play a substantial role in providing adequate fire protection.

The safety professional should help to arrange equipment to allow for quick, easy emergency exit from the room. Operators must be able to quickly shut off the equipment as they exit. This step is especially important when the room is equipped with an automatic sprinkler system.

The electrical system must be adequate for the expected power load. It should be designed to include devices that protect equipment against power surges, brownouts, and power failures.

Specific equipment requirements include (1) emergency electric cutoffs both at the operators' stations and at the exitways (the electric cutoffs should also be labeled); (2) electric service rated at 125% of the total amperage load for the equipment; and (3) covered electric junction boxes at a 15-foot maximum distance apart on flexible electric cords.

All air-handling equipment should be connected to a separate electrical system. Dampers should be provided on all ducts; the dampers should be wired into smoke and heat detectors so they will automatically close when sensing the presence of fire.

The entire equipment complex should have an auxiliary source of electricity in case of a power failure. All electrical equipment should be properly grounded and provided with power surge protection and a manual override system to protect against fluctuations in electric power. All automatic systems should have a manual back-up.

It is to the company's advantage to arrange for emergency use of other facilities so data can be run even if there is an equipment failure. Before the company completes arrangements for emergency equipment use, management should carefully evaluate the rental expense and obtain in writing any agreements reached with other equipment owners. The best time to make emergency use arrangements is before loss occurs.

Waste containers in the computer room should be metal with self-closing lids. If any sound-deadening materials are used, they should be noncombustible. In addition, flammable liquids should not be kept in the room, and liquids with a flash point of less than 300 F (150 C) should not be used for lubrication.

Records Protection

The loss of vital and important records has caused some businesses to undergo such hardships as lowered credit ratings to near bankruptcy. Today, these losses are unnecessary; equipment is available that can adequately protect records from extreme temperatures, smoke, water damage, and other conditions related to fires.

A description of the available equipment and other control measures can be found in the NFPA standard, *Protection of Vital Records*, 232. Only those records that need to be frequently used or referenced should be kept in the EDP room. All other records should be located elsewhere, and protected according to their value. Vital and important records should not be stored in the EDP room.

When appropriate, record duplication is the best control measure. The original document is stored off the premises, and the duplicate is used and stored at the main operating location. To prevent records from being stolen or damaged, management would be wise to investigate the security practices of off-premises storage areas. Both off-premises and in-house record storage areas should have automatic sprinkler systems for the best protection against fire loss.

Periodic in-house records security checks should be conducted and should include assessing records in storage and in use, the EDP equipment, and records of computer time use. In addition, all personnel using the equipment should undergo security screening: reference checks, previous employer checks, salary versus standard-of-living studies, and other relevant verifications.

The EDP equipment can be further protected through internal system safeguards such as error-checking circuitry, redundancy checks, limiting transactions above a stated amount, bound registers, and data encrypting. These security features should be checked periodically to ensure they are working properly.

Safe Storage Practices

Many organizations commonly store extra computer paper in the EDP room. This can be poor practice if the volume of unused and scrap paper is allowed to accumulate since the fire risk is accordingly increased. A minimum supply of paper can be kept in the EDP room, and all scrap containers should be frequently emptied.

Another common practice is to locate fire-resistant file cabinets or media safes in the EDP room. This clutters the work area and can block exits or entryways. Again, only those records needed for daily operation should be kept in the EDP room. This area should not be used as a storage room but should be considered a vital area deserving high-quality fire and security protection. If fire-resistant cabinets or media safes must be located in the EDP room, they should be kept closed, except when actually retrieving or replacing material.

All cabinets or safes used for records storage should be Underwriters Laboratories Inc. (UL) listed. Such equipment is classified by an interior temperature limit and a time limit (in hours). Two standard temperature and humidity limits are used:

1. 150 F (65.5 C) with 85% relative humidity (for photographic, magnetic, or similar non-paper records).
2. 350 F (196 C) with 100% relative humidity (for paper records).

Time limits are 1, 2, 3, and 4 hours. Ratings are listed by class and time: for example, Class 150–1 hour means the internal temperatures will be held to 150 F for 1 hour in normal fire conditions. On UL-listed equipment, the rating is shown on a metal label affixed to the cabinet or safe door.

Utilities Protection

Utility protection also needs to be considered, the prime concern being protection against sabotage. Exposed or overly accessible electrical lines or phone lines present a good target for anyone bent on causing serious monetary damage to a firm through interrupting business operations.

The safety professional should ensure that air-conditioning intake ducts are regularly inspected and maintained to prevent damaging gases, vapors, fumes, or mists being drawn into the EDP room and possibly damaging

or corroding equipment. Particulate matter should be removed from the air by UL-listed filters or by electrostatic precipitators.

Other utility protection measures for EDP rooms are:

- make security checks of all utilities leading into the room
- prohibit electric transformer use in the computer room
- make provisions for protection against lightning surges in accordance with NFPA 70—*National Electrical Code*
- provide an emergency lighting system to allow for safe employee exit from the area in an emergency.

COMPETITIVE EDGE

As mentioned earlier, EDP equipment has become a vital and commonplace tool for business, industry, government, and research groups. The procedural steps (outlined earlier) for installation and protection of EDP equipment are necessary for any business that wants to remain competitive.

When management plans ahead, potential problems can be eliminated at the design stage. The computer room location, equipment, fixed and portable fire protection systems, records and utilities protection, and employee safety must all be considered, and proper safeguards installed. By taking positive steps to minimize exposures to loss, the intelligent businessperson will be able to take better advantage of the competitive edge provided through EDP equipment.

With permission of *Professional Safety Magazine*, the preceding portion of this chapter expanded and updated an article appearing in that publication. (Klonicke, 1983.)

SUMMARY

- Computers are used by safety professionals for a variety of activities from compiling reports on occupational injuries and illnesses to networking with other safety professionals' computers to exchange information, documents, and ideas. These electronic tools help management obtain information necessary to make better decisions, help standardize recordkeeping, assist with regulatory compliance, and help protect employees and organizations.
- Computer systems consist of hardware and software components. Changes in computer technology over the past decade have made personal computers (PCs) faster, more powerful, and more affordable than ever before.
- Many organizations use computer systems to store, analyze, and report on accident experience data, assist in accident investigation, monitor workplace conditions and safety efforts, assist in disseminating information on hazardous substances and employee safety education, maintain files on measures taken to protect employee safety and health, manage employee health and medical records, and provide computer-assisted training programs.
- Computers also provide information on regulatory requirements to ensure compliance with various federal, state, and local ordinances. Operators can program computers to keep track of ongoing events or problems requiring follow-up and schedule physical examinations and other events required in an occupational health and safety system.
- Five steps are needed to develop and implement a successful computer system: (1) understand the organization's operations and needs, (2) identify and review appropriate software and hardware, (3) purchase or develop a system, (4) implement the system, and (5) evaluate the system. Management should tailor computer capabilities to its specific needs as closely as possible.
- Once the system is installed, user training should be conducted on the job. Operators should practice what they learn and receive follow-up training to ensure they are using the system to its best advantage.
- Every system should have a manager or administrator who is responsible for management of the security system, supervision of the data content, problem solving, coordination of changes to the system, archiving of data, and planning for future needs and applications. Periodically, the system should be evaluated for completeness, reliability, user acceptance, costs, improved availability of data, and new capabilities and flexibility.
- Word processing and database management are two of the most common software packages used in occupational health and safety departments. Programs may be purchased off the shelf or customized by programmers.
- Factors influencing a company's choice of computer hardware include corporate standards, telecommunications systems, software, and information needs of users. The key features to look for in computer hardware are speed and capacity. Speed is determined by fast processing chips while capacity refers to computer memory and storage components. Companies must be sure to buy computers whose speed and capacity will enable them to handle an organization's growing workload.
- As computers become more vital to a company's operations, they must be protected. The computer room should be designed to guard against fires and security breaches. The room should be located away from other operations, have locked entryways and exits, have well-protected utility lines, and contain only data processing equipment and supplies. All furnishings and interior finishes should be noncombustible.
- Equipment should be arranged and wired to protect against fires and fluctuations in power supply. Management should provide an auxiliary source of power in case of electrical failure, and all electrical equipment must be properly grounded to guard against shocks and burns.
- Vital and important records must be protected by adequate storage, backup, and security measures. All personnel using computer equipment should be screened by security staff. The equipment itself can be protected through internal system safeguards such as error-checking circuitry.
- Computer supplies and paper should be stored in high-quality cabinets or safes that are fire-resistant and offer security protection. Likewise, all utility lines must be secured against sabotage to prevent loss of vital data or disruption of business operations.

REFERENCES

Best's Loss Control Engineering Manual. Oldwick, NJ: AM Best Co., published annually.

Helander MG, ed. *Handbook of Human/Computer Interaction.* Amsterdam and New York: Elsevier, 1987.

Klonicke DW. Loss control in the computer room. *Professional Safety* 17–20, April 1983.

Miller E and O'Hern C. Microcomputers can make it work for you. *Safety & Health* 35: 28–32, 1987.

National Fire Protection Association, Batterymarch Park, Quincy, MA 02269.

Fire Protection Handbook, 16th ed., 1986.
Standards:
Protection of Electronic Computer/Data Processing Equipment, NFPA 75.
Protection of Records, NFPA 232.

National Safety Council. 444 North Michigan Avenue, Chicago, IL 60611. *Occupational Health and Safety,* 2nd ed., 1992.

Ross DT and Schoman KE. "Structural Analysis for Requirements Definition." In: *Software Design Techniques,* 4th ed. Freeman P and Wasserman AI, eds. Long Beach, CA: IEEE Computer Society, 1983.

PART 4
PROGRAM
IMPLEMENTATION

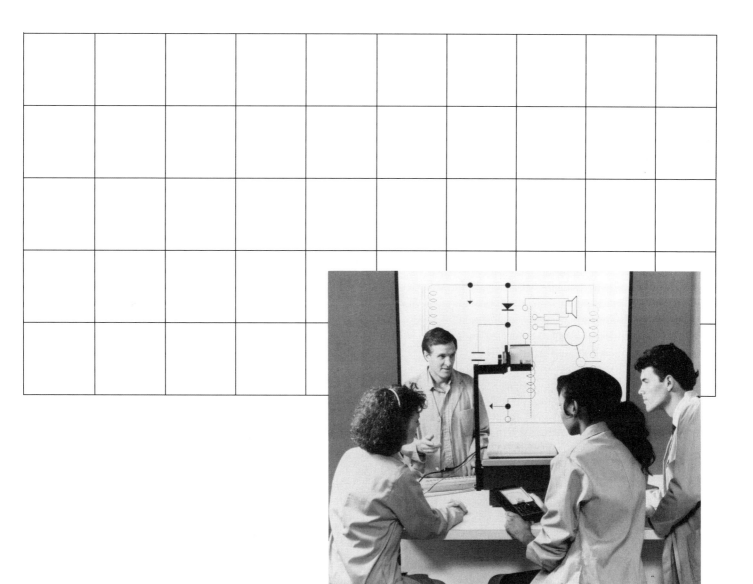

Computerized overhead projection system. (Courtesy 3M Visual Systems Division.)

15

Attitudes, Behavior, and Motivation

The industrial safety and health professional assists line management in achieving maximum production by preventing or mitigating work-related fatal or injury accidents. The occupational environment is composed of interacting components, primarily the worker, materials, and equipment. A comprehensive safety program addresses all aspects of the work environment and recognizes that safe behavior-management stands at the center of the program. Consequently, a major responsibility of line management, with the safety and health professional's assistance, is to motivate workers to change their behavior so that working conditions are safer.

This goal does not mean other facets of a sound safety program are less important. Line management and the safety and health professional are also concerned with safe design and plant layouts, safety devices on machines and use of such devices by employees, the wearing of safe clothing and use of protective equipment—all of which contribute to the reduction of disabling accidents.

However, when despite every precaution on the part of equipment manufacturers, materials suppliers, line managers, and safety and health professionals, accidents still occur, the human element often emerges as an essential factor. If accidents are to be reduced, the safety and health professional must assist supervisors and other line management, as well as individual employees.

This chapter is designed to promote understanding of human behavior in the work environment it complements. It explains why some people have more limitations than others when following safe practices and health and safety procedures.

PSYCHOLOGICAL FACTORS IN SAFETY

The full role and responsibilities of the industrial psychologist are too broad to be covered in this chapter. Some psychology topics and issues play a direct part in the success or failure of sound personnel procedures in industry, but not all of these procedures are directly related to safety, nor do they fall within the assigned duties of the safety and health professional. They can, as part of the regular personnel procedures, indirectly contribute to safety in the workplace.

Psychological factors that influence safety program success include individual differences, motivation, emotions, stress, attitudes, behaviors, and learning processes. These are described below.

Individual differences. Individual differences among employees is an ongoing problem. Although these differences are constant and obvious, some factors are common to all people, a fact that supervisors and safety personnel need to understand when dealing with work groups.

Motivation. Understanding what motivates people is important. When someone has an internal drive to acquire something, it can be said they are motivated to acquire it. It is important to realize that someone can be equally motivated to not want something. To use a guard to protect one's fingers from a saw is, perhaps, indicative of motivation for safe practices, but the desire to ignore a safety device because

347

it might decrease production is also a motivator. Safety personnel and others need to consider the concept of conflicting motivators in any attempt to understand human actions.

Emotion. While emotions can be constructive at times, they also can be destructive—working to the detriment of both the individual and the safety program. Emotions such as anger, fear, excitement also can interfere with the thought processes, resulting in behavior that conflicts with a rational approach to work tasks.

Stress. One definition of stress is factors that accelerate the rate of aging through wear and tear of daily living. Stress is always present and can be both good or bad. How employees choose to react to the stress on or off the job is very important.

Attitudes. Industry has recognized the effect employee attitudes can have on production, plant morale, turnover, absenteeism, plant safety, and the like. As a result, management has spent much time and money trying to determine workers' attitudes. Although workers' attitudes toward safety and health procedures and policies can make a difference between an effective safety program and an empty set of rules, accurately determining an individual's attitude can be difficult. Attitudes are internal and cannot be measured. Accordingly, it is necessary to observe and measure employee behaviors. Change behaviors and you modify attitudes.

Learning processes. Finally, management should be concerned with how people learn on the job. One cannot understand motivation, attitudes, emotions, or even individual differences without some consideration of the learning process involved in establishing each worker's outlook and personality.

Much of the success of a safety program depends upon its acceptance by those to whom it is directed. Program acceptance in turn depends on an understanding of the psychological factors influencing program success.

The safety and health professional and line management must effectively explore and find factors common to the group that can be used to promote safe conduct on the part of all employees. The basic question is, "What factors associated with human behavior can be used to increase the effectiveness of safety programs?"

Five of these psychological factors are discussed below.

INDIVIDUAL DIFFERENCES

When a chemist analyzes a chemical compound, its exact nature and composition can be accurately specified. When this compound is used, its behavior can be predicted. The action and reaction of any one sample are the same for all other samples of the compound.

When studying human behavior, however, the psychologist is not dealing with the same degree of certainty. The psychologist does not know the composition of the person being dealt with—in many instances, there is little knowledge of the subject's past history.

In addition, the behavior of one person is not the same as the behavior of another person. Because behavior can be influenced by both the individual's attitude and situational variables, Person A in the sample is not equivalent to Person B in the same way one cubic centimeter of distilled water equals all other cubic centimeters of distilled water.

The known fact that people differ has been referred to as the "personal equation" or, more commonly, "individual differences." The personal equation presents many problems for both the safety and health professional and supervisor. Yet within the framework of individual differences, certain general patterns common to a group do exist.

For example, human behavior is typically motivated by some belief, need, or drive. Regardless of what the individual does, there is usually some purpose driving the behavior. For example, the purpose may be to resolve or reduce some basic tension. The degree and nature of this tension, or dissonance, will depend in part upon the individual's attitude, values, or perception of situational variables (job expectations, work conditions, physical ability).

Whether or not an employee works safely depends upon (1) the present situation—Is the employee rushed? fatigued? in poor health? (2) past experiences—Were accidents avoided in the past? What amount of training does the employee have? and (3) workplace and methods design—Were the job and workplace ergonomically designed?

The Average-Person Fallacy

Although each person differs from every other person, the important fact is how one deals with individual differences. Figure 15–1 presents the distribution of scores in a normal curve. Many human characteristics (for example, anthropometric characteristics or IQ) are assumed to be distributed according to the normal distribution (often referred to as the "normal curve"). Note that in such a distribution, half the persons are below the average line (called the mean) and half the persons are above it.

Often managers and safety and health professionals realize individual differences exist, but it is physically impossible to deal with each person separately. They may erroneously attempt the next best thing—appeal to the "average person."

Unfortunately, inept use of the "average" as the main characteristic of a population has distorted statistical data and produced a misconception called the *average person.* People, in reality, are highly unlikely to be average. For example, in body dimensions, which can be measured more accurately than emotional, behavioral, or intellectual characteristics, less than 4% of a test group had three common average dimensions. Less than 1% were average in five or more dimensions at the same time. Therefore, when an appeal is aimed at the average, it misses much of the population.

A better approach is to use percentiles. When designing the system, instead of setting the standards to fit the average, design it to fit all but the upper 5% and the lower 5%. The system will then fit 90% of the population.

The use of percentiles (for example, between the 5th and 95th percentile), rather than averages, has long been the technique for relating individual test scores to a group as a

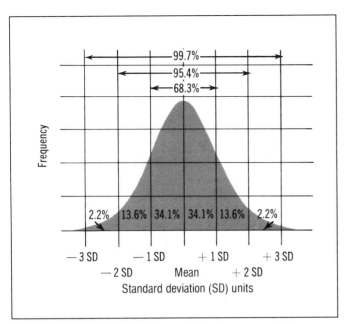

Figure 15–1. Many human characteristics are assumed to be distributed according to the normal curve shown here. Because the normal curve has a known shape, it is possible to state the percentage that lies between +1 and -1, standard deviation (SD), or between any other two points expressed in SD units. Thus, it is easy to describe a normal distribution of data using simple statistics—the mean (average) and the standard deviation (SD). For example, 68% of all the measured values lie within ±1 standard deviation about the mean; 95% are within ±2 SD, and so on. Given the mean and the standard deviation, therefore, the complete distribution in a normal distribution can be determined.

whole. This system can also serve as a guiding rule in the design of human-machine relations.

Fitting the Person and the Job

Very often when replacing or relocating personnel, managers strive to find the individual possessing most of the characteristics deemed necessary. Job specifications and employee requirements are given in terms of a minimum, but very seldom are psychological factors expressed within a minimum and maximum (see Chapter 5, Ergonomics in the Workplace, in the *Engineering and Technology* volume).

Often the ideal candidate is pictured as being the most intelligent, most loyal, most productive, etc. In other words, management tends to look for the perfect person rather than for the right one. For example, a deaf person, properly trained for a task in which appropriate noise levels cannot be controlled, would experience fewer problems than a hearing person, properly trained for the same task, because of the issue of wearing protective hearing devices.

Another example is the person who comes to work under any condition, whether ill or not. This person may be as undesirable as the one who uses every excuse to take off. The worker with a high fever, bad cold, or sore back is a potential hazard to all concerned, especially if on heavy medication.

Physical characteristics of the individual. Reaction time, psychomotor skills (for example, manual dexterity), and visual

abilities seem to have at least some bearing on safe performance. While the extent to which they are directly responsible for accidents is neither clear nor constant, it appears that a certain minimum degree of physical competence is required for successful, accident-free performance. Some types of jobs demand superior physical abilities while others do not.

Most of the research to date has considered these factors individually. More studies have examined the connection between combinations of physical shortcomings and accidents. Such studies were initiated because investigators realized that physical and psychological abilities operate in combination, not separately. For example, one study investigated the relationship between workers' general perceptual skills and their accidents, and not simply the separate factors that make up perception—vision, hearing, etc.

Individual characteristics and the job. Personnel people in industry screen candidates on the basis of the specific characteristics required to perform a particular job. In many cases, both (1) physical characteristics, such as size, visual acuity, and steadiness, and (2) personality characteristics are important. As a result, many individuals are not considered for a job because they do not have all the necessary qualifications.

The same principle applies in other areas of matching the right worker to the right job. When designing equipment, for example, human engineering experts consider physical limitations as well as other human characteristics in an effort to make human-machine interactions as nearly perfect as possible. See Chapter 5 Ergonomics in the Workplace, in the *Engineering and Technology* volume for a discussion of human factors engineering in relation to occupational safety.

Where hazards cannot be eliminated, safeguarding provides protection against the potential failure of people to use equipment correctly. Both safe design and safeguarding minimize the effect of individual differences on accident frequency and severity.

Methods of Measuring Characteristics

Regardless of the technique used to screen, place, and motivate employees, management needs a method to measure safety program effectiveness. Techniques used to obtain feedback range from the across-company accident rates to the within-company approach of safety sampling, or critical incident technique (see also Chapter 13, Recordkeeping and Incidence Rates).

Measuring techniques can be assessed by determining reliability and validity. *Reliability* refers to consistency of measurement. The reliability of a given measurement, such as a test or instrument, can be estimated in various ways. In general, however, the concept of reliability refers to how stable measurements remain. This is the case whether stability is assessed (1) across time or between settings, (2) using the same or a different group of individuals, or (3) for internal consistency or consistency between alternate forms of the same test or instrument.

Underlying the concept of reliability is measurement error. All gages used to assess performance will have, in differing

degrees, some inherent measurement error. Measurement error refers to the estimated fluctuations likely to occur in individual or group performance because of factors beyond anyone's control.

Validity refers to how well the test or instrument measures what it is supposed to measure. The validity of a test or instrument can be assessed in various ways: (1) the relevance or plausibility of items (or the overall instrument) with regard to the given behavior (face validity), (2) whether all aspects within a given concept are adequately covered (content validity), (3) how well the test or instrument measures a specific construct or trait (construct validity), and (4) a demonstrable relationship between performance, as measured by the given test or instrument, and some other related behavior (such as IQ and school aptitude) (Anastasi, 1988).

A measurement can be reliable without being valid, but a valid measurement also must be reliable. To illustrate that a reliable measure is sometimes not valid, consider a yardstick. A yardstick is very reliable—it gives a consistent measurement every time it is used. It is also valid for measuring the length of a table. But if a yardstick is used for weighing the table, it is no longer a valid measure.

In addition to reliability and validity, a measuring technique must be practical. A technique can possess high validity but be so cumbersome and intricate that it can only be used in special situations by highly skilled technicians. In spite of its statistical value, such a technique is almost worthless in most workplaces.

Two sampling techniques used for evaluating potential accident-producing behavior are (1) the critical incident technique, described in Chapter 3, and (2) behavior sampling.

- The critical incident technique involves the following. A random sample of employees is interviewed to collect accident information concerning near-misses, difficulties in operations, and conditions that could have resulted in death, injury, or property loss. Those participating are asked to describe any incidents coming to their attention. This technique can be useful in investigating worker-equipment relationships in past or existing systems, evaluating modifications to existing systems, or developing new systems.

- The behavior sampling or activity sampling technique involves the observation of worker behaviors at random intervals and the classification of these behaviors according to whether they are safe or unsafe place the worker at risk. Calculations are then made to determine either (1) the percentage of time the workers are involved in at-risk practices, (2) the percentage of workers involved in at-risk practices during the observation period, or (3) the percentage of unsafe versus safe behavior observed. Using this technique, management can apply various components of a safety program (such as safety lectures, posters, brief safety talks, safety inspections, motion picture films, supervisory training) and immediately note their influence on workers' unsafe behavior.

Research studies have shown that feedback can be introduced into this process with clearly positive effects on worker behavior and accident rates. Feedback charts are prepared that show the percent of safe behavior observed during each sample. These charts are posted in the workplace and serve as positive reinforcement for safe behavior (see Analyzing and Changing Behavior later in this chapter).

When screening employees or potential employees in regard to their accident potential, human resources personnel must exercise caution. To date, no systematic screening procedures have been developed that meet both reliability and validity criteria and at the same time are adequate for use in all industries or even in a specific industry. While in theory such a screening procedure is possible, the state of scientific knowledge in occupational safety research is too limited to support the development of such screening measures. Therefore, the reduction in human-error potential by means of a good ergonomics program (Chapter 5 in the *Engineering and Technology* volume) is still the best line of defense.

MOTIVATION

Through the interaction of hereditary and environmental factors, each worker develops an individual personality. Line management and the safety and health professional must be continually aware of these individualities when dealing with individuals and groups. Yet, individual differences alone are not all there is to human behavior. People share enough common characteristics that supervisors, safety and health professionals, and managers can influence workers to cooperate for a common cause.

To have all personnel in a company from the president down to the line staff member working together productively and safely is a primary goal of a safety program. Such an effort must be achieved by motivating management and workers to strive for a common goal, a goal that is shared by all parties involved.

It is no wonder, then, that the question most often asked by both safety and health professionals and line managers is, "How do we motivate our employees to work together efficiently and safely?"

One of the best ways of motivating employees is by a joint labor/management safety committee. Such a committee involving the workers and management for safety and health endorsed by all levels of the company can have extremely positive results.

Peer pressure, positive involvement, and recognition combined with utilizing the expertise from every company level for the improvement of safety and health can be a strong motivational tool for the safety and health professional.

Complexity of Motivation

Motivation is perhaps one of the most complex issues in the field of human behavior—people have many needs, all continually competing and influencing behavior. Hersey and Blanchard (1988) called motives the "'whys'

of behavior". Motives and needs create activity and direct the behavior of people.

Psychologists and researchers cannot give clear-cut, concise answers to all the questions that might be asked about other people's motivation. Rather they attempt to describe some basic concepts such as hierarchical motivation, affiliation motivation, and achievement motivation. Understanding these concepts can help managers, safety and health professionals, and others learn about what can motivate workers.

The Hierarchy of Needs was developed by Abraham Maslow. Hierarchical motivation simply means some needs have a higher priority than others in people's lives (Figure 15–2). Once an individual has satisfied the physiological need to sustain life (food, clothing, shelter) the other needs can become dominant. Having satisfied the physiological needs, safety and security needs take precedence over social needs (recognition, affection, social approval). When the safety needs are satisfied, then the social needs become more pressing. For example, a line manager may feel that acceptance and recognition by workers is a major need to be satisfied. Upper management must realize that supervisors who care less about enforcing an unpopular safety rule and more about the affection and recognition of the workers, are not performing their job adequately. Employee safety has the highest priority.

This hierarchy is not static. The strength of certain needs will change depending upon the individual's level of satisfaction. As an individual satisfies a need, that need is no longer as strong a motivator as an unsatisfied need. By satisfying related needs, the individual advances to the succeeding hierarchy level.

A second area of interest to safety and health professionals is the difference between affiliation motivation and achievement motivation. Affiliation motivated people need to be accepted by others and to feel that they belong. They will be motivated by needs which gain them acceptance by the group or retain membership in the group. Achievement motivated individuals are more concerned about the outcome, or results of a task. Achievement motivated people are not extreme risk takers. They will attempt to provide solutions to problems rather than leaving the outcome to fate or chance. Although the achievement motive in a group can be isolated and measured, David C. McCelland (1961) found that while some people have an intense need to achieve, others do not appear to be as concerned about achievement.

People may not be achievement motivated because achievement motivation requires the development of a need. A person may have a high need for achievement, but if the individual is not challenged the need might not be aroused. The individual may lack an understanding of how to control a significant safety problem. By presenting the problem in such a way that it challenges the individual, it increases the likelihood that the person will have an interest in solving the problem. Thus, to develop achievement motivated workers we must create a need that provides clear objectives, a reasonable probability of success, and measurable feedback.

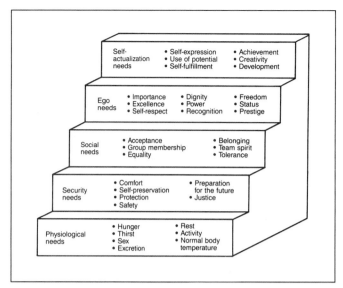

Figure 15–2. Abraham Maslow investigated the various needs that people have. Next he arranged them in a hierarchy; note that tissue needs are near or at the base, and that more intellectually satisfying needs are near the top.

The expectation of success or fear of failure are two important categories of achievement motivation. The employee who works safely because that is the person's goal is success oriented because of the intrinsic value of the goal. The person who works safely because the individual is afraid of the potential negative consequences, such as a serious injury, is doing so to avoid the consequences of failure rather than pursuing the goal for the sake of achievement.

Safety and health professionals and other individuals should never assume that a worker's satisfied need will remain satisfied forever and thus they do not need to continue to give recognition, or affection, or social approval. Situational change at work or home can create motivational changes within the needs of the individual requiring continued feedback.

People strive to satisfy their needs but do not always attain them. Also, individuals will seek to satisfy their needs in different ways at different times. For instance, the need for acceptance may be satisfied through social approval one time while at another time it may be satisfied through affection from family members. Thus, in dealing with people, one must recognize that what was effective yesterday may not work today, although the person's behavior is to some degree similar.

Being aware that people continually seek to satisfy their needs makes it somewhat easier for the safety and health professional or anyone else to deal effectively with people in the work situation. The safety and health professional must develop the ability—and help supervisors to develop the same ability— to provide people with the type of satisfaction they require.

Safety and health professionals and supervisors must spend time together and with workers so they can come to know one another and the workers very well. The safety and health professional must clearly understand what the workers want

from their job. Is it good working conditions, promotion and growth, interesting work, job security, or good wages? They can then determine with a fair amount of accuracy what a worker's behavior at that moment means in terms of need satisfaction. This awareness is based on the little cues in the individual's behavior that the safety and health professional and supervisor can learn over time to recognize, provided they have made the effort to know their people.

Careful observation is required to obtain this type of knowledge. Safety and health professionals should advise line management that any supervisory training program includes effective worker observation methods.

Job Satisfaction

In the interest of further understanding motivation in the work environment, many studies have been conducted to determine what constitutes job satisfaction. Generally, these studies assess the job elements workers claim contribute to their job satisfaction (or dissatisfaction). The results of these investigations suggest that satisfaction of psychosocial needs rather than physiological needs may be the major motivational aspect of job satisfaction.

Table 15–A presents the results of various surveys of job satisfaction conducted over the years. The numbers represent rankings of the factors considered in each study. Because different language and alternatives were used in each survey, the factors have been paraphrased to represent the various elements.

The results of these surveys suggest that high pay is not in itself a primary job motivator. Although workers expect a just and equitable income, they appear to expect only what others would be paid for comparable work. While workers might feel dissatisfied if underpaid, higher pay alone does not guarantee job satisfaction (Clark, 1958; Adams, 1965).

On the other hand, steady work or job security does appear to be a primary job motivator. Workers want the security of knowing if they perform their jobs well, they will have a job in the future. Job security, as a component of job satisfaction, can explain the willingness of a worker to remain in a low-paying, stable job instead of accepting a high-paying, less-stable job.

Other factors appearing important as job satisfiers include type of work, opportunities for advancement, and good working companions. Note that all of these factors seem related to the psychosocial needs of feeling important and belonging to a peer group. Likewise, comfortable working conditions (rated high by a number of employees) are probably associated with a desire to be humanely treated.

The results of these surveys on job satisfaction are important when considering the safety program within the context of personnel policy. Inasmuch as the safety program is designed to ensure the well-being of the employee, it helps to maintain the employee's continued ability to do the work (which, in turn, gives job security).

Likewise, the safety program represents management's interest in the co-workers and working conditions of the employee. All of these aspects of safety programming should

be anticipated and incorporated into the approach which is taken with both supervisory staff and the employees. Honest and sincere positioning of the safety program to enhance employee welfare makes practical sense in light of current knowledge regarding job satisfaction.

Management Theories of Motivation

There are many theories in psychological literature addressing human motivation. Within this literature, specific theories have evolved with special reference to management as it exists in industrial organizations. Although only two such theories are presented here, other equally cogent points of view, such as Blake and Mouton's attitude-based Managerial Grid III or Hersey and Blanchard's behavior-based Situational Leadership, could be discussed (For additional information see the Suggested Reading section at the end of the chapter).

Because theories of human motivation lack sufficient data to support all their tenets, they might best be viewed as philosophies of management. They are important to the safety and health professional because, if accepted, they can influence the direction in which management seeks to develop and implement a safety program.

Theory X and Theory Y

In an attempt to analyze how management regards human motivation, McGregor (1985) discovered two basic ways in which management views workers. Which view a company accepts determines the management practices that are adopted to run the company.

Theory X, according to McGregor, assumes the worker is essentially uninterested and unmotivated to work. In order to change this attitude, management must motivate workers through various external motivators—rewards (bonuses, prizes, etc.) or punishments (docked pay, fewer privileges, etc.). In effect, the worker is motivated to work by virtue of the external reward and punishments offered. Thus, under Theory X policy, management uses control and direction as the means of worker motivation.

Theory Y, according to McGregor, assumes the worker has the potential to be interested and motivated to work. In fact, work is assumed to be as natural and desirable as other forms of human activity, such as sleep and recreation. Under such circumstances, management is confronted with the role of organizing work so the worker's job coincides with the goals and objectives of the organization. Thus, a Theory Y manager views the task as constructively using the worker's self-control and self-direction as the instruments through which work is accomplished.

By emphasizing responsibility and goal orientation, management capitalizes upon the motivation already present within the worker. If conflicts occur between the worker's goals and management's goals, they are resolved through mutual exploration and discussion. It is always assumed, under Theory Y policy, that the worker's inherent motivation is essential to accomplish the organization's goals.

Both Theory X and Theory Y proponents exist, and management systems successfully operating on the basis of

Table 15–A. Summary of Different Surveys on Job Satisfaction in Order of Importance of Different Factors

	Women Factory Workers	Union Workers	Nonunion Workers	Men	Women	Employees of Five Factories
Steady work	1	1	1	1	3	1
Type of work				3	1	3
Opportunity for advancement	5	4	4	2	2	4
Good working companions	4			4	5	
High pay	6	2½	2	5½	8	2
Good boss	3	5½	5	5½	4	6
Comfortable working conditions	2	2½	3	8½	6	7
Benefits		5½	6	8½	9	5
Opportunity to learn a job	8					
Good hours	9	7½	7	7	7	
Opportunity to use one's ideas	7	7½	8			
Easy work	10					

each of these theories can be found throughout American industry. The important point is for safety and health professionals to remember that safety programs can work under whichever system is operating within an organization. While the technique of implementation may differ, Theory X and Theory Y approaches to human motivation can both be used to motivate workers to adopt safe behavior.

Job-Enrichment Theory

Another analysis of human motivation in occupational environments has been developed by Herzberg (1966). Although quite comparable to the Theory X and Theory Y, Herzberg is explicit in both detail and philosophy. His concept of job enrichment, in many ways an extension of Theory Y, has been a major force in the development of management and leadership strategies.

The classic approach to motivation concerns itself with changing the environment in which a person works. This includes the circumstances surrounding the individual on the job (good or poor lighting, an agreeable or offensive supervisor), and the incentives given in exchange for work (money, a pat on the back, a writeup in the company bulletin, etc.).

Herzberg believes worker concerns about environment are important but are not sufficient alone for effective motivation. People are motivated best by using the work itself to satisfy their desires and needs.

Herzberg holds there is no conflict between the classic (environmental) approach to motivation and his approach to motivation through work itself. He regards both as important. The classic approach is called hygiene whereas Herzberg's approach is called "job enrichment." Figure 15–3 presents a contrast of the classic hygiene approach and the job-enrichment approach to motivation.

The hygiene approach may be understood by the following analogy—a person is provided with pure drinking water and waste disposal; both are necessary to keep this person

healthy, but neither makes him any healthier. By extension, good rapport with others may enhance an individual's job satisfaction, but job satisfaction alone will not necessarily motivate the individual to develop safe work habits.

Further, treating a person better does not enrich the job, although the individual can become unhappy if not treated well. Again, a salary increase can keep an employee from becoming dissatisfied for a time, but sooner or later another increase will be required to boost motivation again. Nor does worker protection enrich a job. Even in hazardous industries such as coal mining or bridge building, the worker may regard protection like gas testing and life lines as a part of the job. In such cases, the protection will not be a motivating force; on the other hand, the worker will be very unhappy if no protective effort were made.

Herzberg's idea that work itself can be a motivator represents an important behavioral science breakthrough. Traditionally, work has been regarded as an unpleasant necessity but it has not been thought of as a potential motivator. Although automation is helping to phase out the unstimulating aspect of many jobs, a job should provide an opportunity for personal satisfaction or growth. When it does, it becomes a powerful motivating force.

People, Herzberg further theorized, must be given the opportunity to do work they think is meaningful. Merely complimenting an individual who is doing a routine job or saying that the worker is doing something meaningful accomplishes little. The worker often does not feel this is true. Job rotation is not the answer, either; it does not enrich a job—it only enlarges a worker's responsibilities and tasks.

Herzberg also observed that new technology can make workers feel obsolete. As Herzberg noted, "Resurrection is more difficult than giving birth." Obsolescence must, therefore, be eliminated by continued retraining—not just an occasional session. Jobs should be updated, and people doing the jobs must keep abreast of the latest equipment and techniques.

Hygiene Approach (Classic)	Job-Enrichment Approach
Company policies and administration	Achievement
Supervision	Recognition
Working conditions	Work itself
Interpersonal relations	Responsibility
Money, status security	Professional growth

Figure 15–3. Contrast of the hygiene approach to motivation with the job-enrichment approach.

Even though a company provides the hygiene factors, it also must provide a task that has challenge, meaning, and significance. An unchallenged individual who does not quit but stays on is usually resigned to the work and suffers from poor morale. That, says Herzberg, is the price a company pays for not motivating people.

In summary, seven principles of job enrichment can be itemized.

- Organize the job to give each worker a complete and natural unit of work.
- Provide new and more difficult tasks to each worker.
- Allow the worker to perform specialized tasks in order to provide a unique contribution.
- Increase the authority of the worker.
- Eliminate unneeded controls on the worker while maintaining accountability.
- Require increased accountability of the worker.
- Provide direct feedback through periodic reports to the worker.

FRUSTRATION AND CONFLICT

When a barrier is placed between a goal and the person who is seeking it, the individual experiences frustration. This emotion differs from deprivation or lack of satisfaction when a need is not fulfilled. The thwarting of behavior directed toward a goal can be the result of several factors (Miller and Dollard, 1962).

- The barrier can come from within the individual. The person who sets impossible goals can become frustrated when unable to reach the goals. The solution to this conflict may be an accident.
- The barrier can also arise from the environment. The person who sets a very high personal production goal may be unable to reach it because of faulty equipment.
- A third type of frustration is caused by conflict. If two motives somehow conflict, the satisfaction of one means the frustration of the other. For example, the worker who sets high personal production and safety standards, but is unable to meet both under the present system, must satisfy one and sacrifice the other.

This section addresses the conflict-caused frustration. Ba-

sically, there are three types of conflicts: "approach-approach," "avoidance-avoidance," and "approach-avoidance."

Approach-Approach

As the label implies, the "approach-approach" conflict arises when an individual is faced with two goals that are equally attractive but only one of them can be obtained at any one time. An example is the manager who has been offered a promotion at his company, while another firm has made him an equally attractive offer.

The approach-approach conflict is the easiest to resolve. No matter what goal the individual selects, a need will be satisfied, without much loss to another need. Usually the individual will resolve such a conflict by satisfying one need first and then satisfying the other. If the person is both hungry and sleepy, for example, the resolution might be to eat first and then sleep.

Avoidance-Avoidance

A second type of conflict is the "avoidance-avoidance" or double-negative conflict. This conflict can arise, for example, when an employee is told to wear heavy fire-retardant outer clothing in the summer heat or else risk the possibility of clothes catching on fire. This employee is, as the saying goes, caught between the devil and the deep blue sea.

Two kinds of behavior generally result from such a conflict—vacillation and flight. In vacillation, the individual approaches one goal; retreats, approaches the other, retreats, and so on. As the goal is approached, the unpleasant portions of the goal increase, so the individual withdraws. In the example, the worker may put off doing the job requiring the heavy clothing until just before the quitting bell sounds.

Another result from the avoidance-avoidance conflict is flight. The athlete may leave the field or a worker may quit the job. Quitting a job, of course, has serious consequences in lost prestige and lost income.

A more common type of fleeing, therefore, is in a figurative sense—such as daydreaming and fantasy. The worker faced with an unpleasant task may repress the reality by daydreaming.

The avoidance-avoidance conflict is not as easy to resolve as the approach-approach. No matter which choice the person makes, there will be unpleasant consequences.

Approach-Avoidance

The "approach-avoidance" conflict is the most difficult to resolve. The individual is both attracted and repelled by the same goal (Figure 15–4). The worker who is striving for top production inadvertently may take unnecessary risks to achieve that goal. One way out of such a conflict is by realizing that safety and productivity go together.

Approach-avoidance conflict refers to a single goal having both positive and negative attributes associated with it. Other things being equal, the strength of the approach or avoidance tendency will depend upon how close the individual is to the goal. At point A, for example, the tendency to avoid (or retreat) will be greater than the tendency to approach (since,

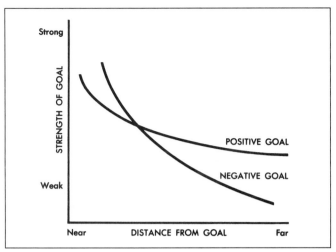

Figure 15–4. The approach-avoidance conflict is complicated by the different goal gradients for positive and negative goals.

as shown, the slope of the negative goal gradient is steeper than the positive gradient at point A). At point B, however, the tendency to approach the goal will be greater than the tendency to avoid it. According to this theory, conflict will lead to retreat, vacillation, or indecision. The approach-avoidance conflict, and more complex double approach-avoidance conflict situations (that is, two goals, each mutually exclusive, having positive and negative attributes associated with each), are often common situations.

Reaction to Frustration

In most cases, the conflicts described will lead to frustration. This frustration often can result in positive, constructive resolution of the situation. For example, the individual who is facing the conflict of top production, but at increased risk, may develop a new system for processing the product at a faster *and* safer rate. On the other hand, the frustration can lead to some form of negative behavior (Miller and Dollard, 1962).

Emotion and Frustration

Negative emotions can have a disturbing influence upon a person's behavior. Anxiety is one emotional reaction that, because of its disruptive influence on a person's normal behavior, can make that person more susceptible to accidents. Anxiety and its disrupting influence can spread to others working in the same area and create an atmosphere in which accidents are more likely to happen.

Anxiety is, however, only one of the emotional patterns that individuals experience in the face of frustration. Some people become angry, or fearful, while others accept frustration as a challenge and attempt to solve the problem. Most people who react negatively to frustration and threat find it difficult to cope with life situations. It is not necessary to discuss the physiological pattern that develops, for everyone is familiar with the feelings experienced during fear, anxiety, anger, shame, and other emotions. Rather, one needs to recognize that over the years people have developed ac-

cepted social expressions for emotions. To some degree, one individual can be aware of how another is feeling. This is not to say that one cannot be fooled or that the outward expression always truly indicates the feelings of the individual. Many people learn to mask their real feelings.

However, after working with people on a daily basis for some time, one can learn to tell when they are holding their feelings in and when they are expressing them. This ability is useful to the supervisor and the safety and health professional in dealing with individuals in the shop. As they go about the plant or establishment carrying out their normal duties, they may discover those individuals having difficulties and take some extra precautions to avoid accidents.

In some cases, however, behavior disorders may be a more serious problem. Individuals whose normal behavior is seriously disrupted or whose ability to handle frustration is seriously impaired may require professional services beyond the scope of the normal supervisor-employee relationship. Although a supervisor should assist in directing the employee to in-plant services or outside agencies, the supervisor should not attempt to deal directly with psychological problems requiring professional care. While the current movement toward company-sponsored psychological services may grow, it is unlikely to be the standard in all industries in the near future. For a detailed discussion of industry programs for drug abuse and alcoholism, see Chapter 4, Occupational Health Programs, and Chapter 7, Employee Assistance Programs.

Non-Directive Counseling

Although no attempt can be made here to train safety and health professionals or supervisors to become personnel counselors, they can do much to help reduce the immediate result of strong emotions. The supervisor is in the best position to evaluate the emotional level of an individual worker and to do something about it.

Everyone is familiar with the emotional outbursts of people who have had a trying and upsetting experience. Such expressions of emotion can take many forms from an outburst of profanity, to a torrent of tears, to physical assault on objects that cannot strike back. The result is relief from the intolerable tension and thus emotional relaxation.

The supervisor can sometimes alleviate work-related tension by providing an employee with an opportunity to talk about whatever is bothering him or her. This may be especially helpful if the problem is work related. Even talking about nothing in particular can be helpful to the tense employee.

The stresses causing emotional upset can occur within the work environment or away from it, but each kind has its effect on the other. A simple statement, such as "How are you today?" or "What's bothering you?" may well be sufficient to let the employee verbalize pent-up emotion. After the discussion, the employee can often evaluate real or imagined problems more clearly and realistically.

To attack the problem at the roots calls for careful listening on the part of the employee's immediate supervisor.

During the normal routine every supervisor has an opportunity to talk with individual employees. Through the use of open-ended questions (those that do not allow for a simple yes-or-no answer), the supervisor can encourage an employee to discuss work-related problems. The supervisor needs to listen not only to words but to the feelings behind them. He or she must pay attention to the nonverbal language of facial expressions, gestures, bodily postures, and the like. This will increase understanding of employee attitudes and the strength of feeling attached to them.

Frequently it is impossible to discuss such matters at length in the work area. If so, the supervisor should set a time and place away from other employees to allow the individual freedom of expression. Even though the supervisor may not have a completely private office, some place can be found for a quiet face-to-face discussion.

This relationship between the supervisor and the worker must be one of the supervisor's normal everyday functions. Diligent observation, and effective and sensitive listening, should be tools of every supervisor.

ATTITUDE, BEHAVIOR, AND CHANGE

Although many existent psychological theories address attitude, behavior, and change, many basic concepts are still controversial. For example, whether or not attitudes indicate future behavior is still a theoretical issue debated in social psychology.

Attitude

Many theories on attitude formation and change suggest that attitudes consist of three components, knowing-feeling-acting. Thus such a theoretical approach distinguishes, yet integrates, such concepts as knowledge, emotions, and behavior or behavioral tendencies. Other theories regard attitudes as the expression of one's values towards something or someone. Most of these theories agree, however, that attitudes (at least in part) reveal a person's tendencies to evaluate objects, persons, or situations favorably or unfavorably (McGuire, 1968).

Katz and Stotland (1959) identified the following three attitude components:

- Affective Components: Affective components are the positive or negative feeling (affects) of attitudes. Although some attitudes can be quite irrational, they involve little more than feelings. People may not like doing something a safe way, but cannot tell you why.
- Cognitive Components: Attitudes can differ due to the extent of knowledge involved in the attitudes. Some people work safely because they have worked through the problem and have decided that working safely is the best way. Others may not work safely because their attitude towards safety is based on incorrect information or knowledge. In fact, misinformation can be a source for many attitudes.
- Action Components: As touched on earlier, attitudes may bear little relationship to behavior. An individual may

express a very strong attitude about working safely, and yet the person fails to wear the personal protection equipment provided for the job. In such a case, the attitude may lack a significant action component.

A person's response or attitude is dependent, in part, upon that individual's previous experiences. For example, if someone sees a person on the street who resembles a friend, the immediate response may be a smile, friendly gestures, and warm voice quality. When that person is perceived to be a stranger, there is an immediate change to another facial expression, gesture, voice quality, and so on. Such reactions can be triggered by many factors—a certain look from another person, a manner of speaking, a mustache or the lack of one, hair color, or style of handshake. All individuals have certain feelings about other people or situations and tend to act according to their attitudes, which have been formed over the years.

Some of these attitudes are latent, or hidden. The responses based on latent attitudes can lie dormant within the individual until triggered by some event. Given the appropriate stimulus, the attitude surfaces and the behavior exhibited is consistent with the individual's feelings. A certain word, for example, connotes different responses in different people. The words "union," "management," "labor," and even "safety" carry with them certain connotations that touch off different attitudinal reactions in individuals, depending upon the kinds of experiences they have had with the subject.

Because attitudes play such an important role in everyday relationships, the safety and health professional should understand how they are developed, their effects on individuals, and what can be done to change them.

Determination of Attitudes

Direct personal experience is thought to be one way attitudes are established. The three components of attitude—feeling, knowledge, and action—can be influenced through personal experience. In other words, experiences that individuals have had in the past, especially those involving strong emotions, determine attitudes. For example, the attitude of a worker toward management or the supervisor may be fearful and hostile, if the worker has been fired several times for what the individual believes are superficial reasons. There might have been sound reasons for the dismissals or layoffs, but to admit this would be a threat to the person's pride; thus, the individual puts the blame elsewhere. Because of the effects these dismissals have had on the employee and the family, such as causing financial stress, the employee may become very angry. Management now is the enemy. The person's present hostile, fearful attitude is set and will be difficult for the next supervisor to change without concerted effort on both parts.

Many people have had experiences that they learned to associate with fear, sorrow, pain, or happiness. All of these emotions tend to make people act the same way when a similar experience occurs.

One note of caution. Although there is only one positive emotion listed among those just mentioned, this does not

mean that most attitudes are negative, for there are many positive ones.

Attitude Change

Numerous factors are associated with attitude change. Commitment and responsibility are core concepts in the formation and changing of attitudes. Mimicking a particular view will have little impact on an individual's attitude. If there is a desire to develop a new attitude or change an existing attitude, the attitude must be learned in a concrete situation where the attitude is tied to action. Attitudes that are acquired at a verbal level cannot be expected to influence behavior. If the safety and health professional plans to influence a change in the safety attitude of an individual or group, the planned for change must include an effective activity component that requires the participants to relate the development of the new attitude to the activity. Without providing individuals with the means or methods to experience change, simple slogans will not create an effective change in attitude.

In summarizing the large body of psychological literature on attitude formation and change, McGuire (1968) suggested three components of the process (Table 15–B). These include (1) types of attitude-change situations (column 1), (2) variables associated with the communication process (column 2), and (3) behaviors associated with attitude change (column 3).

- Relevant situational factors include suggestive situations (where the desired attitude is repeatedly presented), conformity (where social or peer pressure is used to elicit the desired response), group discussion, persuasive messages, and intensive indoctrination. Each situation is associated with varying degrees of attitude change. Also, the implications of each depend on the type of communication variables involved.
- Source variables refer to characteristics associated with the person or represented organization presenting the message. The effect of the message on subsequent attitudes can vary depending on the credibility, attractiveness, and power of the message; the order in which specific issues are presented; and the differences between the sender and receiver of the message. Channel factors consist of ways the message is presented, for example, direct personal experience or mass media. Receiver factors include whether the audience is actively involved in the process and how much the audience can be influenced. Destination factors refer to the degree of post-communication message decay across time, and time latency associated with delayed action.
- According to McGuire (1968), the receiver (or audience) must proceed through five steps for attitude change to occur. These are attention, comprehension, yielding, retention, and action. According to this model, each step depends on the preceding one.

To add to the complexity of this model, communication variables, especially in daily situations, can affect one another and interfere with the message. Also, the sequential-step process may have more intuitive than empirical support.

Nevertheless, this model should alert safety and health professionals and others to the fact that attitude change is not a simple process.

Organizational development. Behavioral scientists, concerned about the amount and rate of change in our technological society, have studied the impact of such change on the industrial environment. These researchers have evolved an approach, called "organizational development," that attempts to assess a corporation's ability to adjust to such conditions as rapid growth, new technology, increasing diversity, and management system problems. Organizational development, which generally has as its goal providing education for employees, is designed to alter the attitude and structure of organizations so they can better adapt to a changing technology.

Effective organizational development requires a strong organizational culture. The development of a strong culture will aid a company's success by providing the employees the means to identify and act on the values of the company (Deal and Kennedy, 1982). The organization's culture provides workers with informal rules on how they are to act. It is through the company's culture that the organization can either reinforce the safety efforts of the workers or create barriers to the development of an effective safety program.

Generally, an organizational development approach gathers information from employees and management to determine the climate and capacity of the organization to adapt its objectives to the new technological environment. Based upon such feedback, an attempt is made to develop organic systems to replace mechanical systems within the organization. Organic systems are characterized by a preoccupation with people as they work together; mechanical systems are characterized by a preoccupation with the structure of a company. Figure 15–5 presents a summary of the differences between mechanical and organic systems. In the final analysis, organizational development rejects bureaucracy as an organizational model and substitutes a model based on interpersonal competence.

To implement the ideas represented by organizational development, management can use a number of procedures to effect changes in the organization. Since the changes are typically people oriented, management should recognize the need to motivate employees to accept the changes within both management and work groups. Typically, people can be educated and trained to learn how to work with the new system. These techniques have merit in implementing change only so long as top management agrees to the changes and provides incentives within the organization for their adoption.

Ultimately, organizational development seems to provide the necessary means for preparing an organization to plan for the future. Included in such an effort should be the recognition of how changes in an organization will affect the current safety program. By involving all levels of managers and workers in developing and managing safety programs, it is likely that the programs will be able to meet the future needs of modern organizations.

Table 15–B. Factors Associated with Persuasive Communication

Attitude Change: Situation Types	Communication Process	Attitude Change: Behavioral Step
1. Suggestive	1. Source	1. Attention
2. Conformity	2. Message factors	2. Comprehension
3. Group discussion	3. Channel factors	3. Yielding
4. Persuasive messages	4. Receiver factors	4. Retention
5. Intensive doctrination	5. Destination factors	5. Action

Source: McGuire (1968).

Structural Change

In the previous discussion of attitude formation and attitude change, no direct relationship between attitude change and behavioral change was assumed. In practice, however, line management and safety and health personnel are interested, in ways of changing employees attitudes only if their behavior also changes; for instances, they not only change their attitude about wearing personal protective equipment, but they also change their behavior by actually using it consistently. While the research literature indicates that many techniques—training and counseling, for example—can change attitudes, there is much less support for a direct relationship between attitude changes and subsequent behavior changes (Ajzen and Fishbein, 1977).

In effect, line management and the safety and health professional must consider techniques other than simple attitude change when considering employee behavior. Evidence suggests it is possible to change behavior by changing the structure of the work environment. Examples of structural changes involve changing situational variables, include changing job contents, modifying the physical arrangements of work, changing worker interaction patterns, and rearranging work procedures.

In each case, it is not necessary to expend the time and effort to change attitudes before changing behavior directly. Rather, the introduction of the structural change can modify behavior and with it, possibly change employee attitudes.

While the structural change model has not been applied extensively in the past in relationship to occupational safety, it warrants consideration by the safety and health professional. Human behavior can be changed by altering the very circumstances under which the individual works. Unsafe practices, for example, cannot occur when the structure of the work and work groups prevents them from happening.

Behaviors Versus Attitudes

Attitudes refer to internal predispositions to behavior; as such, they are difficult to observe and measure. Behavior refers to observable actions, which can be measured. This distinction is critically important because measurement of behaviors lays the groundwork for effective safety management.

While attitudes are difficult to change directly, changing behavior is not as difficult. A good example of this is people's attitudes toward using seat belts. As the behavior of wearing seat belts has changed over the years, so have attitudes to-

ward wearing them. Thus, the most efficient way to change safety-related attitudes is to change safety-related behavior.

Behavioral Management

More attention and recognition have been given to the importance of behavioral management in safety programs. Behavioral management refers to the systematic identification, measurement, and control of safety-related behaviors. This section briefly reviews this approach.

Analyzing and Changing Behavior. Developing and planning for change requires that the safety and health professional or supervisor has the ability to understand the situational variables that influence the behavior outcomes of individuals. Why would a person have a strong desire to work safely, yet continually perform in an unsafe manner? Why do we find discrepancies between the expectations and results of the safety efforts in the work environment?

Often, when there is a discrepancy between the result and what was expected, supervisors may believe that there is a need for additional employee education or training. Ferdinand Fournies (1987) asked over 4,000 managers "Why don't subordinates do what they are supposed to do?" The responses included:

- They don't know what they are supposed to do.
- They don't know how to do it.
- They don't know why they should do it.
- There are obstacles beyond their control.
- They don't think it will work.
- They think their way is better.
- They're not motivated—poor attitude.
- They're personally incapable of doing it.
- There's not enough time for them to do it.
- They are working on the wrong priority items.
- They think they are doing it.
- There's poor management.
- They have personal problems.

Robert Mager and Peter Pipe (1984) developed a useful procedure that provides the safety and health professional and supervisor with the means to systematically and accurately analyze performance discrepancies and create change. Mager and Pipe's first step in analyzing performance is to identify the nature of the discrepancy. What is the issue? The safety and health professional must accurately describe the safety discrepancy. Examples would include driving a forklift with the forks elevated above a specified level, not wearing personal protective equipment, or conducting

Mechanical Systems	Organic Systems
Emphasis upon individual performance	Emphasis on relationships in group
Chain of command concepts	Confidence and trust among everyone
Adherence to delegated responsibility	Adherence to shared responsibility
Division of labor and management	Participation in multimember teams
Centralized management control	De-centralized sharing of control
Resolution of conflict through grievance procedures	Resolution of conflict through problem solving

Figure 15-5. Organization development seeks to implement organic systems in place of mechanical systems.

maintenance without locking out the equipment. The second step is to answer the question, "Is it important?" If the answer is yes, the process continues to the third step, "Is it a skill deficiency?"

If a person drives the forklift with the fork too high because of a lack of skill, the safety and health professional may need to provide formal training if the individual has not had prior experience on a forklift. A second factor could be due to the fact that the person has not driven the forklift very often and requires additional practice. The third factor could be that the person has driven the forklift often, but requires appropriate feedback.

If the person's forklift driving behavior is not due to a skill deficiency—"They could do it if they wanted to"—additional knowledge from education and training may not change the individual's performance. In these types of situations, the safety and health professional must determine if:

- The appropriate performance will result in punishment.
- There is no reward for the appropriate performance.
- The appropriate performance may not matter to the individual.
- There are obstacles to attaining the appropriate performance.

In these types of situations it is important to remove the punishment, arrange positive consequences, or remove the obstacles. Does the plant layout require the forklift driver to constantly keep raising and lowering the forks to clear congested areas, thus reinforcing the driver's decision to drive with the forks elevated? Can the work area be redesigned?

In some situations, such as not wearing personal protective equipment, the safety and health professional may not need to determine if the discrepancy is related to the a skill deficiency. By analyzing the situation, the safety and health professional or supervisor may be able to design a simpler way to do the job that does not require the use of the protective equipment. If a simpler way is identified, the job should be changed or on the job training should be provided.

In addition to redesigning the job, nonperformance may be due to an individual's potential. The individual may be unable to learn the job, lacks the physical or mental potential, or is overqualified for the job. If the safety and health professional or supervisor accurately determines that the person lacks the potential, a decision must be made to either transfer the person to a different job or terminate the relationship. It is important to realize that Mager and Pipe believe that the transfer or termination options are used more than they should be. This could be due in part to the decision maker's lack of understanding regarding human performance or an indication of the decision maker's inability to identify other options.

Because an individual's behavior can be the result of competing factors—created by either internal needs or situational variables—the safety and health professional or supervisor may identify a variety of solutions. By understanding the factors that can influence performance discrepancies, the best possible solution can be selected and implemented.

Evaluating critical behaviors in the work place can assist in changing safety related behaviors. The basic steps of the process are:

1. *Identify critical behaviors.* This means to write, in observable terms, what employees should do to properly perform their jobs. The safety and health professional can list a few critical behaviors or a complete inventory, depending on the scope and results desired.
2. *Conduct measurement through observation.* Trained observers watch the workplace to determine if the listed behaviors are performed safely or unsafely. The total number of observed behaviors is divided into the number of safe behaviors to obtain a percentage figure for safe behaviors.
3. *Give performance feedback.* The percentage figure for safe behaviors is shown on a graph displayed in the workplace. At regular intervals, behaviors are again observed and the new safe behavior figures added to the graph. Studies show this critical feedback will improve safety behaviors. Praise and recognition from supervisors or peer pressure can be effective ways to encourage safe behaviors.

LEARNING PROCESSES

Through the process of learning, people discover their psychosocial needs, establish habitual patterns of behavior and emotional reactions, and develop the attitudes they bring with them to industry. It is important for the safety and health professional to consider learning and the principles that affect it. This is especially vital in any discussion of safety, because training is a major consideration in safety programs.

Often, to change behavior, one must substitute new learning for old habits (Hulse, Deese, and Egeth, 1980). To change behavior in need-satisfaction sequences, one must teach better ways by which employees can achieve their goals. In each case, some new learning must be substituted for the old.

Motivational Requirements

Educational systems must place emphasis on stimulating the individual to want to learn. Materials must be designed to apply to practical situations in students' lives, and instructors have to develop student interest in the materials. To teach a worker something about computers, for example, the instructor might have the person create a simple letter or copy and store files related to their department's work. Instructors try to motivate their students and encourage their learning process. With the changing demographics of the American workforce, effective employee education programs that can provide knowledgeable, safe, and productive workers is now more important than ever before.

The safety and health professional must recognize that if workers are going to learn safe procedures, they must want to do so. To merely point out that accidents cost the company money will not motivate them. The safety and health professional should point out both the risks of improper work habits, and their costs to both the company and the individual. How improper work habits affect the product/service provided. The impact of improper work habits to the employee, the company, the product on a long term basis, if unchecked. Finally, the safety and health professional should seek the worker's aid in addressing these problems in a functional manner for the benefit of all.

It is not wise to assume that, because management sees the value of safe procedures, the workers also will. Management may be motivated to start a safety program because it will lower insurance costs, reduce the amount of waste, and increase the number of units produced. Workers cannot be expected to want the program for the same reasons. To sell a safety program to the employees, one must know what will motivate them. Here the safety and health professional can capitalize on the needs discussed previously, and probably can find many more that are consistent with the aims of the safety program. Everything that will motivate the worker to learn the right procedures should be used. (Chapter 18, Promotions and Campaigns, discusses this in more detail).

Principles of Learning

Some consideration of basic learning principles is valid whether the learning is to be done in a college classroom or in a work area. When training procedures are based on these principles, learning is more efficient and thorough.

Reinforcement. Through experimentation, psychologists have found that reinforcement can often facilitate learning. In practice, reinforcement can, for example, make work more efficient and more productive. When a worker's pay increases because he or she produces more units, the person has received a reward for learning. This type of reinforcement can have negative aspects as well—the worker who figures out a hazardous shortcut, to produce more units may actually be rewarded for an unsafe practice.

Employees must be recognized and rewarded *only* for safe work methods. Higher productivity based on safe work methods satisfies the dual need for achievement and recognition. This fact in itself not only reinforces employee learn-

ing, but also influences other employees who see or hear about it. A supervisor who recognizes the same needs in other workers can reinforce this learning through praise of greater productivity and telling workers how much improvement has been shown. Thus, the publicity given to a safety record, a bonus, a promotion, or anything else that satisfies individual needs would serve as a reward to reinforce whatever learning has brought about the desired behavior.

Reinforcement tends to be more effective if it is given soon after the desired behavior occurs. Praise, for example, should be given at the time the desired behavior happens, or if delayed, associated verbally with the desired behavior. This does not mean it must be instantaneous, but it should be within a reasonable length of time. After a delay, the safety and health professional or supervisor should let the individual know why they are receiving the reward.

Reinforcement often increases the likelihood that workers will repeat the desired behavior. However, hazardous shortcuts or any deliberately unsafe practices must not be rewarded. In fact, in such situations reprimands or disciplinary actions are more appropriate. In short, supervisors should correct unsafe methods by complete explanation and demonstration of safe work procedures, and then reinforce the new, correct work pattern as soon as feasible.

This principle also applies to employee participation in a safety program whether through suggestion systems, safety and health committees, discussion groups, or training sessions. In all these areas an individual gains personal worth if his or her opinions are asked for and graciously received.

A safety program that gathers the ideas of all, either individually or by representation, satisfies the need people have for being "in the know." In this way, the safety and health professional can create a positive atmosphere for the program and instill in employees a sense of obligation and responsibility for its success. Research has demonstrated that when employees feel they helped to create a program, there is more chance for its success.

Research on the effectiveness of disciplinary actions suggests that it can have mixed consequences. Discipline generally is thought to be less effective than reinforcement, perhaps because discipline provides indirect cues or information (what not to do) as opposed to positive reinforcement that provides direct information about what employees should do (Church, 1963).

Recognizing an individual's correct performance will lead to a more positive attitudinal response on the worker's part than will any discipline. Workers need to have a positive attitude toward training procedures. Both the positive and the negative aspects of reinforcement may generalize over the entire work situation. The supervisor's praise for a task well done may influence an employee's attitude toward other parts of the job, including training procedures, safety devices, and the safety program. The proper use of recognition thus can lead to worker acceptance of efficient and safe work methods.

Much of the practice underlying programmed instruction (see Chapter 16, Safety Training) is based upon the rein-

forcement concept as well as the other principles of learning which follow.

Knowledge of results. Closely allied to recognition—in fact, one aspect of recognition—is knowledge of results. People like to know how they are doing on the job. Telling employees how well they are getting along in the training program will likely motivate them to continue learning and to do a better job. To teach a worker and not mention how the person has improved is defeating one's own purpose. It cannot be assumed that the only one who needs to know about the effects of training is the supervisor or the safety and health professional. When workers know that a new procedure is helping to increase their production, they have received reinforcement for their learning and effort.

Knowledge that a job was done right is rewarding to employees. However, the contrary is also true; if the safety part of a job was not done right, then the job was not done right. Employees must know that safety rules and procedures are a part of job performance.

One phenomenon that often occurs in the learning process is known as a plateau. At this point, a worker's progress levels off for some time before he or she shows another increase in learning. Often some individuals become discouraged during this time, and their learning can decline. The trainer needs to understand this phenomenon and tell the individual that such a leveling off is normal and that further progress will come later. In this way, the employee will avoid becoming discouraged, and proceed to keep on learning more easily and efficiently.

Practice. The safety and health professional is interested in developing safe working habits in employees that will become almost automatic on the job. As a result, merely putting a worker through sufficient training sessions is not enough. Despite any apparent mastery of safety methods, the next time through the work, an employee may make one or more mistakes. To be sure habit patterns are firmly entrenched, workers must practice what they learned. The Job Instruction Training (JIT) programs include one of the important aspects of training, namely, follow-up by the supervisor. This is for no other purpose than to ensure mastery of safe working methods. Within a reasonable time, depending on the job complexity, this follow-up will no longer be needed.

Take, for example, a simple task such as bicycle riding. A child, in learning to ride a bicycle, needs to know how to balance, how to get on, how to get off, and how to stop, and must learn all of these skills as part of the total process. Perhaps the process begins with balancing, then how to start and stop, and then how to get on and off. Even after accomplishing all of these, the child should not be left alone. Several more sessions are needed to make certain the task is being correctly done. Each ride the child takes is another practice session. This is reinforcement by practice. So it is with the individual on the job.

In addition to the learning principles covered so far, whole versus part learning has been a knotty problem for industrial trainers for many years. The question is whether trainers should teach a work procedure as a whole, or break it down and teach it part by part. There is no best answer. Both methods have advantages and disadvantages depending upon job complexity, the trainee, and the kind of job breakdown used. Perhaps a combination of the two is best, using the whole method, but with sufficient flexibility to emphasize meaningful parts of the task wherever necessary.

Meaningfulness. Studies in verbal learning have demonstrated how important it is that the material used be meaningful to learners. Meaningfulness to management and workers is understanding the value of producing a marketable product (which contributes to job satisfaction, job retention, profit, etc.) and the negative consequences of failing to engage in appropriate actions, leading to higher production costs, higher product rejection rates, accidents, and their associated losses. Meaningfulness is essential in safety education because the worker needs to understand why a certain procedure is better than another. Adequate explanation, how a given movement or change in position can eliminate hazards with no decrease in production, gives meaningfulness to the procedure. With this understanding, workers will be motivated to learn the safe procedure. Without it, they will be inclined to use their own methods until they learn, perhaps by an accident, how inadequate unapproved work procedures are. The safety and health professional and line supervisor should not forget the advantage of workers' understanding the reason for having protective clothing, safety devices, safety meetings, and discussions, as well as the need for full and complete accident reports.

Selective learning. Out of each day's many experiences, people select those they wish to retain. This selection process probably is related to motivation more than to anything else, and for that reason motivational aspects of a training program need to be considered. Safety trainers must be sure the workers retain the most important facts. Relating subject matter to individual needs will ensure the proper selection and retention.

Frequency. Everyone does best those things he or she practices most. This principle is certainly important to the safety and health professional, for it emphasizes the necessity of frequently applying safety rules and regulations in the training program. When workers hear frequent reference to the kinds of problems, hazards, and procedures that cause and prevent accidents, they will learn more than through a one-exposure routine. Giving the worker a copy of the safety rules and regulations in hopes these will be learned and used is not enough. Instructors should devise ways to bring these rules and regulations to workers' attention frequently and regularly.

A major railroad put this principle into practice when its supervisors discussed with their employees a safety regulation each day before work began. Thus, each day, the staff was responsible for knowing, understanding, and applying this regulation when appropriate or when asked by the supervisor. This company combined the principles of learning, reinforcement, and follow-up in one program activity.

An instructor must insist not only on frequent practice,

but on the trainee's following the correct method. Daily use of safe methods will create safety habits in employees that will become nearly automatic. The supervisor and safety director must make sure the work method practiced is the safe one.

Recall. Closely allied to the principle of frequency is that of recall. What is learned last usually can be most easily recalled. As has been indicated, handing a worker a set of safety rules and regulations does not ensure learning. Those who received printed instructions or a few training demonstrations a long time ago may not be able to recall what the rules and regulations are. Safety and health professionals must devise ways for workers to come into constant contact with these regulations through activities such as contests, reviews of safety regulations, and committee work.

Primacy. The law of primacy must be taken into consideration in two aspects of the safety program. The worker's initial contact with safety procedures should be one of major importance. If this initial contact is of a negative nature—such as being tossed a book of rules and told "Make sure you learn them"—the worker is left with the impression that safety is unimportant. From the first day of employment, the worker must know that not only the safety rules but the whole safety program is of great importance. This will help ensure the positive response desired.

Second, in the training program, habit patterns using safe methods should assume primary importance. The supervisor must be certain the worker does not learn how to work any other way than by the safe method. It is harder to establish good patterns if one has first learned the poor ones. Training must be right from the very beginning. Old habits are hard to break and expensive in terms of training and accident costs.

Intensity. Those things made most vivid to the worker will be retained the longest. Safety programs already use this principle in safety publicity with eye-catching posters, slogans, and the like.

Part of the safety and health professional's responsibility is to help management enhance the worker's interest in the program. In this way, the training material will be remembered for a long time. Otherwise, it might well be forgotten minutes after the training session is over. To some degree, the trainer must position the safety program in the same way advertisers position their commercials to catch the public's attention.

Transfer of training. All new learning occurs within the context of previous learning experiences. The fact is that current learning (or performance) can be influenced by previous learning, a phenomenon known as *transfer of training.* Positive transfer of training occurs when the previous learning experience helps current learning or enhances current performance. Negative transfer of training occurs when previous learning makes the current learning experience more difficult or in some way inhibits current performance. An example would be individuals who have in the past wedged back the guard on a power hand saw experiencing negative transfer

when they are taught that saw guards must be fully operational.

In as much as line management wants to help employees learn correct responses in new situations, the trainer should try to maximize the positive transfer that occurs within the industrial environment. Generally, learning to make identical responses to new stimuli results in positive transfer. For example, learning to drive in a new (or different) automobile is facilitated by the fact that identical responses (such as accelerating or steering) are required in the new automobile, similar to those in the old automobile. If the new stimuli are similar to previous cars (for example, the location of the controls, their texture, and their direction of movement), positive transfer should be high. In such situations, positive transfer increases the more similarities there are between the two learning situations.

Learning to make new responses to what appear to be identical stimuli, but are actually different, results in negative transfer. For example, if the controls of the cars appear identical but in each car the "fifth" gear is in the opposite position, negative transfer can be expected for a time. If the fifth gear position is located at the upper right position on the floor shifter, in one car, the student will experience some difficulty learning to drive a car in which the fifth gear position appears on the lower right. Or, if the OFF position of one toggle switch is the same as the ON position of another, the potential for accidents is serious.

Such negative transfer can cause errors, delayed reactions, and generally inefficient performance on the new task. In such situations, the amount of negative transfer *increases* the more apparently similar the two learning situations seem to be.

With an understanding of transfer phenomena, it is possible to maximize, within the industrial environment, opportunities for positive transfer and minimize those for negative transfer. This requires careful planning of machine purchases and work procedures to make certain the new tasks required of a worker make use of (and do not conflict with) previous learning experiences.

Forgetting

This discussion of learning would not be complete without some consideration of the process of forgetting, which occurs as learning takes place. Never assume that learning and forgetting are mutually exclusive, for such is not the case. As one learns, one also forgets some of what was previously learned. Curves of forgetting indicate that most learning is lost immediately after it has taken place. Depending on the complexity of the job, the amount of learning lost will vary after each day's training session. There should be less forgetfulness or initial mistakes in successive days of training, and more need for systematic refresher sessions to recall what was learned.

To minimize the process of forgetting, trainers must consider the various principles of learning and the motivational aspects associated with them. This needs to be kept in mind from the initial stages of employment through the everyday work situations in the company.

SUMMARY

- The human factor operates at all levels in industry, and is perhaps the most potent factor for success or failure of a safety program. Attitudes are important to safety in a company, but safety-related behaviors are critical. Safety behavior management stands at the center of the safety program. As a result, a major responsibility of the safety and health professional is to manipulate those variables that assist in changing inappropriate behaviors so that working conditions and procedures are appropriate.

- A well-trained joint worker/management safety committee can be a great resource and tool in involvement and motivation of all levels of a company to have a positive safety committee.

- Individual differences among employees present an ongoing problem in industrial settings. Each person is an individual, to some degree different from every other one. Yet within this framework, certain patterns of behavior, such as satisfying basic needs, are common to all people. Whether or not employees work safely depends on (1) current needs, (2) present situation, (3) past experiences, and (4) situational variables including workplace and methods design. By understanding what needs people have and how they go about satisfying those needs, safety and health professionals and supervisors can link company goals to individual goals. Thus, they can motivate workers to achieve both individual and organizational objectives in the safety program.

- People become frustrated when their goals cannot be achieved. There are many ways of reacting to such frustration, but the emotional reactions are of concern to the safety and health professional. These reactions, as well as the attitudes formed during the reactions, can be highly disruptive to safety precautions and procedures.

- Although many theories exist on attitude formation and change, changing workers' attitudes is a difficult task. This is due in part to the influence of past experiences and the emotions those experiences provoke. Because of the questions surrounding the relationship between attitude and behavior, greater emphasis must be placed on the importance of behavioral management in safety.

- Behavioral management's "action"-directed focus involves the identification, measurement, and control of workers' safe work-related behaviors. By accurately analyzing the factors that are creating a discrepancy between safety program expectations and results, the safety and health professional or supervisor can then develop the best possible solution for achieving the expected result. Steps involved in changing safety-related behaviors can include (1) providing training, (2) arranging practice, (3) arranging positive consequences that reinforce correct performance, (4) removing obstacles that prevent satisfactory performance, or (5) finding a simpler way to do the job in a safe manner.

- Understanding the principles of how people learn allows for the establishment of effective employee development and training programs that can influence change and reinforce the concept of safe work-related behavior. In teaching employees, management must understand and apply learning principles such as reinforcement, feedback, practice, frequency, transfer of training, and minimizing loss of learning. These principles can be used to motivate workers to learn and make learning more efficient.

- In no way is this chapter intended to minimize or ignore the work already being carried on to reduce accidents. Safety devices, safer machines, safe work layout, and many other aids and measures all are part of the total program. The human factor is an aspect of the whole system.

- People, machines, and materials are still the three components of industry that can contribute to safety. Machines and materials can be controlled, but the human factor must be guided and behavior channeled in the interest of accident prevention.

REFERENCES

"Accident Causation Theory: The Emerging Pluralism," *Occupational Hazards*, (October 1983):45,10.

Adams JS. "Inequity in Social Exchange." In L. Berkowitz (ed.), *Advances in Experimental Social Psychology*, Vol. III. New York: Academic Press, 1965.

Ajzen I and Fishbein M. "Attitude-Behavior Relations: A Theoretical Analysis and Review of Empirical Research." *Psychological Bulletin*, (1977)84: 888–918.

Anastasi A. *Psychological Testing*, 6th ed. Toronto, Canada: Macmillan, 1988.

Bolton R. *Social Style/Management Style*. New York: American Management Associations, 1984.

Braun WD, PhD. *Welcome Stress. It Can Help You Be Your Best*, 1st ed. Minneapolis: CompCare Publications, 1983.

Chhakar JS and Wallin A. "Improving Safety Through Applied Behavior Analysis." *Journal of Safety Research*, (1984) 15:141–151.

Church RM. "The Varied Effects of Punishment on Behavior." *Psychological Review*, (1963) 70:369–402.

Clark JV. *A Preliminary Investigation of Some Unconscious Assumptions Affecting Labor Efficiency in Eight Supermarkets*. Unpublished doctoral dissertation, Graduate School of Business Administration, Harvard University, 1958.

Cooper CL and Smith MJ. *Job Stress and Blue Collar Work*. Chichester, England: John Wiley & Sons, 1985.

Deal T and Kennedy A. *Corporate Cultures: The Rites and Rituals of Corporate Life*. Reading, MA: Addison-Wesley, 1982.

Fournies, F. *Coaching for Improved Work Performance*. Blue Ridge Summit, PA: Liberty House, 1987.

Dipboye RL and Howell WC. *Essentials of Industrial and Organizational Psychology,* Revised ed. Homewood, IL: The Dorsey Press, 1982.

Glasser W, MD. *Control Theory.* New York: Harper & Row, 1984.

Hersey P and Blanchard K. *Management of Organizational Behavior: Utilizing Human Resources,* 5th ed. Englewood Cliffs, NJ: Prentice Hall, 1988.

Herzberg F. *Work and the Nature of Man.* Cleveland: World Publishing Co., 1966.

Hulse SH, Deese J, Egeth H. *The Psychology of Learning,* 5th rev. ed. New York: McGraw-Hill Book Co., 1980.

Katz D and Stotland E. "A Preliminary Statement to a Theory of Attitude Structure and Change." In S Kock (Ed.), *Psychology: A Study of a Science,* Vol 3, 423–475. New York: McGraw-Hill, 1959.

Mager R and Pipe P. *Analyzing Performance Problems Or You Really Oughta Wanna,* 2nd ed. Belmont, CA: David S. Lake Publishers, 1984.

Marks LN, RN. "OHN's Can Provide Listening Ear Dealing with Employee's Problems." *Occupational Health and Safety,* (May 1986) 55,5.

Maslow AH. *Motivation and Personality,* 2nd ed. New York: Harper & Row, 1970.

McClelland D. *The Achieving Society.* Princeton, NJ: Van Nostrand, 1961.

McGregor D. *The Human Side of Enterprise.* New York: McGraw-Hill Book Co., 1985.

McGuire WJ. "The Nature of Attitudes and Attitude Change." In Lindzey G, & Aronson E, (eds.), *The Handbook of Social Psychology,* 2nd ed., Vol. 3, Reading, MA: Addison-Wesley, 1968.

Miller NE and Dollard J. *Social Learning and Imitation.* New Haven, CT: Yale University Press, 1962.

"Open Your Ears and Learn to Listen." *Safety and Health* Magazine, (July 1990) 44–47.

Rader CM, PhD. and Haber PB, PsyD. "A Psychological Profile of Industrial Injured Workers." *Occupational Health Nursing,* (November 1984) 32,11.

ReVelle JB and Boulton L. "Worker Attitudes and Perceptions of Safety." *Professional Safety,* (December 1981) 26,12.

Weisinger H, PhD. *Dr. Weisinger's Anger Work-Out Book.* New York: Quil, 1985.

"What Does It Take To Change Unsafe Behavior?" *Safety and Health,* (July 1990) 60–63.

Suggested Readings

Baldwin BA. "Professional Burnout: Occupational Hazard and Preventable Problem." *Pace* (January/February, 1981) 8,1:20ff.

Bennis WG and Nannus B. *Leaders: The Strategies for Taking Charge.* New York: Harper & Row, 1986.

Blake RR and Mouton JS. *The Managerial Grid III,* 3rd ed. Houston, TX: Gulf Publishing, 1984.

Deal T and Kennedy A. *Corporate Cultures: The Rites and Rituals of Corporate Life.* Reading, MA: Addison-Wesley, 1982.

Dennis LE and Onion, ME. *Out in Front: Effective Supervision in the Workplace.* Chicago: National Safety Council, 1990.

Drucker PF. New templates for today's organizations. *Harvard Business Review,* Jan.-Feb. 1974.

Fournies F. *Coaching for Improved Work Performance.* Blue Ridge Summit, PA: Liberty House, 1987.

Hersey P and Blanchard, K. *Management of Organizational Behavior: Utilizing Human Resources.* 5th ed. Englewood Cliffs, NJ: Prentice Hall, 1988.

Herzberg F. "One More Time: How Do You Motivate Employees?" *Harvard Business Review,* Jan.-Feb., 1968.

Hinricks JR. "Psychology of Men at Work." *Annual Review of Psychology.* (1970) 21.

Lazarus RS. "Little Hassles Can Be Hazardous to Health." *Psychology Today.* (July, 1981) 15: 48–62.

Machlowitz M. *Workaholics.* New York: Addison-Wesley Publishing Company, Inc., 1980.

Mager R and Pipe P. *Analyzing Performance Problems Or You Really Oughta Wanna.* 2nd ed. Belmont, CA: David S. Lake Publishers, 1984.

Patrick P. *Health Care Worker Burnout: What Is It, What To Do About It.* New York: Inquiry Books, 1980.

Zimbardo PG. *Psychology and Life,* 12th ed. Glenview, IL: Scott, Foresman and Company, 1988.

16

Safety Training

A key element in every successful organization, in any successful accident prevention program, and in any occupational safety and health program is effective job orientation and safety training. Workers must have proper training to do their jobs safely and efficiently.

The responsibility for, and the implementation of, employee training rests with management. Management must recognize that for the organization to achieve its objectives, employees need to perform at a certain level. In order to achieve a satisfactory level of achievement, management needs to establish worker training policies. This in turn will lead to programs for the workers. To help achieve this goal, line managers must:

1. understand how to train employees in the safest and most efficient methods of job performance
2. know and support all company health and safety policies and procedures
3. know how to detect and eliminate or control hazards, investigate accidents, and handle emergency situations
4. be trained in the proper methods of leadership and supervision.

Safety and health professionals should be taught how to identify and employ sound training methods. In the event the company hires training consultants, the safety organization must ensure that the instruction they provide achieves the company's goals and meets all regulatory training obligations.

TRAINING AS PART OF THE SAFETY PROGRAM

A new era in worker safety and health has arrived in the past two decades, and the need for training is being addressed by regulatory and legislative bodies. The responsibility for safety training typically falls under the jurisdiction of the safety and health professional. Whether members of the safety department actually conduct the classes or merely ensure they are conducted depends on the individual organization. Organizations can no longer simply provide a training session, however; the emphasis must be on workers demonstrating satisfactory understanding of how to perform various work procedures using safety equipment, proper guarding, and proper maintenance. With this in mind, it is critical for the safety and health professional to work closely with all levels of management and employees to ensure that training is adequate, accurate, and timely.

Management from top executives to supervisors must recognize the importance of training and support training efforts if these programs are to have any validity. When employees believe that safety training is not important to their supervisors, it will not be important to them.

There are several ways for the safety and health professional to identify the training needs of the organization. The professional can work with company legal counsel and others to stay current with all regulations mandating specific safety requirements. Accident and injury records together with specific investigation reports can reveal areas of weakness or failure in training. Although this approach looks at the past,

it can be turned into a proactive approach designed to prevent similar incidents from occurring. According to Heinrich's theory, past events forecast the possibility of future and potentially more severe incidents (see Chapter 12, Accident Investigation, Analysis, and Costs).

Other ways the safety and health professional can determine training needs include workplace surveys (safety audits), regularly scheduled inspections, and employee opinion surveys. These are preventive methods used to identify training needs so that programs can be established before more accidents occur.

Safety and health professionals must keep in mind that training programs do not eliminate a company's obligation to provide safe working conditions or to remove unsafe materials from the workplace. Training, however, will ensure that employees know how to do their jobs using the proper methods. Reports have strongly suggested that the majority of accidents are caused by workers' unsafe practices and errors rather than by equipment failures or poor engineering. At least in part, training can change undesired behavior and/or reinforce the desired behavior.

Training, unfortunately, is not the answer to all work-related problems. It is only one method of influencing human behavior. Before a training program is developed, management should thoroughly evaluate all problems and expectations related to the operation to determine if training will accomplish the desired goals. It must be clearly understood by everyone that training can be used only to correct skill deficiencies or to improve knowledge. Most training programs cannot correct poor engineering, poor worker attitudes, or substitute for good safety engineering and enforcement.

TRAINING POLICY

In many organizations, written policy statements establish the purpose, goals, and procedures for such areas as the company mission, hiring practices, purchasing, and so on. A written statement should be developed to cover the organization's policy on training. This may be incorporated into an overall policy statement or into separate policies for each issue.

Whichever method management selects, it should keep certain considerations in mind. The first concern is that all policies must be written to work together to achieve the ultimate company objective. Too often policies are developed separately and tend to conflict with one another. When this occurs, supervisors and workers become confused and may show inconsistent compliance. Thus, management must make sure that when several policy statements are used, they do not contradict one another.

A second consideration is the development of training schedules for all employees. A common error, found in many organizations, is to focus on safety training for workers and possibly supervisors and neglect to provide supportive training for management. This oversight can develop into a "conflict of interest" situation because the management team is unaware of the activities of the safety program. When

managers understand the organization's safety efforts, they find it easier to support them.

Therefore, the organization's training policy should define training programs for all levels of management and employees. The importance of this policy cannot be underestimated. Many of the difficulties companies experience in trying to provide safety training are a result of communication problems between different levels of the management team. Likewise, many of these same problems could be resolved by training managers in such areas as their legal obligations, their safety mission, savings potential of safety training, and other concerns.

A strong training policy will avoid pitting the safety organization against management over training issues. The policy should identify the type of training to be provided for workers, line supervisors, and management personnel. Elements of the program, such as the type of training methods used and the location of the training, may also be specified.

A well-planned training program should never interfere with production schedules. The policy statement can help to reinforce the principle that safety training is good business and can actually improve production rather than interfere with it as some managers may believe.

Planning the training program in advance will help to prevent any friction that may otherwise occur between departments. When management staff is guided by a policy statement, the training program becomes part of the company's overall operational standards and will be incorporated into the schedule rather than regarded as an interruption.

TRAINING BUDGETS

To establish an effective safety training program, management must provide an adequate budget. Training is an area where the old adage "you get what you pay for" is decidedly true. The budget for safety training may either be included with other training programs or be divided into separate accounts, depending on the company's financial policies.

When safety training is combined with other training programs, workers more easily accept the message that safety is a part of day-to-day operations. On the other hand, if safety training is separate from, say, job skill training, employees and mangers may tend to regard safety as less important.

The costs of training must be accurately recorded. These costs include but are not limited to:

- physical space
- personnel staffing
- student salaries
- training materials
- downtime (if applicable)
- personnel replacements
- transportation costs
- lodging, meals, etc.

Each organization has its own financial accounting practices. The important element in evaluating program costs is to be sure to identify the complete costs of any training pro-

gram. Even though many of these expenses are only indirectly related, they are still a legitimate part of the program budget.

The decision to train workers in safety is made by comparing the costs of undertaking such training against the costs of no training at all. The cost of not training must also reflect all of the possible expenses associated with worker accidents and injuries. For example, a training program for teaching proper lifting procedures may be estimated at $25,000. However, not training workers in these procedures may result in injuries that can amount to annual costs several times higher than the $25,000 estimated for the safety program. Thus, the company could easily recoup the cost of training by preventing even one injury to only one employee. These figures are only examples. But by reviewing records, the safety and health professional can identify the average cost of a particular class of injury and the number of these injuries experienced annually and compare this figure to the cost of providing a safety training program that will correct the problem.

With the support of the training policy, it is possible to prepare adequate budgets to provide all elements of a strong safety training program. These elements include:
1. planning company training needs *before* an accident or series of accidents
2. considering company growth or expansion
3. meeting all regualtory requirements
4. evaluating past accident causes
5. eliminating or controlling known hazards
6. providing adequate recordkeeping
7. providing for personnel development (trainers, supervisors, etc.).

RECORDKEEPING

Along with the increased emphasis on employee training by regulatory agencies, there has been a corresponding emphasis on company recordkeeping. Besides using records to fulfill legal obligations, organizations are also finding valuable uses for training records within their organizations. For example, an organization may need to demonstrate the comprehensive nature of its training program should a worker or manager file a lawsuit as the result of an accident or injury. Training records will help to identify the strengths and weaknesses of the safety program. In addition, future training budgets can be developed by examining the effectiveness and costs of past training programs.

Both new and current employees require regular safety training. In order to ensure that every member of the organization has received the appropriate training and is scheduled to receive additional instruction when required, the company must maintain adequate records for each employee. If a worker is injured, his or her records may provide valuable information to identify the true cause of the accident. Training records should also document who were the instructors for the training session, the cost of each training session, the content and methods used for instruction, and the results or effectiveness of each training session.

Employees' personal training records should be securely maintained to ensure privacy yet still permit employees to see their files. Accident investigators and regulatory representatives may also need to have access to these records for reasons.

Standard employee records include:
- employee work history
- job descriptions and locations of work performed
- specific task performance (time and duration)
- training programs attended
- dates and costs of training
- location of training sessions (in-house/external)
- purpose of training (initial, refresher, remedial).

Additional information in the records can include:
- course title
- course objectives
- training methods (seminar, on-the-job training (OTJ), home study, etc.)
- training aids or course materials used
- instructor (qualifications)
- methods of evaluation (written tests, performance).

The company must have a written policy identifying who is allowed access to the training records and under what conditions. This policy should be developed under the guidance of legal counsel.

SAFETY TRAINING NEEDS

A safety training program is needed under the following conditions:
1. for all new or reassigned (transferred) employees
2. when new equipment or processes are introduced
3. whenever procedures are revised or updated
4. when new information is available or required
5. when employee performance needs improvement or to provide employee growth
6. when employee interest in safety and efficiency needs a boost.

Many organizations mistakenly assume that transferred employees will not require orientation to their new assignments because they already know the operations. This assumption has been responsible for many worker injuries. In many instances, a transferring worker requires almost the same type of orientation as does a newly hired employee. Information should be presented on the location of safety equipment, Material Safety Data Sheets, MSDS files, proper accident and first aid reporting procedures, evacuation procedures, and other items differ from workstation to workstation. For transferred employees to work safely and efficiently, management must orient them to the new environment.

Certain symptoms will indicate to the alert safety and health professional that training is needed. These symptoms include:
1. proportionately higher accidents or injury rates than other companies that do the same type of work
2. increasing accidents rates and/or insurance rates
3. high employee turnover

4. excessive waste and/or scrap materials from operations
5. recent company expansion or procedural changes
6. increased or high levels of sick days
7. changing regulatory requirements
8. need to upgrade or update employees' knowledge and/or responsibilities
9. poor job satisfaction.

Organizations that have had great success in developing safe working environments understand the importance of the supervisor's role. Therefore, they focus considerable safety training on first-line supervisors. This does not eliminate the need to train all managers and employees but provides sound leaders and good management in the safety effort. (Figure 16–1 provides an example of a course curriculum for a supervisor's training program.)

It is important to note that the training program for supervisors includes human relations and communications skills. The supervisor is the link between upper management and the workforce. With proper training, supervisors can provide guidance, set a positive example, resolve problems, and identify the concerns of employees. Unless an organization provides the necessary leadership training, it cannot expect to have effective leaders.

DEVELOPING THE TRAINING PROGRAM

The first step in developing a training program is to identify the needs of the audience and to develop program goals. Safety should be integrated into all job skill training from the beginning. A successful training program establishes sound course objectives that integrate training with production, quality, and cost-containment issues rather than viewing these as separate concerns.

Further, organizations cannot say that training has been provided if workers do not absorb the skill or knowledge being taught and cannot demonstrate competency. In such cases, the training is either incorrect or incomplete. The employer is responsible for providing appropriate, timely training that covers all pertinent safety issues within the organization.

PROGRAM OBJECTIVES

The first and most important step in developing a training program is to identify those who need instruction and to establish program objectives. The objectives must clearly state what students are expected to gain from the program. In order to do this, objectives must be measurable. Because employee attitudes and awareness generally cannot be measured, they are not considered proper training objectives.

Preparing course objectives requires training and practice. There is too much time and money at stake to allow a "hit or miss" approach. If students are to be taught a particular skill, the objectives must state the measurable level of performance desired. By detailing the program objectives, instructors can develop a training session to meet the company's established goals. Objectives also serve as a means of evaluating the training's effectiveness.

Weak or vague objectives often result in poor or unnecessary training programs. Excessively high objectives may overburden the students with useless information. With good course objectives, the students can focus on the essential requirements of the program.

Prior to stating the objectives, management should review workers' responsibilities. The job description or job safety analysis (JSA) may provide the necessary information (see the discussion on Job Safety Analysis later in this chapter). Supervisor observations and performance testing can also identify areas of discrepancies. For example, a supervisor's job description may include the responsibility of training new workers in proper work methods. Thus, supervisors will need to receive education and experience in training skills. Organizations can make a costly mistake by assuming that everyone who knows how to perform a job also knows how to teach it. Teaching is a skill that requires not only training but practice and performance evaluations. A poor trainer can be the source of considerable problems in safety, quality, and production. As a result, the cost of not training the supervisors would most likely be far greater than the cost of a "train the trainer" program.

As safety and health issues become more complex, organizations must work hard to keep up with current requirements. Therefore, the objectives for training should be reviewed at least annually to ensure that they meet organizational needs. Many organizations are using the team approach to evaluate company training needs. The team may be comprised of members from the workforce, line supervision, safety department, maintenance, and management.

Supervisors and managers must demonstrate their support and approval of the training objectives. Workers are quick to detect when management support of a program is lacking. They will generally follow their supervisor's lead in either disregarding or adhering to safety procedures.

COURSE OUTLINES AND MATERIALS

After defining the training program audience and objectives, the next step is to develop a complete course outline. In many cases, the outline will identify specific text and materials required. The materials must be arranged in a logical sequence. Allocate time to each major topic and subtopic according to the importance of the material in meeting established program goals. The outline (Figure 16-1) also helps to suggest the best methods of instruction, the materials to be provided, and the total time required.

LESSON PLANS

A lesson plan is often referred to as a blueprint for training. Trainers can choose from several types of lesson plans. The selections should be based on organization requirements and instructor preferences. However, once the decision is made, all the company lesson plans should be created in the same format. This approach will make it easier to review the plans periodically. The lesson plan is designed to help the instructor:

A BASIC COURSE FOR SUPERVISORS

The knowledge and philosophy of accident prevention is not just common sense, as some proclaim. It is a specialized body of information accumulated over a period of many years. It is the job of the safety professional to help supervisors gain whatever information is available that will make their safety efforts more productive.

The most direct way to develop the desired attitudes and to impart the necessary information about safety to supervisors is to provide a course of instructions. Courses in safety and other management subjects are conducted for supervisors by many companies. They follow a fairly well-established pattern.

Subject Outline of Course

The following outline, based on the National Safety Council's "Supervisors' Development Program," shows the subjects that should usually be included in a safety course for supervisors. Visual aids are available on all the subjects and should be used at every meeting. Titles of films and other aids are not included here because new ones are being produced constantly. The person conducting a course should secure and use those best suited to his needs. (See the latest General Materials Catalog published by the National Safety Council.)

Session 1
Loss Control for Supervisors
    Accidents and incidents, areas of responsibility, the cost of accidents, and a better approach to occupational safety and health.
Session 2
Communications
    Elements of communication, methods of communication, and effective listening.
Session 3
Human Relations
    Human relations concepts, leadership, workers with special problems, and the drug and alcohol problem.
Session 4
Employee Involvement in Safety
    Promoting safe-worker attitudes, employee recognition, safety meetings, and off-the-job accident problems.
Session 5
Safety Training
    New employee indoctrination, job safety analysis (JSA), job instruction training (JIT), and other methods of instruction.
Session 6
Industrial Hygiene and Noise Control
    General concepts, chemical agents, physical agents, temperature extremes, atmospheric pressures, ergonomics, biological stresses, Threshold Limit Values (TLVs), and controls.
Session 7
Accident Investigation
    Finding causes, emergency procedures, effective use of witnesses, and reports.
Session 8
Safety Inspections
    Formal inspections, inspection planning and checklists, inspecting work practices, frequency of inspections, recording hazards, and follow-up actions.

Session 9
Personal Protective Equipment
    Controlling hazards; overcoming objections; protecting the head, eyes, and ears; respiratory protective equipment; safety belts and harnesses; protecting against radiation; safe work clothing; and protecting the hands, arms, feet, and legs.
Session 10
Materials Handling and Storage
    Materials handling problems; materials handling equipment; ropes, chains, and slings; and material storage.
Session 11
Machine Safeguarding
    Principles of guarding, safeguard design, safeguarding mechanisms, and safeguard types and maintenance.
Session 12
Hand Tools and Portable Power Tools
    Safe working practices, use of hand tools, use of portable power tools, and maintenance and repair of tools.
Session 13
Electrical Safety
    Electrical fundamentals review, branch circuits and grounding concepts, plug- and cord-connected equipment, branch circuit and equipment testing methods, ground fault circuit interrupters, hazardous locations, common electrical deficiencies, safeguards for home appliances, and safety program policy and procedures.
Session 14
Fire Safety
    Basic principles, causes of fire; fire-safe housekeeping; alarms, equipment, and evacuation; and reviewing the supervisor's fire job.

Tips on Running a Course

All supervisors should be given a basic course of the type just outlined. It should be repeated from time to time for new supervisors and prospective supervisors.

The format should be formal; attendance records should be kept and certificates or diplomas issued upon completion. Some companies give a diploma that can be framed and displayed; others favor a pocket-sized card, and still others present both to their graduate supervisors.

It adds to the dignity of the training program if, at the beginning of a course, a company executive meets with the group and explains the importance of the supervisor's role in the safety program and the value of the training. A formal graduation ceremony is often held upon completion of the course.

Some large companies with widespread operations personalize a course and give it throughout the organization. The preparation of such a course can be a major and costly project. It is not recommended except for very large organizations with proportionately large safety and training staffs.

Prepared courses are available, such as the National Safety Council's "Supervisors' Development" course, which has a text book, the *Supervisors' Safety Manual,* Instructors' Guide, student kit, certificate of completion, and visual aids. The Council's courses can be adapted to individual needs. They have been thoroughly tested by use in hundreds of companies and many kinds of operations. Use of a prepared course saves a great deal of preparation time and effort. Let the instructor give it the company slant.

Figure 16–1. A basic course for supervisors.

1. present material in the proper order
2. emphasize material in relation to its importance
3. avoid omission of essential material
4. keep classes on schedule
5. provide trainee participation
6. increase trainees' confidence, especially new employees.

The order of information within the lesson plan may vary with the style of each instructor, but should contain the following:

1. Title: Should clearly identify the subject matter
2. Objectives:
   a) Indicate the requirements for successful completion
   b) Establish level for course material
   c) Stimulate thinking about the subject
3. Training aids: Identify materials used in course presentation such as videos, charts, equipment, or tools
4. Introduction:
   a) Establish the scope of the subject
   b) Demonstrate the value of the information to the student
   c) Focus the student's thought on the subject material
5. Presentation:
   a) Provide the plan of action in a logical sequence
   b) Describe the method of teaching to be used, such as lecture, demonstration, class discussion, or any combination of these
   c) Provide direction for the instructor activity, for example, information to be written on chalkboard or use of visual aid
6. Application: Indicate how this material or information will be utilized by the students
7. Summary:
   a) Restate the main points of session
   b) Tie up any loose ends or confusing issues
   c) Allow for questions and answers
8. Test (if applicable): Determine the level of understanding and/or performance to ensure the course objectives have been met
9. Assignment: Provide references or instructions for preparation of future classes
10. Type of certificate, diploma, etc., to be given upon completion.

A self-checklist can help instructors improve their teaching techniques (Figure 16–2).

TRAINING METHODS

Instructors can select one of several different training methods when preparing a program. Each method has strengths and weaknesses. The technique selected should be determined by the objectives to be met, type of student participation, time allocated, facility being used, and equipment available. Everyone learns at different speeds, and through different methods. Instructors must have the training and teaching skills to address these elements of human behavior. The following are the most common types of training techniques used in industry:

1. On-the-job training (OJT)
   a. Job instruction training (JIT)
   b. Over-the-shoulder coaching
2. Group techniques
3. Conference method
4. Individual techniques
5. Computer-assisted training (interactive).

On-the-Job-Training

On-the-job training (OJT) is widely used because it allows the worker to produce during the training period (Figure 16–3). However, this method has several drawbacks. As a result, when OJT is used, the following concerns must be addressed:
1. The trainer must possess proper training skills.
2. The training program should be developed to ensure that all workers are trained in the same way to perform their tasks in the safest and most productive fashion.
3. Adequate time is allotted to the trainer and trainee to be sure the subject is well covered and thoroughly understood.

Job Instruction Training

A variation of OJT is known as job instruction training (JIT), also referred to as the four-point method. Instruction is broken down into four simple steps:
1. preparation
2. presentation
3. performance
4. follow-up.

This method of OJT has been highly successful. Workers are taught each job skill from a prepared schedule of training. The program is adjusted to each student so that workers learn one task before beginning the next.

In all training programs, the selection of the trainer is critical to program success. The one-on-one relationship between trainer and trainee allows for better control and communication. In the four-point method, this trainer-trainee relationship works in the following ways:

STEP 1.   Preparation. During the preparation stage the instructor puts the workers at ease. He or she explains the job and determines what the worker currently knows about the subject. This stage also includes the preparation of the proper learning/ working environment.

STEP 2.   Presentation. In the second step, the instructor demonstrates the work process. The student watches the performance and asks questions. The instructor should present the steps in sequence and stress all key points.

STEP 3.   Performance. In the next step, the worker performs the task under close supervision (Figure 16–4). The instructor should identify any discrepancies in the work performance and note good performance. The worker should explain the steps being performed. This ensures that the student cannot only perform the task but understands how and why the task is done. This stage continues until the instructor is satisfied with the worker's competence at the job.

SELF-CHECK TEST FOR INSTRUCTORS

A good instructor should be able to answer "Yes" to at least 20 of these questions. Under 15 would be below average.

	Yes	No			Yes	No
1. Do you check your classroom before the session for proper ventilation, lighting, and seating arrangement?	☐	☐	12. Do you refrain from using sarcasm or vulgarity in class?	☐	☐	
2. Do you preview all films or slides before showing them?	☐	☐	13. Do you stay on the subject during each session?	☐	☐	
3. Do you prepare and test equipment before class begins?	☐	☐	14. Do you cover all the material in the lesson for each class period?	☐	☐	
4. Do you keep the classroom clean and orderly?	☐	☐	15. Do you allow time for questions and for clarification of the material?	☐	☐	
5. Do you introduce yourself to each class?	☐	☐	16. Do you encourage student participation or use group discussion methods, if appropriate?	☐	☐	
6. Do you attempt to know your students and learn their names at the beginning of each course?	☐	☐	17. Do you use audiovisual aids whenever possible?	☐	☐	
7. Do you state and clarify the course objectives in the first session?	☐	☐	18. Do you demonstrate techniques or work procedures clearly and ask students to repeat the steps as you showed them?	☐	☐	
8. Do you introduce each subject and explain its importance to course objectives?	☐	☐	19. Do you summarize each lesson at the end of the class time?	☐	☐	
9. Do you vary your teaching methods according to the material you are presenting and the students taking the course?	☐	☐	20. Do you give regular assignments with clear instructions?	☐	☐	
10. Do you speak directly to the class and avoid distracting mannerisms such as chewing gum, pacing, and juggling change or chalk?	☐	☐	21. Do you keep to the class schedule throughout the course?	☐	☐	
11. Do you use words easily understood by the students and explain all technical terms?	☐	☐	22. Do you test students to identify your weaknesses as an instructor, and do you use the results to improve your skills?	☐	☐	

Figure 16–2. A self-check test can help a supervisor rate teaching methods.

STEP 4.   Follow-up. The final step is the follow-up of performance. The instructor and/or supervisor must monitor the worker's performance to be sure the job is being performed as instructed and to answer any questions the worker may have.

Of the various OJT methods, JIT is probably the most flexible and direct. Through coaching, the trainee is expected to develop and apply the learned skills in a typical work environment while under the guidance of a trained worker/trainer. The trainer must know the job thoroughly, be a safe worker, and have the patience, skill, and desire to train.

The advantages of training in the JIT method are:

1. The worker can be more easily motivated because the training/guidance is personal.
2. The instructor can identify and correct deficiencies as they occur.
3. Results of the training are immediately evaluated because the student is performing the actual job on actual equipment. The work performed can be judged against reasonable standards.
4. Training is practical, realistic, and demonstrated under actual conditions. Workers can easily ask questions.

Timing is an important element in this type of training. The trainee can receive help when needed, and the instructor can provide feedback as the training progresses. The instructor can also determine when the trainee is ready to move on to new levels of training.

It may be helpful to prepare a chart of tasks or subjects for which workers must receive training. This chart will make it easier to keep track of workers training progress and the levels of competence they achieve.

Of the other methods of OJT that can be used, the most common is known as over-the-shoulder coaching or the "buddy system." This system is considered effective in some situations, but has certain disadvantages:

1. Instructors may be selected for their availability rather than for their training skills. A trainer who lacks basic teaching skills can undermine the entire orientation of a new worker.
2. Each instructor may have his or her own way of performing the tasks being taught. This lack of continuity can make it difficult to control hazards in the workplace and lead to many accidents.
3. Key elements of orientation may be overlooked in the training program; many are not realized until an incident occurs.
4. Poor techniques or bad habits may be spread from one worker to another. Short cuts or safety violations are often demonstrated to new workers as the "way we do it."
5. Safety performance may not be emphasized during the training. Job performance should never be separated from safety standards in any training provided to workers.

Group Techniques

Group techniques encourage participation from a selected audience. These methods allow trainees to share ideas, evaluate information, and become actively involved in the planning and implementation of company policy. Several types of group training are used, but all require skilled facilitators to be successful.

Conference Method

The conference method of teaching is widely used in business and industry. The reason this method is so readily accepted is because of the knowledge each participant brings to the group. In this process, the instructor controls the flow

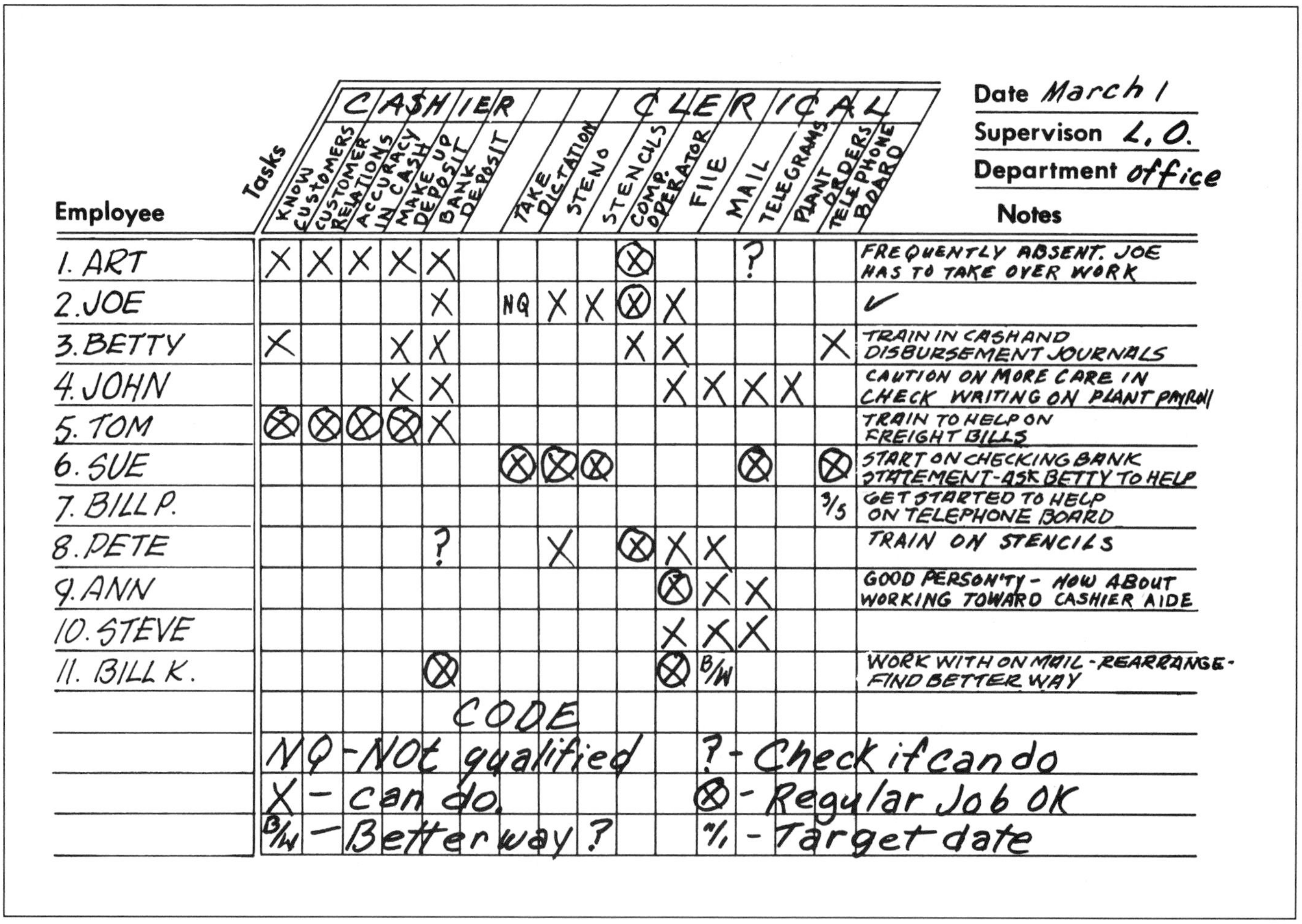

Figure 16-3 (Training chart):

Employee	Know Customers	Customer Relations	Accuracy in Cash	Make up Deposit	Bank Deposit	Take Dictation	Steno	Stencils	Comp. Operator	File	Mail	Telegrams	Plant Orders	Telephone Board	Notes
1. ART	X	X	X	X	X				⊗			?			FREQUENTLY ABSENT. JOE HAS TO TAKE OVER WORK
2. JOE				X		NQ	X	X	⊗	X					✔
3. BETTY	X			X	X			X	X					X	TRAIN IN CASH AND DISBURSEMENT JOURNALS
4. JOHN				X	X					X	X	X			CAUTION ON MORE CARE IN CHECK WRITING ON PLANT PAYROLL
5. TOM	⊗	⊗	⊗	⊗	X										TRAIN TO HELP ON FREIGHT BILLS
6. SUE						⊗	⊗	⊗			⊗		⊗		START ON CHECKING BANK STATEMENT-ASK BETTY TO HELP
7. BILL P.													3/5		GET STARTED TO HELP ON TELEPHONE BOARD
8. PETE				?			X	⊗	X	X					TRAIN ON STENCILS
9. ANN									⊗	X	X				GOOD PERSON'TY - HOW ABOUT WORKING TOWARD CASHIER AIDE
10. STEVE									X	X	X				
11. BILL K.				⊗					⊗	B/W					WORK WITH ON MAIL - REARRANGE - FIND BETTER WAY

Date: March 1
Supervison: L. O.
Department: office

CODE
NQ - Not qualified   ? - Check if can do
X - Can do.   ⊗ - Regular Job OK
B/W - Better way?   3/1 - Target date

Figure 16-3. Training chart helps keep track of who needs, and who has, what training

of the session as the participants share their knowledge and experience. The skill of the facilitator can mean the success or failure of these sessions. Facilitators must use various techniques to draw information and opinions from members. The number of people involved should be limited to allow open discussion from all participants. The opinions of each member should be recorded, and a summary of the group conclusions provided (Figure 16-5).

The conference technique is a valuable tool in problem solving. There are several situations where a safety and health professional may use the knowledge and expertise of members of the organization to address safety and health issues. It is important that at the beginning of the conference, members identify their goals and expectations for the session. For example, if a conference has been called to identify possible solutions to a safety and health concern, the group must understand that they are to make recommendations only, not to establish policy or procedures. By defining the actual role of the conference at the start, the groups can avoid misconceptions and misdirection. When a group is asked to make recommendations, they should be kept informed of the results of those recommendations.

If management fails to establish these ground rules and to provide follow-up information, the members may feel their efforts are ineffective. On the other hand, proper control and guidance of a conference can ensure its success and make it a gratifying experience for the participants.

Brainstorming

Brainstorming is a technique of group interaction that encourages each participant to present ideas on a specific issue. The method is normally used to find new, innovative approaches to issues. There are four ground rules:
1. Ideas presented are not criticized.
2. Free-wheeling creative thinking and building on ideas are positively reinforced.
3. As many ideas as possible should be presented quickly.
4. Combining several ideas or improving suggestions is encouraged.

A recorder should be selected to write down all the ideas presented. The moderator must control the flow of suggestions, cut off negative comments, and solicit ideas from each member.

Brainstorming allows ideas to be developed quickly, encourages creative thinking, and involves everyone in the process. The group can go beyond the old stereotypes or the "way it's always done" limitations.

Figure 16–4. This supervisor is explaining an enclosed operation to one of his workers. He is instructing him in the correct and safe operation of the machine and will make frequent followups on his progress until certain of his proficiency.

Case Study

A case study may involve an actual situation or be fictitious. The goal of the session is to develop group members' insight and problem-solving skills. The study normally is presented by defining what happened in a particular incident and the events leading up to the incident. The group is then given the task of determining the actual causes or problems, the significance of each element, and the solutions acceptable.

Incident Process

This is a type of case study in which the group works with a written account of any incident. The group is allowed to ask questions about facts, clues, and details. The instructor provides the answers to the questions, and the group must assemble the information, determine what has happened, and arrive at a decision. The instructor must guide the group to prevent arguments and to prevent one or two members from dominating the discussion. This is a useful method of training that encourages employee participation in the accident prevention program. Situations can be real cases in the company or developed from potential hazards that exist.

The moderator must be capable of controlling the group flow and of preventing the group from missing the true or root causes. For example, an employee was struck in the eye by a foreign body and an investigation revealed the employee was not wearing safety glasses while operating a bench grinder. The group must seek the root cause of the accident and not settle for the common conclusion "the worker failed to follow procedures" or "the worker failed to wear eye protection." They must specify why the supervisor or management failed to enforce the proper procedures (assuming there was an established procedure). Why did management allow this lax supervision?

Other Group Methods

Discussion. When used with small groups, the discussion method allows for the open exchange of ideas and concerns among workers. The leader must control the flow of the discussion and be aware of the time involved for adequate exchange of information.

Role playing. This training method is effective for evaluating human relations issues. Members attempt to identify the ways people behave under various conditions. Although this technique is not an effective method of prob-

Figure 16–5. In a participation class, the number of people in the conference group should be small enough so open discussion can occur.

lem solving, it can uncover issues not previously considered. This method is particularly helpful in identifying and changing personnel issues such as poor morale or negative attitudes.

Lecture. With this method, a single person can impart information to a large group in a relatively short time. This method is normally used to communicate facts, give motivational speeches, or summarize events for workers. There is little time or opportunity for interaction by the attendees. Follow-up for these sessions must be well planned in advance to be successful.

Question and answer (Q and A) sessions. Normally Q and A sessions follow training periods after the instructor has summarized the material presented. Workers can use this method to clarify individual concerns or points of fact. However, students often will need time to prepare and organize their thoughts before they can ask questions. In situations where they must absorb a large amount of material, allow them time to reflect on or apply the knowledge and to formulate questions. The instructor can plan to have a follow-up session or allow students to present their questions personally. Question-and-answer sessions are helpful in clarifying issues of policy or changes in schedules or events.

Simulation. When actual materials or machines cannot be used, trainers can use a simulation device. This method is used effectively in aircraft pilot training, railroad engineer training, and other applications. Various methods are employed in management training programs as well, such as the "in-basket technique," "war games," and others. One simulation demonstrates the loss of eyesight to workers to encourage them to wear safety glasses on the job. The only limit to this technique is the trainer's creativity.

Simulation is most effective when the students can participate. Careful planning and attention to detail are required. The initial costs of these sessions can be high because of the equipment required and time involved in conducting training.

Individual Techniques

Drill. Using the elements of practice and repetition, this method of instruction is valuable for developing worker skill in fundamental tasks and for performing under pressure.

Workers required to perform in crisis situations should be trained under conditions that resemble the crisis as closely as possible. For example, when instructing workers in cardiopulmonary resuscitation (CPR), the trainer must try to instill the tension that workers will experience when they attempt to resuscitate a real victim.

Demonstration. As discussed in the section on JIT, the

method of demonstration allows the instructor to perform the actual task and then have the worker repeat the performance. Instructors must be sure the job is performed exactly as required to prevent workers from developing poor habits and performance standards (and supervisors must see that employees follow the designated procedure). If the conditions used in demonstrations are not similar to the actual workstation or equipment on the job, this method will yield little useful results.

Quiz. This technique is normally used to determine if students understand the necessary information and can apply the knowledge when required. Developing good tests is a skill that requires constant review to ensure that training objectives are being met. Poor tests can reduce workers' morale and undermine training objectives.

Television. An increasing number of training programs are designed as videotape presentations. Tapes are available on nearly any subject. This method of instruction is effective if properly applied. The use of videotapes does not eliminate the need for professional instruction, but can enhance a classroom presentation. Videos are available from the National Safety Council and numerous private companies. Instructors should screen the tapes to make sure they fulfill the needs of the training program.

Computer-assisted training. Interactive computer programs are being developed for many areas of employee training. They allow the students to receive information by reading or watching a video presentation and then respond to questions. If the correct response is entered, the computer will advance the program to the next section; if the wrong response is entered, the program will repeat the information and retest the student. The system is valuable for several reasons:

1. The students can work at their own pace.
2. Records are automatically kept of all training. The amount and type of records maintained can be modified to meet the company requirements.
3. Correct answers are required before a student can proceed to the next lesson.
4. Students receive training as time is available in their schedule, rather than having to meet training schedules.

With computer-assisted training programs, instructions can guide students step by step through a curriculum designed to meet the goals of the individual, the company, and/or any regulatory obligations. The company can keep records that include the amount of time each student spends in training, the type of material or information presented, and the success of the training. This program works extremely well for organization with small workforces or those that cannot remove large groups of workers from their jobs at any one time. (See also Chapter 17, Audiovisual Media, in this volume.)

Reading. Companies should provide employees with written safety materials such as monthly newsletters, supervision and safety magazines, and poster. In addition, organizations may establish a library where employees can research information on subjects such as work procedures, safety, leadership, health care, family or home safety, and other subjects

of concern. It is important, however, that management does not assume everyone has the ability to read and comprehend all of the written material provided. Companies cannot replace instruction or training programs simply by handing an employee a training manual. Written material is meant to supplement or to serve as a reference for training.

School Courses

Independent study. Home-study courses or correspondence courses are used by many companies. They can help workers to advance within the organization or to improve their knowledge of their jobs and industries. A major advantage of this method is that the worker does not lose any time from work and can complete the course at his or her own pace. Another advantage is the low cost of home-study programs. Normally they are centered around a textbook assignment, followed by self-tests using multiple choice, true/false, fill-in, or essay questions. Several courses also provide laboratory or performance materials such as television, radio, or computer repair programs that work on actual equipment. Some home-study programs come with videotaped presentations for students to view.

The National Safety Council offers home-study programs for supervisor training—"Supervising for Safety" and "Protecting Workers' Lives."

Seminars and short courses. Seminars for safety and leadership skills are offered by many colleges and universities as well as by insurance companies and private organizations. The Safety Training Institute of the National Safety Council offers on-site instruction for workers, supervisors, and management. Seminars range from one-hour sessions to several days.

POLICIES AND ATTITUDES IN WORKER TRAINING

Teaching at the Adult Level

Instructors must always keep in mind that workers are adults and must be treated accordingly. It is generally agreed that the most accepted method for teaching adults is through student interaction. As a result, organizations should limit group size to allow for student participation and response. Training must emphasize that safety is not a separate job to be done if and when other responsibilities permit the time. Rather, the supervisor must always consider how to maintain the production schedule with safe work practices.

In this regard, workers most likely will reflect the attitudes of their immediate supervisors. When supervisors make safety an important element in job performance, it will be treated as such by the workers. On the other hand, if a supervisor sends a message that production is the only important consideration, workers will tend to ignore safety policies to maintain the production schedule. These messages are communicated more often by a supervisor's actions than by words or written instruction. For this reason, supervisor's training should include human relations

and communication skills to make the person a more effective leader and role model.

Integrate Safety into All Training

When an organization has a formal training program, safety training can easily be incorporated into the curriculum. To do so, the training professional and the safety and health professional should work together closely to plan the activities. Training supervisory staff in safety issues is normally the primary concern of the safety professional. The mission of the safety professional is to educate the members of the organization in all elements of the safety and health program.

In many instances, such a program may require several years to educate and modify workers' existing attitudes toward "safety." However, integrating safety into production training will help to accomplish this goal. When supervisors are educated in leadership techniques, they become a strong ally for the safety and health program. This is because almost everything taught about good supervision helps to promote the safety and health program. In the same sense, almost everything taught in safety and health program has a positive effect on production.

Safety training, like performance training, is most effective when it is provided on a continuing basis. This will test the creativity of training department staff. The dynamic, changing nature of the working world requires a constant review and updating of training programs. When instruction lacks imagination or stimulating materials, workers quickly lose interest in both the medium and the message.

TRAINING NEW EMPLOYEES

Safety training begins when a new worker is hired. The person is usually open to ideas and to information about the way the company operates. From the first day, the new employee formulates opinions about the management, supervisors, fellow workers, and the organization. Personnel managers may say they have never hired a worker with a bad attitude. This may not be entirely true, but most employees' negative attitudes are developed after they have been on the job.

Timeliness of instruction can become a key issue in the orientation program. For example, although statistical data differ, it is generally agreed that new employees are significantly more prone to work-related accidents. This fact is attributed to the inexperience of new workers, their unfamiliarity with procedures and facilities, and their over zealousness to do the work. There are also a significant number of workers injured between the four- to six-year range. This is generally attributed either to a change in work duties (new or transferred employee) or worker complacency. These issues can be covered in an ongoing training program and help to reduce the number of accidents.

The following subjects are suggested as part of the orientation program:

- company orientation: history and goals
- policy statements
- benefit packages

- organized labor agreements (if applicable)
- safety and health policy statement (if separate)
- acceptable dress code (as required)
- personnel introductions
- housekeeping standards
- communications about hazards
- personal protective equipment
- emergency response procedures: fire, spill, etc.
- accident reporting procedures
- near-miss accident reporting
- accident investigation (supervisors)
- lockout/tagout procedures
- machine guarding
- electrical safety awareness
- ladder use and storage (if applicable)
- confined space entry (if applicable)
- medical facility support
- first aid/CPR
- hand tool safety
- ergonomic principles
- eye wash and shower locations
- fire prevention and protection
- access to exposure and medical records.

This is by no means a complete list of the training needed to properly safeguard new workers. However, the subjects mentioned cover essential information that is far too important to overlook or leave to casual learning. A formal program should be developed not only to provide the worker with this information, but to create a strong link between employees and the safety and health organization.

Many of these subjects are left to the supervisor to cover with employees when they arrive at the worksite. The supervisor must have time cover these points with the employee. Unfortunately, the importance of going over these subjects with the employee is not recognized until an accident occurs.

The supervisor should be current with the information presented to new workers to avoid the age-old problem of contradicting the training material. Supervisors should also become involved in the development of training programs to keep the information current and practical in relation to work demands. When supervisors start disregarding or contradicting the training program, the entire program—and the company's image—lose credibility with the workers.

The supervisor should re-enforce the training program and enhance the information by demonstrating how it will apply to the worker's specific job assignment. For example, the new employee may have been taught how to read the warning labels on chemical containers. The supervisor can then enhance this information by conducting a tour of the department, pointing out all hazardous materials, and identifying the protection provided through personal protective equipment and control measures.

Manuals and Rule Books

The use of written policies and procedures has been a traditional means of providing information to new employees. The company should exercise care in preparing the manuals

Table 16–A. Index to OSHA Training Requirements.

Hazard	Part	Subpart	Section
Blasting or Explosives	1910.109	H	(d)(3)(i)(iii)
	1926.901	U	(c)
	1926.902	U	(i)
	1915.10	B	(a) thru (b)
	1916.10	B	(a) thru (b)
	1917.10	B	(a) thru (b)
Carcinogens			
4-Nitrobiphenyl	1910.1003	Z	(e)(5)(i) thru (ii)
alpha-Naphthylamine	1910.1004	Z	(e)(5)(i) thru (ii)
4, 4'-Methylene bis (2-chloroaniline)	1910.1005	Z	(e)(5)(i) thru (ii)
Methyl chloromethyl ether	1910.1006	Z	(e)(5)(i) thru (ii)
3, 3'-Dichlorobenzidine (and its salts)	1910.1007	Z	(e)(5)(i) thru (ii)
bis-Chloromethyl ether	1910.1008	Z	(e)(5)(i) thru (ii)
beta-Naphthylamine	1910.1009	Z	(e)(5)(i) thru (ii)
Benzidine	1910.1010	Z	(e)(5)(i) thru (ii)
4-Aminodiphenyl	1910.1011	Z	(e)(5)(i) thru (ii)
Ethyleneimine	1910.1012	Z	(e)(5)(i) thru (ii)
beta-Propiolactone	1910.1013	Z	(e)(5)(i) thru (ii)
2-Acetylaminofluorene	1910.1014	Z	(e)(5)(i) thru (ii)
4-Dimethylamino-azobenzene	1910.1015	Z	(e)(5)(i) thru (ii)
N-Nitrosodimethylamine	1910.1016	Z	(e)(5)(i) thru (ii)
Vinyl chloride	1910.1017	Z	(j)(1)(i) thru (ix)
Cranes and Derricks	1910.179	N	(m)(3)(ix)
	1910.180	N	(h)(3)(xii)
Decompression or Compression	1926.803	S	(a)(2)
	1926.803	S	(b)(10)(xii)
	1926.803	S	(e)(1)
Employee Responsibility	1910.109	H	(g)(3)(iii)(a)
	1926.609	U	(a)
Equipment Operations	1910.217	O	(f)(2)
	1926.20	C	(b)(4)
	1926.53	D	(b)
	1926.54	D	(a)
	1910.252	Q	(c)(6)
Fire Protection	1916.32	D	(e)
	1917.32	D	(b)
	1926.150	F	(a)(5)
	1926.155	F	(e)
	1926.351	J	(d)(1) thru (5)
	1926.901	U	(c)
Forging	1910.218	O	(a)(2)(i) thru (iv)
Gases, Fuel, Toxic Material, Explosives	1910.109	H	(d)(3)(i) and (iii)
	1910.111	H	(b)(13)(ii)
	1910.266	R	(c)(5)(i) thru (xi)
	1910.106	H	(b)(5)(vi)(v)(3)
	1916.35	D	(d)(1) thru (6)
	1926.21	C	(a) and (b)(2) thru (6)
	1926.350	J	(d)(1) thru (6)
General	1926.21	C	(a)
Hazardous Material	1915.57	F	(d)
	1916.57	F	(d)
	1917.57	F	(d)

Hazard	Part	Subpart	Section
Medical and First Aid	1910.94	G	(d)(9)(i) and (vi)
	1910.151	K	(a) and (b)
	1915.58	K	(a)
	1917.58	F	(a)
	1926.50	D	(c)
Personal Protective Equipment	1910.94	G	(d)(11)(v)
	1910.134	I	(a)(3)
	1910.134	I	(b)(1), (2) and (3)
	1910.134	I	(e)(2), (3) and (5)
	1910.134	I	(e)(5)(i)
	1910.161	K	(a)(2)
	1915.82	I	(a)(4)
	1915.82	I	(b)(4)
	1916.57	F	(f)
	1916.58	F	(a)
	1916.82	I	(a)(4)
	1916.82	I	(b)(4)
	1917.57	F	(f)
	1918.102	J	(a)(4)
	1926.21	C	(b)(2) thru (6)
	1926.103	E	(c)(1)
	1926.800	S	(e)(xii)
Pulpwood Logging	1910.266	R	(c)(5)(i) thru (xi)
	1910.266	R	(c)(6)(i) thru (xxi)
	1910.266	R	(c)(7)
	1910.266	R	(e)(2)(i) and (ii)
	1910.266	R	(e)(9)
	1910.266	R	(e)(1)(iii) thru (vii)
Powder-Actuated Tools	1915.75	H	(b)(1) thru (6)
	1916.75	H	(b)(1) thru (6)
Power Press	1910.217	O	(e)(3)
Power Trucks, Motor Vehicles, or Agricultural Tractors	1910.109	H	(d)(3)(iii)
	1910.109	H	(g)(3)(iii)(a)
	1910.178	N	(1)
	1910.266	R	(e)(9)
	1910.266	R	(e)(6)(viii)
	1928.51	C	(d)
Radioactive Material	1916.37	D	(b)
Signs—Danger, Warning, Instruction	1910.96	G	(f)(3)(viii)
	1910.145	J	(c)(1)(ii)
	1910.145	J	(c)(2)(ii)
	1910.145	J	(c)(3)
	1910.264	R	(d)(1)(v)
Tunnels and Shafts	1926.800	S	(e)(xiii)
Welding	1910.252	Q	(b)(1)(iii)
	1910.252	Q	(c)(1)(iii)
	1915.35	D	(d)(1) thru (6)
	1915.36	D	(d)(1) thru (4)
	1916.35	D	(d)(1) thru (6)
	1917.35	D	(d)(1) thru (6)

to ensure that the information is easy to understand, complete, and that all rules are enforced. Rules should cover items such as first aid, personal protective equipment, electrical safety, and housekeeping, just to name a few.

Regulatory Obligations

There are many regulatory agencies which affect the employers responsibility to provide training to their employees. All regulatory agencies have published requirements for the employer. There may still be difficulties in the final interpretation, and the best solution for these problems is to communicate with the agency involved, if necessary, and the organization's legal counsel. Table 16–A lists the references published by the Occupational Safety and Health Administration (OSHA) that reflect training requirements. Table 16–B identifies the requirements for training as defined by the

Table 16–B.

OSHA AND MSHA TRAINING REQUIREMENTS

The continued importance of training is evidenced by the requirements of both the Occupational Safety and Health Administration (OSHA) and the Mine Safety and Health Administration (MSHA).

OSHA requirements

Listed next are the major parts of the OSHA regulations (Title 29—Labor, Code of Federal Regulations) covering training requirements .Table 9-A gives a convenient index indicating the type of hazard and the parts of the regulations requiring training to protect against the hazard.

Part 1910, Safety and Health Training Requirements for General Industry

Part 1915-18, Safety and Health Training Requirements for Maritime Employment

Part 1926, Safety and Health Training Requirements for Construction

Part 1928, Occupational Safety and Health Requirements for Agriculture.

MSHA regulations

The following is a summary of the training requirements under the MSHA Regulations, published in the Federal Register, Vol. 43, No. 199, October 13, 1978.

Subpart B— Training and Retraining Miners Working at Surface Mines and Surface Areas of Underground Mines

§48.21 Scope

Subpart B sets forth the mandatory requirements for submitting and obtaining approval of programs for training and retraining miners at surface mines and surface areas of underground mines. It also includes requirements for compensation for training and retraining.

§48.22 Definitions

(a) "Miner"—Any person working in a surface mine or surface area of an underground mine and who is engaged in the extraction and production process or is regularly exposed to mine hazards, or who is a maintenance or service worker (whether employed by operator or contractor) working at the mine for frequent or extended periods.

Short-term, specialized contract workers (drillers, blasters, etc.) who have received training under 48.26 (training of newly employed experienced miners) may be trained under 48.31 (hazard training) in lieu of other subsequent training.

Excluded from definition

   (i) Construction, shaft and slope workers covered in Subpart C, Part 48.

   (ii) Supervisory personnel (covered under MSHAapproved state certification requirements).

   (iii) Delivery, office or scientific or short-term maintenance workers and any student engaged in academic projects.

(b) "Experienced miner." A person currently employed as a miner; or a person who received training acceptable to MSHA from an appropriate state agency within the preceding one month; a person with 12 months experience working in surface operations during the preceding 3 years; a person who received new miner training (48.25) within the past 12 months.

(c) "New miner." Not experienced.

(d) "Normal working hours." Regularly scheduled workhours.

(e) "Operator." Owner, lessee, person that controls or supervises the operation; or any contractor performing similar function.

(f) "Task." Regular work assignment which requires physical abilities and job knowledge.

(g) "Act." The Federal Mine Safety and Health Act of 1977.

§48.23 Training Plans

(a) Each operator shall have a MSHA-approved plan for training:

New minersNewly employed experienced miners

Miners for new tasks

For annual refresher

For hazard.

   (1) Existing mines shall submit the training plan to MSHA for approval within 150 days of the effective date (October 13, 1978). The plans must be filed by March 11, 1979.

   (2) Unless extended, MSHA shall approve the operator's plan within 60 days.

   (3) New mines—reopened mines. Must have an approved plan prior to (re)opening.

(b) Training plan shall be filed with the Chief of Training Center, MSHA, for the area in which the mine is located.

(c) Information to be filed:

   (1) Company nameMine name MSHA I.D. number.

   (2) Name and position of person responsible for health and safety training.

   (3) List of MSHA-approved instructors along with the courses they are qualified to teach.

   (4) Location of training site.

   (5) Description of teaching methods and course materials which are to be used.

   (6) Number of miners. Maximum number of miners to attend each session.

   (7) Refresher training—a schedule of time or period of time when such training will be given. To include titles of courses, total number of instruction hours for each course, and predicted time and length of each session.

   (8) New task training for miners.

     (i) Submit complete list of task assignments to correspond with the definition of "task."

     (ii) Titles of the instructors.

     (iii) Outline training procedure for each work assignment.

     (iv) The evaluation procedures used to determine the effectiveness of training.

(d) Two weeks prior to plan submission, a copy of the plan shall be given to the employees' representatives.

Should there be no employee representative, the plan shall be posted on the mine bulletin board two weeks prior to submission.

All written comments from employees must be delivered to MSHA. Miners may deliver such comments directly to MSHA.

(e) The training plan is subject to review and evaluation by MSHA. Course materials, including visual aids, handouts, etc., must be available to MSHA. At the request of MSHA, the operator must alter, change or modify the plan.

A schedule of upcoming training must be given to MSHA.

(f) A copy of the approved plan must be available at the mine for MSHA, miners, and miners' representatives.

Table 16–B. *(Continued.)*

(g) All courses shall be conducted by MSHA-approved instructors except as provided for in the "New Task Training of Miners" and "Hazard Training" sections.

(h) Instructors are approved in the following ways:
  (1) Receive instructor training from MSHA or from a person designated qualified by MSHA.
  (2) Instructors may be approved by MSHA to teach specific courses based on written evidence of qualifications and teaching experience.
  (3) MSHA may approve instructors based on the performance of the nstructors while teaching classes are monitored by MSHA. This program must be approved in advance by MSHA.
  (4) Cooperative instructors, designated by MSHA to teach approved courses within the past 24 months, shall be considered approved.

(i) MSHA can revoke the approval of an instructor for good cause. There is a specific appeal procedure provided.

(j) MSHA shall notify the operator and miners' representative in writing the status of the MSHA approval within 60 days from the date the plan is submitted.
  (1) Any revision to the plan required by MSHA in order to gain approval shall be given to the operator and employees' representative. The operator and the employees' representative have the right to discuss alternative revisions or changes with MSHA—within a specified period of time.
  (2) MSHA can approve portions of a plan and withhold approval on the balance.

(k) Training shall begin within 60 days after approval of the plan.

(l) The operator shall submit any proposed changes or modifications to an approved plan to the employees' representative and MSHA prior to making such changes. MSHA must approve such changes prior to implementation.

(m) MSHA must notify, in writing, the operator and employee representative of the disapproval or recommended changes to the submitted plan. Such notification will include:
  (1) State specific changes or deficiency.
  (2) Action needed to bring plan into compliance.
  (3) MSHA will take punitive action against the operator should remedial action to effect compliance be delayed or ignored.

(n) All MSHA-recommended changes shall be posted on the bulletin board and a copy of same delivered to the employees' representative.

§48.24 Cooperative Training Program

(a) Training programs may be conducted by the operator, MSHA, MSHA-approved programs conducted by state or other federal agencies, or associations of operators or miners' representatives, private associations, or educational agencies.

(b) Instructors and courses shall be approved by MSHA.

§48.25 Training of New Miners; Minimum Courses of Instruction; Hours of Instruction\

(a) Each new miner shall receive not less than 24 hours of training. Unless otherwise stated, this training shall take place before they start work duties. At the discretion of MSHA, a new miner may receive a portion of this training after he starts his work duties. Provided that not less than 8 hours of training shall be given before the employee starts work. This first 8 hour training shall include:

  (1) Introduction to work environment
  (2) Hazard recognition
  (3) Health and safety aspects of the tasks assigned.

The remainder of the 24 hours training or up to 16 hours will be given within 60 days. This program must be approved by MSHA. Certain conditions at a mine, such as employee turnover, mine size or safety record may cause MSHA to require the full 24 hour training prior to the start of work.

(b) New miner training program shall include:
  (1) Instruction in the statutory rights of miners and their representatives.Authority and responsibility supervisors A review and description of the line of authority of supervisors and miners' representative Introduction to mine rules and procedures for reportihazards.
  (2) Self-rescue and respiratory devices—instruction and demonstration in the use, care, and maintenance (where applicable).
  (3) Transportation controls and communication systems—instruction on the procedures in effect for riding on and in mine conveyances where applicable; the controls for the transportation of miners and materials; use of mine communication systems, warning signals, and directional signs.
  (4) Introduction to work environment. Includes tour of mine and a description of the entire operation.
  (5) Escape and emergency evacuation plans, fire warning and firefighting. Review mine escape system and emergency evacuation plans and instructions in fire warning signals and firefighting procedures.
  (6) Ground control; working in areas of high walls, water hazards, pits and spoil banks; illumination and night work.
  (7) Health—instruction includes the purpose of taking dust measurements and noise and other health measurements, and any health control plan at the mine shall be explained. The operator shall explain the health provisions of the act and warning labels.
  (8) Hazard recognition—course includes recognition and avoidance of hazards present in the mine.
  (9) Electrical hazards—includes recognition and avoidance of electrical hazards.
  (10) First Aid—must be a MSHA-approved course.
  (11) Explosives—includes a review and instruction on the hazards related to explosives. This course can be omitted if no explosives are used or stored at the mine.
  (12) Health and safety aspects of the tasks to which the new miner will be assigned. The course includes instruction in the health and safety aspects of the work, the safe work procedures, and the mandatory health and safety standards pertinent to the work.
  (13) Any other courses deemed necessary by MSHA based on special circumstances and conditions at the mine.

(c) The training plan shall include oral, written or practical demonstration methods to determine successful completion of the training. These methods shall be administered to the miner prior to assignment to work duties.

(d) A newly employed miner who has received the full 24 hours training within the 12 months preceding employment need not go through the operator's new miner training program. However, the miner will have to receive and complete the instruction for the "newly employed experienced miner" and "new task training of miners before commencing."

Table 16–B. *(Continued.)*

§48.26 Training of Newly Employed Experienced Miners, Minimum Courses of Instruction

(a) The newly employed experienced miner shall receive and complete the training listed below before being assigned to work duties.

(b) The training program includes the following:

(1) Introduction to work environment. Includes a tour of the operation and a description of the total operation.

(2) Mandatory health and safety standards. Includes those standards pertinent to the tasks assigned.

(3) Authority and responsibility of supervisors and miners' representatives. Includes a review of supervisors and miners' representatives line of authority, and the responsibility of such persons. Also, an introduction to the operator's rules and procedures for reporting hazards.

(4) Transportation controls and communication system. Includes instruction on the procedures for riding on and in mine conveyances; controls for the transportation of miners and materials; and use of the mine communication system, warning signal, and directional signs.

(5) Escape and emergency evacuation plans; fire warning and firefighting. Includes review of the mine escape system; escape and emergency evacuation plans and instruction in the fire warning signals and firefighting procedures.

(6) Ground controls; working in areas of high walls, water hazards, pits and spoil banks; illumination and night work. Includes introduction and instruction on the high wall and ground control plans; procedures for working near high walls, water hazards, pits and spoil banks, illuminated work areas, and procedures for working during hours of darkness.

(7) Hazard recognition. Includes recognition and avoidance of hazards, particularly any hazards related to explosives where explosives are used or stored at the mine.

(8) Any other courses MSHA deems necessary based on special mine circumstances and conditions.

§48.27 Training of Miners Assigned to a Task in Which They Have Had No Previous Experience; Minimum Courses of Instruction

(a) A miner shall be trained to safely perform any new work task prior to starting such work. The exceptions to this rule are:

(1) If the miner received such training within the preceding 12 months and can demonstrate knowledge of the safe procedures.

(2) If the miner performed the work within the preceding 12 months and can demonstrate knowledge of the safe procedures.

The training program shall include the following:

(i) Health and safety aspects and safe operating procedures for work tasks, equipment or machinery. Includes instruction in the health and safety aspects and safe operating procedures and given on-the-job.

(ii) Supervised practice during nonproduction. Practice training in the work task will be conducted at times or places where production is not the primary objective.

Supervised operation during production. Training will be conducted while under supervision and during production in the operation of equipment and performance of work task.

(iii) New or modified machines and equipment. Where new or different operating procedures are required as a result of new or modified equipment, the miner will be fully trained in the new procedures.

(iv) Any additional courses MSHA may deem necessary as a result of special conditions or circumstances at the mine.

(b) Miners shall not operate equipment or engage in blasting operations without direction and immediate supervision until the miner has demonstrated safe operating procedures for equipment or blasting operation.

(c) Miners assigned to a new task not covered in the paragraph shall be instructed in the safety and health aspects and safe procedures of the task prior to starting such task.

(d) All training and supervised practice shall be given by qualified trainers or experienced supervisor or other person experienced in the new task.

§48.28 Annual Refresher; Training of Miners

(a) Required—8 hours of annual refresher.

(b) Refresher shall include the following:

(1) Mandatory health and safety standards. The standards relating to the miner's task.

(2) Transportation controls and communication system. (Same as paragraph 48.26, b, 4.)

(3) Escape and emergency evacuation plans; fire warning and fire fighting. (Same as paragraph 48.26, b, 5.)

(4) Ground control; working in areas of high walls, water hazards, pits and spoil banks; illumination and night work. (Same as paragraph 48.26, b, 6.)

(5) First aid—method acceptable to MSHA.

(6) Electrical Hazards—recognition and avoidance of electrical hazards.

(7) Prevention of accidents—review of accidents and their causes and instruction in accident prevention in the work environment.

(8) Health—explain purpose for taking dust, noise and other health measurements, and any health control plan in effect at the mine. Further, explain warning labels and health provisions of act.

(9) Explosives—review and instruct on the hazards related to explosives: This course is not needed when explosives are not used or stored at the mine.

(10) Self-rescue and respiratory devices. Instruct and demonstrate the use, care and maintenance of self-rescue and respiratory devices.

(11) Any additional courses MSHA may deem necessary.

(c) All experienced miners will receive refresher training within 90 days after the training plan is approved by MSHA.

(d) Annual refresher training sessions shall not be less than 30 minutes of actual instruction time and miners shall be notified that the session is part of annual refresher training.

§48.29 Records of Training

(a) Upon completion of the MSHA-approved training, the operator shall record and certify on MSHA form 5000-23 that the miner has received the specified training. A copy of the training certificate is given to the miner. A copy of the certificate is filed at the mine site to be available to the various government agencies and the miners.

(b) False certification that training was given is punishable under Section 110 (a) and (f) of the Act.

(c) Copies of training certificates for current employees shall be retained at the mine site for two years and for 60 days after a miner terminates.

Table 16–B. *(Continued.)*

§48.30 Compensation for Training
(a) Training shall take place during normal working hours and the miner receives the rate of pay as though working at the work task.
(b) Should the training by given at a location other than the normal workplace, miners shall be paid for additional costs, such as mileage, meals, and lodging, they may incur in attending the training.

§48.31 Hazard Training
(a) All miners shall receive hazard training before starting work duties. Such training shall include the following:
    (1) Hazard recognition and avoidance;
    (2) Emergency and evacuation procedures;
    (3) Health and safety standards, safety rules, and safe working procedures;
    (4) Self-rescue and respiratory devices; and,
    (5) (i) Any additional courses MSHA may deem necessary.
        (ii) Miners will receive training at least once every 12 months.

    (iii) The hazard training program will be submitted to MSHA along with the other training programs.
    (iv) Recordkeeping and completion certification shall be maintained in the same manner as the other training plans.

§48.32 Appeals Procedures
The operator, miner, and miners' representative can appeal any decision of the MSHA Training Chief.
(a) Appeals to MSHA shall be in writing to:
Director of Education & Training
MSHA
4015 Wilson Blvd.
Arlington, Va. 22203
The appeal must be within 30 days after notification of a MSHA decision.
(b) The Director can request additional information from all parties.
(c) The Director shall render a decision on the appeal within 30 days after receipt of the appeal.

Mine Safety and Health Administration (MSHA). Remember these regulatory obligations only represent the minimum requirements and are rarely comprehensive enough to provide a truly safe and healthful environment for our employees.

SUMMARY

- A key element in successful accident prevention or occupational health and safety programs is proper training to help workers to do their jobs safely and efficiently. Safety and health professionals, supervisors, and other managers must develop training skills to teach workers how to perform competently and safely on the job.

- Safety training must have the wholehearted support and financial backing of top management and must be integrated into all other types of skills training. If management regards safety as a major goal, workers will take it seriously as well.

- The safety and health professional can identify training needs in an organization by keeping up with government regulations and by conducting workplace surveys, inspections, and employee opinion polls. However, training programs do not relieve a company of its obligation to provide a safe work environment or to remove unsafe materials from the workplace.

- Training should be provided for all levels of employees and management. This will avoid conflicts between the safety department and the management team over training issues. However, a well-planned training program should never interfere with production schedules.

- When developing a training program budget, companies must not only factor in all costs of a program but estimate what it might cost *not* to do safety training in terms of medical expenses and lost time due to injuries and accidents. If the program eliminates even one major accident or worker injury a year, it often can pay for itself.

- Accurate records of safety training can be used to show a company's compliance with regulations, to defend against liability claims in court, to show the strengths and weaknesses of the program, to monitor worker training schedules, and to develop future training budgets. Employee records should be regarded as confidential—available only to employees or to those with a legal right to view the information.

- Safety training programs are needed for all new or transferred employees, when new equipment or processes are introduced, when procedures are revised/updated, when new information is available, whenever employee performance needs upgrading, and when employee interest in safety needs a boost.

- Symptoms of safety problems that can be corrected by training programs include higher rates of accidents or injuries, increasing insurance rates, high employee turnover, excessive material waste, and changing legal requirements.

- The first step in developing a training program is to identify the target audience and to create program goals. The company must develop a written policy to establish the organization's purpose, goals, and procedures for its training programs. All company policies should work together to achieve company goals.

- Program objectives should clearly state what students are expected to gain from the program and be able to measure student compentence. Objectives should be based on a realistic assessment of the job responsibilities and tasks that workers are expected to perform.

- The next step is to develop a complete course outline, specifying the text and materials to be used. The out-

line serves as the basis for individual lesson plans. The lesson plan helps the instructor present material completely and in the proper order, keep classes on schedule, provide for student participation, and increase student confidence.

- Instructors can use a variety of training methods including on-the-jog training (job instruction training and over-the-shoulder coaching), conference methods, group techniques, individual techniques, and computer-assisted training. Instructors should adapt their teaching methods to the needs of the students and content of each course.
- Training must always emphasize that safety is not a separate job but an integral part of each task. This attitude is particularly important to communicate to new employees and first-time workers in an orientation program. Stressing safety issues can create a strong link between employees and the safety department and enhance workers' well-being on the job.

REFERENCES

Programmed Instruction

"A Bibliography of Programs and Presentation Devices." C Hendershot, 4114 Ridgewood Drive, Bay City, MI 48706.

"Library of Programmed Instruction Courses." EI du Pont de Nemours and Company, Inc., Education and Applied Technology Division, Wilmington, DE. 200117

National Society for Programmed Programmed Instuction, PO Box 137, Cardinal Station, Washington, DC 20017

Management Games

"A Catalog of Ideas for Action Oriented Training." Didactic Systems, Inc., PO Box 4, Cranford, NJ 07016.

"The In-Basket Method." Bureau of Industrial Relations, Department of Training Materials for Industry, The University of Michigan, Graduate School of Business Administration, Ann Arbor, MI 48104.

"Simulation Series for Business and Industry." Science Research Associates, Inc. Department of Management Services, 155 North Wacker Drive, Chicago, IL 60606

Zoll AA. "The In-Basket Kit." Addison-Wesley Publishing Co. Reading, MA 01867

Books

Dennis L and Onion M. *Out in Front: Effective Supervision in the Workplace.* Chicago, IL: National Safety Council, 1990

Mager RF. *Preparing Instructional Objectives.* 2nd ed. Belmont, CA: David S. Lake Publishers, 1984.

Mager RF and Beach KM Jr. *Developing Vocational Instruction.* Belmont, CA: David S. Lake Publishers, 1984.

Mager RF and Pipe P. *Analyzing Performance Problems of 'You Really Oughta Wanna'* 2nd ed. Belmont, CA: David S. Lake Publishers, 1984.

ReVelle JB. *Safety Training Methods.* New York: John Wiley and Sons, 1980.

Supervisors Safety Manual. 7th ed. Chicago, IL: National Safety Council, 1990.

"Training Requirments of OSHA Standards," OSHA No. 2254. U.S. Department of Labor. Available from U.S. Government Printing Office, Washington, DC 20402, or local OSHA regional office.

17

Audiovisual Media

Audiovisual (AV) communication tools have become an essential part of occupational safety, health, and environmental instruction. Those who plan a safety talk or lecture often give much thought to how AV materials can reinforce their message. Research shows that when audiovisuals are used, meetings tend to be 28% shorter and 43% more productive.

EFFECTIVENESS OF AV COMMUNICATION TOOLS

AV media take advantage of the two senses people use most when learning—sight and hearing. Human beings obtain 83% of what they learn through sight, and 11% through hearing. Only 1.0% is learned through taste; 1.5% through touch; and 3.5% through smell. Further, extensive research shows that people retain 50% of what they both see and hear, but only 20% of what they learn through hearing alone and only 30% of what they learn through sight alone (Figure 17–1).

Range of AV Media

The full range of AV media includes posters, flip charts, slides and other projected transparencies, videos, computer disks and tape, motion pictures, recordings, models, and fully operating equipment. Teaching aids include books and workbooks, chalkboards and marker pen boards and tablets, bulletin boards, display cases, and flannel boards. AV equipment (hardware) covers projectors, simulators, tape and cassette players for sound and video, cameras for stills and video, monitors, television equipment, and equipment for computer-interactive training, such as personal computers, optical laser disc players, and CD ROM players (Figure 17–2).

Live demonstrations and experiments also are classified as audiovisual events. Health and safety professionals who want to make full use of the power of AV media must first understand employee training needs and how AV media can help to meet these goals.

DEFINING TRAINING AND COMMUNICATION NEEDS

Regulatory agencies and legislative bodies are mandating safety training in far more detail than ever before. For an in-depth discussion of safety training, see Chapter 16. In general, employers can no longer rely on class rosters as sufficient evidence to prove that workers received adequate training. Instead, they must train workers in specific competencies and document that such training has been effective. For example, new regulations or changes in existing rules (see the Lockout/Tagout section in Chapter 13, Safeguarding, in the *Engineering and Technology* volume) will require additional training time, even for veteran employees.

Accident Evaluation

Accident investigations are beneficial in determining areas of weaknesses or failures in training. Although identifying high accident areas and analyzing possible training defi-

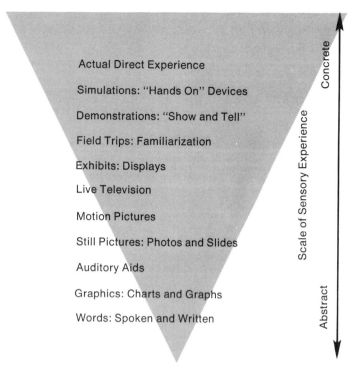

Figure 17–1. The more concrete the medium of communication, the more effective it is.

ciencies are reactive measures, such an approach can be proactive if it helps management to prevent similar incidents. If accident reports like those shown in Figure 13–4, "Supervisor's Accident Investigation Report," are used, each form filled out after an accident should indicate whether training for management and workers is helpful in preventing recurrence. By compiling and studying those forms, management can pinpoint areas where the training program should be reviewed and possibly strengthened or where a new training program may be needed.

Personnel Changes

New employees. New employees need training in their required job skills and safety practices. When workers are cross-trained on several machines—so they can fill in for other employees when needed—they require even more safety training while learning a variety of procedures.

Veteran employees. Management should also provide refresher courses; training for reassigned employees; and training on new equipment, processes, and procedures to ensure that workers follow safe practices at all times.

Personnel who have been promoted or transferred. One of the best ways to help management acquire a safety orientation is to make sure that personnel who have been promoted are thoroughly trained in their new safety responsibilities. A supervisor who has come off the line, for example, needs training in his or her new role—taking charge of a work group. Someone who has been transferred does not automatically know where first-aid equipment is kept at the new location, what to do in case of an emergency, and what the housekeeping practices should be at the new workstation.

This is true even though the worker may be familiar in general terms with company policies.

Meeting Presentations

Effective communication of safety and health issues to higher levels of management is an important aspect of the safety and health professional's job. Delivery of facts using AV materials can help clarify the message and improve the relationship between the safety and health professional and management.

Establishing Objectives

For both management and employees to get the most out of training, managers must establish clear objectives early in the development of training programs. Once trainers know what they need to accomplish, they can begin to concentrate on the details of how to get the job done.

Size and composition of audience. A training presentation for a group of 200, in an auditorium setting, will require a different approach than a hands-on, "how-to" session given by a supervisor to a small group of employees. The larger, more formal session can be handled best by a lecture format, with appropriate audiovisual tools for reinforcement: slides, video, or film. The factory floor discussion, which may take place in a small conference or training room, probably can benefit from a give-and-take format. A flip chart, a chalkboard, or a large pad of paper on which the supervisor can list ideas the group contributes are appropriate audiovisual tools for this setting.

Training materials—and the accompanying audiovisuals—should be designed with the target audience in mind. Language levels should match the audience comprehension. For example, if an instructor uses too many technical terms without explanation, the talk will be difficult for a general audience to understand. On the other hand, too simple a presentation is not appropriate for a group of technical engineers.

Trainers should ask the following questions about program materials. If more than one language is spoken in an office or factory, has the company provided translations for key audiovisuals? Is the translation not only accurate but sensitive to cultural differences that may exist in the workforce? Do the visual messages, as well as the text, reflect this sensitivity?

Type of message to be communicated. To be successful, training should correct deficiencies in workers' performance, skills, or knowledge. Although safety trainers would like to encourage good safety practices and worker participation in safety goals, research has revealed that motivational messages, even when reinforced through audiovisuals, are not very successful. Furthermore, the effectiveness of motivational messages cannot be measured accurately or correlated with training procedures.

Good training philosophy calls for *behavioral* objectives, that is, specific changes in workers' behavior that can be measured by pre- and post-training testing or performance. Behavioral objectives have several components:

Figure 17–2. One of the new developments in audiovisual media, a motion analyzer, enables industrial engineers to scrutinize machinery that moves too fast for the human eye, to pinpoint any problems, and to correct them immediately. (Courtesy Eastman Kodak Company.)

- demonstrable behavior for each objective
- time frame following the training in which the behavior *change* should take place
- agreed-upon standard for measuring the competency or performance of the workers after the training.

A good example of a training objective might be: "After completing the program, the worker will respond to a simulated hazardous materials emergency by obtaining the proper MSDS and responding to the emergency according to the MSDS directions with less than 10% error."

In contrast, the following is not a good training objective: "After viewing this training tape, workers will be better motivated to follow recommended procedures for wearing appropriate eye protection when working with lasers." "Feel," "believe," "support," and similar words do not make good training objectives because managers cannot measure attitudes and motivation objectively.

Veteran trainers know that training is never a completed task. Teaching, reinforcement, and practice are necessary if workers are to retain and assimilate information. Carefully chosen audiovisuals help employees to understand the information they receive, shorten the learning curve, reinforce concepts and skills, and reduce the time needed to present the material.

SELECTING AV MEDIA

Each type of audiovisual material available has an appropriate role to play in training. Advantages and limitations of many of the more widely used AV tools are given in Table 17–A.

Table 17–A. Major Features and Limitations of Various Audiovisuals

Type and Popular Size	Audience Size	Limitations	Strong Points	Comments
MOTION PICTURES 16 mm	M/L	Camera and projector expensive; require trained operator, except for self-threading models. Film not easily changed or updated.	Effective for training and motivating. Uniform professional message. Optical sound nonerasable. Sharper image than 8 mm for given projection size. Single-frame, stop-motion projectors are available.	Silent verson less costly, but less effective.
SLIDES 2×2 in. (35 mm, 126, or 127) 2¼×2¼ in. (120 film) 3¼×4 in. (super-slides) (theater projector)	S/L	Slides may get out of sequence, reversed, etc. Cardboard mounts not durable.	Effective for training and motivating. Less of a "canned" show since slides may be rearranged. Slides can be made and processed quickly. Color inexpensive.	Taped message or reading script easily added or changed. Remote control and multiple projection possible.
FILMSTRIPS 35 mm sound	S/L	Strips and records not easily updated. Message might not be effective or suitably paced for user's needs.	Effective for training and motivating. Message uniform. Sounds easily added to tape or disk.	Silent strips with scripts less expensive, but still effective. Seldom used any more.
OVERHEAD PROJECTORS 10×10 in. 7×7 in.	S/L	Transparencies positioned by hand. Projector close to screen; it or user may block view unless screen is raised or set at an angle. Ready-made material not widely available.	Effective for training. User can write on transparency while facing audience. No need to darken room Transparencies easily made and filed. Presentation informal and flexible.	Color transparencies or overlays easily made.
OPAQUE PROJECTORS 10×10 in. max.	S/M	Projectors require manual operation. Material in books may be difficult to store or ship. Room must be darkened. Copy may be too small.	Effective for training. No transparencies required; small objects, printed material, drawings, and photographs used "as is."	Copies or originals can be hinged or put on rolls to maintain sequence.
CLOSED-CIRCUIT TELEVISION VIDEOTAPE CASSETTE	S/M	Initial investment expensive. Requires adequate lighting. In color or black and white. Copies must be be made one at a time unless duped by lab.	Instant replay. Excellent for training situation where trainee must "see himself in action." Has relatively low operating cost. Can be shown in lighter room.	Small number of people can view screen. TV is a culturally natural transmission medium.
COMPUTER-BASED INTERACTIVE TRAINING	S	High initial expense. May be difficult to copy. Difficult to produce in-house, requires VCR or laser disk equipment to generate some images.	Low operating cost. Does not require a darkened room. High retention of material presented, user paced. Can automatically record and verify training and test scores for individual employees. Presentation and message uniform, does not require presence of instructor.	Procedure without VCR-Laser, disk effective; but less interesting to student.
FLANNEL, HOOK AND LOOP, MAGNETIC 12×36 in. to 48×72 in.	S/M	Presentation requires advance preparation. Few ready-made presentations available. Flannel board material may fall off if not applied correctly or if board too nearly vertical.	Effective for training. Message easily changed, yet can be filed and reused. Permits informal presentation with desirable audience contact. Dramatic, "slap-on" effect builds interest.	Boards suitable for heavier displays; cost slightly higher than cards or pads.
FLIP CHARTS AND CARDS 38×48 in. 18×24 in.	S/M	Limited to small groups. Limited as to amount of copy. Good lighting necessary.	Effective for training and informing. Prepared material can be arranged in sequence. Good audience contact. Material easily prepared; can be added during talk and can be saved.	Ready-made letters, color, sketches cut-outs easily added. Colored paper effective.
PAPER SHEETS AND PADS 28×36 in.	S	Speaker must print legibly. Good lighting necessary. Ink from felt markers may bleed onto adjacent sheets.	Effective for training or discussion; informal. Permits reference to other sheets both during the discussion and for later writing of minutes. Low-cost pads easily obtainable.	Used in place of chalk boards, no erasing.
CHALKBOARDS (portable and wall mounted) 36× 48 in., larger for wall mounted	S	Board must be erased before reuse and recall not possible. Good lighting necessary. Ordinary chalk marks hard to see. Dust from chalk and erasers annoying.	Effective for training or for discussing a limited number of points. Presentation informal. Portable chalkboards also useful for holding charts or displays.	Colored or fluorescent chalk adds life to talk. Magnetic boards available.
POSTERS AND BANNERS 8½×11½ in., 17×23 in., and larger	S/L	Only one or two ideas can be presented at a time; considerable time needed for changing.	Effective for motivating; support training. Specific messages can be posted at points of hazard or to meet timely situations. Ample posters available.	Homemade posters supplement general posters.
WORKING MODELS, EXHIBITS, AND DEMONSTRATIONS	S/L	May require special training to use. Live action is subject to errors.	Action can closely simulate actual conditions. Permit group participation.	

Table 17–B. Some Safety Uses for Videotape

Videotape is a versatile medium. Here are a few ways it is currently being used to put across the industrial safety message:

- **Security surveillance**—Mounted cameras observe distant gates, loading docks, and payroll departments.
- **Job review**—Used in conjunction with Job Safety Analysis, Task Safety Analysis, and Step Safety Analysis for observation and review of both good and bad procedures.
- **Management training and development procedures**—First train the trainer, then have him train employees down the line. Playbacks assure that procedures are practiced correctly by the instructor before others are taught to do as he does.
- **Incident investigation**—When brought to the accident scene, a recording is made of the

physical set-up, personnel present, time, weather, lighting, and other conditions pertinent to the accident.

- **Motivation and enforcement**—Showing the employee how his performance can lead to an accident is an effective way of gaining his cooperation. OSHAct violations can be spotted and corrected. Before-and-after scenes can prove dramatic.
- **Training in sophisticated equipment, complex procedures**—Videotapes can show and repeat processes a step at a time, permitting interruptions for questions.
- **Informing distant audiences of a procedure or announcement affecting all units**—Duplicates of a tape can be made and mailed anywhere so that all personnel are informed simultaneously.

Role of Trainer

One factor sometimes neglected in the selection of AV media is the role a trainer expects to play in a particular training session and how he or she perceives training responsibilities. Good trainers continually ask themselves these questions: "Is what I want to do appropriate for the training objective in this particular session? Does the audiovisual tool I'm considering enhance my training role, as I see it?" (Table 17–B.)

Trainer as expert. Trainers can choose the "expert" role, communicating information in a more formal presentation. The trainer answers questions from the audience but allows little opportunity for give-and-take discussion. However, such a choice may be appropriate when trainers need to convey information to a large audience, for example, explaining new or revised regulations or policies to the workforce. Appropriate AV media would include projected visuals, such as slides and overheads, and a take-home handout.

Trainer as skill-giver. At times, trainers can assume the role of a skills teacher. Supervisors who have come up from assembly-line work or who have had experience on specific equipment often are called upon to teach their skills to new employees or to promoted or transferred workers. Hands-on film demonstrations, either those commercially available or company-produced videos, would be appropriate AV media. If the organization uses interactive training, a laser disc module—if one is available on the subject—is another choice. For example, such a module might show a worker how to use a full-face mask respirator: the respirator parts and their functions; how to put on and take off the respirator; and how to test, maintain, and store the equipment.

Trainer as facilitator. A third role the trainer can play is that of a resource person, who helps to match individual student needs with appropriate materials. Such a role assumes the trainer is flexibile and has a broad knowledge of available teaching aids, including various AV media. Trainers who function successfully in this role work to create a training environment in which students are not afraid to ask questions or to admit their own need for extra help.

Student workbooks or worksheets, which can be completed at the employee's own pace, can serve as the basis for individualized discussions. The trainer uses the material to help employees understand their shortcomings and to suggest remedies. Management also can provide audiotapes and videotapes that employees can take home, enabling students to go over the material until they understand it. Interactive laser disc training also may be an option if the organization has such equipment. Workers can practice privately without feeling pressured to keep up with faster learners in the class.

Audience Size

Some AV media work best in small-group learning situations. Interactive training—by definition—is a one-on-one experience, even though a hundred students may be using terminals at the same time. Large audiences are usually too far away to read flip charts, flannel boards, chalkboards, or paper charts easily. Overheads and slides, on the other hand, can be made readable for audiences up to several hundred, depending on the screen size, graphic quality, and projector used. Likewise, films can be viewed by one person or several hundred. Videos—if shown simultaneously on round-the-room monitors or on large projection screens—work for large audiences as well as for individual viewers.

Cost Factors

To determine whether the cost of an AV medium is justified, management must consider what the expenditure will buy, how quickly the material will be outdated, how the audiovisual will be used, and how many students will view it. A bottom-line cost for a single AV never tells the whole story; management must always rely on personal judgment to determine if the return on investment (ROI) is appropriate for the organization and its needs.

For example, interactive laser disc simulation of an airplane cockpit may seem expensive. However, by allowing a pilot to practice correct responses to emergency conditions, such simulation may prevent an airplane crash, with its potential loss of life and property damage. Because the upfront expense of interactive laser disc training is in design and production, costs per student for instruction are affordable if many students will use the material. One major company estimates that interactive training technology is saving $100 million in training costs.

Even in small companies, a training department can easily justify purchasing a computer graphics package to produce attractive slides, overhead transparencies, and posters if the company conducts worker training frequently. The training department can make the case that *effective* training will help cut down on employee turnover as well as accidents, thus lowering company costs. Inexpensive teaching aids that also

can be used for in-the-field training or mobile classrooms (e.g., portable chalkboards, flip charts, audiotape players, and even VCRs) make it possible for organizations with limited resources to present information with the added impact of audiovisual reinforcement. On the high end of the cost spectrum, custom-produced videos, commissioned by a trade association or other industry group, may give users access to the latest in technology and training materials that they would be unable to afford on their own.

Commercial versus In-house Production

Today's workers are far more experienced and sophisticated as viewers than employees were in the pretelevision era. According to the American Academy of Pediatrics, by the time the average person reaches age 70, he or she will have spent approximately seven years watching television. Data from A.C. Nielsen, a market research organization, indicate that adolescents 12 to 17 years of age watch more than 23 hours of television a week. As a result, viewers expect that company AV materials will equal commercial productions in quality.

Companies with in-house production organizations can consult with trainers about their needs for slides, transparencies, posters, and videos. The AV department can provide cost estimates, show trainers what has already been done for other company departments, and—should a custom-designed presentation be required—work with the trainers from conception to execution. If trainers want a second or third opinion (and they should always get more than one bid on any major film or video project), they can include their in-house organization as one possible alternative. They also should look outside the company for others, such as independent production houses or organizations that specialize in custom training materials.

State-of-the-art technology and equipment have become so complex that the average training department's personnel cannot match the skills and expertise of the AV pros. The trainer is the expert on what should be communicated and on the required training methods. The audiovisual experts know how to manipulate the media to get the best effects and quality. (See the discussion on video later in this chapter for an easy-to-use guide on how to select the appropriate level of video quality.)

Does this mean trainers should never use home-made AV materials? Not at all. Trainers can easily make inexpensive flip charts, flannel boards, paper and pencil charts, chalkboard presentations, instant photos, and audiotape interviews. As described in the section on computer graphics, some software packages are fairly easy to use and will produce acceptable AV materials.

The choice between professionally developed or home-made AV materials always depends on a number of factors that the trainer must weigh according to their relative importance for his or her organization. These factors include the number of showings, size and composition of the audience(s), degree of customization (is this AV aid for general use, or is the information for a particular

department or plant site only?), and the importance of the message.

COMPLEMENTARY INSTRUCTIONAL TOOLS

See Chapter 16, Safety Training, for a thorough examination of live presentations, including lectures, discussion, demonstrations, and performances.

Exhibits and models make effective three-dimensional displays for use in instruction. Cardiopulmonary resuscitation (CPR) training, for instance, requires a mannequin on which students can practice. A cut-away model showing layers of muscles or a skeleton showing vertebrae can be effective in demonstrating how injuries can occur and what body parts are affected (Figures 17–3a and 17–3b).

Trainers can use exhibits to demonstrate the safe operation of a machine or process. Exhibits can also feature examples of first aid equipment, protective clothing, rescue equipment, respiratory protective equipment, and fire protection appliances.

Companies at times may organize demonstrations of fire prevention and suppression with the assistance of the local fire department (Figures 17–4a and 17–4b). Trainers also can arrange demonstrations of good and bad lighting; impacts on safety hats, safety shoes, and eye protection devices; and first aid treatment and transport of injured workers.

SPEAKERS

Speakers, to be most effective in a training session, should take full advantage of AV attention-getting devices, such as pointers, easels, charts, TelePrompters, props, and aids.

Pointers. A pointer, usually made of wood or aluminum, is a necessary item when a speaker is using a chalkboard or charts on an easel. A pencil or a finger is a poor substitute. However, for overhead transparencies projected on a screen, a speaker can use a pencil or pen, laid on the transparency itself, to call attention to a word, phrase, or portion of a chart. The shadow of the pencil or pen shows up on the screen, and focuses the audience's attention on a particular fact. Using a pointer allows the speaker to face the audience, while at the same time referring to the visual material. A speaker should not fidget with the pointer, however, since that distracts from the presentation.

If visibility is a problem, use a pointer with a fluorescent tip. In a dark or semi-dark room, a battery-operated or 110-volt flashlight-type pointer will project a spot of light or a bright arrow onto a screen from a considerable distance. If the speaker must stand at some distance from the chart or screen, a high visibility electric pointer is useful, even in a fully lighted room. A telescoping, pocket-sized pointer is handy for speakers who must carry a pointer with them.

Easels. Portable aluminum easels are helpful for holding flip charts and paper pads. Be sure the easel is sturdy enough to hold the audiovisuals and will not collapse during a speech. Some easels fold for use on a table.

Easels are effective when used close to the audience.

Here Is How to Use It
There is a correct and an incorrect way to pick up a heavy object. The correct way is to keep the back straight, the knees bent and spread, and the load close to the body. The incorrect way is to reach way over and lift. (Twisting the back complicates the bad effect.)

To demonstrate this effectively, a special model can be used. If used with its block spine locked, the model simulates lifting with strong leg muscles; the ribbon on the spine remains limp, indicating very little tension of the back muscles. It demonstrates that the back cannot be kept straight without bending the knees.

To demonstrate improper lifting, the model is used as is shown in the sketch. The legs are bent only slightly (or held straight). One hand lifts the handle just ahead of the fulcrum at the hips in order to lift the weight in the hands. The back arches under the strain and visibly pulls each block apart.

The model can also be made to show proper foot placement (see drawing at left).

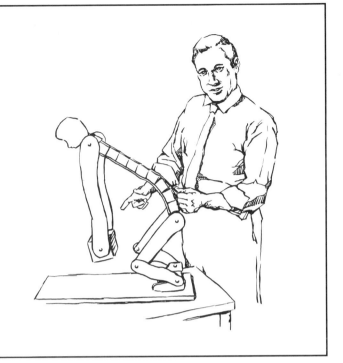

Figure 17–3a. A model is used to demonstrate lifting techniques.

However, the impact of the display is often lost when audiences are large or are more than 20 to 30 feet away from the board.

Charts. Well-chosen and well-prepared charts and diagrams allow the audience to "see" what the speaker is saying. Charts tell a story faster and more clearly than can an oral presentation alone. Speakers who use charts (rather than projected visuals) should practice showing them before their talks to make sure the charts appear in the correct sequence. Trainers who fumble with the charts or switch them haphazardly distract from their message, losing any impact the information may have had. If speakers discuss material other than topics illustrated by the charts, they should cover the charts temporarily to focus audience attention on what they are saying.

TelePrompters. TelePrompters or other cueing devices can be used for dramatic or formal presentations in which speakers are required to follow a prepared script. Tele-Prompters require technical help to set up. Speakers must know how to use the equipment so their presentations will appear natural.

Rehearsals

Successful speakers never give presentations without preparing well in advance. The smooth, seamless talk that looks so effortless was undoubtedly rehearsed ahead of time, often more than once. By practicing their presentations, speakers are able to stay within their time limits and to deliver a more effective talk. When a speaker is confident in what he or she is doing and handles accompanying audiovisuals well, audiences are free to concentrate on the message being delivered.

If a presentation includes several speakers, it is a good idea for all participants to show their speeches to one another in advance. This is especially important if company executives or important visitors will be in the audience. This approach will help speakers to avoid duplicating information or audiovisuals. Individuals may wish to revise their delivery or practice certain skills, depending upon suggestions they receive from other speakers.

Props and presentation aids. As a first step in preparing for a speech, go through the talk and the audiovisuals well ahead of time. Make a duplicate copy of the outline and in the margin mark items needed for the talk. Make a separate list of those items to ensure that nothing is left out.

Use a simple notepad or a preruled accounting notebook page with multiple columns to keep track of all requirements. Head the columns: "Have," "Need to reserve," "Need to buy," "Need to rent." Down the left side, list all required audiovisual props and presentation aids in the order in which they will be used during the talk. Don't forget to include items like chalk, erasers, markers, extra pencils or pens, etc. Next, check the appropriate column for that particular presentation aid. In this way, a speaker can easily see what has to be acquired.

Revise the list at least once before the presentation is scheduled, updating the status of each item. Use the far right column to indicate who will be responsible for seeing that the item is on hand. However, even if some tasks are delegated, the speaker is still responsible for double-checking that everything is ready and in the proper place well before the presentation.

Speaker Aids

Chalkboards, marker boards, and paper pads. Chalk-

Construction Specifications

The spine is made with 9 blocks, each 2 in. square and 1⅝ in. high. Each is drilled at the center to accommodate a standard screen door spring. The T-shaped head and shoulder piece is about 8 in. long, 2 in. thick, and supports the arms on shoulders about 5 in. apart. The hip block is 2 in. square, with sloping sides so the legs will spread open in front. Add a ⅜-in. spacer between the hip block and each leg.

Assemble the body by using wood screws to attach ends of spring to the head and hip pieces. Tack a piece of 2-in. wide belting to the *front* of the spine blocks to hold them in alignment.

The arms and legs can be shaped from ¼-in. plywood.

A block, approximately 5 in. square and 3½ in. high, represents the lifting weight. Elastic tape or multiple rubber bands are stretched from the shoulder to the hip, along the *back* of the spine blocks to represent the spine muscles.

A metal handle can be secured to the lower end of the hip block, as shown in the drawing.

The model can be mounted on a 1×12-in. board, 18 to 24 in. long.

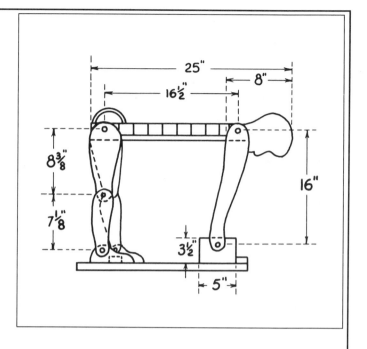

Figure 17–3b. Ergonomic drawing of the model.

boards, marker boards, and paper pads lend themselves to small group instruction and provide inexpensive visuals. Each is an excellent tool for stimulating discussion. Students can comment on points the speaker previously listed, and the speaker can quickly record student responses or suggestions.

Although chalkboards come in several colors, light green is considered standard. A dustless chalk should be used, preferably a strong, bright color for maximum visibility. Fluorescent chalks are particularly effective.

Marker boards have replaced many chalkboards. Colored felt-tip markers are used to write or draw on the polished white surface. Marks are erased with an ordinary cloth with little or no dust. One advantage to these boards is that they serve a dual purpose as a projective screen and marker board for small audiences.

Large paper pads, available at art stores or office supply stores, are easy to use. Clamped on a lightweight or folding easel, such pads are portable and always ready for use. A speaker can keep training material written on the pad or discard it after the session. Feedback or questions from students, recorded and saved on the paper, can serve as a nucleus of ideas for a future training session.

In brainstorming sessions, where the leader's objective is to stimulate quickly as many ideas as possible without detailed analysis or criticism, pads are of real value. As each chart or page is filled, it can be tacked to a bulletin board or clipped to a wire running along one side of the conference room. Group members then have a continuous record of what has been discussed. Latecomers can use these sheets to catch on to the discussion, and the speaker has a good record of the meeting.

Inexpensive rolls of white paper can be used for charts;

occasionally newsprint "ends" are available at small cost from a local publisher. Even brown wrapping paper will do if bright-colored contrasting markers are used to write down the ideas. The speaker can write and read material while unrolling the chart like a scroll.

Flip charts, similar to paper pads, but usually prepared in advance, often are used in more formal meetings. Blank pages can be provided for on-the-spot notes. Portable units are easy to make and transport to different locations.

Flannel boards. A flannel board is a plywood board, commonly 3 x 4 ft (0.9 x 1.2 m), covered with dark flannel. A number of frames can be attached or placed close together to form a larger surface. Speakers can use roll flannel to cover the board(s). Make artwork or lettering out of a special velour-backed paper, lightweight cardboard, or heavyweight paper and affix flocking paper with felt adhesive or strips to the back. If necessary, use sandpaper instead of the special flocked paper. The flocking or sandpaper grips the long-napped flannel firmly enough to bear lightweight cards (Figure 17–5).

Most speakers and trainers find that flannel boards are easy to make and can be used in field training or mobile classrooms when other, more sophisticated audiovisual equipment is not available. Attaching words, designs, or messages on the boards creates a dramatic effect not possible with other visuals.

Like posters, flannel boards attract attention when used in safety displays. Their design should be striking enough to make workers stop and read the message.

Speakers often use flannel boards as an audiovisual aid in a presentation and change the cardboard cards to emphasize different safety messages. If so, the speaker should practice

Figure 17–4a. Railroad safety officers are receiving hands-on training in firefighting. (Courtesy of Association of American Railroads.)

Figure 17–4b. As part of hazardous material response training, these students extinguish a liquified petroleum gas fire at the AAR Transportation Test Center. (Courtesy Association of American Railroads.)

making the switch smoothly without interrupting the effectiveness of the presentation.

Hook and loop boards. A hook and loop board is a heavy-duty "slap board." The material covering the board contains countless nylon loops. These loops are caught by the almost invisible nylon hooks on the small pieces of tape that are mounted on the backs of signs or other display objects. A tiny patch of the hook material fastened to a heavy or large object will hold it securely on the board.

Hook and loop boards are available commercially in sizes ranging from 18 x 24 in. (45 x 60 cm) to 48 x 72 in. (1.2 x 1.8 m).

Magnetic boards. Trainers can make a magnetic board of either a spray-painted sheet metal plate or a steel-backed chalkboard. Magnetic boards are also available commercially. Small objects or cutouts mounted on small magnets or on magnetic tape can be placed on the board and moved at will.

Instructors often use this type of visual for training operators of vehicles such as forklift trucks. The mobility of the objects—toy vehicles or cutouts of trucks, together with the cutouts of aisles and loads—enables the instructor to give a realistic demonstration of safe practices.

DESIGNING EFFECTIVE AUDIOVISUALS

Designing effective audiovisuals is easy to do if trainers remember the basic purpose of AV media: to reinforce a message. When developing an effective meeting, trainers must select a training objective, carefully consider how to enhance their presentation with audiovisuals, and organize the audiovisuals beforehand. Even a four-walls-and-a-chalkboard, 15-minute session can be more effective when it is well prepared.

Step one in effective training is to analyze the target audience. What do they need to know? Why do they need to know it? How much knowledge do they already have about the topic? What is their level of vocabulary and experience? Are there special conditions, such as cultural or language sensitivities, that the trainer needs to consider in planning the training session?

A necessary corollary is to analyze all training conditions. How much time will the trainer have? Where will the training be conducted?

Organizing the AV

Every presentation begins with an idea or concept and builds from the basics. Here are several ways of organizing a session:

- Start with a question or problem, and move toward an answer or solution.
- Present a problem, fill in the details, and end with a solution (a good way of using case histories effectively).
- Describe "what if?" possibilities; after discussion, reach a workable solution.
- Do a time sequence, developing visuals to illustrate chronological order.
- Present information, using a near-to-far or far-to-near sequence.
- List "for" and "against" points in support of, or opposed to, a main idea.

Figure 17–5. Dramatic effects not possible with other visuals can be created using a flannel board.

▪ Explain the features and benefits of an idea or problem.

Once the trainer has decided on a sequence (*one* of the above, or a different choice), he or she should write down the main ideas that will be presented in the training session. If 3 x 5 index cards are used, the trainer can shift the order of ideas around to find the best arrangement. Trainers do not necessarily need a formal script to organize their visuals, but they do need to go through certain steps after they have analyzed the target audience.

Develop a simple outline of the visuals to be shown in the meeting. Work through, in advance, key training points that visuals can help to explain. Visuals should clarify spoken information or dramatize an important point. The message is what counts, not slickness or glitz.

The most important thing to remember about audiovisuals is to keep them simple. Visuals should summarize, not repeat, what the speaker will be saying. Streamline the visuals to a few phrases or words, rather than complete sentences. Use 3 x 5 cards to write down the visuals needed for the talk. Try several arrangements of the cards to discover the right sequence.

Trainers should ask themselves the following questions. Does the outline flow logically? Does it match the order in which the speaker presents ideas? Although trainers will want to build from simple thoughts to more complex concepts, they should not forget to put at least one visual early in the presentation to communicate clearly what topics will be explained. Such a visual focuses attention on succeeding material, because it lets the audience know what is coming.

Chart Formats

Audiovisuals work best in training when speakers use easily understood charts and graphs that provide accurate information. For projected visuals, such as overhead transparencies and 35mm slides, trainers have a choice of formats.

For most projected images, use the horizontal format (called "landscape" by many computer graphics software packages). Like a landscape, such visuals are wider than they are long. Because fewer lines fit on a horizontal visual than on a vertical one, speakers are forced to limit their content, resulting in a clearer visual.

Long lists, forms, and detailed flow charts often project better when designed for a vertical format (called "portrait" in computer graphics). Like a portrait, such visuals are longer than they are wide. One disadvantage of portrait format is that viewers' heads sometimes get in the way, making it harder for most of the audience to see text at the bottom of the screen.

Text Charts

Text charts are one of the most commonly used audiovisuals. They are easy to create, either by keying copy into the computer, using a typewriter, hand-lettering, or using press-on letters available at art stores or office supply stores.

Text charts are made in four styles:

Titles. Title charts are good for introducing presentations. For maximum effectiveness, include no more than eight lines. As a general rule, use the top one-third of the chart for titles and the center of the chart for the message or announcement of the topic.

Lists. Lists are a good way of helping people organize their thoughts about a central idea. They can show sequential steps in a process, such as Lockout/Tagout instructions for a particular machine, or can present ideas that then serve as a springboard for discussion. In a simple list, each idea is emphasized equally. Speakers may want to highlight ideas by placing a star (*) or a bullet (·) in front of each one.

A good way to stimulate discussion on each bulleted idea is to present them as single thoughts. The trainer can make the first visual using only the first line of the list; the second visual, with two; the third visual, with three, and so on, building the list as the talk continues. If the trainer is using projected visuals, another easy way to control what the audience sees, bit by bit, is to create a single visual with three or four ideas on the list. Then place a piece of paper under the visual to block out all but the area viewers should see. Pull the paper down, revealing the next idea on the list, when ready to begin discussion on that topic. This technique is especially good for ideas that follow a logical sequence.

Organization charts. Many computer graphics programs have software that will automatically do the drawing, once the trainer has keyed in the names and titles of the staff and indicated the reporting relationships. Speakers can make these charts by drawing boxes and lines by hand. An organization chart is a quick way to illustrate lines of authority and corresponding responsibilities. Use an organization chart, for instance, to indicate safety committees for each shift.

Tables. Tables are good for hand-outs, but difficult to follow on projected visuals unless they are extremely simple. Resist the temptation to put too much information on a table. Viewers cannot absorb details that quickly. A variation of a table is a column chart, with related data displayed in two or three columns. "Days Without Reportable Accidents" could be the overall title, with *June, July,* and *August* listed as column headers, and number of days under each month. Column charts are also useful for listing ideas or problems, with potential benefits in one column and drawbacks in another.

Type Selection

As a rule, the simpler the posters, slides, and overheads are, the more memorable they will be. Thus, avoid selecting type that is too complicated, placed unattractively on a page or on the screen, or too small to be read comfortably.

Choosing type styles. Typeface is a term for a whole family of letters with a particular design. Some typefaces have serif lines or curves that project from the ends of letters. Times Roman, Bookman, and Bodoni are examples of serif typefaces. Serif typefaces may be easier to read when used for the body of text charts because they are widely found in newspapers and magazines.

Typefaces without these lines are called *sans serif.* Helvetica, News Gothic, and Futura are three sans serif typefaces. They look more modern, but some studies show that

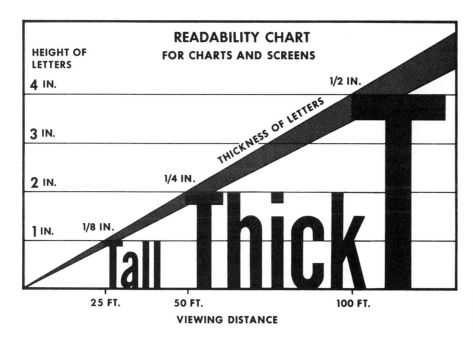

Figure 17–6a. This chart shows what heights and thicknesses letters and symbols should be for easy viewing of projected and nonprojected visuals.

multiple lines of sans serif type may be harder to read, especially for older people. Most designers recommend restricting their to use to headlines and outlines rather than to the body text.

For big, bold lettering, Helvetica and Gothic Bold (both sans serif) have long proved popular. They make attractive headlines and display type to accompany body text set in Times Roman or Bookman (both of which are serif typefaces).

Script typeface, which looks like handwriting, is a third style of typeface, but is the hardest of all to read. If another, less common typeface is desired, trainers should consult with a graphic designer or typesetter to make sure typefaces and graphic images do not clash, distracting from the importance of the safety or health message being conveyed.

Aligning type. Type set on the page is aligned in straight or ragged (uneven) margins. Type said to be "flush left" has a straight left margin and a ragged, or uneven, right margin. Type set "flush right" has a straight right margin and a ragged left margin. When type is "justified," it has both left and right straight margins.

Text charts are generally easier to read, and easier to produce, if designers set the text flush left. This is because viewers are accustomed to looking first at the left-hand side of a page. For clarity, double-space between lines on a text chart and keep lines short. Try to put no more than eight lines on a single text chart.

Capitalization. Studies show that viewers find it easier to read text set in upper and lower case than text set in all capital letters, or "all caps." All caps works for short headlines, but not for body text.

If trainers want to emphasize certain words, boldface (dark) type works better in headlines. Inside a paragraph, however, italic type works better than either boldface or all capital letters.

Type size. Any visual is ineffective when lettering is too small to be read comfortably. People who wear prescription lenses, especially bifocals, and older workers may need to have larger type. Design lettering so it can be read easily from the back row of the audience; the message will have a much better chance of getting across.

Use Figures 17–6a, 17–6b, and 17–7 as a guideline for recommended heights and thicknesses of lines and symbols. Block letters show up better than handwritten copy. To be easily read at a distance of 50 ft (15 m), printed or projected letters should be at least 2 in. (5 cm) high and ¼ in. (8 mm) thick.

For use with overhead projectors, material prepared with characters at least ¼ in. high, with spacing of at least ¼ in. between lines, will give a letter height of 2 in. on a screen 6 ft (1.8 m) wide. Text should be clearly visible at a distance of six times the screen width.

Plenty of white space in visuals makes them more effective. Double-space text for high readability. For projected visuals, use no more than nine lines of text in a title chart, 24 bulleted lines for a bullet list, 23 lines for a two-column chart, and 14 lines for a three-column chart.

Computer graphics and desktop publishing software programs make it easy to create text charts and legends for graphs by letting users change and control type. Traditionally, type size is measured vertically, in points, and line lengths are measured in units called picas. As a rough rule, 12 points = 1 pica, and 6 picas = 1 inch. The conversion from picas to inches is slightly different on a type gauge (a special, inexpensive ruler available in most art supply stores) because the computer software rounds off the conversion.

Instruction manuals for software packages whose purpose is to help users create overheads and slides for presentations often suggest appropriate type sizes for varying usage. Some even specify a particular size as the default, unless the user deliberately overrides the program. Usually, the designer can preview how a chart will look and make any adjustments before printing out the final product.

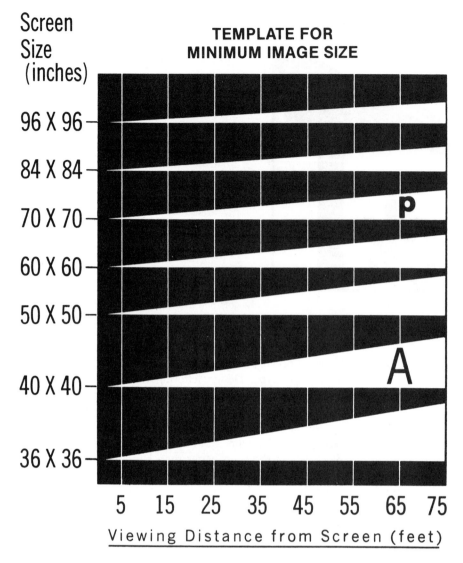

Screen Size (inches)

TEMPLATE FOR MINIMUM IMAGE SIZE

96 X 96

84 X 84

70 X 70

60 X 60

50 X 50

40 X 40

36 X 36

p

A

5   15   25   35   45   55   65   75

Viewing Distance from Screen (feet)

Figure 17–6b. The correct type size for overhead projection can be scaled by placing the transparency over this template. For example, lettering for a transparency to be projected on a 70- x 70-in. screen must be at least as large as the "p" ($^1/4$ in. high) if it is to be viewed from a distance of 65 ft. For this same viewing distance, but with a 40- x 40-in. screen, a letter of minimum size "A" ($^1/2$ in. high) should be used when making the transparency.

Computer instruction manuals and "how to use" books about popular programs available at most computer stores are good sources of helpful information. So are typeface specification sheets that show different styles and sizes of type. If trainers consider working with an outside AV service bureau to have overheads, transparencies, or even posters made, the bureau's designers can give trainers a quick overview of what will look best for their particular purposes.

In general, use 12 point or larger type for projected visuals to make them easy to read. Remember, larger is better. Keep at least a 2-point difference between text and headline, or text and subheads, to make a visual distinction between the two. For example, if your text type is 12 point, make a headline in 14 point type. For projected visuals to be read comfortably, the designer should be able to read the smallest line of type on the visuals at arm's length without a magnifying glass.

Because each typeface has its own personality, the number of characters per line may vary even when the typefaces are the same size. Some typefaces have a relatively high character count per line, and look smaller than others. Commercial printers and service bureaus can help trainers choose a typeface that will give visuals the clearest, cleanest look.

Graphs

Graphs are a visual summary of data. To make them easier to understand, always put a legend or short title alongside the graph to tell viewers what the information is about. Graphs work most effectively when they are matched to the information presented. Although there is no rule about which type of graph to use, you'll find the following suggestions helpful:

Pie charts. A pie chart, so-called because of its circular form and pie-shaped wedges, is useful to show how much of a total is represented by each subdivision. For example, a pie chart can show that out of 1,000 reportable accidents, 250 of them took place at only one facility. This would clearly demonstrate the need to improve safety at this plant. Pie charts also help to illustrate good safety records. Of those 1,000 reportable accidents, for example, perhaps only 16 took place in the shipping department.

Remember, use a pie chart to show data from only one time period. Do not compare two years' accident records within a single pie. Instead, use two separate pies, one for each year.

Creating a pie chart is fairly easy. Most computer graphics software packages will make a pie chart automatically if a user merely keys in the data. If drawing the chart by hand,

remember that the total pie contains 360 degrees. The angle needed for each component of the pie chart must correspond to the correct percentage of the pie. In the case above, the 250 accidents represent ¼ of the 1,000 accidents. Consequently, the pie "piece" drawn must represent ¼ of the 360 degrees of the pie, or a 90-degree angle.

250/1,000 = 25%/100% x 360 degrees = 90 degrees

All the components in a pie chart should add up to 100% of the total amount represented, just as the angles for each

Figure 17–7. Rule-of-thumb for lettering to be shown on a 2-x 2-in slide. (Printed with permission from Eastman Kodak Company.)

of the pie wedges should total the 360 degrees.

To break down data further, create a chart within a chart. For example, a pie wedge is only part of the whole pie, but can become its own series of data. Pull the wedge to one side of the visual for extra emphasis, and turn the information contained in that wedge into a column chart.

For example, assume that of those 250 hypothetical accidents, 50 occurred in the warehouse, 175 happened in the factory, and 25 took place in the plant office. For the column chart, which is basically a vertically stacked "pie," 250 becomes the 100% total. Consequently:

50/250 x 100% = 20% warehouse accidents
175/250 x 100% = about 70% factory
25/250 x 100% = 10% plant office

Either draw the column by hand or use a computer graphics package that calculates the percentages and draws the column automatically. The column will look more balanced if the largest section (with the biggest percentage) is placed on the bottom. Emphasize that section's importance by giving it a dark color or heavy pattern. Link the pie and the column for an effective visual (Figure 17–8).

Visual techniques also can make a pie chart more effective. Since most people look at the right-hand side of a graph first, place the most important "slice" of the pie at 3 o'clock on an imaginary clock face. As you work your way counterclockwise (backwards) around the clock, each slice following

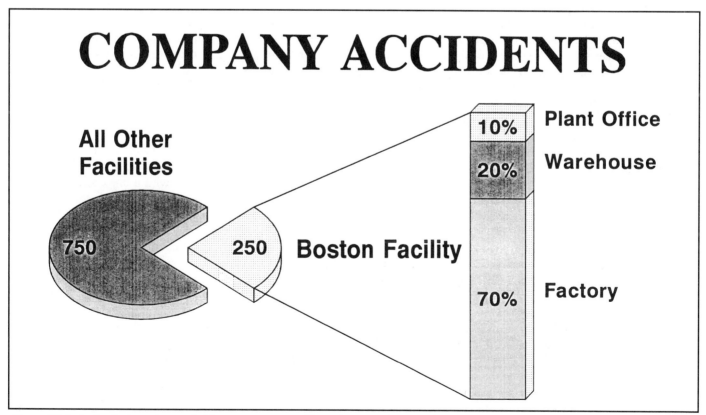

Figure 17–8. A pie chart can be created by a computer graphics software package to show accident (or other) data at a glance. (Courtesy Karen Zmrhal.)

COMPARISON OF ACCIDENTS

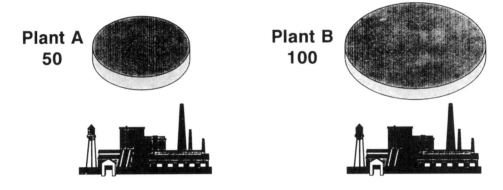

Plant A Accidents

Plant B Accidents

Plant A
50

Plant B
100

Figure 17–9. Two pies of different size may be clearer than one pie with several slices. In this computer-graphics-generated chart, the area of the pie on the left is 50% of the area of the pie on the right. (Courtesy Karen Zmrhal.)

the most important one should be a slightly lighter color. If the chart is in black and white, rather than color, give the most important segments black or dark gray colors. Make the least important elements light gray or white.

Avoid putting too many slices in a pie, so as not to confuse the audience; six or seven slices is about the limit an average viewer can comprehend. When there are too many slices, it is hard to compare their sizes.

Computer graphics packages allow trainers to make proportional pies, or three-dimensional pie charts. Just because they can, however, doesn't mean they should. Two pies on a single visual, equal in size but representing different amounts of data, are often difficult for viewers to interpret. For example, suppose that Plant A has only half as many accidents as Plant B. One pie could represent the number of accidents at Plant A, and a pie twice as large could represent the number of accidents at Plant B (Figure 17–9).

Although the proportional pies shown in Figure 17–9 are statistically accurate, they may mislead the audience. Most people find it hard to judge circular proportions quickly, because they tend to think in terms of line lengths.

Three-dimensional pies (which show the "depth" of the pie) also can be misleading. If the top of the pie is made lighter to show perspective, that section may appear to be too large to represent its data correctly.

Bar charts. Bar charts are one of the simplest ways to display data to an audience. They are used to illustrate comparisons of volume over time. For example, a bar chart can show the dollar amounts a company has paid in Workers' Compensation over the past five years. Designers can set the

bars horizontally or vertically, provided there are only 10 to 12 bars on a single visual. If bar labels are long, a horizontal bar chart is a good choice. When the data permit, arrange the bars in sequential order, either low to high, or high to low. Ranking lets viewers compare data more easily.

Computer graphics software packages allow users to choose among several designs, such as overlapping bars, stacked bars, or paired bars. When choosing an overlap option, start the series with the smaller values first, so that the shorter bars in the overlapping groups come before larger ones.

Line charts. Line charts are plots of points that trace changes in data over time. For example, "How many accidents did we have in the machine shop from 1982 to 1991?" is a good question to answer with a line chart. It is easy to draw such a chart by hand; the years, which are the independent variable, go on the X, or horizontal, axis, and the number of accidents per year go on the Y, or vertical, axis. Plot the points, and draw a line connecting them or use a computer graphics package to draw the line chart.

Area charts. Area charts are a good way to show volume, especially to emphasize changes in data. These charts use lines and patterns to separate different kinds of information. Trainers can use area charts as an alternative to a line chart if they are working with a single series of data. For example, to show dramatically how accidents have declined at a factory, an area chart can make the information easy to remember. Designers can combine data series by constructing an area chart with several layers: e.g., the total number of accidents and the number of accidents in the office, the factory, and the warehouse.

To make a multiple stacked chart, put the darkest pattern on the bottom, and work from dark to light, with the lightest pattern on the top of the stack. Either shade the chart by hand, or use a computer graphics software package to create the chart from data values keyed in.

Drawings and sketches. A simple drawing, done by hand or with a computer program, can often help make a process or procedure clearer. Keep the illustration as simple as possible, and always include a brief title or explanation next to the drawing.

For consistency within a single presentation, try to keep elements of the visuals, if they are repeated, in approximately the same place for each of the slides or transparencies. If a company logo is used on each visual, always place it in the same location on each slide or transparency. Do the same on all drawings for titles or explanations.

Color

The skillful use of color will add greatly to the impact of visuals. On charts (including those that are all text) a too-light background makes information difficult to read. Overheads with lettering on a clear background project well; so do transparencies and slides that use "reverse lettering," or white text on a darker background.

Remember that approximately 10% of the population is red-green color blind or color poor. If lettering or graph markings are in either color, such persons will see them only as varying shades of gray. The impact of the message will be lost.

If using a computer graphics program to generate text charts and graphs—whether print materials, overhead transparencies, or slides—control the impulse to try every color combination the program offers. Experts suggest using no more than four colors in a projected visual.

Experiment with color schemes and test them for visibility and legibility. A good color concept on paper may not look good on screen, especially for those viewers sitting in the back of the room. Make sure lettering and background offer sufficient contrast; green lettering on a gray background, for instance, blurs at a distance if the values of each color are too similar.

COMPUTER GRAPHICS

Most computer graphics software packages use templates, that is, predesigned styles that specify type size, justification, and color schemes. Users need only to key in their data. The automatic layout features ensure consistent, attractive results. Software packages usually include a predefined color palette, used when the program displays charts on the computer monitor or sends chart output to a film recorder to create slides. If the trainer does not like the standard color palettes, he or she can modify them to create a customized color combination.

Some software can read data directly from existing spreadsheets and worksheets, including text, titles, labels, and graph settings, thus eliminating time-consuming double entry of data.

Low-end computer graphics software and high-end word

processing software have document-formatting capabilities. After the user has created text, he or she can adjust the type style and type size, and print the output on a dot-matrix or laser printer. The computer will produce sheets that can be run through a plain paper copier to make black-on-white overhead transparencies.

High-end computer graphics software has several options for output. If the company has a film recorder, the trainer can make slides instantly. Print the output and make overhead transparencies on dot-matrix, laser, or ink-jet printers. Some graphics software offers options that allow sequence presentation over a monitor. Although the monitor limits the size of the audience, hardware can be purchased or leased to project the sequence on a screen. The computer output is diverted from the monitor to the window-like device that is set on or built into an overhead projector. You can also send your computer images to a service bureau, which will turn them into 35mm slides, color overhead transparencies (mounted or unmounted), printed colored handouts, or even colored posters. Either mail the files on computer disk to the service bureau or transmit them directly over the telephone lines by using a modem and special communications software.

Choosing a Computer Graphics Package

A number of factors must be considered before deciding if a computer graphics package is cost- and time-efficient. Here are suggested questions to ask:

- Have the trainers analyzed their training schedule? How many presentations or displays do they or their departments expect to put together within the next 12 months or the next budget cycle? What percentage of them could benefit from better visuals?

- What are the minimum computer hardware requirements for the graphics package? Today's sophisticated software packages require a computer with a hard disk drive. Even if the company has a laser printer, users may not be able to print a full page of high-quality graphics for complex charts if the printer has less than 1.5 megabytes of memory.

- Do users, or staff in the organization who will make the visuals, have unrestricted access to a computer and printer that meet those requirements, or will they have to schedule computer time? How far ahead will they have to schedule time to get the visuals made? How long, realistically, will it take to make them?

- Who within the organization actually will use the software: the trainer, a supervisor, a secretary? How enthusiastic are those people about taking on a new task? How proficient (or willing to learn) are they already with computers? Can they read the manuals, or are they intimidated by the documentation? Can their work schedules be adjusted to allow for the new time demands? Will they be asked to prepare visuals for other departments, now that the company has the graphics package?

- What training is available on how to use the software, and what will it cost? Reading a manual or "how to use" text will not be enough to teach the necessary skills. Is

training given in a lecture format only, or is hands-on training possible? Some computer graphics packages are more complicated to learn than others and require considerable time and repetition before a user is proficient. Users who make visuals infrequently may have to relearn parts of the procedure.

- Where does the user have to go for training? Is the vendor willing to train on site if enough workers from an organization use the software? Some community colleges offer training in popular packages. If so, can employees attend? Can management arrange an in-district tuition rate or discount?
- Can users' work schedules be rearranged to permit them to attend a training session? Are there fees for the training? Can more than one person from a company attend training sessions at a reduced rate? After the training session, what telephone support is available, and from whom? Is there a charge? Will the company have to pay long-distance tolls to talk to the software manufacturer's technical department?
- Will the vendor or software manufacturer supply names and phone numbers of others within an industry who are using the software for similar purposes? If so, users should call them and ask for their opinion of the software. Would they be willing to buy the package again? Why or why not? If they hedge, ask for details; then ask the vendor or manufacturer more specific questions.
- If service bureaus exist in a firm's area, users should ask two or three of them to recommend software that will produce the type of visual output that could be processed by the bureaus. Make sure the bureaus are already using the software for customers and ask for references that the company can contact.

When using computer graphics software to create projected visuals, trainers need to consider how they will present the information to an audience. They may have to trade off top-of-the-line (but costly) quality for budget considerations. For example, high-resolution projected visuals look wonderful on screen. However, unless the company has an in-house graphics department willing and able to produce them, management almost certainly will have to order the visuals from an outside service bureau.

Using a Service Bureau

Virtually every major city has at least one service bureau that specializes in producing slides, overheads, prints, and posters from the computer files created with a computer graphics software package. Companies not close to a major city can send their files directly to a bureau by modem over the telephone line. They also can mail the computer disk or send the disk via overnight air express. Turnaround time is usually 24 to 48 hours, plus transit time to get the slides or transparencies back.

Costs for this professional service may be a significant factor. If a firm is training a small group of people in a half-hour meeting that will not be repeated, trainers can almost certainly get by without expensive audiovisual aids. They can use the word-processing or computer graphics program to create overheads, even if they project them in black and white, with shaded patterns to indicate varying data. On the other hand, if users are showing safety data at a stockholders' meeting, giving a presentation to a top executive, or speaking at a conference, they will want the best audiovisuals possible, regardless of cost.

Look closely at the services such a bureau can provide before hiring them. Ask for names of clients and companies, and check references. One question to ask: does the bureau deliver what it promises on time? No matter how professional a bureau's output may look, if users are depending on slides or transparencies for an important presentation, they cannot afford to have their visuals arrive late. Some bureaus will produce a sample slide or transparency from a company's computer files for a nominal fee, giving users a chance to evaluate the bureau's work.

PROJECTED VISUALS

Projected visuals include slides, transparencies for overhead projectors, objects used in opaque projectors, motion pictures, television, and video.

This chapter will not go into great detail regarding slides and movies, because excellent information is available from the manufacturers of film and projectors, camera stores, libraries, schools, and publications. The Council's *Safety and Health*, for instance, frequently carries detailed articles on producing slide shows, using slides, and related topics. The National Safety Council also publishes Data Sheets on the subject (see References).

Overhead Transparencies

Overhead projection offers a number of advantages to speakers (Figure 17–10):

- Visuals can be shown while room lights are left on, so audience can take notes.
- Speakers can project to any size audience.
- Speakers can face the audience at all times.
- Speakers can use a pen or pencil as a pointer to indicate an item on the transparency. The shadow of the pointer shows up on the screen, helping to focus audience attention.
- Overhead projectors are easy to operate and usually available.

Overhead transparencies are easy to make when you use a computer graphics software package (Figures 7–8 and 7–9). However, if you can type, draw, use transfer lettering (available at art stores or office supply stores) or clip art, (available in purchased books or on computer disk), you can create overheads easily and inexpensively.

You can type, draw, or letter your message with dark ink on plain paper, send it through a plain paper copier in your office or at an office supply store, and get a black-on-clear transparency. Or make a reverse transparency, using white lettering on dark background. Using a service bureau or outside provider, you can take a 35mm slide or photo and have color transparencies made professionally.

Figure 17–10. This computerized overhead projection system assists the trainer in an explanation of electrical circuitry. (Courtesy 3M Visual Systems Division.)

Special effects. You can highlight certain words or phrases on your overheads by spot-coloring them ahead of your presentation with permanent color pens. Or, if you're using the same visuals for different audiences, use washable color pens, and write directly on the transparencies while showing them. Afterwards, wipe off the markings.

Overlays help explain step-by-step concepts and build the visual's story in a meaningful way. Two or more imaged transparencies are used in sequence over each other. Overlayed visuals are hinged to the frame on one side with tape to allow the base visual to be presented first. Then the overlay is flapped over it to complete the message.

A technique good for title visuals is an overlay. Make two transparencies from the same data; one, in a positive color; the other, in a reverse image. Frame the two transparencies together, but slightly offset, so the net effect is that of shadowed lettering.

Framing transparencies. Because frames help to block out distracting light from the screen, add rigidity for handling and storage, and provide a convenient border for writing notes, most speakers do frame transparencies, even though you can use them unframed. Frames also provide a space to number and date visuals so you can keep them in order.

Traditional cardboard frames, preframed films, and flip-frame transparency protectors (clear plastic sleeves) are available. The latter two are lighter in weight than conventional cardboard, and take less room in storage.

Projection panels. One option for presenting overhead transparencies is to use a projection panel to display the output of your computer. Such panels are available with color or monochrome displays. Used together with an overhead projector that shows the computer images "live," the panel lets you make changes in your data and immediately show those changes on the screen. The technique is often used to dramatize "what if" data or spreadsheet figures.

Changing data can be used in combination with overhead transparencies for even more dynamic impact. Set up two projectors and two screens. On one set, use transparencies for "static" information, such as safety goals or workdays lost to accidents. Combine the second projector and a production panel to display the "what if" information. "What's the dollar impact if we reduced the number of accidents by 5%?" by 10%?" etc. As you manipulate the computer spreadsheet to show the changing figures, the audience sees the changes being made, because they're projected on the second screen.

The two-projector/projection panel technique also works well as a discussion springboard for problems. Display the problem on the first screen, and show the range of solutions on the second. As students talk about each alternative, adjust the computer data to show the impact of each possibility, and project the display on the second screen.

Slides

Slide shows, whether developed by computer graphics, photography, or a mix of the two technologies, are another technique for presenting audiovisuals. If sound is added, programs can be recorded, and sound can be synchronized to slides on a cassette recorder, which can automatically advance the slides with the narration.

If cassette recorders are equipped with a stop button on the visual synchronizing mechanisms, the speaker can stop the show at appropriate points. He or she can pose questions, get a class response either in writing or verbally, and can start the show again, explaining and discussing correct answers.

Slides offer certain advantages:
- easily updated
- inexpensive
- easy to make
- produced by photography, computer graphics software packages, or a combination of techniques
- commercially available
- targeted to audience or training needs, since specific slides can be substituted in a series
- can be coupled with additional slide and/or film projectors for multimedia impact
- audio tape for automatic advance available.

Slides, however, have some potential disadvantages:
- easily get out of order
- can stick in the holding tray
- can easily be projected upside down or backwards
- if poorly made, can be distracting.

AV Preproduction

Creating a storyboard is one effective way of visualizing and assembling a slide show. On index cards, or on paper on which you've drawn frames about 2 x 3 in., sketch in what you want your first visual to be. When you've made a quick sketch of what audiences will see (just as the camera will see it), jot down a brief description of the scene on the same card.

Arrange the cards in order on your storyboard. An 80-slide show needs somewhere between 70 and 80 cards mounted on your storyboard, running in order from top left to bottom right. Edit the cards until you're pleased with the sequence. Then number them. A well-thought-out storyboard will help you fit images together in a sequence that helps your presentation.

If you're shooting (or commissioning) slides that some day might be transferred to film or videotape, there are several technical points to keep in mind as you plan the photography. That's because viewers will not remember slight visual differences in a series of projected slides. Film and video cameras, however, pick them up. The following suggestions come from the Association for Multi-Image International, Inc.

Aspect ratio. The ratio of width to height of a projected picture image (W:H) is called the aspect ratio. The standard in slide production is 3:2, but the aspect ratio of film and videotape is 1.33:1. If you want the full image of your slide transferred to film and video, you will have a dark band at the top and bottom of the image unless you keep your critical information within the film and video aspects. Many camera reticles (the lines you see as you look through the camera to focus it) are marked with these areas. Be sure that everything you are shooting that you want to appear on screen falls within the "lines" as you snap the photograph.

Type and character sizes. Slides have a higher resolution (the clarity or fineness of detail visible on-screen) than 16mm film or videotape. Consequently, if you're planning a transfer to film or video, make your type and character sizes larger than normal on the slides. Be especially sure that complicated typefaces such as serif or extended italic type are larger because of the lack of definition of video.

Color choices. Limit the graphics colors you select for your slides if you plan eventually to transfer them to video. Video has a reduced contrast range, causing poor reproduction of dark, intense colors or of highly-saturated colors, especially reds.

Differences in brightness. If the brightness of any one slide varies greatly from the average of the entire show, color-correct when slides are transferred by having the lab use neutral density filters.

Don't underestimate the time, money, and effort it takes to produce a good slide show. Check into purchased slide packages, computer graphics software, your organization's in-house visual production unit (if one exists), or an outside production company before deciding which route to go.

Safety Considerations

Photographers—whether you, your company professionals, or outside photographers from a production company—should observe these safety precautions when on location.

Safe background. It's not enough simply to make sure an object or person who's the subject of a photo is in complete compliance with all pertinent safety regulations. Persons in the background who might also show up in the picture also need to be observing safety precautions and wearing all the personal protective equipment required for the job they're doing.

NOTE: A safety violation or evidence of poor housekeeping showing up in a photo that's supposed to be promoting safety and health is likely to attract a lot more attention than the main safety message.

Safety equipment. Photographers, their helpers, and everyone else who goes into an area where personal protective equipment is required for the purpose of taking safety photos must wear all the equipment specified, even though it might appear as if they personally face little risk of injury.

It's essential that persons promoting safety follow all the rules of personal protective equipment, as well as all safety rules for that plant or work site.

Safe practices. Photographers, crew members, and visitors must observe all safety rules, including no-smoking rules and rules that prohibit certain types of electrical equipment in areas where sparks, radio frequency (RF), or electromagnetic interference (EMI) could be hazards. Rules must be observed, both for the protection of the visitors, and to demonstrate

respect for safety principles.

Lighting equipment. Photo floods and other special equipment used on a photo shoot should be in good working condition, and of the 3-wire, automatically grounded type. Electrical cables must be strung in a way so they won't cause a fall or some other accident. When appropriate, a plant electrician or safety personnel should be on hand to help a photographer avoid mistakes or accidents. All the equipment used on a safety assignment should be UL-listed or approved by another equivalent agency.

Fall protection. Photographers should use sturdy ladders and work platforms to climb on when going after elevated photo angles. They shouldn't climb on machinery or tables. Fall-arrest harnesses and lifelines ought to be used when working from heights.

Opaque Projectors

Opaque projectors are useful for showing printed material and even three-dimensional objects. Printed material up to 10 in. (25 cm) square or an object up to 2½ in. (6.5 cm) thick is placed in the machine, and the image is projected onto the screen by means of a powerful light and a mirror. The room must be darkened for effective viewing.

With an opaque projector, material that cannot conveniently be transferred for overhead projectors or photographed for slide projectors can be quickly shown. An example of such material would be a safety catalog or a safety poster in full color. Of course, to be suitable for use in an opaque projector, printed material must have type large and clear enough to be easily read by the audience when the material is projected.

Another use of the opaque projector is to project a picture, map, or other shape onto a pad or chalkboard so it can be traced. The size of the projection can be varied to the exact size required, for a neat, professional-looking drawing that can also be used as a flip chart.

MOTION PICTURES

Films are excellent tools for training and motivating workers. Because of films' higher costs compared to other visuals, trainers should plan their use with special care. The need for a quality product, and the sophisticated equipment and technology required to produce a film, are reasons enough to justify film production by a company's in-house audiovisual department or by an outside production company.

In addition, a wide selection and variety of 16mm safety films are available from insurance companies, local safety organizations, commercial film libraries, trade associations, industrial producers, and the National Safety Council. Many films can be rented or purchased, or previewed before a purchase decision is made. Because design and production costs (the largest cost components of films) have been absorbed by the producer and spread out over many renters or buyers, such films may not be personalized to each company's situation. However, they are almost always more economical than those your firm's AV department can produce.

VIDEO

Technical training by video has become extremely common, because videos offer a number of advantages:

- Job site brought to the classroom. A major advantage of video is its ability to show work-site situations. Students easily identify with on-screen visual representation of conditions or problems.
- Familiar format. Almost everyone is used to receiving information from a television screen, and seven out of 10 U.S. households have video cassette recorders (VCRs).
- Instant playback. If a trainer desires, he or she can replay a sequence, pointing out specific techniques or incidents. Students viewing the video can see material over again until they feel they have mastered the content.
- Availability of easy-to-use equipment. It is much simpler to pop a prerecorded video cassette into television playback equipment than it is to locate a 16mm projector and screen, thread the projector properly, and show the film. Virtually no special training is needed to run the video cassette player. Some companies allow workers to check out safety videos for home viewing and self-study, much as commercial chains rent prerecorded tapes.

Showing prerecorded tapes. Safety videos on a wide variety of subjects are available, either as single topics or as part of a series. Many videos, including a number of those offered by the National Safety Council, come with additional learning material designed for reinforcement: textbooks, instructor's guides, transparencies, and program guides.

Shooting your own videos. Photographing short onsite videos with a camcorder is not difficult to do. Often, camera stores run free or low-cost classes in "how-to" techniques. Local television studios may offer cable access classes in video production; high schools and community colleges frequently have training classes.

Be sure, before investing time or money in making videos, that the video can be played on company-owned equipment, or that management can acquire appropriate equipment on a cost-efficient basis.

Trainers who decide to shoot their own videos should make them short and simple. They may want to have a colleague videotape their training sessions before giving presentations to students in order to review, analyze, and improve their performance.

For onsite shooting, plan a script carefully, using storyboard techniques previously discussed in this chapter. Be sure to obtain appropriate releases from those who appear in the video.

Working with a video production company. Because people are so familiar with television programs and videos, they are highly conscious of production quality. Unless a company's staff are experienced AV professionals, the firm should consider using an out-of-house training organization, the company's own audiovisual production department, or an outside video production company to achieve the visual quality and impact needed for effective training.

Before trainers commission a custom-designed training video, they should answer the following questions, which

will be asked by firms who may bid for the project:

- Benefit/impact: How will the project save the company money, improve productivity, safety, or efficiency? Can the trainer provide "hard data," with potential benefits?
- Background. Why does the trainer believe a video is necessary? Can he or she describe the situation or problem that needs to be communicated? What solutions has the company aleady tried? What solutions, other than video, are they considering?
- Objectives. Can the trainer describe the session objectives in behavioral terms and list them in order of priority? Should the video primarily train (impart specific skills); inform (give background, ideas, facts); or motivate (difficult to measure effectiveness or results)? What should the audience be able to do—or stop doing—after they see the video?
- Audience demographics. What are workers' ages, years with the company, gender, education, etc.?
- Audience interest/need. Why should the audience be interested in the situation shown in the tape or in solving a work-related problem? How will they benefit from the video?
- Audience knowledge/experience. What does the audience know about the situation or problem? What is their past experience or involvement? (Before video production is started, the producer should poll the target audience for their input.)
- Audience attitudes/prejudices. What attitudes do the audience members already have about the situation or problem? What motives might the audience have for paying attention to, or ignoring, the video?
- Company goals and strategies. How does the intended video support the company goals and strategies?
- Generic value. Can the proposed video be used in more than one facility? Can it be written generically and still meet company objectives? Can alternate scenes be shot to make the program generic? Would other organizations be interested in the program?
- Use. How will the program be used? What will the viewing environment be? Will the video be used as a "stand-alone" program or as part of a course? Will any support materials (leader's guide, student study guides, handouts, booklets, etc.) be needed to go along with the program? Who will develop them?
- Evaluation. Will the video be evaluated formally, with specific questions, or informally through comments? Who will do the evaluation, and what is the evaluation strategy? Will the video be previewed with a sample from the target audience before production is finalized?

Choosing a production company. If management decides to produce a custom-designed video, trainers will want to get several bids before selecting one firm. The organization's audiovisual production company, if it exists, should be one of the candidates.

As a basis for bid comparisons and to help manage client expectations, one major company has developed a program classification matrix describing four levels of video production. (Table 17–C). These levels range from a "talking head"

video (Class I) to a comprehensive, top-of-the-line program comparable to broadcast television (Class IV), using professional talent, studio sets, extensive editing, and custom music.

Use the program classification matrix as a tool in selecting the right video production. Choose blocks across the rows that represent trainer expectations, total the columns, add the answers, divide by 10, and round up or down to the nearest whole number.

In 1990, production costs for an 8- to 12-minute video at Class I were approximately $5,400; Class II, $13,800; Class III, $40,000; and Class IV, $90,000.

Trainers may want an outside consultant to help them evaluate the bids, check references, and evaluate past productions of any company they are considering. Remember, too, that production costs do *not* include distribution. Get bids on duplicating and distributing the video from several vendors.

If a company has a culturally diverse workforce, trainers should consider using a production firm specializing in translations to do voice-overs for the video. For maximum effectiveness, use a competent, experienced organization that is sensitive to nuances of language and cultural differences.

COMPUTER-BASED TRAINING

Computer-based training is a viable, high-tech audiovisual option that a number of companies have used with good results. Benefits include shorter training time, because students proceed at their own learning rates; more timely training, as a training module can be given to a student whenever he or she needs it; and an increased student-to-instructor ratio. One instructor can supervise more students who are using computers than is possible in a conventional classroom setting.

Because concepts and "how-to" instruction are standardized, computer-based training can be more effective. All students receive the same information, regardless of their location or their teacher. Interactive computer-based training provides standardized feedback, giving the same responses each time a student answers a question. Also, because students have the opportunity to practice until they master the skills taught in the training, they become more proficient (Figure 17–11).

If individualized instruction is desired, interactive computer-based training can pretest and posttest students, tracking and recording their performance as they complete various segments of the program. Some interactive computer-based training programs can be customized to adjust the work within a module, based on a student's performance.

Drawbacks to computer-based training include the cost and time necessary to develop high-quality instruction, based on specific training needs. Generic computer-based training packages are difficult to match with an individual company's objectives. Organizations that wish to develop their own programs (or work with an outside organization that will do so) must be prepared to invest substantial time and money. Programs often need to be tested and refined several times during development.

A second drawback to computer-based training lies in its standardization. Not all people learn in the same way or think

Table 17–C. A Video Program Classification Matrix. (Courtesy GTE.)

Classification	Class I	Class II	Class III	Class IV
Description	Simple Production	Intermediate Production	Advanced Production	Complex Production
Definition	Little or no pre- or post-production.Talking head (field or studio), event documentation or file footage production, with little or no graphics or switching. Script, if any, provided by client.	Moderate pre- or post-production. In studio or field, with minimal research, scripting, graphics or video inserts, with in-house talent.	Considerable pre- or post-production. Studio and/or field, with professional talent. Moderate research, scripting, graphics and use of music with basic set.	Extensive pre- and post-production. Complex field and/or full studio using elaborate sets, professional talent, in-depth research, fully scripted, special video effects, sophisticated graphics, sound effects, and music track.
Matrix Point System	1	2	3	4
Writer	0–1 Day	2 Days	3–6 Days	7 or more days
Director	0–1 Day	2 Days	3–6 Days	7 or more days
Production Days	0– ½ Day	1	2–4	5 or more days
Edit Days—Off Line	0– ½ Day	1	2	4
Edit Days—On Line	0–1 Day	2	4	5
Production Style Studio	Simple existing set pieces.	Combining available on hand set pieces with purchase of small set items.	Combining available on-hand set pieces, and limited simple set construction.	Major new set construction
Remote	ENG shoot at unaltered remote location.	ENG shoot requiring only minor location alterations	EFP shoot requiring only minor location alternations.	EFP shoot requiring staging and possible disruption of area work flow and schedules.
Talent	None, non-professional, talking head, voice over.	Simple on-camera or voice-over narration. Using either non-professional or professional talent. Requiring little direction.	Usually professional talent. Requiring moderate direction, several scenes and/or locations.	Requires multiple professional talent, or single talent requiring complex direction with multiple scenes on numerous sets or locations.
Music	Basic sound track with little or no music. Selections from blanket fee music library.	Simple sound track usually with blanket fee music, or limited use of library selections.	Advanced sound track with multiple music selections and sound effects. Selections from multiple libraries.	Complex sound track with multiple music selections and sound effects, and requiring separate audio edit sweetening session. Possibly the production of original sound track.
Graphics	Little or no graphics.	Simple graphics or titles from character generator.	Type set or art card graphics combined with character generator and basic switcher effects.	Production or purchase of complex graphics or special effects.
Editing	Little or no post-production editing. Very basic machine-to-machine editing.	Simple cuts only editing, possibly using some basic switcher effects, and on-line audio board mixing.	Advanced audio/video editing, requiring split audio tracks, advanced switcher effects, but all created during the on-line session.	Complex audio/video editing, requiring a separate edit session to create multi-layered special effects for both video and audio.

Each block in a column represents a numeric value of 1, 2, 3, or 4. To determine the program level, total columns and divide by ten. Round up or down to the nearest whole number.

Example: Total points = 35 ÷ 10 = 3.5 or Clas IV Program.

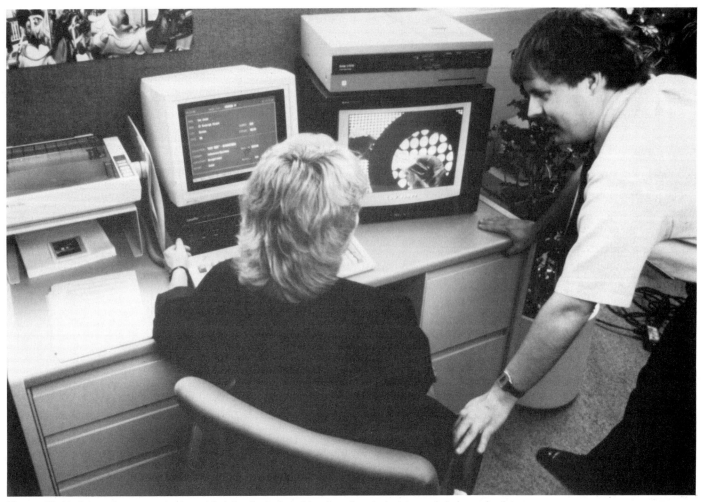

Figure 17–11. Interactive, computer-based training is designed to allow the student to work independently, correct mistakes, and reinforce correct responses. (Courtesy Eastman Kodak Company.)

alike. If computer-based training is used by itself, without an instructor to ask questions, students may learn to respond with the "correct" answers but may not understand those answers or why they are "correct." When faced with a situation on the job that is slightly different from the computer version, they may not know how to handle the problem.

Because computer-based training is impersonal, keeping students "on target" and motivated can be difficult. Well-designed training materials can help, but the student's attention span may be shorter than would be the case in a more conventional training class. (See also Chapter 14 and Figure 14–2.)

Assessing Needs

Deciding whether or not computer-based training is appropriate for an organization is easier if managers use a systematic approach. They will need to evaluate their training requirements carefully. Are they training for specific skills needed to operate certain equipment? If the problem is shared by many companies across a variety of industries, computer-based training software packages may have already been developed.

Programs in widespread use in business, such as word processing and spreadsheets, often come with self-paced on-screen tutorials. Instruction on how to use such software

may be available on audiotape or video. On the other hand, if managers want to show new employees how to change machine tool set-ups by reprogramming a computer-controlled machine, the company will almost certainly have to develop its own computer-based training, or find a different vehicle for instruction.

If company training programs must meet regulations that require documented employee instruction, trainers may find that commercial packages already have been developed on these topics for the industry. Before investing in any packages, and the necessary hardware to run them, however, trainers should ask if vendors will release names of companies that have used the computer-based training successfully, or that can provide evidence that the training, in fact, does the job it is designed to do.

A substantial investment in computer-based training also may be hard to justify if new procedures, new equipment, or new regulations make the instruction outdated soon after its production and distribution.

Computer-Managed Instruction

Trainers can choose to use computer-managed instruction with computer-based training. Under computer-managed in-

struction, trainers monitor students on their training time, attendance, and participation; the training materials students have used; tests taken; and the final results achieved. The computer can keep track of how a student performs and suggest additional materials or extra drill, if the performance needs improvement. If some students complete the modules ahead of schedule and are ready for a higher level of instruction, the computer can generate a list of names for the instructor's review.

Interactive Training

Interactive training, a technology that uses a laser disc video and CD-ROM (laser-operated) based technology in combination with a computer video display terminal screen, allows a person to be trained systematically by simply touching the screen to choose from various options. The user does not have to know anything about computers, nor key in the answers.

The system combines sight, sound, and kinesthetics in a single presentation, thus reinforcing learning. Because training is one on one and requires touch-screen action by the worker before the module continues, employees stay motivated and involved in learning.

Interactive technology offers a number of advantages, including:

- Reduced learning time. In some studies, learning time can be reduced up to 50%. Immediate feedback provides constant reinforcement of concepts and content.
- Reduced delivery costs per student. In interactive instruction, the largest costs cover design and production, not duplicating or distributing the film. Delivery cost per student decreases as more students use the same program. In traditional training systems, which depend heavily on teachers, delivery cost remains constant or rises as numbers of students increase. Even for a custom interactive program, the cost-per-student breakeven point occurs with a relatively small number of students. For large organizations with many students who need to learn the same material, savings can increase dramatically.
- Reduced risk. Interactive systems can allow students to explore potentially dangerous subjects without compromising safety. For example, a course on basic electronics and maintenance lets students accidentally "touch" the wrong parts without risking electrocution, yet "see" the consequences on screen.

Because interactive systems do not allow students to continue on to new material until they have demonstrated their mastery of fundamentals, trainees gain a strong foundation on which to build further skills. Yet, because instruction is self-paced, students can repeat and review materials until they are confident they have mastered the information.

Interactivity formats

Interactivity is achieved by a question-and-answer, or challenge-and-response, format. These formats require the student to participate actively in the training. Two types of interactivity are often stressed: conceptualization and simulation.

Conceptualization. Primarily used for training that depends on theory or other abstract ideas, conceptualization is a good choice when employees must learn "soft skills," such as human relations or management training. This type of training helps to communicate psychological or philosophical viewpoints.

Simulation. The use of simulations that duplicate the look and feel of real experiences has proven to be a valuable learning tool. The most sophisticated simulators are probably those used by the aerospace industry to train pilots. To the extent possible, the simulated cockpit replicates visual, auditory, and tactile experiences of real flight. The primary advantage is to place the pilot in emergency situations under controlled conditions so that safety procedures can be practiced. Trainers can evaluate both the procedures and the pilot's performance.

Simulated electrical panels are used widely for electrical training. The electronic components for a machine or systems are installed on a board, so that trainers can create problems requiring students to practice troubleshooting skills in a classroom setting.

Fire training institutes and some fire departments use simulations to create various firefighting and rescue situations. Trainees then practice the recommended techniques under controlled conditions.

The National Aeronautics and Space Administration (NASA) uses special water-filled tanks and weighted suits for astronauts to simulate working in a weightless environment. Some mining companies have replicated underground conditions in surface training labs to enable miners to practice machine operations under controlled conditions.

SUPPLEMENTAL MATERIALS

Supplemental materials can enhance and support a company's audiovisuals. Whether supplemental materials are stand-alone safety reminders or designed as part of a comprehensive presentation package, they can increase safety awareness and reinforce training messages.

Posters. A number of commercially available safety posters on various topics are offered by a variety of organizations, including the National Safety Council. State-of-the-art procedures, single-topic reminders, and compliance/regulatory warnings are good visuals.

Displayed at various locations throughout an organization, posters can help employees avoid hazards and can be a useful part of a company's comprehensive safety program. For more detail, see the discussion of posters and displays in Chapter 18, Promotions and Campaigns.

If a trainer or someone in the department is making posters, keep them simple and short for maximum impact. Each poster should convey only one main idea, with no more than two or three additional details for reinforcement. Eye-catching visuals can be used; for example, a drawing of wiggling toes sticking out of a cast that is decorated with signatures of fellow workers, and the message, "Wearing Safety Shoes Might Have Prevented This," makes the point graphically.

Ideally, posters should be large: 15 x 20 in. (38 x 51 cm) is a minimum size, although 22 x 28 in. (56 x 71 cm) is better. If posters are too small, people will pass by without noticing them. Vivid colors such as golds, yellows, hot orange, or brilliant blues grab attention immediately. For even more emphasis, highlight with a splash of contrasting color.

Type should be as large as possible—38 to 72 points, or letters at least one inch high for headlines. Trainers can use markers, transfer letters, cut paper, or India ink for lettering.

Pictures. A good safety photograph also attracts attention. Photos can show the safe way to perform an activity, such as lifting a package, putting on a respirator, or using a computer keyboard correctly to minimize the risk of carpal tunnel syndrome.

Like posters, photographs should convey one main idea to be most effective. Pictures of people are always more interesting than those of equipment alone; close-ups are more attention getting than long, "establishing" shots. If the photographer is shooting a series of photos, however, he or she may want a long "establishing" shot to set the location and orient the viewer, plus some medium shots and close-ups for variety. Someone skilled at focusing may want to use a tighter close-up by including only the important details, such as focusing on a worker's hand to illustrate proper tool use.

To ensure top-quality photos, match the film to the environment. Daylight film works best with daylight or flash. while tungsten film works best with tungsten light. If the photographer has to shoot with the "wrong" film, he or she should use the correct color-compensating filter. Color-correction filters can also help maintain the correct color balance with fluorescent lighting, found in many offices or factories. Fluorescent lighting tends to produce greenish tones. If subjects wear bright, strong colors, the green cast will not be as noticeable.

When shooting outside or in a well-lighted room, film with an ASA (ISO) reading of 400 can take advantage of available light. The higher the ASA (ISO) number, however (check the box), the grainier the photograph when enlarged or projected.

To improve the chances of getting successful photographs, overshoot; that is, take substantially more pictures than needed. Film is the cheapest commodity; the time of the people involved is far more costly. Taking four or five photos of a single set-up to get one usable picture makes economic sense. Bracket exposures; adjust the camera meter and settings for what will be the ideal exposure. After snapping the picture, take at least two more, one f-stop under and one f-stop over, the recommended setting. If trainers want to brush up on photography, a camera store, an equipment manufacturer, a high school, or community college will often offer free or low-cost classes on simple techniques.

On the other hand, if the project calls for a number of photographs, particularly in a factory or large office setting, or if the upcoming event is a major one, it is almost always cost-efficient to use a professional photographer. AV professionals have the lighting equipment and know-how to make a shoot run smoothly. Clear the way with the supervisor and with whatever authorities might be involved, including the facility's manager, and any governmental body that might have jurisdiction. Get releases from all persons photographed (including employees and managers) before including them in pictures. The company's legal department can draw up the correct form. Make sure the release is open-ended and nonspecific enough so that if the company wants to recycle the slide into another show, it will not be necessary to acquire a second permission signature.

Handouts. Handouts represent another supplemental way of reinforcing training concepts or of testing specific skills taught in the presentation. Employee safety booklets, guidebooks, manuals, compilations of case studies, and "hard copy" of overhead transparencies and slides are good learning aids. Many are available commercially from various sources, including trade associations and the National Safety Council.

Service bureaus can turn overheads or slides into hard-copy handouts. However, the projected visuals will be summaries, with few words and lines on each overhead or slide. Handouts give the speaker a chance to expand on the information shown on screen, including more details or cases.

When trainers plan to use such handouts as part of a comprehensive program that includes other AV media, they should make sure not only to preview visuals, but to check the printed handouts carefully. Both media should give the same message, not confuse students.

Industrial data sheets. Industrial data sheets, such as those available from the National Safety Council, can reinforce safety concepts by providing supervisors and workers with facts, diagrams, photos, and analyses in one convenient handout. Because such data sheets are periodically revised, they can help trainers, supervisors, and others keep up to date on the lastest regulations, compliance issues, or correct factory procedures.

Audio cassettes. Used as stand-alone training or as components in a comprehensive package, audio cassettes can promote safety effectively. Purchased cassettes—some available with response booklets and and self-tests—are valuable for individualized instruction or refresher training. Some are available commercially with instructions in a choice of languages. Because students can replay audio cassettes, they can go over material until they have mastered it. Cassettes, also can be part of classroom presentations.

If trainers make tapes themselves, they should use a hand-held microphone to filter out ambient room noise. Sixty-minute tapes (even if the recorded material is not that long) work better in a recorder than 90-minute tapes. Trainers can do brief on-the-job interviews that may prove useful later for bulletins, newsletters, and future meetings. They also can dictate a running commentary while photographing safe or unsafe operating conditions, new processes, or equipment that require subsequent study. They can record minutes in

safety meetings or valuable discussions in training sessions.

In addition, trainers can tape safe, efficient job procedures, provided the trainer allows sufficient time for performing each task. The trainee then listens to the tape and follows the instructions while performing the work.

PRESENTATION OF AUDIOVISUAL MEDIA

Most modern meeting rooms and classrooms are well designed for AV presentation. Occasionally, trainers may find themselves leading a training session in a less-than-ideal environment. Here are some tips that will help them either way:

- Make sure the room in which they expect to lead the presentation is, in fact, available. Know in advance what they will do and where the class can go if forced to switch rooms at the last minute. Will the equipment they have planned to use work in an alternative location? Can they get three-pronged adapters or extension cords fast, if necessary? Is a blackboard on wheels available? Is there chalk and an eraser? Will the number of tables and chairs needed fit in the substitute room? How will they get there?
- Collect everything trainers could possibly need for their presentation—extra chalk and eraser, if they are using a blackboard, extra batteries as well as an AC adapter cord, if they are using an audiotape cassette player. Don't forget to have back-up for essential items such as spare projector bulbs and fuses, and make sure trainers or their staff know how to replace them.
- Be sure the person responsible knows how to operate all the equipment to be used. Have trainers coach and rehearse their talk with a substitute, should the first helper be unable to attend the meeting.
- Make an agenda of the meeting with approximate times listed for each part of the presentation and give it to their helpers well ahead of the session. Have an extra copy of the agenda on hand should a last-minute substitute helper become necesary.
- Staff should know when the speaker wants the lights on and off, or adjusted to a particular level. Will the staff handle lighting controls?
- Check the equipment a day ahead of time to be sure it is in good working order. If equipment such as projectors and screens is shared with other users, trainers should make sure they have signed up for the required equipment well in advance. Double-check the day before the presentation to make sure no one has preempted the reservation. Does all equipment "match?"—i.e., is the VCR player the correct size for the videotape? Will the slide tray fit the projector? Don't take anything for granted; assume that if something can go wrong, it will, and prepare accordingly.
- Make sure that prerecorded media (i.e., videotapes, films, and audiotapes) are of acceptable quality for presentation. *Preview everything* well in advance, if possible, in case alternatives are needed. Time any prerecorded media in advance to be sure the presentation will run smoothly and on schedule.

Also preview projected visuals made in-house. If possible, use the same room, equipment, and lighting chosen for the presentation. Slides that look good on a slide sorter may be under- or overexposed when seen on the actual screen to be used at the meeting. Leave slides or transparencies on screen, and sit in the furthermost chair. Can the visuals be read comfortably? Are they sharp and in focus?

- Before the presentation begins, preset the brightness and audio controls on projectors, audio equipment, and video monitors to avoid disrupting the program later to make adjustments.
- Make sure all the slides to be used in a presentation are in order and good condition. When the slides are in proper sequence, draw a line diagonally across the top of the pack. In this way, a slide that happens to be turned accidentally on its side or moved out of sequence will be noticeable at a glance.
- Use duct tape to tape down electric cords, so they won't pose a fall hazard.
- For effective communication, presenters should face the audience or keep their faces turned in profile while discussing slide shows, movies, and videos.
- If a loudspeaker is used, place it near the screen and slightly elevated for realistic-sounding audio reproduction. If a clip-on microphone is used, test it ahead of time to be sure it actually works and that the volume is set at the correct level for the meeting room conditions.
- Make sure microphones are placed at convenient locations if audience participation is planned so that everyone can hear what is said.

Seating. Check the presentation room for safety features and for seating arrangement. Aisle space should be adequate, and exits should be available for emergency use.

The seating arrangement should give every member of the audience an unobstructed view of the visual. White matte screens do better at an angle of 40 degrees or less, while beaded screens are not as effective for viewing at an angle. To view a projected visual, the most desirable seating area is within a 30-degree angle from the projectors for a matte screen and within a 20-degree angle for a beaded screen (Figures 17–12a and 17–12b).

The recommended minimum viewing distance is at least twice the screen width; the maximum should be no more than six times the screen width. A 4-foot-wide (1.2 m) image, therefore, could be viewed at a maximum distance of about 24 ft (6.3 m); a 6-foot-wide (1.83 m) image could be viewed comfortably at a maximum distance of about 48 feet (14.6 meters).

For meeting room or auditorium seating, allow 21 to 24 in. (53 to 61 cm) width for each chair and 36 in. (91 cm) for each row of chairs, or about 5 to 6 sq ft (0.46 to 0.56m²) per viewer. In classrooms or conference rooms, allow twice as much space.

Various seating arrangements work better for certain kinds of meetings. If a conference table is not available and the

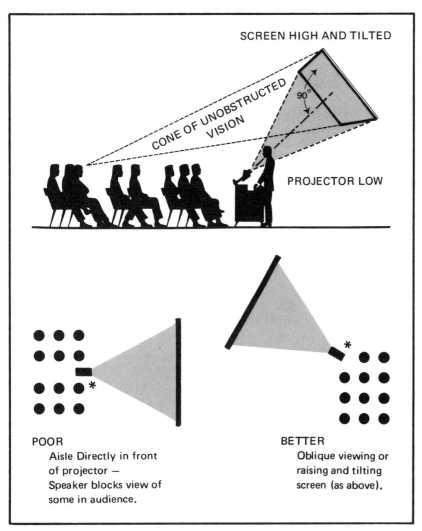

Figure 17–12a. To obtain optimum viewing area, especially when instructing large groups, tilt the overhead screen *(top)* or place it toward the corner of the room *(lower right)*. Both placements reduce the possibility of the instructor and equipment blocking the vision of some viewers, as might be the case when the projector is located in front of the group *(lower left)*.

group consists of five persons or less, set up an overhead projector on a desk and project visuals onto a portable screen or against light-colored walls. Meetings with six to 12 people work well with a single center table. For groups of 20 and under, use a U-shaped table to encourage participation and discussion.

Tables arranged in a herringbone pattern are good when the presentation is mostly a lecture format, and the speaker wants to focus attention on talk and visuals. The tables provide work space, yet the participants can still see each other and interact. A more formal auditorium setting, or an amphitheater with elevated levels, probably provides little or no audience participation. In such settings, the speaker almost certainly needs a microphone and a screen and visuals large enough so the audience can see them clearly.

In some extremely wide rooms, the seats cannot be arranged so the audience sits within the recommended viewing angle. For these situations, project slides or overheads simul-taneously on two or three screens, separated by at least 25 ft (7 m).

EVALUATION

Evaluation helps trainers get the maximum benefit from the time and effort they have spent in creating and using audiovisuals. They need to know how they are doing and what areas require improvement.

Effective evaluation involves more than handing out forms asking audiences how they liked a speaker or presentation. A well-designed evaluation strategy can yield a great deal of useful information. Such a strategy, planned in advance, should consider:

- What will be evaluated? At times, speakers want their audiences to evaluate audiovisual quality. Other times, they will ask for feedback on how well the AV materials were used. A third question to consider is, "Did the au-

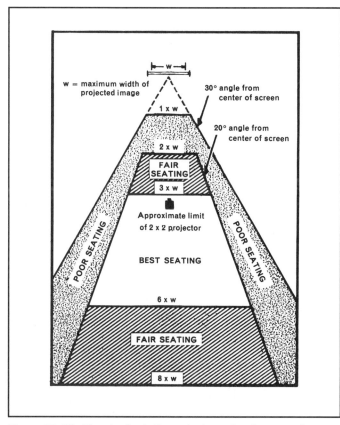

Figure 17–12b. The shading indicates both good and poor seating area when ordinary (beaded) projection screens are used. A lenticular screen would widen the angle, which is measured from the center-line. The best viewing area is within 20 degrees of the projection axis for a beaded screen, and within 30 degrees for a matte screen.

diovisuals and training session influence worker motivation or behavior?"

- Who will do the evaluating? The speaker? Audiences? The supervisor? Executives? Outside consultants? Establish who will evaluate the sessions.

- When will the evaluating be done? Speakers can lose thoughtful comments and feedback if evaluating becomes a routine fill-out-the-form exercise after each training session. Instead, consider polling individuals in the audience to keep interest alive.

- How will the evaluating be performed? Will formal rating sheets be used? Will the audience fill out 3 x 5 index cards and drop them in a suggestion box? Do speakers want to be ranked "by the numbers" or are they looking for descriptive comments?

No one strategy or form is "right" for a particular organization or situation. But by working together with others in the facility, trainers should be able to devise an effective evaluation strategy for their organization's needs. Also, they should evaluate the strategy itself, at least once a year, to see how it can be improved.

Creating an Evaluation Climate

If speakers want honest comments, they need to create a climate that encourages feedback and respect for each other's ideas. If speakers become defensive when criticized, or if workers believe those who make suggestions will be penalized, speakers will not get the thoughtful appraisal they need.

One useful approach which often works involves asking employees who say they don't like something to tell you how they would make it better. By forcing them to focus on the positives, you not only get suggestions that might work, but you reinforce the team feeling of "we're in this together."

Just as you do in a brainstorming session (see Chapter 16, Safety Training), accept all comments without allowing negative discussion on any ideas presented. You want to reinforce creative thinking and involvement, weaning the group from the "we've always done it this way" limitations.

Evaluating the Audiovisuals

Here are some criteria you can use as a starting point to judge the quality of audiovisual tools:

Appearance. Were diagrams and graphs appropriate for the information being displayed? Was an easy-to-read typeface chosen? Was lettering large enough to be seen easily by the audience, no matter where a viewer was sitting or whether he or she wore glasses? If more than one color or pattern was used, were they distinct enough so viewers could distinguish information easily?

Were the visuals in focus? Were they appropriate for the message? Was the information they contained up-to-date? If "how-to" skills were shown, was the instruction paced slowly enough so viewers could understand and follow the teaching?

Audio. If sound was part of the audiovisual, was it clear and distinct? Was the volume level comfortable, but adequate? Was the sound consistent throughout the session? Was the sound appropriate for the visuals? Was the level of language appropriate for the audience and message? If technical terms were used, were they defined immediately?

For live presentations, were microphones free of annoying howl or distortion? Were there enough microphones, correctly placed, for all who spoke?

If a soundtrack was used in a second language, was the message accurately translated? Did the translation take into account any cultural differences or sensitivities?

Props. If chalkboards, flannel boards, hook and loop boards, or easels were used, were they placed so the audience could see them? Could each member of the audience see the material they contained easily?

Evaluating the Use of Audiovisuals

Did the audiovisuals help make the presentation message clear? Did they reinforce what was being taught? Were audiovisuals well-suited to the presentation format, whether it was individual or small group instruction, computer-based training or simulation, discussion, lectures, or a presentation by a panel? Were audiovisuals integrated into the presentation appropriately, that is, at appropriate places, and for an appropriate length of time?

Had the speaker rehearsed enough with the audiovisuals to use them well, and with confidence? Was the room

comfortable? Was the seating arrangement satisfactory? Was lighting appropriate for the audiovisuals used? If projected visuals were used, (overheads, slides, films, videos), did the person responsible for the projection handle the equipment properly and efficiently?

Evaluating Audience Reaction

- Was the objective of the presentation motivational? If so, did the session stimulate or reinforce motives? Did the audiovisuals make a direct link to the core desire or identified need?
- Was the objective of the presentation skills training? If so, did the audiovisuals provide illustrations and examples of application? Did the presentation reinforce the message by also using learning activities?
- Was the objective of the presentation informational? If so, did the audiovisuals present facts and supporting material in a clear, logical structure? Was the pacing of the total presentation appropriate, so that the audience could acquire new knowledge?

SUMMARY

- Audiovisual (AV) media have become indispensable tools in occupational safety, health, and environmental training. They take advantage of the fact that people learn and retain material better when they use both their eyes and ears. AV media includes posters, films, videos, recordings, books, flipcharts, bulletin boards, computers, CD ROM players, and live demonstrations and experiments.
- To make the most of AV media, trainers must not only define their training needs but document that such instruction has been effective. Most companies focus their training efforts on high-risk areas and on workers who have recently been hired, promoted, or transferred. Management must establish measurable training objectives tailored to a specific audience and message.
- The selection of AV materials depends on the role the trainer must play (expert, skill-giver, or facilitator), audience size, cost of the materials, and whether in-house or outside personnel will make the materials. Management must consider how often the materials will be used and whether the message to be communicated justifies the time and expense involved.
- Speakers can be more effective if they take advantage of attention-getting AV devices such as pointers, easels, charts, TelePrompTers, props, and other aids. The speaker should make sure that such equipment is appropriate for the training session, easy to use, and sturdy enough to withstand repeated use. The best speakers are well organized and rehearse their presentations thoroughly before the actual session.
- Designing effective AV materials is easy if trainers remember the basic purpose of AV media: to reinforce a message. A general guideline is to keep audiovisuals simple; this principle applies to charts, text charts, type selection, colors, and graphs. Visuals must summarize, not repeat what the speaker says.
- Charts use either a landscape or portrait format. Text charts are made in four styles: title, list, organization, and tables. Graphs such as pie charts, bar charts, line charts, and area charts are effective for comparing changes in information either between categories or over time. For all charts, text, and other visuals, choose simple, attractive type styles for headlines and text, and use strong, contrasting colors. Make sure visuals can be read from all angles of a meeting room and from the back of the seating area.
- Most computer graphics software packages use templates to specify type size and style, justification, and color schemes. In choosing the right software, trainers must consider company training schedule, minimum hardware requirements for the package, who will use the software and how they will be trained, references or names of current users, and whether the software can be used by service bureaus to create visuals.
- Movies and videos are two of the most commonly used AV materials in training. Videos are particularly valuable because they are easy to use and make, show on-site situations, and allow workers to replay key procedures or practices until they have learned the material. A company can make its own films or videos or hire a service bureau or production company to create them.
- Computer-based training is a high-tech audiovisual option whose benefits include shorter training time, more timely training, an increased student-to-instructor ratio, and better training results. Disadvantages include the time and cost required to develop, test, and refine instruction materials and the lack of individualized training approaches.
- The most common computer-based systems include computer-managed instruction and interactive training, which uses conceptualizations and simulations to teach workers. Interactive training combines sight, sound, and kinesthetics to reinforce learning and to reduce learning time, cost, and risk. Simulations allow students to practice high-risk procedures and safety techniques without incurring injury.
- Supplemental materials can enhance and support a company's audiovisuals and make a speaker's presentation more effective and memorable. These materials include posters, photographs, handouts, industrial data sheets, and audio cassettes. Trainers can either make these materials themselves or have them developed by professional AV staff.
- The key to effective audiovisual presentation is preplanning. Trainers should confirm meeting room reservations, collect all supplies and materials, draw up an agenda, check equipment beforehand, preview all visuals and recorded media, arrange slides or transparencies in the right sequence, test the sound system, determine the best seating arrangements, and eliminate all safety hazards in the meeting room.
- The final step is evaluation of the training session. An

evaluation strategy involves answering such questions as what will be evaluated, who will do the evaluating, and when and how will it be done? Trainers need to establish a climate that encourages open, honest feedback from participants. Audiovisuals should be evaluated on the basis of their quality and their use.

REFERENCES

Films/Videos

National Safety Council, 444 North Michigan Avenue, Chicago, IL 60611. "How-to" techniques, taught through short visuals. Available in 16mm film or $^1/_2$" VHS videotape.
Communication Skills.
Listening Skills.
Presentation Skills.

Periodicals

Audio-Visual Communications. Media Horizons, 50 West 23rd Street, New York, NY 10010.

Fast Forward. Association of Audio-Visual Technicians, Box 9716, Denver, CO 80209.

In Plant Video/AV Communicator. (bi-monthly, February through December), PTN Publishing Corp., 210 Crossways Park Drive, Woodbury, NY 11797.

Technical Photography. PTN Publishing Corp., 2210 Crossways Park Drive, Woodbury, NY 11797.

Books

Bruccoli M, ed. *Audio Visual Market Place.* New York: R. R. Bowker Co. (annual).

Bunyan, JA. *Why Video Works: New Applications for Management.* White Plains, NY: Knowledge Industry Publications Inc., 1987.

Carlberg S. *Corporate Video Survival: A Book of Strategies.* White Plains, NY: Knowledge Industry Publications Inc., 1989.

Cole M and Odenwald S. *Desktop Presentations.* New York: AMACOM, Division of American Management Association, 1990.

Hansell KJ. *The Teleconferencing Manager's Guide.* White Plains, NY: Knowledge Industry Publications Inc., 1988.

Kerlow IV and J Rosebush. *Computer Graphics for Designers & Artists.* 2nd ed. New York: Van Nostrand Reinhold Co., 1987.

National Safety Council, 444 North Michigan Avenue, Chicago, IL 60611.
Industrial Data Sheets.
Non-projected Visual Aids, 12304–0564.
Photography for Safety, 12304–0619.
Posters, Bulletin Boards, Displays, 12304–0616.
Projected Still Pictures, 12304–0574.

Wiese M. *Film & Video Budgets.* rev. ed. Studio City, CA: Michael Wiese Productions, 1990.

Associations and Government Agencies

American Society for Training and Development, 1630 Duke Street, Alexandria, VA 22313.

Association of Audio-Visual Technicians, 2378 South Broadway, Denver, CO 80210.

Association for Multi-Image International, Inc., 8019 North Himes Avenue, Suite 401, Tampa, FL 33614.

International Association of Business Communicators, 1 Hallidie Plaza, Suite 600, San Francisco, CA 94102.

International Television Association, 6311 North O'Connor Road, LB 511, Irving, TX 75039.

National Safety Council, Safety Training Institute, 444 North Michigan Avenue, Chicago, IL 60611.

U.S. Office of Education, Bureau of Adult and Vocational Education, 400 Maryland Avenue SW, Washington, DC 20036.

Women in Communications, Inc., National Headquarters, 2101 Wilson Boulevard, Suite 417, Arlington, VA 22201.

(Other data books and pamphlets are available from distributors and manufacturers of cameras, films, computer graphics software, and other visual media equipment. Trade journals in the fields of computers, audiovisuals, photography, education, sales management, advertising, and training also contain excellent information.)

18

Promotions and Campaigns

This chapter deals primarily with promoting and maintaining managers', supervisors', and employees' interest in safety. The chapter covers promotions (what an organization does to promote safe behavior and practices within its workplace) and campaigns (what it does to publicize safe behavior and practices to workers' families and areas served by the organization).

To promote and maintain interest in safety, the safety and health professional must inform key executives and managers about the safety program's objectives and how to achieve them. However, management must also demonstrate its commitment and interest and actively support a solid safety program. Only with top management leadership and support will activities to promote employees' interest be successful. A well-promoted program involves both management and employees in safety and health activities. (See Meetings of Executives, later in this chapter.)

If top management seems uninterested in such issues, the safety professional should not assume that executives are indifferent or opposed to safety. Many times they may simply be unaware of the importance and basic benefits of an organized safety program.

REASONS FOR MAINTAINING INTEREST

Why is it necessary to maintain interest in safety (1) if the workplace has been designed for safety, (2) when work procedures have been made as safe as possible, and (3) after supervisors have trained their crews thoroughly and continue to enforce safe work procedures? The answer is simple: even with these optimum work conditions, accident prevention basically depends upon the desire of people to work safely.

All hazardous conditions, unsafe practices, and loss-control problems cannot be anticipated. Employees frequently must use their own imagination, common sense, and self-discipline to protect themselves; they must think beyond the immediate work procedures to act safely in potentially hazardous situations when on their own.

Indications of Need for a Program

Various yardsticks indicate supervisor and employee attitudes toward accident prevention. These measures also indicate problems that the safety professional needs to address by creating a program to stimulate and maintain employee interest in safety.

- An increased rate of injuries, accidents, and near-accidents may signal that a program is needed. If such an increase cannot be explained through engineering methods, training, or supervision, then perhaps employees are forgetting or ignoring work rules, failing to stay alert, or taking chances. A program to develop and maintain their interest in safety will help to reverse this trend.
- If housekeeping is deteriorating, protective equipment is not being used, and guards are not being replaced, it is time to improve supervisory safety awareness and to promote renewed and increased interest in safety.

- Incomplete or missing accident reports also indicate decreasing supervisory interest and, perhaps, even a failure of employees to report minor accidents and injuries. Motivation to ensure better reporting is then in order.

Program Objectives and Benefits

A well-planned program, although it cannot be expected to do everything, can help workers become more committed to safety. For example, the program can:

1. Help to develop safe work habits and attitudes—but it cannot compensate for unsafe conditions and unsafe procedures.
2. Focus attention on specific causes of accidents, although by itself it cannot eliminate them.
3. Supplement safety training, yet it cannot be considered a substitute for a good training program.
4. Give employees a chance to participate in accident prevention activities, such as suggesting safety improvements in job procedures.
5. Provide a channel for communication and cooperation between workers and management, because accident prevention is certainly a common meeting ground.
6. Improve employee, customer, and community relations, because it is evidence of management's commitment to accident prevention (Figure 18–1).

The major objective of a program to maintain interest in safety is to involve employees in preventing accidents. Usually, though, it is difficult to determine the degree of success such a program achieves. This is because companies with these activities also have sound basic safety programs: working conditions are safe, employees are well trained and safety minded, and supervision is heavily involved in promoting safety measures.

However, one company that already had a good basic program attributed a reduction in its work injury rate to stepped-up efforts to maintain interest in safety. It was based on an idea submitted by an employee: Each month candy bars were distributed to injury-free employees. Wrapped with some of the candy bars were slips that could be traded for free pairs of safety shoes. Another company gave each employee who had an injury during the month a package of gum with the slogan "Something to chew on" along with a friendly safety message and wishes for an injury-free future.

One poultry-processing company was able to boost its safety program by giving away "big ticket" items. Prizes ranged from appliances up to an all-expense paid trip to Hawaii for eligible employees and their guests. An expenditure of $40,000 netted the organization a 29% reduction in lost workday cases and a reduction of $450,000 in workers' compensation costs.

SELECTION OF PROGRAM ACTIVITIES

Modern advertising and marketing techniques have much in common with those used to "sell" safety. Just as steady and imaginative sales promotion is needed to sell

Figure 18–1. A sticker like this one (actual size shown here) can be placed on surfaces to remind everyone of a company's continuous efforts to work safely.

most products and services, safety also requires constant, skillful promotion. This approach makes the basic elements of accident prevention easier for workers to understand and accept.

Many safety and health professionals make the common mistake of instituting promotional activities with no preliminary planning or objectives in mind. They may use films, videos, or posters solely because they are available at low cost. These managers are doing their employers and co-workers a disservice. The same is true of safety and health professionals who spend a disproportionate amount of time on committee work, contests, or "homemade" visual aids because of their bosses' or their own personal interests. This haphazard approach cannot be used to develop an entire safety program. Such cosmetic activities are only part of safety in the workplace. They offer no substitute for a sound, well-planned program based on managerial involvement, responsibility, and accountability.

Basis for the Program

To be effective, a program to maintain interest in safety must be based on employer and worker needs. Management should select activities not simply because they will be popular, but because they will yield results. To develop suitable activities and promotional materials, management must know the needs of supervisors and employees.

To find out what employees really thought of their safety program and just how interested they were, one company inaugurated a safety inventory plan. Each year after the regular stock inventory had been taken, safety inventory cards were distributed to all employees—salaried workers as well as hourly.

The cards were distributed by supervisors who asked employees to take stock of their jobs and environment with regard to safety. The following year's safety program was planned on the basis of the questionnaire returns, which ran better than 90%. Many suggestions for improving the safety program were received and subsequently put into practice.

Factors to Consider

Company policy and experience. If a company ordinarily uses activities such as committees, mass meetings, award programs, and contests in areas other than safety, then these should be used for the safety program. It is generally unwise to spend much time on activities that are unknown to company policy and experience, unless it is believed that a new "sales pitch" is justified.

On the other hand, if supervisors and employees are overinvolved in committee work and activities such as sales promotion, quality control, and tool damage programs, similar activities for safety may be lost or considered burdensome. Other approaches might prove more effective. However, if management believes safety is important, effective measures will be taken regardless of the burden.

Once a promotional program is under way, it should not impair other aspects of the accident prevention program or other company activities. For example, safety meetings should not take much more time than meetings for quality control, sales, or industrial relations.

Budget and facilities. Budget considerations always affect plans for a safety promotion program. At first the program will require extra effort, time, and money. The safety and health professional can justify this expenditure, however, as an investment that will produce direct and indirect benefits from both a financial and employee-relations standpoint. If the program is to be successful, management must allocate sufficient funds to carry it out.

In selecting program activities, line management and the safety and health professional should consider the available facilities. Films, for instance, require not only a projector and screen, but also a darkened room free of background noise. In many companies, safety-oriented facilities, publications, or services may be obtained through the industrial or public relations departments. Public relations also may be a valuable source in helping to plan promotional activities. The Safety and Health Department or the Customer Service Department can provide information as well.

Types of operations. The nature and organization of company operations affect the choice of activities and materials for maintaining interest in safety. When operations are widely scattered and diversified, as in the construction, railroad, marine, motor transport, sales, and air transport industries, the job of selecting and disseminating safety information becomes more complicated.

In decentralized operations, the safety and health professional must help management choose materials, such as publications, videos, and films, that can be used easily in the field. Local supervisors must be able to conduct meetings, present material, and handle posters.

The safety and health professional must also consider the needs of employees involved in widely different kinds of work at far-flung locations. For example, a poster program may be used as one means for maintaining interest in safety at each outlying location. If so, the supervisor can designate a trustworthy employee to receive posters and take care of their distribution and posting. This individual can also be

sure that poster boards and display cases are kept clean, attractive, and free of extraneous paper.

Another method is to equip a trailer with permanent displays and transport the company safety story to remote locations. Rear-screen projectors (35mm) and videotapes are also well suited for use in scattered locations and in the field. A relatively new and inexpensive method of communication is the video camera. Videos can be used to communicate a variety of short subjects to distant work sites. This method will ensure that the same message is communicated in the same way to all employees. In addition, video cassette players are now available with a continuous play mode. After a tape is shown, the VCR automatically rewinds the tape and replays it as many times as desired.

For many organizations, the types of educational materials used for different groups of employees may vary considerably. For instance, movies may be suitable for the day shift when many employees can be taken off the job, but would not be appropriate for a small night shift when no one can leave the job site even for a short time.

However, night crews and maintenance employees should not be overlooked. Their work is as vital as that of other employees to the overall accident prevention effort. Programs for them will have to be planned to fit their specific requirements and time constraints.

Types of employees. The types, backgrounds, and educational levels of employees must be considered when choosing safety promotion activities. For example, migrant workers frequently do not receive sufficient job training. Material for these employees should present basic safe practices for their jobs in brief, easily-understood form and in the appropriate languages when their English skills are still poor. Management can use local colleges or universities as a resource to assist in translating material into other languages. Be aware that different dialects, even in the same language, may result in the safety message being misunderstood. A similar approach should be used with temporary workers or those assigned from union halls. Employees who have difficulty understanding or reading English need visual material. Material in Spanish is available from the Inter-American Safety Council (see the Appendix, Sources of Help).

Basic human interest. If employees seem bored or uninterested in safety activities, an extra push is needed to involve them. One approach is to base promotional activities on employee interests, such as bowling or fishing, as an incentive.

The wise choice of these activities depends on an understanding of basic human needs and emotions, such as those listed in Figure 18–2. The safety activities suggested for employees should have general appeal.

Other considerations. The tasteful use of employee human interest materials featuring children, animals, and cartoon figures along with activities and contests fostering good-natured competition play a big part in most companies' safety promotion programs. These elements can be as effective in safety promotion as they are in sales and marketing. Their wise use should not compromise company policy or offend anyone.

BASIC HUMAN INTERESTS AND CORRESPONDING ACTIVITIES

Basic Interest Factors	Ways to Use These Factors
Fear of painful injury, death, loss of income, family hardship, group disapproval or ridicule, supervisory criticism.	**Visual material:** emotional or shocker posters, dramatic films, pictures and reports of serious injuries on bulletin boards, in company papers.
Pride in safe workmanship, in good records, both individual and group.	**Recognition** for individual and group achievement; trophies, personal awards, letters of appreciation.
Recognition: desire for approval of others in group and family, for praise from supervisors.	**Publicity:** photos and stories in company and community papers, on bulletin boards.
Participation: desire to be "one of the gang," "to get in the act."	**Group and individual activities:** safety committees, suggestion plans, safety stunts, campaigns.
Competition: desire to win over others, such as shown in sports.	**Contests** with attractive awards.
Financial gain through increased departmental or company profits.	**Monetary awards** through suggestion systems, profit-sharing plans, promotions, increased responsibility.

Figure 18–2. Ways to put six basic human interest factors to effective use in promoting safety. Basic needs and desires that motivate people are shown on the left. The right column lists direct appeals safety promotional programs can make.

In one instance, an Ohio company capitalized on the interest of many people in wagering. Any employee in a carpool who forgot to buckle his or her safety belt had to pay for lunch or dinner. The idea proved so popular the company decided to expand the program by using posters and announcements.

Ideas for maintaining interest often use humor. The "light touch" is essential and should be good-natured. Ridicule should never be used; it only arouses resentment (Figure 18–3).

A positive, constructive approach is generally better than a negative approach. However, the latter sometimes is preferable if it is more dramatic. A picture showing the consequences of an accident, such as a fall on a slippery floor, may have a greater impact than one depicting the safe practice that could have prevented the accident.

Variety is essential. Often a simple change, such as a different type of contest, a bulletin board redesign, or revising the format of safety meetings, can renew workers' interest. The activity itself may not be more effective, but its new form stimulates thought and discussion. Although safe work practices should become routine, their presentation should not.

Activities that require employee participation generate more interest than do those that involve only seeing and listening. Many companies have worked with the National Safety Council to produce films, videos, and movies in their plants. These firms report an upsurge in safety interest when some employees are asked to act in a video or film emphasizing safety procedures or practices that may have seemed routine.

In some instances, management urges employees to submit suggestions for equipment guards or to help in the selection of personal protective equipment. Companies find that these workers are more inclined to use the guards and the personal equipment than they would be if they had not been asked for their opinions.

For the same reasons, including employees in the process of drafting safety rules encourages compliance on the part of those workers who participate in the project. Companies have also found that workers who serve on a safety and health committee tend to develop increased awareness of safety responsibilities.

STAFF FUNCTIONS

Creating and maintaining employee interest in a line management responsibility falls to the safety and health professional. This individual is responsible for assisting line management in planning the safety program. Line managers and supervisors are responsible for planning, organizing, and executing the program.

Role of the Safety and Health Professional

Safety and health professionals coordinate the program, supply the ideas and inspiration, and enlist the support of management, supervision, and employees. Programs should be designed to involve both management and employees. Employees like to receive awards; managers like the public-relations aspects of presenting them.

Safety and health professionals may work with local safety councils, chapters of the American Society of Safety Engineers, and other civic or technical groups interested in accident prevention. They can gain much by attending the National Safety Congress and Exposition and regional safety conferences. Here they learn what other companies are doing and how those ideas can be translated into practical activities for their own organizations. After participating in roundtable discussions, listening to speakers, and meeting many people with similar interests, safety and health professionals often return to their jobs with renewed enthusiasm.

Because they frequently are called upon to address groups,

Figure 18–3. The return on investment for public safety advertising is high, according to the Salt River Project. This billboard is one of 22 in the area that carry a new message each month and reach an estimated 76% of adult Valley residents. The effort costs between $50,000 and $70,000 and has been in effect 10 years. The advertising contributes to halving of the insurance premium in five years. (Reprinted with permission from Salt River Project, Arizona.)

safety and health professionals should be able to present their ideas clearly, effectively, and convincingly. Polishing public speaking skills will also help when dealing with people on a one-to-one basis. Safety professionals should know when and how to use visual aids.

Showmanship tactics, however, should be used only with considerable discretion. If they backfire, they can hurt the credibility of the safety and health professional, the company, and the safety program in workers' eyes.

Safety and health professionals should help educate line management in two areas. First, managers should learn how to keep working conditions as safe as possible. Second, they must know how to motivate workers to follow safe procedures consistently, as a part of good job performance.

A vast amount of program material useful to the safety professional is available through National Safety Council publications. Safety and health professionals also are invited to submit interesting data, difficult problems, or "gimmicks" of any nature to the Council for help in solving safety problems and to share ideas and solutions with others. The Council has information on every phase of safety, gleaned from the experience of members in various industries. One company's solution to a problem may prove invaluable to other firms.

Role of the Line Manager

The line manager is the key person in any program designed to create and maintain interest in safety. This is because the manager is responsible for translating management's policies into action and for promoting safety activities directly among the employees. How well this responsibility is met will determine to a large extent how favorably employees will view safety activities.

Ranking managers must learn that under existing laws they are directly accountable to their organization and society for their employees' safety. Management, with the safety professional's assistance, must see that supervisors receive adequate safety training.

The supervisor's attitude toward safety is a direct reflection of upper management's attitude towards safety and is a key factor in the success not only of specific promotional activities but also of the entire safety program. Supervisors look to upper management and employees look to the supervisor for leadership. Line managers, including supervisors, who are sincere and enthusiastic about accident prevention can usually maintain employee interest. Conversely, if line managers give only lip service to the program or ridicule any part of it, their attitude can cancel any good that the safety and health professional may be able to accomplish.

Many supervisors are reluctant to change their modes of operation or to accept new safety engineering ideas, much less to enthusiastically support contests, safety stunts, committee projects, and other activities used to promote and maintain interest in safety. It is line management's task to sell these supervisors on the benefits of accident prevention. They

must be convinced that promotional activities are not "frills" but important projects that can help to prevent injuries. Line management's wholehearted cooperation is essential to the success of the entire program. It can be pointed out that a successful program makes management's job easier and less time consuming.

One way line managers can support safety is by setting a good example and educating their workers in safe practices. They must follow safety rules and procedures, and wear safety glasses and other personal protective equipment whenever they are required. Teaching safety is also an important function of supervisors. To be successful in this area, they cannot depend upon safety posters, a few warning signs, or even general rules to do their teaching for them. Supervisors themselves must first be trained to teach if they are to be competent in this area. (See Chapter 16, Safety Training, for details on training courses and techniques.)

When enforcing safety and health rules, line managers should not suddenly adopt a "get tough" approach after being lax. They should be consistently firm and fair all along. Otherwise, workers who have the impression the supervisor cannot recognize unsafe conditions and practices or simply does not care will not take the new tougher line very seriously.

Supervisors should be encouraged to take every opportunity to exchange ideas on accident prevention with workers, to commend them for their efforts to do the job safely, and to invite them to submit safety suggestions. They can also be most effective in relaying personal reminders on safety to employees. Such reminders are particularly appropriate in the transportation and utility industries, where crews are on their own from terminal to terminal.

Line managers, including supervisors, should request and receive help from the safety department. Help may also come through correspondence, educational materials for distribution, and personal visits from safety personnel. Supervisors should also receive adequate recognition for independent and original activity.

SAFETY COMMITTEES AND OBSERVERS

Various types of safety and health committees have many different functions. (For further details, see the Council publication, *You Are the Safety and Health Committee.*) However, the basic function of every safety committee is to create and maintain interest in safety and health and thereby help to reduce accidents.

Some organizations prefer other types of employee participation to formal safety committees. This is because they feel safety and health committees require a disproportionate amount of administrative time, generally tend to pass the buck, may stir up more trouble than they are worth, and may be a scapegoat for supervisors who want to unload their responsibilities.

The answer to these objections is not to abolish the committees but rather to reexamine their duties, responsibilities, and methods of operation. Such analysis often can lead to constructive changes enabling a committee to fulfill its original objective—that of stimulating and maintaining interest in safety.

Committee membership should be rotated periodically. This ensures a fresh perspective and also increases the number of employees who are trained to look at operations with safety in mind.

Involving employees in safety inspections, either alone, as observers, or as part of a formal safety and health committee has the same basic objective: to get more employees actively involved and interested in the safety and health program. Planning, organizing, publicizing, and following definite procedures will streamline the work of both committees and observers and help ensure effective results.

QUALITY CIRCLES AND SAFETY CIRCLES

Hazard recognition and control is one area in which employee involvement produces substantial results. The following discussion is taken from the Indiana Labor and Management Council's booklet *Worker Involvement in Hazard Control* (see References), used with permission.

There are two reasons for the popularity of involving employees in hazard recognition and control. The first is management's desire to use all available resources to increase productivity and quality in the face of growing competition. The second is management's understanding that employees want to accept new challenges and to participate in activities that affect their work life.

Popular forms of worker involvement programs are quality and safety circles. A quality circle is a group of employees, performing similar work or sometimes varied work, who meet weekly. In the meetings, they learn about and apply basic techniques to identify problems within their area(s), analyze them, and recommend solutions to management. In some instances, circle members discover hazards during their analysis of other plant problems and make excellent recommendations for their control.

Safety circles are a type of quality circle used by some companies to reduce the number of accidents and injuries by keeping safety and all its important features foremost in the minds of the employees. This implies a change in the employee's role from passive to active, while management's role becomes less negative (fewer don'ts) and more positive (more do's).

In many situations, safety circles are established on a plant-wide and departmental level. Safety circle meetings are held monthly for approximately one-half hour to one hour. Each circle meeting is usually preceded by a presentation explaining the successes and problems that the safety circle team experienced during the preceding month. The circle reviews a breakdown of all injuries, including first aid cases, as a means of measuring safety progress and as a means of pinpointing trouble areas. Usually each member of the safety circle is assigned a specific responsibility to review. Companies employing the safety circle concept report an improvement in their accident and injury experience.

MEETINGS

Safety and health meetings may be conducted for managers, supervisors, employees, or other groups. In every case, the purpose of the meeting is to stimulate and maintain interest in safety and health issues. If meetings fail to achieve this, their format or content should be changed to make them effective or they should be discontinued and a new approach taken.

A formal agenda for all attendees should be distributed prior to the safety meeting. An agenda helps to focus attention on specific issues, and avoids wasting time on nonrelevant topics.

Types of Meetings

The following types of safety and health meetings commonly arouse and maintain interest in accident prevention (see also Chapter 16).

1. Meetings of operating executives and supervisors to formulate policies, initiate and maintain a safety and health program, or plan special activities.
2. Mass meetings of all employees, sometimes including families, or even the entire community to serve special purposes such as launching a major new program or contest and "selling" safety to everyone affected by accidents and injuries.
3. Departmental meetings to discuss special problems, plan campaigns, or analyze accidents.
4. Small group meetings to plan the day's or week's work so that it can be done safely, to discuss specific accidents, or to review safety instructions.

Meetings of executives. When a safety and health program is inaugurated, it is especially important that the chief executive officer (CEO) of the company or top plant manager should call a meeting to announce the general accident prevention plans and policies to all line managers and other operating executives. Under some labor contracts this must be, or should be, a joint union-employee announcement. If these persons meet at regular intervals to discuss operating problems, this announcement can be made at a regular gathering. Otherwise, the CEO or manager should call a special meeting. Periodic safety inspections by executives help to lend credence to management's interest in safety.

After this first meeting, the group may hold sessions periodically (usually monthly) to evaluate and coordinate the safety program, to check on the progress being made in accident prevention, to appraise proposed activities, to set policies, and to make decisions. It is also desirable for this group to review and/or investigate all fatalities, multiple amputations, and other serious injuries.

Departmental meetings serve many safety and health purposes. They may be used to discuss the company safety program so that employees will better understand its policies and procedures, to provide information about accident causes and accident types, or, in a purely inspirational manner, to create an awareness of hazards and a desire to prevent accidents. Departmental meetings also can be used to review and/or investigate all injuries involving lost workdays or restricted work.

Many departmental safety meetings are held monthly and most are conducted by the supervisor. The safety department usually assists in planning and provides materials, such as visual aids.

The program for a departmental meeting may include:

1. Reports on injuries in the department since the last meeting, a safety inspection in the department, and the department's standing in a contest. (The total time spent on reports should be limited so that this part of the meeting does not become tiresome.)
2. Discussion by the supervisor of specific safety practices or unsafe conditions that need to be improved.
3. Talk, demonstration, or audiovisual presentation on an appropriate accident-prevention subject. The speaker may be the supervisor, a department employee, the company safety and health professional, an outside expert, or a company executive.

Departmental meetings give the supervisor an opportunity to point out the dangers of particular unsafe practices or conditions. By condemning those practices, the supervisor sets a good example and lets workers know they are to follow the same rules and procedures. Most workers welcome an opportunity to share their safety ideas in these meetings.

At the conclusion of departmental meetings, the supervisor may prepare written reports for the plant (or company) safety committees and managers.

Supervisors also can hold small group meetings with people doing similar work at or near the workplace. The supervisor may discuss the causes of a recent accident workers have witnessed or learned about. Employees should be encouraged to join in the discussion with the goal of reaching some conclusions about how the accident might have been prevented.

The supervisor may present a problem that has developed because of new work or new equipment. Again, all should participate and offer their views.

At times the supervisor may present a film or chart talk on a subject related to the work of the group members. Other audiovisuals such as models or exhibits may be used. Safety devices or pieces of equipment or material may be brought in and discussed.

"Production huddles" are instruction sessions about a specific job that include safety information. Such meetings are particularly useful with maintenance crews when an unusual job is about to start. The plans for doing the job safely and efficiently are discussed and a procedure is agreed upon. Public utility line crews who use this type of meeting call it a *tailboard* or *tailgate conference*. Before starting a job, the crew gathers around the truck and discusses the job, laying out the tools and materials they will need and agreeing upon each person's tasks.

A particular advantage of small group meetings is that they provide excellent opportunities for presenting all types of information, including safety information, directly to employees. They also stimulate an exchange of ideas that

can benefit the accident prevention program. To be successful, each safety and health meeting must have a tangible message, imaginative presentation, opportunity for audience participation, and a conclusion that spurs action toward an attainable goal.

Mass meetings. Mass meetings are held for special purposes, such as launching a contest or an award program, presenting awards, introducing new equipment, explaining a change in company policy, or celebrating an exceptionally fine safety record with an event such as "safety day."

In companies with plants in different cities, a top executive may call a meeting of employees during a plant visit. The talk may cover safety as well as other subjects. One company president makes an annual round of plants with the safety director and speaks at a safety rally of all employees at each plant.

Under certain conditions, particularly in smaller communities, large meetings can be held in a local theater or public hall. These meetings require fairly elaborate arrangements and considerable publicity to ensure good attendance on employees' part.

A mass meeting in a public hall using the "family safety night" theme allows not only employees to attend but also their families and friends. In addition to a presentation or speech targeted at the program theme, management should provide for some other entertainment. Often good talent can be found right in the plant or shop.

Mass meetings afford an excellent opportunity to use an outside speaker who can talk with authority and in a crowd-pleasing manner on general accident-prevention work. Such speakers can be found in nearby plants, insurance companies, city administrations, automobile clubs, or community safety councils.

If management wants to include movies, videocassettes, or slide shows relating to accident prevention at the meeting, they can find a suitable selection from those available through the National Safety Council or regional film service organizations, whose locations may be obtained from the Council.

Planning Programs

Making the safety meeting interesting is of the utmost importance. Speakers should not complain about or scold workers on their job performance. Talks should be brief and start and end on time. The subject matter should be studied in advance to make sure it is pertinent and is not a repeat of other recent talks.

Large occasional meetings need even more preparation and timing than do small meetings. Speakers, including company executives, should review their remarks with the meeting planner to ensure they will add to the desired purpose. Films, videos, and other visuals should also be reviewed in advance.

Persons responsible for employee meetings should analyze them periodically to determine whether they are accomplishing their purpose. It is all too easy for meetings to degenerate into dull routine. Only continual effort and planning will prevent this trend.

A plan of action to develop a successful safety and health meeting includes these points:

- Prepare in advance. The preliminary arrangement determines the results. Do not conduct a meeting without preparation.
- Select a major topic. Make it timely and practical—one that the group can discuss.
- Obtain facts and figures. Be sure they are correct and complete. Prepare a visual, such as a simple chart or table, whenever possible.
- Map the presentation. Decide on the best way to present the subject of the meeting. Try to anticipate the group's reaction and questions. Outline results you hope to accomplish.
- Set a timetable. Allow adequate time, but set a reasonable limit.
- Have an agenda distributed in advance.
- Be sincere. Managers' sincerity and interest in employees' welfare must be unmistakable.
- Introduce the topic. Tell in simple terms what the meeting is all about. Use a punch line or other short, to-the-point lead-in.
- Present facts, arouse interest. State pertinent facts in an interesting manner.
- Promote group discussion. Ask questions that cannot be answered "Yes" or "No." Prompt members of the group to think individually and collectively. Let them talk.
- Agree on some action. Try for group agreement on methods of correction and improvement. Write these down.
- Summarize the meeting. Review briefly what has been discussed and decided. Follow up in the various departments in writing, if possible.

CONTESTS STIMULATE INTEREST

Management and supervisors should always keep in mind that contests (such as housekeeping contests, interdepartment contests, etc.) are not substitutes for management interest, safe procedures, and "built-in" safety. Nevertheless, although a successful accident prevention program depends on good management, good training, and efficient operations, some special effort may be needed to maintain worker interest in safety. Moreover, the interest value of contests has direct bonus values in good publicity and improved employee morale.

Therefore, a competition should be held *only* after management has taken the basic steps in a safety program: developed a policy statement, adopted a recordkeeping system, safeguarded equipment, and installed a first aid department. Such substantial demonstrations of management's interest, sincerity, and responsibility greatly help win the active participation of supervisors and workers in a contest. However, a contest serves only as part of a safety program, not as a substitute for one.

Purpose and Principles

Safety contests are operated purely for their motivational value. A contest that does not motivate workers to be more

safety minded is worthless. In most contests, the competing groups are departments of the same plant or divisions of the same company. This type of contest is useful when investigations of accident injuries reveal that most of them are caused by employees' unsafe practices. If accidents are the result instead of unsafe conditions or a combination of unsafe practices and unsafe conditions, the contest should be reconsidered. They might well focus negative attention on one of a company's most conscientious, safety-minded employees injured through no fault of his or her own.

Generally, contests are based on accident experience and are operated over a stated period. A prize is offered to the group attaining the best record according to the contest rules. Contests have been important almost from the time of the first safety programs. Over the years, companies have developed a fairly well-established set of operating principles:

1. A contest should be planned and conducted by a committee representing all competing groups.
2. Competing groups should be natural units, not arbitrary divisions.
3. Methods of grading must be simple and easily understood.
4. The grading system must be fair to all groups.
5. Awards must be worth winning and generate interest among employees.
6. Good publicity and enthusiasm are important.

Contests may run for various periods—from a few months to a year. Those who recommend longer periods believe that if workers are kept on their toes for a longer time, safe working practices are more likely to become a habit. Some safety professionals, however, believe that greater interest can be aroused and maintained during a short period and therefore prefer short and more frequent competitions. Management should devise contests of different lengths to see which is most effective.

A safety contest stock certificate idea was developed by the Maxwell House Division of Kraft General Foods and ran for one year. For each week of the contest that a department worked without a disabling injury, employees received a stock certificate worth 50 cents. Dividends were paid on this stock at the rate of 10 cents for the first 1,000 consecutive safe hours worked; 25 cents for the first 10,000 consecutive safe hours worked; 50 cents for the first 50,000 consecutive safe hours worked; and one dollar for the first 100,000 consecutive safe hours worked. This meant that if a department worked 100,000 consecutive workhours without a disabling injury it received a total of $1.84 in dividends for each share of stock held.

There were penalties, however. If a disabling injury occurred in a department, it was penalized one month's stock earnings. This meant that during that particular month the department could not be awarded any stock or dividends. It also meant that if an accident should occur in the latter few days of the month the department would lose all dividends and stock certificates previously issued for that month.

Injury Rate Contests

In a contest based on injury rates, the measure of safety performance is the Occupational Safety and Health Administration (OSHA) incidence rate (described in detail in Chapter 13, Recordkeeping and Incidence Rates).

Injury rate contests carry with them an inherent risk of possible abuse and must be closely monitored to ensure honesty. All injuries must be reported. These contests tend to put peer pressure on employees not to report their injuries, to bring administrative pressure to bear on those doing the recording not to count certain injuries, and to tend to make management focus attention on the contest rather than on the safety program. If these conditions arise, the entire safety program is discredited.

Contests should not be based upon injury severity, because severity data cannot be determined promptly. Also, the classification is frequently a matter of chance and contributes little to devising accident prevention measures. Nor is using a combination of frequency and severity classifications helpful, for the same reasons. Finally, contests should not be based on reducing the number of reported first aid cases; people will simply stop reporting such injuries.

In-plant contests. To be effective, a contest must run for a specific length of time. The contest will be more successful if the campaign is launched with advance publicity and fanfare. Often the president or other high official makes the original announcement, presents awards, and otherwise lends prestige to the contest.

Competition, if properly organized, can do much to develop teamwork in safety. Some employees who may give little thought to their own work habits can be influenced to cooperate with their fellow workers if they know that their actions can either discredit their department or "team" or result in a payoff for safe work.

Council award programs. The National Safety Council firmly believes in the value of award programs for maintaining interest in employee safety and health. The award programs are available to employer members, and some include participation of nonmembers through cosponsor agreements with trade associations.

The Occupational Safety/Health Award Program is a noncompetitive recognition program. Awards can be earned by organizations achieving a perfect safety record—no deaths or cases involving days away from work. Organizations with nonperfect records may also earn recognition if they achieve the criteria for a significant reduction of their incidence rates. There are four levels of awards:
- AWARD OF HONOR
- AWARD OF MERIT
- AWARD OF COMMENDATION
- PRESIDENT'S CITATION LETTER

In addition, participants earning one of the above awards for the entire organization may also earn the Council's highest award, the Distinguished Achievement in Occupational Safety and Health Award. Only three of these awards are given annually, one each for small, medium, and large-sized employers.

The Occupational Safety/Health Contest is a competitive award program. Participants are able to compete in one or

more of 20 different industry contests. Each industry contest is subdivided into divisions according to the products being produced. Whenever possible, organizations of like size are grouped together.

Awards are given based on rank within each division. The site having the lowest incidence rate for cases involving days away from work is ranked first. First, Second, and Third Place awards can be earned.

The Council designated the Safe Worker Award Program to reward and encourage the accident-free performance of individual employees. Each employee is awarded a Safe Worker lapel pin or other incentive for each 12-month period during which no cases of days away from work are incurred. The employee continues to earn an incentive item for each additional year of safe work achieved, striving for the pinnacle of 45 years.

The National Fleet Safety Contest is a competitive award program comparing fleets from the same industries, having similar vehicles, or of the same size. The program is open to all National Safety Council members and members of co-sponsoring associations whose fleets operate in the United States and Canada.

Data from contestants are collected on a monthly basis from January 1–December 31 of each year. This data includes the cumulative number of miles driven, the cumulative number of accidents experienced, and the number of vehicles in the fleet. From this data, an accident rate is developed and used as an intitial means of comparison.

Outlined by the National Fleet Safety Contest Rules, the number of fleets within the following awards are presented: First Place plaques, Second Place plates, or Third Place plates. Two noncompetitive awards, Perfect Record and Certificate of Achievement, can also be earned by participating fleets.

The National Safety Council Safe Driver Award is the recognized trademark of professional drivers designed to prevent accidents and promote safety in the professional driving industry. Member organizations of the National Safety Council enrolled in the Motor Fleet Safety Service, or the School Transportation Safety Service, are eligible to participate in this program.

Companies certify drivers who have driven a continual number of years without a preventable accident. Through this certification, the driver receives a wallet-sized certificate denoting the number of years of safe driving service. Other incentive items, such as lapel pins and patches, may be purchased to signify this accomplishment. (See page 425.)

The Safe Driver Award rules spell out what is expected of professional drivers in the way of safety performance—driving a motor vehicle without having a preventable accident. The award rules constitute a yardstick by which drivers can measure their own performance and by which supervisors can meassure the performance of individual drivers.

The 'XYZ Company' Contest Explained

The "XYZ Company" is engaged in steel fabrication and erection. Rules for one of its safety contests are given in Figure 18–4, but some explanation is needed.

Rule 1. Although contests can run for any period of time, the company chose a one-year period. This was done for two reasons. First, the period allows development of safe working habits—the useful objective of the contest. Second, because some of the departments were relatively small, the one-year period eliminates random (chance) factors from influencing an individual department's experience record.

Rule 2a. Hazards in steel fabrication and erection vary greatly; therefore, fabricating units compete in one division and erection units in another. Operations are similar enough to have a common basis for determining standings—see Rule 4.

Other companies may find that hazards differ sharply from one plant or operation to another, and the similar units may be too few to group. Such plants or operations may compete on an equitable basis in several ways.

- The participant achieving the largest percentage reduction in its frequency rate in comparison with a base period, such as the previous year, may be the winner. Because each unit competes only against its past record, this method provides a fair basis for comparison of rates.

- The workers' compensation insurance rates for different types of units in the same state have been used to establish a handicap factor to compensate for differences in hazards. If the rates per $100 payroll are $3.00 for Plant A, $2.00 for Plant B, and $1.50 for Plant C, factors for the units are 3, 2, and 1.5, respectively. The frequency rate of each plant is adjusted by dividing the rate by its factor. Plants are ranked from the lowest to the highest on the basis of the adjusted rates.

- The national average rates for different types of units may be used similarly for determining standings. If Plant A achieved a frequency rate of 10.0 and the national average for units in this industry was 12.0, Plant A's rate is 0.83 of the national average. Plant B's rate in comparison with its national average rate is 0.75. Therefore, Plant B would rank nearer the top than Plant A. Average rates for most industries are published annually in the National Safety Council's *Work Injury and Illness Rates.*

Rule 2b. Separation of large and small units is essential. Generally, a small unit finds it easier to go through an entire contest period without a disabling injury than does a larger unit. If the number of units is sufficient, management can set up three size classifications. Unequal size groups can compete if they include units that can attain no-injury records. For this reason, the erection departments of the company in this example were divided into Group A—the five largest departments—and Group B—the remaining units. Five contestants in a group are usually considered minimum.

Rule 3. Awards should be specified. "XYZ Company" follows the general practice of giving first, second, and third place awards.

Rule 4a. The frequency rate is most often used as the basis for determining standings because it can be computed promptly and easily.

Rule 4b. There should be a satisfactory method of determining the winner between two or more units having perfect records to give the smallest contestant a fair chance. For

---

RULES FOR THE 'XYZ COMPANY' SAFETY CONTEST

Rule 1. The contest shall begin January 1, and end December 31, 19. . . .

Rule 2. The contest shall consist of two divisions.
   a. Fabricating units shall participate in Division I.
   b. Field erection departments under the direction of a superintendent shall participate as separate units in Division II, which shall be divided on the basis of size into two groups: Group A shall consist of the five erection units working the largest number of man-hours, and Group B shall consist of all other units. Units shall be tentatively grouped by size during the first three months, and the final classification shall be made on the basis of total man-hours worked at the end of four months. No further changes in size groups will be made after April 30, 19. . . .

Rule 3. Recognition awards shall be:
   a. Trophies to the winners in Division I and Groups A and B of Division II.
   b. Engraved certificates to plants and erection units ranking second and third in Division I and in Groups A and B of Division II.

Rule 4. a. The winners shall be the contestants having the lowest weighted frequency rates.
   b. In the event that two or more contestants in any classification have had no chargeable injuries during the contest period, the winner shall be the contestant who has worked the largest number of man-hours since the last chargeable injury.

Rule 5. All injuries and illnesses resulting in death or days away from work shall be counted, as defined by OSHA record-keeping requirements.

Rule 6. A sum of $50.00 shall be presented to the units that have had no disabling injuries during the first six months of the contest or have reduced their average frequency rates for the first six months 50 per cent in comparison with the average rate for the preceding six months' period. The award shall be divided into prizes of $12, 10, 8, 6, 4, 2, and eight $1 prizes and raffled to employees. No employee may win more than one prize.

Rule 7. Standings shall be compiled monthly and published in a bulletin that will be distributed to all managers, superintendents, and foremen.

Rule 8. All questions pertaining to the definitions of injuries and rules shall be referred to the Contest Committee, whose decisions shall be final.

Rule 9. Awards shall be presented at an appropriate ceremony to be announced at the end of the contest.

Figure 18–4. A typical set of rules for a safety contest. See text for a point-by-point discussion.

example, selecting the winner on the basis of the largest number of employee-hours worked since the last chargeable injury is fair to all units, regardless of size.

Rule 5. A standard method of counting injuries is essential to avoid controversies and maintain confidence in standards. Use the incidence rate described in Chapter 13, Recordkeeping and Incidence Rates.

First aid and other minor injuries should be excluded for purposes of the contest. If they are counted, workers may fail to obtain treatment so that cases will not be recorded. The result could be an increase in infections.

Rule 6. Various "special rules" may be used in a contest that runs six or more months to maintain employee interest.

Rule 7. Frequent contest bulletins keep everyone informed about standings. Supervisors can post bulletins and discuss standings at safety meetings. Safety personnel can announce contest results in plant publications and in other ways to build and maintain interest.

Rule 8. Questions about rule interpretation will arise and must be settled fairly. This is an important function of the contest committee, which, in turn, may refer decisions about

disputable injuries to outside judges. A committee of the American National Standards Institute has been set up to interpret standard definitions.

Rule 9. (See the section Meaningful Awards, later in this chapter, for award ideas.)

It is important that competing plants or groups be informed of the latest findings. Figure 18–5 shows a typical contest bulletin.

Interdepartmental Contests

Interdepartmental competitions get "close to home" and place responsibility for a good showing on supervisors. Because workers have a greater personal interest in the standing of their own department than in the record of the entire plant or other unit, this type of competition has proved popular for creating interest in safety among both supervisors and workers.

An interplant contest plan often may be adapted to an interdepartmental competition. A company operating a number of similar facilities may take advantage of workers' interest in their departmental records by conducting a

---

STANDINGS IN THE 'XYZ COMPANY' SAFETY CONTEST
JANUARY–JUNE

Oakland leads at the halfway mark!

Tulsa moved into second place.

Chicago slipped from second to third place in June.

Corbin, in last place, had no disabling injuries during June. Good work!

Tulsa still leads for the President's Award for largest improvement over the previous year's record.

Plant	Rank	January–June Incidence rates*	% Increase + or decrease – over last year
Oakland	1	2.8	−30%
Tulsa	2	2.9	−40%
Chicago	3	3.5	+ 5%
Cincinnati	4	4.4	−20%
Corbin	5	5.0	+22%

*Incidence rate is number of cases resulting in death or days away from work per 200,000 employee-hours.

Tips on How to Win, No. 6

One-sixth of our disabling injuries occur in the use of cranes and hoists. Have foremen hold a safety meeting on safe practices in hitching loads and other crane operations for shop men. We'll furnish a film. Use posters on the subject. Require safe methods. See the enclosed bulletin for suggestions on how to solve this major accident problem.

Figure 18–5. Simple monthly contest bulletin gives essential information about current standings and also includes a suggestion for improvement in "Tips on How to Win."

competition among the same departments in various organizations. Public utilities may conduct a contest among districts, and other nonmanufacturing companies may follow a similar plan. Because hazards in similar operating units are about the same, the basis of standings may be the frequency rate. Departments or other units may be grouped according to size.

Most departmental contests, however, are conducted among dissimilar departments in one plant; the departments often differ greatly in size. The difficulties due to variations in hazards and numbers of employees from one department to another may be overcome using the same techniques discussed above regarding differences between plants.

Always measuring the numbers of accidents of this year versus last year can have unwanted results. *Too much* emphasis on negative measurements can result in underreporting because participants do not want to look bad or lose the contests.

Another approach can be considered; that is, measuring the success of a safety program by using *positive* numbers. Here are a few examples:

1. Count the number of safety and health suggestions submitted by a department and count the number of positive ideas and corrections made because of these suggestions. This effort could be tied into production and quality suggestion programs.
2. Give credit for accident investigation reports that are submitted complete, on time, and with positive recommendations for preventing an incident from happening

again (avoid accepting reports recommending "I told Joe to be more careful next time!").
3. Encourage employees to report all accidents, including near-accidents, first aid cases, property damage without injuries, and accidents without injuries.
4. Give credit for safety meetings that are planned, outlined, and run by supervisors and managers.
5. Use safety and health inspections in a system that counts the number of corrections that were made because of these inspections.

Intergroup Competitions

Intergroup contests are particularly suitable for units that employ fewer than 400 people and have small departments in which hazards vary sharply. Employees are divided into teams of from 20 to 50 workers. To equalize factors of size and difference in hazards, each team has a proportionate number of employees from the most and least hazardous occupations. Each group is led by a captain, whose principal duty is to contact members of the team and create interest in winning. Team members may be members of contest committees.

Interest is often promoted by naming teams after prominent baseball, football, or other outstanding sports teams, and the entire competition may be named after a league or other sports organization. Workers can draw their team names from a hat and show their membership by wearing colored pin-on buttons. Supervisors can place contest signs in competing departments.

Figure 18–6. A National Safety Council wallet certificate designating the number of years of safe driving service would appeal to the basic human feeling of pride in a good safety record.

Using colored buttons to identify workers helps supervisors or others to keep score more accurately. Because members of different teams often work together, the buttons make it easy to ensure that every case involving a competitor is charged against that group's record.

Intraplant or Intradepartmental Contests

With set standards of performance, an intraplant or intradepartmental safety contest helps to emphasize the supervisor's and employees' responsibilities for avoiding accidents. These standards often include no-injury records for varying periods, achievement of lower injury rates compared with a previous period, and improvement over the average injury rates of similar units or of the industry. In this type of contest, units of an organization do not compete with one another; rather, each unit attempts to match or surpass the established standards.

Personalized Contests

Safe driver award program. This award program is the recognized trademark of the professional driving industry for those who have proven their skills as accident-free drivers. Over 6 million drivers have earned this prestigious award since 1930. Enrollment is open only to National Safety Council members receiving the Motor Fleet Safety Service. Employees of these eligible organizations must be certified by their company in order to receive wallet certificates designating the number of years of safe driving service (Figure 18–6).

Safe worker award program. Recognize your safe workers with distinctive incentive items that encourage continued safe performance. Recognition items indicating one or more years of accident-free performance are part of the

program. All full-time employees of Council member organizations are eligible to participate in this program. Registration for the program is available through the Council's Safety and Health Motivation Programs Department.

Some plants single out for special recognition employees who have safe records. Managers give certificates to those who have worked one, two, five, and ten years without an injury. Holders of certificates have found them useful in obtaining promotion and even in seeking employment with other organizations.

Periodic raffles of merchandise or cash, or the use of a new car for three or four months, regardless of the injury record of a department or plant, also have been used successfully to encourage and acknowledge the efforts of safety-conscious workers. Employees who are involved in disabling accidents (or drivers who have had a "preventable accident") during the period become ineligible for drawings.

Various sweepstakes plans have proved popular and effective in maintaining interest in a good record from month to month. One such plan was operated successfully by a branch plant of a well-known paper company. Here's how it worked.

On the payday prior to the beginning of each month all hourly employees received cards with serial numbers at the pay office window. The workers wrote their names and departments on the cards, tore off the stubs and put them in a box, and retained their portions of the cards. Names of employees in departments having no disabling injuries during the month then were drawn for prizes ranging from $5 to $25. The company contributed a total of $75 per month.

Since eligibility for the drawing depended on a perfect departmental record, each worker had to be careful about all job-related actions. Workers frequently corrected others for unsafe practices that might spoil chances for prizes.

If no accidents occurred during a three-month period, supervisors participated in a drawing for their own prizes ranging from $5 to $15. This feature proved helpful in enlisting their cooperation.

Some companies give awards like first aid kits or trading stamps to those in every department who have worked during a given period without an accident. This approach is effective only if going a month, for instance, without an accident is unusual.

Overcoming Difficulties

Although there are several difficulties in operating contests, they can be overcome rather easily. One of the most common difficulties is that some departments are inherently more hazardous than others. To overcome this problem, management can establish handicaps based on annual rates set by insurance companies or on average accident frequency for the different kinds of work.

Another method is to base standings on improvement over past records. Thus, a department with a past average accident rate of 20 and a current rate of 15 would be acknowledged as achieving a 25% improvement. They would win over a department with a past frequency of 15 and a

current frequency of 12—only a 20% improvement. Usually, in both methods an average of rates over three to five years is used as the base.

Another problem is that a department may have so many accidents at the beginning of the contest period that it is out of the running and loses interest. Having shorter contest periods helps to overcome this difficulty. Another remedy is to have different awards for different achievements. An award for the department having the longest run of injury-free workhours is one example.

Division of responsibility for the cause of an accident may become a problem. For example, the unsafe practices of one supervisor's worker (or an unsafe condition in this supervisor's department) can result in an accident involving an employee from another department. This would affect another supervisor's record. These situations must be anticipated and the contest planned to deal with them fairly.

Noninjury-Rate Contests

The use of noninjury-rate contests is gaining favor among safety professionals to motivate program performance. They are used to target problem areas and are not subject to the same potential misuse as are injury-rate contests.

For example, an employee contest can be based on the safe worker. The contest uses either a monetary award or a status award, such as a drawing for a television set or a private parking space. All employee names are included in the competition. During the contest period each worker spotted performing an unsafe practice or not wearing required protective equipment has his or her name removed from the contest. To maintain credibility, those experiencing recordable injuries would also be eliminated.

Supervisory contests, such as a "Supervisor's Safety Club," generally are based on how well supervisors achieve program objectives. These objectives range from the percentage of a supervisor's employees receiving safety talks to the number of self-inspections conducted and deficiencies promptly corrected. Here again, to maintain credibility the recordable injury of an employee would remove the supervisor from the contest. Awards range from steak dinners and sports events to a Supervisor's Safety Club jacket. Managers overseeing several supervisors or departments would have similar competitions and similar awards.

Contests at the plant level generally cover multiplant performance in a corporation. In addition to how well corporate safety program objectives are met, injury experience should be included in the contest criteria. Such contests usually have a corporate trophy to be displayed at the winning plant rather than providing any monetary consideration.

Other noninjury-rate contests include safety slogan, poster, housekeeping, and community contests. No matter what the contest, the objective is to get the maximum number of people talking, thinking, and participating in safety.

Slogan, limerick, and poster contests. Safety slogan contests vary. One can be for the best safety slogan submitted by an employee. Another may ask the employees or their spouses if they can repeat the "slogan of the week" or the message on a certain safety poster.

Company magazines may conduct contests to "finish the limerick" or "write a rhyme" or write "twenty-five words on the best way to be safe." Often these are open to both employees and family members.

The value of homemade posters is in their special application to a particular industry or company. If the posters are the result of an employee contest, their motivational value will be increased, possibly exceeding that of the industry-made variety of posters. An important ingredient of such a contest is to get employees and their families participating in the planning and judging stages as well. Frequently, employee poster ideas—aside from the quality of the artwork—are so good that companies even submit the winning contest entries to the National Safety Council for possible conversion into printed safety posters.

Departments often conduct housekeeping contests. This type of competition is aimed at fundamental causes of accidents and usually tries to eliminate unsafe practices and conditions. Housekeeping contest plans differ from one company to another. The following plan has been used successfully by a metals firm.

Once a week a committee of three management representatives inspects each department and reports unsafe conditions to the superintendent. A copy of the report is furnished to the works manager, and another is kept for the use of later inspection committees. A demerit for each unsatisfactory condition is charged to the department. If the condition is not rectified within one week, an additional demerit is added.

At the end of the month the demerits are totaled, and departments are rated on the basis of the proportion of demerits to the total number of employees in the department. If Department A employed 175 people and had 25 demerits, its rating would be 85.7. This figure is obtained by dividing 25 (number of demerits) by 175 (number of employees), multiplying by 100, and subtracting the product from 100. Standings are posted monthly on the bulletin boards in each department.

Awards are made at a mass meeting of the employees and guests. Names of employees in winning departments are placed in a box from which is drawn the name of the winner for the month. The winner's picture is posted on a special bulletin board, and a short talk on safety by a representative of management is broadcast throughout the plant. (A general rule prohibits an employee from winning more than one award during the contest.)

The name of the winning department is inscribed on a plaque each month. The head of the winning department receives the plaque from the previous month's winner at the mass meeting. At the end of the year, the department that has won the plaque the greatest number of times receives it permanently.

Community and family contests. Many companies have stimulated interest by sponsoring safety essay or poster contests for children of employees, local school children, or young art students. The publicity before and after such

contests, plus the interest generated by the posters themselves and the judging, not only stimulates the interest of employees, but also promotes the company's community and public relations.

More than one company has launched a safety poster or essay contest for the children of employees. Management knows full well that employees will give their children considerable help and that there will be much favorable discussion about the contest in lockerrooms, lunchrooms, and car pools.

To promote greater safety at railroad crossings, Texas railroads sponsored a contest at the Texas Press Association Convention. The journalist who made the closest guess of the total number of grade crossings in Texas won a prize. Because of the contest, safety at grade crossings received considerable attention in the press.

Miscellaneous contests. There is an endless number of different types of contest possibilities. Often they can be combined with injury reduction contests. Contests can be held for attending safety meetings, for wearing safety shoes, for reporting unsafe conditions or unsafe practices, or for off-the-job or public safety activities of individuals, departments, or branch plants.

Safety contests are popular with both management and employees. However, the safety and health professional should always determine before recommending one whether the contest will take valuable time and effort away from providing safer equipment or better training for supervisors and employees.

Contest and Other Publicity

Management should publicize all stages of a contest as fully as possible. It can announce the contest through placards and stories in employee newsletters. Special signs, banners, or posters should publish standings at frequent intervals. Supervisors can hand out bulletins to employees urging them to keep the record perfect. Trade journals like *Safety and Health* and National Safety Council newsletters are other publicity outlets. In a smaller community, outstanding safety performance by a well-known company deserves—and usually gets—excellent publicity in the local newspaper and on the radio station.

The publicity value of a successful contest is considerable, although difficult to estimate. It should be commensurate with the significance of the occasion.

The presentation of an award to an employee who has gone 25 years without a disabling injury has human interest value for both internal and external communications resources. Recognition of an exceptionally fine "no-injury" record by a plant or company, or presentation of a National Safety Council award, would also call for widespread publicity.

Some companies purchase radio or television time to announce the results of a contest. One company had large campaign-type buttons made, and photos of children wearing them appeared in the local press.

Publicity (including photographs) can be sent to the local newspaper. The information should be prepared as a press release, indicating the nature of the contest or award. See Communications by the Safety Office later in this chapter for details of preparing a news release.

Meaningful Awards

An award serves several purposes: an inducement, a goodwill builder, a continuing reminder, and a publicity tool. To serve these purposes, however, an award must be meaningful to those participating in the contest.

Employees sense when management is giving awards only for "sales" or publicity purposes and is providing little or no safety effort. If a number of awards are given, the sheer volume can detract from the value of the program.

The value of awards lies in their appeal to basic human interest factors, such as pride, need for recognition, urge to compete, and desire for financial gain (Figure 18–2). Some organizations present monetary awards as a bonus for making an extra safety effort. The distribution of U.S. Savings Bonds for safe records gets away from the direct monetary nature of a financial award. Other organizations, however, prefer gift or plaque awards appealing to the other basic human interest factors.

The originality or cleverness of an award or of its presentation is an important factor. Select awards that are worthy of good publicity, photograph well, and provoke employee conversation. Refreshments for all employees in a department after the completion of a set number of injury-free hours probably will create more favorable comment than would the presentation of a fancy plaque to the department supervisor. Drawing for a small cash prize or a grab bag prize would attract more interest than a routine presentation of the same award. An award to an employee's spouse for completing a home safety checklist, for writing a safety slogan, or for contributing to the company paper also generates more interest than would the same award given on the job.

- One Council member reported an award idea that received an unusual amount of publicity, both within the plant and locally. A local automobile agency loaned the company a new car that an injury-free employee, whose name was drawn from a hat, drove for a week. The employee also had a special "reserved for John Doe" parking space in the company lot. The only expense for the employer was a few dollars to cover special insurance. Even a special parking space awarded on a rotating basis can be effective.

- Another way to gain interest is to let the employees participate in selecting the award, planning its presentation, and helping with publicity. Frequently, they will suggest a humorous or novel award or publicity approach that may attract more interest than one planned by management.

- Payment of bonuses as awards for good safety records evokes considerable differences of opinion. Some managements and safety people feel that this approach is unwarranted, because all employees are paid to work

safely. Others believe it can enhance an already successful program.

- Some companies raffle household or sports merchandise. Interest in safety among supervisors and workers often is developed by personal awards like wallets, knives, or key cases, often suitably inscribed.
- Many employees value attractive pins or engraved cards commending them for years of employment without an accident. One company places a safety record sticker on the employee's hard hat. Others provide special badges, pins, or shoulder patches to recognize safety achievement or service on a safety committee.

In addition to contest awards, recognition should be given to those who have saved lives, served on safety committees, submitted valuable safety suggestions, or made other significant contributions to accident prevention.

Award Presentations

To make an award presentation something special, one company rented an auditorium and invited civic and labor leaders, company executives, other dignitaries, and employees' families to a large celebration party. Another approach is for the president of the company, or some other high official, to present the awards at a general meeting, a picnic, or a dinner (or breakfast) that may even include entertainment. The reason for inviting VIPs is not only to add prestige to the presentation, but also to promote their interest and commitment to safety.

Such presentations require planning and must be in keeping with the importance of the occasion. The location chosen for the event should be appropriate and comfortable, not noisy or crowded. An award to an individual might be made in an executive office; groups might receive awards in a conference room, private dining room, or company lounge or cafeteria (during a nonrush period).

Inform participants about the agenda. Familiarize those who make the presentation with the significance of the award, the achievement it recognizes, and the background of the individual(s) who earned it. Ask the press to cover the event. Managers can post press photographs after the actual presentation to take advantage of desirable backgrounds, such as the plant or company name, some prominent trademark, or another eye-catching effect. Feature the award itself prominently in the picture. (See the sections following Campaigns later in this chapter and the Council Data Sheet 619, *Photography for Safety,* for other ideas.)

For a group award—such as a company, plant, or department completing an injury-free year—management can provide free refreshments for a specified time ranging from one coffee break (per shift) to a full 24-hour period. Another, more elaborate award was given by a company employing about 300. At the end of a year in which there had been no disabling injuries, the president took the group to a major league baseball game. The following year, after the company maintained its no-injury record, the employees, plus their families, were invited to an all-day picnic and cruise.

Such activities help build better employee relations, as well as promote more interest in safety on the job and within their communities.

POSTERS AND DISPLAYS

Posters and displays can reach a large audience with brief, simple messages designed to accomplish one or more missions: to convey information, to change attitudes, and/or to change behavior. They are used to communicate with people who are going about their normal activities. As a result, the audience must notice, read, and take in the message in a very short time—often in a matter of seconds (Figure 18–7).

Safety posters are one of the most visible evidences of accident prevention work. Because of this, some companies have mistakenly assumed that posters alone can accomplish safety instruction and have neglected such essentials as real management support, guarding, and job instruction. In fact, hit-or-miss use of posters in a plant where no other safety work is done is likely to have a negative influence, making employees feel that the company is not sincere. (See National Safety Council Data Sheet 616, *Posters, Bulletin Boards, and Safety Displays.*)

Purposes of Posters

Posters properly used have great value in a safety program through their influence on attitudes and behavior. One has only to see the efforts that commercial advertisers make to acquire space in business areas or near factory gates to appreciate the value of posters inside the workplace.

When selecting posters, keep their specific purposes in mind. They are used to:

1. remind employees of common human traits that cause accidents
2. impress people with the value of working safely
3. suggest behavior patterns that help prevent accidents
4. inspire a friendly interest in the company's safety efforts
5. foster the attitude that accidents are mistakes and safety is a mark of skill
6. remind employees of specific hazards.

Posters also help to support special campaigns, for instance, using guards, wearing eye protection, maintaining good housekeeping, offering safety suggestions, or driving carefully. They promote traffic, home, and even pedestrian safety by encouraging safe habits both on and off the job.

Effectiveness of Posters

A number of studies have underscored the effectiveness of posters for training and motivating.

- One study, conducted by the British Iron and Steel Research Association, used three posters that reminded workers to hook cable slings. The posters were displayed in six plants over a period of six weeks. A seventh plant was used as a control. Tallies made in the six plants before and after display of the posters showed about an 8% increase in compliance with the rule. The seventh plant, in which the posters were not used,

Figure 18–7. Posters convey safety messages forcefully and often humorously, and help emphasize a safety message in a memorable way.

showed a very slight decrease in compliance.

Although use of the posters merely supplemented previous training, the plants that originally had the lowest rates of compliance showed some of the best gains. Using the three test posters separately, on a biweekly basis, proved slightly more effective than simultaneous use of all three during the entire six-week test.

- In a survey conducted by a prominent casualty insurance company, over 200 employees were interviewed in depth on the effectiveness of safety posters, films, and leaflets. Results indicated that all the media were instrumental in bringing workers to a higher level of safety awareness and that all were effective in sustaining that awareness. Employees preferred posters to leaflets, although they acknowledged the value of leaflets for more detailed coverage of safety issues.

- A survey of Council members indicated that about three-fourths of the nearly 800 respondents use a variety of poster subjects with one-third preferring cartoons and all-industry posters. Horror or shocker posters were least preferred.

Fifty-four percent of the respondents used posters to influence general attitudes; 27% to cover special operations;

18% to meet special or seasonal problems; and 14% to promote off-the-job safety.

Sixty-nine percent of the respondents in the Council's poster survey preferred posters of 8 x 11 in. and 17 x 23 in. sizes (21.5 x 28 cm and 43 x 58 cm). Of these two sizes, the smaller was preferred by a six-to-one margin over the larger.

Frequently, an unusually striking photo or overly elaborate (and expensive) artwork actually detracts from the safety message. This is not to say that "eye-popping" illustrations should never be used; they may serve to attract attention to the more conservative, serious messages on other posters. However, be sure photos are related to the message being presented.

Types of Posters

Industrial posters available from the National Safety Council, insurance companies, associations, and other sources fall into three broad categories: general, special industry, and special hazard.

- General posters are concerned with such subjects as risky behavior, disregarding safety rules, forgetting to replace guards, and other human failures.

- Special industry posters, as the term indicates, have

application only in specific industries, such as mining or logging.

- Special hazard posters, for example on lifting, ladders, or the storage or handling of flammable liquids, are useful in every industry where the particular hazards are encountered. In some cases, a hazard is so serious that a special poster is developed based on the severity and not the frequency of exposure.

The National Safety Council carries more than 800 different safety posters in stock at any one time. These range from pocket-sized, pressure-sensitive stickers to billboard-sized jumbo posters.

Subject matter is roughly in proportion to the frequency rate of certain types of accidents. For example, more posters are concerned with materials handling than with chemicals and gases. (Call the National Safety Council for illustrations of the many posters available.)

Other locations. Posters and stickers can be mounted on delivery trucks, buses, industrial trucks, mail trucks and carts, in elevators, and even on doors. Pocket cards or plastic pocket protectors, such as those available from the National Safety Council, might be called "walking safety posters."

Other materials. Safety messages need not be limited to printing and artwork. They can be very effective when used in illuminated or changeable signs. One company paints a safety message on a plywood welding screen.

Homemade posters. A company can develop its own posters to deal with special hazards not covered by posters available from outside sources. Even the smallest company can make an occasional special poster inexpensively—using colored paper, crayons, or felt marking pens—to call attention to a special hazard, to commemorate the winning of a safety award, or to point up a problem not likely to be covered by a commercial poster. (See Chapter 17, Audiovisual Media, for details.)

Safety personnel or supervisors can make effective posters using photographs of local conditions or accidents, even if the situations must be posed. A common type of homemade poster is the "testimonial," featuring a photo of an employee and a close-up of damaged safety glasses or safety shoes. The poster carries a brief statement explaining how this equipment protected the wearer.

Homemade posters on new processes, new guards, or new rules personalize the safety program and augment even the best selection of commercial posters. An excellent source of this type of poster is an employee contest, which is easily administered at a nominal cost.

Changing and Mounting Posters

No specific rule can be given for how often posters should be changed because of varying definitions of the term "poster." Some types probably should be mounted permanently. For example, a poster on artificial respiration can be kept in the first aid room, or one on the use of a certain kind of fire extinguisher can be posted near it.

Most companies change general interest posters at definite intervals, usually weekly. They may rotate them from one area to another or file and reuse them after a year or so.

Management should vary the type of posters displayed. Consecutive posting of several health-related messages, for instance, or of new machinery posters is not desirable unless the company is conducting a special campaign on the topic. It is better to mix poster topics; for example, present an eye safety poster, then a machinery guarding poster, followed by a poster warning against wound infections, and so on.

For maximum effectiveness, posters not only must be selected carefully and changed on a definite schedule, but also displayed attractively in well-lighted locations where they will be seen by the greatest number of people. They should be placed on safety bulletin boards, near time clocks, in cafeterias, and at points of special hazard, such as paint storage rooms, rubbish cans, hazardous machinery, or dangerous intersections. Consider placing safety posters in unique places such as in the washroom above urinals, on the back of stall doors, on mirrors, or by paper towel dispensers.

Bulletin Boards

Bulletin boards should be large enough to allow convenient change of posters and should be placed where employees can see them during break times, such as near drinking fountains. They should be centered at eye level, about 63 in. (1.6 m) from the floor. Place them in a well-lighted location, or provide special lighting. A good size for a bulletin board is about 22 in. wide x 30 in. long (56 x 76 cm).

Boards should be painted attractively and glass-covered. Although a single board in the workplace is usually enough, several panels may be better in lunchrooms or lockerrooms. Flashing lights or other eye-catching extras may be suitable for nonproduction areas but are likely to distract and annoy employees in workplaces.

A bulletin board should be used for only one display at a time, but need not carry safety posters exclusively. Any program of mutual interest to company and employees can be displayed on the bulletin boards. In fact, safety posters may have a stronger appeal if they appear along with other displays and announcements.

Bulletin boards in the same company may range from large, enclosed, illuminated boards with special sections for posters, safety bulletins, and other messages to a number of small frames or other inexpensive poster mounts installed at strategic points.

Poster frames to which clip-on literature racks can be added are available from the National Safety Council. This arrangement allows the company to conveniently distribute leaflets and other pickup literature that supports the safety message.

Displays and Exhibits

The company can also use displays and exhibits to promote off-the-job safety as well as workplace safety. For example, many companies try to educate their employees to understand that a safe vacation is a happy one.

Personal protective devices, tools, and pieces of fire-fighting equipment can be used to make up displays or

exhibits, with or without corresponding posters (Figures 18–8 and 18–9). Another good eye-catcher is a seasonal exhibit combining a Council poster and a safety display featuring such items as proper footwear for winter walking or skin and eye protection items in the summer to guard against sunburn and eyestrain. Management can purchase signs with changeable letters, electric tape messages, or unusual lighting for safety topics.

Companies and associations have devised many simple and attractive displays for presenting statistical data to workers. One is a safety clock, whose face is marked off to indicate the frequency of disabling injuries. Twin clocks or dials often are used, one recording the present injury rate and the other the rate for the corresponding period of the past month or year.

One company used large thermometer-like boards, placed at every gate and clock-house. Arrows indicated the present and previous month's records. The comparative standings of departments appeared below the symbols.

An auto race was the theme of another display. Each car represented a department, and the cars moved daily to denote progress being made. Airplanes can be used similarly. Another exhibit featured racehorses participating in a "Safety Derby"; the horses were named after items of personal protective equipment. A display of highway signs with photographs of accidents below them was placed near the plant entrance of another company.

OTHER PROMOTIONAL METHODS

Other methods that can be used effectively to arouse and maintain interest in safety are campaigns, safety stunts, courses and demonstrations, publications, public address systems, and suggestion systems.

Campaigns

Campaigns focus attention on one specific accident problem. They are additions to, and not substitutes for, continuous accident prevention efforts the year round.

Campaigns may promote home safety, vacation safety, fire safety, or the use of safety equipment such as safety shoes. Companies can sponsor a "Clean-Up Week" or run a "Stop Accidents" campaign to promote the development of safe attitudes both on and off the job.

The National Safety Council's nationwide campaigns include posters, films, booklets, and specialty items to promote and maintain employees' interest. In addition, the Council has developed special materials in collaboration with trade associations, often to support special campaigns such as "Stop Shock" and "Fight Falls." Information on current campaign material is found in Council catalogs and other publications.

Many large corporations also have conducted extensive campaigns to promote safety on and off the job. Much of their safety awareness material is aimed at employee families, local citizens, and even groups outside the United States.

Companies must plan suitable publicity for the campaign from kickoff to conclusion, similar to that discussed earlier in this chapter for safety contests. Signs, flags, desktop symbols, and other items can be used to dramatize the campaign. To wind it up, schedule a special event such as giving each employee an inexpensive novelty item, free coffee for a week, or a free breakfast or dinner.

Many of the same promotional ideas that help to maintain interest in contests also can be effective in special campaigns—for example, a first aid drill or a demonstration of cardiopulmonary resuscitation (CPR). Some companies use safety parades, exhibits of unsafe and safe tools and equipment, pledge cards, and other such features.

Timeliness may be an important factor in the way employees respond to a campaign. Successful safety campaigns have been linked to elections, the World Series, the football season, Thanksgiving, and other special events. See Council Data Sheet 616, *Posters, Bulletin Boards, and Safety Displays,* for more ideas.

Unique Safety Ideas

Unique safety ideas or "stunts" capitalize on all the effective aspects of showmanship. Companies can develop them as separate devices for maintaining interest or use them to supplement contests and campaigns. *Safety and Health* and other publications regularly give details on various stunts.

Most companies agree that constructive stunts help inspire employees to high standards while stunts that ridicule usually do more harm than good. Moreover, employees and supervisors who are the objects of ridicule may have just cause to blame management for not setting up safe procedures or providing safe facilities and equipment.

Safety stunts can involve an entire company, a department, a small group, or just the individual. A stunt may be humorous, novel, or dramatic, and occasionally even shocking.

A simple stunt is often the most effective. A pivoted hammer, mounted over a pair of safety glasses in a display case, can be operated by a string to demonstrate the impact resistance of the glasses. To dramatize the importance of eye protection, the "let's pretend" stunt can be used. Several volunteers are blindfolded and then asked to eat, write, and move around.

Stunts developed for the company safety program often serve equally well at company open houses or safety picnics and in community safety projects as well. Such stunts, when supported by visual aids, signs, and printed material, demonstrate the company's interest in accident prevention. They also give employees a chance to participate in programs that help create safer attitudes on the job.

Some firms use a somewhat unique poster or card as an attention getter. For example, safety cards handed out by employees can alert co-workers that they have been risking an accident by their behavior.

Courses and Demonstrations

Most safety and health professionals agree that courses in first aid, lifesaving, water safety, civil defense, and disaster

Figure 18–8. Personal protective equipment is the theme of this display at NRPC's Beech Grove, Indiana, maintenance shop. The display is lighted at night and is in Amtrak's impressive 15-foot display case.

control have bonus values that help prevent work injuries. The worker who has completed a course in first aid and has learned to do CPR will be more aware of the hazards of electric shock and more likely to help maintain electrical equipment in safe condition. Likewise, the employee who learns how to stop arterial bleeding better appreciates the serious consequences of using an unguarded saw, power press, or cutter.

Home study and extension courses also serve to stimulate and maintain interest by giving the employee a better understanding of hazards involved in the job. Most safety training courses are designed specifically to improve the attitudes of both supervisors and employees. Using appropriate videotapes and other visual aids will enhance the effectiveness of the courses. (See Chapter 17, Audiovisual Media.)

Public participation in courses involving employees promotes community goodwill. Many industrial safety people are doing an excellent job of promoting safety and fire prevention by arranging or teaching courses on these subjects for the Boy Scouts and Explorer Scouts, the Girl Scouts, Camp Fire groups, Junior Achievement companies, and other youth or school groups (Figures 18–10, 18–11, and 18–12).

The National Safety Council's "Defensive Driving Courses" provide an excellent means of promoting good employee and public relations. They stimulate safer attitudes both on the job as well as off.

Fire equipment and fire safety demonstrations have a practical value beyond that of teaching employees how to react in an emergency. The fact that the equipment is provided for their use reminds them of management's concern about their welfare. Moreover, the demonstrations make employees more aware of the dangers of fire and point up the need for obeying fire prevention rules.

Such demonstrations are easily arranged through local fire departments or fire equipment distributors. Many companies conduct their own demonstrations, using extinguishers that require recharging, or "not-in-service" extinguishers kept specifically for this purpose.

Publications

Reports. Safety program progress reports should be written in an interesting and concise manner for superiors and supervisors. Visuals can be used effectively where appropriate. (See Chapter 17, Audiovisual Media.)

The safety and health professional can present the cost of

Figure 18–9. Local school children are led on a tour of the Canadian Pacific Railroad station in Winnipeg, Canada, by clerk Jim Fisher. This excellent community public relations effort includes distribution of a safety booklet and the viewing of a safety film.

accidents and, perhaps, the cost of prevention in terms significant to management. These costs include medical and compensation costs, production losses, sales losses, increased maintenance costs, and the less tangible, but perhaps more important, hidden costs involved in administrative problems and in impaired public, customer, or employee relations. Reports need not be dull. Photographs, for example, can pinpoint a company's major sources of disabling work accidents.

A statement of accident losses and safety achievements may be included in the company's annual report or in a special annual or monthly safety report issued to top executives and supervisors. If departmental accident losses, like incidence rates, can be charged on an equitable basis, such as "per hundred thousand dollars of sales" or "per one thousand employee-days of production," the comparative standings of departments and improvement in departments or units are easy to evaluate. (See the discussion in Chapter 13, Recordkeeping and Incidence Rates.)

The fact that the company records and publicizes such information is in itself an incentive to supervisors. It reminds everyone concerned that accident costs are just as much an integral part of profit and loss as production, sales, maintenance, distribution, and advertising.

Special charts, graphs, and statistical reports also can visually show facts about accidents. One chart can track the number of disabling injuries, others the number of days lost, injury causes, accident causes, or type of injuries. It cannot be too strongly emphasized that unless such charts are kept up to date they can do more harm than good. An outmoded chart indicates management is losing interest in maintaining a strong safety program.

Annual reports. In recent years, companies have worked hard to make their annual reports to stockholders interesting and clearly understandable. In many cases, annual reports are also distributed to employees so that they can become better acquainted with the company's purposes and problems. A section on aims and accomplishments in accident prevention attracts employee interest and further serves to emphasize the interest of management in the safety of its employees.

Newsletters. Monthly or weekly newsletters or videos are especially important as a means of maintaining interest. They keep employees and supervisors informed, particularly in decentralized or field operations where bulletin boards are not practical. Newsletters can give detailed information on standings in a safety contest and publicize unusual accidents or serious hazards. They can help explain safety rules, remind

Figure 18–10. In this scaled down trailer-mounted house, children learn to escape from fires. They learn to use the proper procedures as they exit the bedroom when the smoke detector goes off (Operation EDITH: Exit Drills In The Home). (Courtesy Springfield, IL, Fire Department.)

employees of safe work practices, and support the safety program in general. If workers can serve as "reporters" or help to produce such a newsletter, so much the better (see details later in this chapter). A case history of a particularly unusual or spectacular accident can sometimes be featured.

Booklets, leaflets, and personalized messages take many forms: safety rule booklets, NSC's "New Employee Booklets," special "one-shot" leaflets, monthly publications such as the National Safety Council's *Today's Supervisor* magazine and its *Safe Worker* and *Safe Driver* for employees, and letters from management. The content of an employee rule booklet, except for material involving company policy, may be developed with the help of safety committees or selected workers to stimulate interest in the topic and help to ensure compliance with the rules.

Larger companies may have their own editors and artists, even their own printing facilities. Smaller companies, however, also can issue attractive booklets, leaflets, and personalized messages, and at negligible expense by using local consultants and printers or their desktop publishing systems.

The National Safety Council, trade associations, and professional organizations publish a wide variety of booklets

and leaflets that are authoritative, attractive, and relatively inexpensive. They cover a wide range of subjects—material handling, first aid, housekeeping, fire prevention, vacation safety, safe driving, and the like. Such materials, carefully selected and regularly distributed, effectively supplement company-prepared publications.

Letters commending meritorious service, signed by the manager and addressed to individuals, also make an excellent impression upon workers.

Safety calendars, published by the National Safety Council, together with Christmas letters from the manager have a direct appeal that reaches the workers' homes. Such mailings should include all employees. *Family Safety and Health,* a quarterly magazine, is sent to more than two million homes by organizations who are interested in the welfare of both employees and their families.

Buttons, blotters, book matches, pencils, and other small novelties all conveying a safety message also may be used. For example, packets of silicone tissues for cleaning glasses, imprinted with brief messages or safety rules, serve to remind employees of the rules and to encourage proper use of safety equipment.

Figure 18–11. Management, employees, and their families participate in a safety fair. Before the children can slide down this authentic firefighters' pole, they are given a safety question to answer.

Public Address Systems

Public address systems often are used to broadcast announcements and page employees. Many companies have taken advantage of these installations to broadcast safety information as well.

Such messages should be planned carefully. Employees might readily lose interest in long speeches or too-frequent safety reminders. When the public address system is used for broadcasting music, safety announcements can be made between numbers.

Suggestion Systems

Because accident prevention is closely associated with efficient operation, many suggestions not only help to prevent

Figure 18–12. This truck contains a mobile puppet show and sound system. The current repertoire consists of four shows: Burn Prevention, Fire Prevention, Substance Abuse, and Safety on Wheels.

accidents, but also lower production costs, improve manufacturing conditions and methods, and change the outlook of workers.

If a company does not have a general suggestion system, it is probably better not to establish one for safety suggestions alone. Setting safety apart from ordinary operating procedures may deemphasize its importance.

Getting good suggestions is essential and must be encouraged by all levels of management. Company posters, contests, campaigns, merchandise incentives, direct mail, printed handouts, personal appeals, supervisory training, safety clubs, and press releases can motivate employees to submit safety suggestions.

Many effective suggestion systems offer money for ideas, and companies may pay considerable sums for employee suggestions. One firm awards more than $10 million annually, but considers the money well spent on valuable suggestions.

To merit an award, a safety suggestion, like a production suggestion, should be substantial and practical, and offer a real solution to a problem. For example, suggestions worthy of awards include changing a method or material, guarding

a hazard, or inventing a safety tool or device. Suggestions that would not merit awards include erecting a sign, cautioning workers, or publishing slogans.

Safety suggestions generally are regarded as a highly desirable way to avoid safety grievances. Improved employee interest and personal involvement are additional benefits.

Suggestion awards. It is easy to measure the monetary value of suggestions that result in greater efficiency, lower material cost, decreased labor cost, or reduced waste. Usually awards for suggestions in these categories are in proportion to the savings derived by the company. In other instances safety suggestions may have a monetary value but be harder to evaluate. As a result, payments for suggestions that contribute to the welfare of the employees but result in no direct savings to the company are most often estimates or composite judgments. Some firms have developed guidelines that consider such factors as the degree of hazard, originality, extent of application, etc. One company has also developed an award guide based upon disability cost experience.

If management downplays "safety awards" and gives more weight to other awards, employees will believe that the company regards safety as an unimportant sideline. Payment

for a safety suggestion must be equal to awards given for other suggestions—it should be based upon its real worth if it can be determined. If a suggested safety device enables an operation to be run at a speed that would be dangerous without the device, a saving may be measured. If a number of accidents have occurred in an operation and a suggested device will eliminate them, the cost of those accidents can be projected and a saving calculated.

Awards should reflect the merit of the suggestions. Most companies award cash and/or bonds. Some award merchandise, all-expense-paid trips, company stock, certificates of merit, medals, gifts for the suggester's spouse and family, or attendance at a recognition luncheon.

Some companies exclude superintendents, supervisors, designers, methods and systems personnel, and other supervisory or technical personnel from receiving awards, so that the other workers will have someone to whom they can go for assistance. Some companies feel supervisory personnel should not be excluded; several firms have separate plans and award schedules for salaried, supervisory, technical, and management personnel.

Suggestion committee. If necessary, a special subcommittee can be set up to determine the monetary value of safety suggestions so that employees will be rewarded for them exactly as they would be for other money-saving suggestions. However, some firms believe such special treatment sets safety ideas apart from other ideas.

Many established suggestion systems now in operation are producing excellent results in monetary and "people" savings. No company should consider a suggestion plan without first studying carefully the plans now in existence.

Boxes and forms. Suggestion boxes should be attractive and well placed, and stocked with special blank submission forms. Commercial suggestion forms also are available. To increase the interest of the employees and establish a spirit of cooperation and importance, it is essential that management acknowledge and resolve all suggestions promptly.

Recognition Organizations

Some organizations recognize people in the United States and Canada who have eliminated serious injuries, or who have minimized them by using certain articles of personal protective equipment. The award provides an excellent opportunity for publicity. Three of these organizations are:

Wise Owl Club. Founded in 1947, this club honors industrial employees and students who have saved their eyesight by wearing eye protection. Address inquiries to Wise Owl Club, National Society to Prevent Blindness, Schaumburg, IL 60173.

The Golden Shoe Club. Awards are made to employees who have avoided serious injury because they were wearing safety shoes. Address inquiries to Golden Shoe Club, 2001 Walton Road, P.O. Box 36, St. Louis, MO 63166.

The Turtle Club. Founded in 1946, the Turtle Club recognizes and honors workers who have escaped serious injury because they were wearing a hard hat at the time of an industrial accident. Although OSHA mandated the use of head protection in many U.S. industries in 1971, the Club's goal—to save lives and prevent injuries through promoting the use and acceptance of hard hats—and its purpose—to help business understand the benefits of industrial head protection—remain important today (Figure 18–13). The Club's sponsor is E.D. Bullard. Address inquiries to Mr. E.D. Bullard, International Sponsor, The Turtle Club, P.O. Box 9707, San Rafael, CA 94912–9707.

CAMPAIGNS

The preceding sections covered internal publicity within a company or organization. The remainder of this chapter discusses how to influence the way a company looks to people on the outside. Favorable publicity is an unmistakable bonus to a good safety program. Yet many organizations often disregard it.

Any company likes to have someone—especially a prospective customer—say, "I like what I hear about this company. I understand that it really takes care of its employees. So I figure it must treat its customers right; therefore, I'll be treated right." One good way for a company to get a reputation for taking care of its employees is to be known as a safe place to work. Surprisingly, an amazing number of companies do little or nothing to let their public—customers, stockholders, the community—know that the safety and health of their employees are important to them. This chapter presents a simple and sensible formula for letting people know that "at my company, the health and safety of the workers are important."

Most companies have a professional public relations (PR) department to handle the public communication program. In smaller companies, the safety and health professional may have to handle public communication. In both cases, the information in the following pages should prove useful. The safety professional should be aware of overall company policies and programs and know when to turn to specialists and when to ask for creative help.

PUBLIC RELATIONS

Public relations is the "management function which evaluates public attitudes, identifies the policies and procedures of an individual or an organization with the public interest, and plans and executes a program of action to earn public understanding and acceptance," according to the magazine *Public Relations News.* Abraham Lincoln knew about PR. Speaking at Ottawa, Illinois, in 1858, Lincoln said: "Public sentiment is everything. With public sentiment nothing can fail; without it, nothing can succeed. He who molds public sentiment goes deeper than he who enacts statutes or pronounces decisions. He makes statutes or decisions possible or impossible to execute."

Lincoln's classic quotation on public sentiment can be traced to Jean-Jacques Rousseau, the 18th-century French philosopher who is generally credited with developing the term *public opinion* as we use it today. At its simplest, PR is

Figure 18–13. The Turtle Club honors people who have escaped serious injury because they were wearing a hard hat. Displayed here are the items new members receive: a hard hat with the club insignia, a membership certificate, a wallet card, and a lapel pin.

the promotion of good public opinion about a person, a company, a government, or other entity. Thus, every employee, every activity, every facility of a company contributes in many ways to the overall opinion that people outside the company have about that company. This is true PR.

Anyone concerned with accident prevention—safety and health professional, supervisor, member of the plant safety and health committee, or officer of the school or community safety council—should realize that any time they communicate with someone outside the committee, department, or company, they are involved in PR. Even an event such as a company-sponsored family picnic can strengthen PR.

To be successful, a PR program must be backed up by a sound management team and a quality organization. Public relations can then reflect the nature of that organization. A strong safety program merits good publicity, but false publicity or publicity based upon inflated facts or specious statistics will be recognized for what it is—company propaganda—and can do more harm than good.

Public relations information, to deliver an effective message, need not always be red-hot news. However, it must have an element of spot news, human interest, or self-help. These qualities will give it feature value for the media. Activity—real, honest, legitimate activity—makes news. Of course, dramatic circumstances such as a rescue or near-miss make news. But until they turn up, the best stories are those that describe a company's constructive and creative efforts to promote and maintain a worker health and safety program.

Basis for Success

The basis of a successful PR program is a successful management. Such a management team makes sure that staff and employees produce good products safely and efficiently, that they cooperate with each other and with the customers, and that all give the best and friendliest service humanly possible at all times.

The plain fact is that poor PR is costing individuals and organizations in this country millions of dollars each year. The remedy is simple: a better understanding and use of fundamental PR on the part of everyone—and a sincere effort to put it into practical use. For lack of good PR, many a worthy cause has failed to get the support it deserves, and many an organization has failed to gain the business it needs. A good PR program need not cost a great deal of money. But it is worth all the time, effort, and reasonable expense allocated to support its activities.

THE VOICE OF SAFETY

In any genuine, effective PR program, emphasizing safety can be a real help. In fact, it is hard to imagine a PR program where sincere and effective concern for the protection of employees from accidents is not a top priority.

If a company does not have a safety program, it misses vital opportunities for good PR and dramatically increases the chance for adverse public attitudes, not to mention the risk of an accident. A safety program, properly managed and communicated to internal and external audiences, can help offset or at least mitigate adverse news should accidents occur. In addition, the safety program is long-range and sustained, whereas "news" is here today and replaced by other news tomorrow. The time to implement a safety PR program is not when trouble strikes. The program must be in place and functioning, and the public must know about it.

The safety and health professional should not only welcome publicity for safety efforts but should energetically seek it.

Working within the Company

In evaluating a company's PR function, one can begin by asking two questions:
1. Does the company have a PR department?
2. Is there an employee publication in the company?
If both answers are "yes," the safety professional should get in touch with both these units before doing anything about publicizing the safety program. This step is important. It not only ensures professional skill and consistency of efforts to publicize safety activities, but it will save confusion, avoid duplication, and possibly prevent conflict between safety personnel and the PR department.

Why does such an obvious step have to be mentioned? The reason is that too often there is little, if any, communication between a safety and health professional and the PR department and publications editor. Communication between the safety and health professional and the PR staff and editor is indispensable, for these three must work together, or safety will not receive the attention it deserves and needs.

The safety and health professional should explain to the PR staff and the editor, if this has not been done before, that positive safety motivation and accident reduction at the company are priorities and that their communications help and advice are needed to reach these goals. They need to understand the necessity of employee and public acceptance

of the safety program and how much the safety and health professional is depending upon them to help earn employee acceptance.

Safety activities contain a wealth of real news. It may be that everyone in the safety business has assumed that safety activities and programs by their very nature must be on the dull side. In recent times, however, more writers and others have begun to recognize that safety can be made interesting. It takes the combined efforts of safety and health professionals, PR people, and editors to do the job.

Sixteen Ways to Make Safety News

It might be useful to list some of the items and activities that can make safety news in an organization and that editors and PR staff ought to know.

1. No-accident records for the entire company or for any one unit—in terms of either days or worker-hours (Figure 18–14).
2. Improved safety records for the company or for any one unit, even if no prolonged no-accident period is involved.
3. An interplant safety contest, or an intercompany contest—especially if anyone has dreamed up an unusual angle or prize.
4. Any unusual safety record for safety performance by an officer or employee of the company—either in length of time or character of the job done.
5. Innovations in safety programs that will prevent accidents. An invention, too, has special news value if the company has been plagued with accidents that the new gadget may prevent.
6. An unusual or highly valuable safety suggestion by an employee.
7. Safety conventions or meetings, either those held by the company or those held elsewhere, to which company representatives will go. A digest of such meetings should be publicized.
8. Other special safety events besides conventions—a safety banquet, a safety training course, fire and first aid demonstration, a special meeting, or an award ceremony.
9. Some unusual event intended to get the employees to take their safety training home to their families, or something the company is doing directly with workers' families to promote around-the-clock safety. Open-house tours, local water safety shows, or public showings of safety films are examples.
10. Some pronouncement or statement by the president or other high officer of the company on some unusual or new safety device or company safety service, such as free inspection of employees' cars.
11. A speech by the head of the company or the safety and health professional at a local, regional, state, or national safety convention or conference. The editors and PR staff should have advance copies of it. The person making the speech should be sure to say something worthy of public attention.
12. The company's annual report is the foundation for corporate communications. Stockholders do read these reports. A paragraph or two on the good safety record for the past year will go a long way in achieving sound publicity within the corporate family, as well as to inform the analysts, who recommend stocks, what the company is doing beyond its financial performance.
13. Any act of heroism by someone in the company. This is a sure-fire story for local papers as well as company publications. Maybe this type of news is not pure safety, but news media regard it as part of safety, and it can always be tied in with an indirect safety message (Figure 18–15).
14. A survey or study of some phase of accident prevention in the company. If the investigators discover that married people who own their own homes are safer than their single counterparts, they have provided a ready-made story.
15. Election or appointment of a company officer or safety professional to an important post as a volunteer officer of the National Safety Council, American Society of Safety Engineers, Board of Certified Safety Professionals, local safety organization, or governmental agency.
16. A company or employee winning an award in a National Safety Council award program or contest. Winners, not losers, are publicized.

A good rule of thumb is to stress the positive, rather than the negative side of an event. For example, instead of a story that says "single people are less safe," it could say that "a company study shows a need for special safety efforts by singles." Another way to stress the positive is to report "Last year 290 employees worked without an occupational injury. Only 10 did not."

SOME BASICS OF PUBLICITY

If a company does not have a PR department, this fact need not prevent its getting information into local papers or on the air. Without PR personnel, the task may be more difficult, because PR people are more experienced and know the media game better than the safety and health professional does. However, even in the absence of a company PR department, the safety and health professional can get publicity for safety activities by going directly to the newspapers, magazines, and radio and TV. It helps to solicit and secure PR advice from local safety organizations, or even business or trade associations with such service. The safety and health professional should always keep the company management informed of what is going on at all times.

The safety and health professional should not pretend to be a PR expert. Instead, he or she should be sincere in approaching editors and TV/radio program managers with sound, worthwhile information and not worry about whether it is packaged perfectly for media acceptance. A good story is a good story. Most editors and reporters are glad to have the material and often will use it. News value and reader interest are what count the most.

Editors and program people are usually not difficult to approach, provided the safety and health professional is courte-

Figure 18–14. Twenty-seven years of operation without a lost workday injury was celebrated with a large billboard. (Courtesy El du Pont, Wilmington, DE.)

ous and friendly and admits to a lack of specialized knowledge of PR techniques. Naturally, no editor or anyone else likes to have someone come charging in and claim to be doing a big favor by delivering the story the world has been waiting for.

In general, the common sense and salesmanship needed for success in the safety field are enough to enable the safety professional to present his or her case clearly and effectively to the media. Nevertheless, the person should be willing to accept advice from PR professionals on how best to tell the story. The safety and health professional needs to be careful about being quoted out of context.

Select the Audience
"Who must be reached with safety information, and is it of interest to them?" These seem like fundamental questions, but many companies never think to answer them. Because publicity cannot reach everyone, the company must chose its audience carefully, especially if the budget is tight or time is limited. In that case, it is logical to assume that—in addition to the in-plant (company) audience—the company would prefer to reach people who might buy the product or help the company in some other direct and profitable manner.

Use Humor and Human Interest
It is worthwhile to try to brighten safety, to make it positive,

rather than ponderous and dreary. It is even possible to evoke a chuckle now and then (Figure 18–16).

Editors are familiar with the solemn pronouncement that "Safety is a serious subject, and must be taken seriously—safety is no laughing matter." No one can argue with that position. But does it follow, therefore, that no one can put into safety some of the same techniques, the same sales appeal, the same sparkle that are used so successfully to sell commercial products? If those techniques can sell shampoo or a personal care product or an automobile, is it unreasonable to expect they can also sell safety?

Or how about a cartoon treatment? This just may brighten what might otherwise be a slightly dull and drab presentation. The safety and health professional should not be too disturbed if someone points out that a cartoon has shown a negative aspect of a safety message—that is, what *not* to do. It may well have done just that. This is the very thing that gives a cartoon its punch. A pratfall cartoon will draw attention to the slippery, icy sidewalk in a way that cannot be shown by a person walking and not falling. However, care should be taken to avoid ridiculing or negatively portraying victims of accidents or the negative results of accidents.

It is even possible to get a child or baby into the act, or even a faithful, shaggy dog, in order to get the spark that

Figure 18–15. Rescuers carry a concrete worker whose small boat capsized, throwing him into the chilly, turbulent waters of the Mississippi River. He went over two spillways, one of them 30 feet long, before he was rescued. He was wearing a U.S. Coast Guard approved Type III work vest, which also had OSHA-required safety features and hypothermia protection. (Reprinted with permission from the Minneapolis Tribune.)

lifts safety activities out of a dreary, impersonal rut and presents them in terms of human interest.

Names Are News

Remember that facts and figures about injuries and their frequency and severity are made more interesting by injecting human interest elements into safety news. Because human interest means people, the safety and health professional should include more information on *people* and safety and less on *things* and safety. Mention which managers or employees accomplished safety goals, contributed ideas to the program, and other such activities.

Friendly Rivalry

Safety awards, safety records, safety contests, safety inventions, and gimmicks—these are only a few of the many things that make good safety news. For example, if the company has reached a new injury-free record in its industry, it is

headed for headlines. The editor, the PR department, and senior management must be kept informed of safety progress all along the way. They will help arouse public interest in the performance and also stimulate greater interest and greater effort among the employees themselves.

An award loses its motivational and publicity value if kept a secret. The safety professional must publicize the event. Photos of award presentations are sometimes used, but they should be interesting and even unusual to attract special attention. Editors dislike the typical "grip and grin" award photos, so be inventive in trying to get a picture worthy of publication. For example, a representative from the organization conferring the award on the organization could be shown handing the award to a foreman or worker in the workplace setting.

In some instances, top safety awards have been accepted by some companies as if they were a "dime a dozen." On the other hand, companies that have made similar awards

Figure 18–16. Awards can be displayed against meaningful backgrounds. Clyde Nyquist, a senior warehouse specialist at Abbott Laboratories, North Chicago, Ill., poses with the trophies he won at the 10th annual International Materials Management Society Fork Lift Truck Rally. The news release that Abbott sent out stressed the company's lift truck operator training programs, refresher courses, and good safety and production records. (Reprinted with permission of Abbott Laboratories.)

the occasion for a major bell-ringing celebration have seen their "safety stock" rise sharply among employees.

A public utility company in Michigan, for example, made so much of its award-winning safety records of the various units throughout the state that an outsider might have thought the company had won the World Series. At one celebration, more than 1,000 employees, from top management on down, jammed the closed-off street in front of company headquarters for presentations, followed by an all-company picnic. This celebration got coverage in the papers and on the air throughout the state of Michigan. Here was public relations that any organization would welcome. Safety had made news.

Techniques

These pointers are offered to the safety and health professional who wants to make the most of PR opportunities:

1. Be honest in what you say. Never exaggerate.
2. Deliver what you promise. If you say to the media that something is going to happen, make certain it happens—

as you said it would when you said it would. This often calls for a "run-through" in advance.

3. If for any reason there is a change in plans from what you have announced, notify the papers, radio, and TV stations at once.
4. Be scrupulously accurate with all names, dates, places, and other facts. There is no such thing as being too careful in this respect. If an editor misspells names, the only thing to do is to resubmit the names correctly spelled and hope for the best.
5. Offer ideas but do not ask for specific space or time. Complaining about your company PR department, or complaining that the local paper or station has treated your company shabbily will not only be fruitless but will create a strained relationship between you and the press.
6. Do not alert your PR department or publications editor (or put out anything yourself) unless you have real news or features to offer. You must not issue material just to be issuing it. Be reasonable with the amount of material you send out. You can wear out your welcome.

7. Tip off your local or industry association, safety councils, and your PR department (or if you do not have one, the papers or radio and TV stations) to anything worthwhile you run across that might make an interesting news story. Do so even if it has no relation to you or your company; the media will appreciate it and be more inclined to view your own materials favorably.

8. Above all, do things that make news. Almost every routine safety item can, with a little extra effort by the safety and health professional and the editor, become a more readable, more constructive piece. News will be published only if something is being done for safety that makes news. News can always be heightened by intelligent, imaginative treatment, but the story must be worth telling. Advertising space can even be purchased for special items.

9. Use good sense and an honest approach if the news is bad. Prove to press representatives and the public that you can roll with the punches. Know your organization's policy and procedure to follow in all situations.

Communications by the Safety Office

If a safety and health professional must handle his or her own publicity with the local media, here are some tips. Many stories have died because someone failed to observe them.

1. Editors and news directors are busy people. Unless a situation is an emergency, media people prefer to receive possible story ideas in writing. Send them a brief release or a letter outlining what you have to offer. Include facts and by all means make your material interesting. Timing is vital: Don't notify the media today of something you are planning for tomorrow.

    Never suggest that a reporter or news crew be sent out—this is the editor's or news director's province, not yours.

2. Generally, the person to write to is the city or business editor of the paper or the news director of the radio or TV station. Of course, if an item is specifically written for a certain columnist or commentator, it is better to make direct contact. If it applies only to a specialized area (finance or sports, for example), it should be brought to the attention of that editor. Know the publication, whether it is general news or a trade journal. If you are working with TV or radio, don't waste their time on information or items they'll never use. (Example: broadcast media seldom mention personnel changes.)

3. Write not for your boss, but for your reader or listener. Answer objectively the questions: who, what, when, where, how and why? Do not load your releases with propaganda for the company. There is no surer way to kill your positive relations with the media.

4. Make your releases as professional in style, appearance, and general quality as you possibly can.

5. In writing a release, be brief and to the point. Printed space is limited and costly. Try to keep the piece to one page. Publications receive thousands of releases each month. These are skimmed, and only the best get into print. Likewise, remember that TV and radio broadcasts usually are measured in seconds, not minutes. Keep your material short and factual.

6. If you are sending a picture with the release, the caption should be typed on a piece of plain white paper and taped to the back or bottom of the photo. Never use paper clips, and never write with a pen or pencil on the back of the picture. Either of these will likely damage the photo and make it difficult to reproduce clearly. Think visuals when working with TV—what can they show that will help to tell or illustrate your story?

7. Leave script writing to the professionals, but check facts. If a radio or TV station requests material, send them the facts, figures, and whatever narrative is necessary. The people at the station will put it into the proper form.

Hints for TV Interviews

The safety and health professional must often be the spokesperson for his or her company, not only for newspaper coverage, but also for radio and television. Although getting the facts correct may be adequate for a newspaper interview, a television or radio interview reflects more of the company than merely what the facts show. It projects a company image through the company spokesperson. If the safety and health professional does not feel that he or she projects a good image over the radio or television, then someone who does should be picked from the safety department or from the PR department.

Here are some tips that will help the safety and health professional give a better radio or TV appearance.

1. Remember that you are being interviewed for your knowledge, not for your personality, entertainment value, or good looks. Be yourself. Don't put on a special voice or worry how the lavaliere microphone looks with your clothes. Don't wave your hands or touch your face or hair. Don't jingle coins or play with jewelry. Keep your hands down.

2. If your TV appearance is preplanned, dress conservatively. Avoid loud clothing or busy patterns that can affect the camera or overpower your message.

3. Review with the interviewer in advance what areas will be discussed. You can steer away from areas you cannot discuss, and you can get help on questions that you might not be able to answer. If you don't know the answer to a question, say so. But add that you will check and get back with the answer in a specific time period.

4. Be natural and cordial. Smile, if the situation is friendly. Be sure to maintain eye contact with the interviewer or camera. Do not memorize a statement and rattle it off. You will waste everyone's time. It is better to be well-prepared and let the on-air material be a question and answer or discussion between you and the interviewer. Don't rush, and don't try to cram a lot of information into a TV slot. Keep it short and light. Avoid statistics.

Handling an Accident Story

In any PR program, it is just as important to know what *not to* do as to know what to do. In fact, it can be even more important.

The foremost warning is this: do not cover up bad news. Good media relations are of utmost importance. It is at such a time that a sound PR program pays off.

Every safety and health professional hopes the day will never come when a serious accident damages a company's safety record, but it happens. In some instances, the repercussions of how the accident reporting and news coverage are handled have been even more unfortunate for the company than the accident itself.

Here is an example of how *not* to handle a press representative: In a midwestern city some time ago, two workers were killed by a crane. This company enjoyed a first-rate relationship with the newspapers and radio and TV people in that city. It had worked hard at safety and at PR. It was good to its employees and had a fine reputation for honest and open communication.

On this particular occasion, however, someone in the company's top management became anxious over the deaths. As a result, when a reporter came out to the plant to get what was to her paper a routine story of the accident, she ran into a maze of censorship.

First, the safety and health professional turned the reporter over to the personnel manager. This person switched her to the general manager, who gave her some "double talk" about the accident and told her the company doctor was the one to see. The doctor said the safety and health professional was the person to talk to.

By this time the reporter's anger and suspicions were rising. What had started out as a routine assignment now had become a challenge to discover what everyone in management appeared to be covering up.

It is important to remember that a reporter cannot lose in a contest like this. Since the workers had been killed, the coroner would have all the facts. If they had been badly hurt, one of the hospitals would have the information. If the workers had not been killed or hurt badly, then there would have been no story in the first place.

The reporter obtained the facts regarding the workers' deaths from the coroner's office. She then wrote a story that was as harsh toward the company as it could be without committing libel. The story was edited, headed, set in type, and lay in the composing room, awaiting its turn to get into the paper.

Ordinarily, newspapers set more news in type than they can print. Each day dozens of items get left out—the "overset," as it is called in newspaper parlance. Under normal circumstances, the story of the accident might well have ended up as "overset" or, if printed, might not have received much attention. However, the coverup and runaround the reporter had received at the plant changed all that. As a result, the story was marked "must" when sent to the composing room and ended up as front-page news.

When there is bad news to report, the safety and health professional must back up the PR department 100% to help get out the news as quickly as if the tidings were all in the company's favor. Unless directed to deal personally with the media, the safety professional should stay in the background and provide the proper information to management and public relations.

Along with the bad news, the safety professional's material should mention the good things. It may be that this is the first accident in months or years, that the company has a safety record far better than the national average for its type of operation, and that it has won a number of safety awards. Such counterbalancing information will help to take the sting out of the story. Reporters are usually willing to include these facts, too.

It is not only good policy but a wise precaution to be honest and fair to press representatives whenever there is news—whether pleasant or unpleasant. This principle is vital to a good PR program and will pay considerable dividends in terms of getting fair treatment from the media.

It would be wise for a safety and health professional to anticipate that some day he or she may have to serve as a company spokesperson at an accident or disaster scene. It is imperative to follow company policy and procedure when acting as a company spokesperson. This needs to be planned ahead of time.

News media can actually help during a big emergency. Families, friends, and neighbors will be clamoring for news and the media can get it to them fast.

WORKING WITH COMPANY PUBLICATIONS

If the safety and health professional thinks the company publication has neglected safety, it is time to correct the situation. He or she should ask the editor how more news value and human interest can be worked into safety stories. With some editorial assistance, the safety program can be just as newsworthy as other stories. The safety professional should offer to provide details and descriptions of events. The idea is to give the editor plenty of good, current information. The editor is just as eager as anyone to publish interesting news and features, and will go more than halfway to think up ways to put news value and reader interest into safety doings.

Here is an example of how the safety and health professional and the editor can team up to make a routine safety happening more newsworthy. Suppose one of the employees, Oscar B—, reaches his 25th anniversary of steady work without a day's lost time due to an injury or occupational illness. This achievement probably entitles Oscar to a button or badge or some other award. The PR-conscious safety and health professional can ask, "Instead of just pinning this button on Oscar with a hearty handclasp and a few words of commendation, why not make a real event out of it? Take the occasion to tell Oscar—and all the world—that this plant places top priority on recognizing the contribution employees make to a safer, better workplace. Oscar has made such a contribution through his personal example of safe practices over the years."

As a result, the company publications editor does not merely publish a picture of Oscar and his award along with a two-line caption. The editor finds Oscar, sits down with him over a cup of coffee, and asks him a few questions

about his career, about his opinions on safety "way back yonder" and now, and about any ideas he may have for making things even safer at the plant.

This example is only one idea of what can happen when the safety and health professional and the editor of the company publication get together to do a more imaginative and energetic job of publicizing the safety program. For instance, they can develop a system to be used when one department of the company wins an interdepartmental safety contest. Instead of merely recording the results of the contest, the editor can dig into the program of the winning department, interview the people responsible for its success, and perhaps come up with a magazine story that will give every department in the organization some hints on how to improve its safety activities.

Producing a Publication

Materials, such as safety newsletters, instruction cards, bulletins, broadsides, booklets, and manuals for communicating safety rules, information, and ideas in print, require careful planning and preparation. Among steps to be taken in planning both internal (to a company audience) and external (to the general public or other out-of-company groups) publications are:

1. Clearly define the objectives of the publication. Consider the type of audience to be reached by those objectives.
2. Determine how general or how restricted the message is to be.
3. Decide what form of publication will best convey the message.
4. Estimate cost of preparing and printing the publication in whatever forms, sizes, and quantities needed. An expenditure for a new publication must, of course, be provided for in the budget, whether or not the item is produced in-house.

If the objective is to give the worker specific rules to follow in doing a job safely and efficiently, an instruction card may be suitable. To stimulate general safety-consciousness, a broadside (single-sheet printed on one side) may be effective. If a series of short reminders, for example, on fire prevention, is needed, posters may be the answer. To treat a topic of general interest, such as methods of materials handling, a leaflet can be the best choice. Here, posters or leaflets from the National Safety Council, insurance company, or other organizations may be more effective, and more economical than in-house materials. Manuals may be required for highly technical jobs or for more thorough coverage of a plant's safety policies and rules. Even a company-wide (or plant-wide) public address system, closed-circuit TV network, or videotapes would be appropriate.

When the safety professional and editor are deciding on the form of the publication, they should keep in mind that there is a direct relationship between the appearance of a printed piece and the degree of interest it arouses. Most readers will react unfavorably to a bulletin, newsletter, or booklet with text in small type, few or no illustrations, narrow margins, and long paragraphs.

Reasonably large type (10 point or 12 point), selected to fit the size of the page and, of course, to accommodate the volume of material, will make the material more readable. The Council newsletters, published by various sections of the Industrial Division of the National Safety Council, are set in 10-point type in about 2-in.-wide columns. In addition, judicious use of white space and variety in size and placement of illustrations help make a publication both pleasing to the eye and easy to read. In safety, as in other fields, ideas conveyed in print are best received and absorbed if they are well organized and attractively presented.

Illustrations break up the text and help to convey information to the reader. Photographs that show action described in the text add realism to instructional materials such as manuals. Human interest photos are desirable in newsletters. Line drawings and sketches are valuable to clarify technical points on instruction cards, in manuals, and in other training materials. Awards can also be publicized.

If the printing process permits reproduction of photos and other illustrations, pictures of award winners, safety devices, and safe and unsafe practices can be used. To avoid embarrassing or ridiculing employees who have been injured or caught in an unsafe procedure, their features can be blocked out or the pictures can be posed (and so identified) to reproduce the event. Or cartoon illustrations can be made up depicting the unsafe procedures from the photograph.

Some state laws forbid publication of a person's photograph without their written permission. As a result, the editor should obtain a signed release from every person who appears in recognizable form in any picture. Often having a new employee sign a photo release is part of the employment routine. Asking for a photo release is just good manners. Details on illustrations are given in Chapter 17, Audiovisual Media.

Preparation of material. Once the objectives, scope, and form of a publication are determined, the person preparing it should make an outline of the subject or subjects to be covered. For most types of material, the outline need not be elaborate, but it should be logical and complete, showing how each topic is a part of the overall plan.

Before gathering material, the writer should study the audience for whom the message is intended in order to get some idea of the readers' knowledge and level of comprehension. In the interests of accuracy, completeness, and balance, material should be gathered from several sources—including articles, books, and especially supervisors, workers, and others in the company who have had experience in the matters to be treated. To ensure technical accuracy, it may be necessary to solicit help from specialists in specific areas.

No matter what the form of the publication, the writer should keep in mind certain basic rules of good writing. To get ideas across quickly and easily, use short sentences, simple words, and brief paragraphs.

In a piece of some length, such as a booklet, a system of headings, kept as informal as possible, will both arouse the reader's interest and guide his or her thinking. In a piece designed to instruct, numbered lists of job steps, for instance,

will prove helpful. In any case, the writer should follow closely the line of logical thought developed in the outline.

Copy should be written in a positive, constructive style. When the nature of the material and the form of publication permit, a friendly—but never condescending—tone can be used effectively. Personal references and names, as in a newsletter, will increase readership. Humor tied to the message and pitched to the employees' sense of what is funny can add a great deal to some types of publications. For instance, cartoon illustrations and a light touch may be particularly effective in a rule booklet. The proposed publication's readability can be gauged by having a few people from the intended audience test-read the piece for understanding.

Production of publications. For the technical details of printing, the advertising department or experts in the publishing field can be consulted. In terms of layout and typography, readability should be the first consideration.

How the piece is to be used will determine its size, paper, binding, cover, and similar details. If the materials are to be filed or if revised pages will be inserted, loose-leaf binders may be used. The in-company or outside editor or printer who will handle the job should be asked for technical advice.

Getting ideas. Everyone in the PR business runs out of ideas now and then. Anyone suffering from this affliction should not hesitate to call on others for help. Employee publications do not compete with one another, so ideas can be borrowed freely from them. Some national agencies produce and supply safety materials that may serve as inspiration. (See listings in Chapter 11.)

The National Safety Council publishes in *Safety and Health* a "Safetyclips" page, which contains stories and illustrations for editors of employee publications. Council newsletters and other publications contain a wealth of interesting and informative material. Council poster miniatures and other materials usually are released for general use if the customary publication credit is given.

Volunteering to serve as an editor of a National Safety Council section newsletter is good practice.

SUMMARY

- Companies must work to maintain management and employee interest in safety and health programs because accident prevention depends on the desire of people to work safely. Thus, the major objective of a program designed to maintain safety awareness is to involve employees in accident prevention.
- Indications of the need for such a program include an increased rate of accidents and injuries, deteriorating housekeeping, and decreasing interest on the part of supervisors in enforcing safety and health regulations and procedures.
- A well-designed program to maintain interest in safety and health issues attempts to "sell" safety to employees. It must be based on management and worker needs and opinions. When developing the program and its related activities, the

safety professional will need to consider company policy and experience, budget and facilities, types of operations, types of employees, basic human interest factors for promotional activities, and other considerations. Activities and promotional efforts should emphasize a positive, constructive approach to safety and be changed often to catch employees' attention.

- Safety and health professionals are responsible for assisting line management in developing and coordinating the program and for encouraging the support of management, supervisors, and employees. Safety and health directors also work with local, community, state, and national groups interested in promoting accident prevention. They must be able to communicate their ideas clearly and effectively and to educate supervisors to maintain high levels of safety performance in their work areas.
- The ranking managers play a key role in any program designed to create and maintain interest in safety. Under existing laws, they are directly accountable to management and society for their employees' safety. Line managers, including supervisors, must (1) set a good example for their workers, (2) teach safe practices and procedures to employees, (3) enforce the rules fairly and consistently, and (4) keep abreast of the latest developments in safety and health issues.
- Programs designed to create and maintain interest in safety can include safety committees and observers along with quality and safety circles. Committees should consist of management and worker representatives, and membership should be rotated periodically. Quality and safety circles involve employees in hazard recognition and control. These groups meet regularly to discuss safety and health-related issues.
- Safety and health meetings can be conducted for managers, supervisors, employees, or other groups within a company. These gatherings can be of top executives, mass meetings of all employees, departmental meetings, or small group meetings. The gatherings must be carefully planned to focus on specific problems and solutions, or their effectiveness will be diluted.
- Various promotional activities such as contests can help to stimulate worker interest in safety and health policies and procedures. Contests should be undertaken only after management has established a sound safety program in the company—they must not be viewed as a substitute for such a program.
- To be successful, contests must be planned and conducted by a committee representing all competing groups; competing groups must be composed of natural units; methods of grading must be simple and fair; awards must be worth winning, and the contest should be well publicized and promoted. Types of contests include injury-rate contests, Council contests and awards, associate contests, intergroup competitions, personalized contests (raffles, sweepstakes), noninjury-rate contests, community and family contests, and miscellaneous contests.
- The value of contest awards lies in their appeal to basic

human interest factors such as pride, need for recognition, urge to compete, and desire for financial gain. Awards must be meaningful to workers to be properly motivating. Many companies allow employees to select their own awards. Award presentations should be staged to provide maximum recognition for the recipients and maximum publicity for the company.

- Posters and displays can reach a large audience with brief, simple messages designed to convey information, to change attitudes, and to change behavior. Properly used, posters and displays have great value in safety programs. They are used to remind employees to work safely, suggest behavior patterns that prevent accidents, inspire interest in safety efforts, foster attitudes that accident-prevention is a skill, and remind employees of specific hazards.

- There are three types of posters: general posters concerned with broad safety topics, special industry posters that apply only to specific industries, and special hazard posters emphasizing particular hazards.

- Posters and displays should be set up where employees are most likely to see them. They should be prominently visible and changed often to maintain worker interest in their messages. Companies can use professionally designed posters or ask managers and employees to design their own. Posters can be mounted on bulletin boards or hung in their own frames.

- Other promotional methods include various campaigns to promote home safety, vacation safety, fire safety, and the like. These campaigns can be timed to coincide with elections, sports events, holidays, and so on. Unique safety ideas or stunts can be effective if they are not overdone. Courses and demonstrations on safety and health topics serve the dual purpose of stimulating interest in safety and teaching workers valuable skills and techniques. Finally, various publications such as reports, annual reports, newsletters, and booklets, leaflets, and personalized messages can help generate interest in safety among workers.

- Company public address systems and suggestion systems are also effective ways to maintain workers' interest. Public address messages should be brief and well timed, or employees tend to tune them out. Suggestion systems must be carefully planned and administered to reward good ideas and demonstrate the company's interest in safety and health policies.

- The Wise Owl Club, Golden Shoe Club, and Turtle Club are three of the most well-known safety recognition organizations. These groups offer assistance to companies that wish to improve their safety record or who want to recognize workers who have escaped serious injury through the use of personal protective equipment.

- Many companies do little or nothing to publicize their internal safety and health programs and activities. Yet the best way for a company to develop a reputation for taking care of employees is to be known as a safe place to work.

- Public relations is defined as the management function that evaluates public attitudes, identifies the firm's policies and activities with the public interest, and brings company news to the public's attention. Most organizations have either a professional public relations department or delegate public relations responsibilities to the safety director.

- To be successful, a public relations program must be actively supported by top management, a strong safety program, and newsworthy stories supplied by the safety professional. Lack of a safety program and poor public relations efforts can cost companies millions of dollars each year in lost productivity and potential business.

- The safety professional can conduct publicity work within a company by coordinating efforts with the public relations department and with the employee publications team to publicize the safety program. The safety professional can point out how the PR director and publications editor can help to gain employee acceptance and support of the safety program.

- Safety activities contain a wealth of real news. Items such as no-accident records, interplant safety contests, safety suggestions by employees, safety award presentations, and the like can be the focus of a company media event.

- In companies without public relations departments, the safety and health director can still obtain publicity for the firm's safety activities. Stories should be selected for a specific audience, contain human interest and appropriate humor, include graphics (cartoons, photographs, posters, etc.), and focus on people not things.

- Safety professionals who want to make the most of public relations opportunities must (1) be honest in what they say, (2) deliver what they promise, (3) notify the media of any changes immediately, (4) supply accurate information, (5) offer ideas not demand space or time, (6) feature only real news, (7) tip off local media to any worthwhile story, (8) give stories news value and (9) be forthright about bad news (accidents, injuries, etc.).

- In working with local media, the safety and health professional should keep in mind the following tips: (1) keep press releases brief, timely, accurate, and complete; (2) be familiar with the publication or radio/TV station and know which editor or news director to contact; (3) tailor press releases for the audience; (4) include captions with photographs; and (5) leave script writing to the professionals but make sure all facts are accurate.

- When the safety and health professional must give a radio or TV interview, the individual should (1) be natural and refrain from distracting gestures or adopting an "official" tone, (2) dress conservatively, (3) review in advance what topics will be covered, and (4) respond naturally to the interviewer's questions without attempting to cram information into the time allotted.

- When a serious accident occurs at a plant or company, the public relations and safety professional must work together to provide accurate, honest information about the incident. Bad news about the accident should be balanced by pointing out the company's overall safety record or safety achievements.

- Company publications represent excellent opportunities

to publicize safety programs, activities, and issues. In planning to start internal and/or external publications, the company should take the following steps: (1) clearly define the objectives and purpose of the publication, (2) determine how general or restricted the message will be, (3) decide what form of publication is best, and (4) estimate costs of preparing and printing the publication.

- Eye-catching illustrations and concisely written, lively copy add interest to publications. The more audience-based the publication is, the more readers are likely to accept the messages they deliver. Companies should also seek the advice and expertise of professionals in the public relations field to help ensure the success of their publications.

REFERENCES

Alliance of American Insurers, 1501 Woodfield Road, Schaumburg, IL 60195. *Fire Prevention and Control, A/V.*

Culligan MJ and Crewe D. *Getting Back to the Basics of Public Relations and Publicity.* New York: Crown Publishers, 1982.

Forrestal D. *Public Relations Handbook,* 2nd ed. Chicago: The Dartnell Corp., 1979.

Indiana Labor and Management Council, Inc., 2780 Waterfront Parkway, Indianapolis, IN 46214. *Worker Involvement in Hazard Control Booklet,* 1985.

Lesly P. *Lesly's Public Relations Handbook.* Report. Englewood Cliffs, NJ: Prentice-Hall, 1983.

Moore HF and Kalupa FB. *Public Relations Principles, Cases and Problems.* Homewood, IL: Richard D. Irwin, 1985.

National Safety Council, 444 North Michigan Avenue, Chicago, IL 60611.
> *Accident Facts* (annual).
> > Catalog and Poster Directories.
> *Family Safety and Health Magazine.*
> > Industrial Data Sheets.
> *Nonprojected Visual Aids,* 12304–0564, 1980.
> *Photography for the Safety Professional,* 12304–0619, 1982.
> *Posters, Bulletin Boards, and Safety Displays,* 12304–0616, 1986.
> *Projected Still Pictures,* 12304–0574, 1980.
> *Writing and Publishing Employee Safety Regulations,* 12304–0664, 1982.
> *Section "Newsletters."*
> *Today's Supervisor Magazine.*
> *Safety and Health Magazine.*
> *101 More Ideas that Worked.*
> *Safe Driver Magazine.*
> *Safe Worker Magazine.*
> *You Are the Safety and Health Committee—Booklet.* rev. 1986.

Nolte LW and Wilcox DL, eds. *Fundamentals of Public Relations: Professional Guidelines, Concepts, and Integrations,* 2nd ed. Elmsford, NY: Pergamon Press, Inc., 1979.

Parkhurst W. *How to Get Publicity.* New York: Times Books, 1985.

19

Office Safety

Many large organizations today, such as insurance, governmental, and financial companies, consist almost entirely of office workers. Accidental injuries are just as painful, severe, and expensive when they occur to office workers as when they happen to production workers. Unless an organization develops an effective office safety program, accidental injuries are far more likely to occur.

It is true that the risk of an occupation-related accident or injury to the office worker is lower than the risk to those employees involved in manufacturing or transportation. However, office risks often go unrecognized and unmanaged, and some could eventually lead to serious injuries and property loss.

A company safety program cannot be fully effective if there is only partial participation by employees and management. A safety program that is not vigorously pursued in company offices probably will not be vigorously pursued in the factory, shop, or plant. If office workers are exempt from safety policy, then production workers often feel that following rules to avoid hazards is an unnecessary burden, and, perhaps, an unfair exercise of authority by management. Exempt office workers seldom understand the importance of safety and may scoff at or criticize production-oriented safety activities, especially in front of production employees.

The health and safety professional who expects to sell safety to management must get management involved in a total safety program. The emphasis must be on preventing office accidents as well as production accidents.

SERIOUSNESS OF OFFICE INJURIES

One reason office safety programs are not more widespread is that many people believe office injuries are minor. This is a serious mistake. One aerospace firm, for example, paid out $102,000 over eight years at one facility just for injuries incurred by people falling out of chairs. Approximately 25,000 people worked at the plant, and more than half were office workers. Of the 14 chair accidents that occurred, the two worst cost the company a total of $97,000. Not only are medical and wage replacement benefits expensive for disabling accidents, but there also are hidden costs such as the loss of productivity. (See Chapter 12, Accident Investigation, Analysis, and Costs.)

Studies made by the State of California Department of Industrial Relations and the Equitable Life Assurance Society of the United States (see References) show that this is not an isolated incident (Figure 19–1). The California State Department of Industrial Relations analyzed reports filed by more than 3,000 California employers on disabling injuries to employees. These employers together employed more than one million office workers. "Office worker" was defined in the California study as "a person primarily engaged in performing clerical, administrative, or professional tasks indoors in an office at the employer's place of business." This definition did not cover salespersons, claims adjusters, social workers, medical and teaching personnel (other than clerical or administrative), and certain stock, order, and inventory clerks.

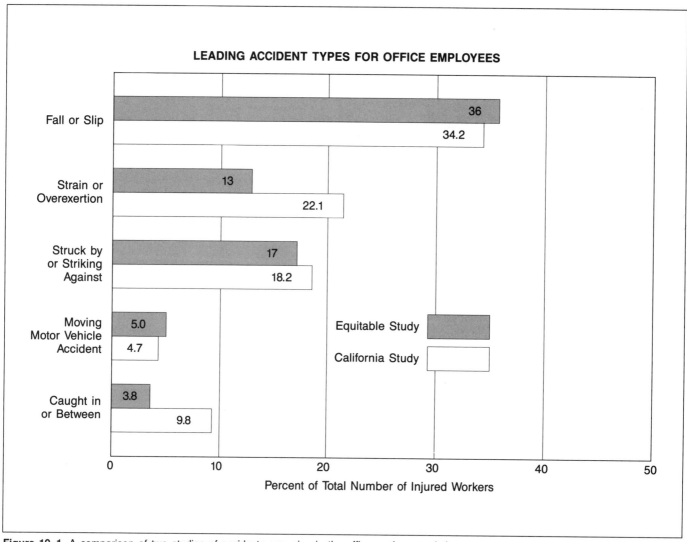

Figure 19–1. A comparison of two studies of accidents occurring in the office worker population.

Table 19–A. Disabling Accidents from Falls in Offices

Equitable Life—8,000 employees (an eight-year study)	Disabling Accidents	Days Lost
In hallways and work areas, caused by running, slipping, tripping over wires, desk drawers, file cabinet drawers, etc.	53	553
From chairs	21	120
Stairs	16	117
Escalators or elevators	8	55
Total	98	845

California Survey—1,000,000 employees (a one-year study)	Disabling Accident Totals
Falls (all categories)	4,360
Falls or slips on stairs or steps	752
Falls from other elevations	370
Falls on the same level	3,238

When the results of the study were extrapolated nationwide, investigators found that on-the-job office accidents amounted each year to about 40,000 disabling injuries at a direct cost (indemnity benefits and medical expenses) of about $100 million. This figure does not include any indirect costs for employers, workers, or the nation.

Of the many accidental fatalities occurring to office workers, approximately half are work-connected automobile ac-

Table 19–B. Percentage of Disabling Work Injuries* and Illnesses to Office Workers by Occupation and Length of Service

Occupation	Total Disabling Work Injuries	Length of Service					
		1 mo.	2 mo.	4–6 mo.	3–5 yr.	6–10 yr.	11–20 yr.
Clerical and kindred	12,858	2.1	4.3	4.0	17.1	15.7	9.2
Professional, technical, and kindred	1,418	1.5	2.5	7.5	17.7	15.7	12.1
Managers and administrators	2,000	1.4	2.4	3.4	18.1	16.0	18.9
Totals	16,276	2.0	3.9	8.2	17.2	15.7	11.0

*These figures are true only for the California Study.

cidents. However, this amount can be significantly reduced by using a defensive driving program (DDC) such as that developed by the National Safety Council.

The California study did not cover employees of the federal government, maritime workers, and railroad workers in interstate commerce. The Equitable Life Assurance Society study, on the other hand, did include salespeople, claims adjusters, medical personnel, and supply and warehouse personnel. The insurance company study covered approximately 8,000 employees of one company working in one building, about 5% of whom were maintenance personnel. During the eight-year study period, the injury frequency rate was 2.3, and severity rate was 15 for the office workers (see Chapter 13, Recordkeeping and Incidence Rates). The average days charged per disabling injury was 5.9. See Tables 19–A and 19–B for details of the studies.

Complacency—Primary Cause of Injury

Complacency—the attitude that office accidents do not amount to much—is one of the prime causes of office accidents. The average office worker gives little thought to safety because office work is not perceived as being hazardous. Office injuries may seem to be inconsequential because they often lack dramatic impact. For example, the person who is injured falling backward from a chair seems to merit little sympathy or attention. The image of an office worker slipping and crashing to the floor on his back may even seem amusing to some people.

Industrial accidents are commonly considered to occur more frequently and with greater severity—amputations, lost eyes, and broken bones—than office accidents. But who is likely to be absent from work longer: the office worker who sustains a severe compression fracture of one or more lumbar vertebrae from a bad fall, or the production worker whose hand is amputated? Probably, it would be the office worker.

On the other hand, the well-instructed worker in a plant manufacturing flammable, toxic chemicals knows that it is a risky business—and understands why safe procedures must be used and safety equipment worn. The worker's safety behavior and training are the best defense. As a result, the production department may have fewer accidents than the plant office.

The office worker, therefore, has to be informed of the hazards and safe work procedures that apply to office jobs.

The employee must also be shown that management provides a safe environment and safe equipment to encourage proper safety behavior. The worker must be willing to adopt safe procedures, and be encouraged to do so. Even more important, office supervisors must understand the nature of office hazards and unsafe practices and take necessary measures to reduce and manage these hazards.

Who Gets Injured?

The California and the Equitable Life studies pointed to whom most injuries occurred, and how they occurred.

New surroundings. These studies showed the importance of teaching office employees when working in new surroundings to look for new hazards and to correct them. Studies revealed a substantial increase in the number of injuries in the first year a company moved into a new office building. The change upset established routines and presented unknown hazards. Even going to and from work became more hazardous as employees had to explore new driving and transportation routes.

New or young employees. The California study found that the new or younger employee does not have a higher accident rate than the longer employed and older worker—at least this was true for those employed in California (Table 19–B). In California, office employees who had been on the job for one and two months had an injury percentage rate of 2.0% and 3.9%, respectively. At the same time, those who had been employed for three to five years had the highest accident percentage rate (17.2%). As for age, the study showed that only 3% of the injuries occurred in the 18-to-19-years bracket.

Sex of employee. In the California study, 70% of the disabling injuries occurred to women, who comprised 68% of the office labor force. The study also shows that for office occupations, the estimated disabling work injury rate for women was very close to the injury rate for men. The rate was approximately 7.8 disabling injuries per 1,000 women employed in office work compared with 7.1 for the same number of men.

The injury statistics compiled from the California study were similar to the Equitable study. The rate of injury accidents per thousand male employees was about the same as it was for female employees. However, the rate of total days lost from disabling injuries was two and one-half to three times higher for men than for women.

Figure 19–2. Desk and file drawers should always be closed when not in use. Labels on drawers help relieve stress and strains. (Courtesy Zee Medical.)

Types of Disabling Accidents

Falls are the most common office accident and account for the most disabling injuries, according to both surveys (Figure 19–1). They cause from two to two and one-half times as high a disabling injury rate among office as among nonoffice employees. Falls were the most severe office accident and were responsible for 55% of the total days lost because of injuries.

Most chair falls came when a person was sitting down, rising, or moving about on a chair. A few were caused by people leaning back and tilting their chairs in the office or cafeteria, or putting their feet up on the desk. Although stairs seem more hazardous than chairs, people seem to recognize the stair hazard and are more cautious. Furthermore, people are not as often exposed to the stair hazard as they are to a chair.

Another major accident category is falls on the same level. This includes slipping on wet or slippery floors and tripping over equipment, cords, or litter left on the floor. Good housekeeping procedures can reduce accidents in this category.

A final category was falls from elevations, caused by standing on chairs or other office furniture and by falls from ladders, loading docks, or other miscellaneous elevations. These falls (not including stairs) accounted for approximately 2% of the disabling injuries in the California study, whereas falls on stairs accounted for almost 5%.

Overexertion. Almost three-fourths of the strain or exertion mishaps occurred while employees were trying to move objects—carrying or otherwise moving office machines, supplies, file drawers and trays, office furniture, heavy books, or other loads. Often, the employees were moving office equipment or furniture without authorization from their supervisors. A significant number resulted when the employee made a sudden or awkward movement and did not involve any outside agency. Reaching, stretching, twisting, bending down, straightening up, and cumulative trauma were often associated with these injuries.

Objects striking or struck by workers. Objects striking office workers accounted for about 11% of the injuries in both studies. Most of these injuries were sustained when the employee was struck by a falling object—file cabinets that became overbalanced when two or more drawers were open at the same time, file drawers that fell out when pulled too far, office machines and other objects that employees dropped on their feet when attempting a move, or typewriters that fell from a folding pedestal or rolling stand. In addition, a number of employees were struck by doors being opened

from the other side. Office supplies or other material and equipment sliding from shelves or cabinet tops caused a few injuries in this classification.

Striking against objects caused approximately 7% of the office injuries in both studies discussed here. Two out of three of these injuries were the result of bumping into doors, desks, file cabinets, open drawers, and even other people (Figure 19–2) while walking. Hitting open desk drawers or the desk itself, while seated at a desk, or striking open file drawers while bending down or straightening up caused most of the rest of these injuries. Other incidents included workers bumping against sharp objects such as office machines, spindle files, staples, and pins. This category also comprised infected cuts incurred when employees handled paper, file drawers, and supplies.

Caught in or between. The final major classification was accidents where the worker was caught in or between machinery or equipment. Mostly, this was getting caught in a drawer, door, or window. However, a number of employees got caught in duplicating machines, copying machines, addressing machines, and fans. Several got their fingers under the knife edge of a cutter.

Miscellaneous office accidents included foreign substances in the eye, spilled hot coffee or other hot liquid, burns from fire, insect bites, electric shocks, and paper cuts. Eyestrain or muscle aches from extended CRT/VDT use and cumulative trauma injuries also are to be considered hazards to office workers.

CONTROLLING OFFICE HAZARDS

Office accidents can be controlled by eliminating hazards or reducing exposure to them. Management can eliminate or reduce hazards most easily when the office is in the planning stage, when equipment is purchased, or when new office procedures are set up.

Layout and Ambience

Offices should be laid out for efficiency, convenience, and safety (Figure 19–3). The principles of workflow apply to offices as well as to factories. Location of the office should take into account the inherent hazards of adjoining operations.

Stairways and exits (including access and discharge) should comply with NFPA 101, *Life Safety Code;* floor and wall openings should comply with ANSI *Construction Safety Requirements for Temporary Floor and Wall Openings, Flat Roofs, Stairs, Railings, and Toeboards,* A10.18. Handrails, not less than 30 in. or more than 34 in. (0.8 and 0.9 m) above the upper surface of the tread, are specified for one side of stairs up to 44 in. (1.1 m) wide, and both sides for stairs wider than 44 in. For stairs wider than 88 in. (2.2 m) add an intermediate (center) rail.

Exits should be checked frequently to be sure that stairways are unobstructed and well illuminated. Exit hallways or paths should have emergency lighting. Exit doors, if locked, should not require the use of a key for operation from inside the building.

Doors are another frequent source of accidents in offices. Glass doors should have some conspicuous design, either painted or decal, about 4 ft above the floor and centered on the door so that people will not walk into it (see Chapter 21 for details). Safety glass complying with ANSI standard Z97.1 should be installed, particularly in doors, rather than plate glass. Sometimes local codes specify the type that must be installed.

Frosted safety glass windows in doors provide vision to prevent accidents while preserving privacy. Solid doors present a hazard because they can be approached from both sides at the same time, and one person can be struck when the door opens. Employees should be warned of this hazard and instructed (1) to approach a solid door in the proper manner, that is, away from the path of the opening door; (2) to reach for the door knob so that, if the door is suddenly opened from the other side, the hand receives the force of the impact rather than the face; and (3) to open the door slowly, especially if the door opens outward.

Another hazard is the door that opens directly onto a passageway. If the door opens directly into the path of oncoming traffic, somebody might bump into the edge of the door. If doors that open onto hallways cannot be recessed, they should be protected with short-angled deflector rails or U-shaped guardrails that protrude about 18 in. (46 cm) into the passageway. As an alternative, the area they swing over can be marked as suggested in the following paragraph. Another procedure is to place storage lockers or benches along the wall near the door, which provides the safety of a recessed door.

Some offices having tile floors will paint white or yellow stripes or apply tape to mark traffic flows or to guide people away from a rapidly opening door. The floor in front of a swinging door also can be marked or painted as a warning or a warning sign could be posted. As a final precaution, it is good practice to have the door hinges on the upstream side of the traffic; that is, on the righthand side as one faces the door from the hallway. Doors are covered in the *Life Safety Code,* NFPA 101.

Lighting. Adequate light, ventilation, and other employee services have an important influence on employee morale. Business growth often requires installing more desks and other equipment than originally planned. Overcrowding is undesirable both for appearance and the physiological effect on employees, especially if it overtaxes ventilation facilities. Smaller offices can be made to appear larger and less crowded if walls, woodwork, and furniture are the same color.

Illumination levels recommended by the Illuminating Engineering Society for an office are listed in Table 19–C. Also, see Chapter 1, Buildings and Plant Layout, of the *Engineering and Technology* volume of this *Manual.*

Some accidents can be attributed to poor lighting. However, many less-tangible factors associated with poor illumination are contributing causes of office accidents. Some of these are direct glare, reflected glare from the work, and harsh shadows, all of which hamper sight.

Excessive visual fatigue can be an accident-causing ele-

Figure 19–3. Office work-stations should be arranged so that required equipment is easily accessible. Provide wheeled carts for transport of light loads, such as computer printouts or books. (Courtesy Zee Medical.)

ment. Accidents also can be prompted by delayed eye adaptation when moving from bright surroundings into dark areas and vice versa. Some accidents attributed to an individual's carelessness can actually be traced to difficulty in seeing.

Office design can facilitate good lighting. If offices depend primarily on daylight, for example, employees engaged in visual tasks should be located near windows. North light is preferred by drafters and artists. However, employees generally should not face windows, unshielded lamps, or other sources of glare. Indirect, shielded fluorescent lamps can produce high levels of illumination without glare. Furthermore, walls and other surfaces should conserve light while avoiding annoying reflections. Ceiling, walls, and floor act as secondary large-area light sources and, if finished with the recommended reflecting paints or wall coverings, will increase light and reduce shadows. Finally, modern office lighting must be designed to counteract potential stress and eye fatigue in CRT/VDT users.

As a uniform means of evaluation, a standard procedure, entitled "How to Make a Lighting Survey" has been developed in cooperation with the U.S. Public Health Service. (This publication is not part of the standard, *Practice for Office Lighting*, ANSI/IES, RP1-1982, but is presented as background material for the use of the standard.) Table 19–C shows the recommended illumination levels.

Ventilation. For large interior spaces, forced ventilation is needed if the space is to be used as an office area. All mechanical ventilation and comfort conditioning systems require careful planning and installation by qualified specialists. Private offices installed around the outer walls of a large office space should not cut off light and ventilation to other employees. If fans are used in an office, they should be guarded, secured, and placed where they cannot fall.

The designs of many office buildings seal in office air—odors, smoke, office solvents and chemicals, molds, fungus, and other contaminants. In these buildings, it is particularly important to design and install adequate ventilation systems. Management should regularly inspect the systems and update them if problems appear.

Electrical. Management must protect workers against electrical equipment hazards in an office. In some cases, the hazard can be avoided completely, such as not using electric-key switches. In other cases, hazards can be reduced by using UL-listed equipment, use of ground-fault circuit interrupters (GFCI), installing sufficient well-located outlets (receptacles), and arranging cords and outlets to avoid tripping hazards.

Employee should not use poorly maintained, unsafe, or poor-quality, non-UL-listed electrical equipment, such as coffee makers, radios, and lamps. Such appliances can create fire and shock hazards.

Table 19–C. Levels of Illumination for Offices

	Recommended Illumination* (Footcandles)
Cartography, designing, detailed drafting	200
Accounting, auditing, tabulating, bookkeeping, business machine operation, reading poor reproductions, rough layout drafting	150
Regular office work, reading good reproductions, reading or transcribing handwriting in hard pencil or on poor paper, active filing, index references, mail sorting	100
Reading or transcribing handwriting in ink or medium pencil on good quality paper, intermittent filing	70
Reading high-contrast or well-printed material, tasks and areas not involving critical or prolonged seeing such as conferring, interviewing, inactive files, and washrooms	30
Corridors, elevators, escalators, stairways	20 (or not less than ⅕ level in adjacent areas)

*Minimum on task at any time.

The company should make sure that a sufficient number of outlets are installed to eliminate the need for extension cords. Those that are necessary should be clipped to backs of desks or taped down. If cords cannot be dropped from overhead and must cross the floor, cover them with rubber channels designed for this purpose. Cords should not rest on steam pipes or other hot or sharp metallic surfaces.

Outlets should accommodate three-wire grounded plugs to help prevent electric shock to operators. Floor outlets should be located, if possible, so they are not tripping hazards and cannot be accidentally kicked or used as a foot rest. A floor outlet protruding above floor level is frequently shielded by a desk or some other piece of furniture. However, when the desk is moved, the outlet can become an immediate tripping hazard unless it is appropriately covered. Such floor design is not common any longer; underfloor or cellular floor raceways are usually used in new construction.

The National Electrical Code, ANSI/NFPA 70, requires GFCI in restroom areas. In all areas, wall receptacles should be so designed and installed that no current-carrying parts will be exposed, and outlet plates should be kept tight to eliminate possibility of shock or collision injury.

Cords for electrically operated office machines, fans, lamps, and other equipment should be properly installed and frequently inspected to see if there are any defects that can cause shocks or burns. Switches should be provided, either in the equipment or in the cords, so workers do not have to pull the plugs to shut off the power. The fact that office equipment is operated on 110–volt circuits is no assurance that serious injury will not occur. Fatalities can be caused by current (amperes) as low as 100 milliamperes if it passes through the vital organs (current is more significant than volts). (See Chapter 15, Electrical Equipment, in the *Engineering and Technology* volume.)

Office electrical service should be designed to accommodate changes in equipment and technology. Electrical, co-axial, computer, and telephone cables are easily serviced and moved when run through modular channels dropped from the ceiling.

Installation or repair of any electrical equipment should be done by qualified workers using only approved materials. Because defective wiring may constitute both shock and fire hazards, management and workers must observe all recommendations of the National Electrical Code, ANSI/NFPA 70.

Equipment. Workers should never place an office machine on the edge of a table or desk. Maintenance staff or manufacturer representatives should secure machines that tend to creep during operation. They can either affix the machines directly to the desk or table or place them on a nonslip pad. In particular, typewriters on folding pedestals should be fastened to the pedestal.

Workers should place heavy equipment and files against walls or columns; files also can be placed against railings. File cabinets should be bolted together or fastened to the floor or wall so workers cannot tip them over.

Floors. Floor surfaces are one of the major causes of office accidents. They should be as durable and maintenance free as possible. Management should select floor finishes for slip-resistant qualities. Well-maintained carpet provides good protection against slips and falls. Maintenance staff must repair defective tiles, boards, or carpet immediately. They should also replace or repair worn or warped mats under office chairs and rubber or plastic floor mats with curled edges or tears. These conditions can create tripping hazards.

Highly polished and extremely hard but unwaxed surfaces such as marble, terrazzo, and steel plates represent slipping hazards. Slip-resistant floor wax can give these materials a higher coefficient of friction and can reduce their slipping hazard. However, maintenance staff must not apply wax so thickly that a smeary coating results. Also, they should not use an oil mop on a waxed floor because it creates a soft, smeary coating that could become a slipping hazard (see Chapter 21, Nonemployee Accident Prevention).

Special slip-resistant protection should be used on stairways and at lobby or elevator entrances. Be sure these es-

pecially hazardous areas are always maintained in the best possible condition. Floor mats and runners often provide a better, more slip-resistant walking surface. Their use is discussed in the National Safety Council's Industrial Data Sheet 12304-0595, *Floor Mats and Runners*. A well-planned routine maintenance program is needed to keep entrance, cafeteria, and vending area floors dry. Refer to Chapter 21, Nonemployee Accident Prevention.

Outdoor areas. Parking lots and sidewalks represent major problem areas. Many slipping and falling accidents occur in the company parking lot. To reduce the hazards, maintenance workers must keep the lot clean, remove debris, fill potholes, and correct uneven surfaces. In colder climates, effective snow and ice removal controls should be used during the winter months. Again, Chapter 21 covers the subject in greater detail.

Aisles and stairs. A suggested minimum width for aisles is four feet. All aisles and passageways must meet the width requirements of NFPA 101, *Life Safety Code*. Keep passages through the work area unobstructed, and place wastebaskets where people will not trip over them. Be sure to install telephone cables and electrical outlets so their wires do not create a tripping hazard in passageways. These and other obstructions, such as low tables and office equipment, should be placed against walls or partitions, under desks or in corners. Avoid building stepoffs from one level to another in an office; if one exists it should be well marked and guarded with a railing.

File drawers should not open into aisles, particularly narrow ones, unless extra space is provided. Pencil sharpeners and typewriter carriages must not jut out into aisles.

Storage. Materials stored in offices sometimes cause problems. In general, materials should be stored in areas specifically set aside for the purpose. The storage area should be located so general traffic patterns do not have to be crossed to reach the stored items. Also, workers should not store or leave anything on the floor in a passageway where it could become a tripping hazard.

Train workers to stack materials in stable piles that will not fall over. They should put the heaviest and largest pieces on the bottom of the pile. When materials are stored on shelves, the heavy objects should be on the lower shelves. Workers should not stack objects on window sills if there is a danger the objects may break the window or fall through it.

Supervisors and managers should plan storage areas to make items easily accessible. Appropriate stepladders should be provided where necessary. Office falls can occur when workers stand on chairs, counters, or shelves to reach inconveniently stored items. Rolling ladders are discussed in the next section, Safe Office Equipment.

Companies should prohibit smoking in mailing, shipping, print shops, or receiving rooms. They should also ban smoking in other areas where large quantities of loose paper and other combustible material may be stored and in areas where flammable fluids are used, such as duplicating rooms or artists' supply areas.

Workers should store flammable and combustible fluids and similar materials in safety cans, preferably in locked and identified cabinets. Only minor quantities should be left in the office, and bulk storage should be in properly constructed fireproof vaults. (See Chapter 16, Flammable and Combustible Liquids, in the *Engineering and Technology* volume of this *Manual*.)

Safe Office Equipment

Good quality office furniture not only contributes to the safety of the office but also enhances its appearance. This, in turn, improves the attitudes of both employees and visitors.

Chairs, especially, should be comfortable and sturdily built with a wide enough base to prevent easy tipping. Five-legged chairs are more stable and discourage employees from tilting back on their chairs. (See Chapter 5, Ergonomics in the Workplace, in the *Engineering and Technology* volume, for further details.) The casters on swivel chairs should be on at least a 20-in. (0.5 m) diameter base, but a 22-in. base is preferred. The casters should be securely fixed to the base of the chair and well constructed because loose or broken casters are a frequent cause of chair falls. About 20% of the chair falls in the California study were due to chair defects.

Companies should purchase chairs with easy-to-adjust seat heights and back supports. Show employees how to properly adjust their chairs. The correct fit will make the employee more comfortable and help to reduce acute and chronic back strain—enabling office workers to work more safely and productively.

Desks and files. Spring-loaded typing desks should be carefully selected. If a worker opens some models without due care, the typewriter table will snap out and could bruise or cut the person.

Also, even if good quality desks and file cabinets are purchased, it is still possible that occasionally one will have a sharp burr or corner on it. Supervisors or maintenance staff should inspect office furniture when it is received and remove such burrs or corners immediately. Drawers on desks and file cabinets should have safety stops to prevent workers from pulling them out of the drawer slot.

Other safety tips can help to prevent accidents and injuries. Purchase office machines, such as rotary files, copying machines, paper cutters, and paper shredders, with well-designed guards. Glass tops on desks and tables can crack and cause safety hazards. Durable synthetic surfaces are safer. Make sure workers have enough noncombustible wastebaskets. If smoking is permitted in the area, provide safety-type ashtrays that are large and stable enough to safely contain smoking materials.

Office fans should have substantial bases and convenient attachments for moving and carrying. If located less than 7 ft (2.1 m) from the floor, they must be well guarded, front and back, with mesh to prevent workers' fingers from getting inside the guard. Many cut fingers result when people try to move fans by grasping the guard or try to catch falling fans. Train workers not to handle fans until they shut off the power and the blades stop turning.

Computers. If a computer is to be installed in a building with overhead sprinklers, keep sprinkler protection in service throughout the area, but get advice on necessary protection against both fire and water damage to computer hardware. Actually, water damage is not to be feared as much as previously thought. For one thing, most new computers are less susceptible because of their solid-state circuitry. In addition, tests have shown that water does not harm magnetic tape. Most of the damage suffered by computers in a fire results from the heat. One of the best ways to prevent a damaging fire is to keep combustible materials such as paper, tapes, and cards at an absolute minimum in the room with the computer. When safeguarding such an investment, call in a fire protection adviser as well as a computer installation expert (see Chapter 14, Computers and Information Management).

Rolling ladders and stands used for reaching high storage should have brakes that operate automatically when weight is applied to them. All stepladders should have nonskid feet.

Chemical products. Organizations often underestimate the number and types of hazards represented by office chemicals. The safety professional should assess all chemical products used in copying and duplicating machines and in print shops and should assess all adhesives and cleaning materials. Workers must be informed of any dangers and instructed in the safe use of hazardous chemical products.

If possible, substitute nontoxic and nonflammable solvents for those used in printing and duplicating or other operations. (Details are given in *Fundamentals of Industrial Hygiene,* 3rd edition, part of this *Occupational Safety and Health Series.*) If chlorinated bleaches are purchased for cleaning purposes, make sure that they will not be mixed with strongly acidic or easily oxidized materials. Purchase a good grade of slip-resistant floor wax.

Purchasing equipment. As discussed in Chapter 3, the company health and safety professional should work with the purchasing agent in buying office furniture and equipment. Both should be aware that although advertisers sometimes stress the safety features of office equipment, the machines may be delivered without these important safeguards. Mechanical hazards of heavy office equipment can be determined by careful, expert inspection before purchase. These hazards can almost always be eliminated or minimized, although sometimes at substantial expense.

The safety professional should also inform the purchasing department of precautions to be taken in connection with chemicals, dyes, inks, and other supply items. Particular attention should be paid to toxic, irritant, or flammable properties. Where hazards are unavoidable, manufacturers, suppliers, or the safety department should supply labels and specific instructions for careful handling or issue instructions when workers receive the material.

The purchasing department should gather all pertinent information from the manufacturer on equipment design and electrical and space requirements, and should try to determine the composition of proprietary compounds. They can forward this information to the safety professional (or safety department) for an opinion concerning inherent safety hazards before purchasing new equipment or supplies.

Office machines. All machines having external moving parts that could be hazardous should have enforced safety procedures and constant training and retraining of operators as necessary. Tell employees that if any office machine gives a shock, appears defective, sparks, or smokes, they should turn it off, pull the plug, and inform the supervisor.

Some office machines are noisy, especially the telex, computer printers, and printing equipment. This noise is usually more of an annoyance than a health hazard, but may need to be evaluated by a professional for possible adverse effects. Various covers help to dampen machine noise.

Printing Services

Larger offset presses. Check the operation of offset presses. Is the operator putting his or her fingers on the blanket while the press is in motion? One offset press department had seven finger-injury accidents in the first two weeks of operation, all caused by press operators who put their fingers in the running press to remove dirt or other particles from the plate. Presses should meet all guarding regulations imposed by local, state or provincial, and federal agencies.

Make sure the area around the presses is free from clutter and well lighted. The flooring should be resilient, or rubber mats should be provided to minimize operator fatigue and to prevent slipping.

Only qualified operators should operate presses. Loose clothing and long hair are hazardous around these machines. Train workers to use a safe, nontoxic substance to clean the presses; office supervisors and press operators should understand the fire and possible health hazards involved and follow all instructions for safe use and storage. The disposal of cleaning materials should be planned so that the procedure is an acceptable practice.

Gathering and stitching machines. Supervisors should make sure guards are installed on open sprockets and collector chain drives of gathering and wire stitching (or stapling) machines to protect employees from hand and body injury. The operating arm on the end of the gathering machine should be guarded.

Install hinged drop-guards to cover any exposed operating mechanism that creates nipping hazards under the machine and along the working area where operators fill the pockets. Nonskid material should cover the floors and work platform at this area. Supervisors should train operators to open signatures in the middle and place them on the saddle or rod between the hooks on the moving chain. If the hook is not put on the rod or chain correctly, workers must shut off the machine before attempting to straighten the hook out. Operators should also shut down the machine when threading stitcher heads, making any adjustments, or removing jams.

Folding machines. Here are points to be stressed for safe operation of folding machines:

1. Before jammed paper is pulled from the machine, shut the

motor off to avoid getting hands in the feed rollers.

2. Finger clearance at the folding knife should be checked before pulling out paper, putting tape on rollers, or adjusting plates and roller pressure.
3. Workers should walk down the steps of folder feeder platforms facing forward, never backward.
4. On large-sized folders, all steps and platforms should be protected by railings.

Defective staples protruding from reports or booklets should be removed to avoid cuts from them while books are being jogged, trimmed, or wrapped. Workers should be trained to cup their hands over the work when removing defective wire staples. Employees engaged in this operation should wear eye or face protection, and passers-by should be protected against flying staples by screens or by isolation of this work.

Enforced Safety Procedures

Because the major category of office accidents is slips and falls, employees should never run in offices. Also, a number of office accidents can be prevented if everyone walking in passageways would keep to the right. Convex mirrors should be placed—and used—at corners and other blind intersections. As discussed under Doors earlier in this chapter, collisions at doors also can be prevented if people do not stand directly in front of the door, but away from the path of its swing when they go to open it.

People carrying material must be sure they can see over and around it when walking through the office. They should not carry stacks of materials on stairs but use the elevator instead. If one is not available, the person should make two trips. People should not have both arms loaded when using stairs; one hand should be free to use the handrail.

When using stairs outside at night or in a dimly lit area, workers should be instructed to go single file, to keep to the right, and to always hold the handrail. People should not crowd or push on stairways; they should pay attention to where they are going. Falls on stairs often occur when the person is talking, laughing, and turning to friends while going downstairs. Other safety rules for stairs include: do not congregate on stairs or landings, and do not stand near doors at the head or foot of stairways.

Good housekeeping is essential to prevent falls. Employees should not be permitted to litter in their work areas and should wipe up all spilled liquids immediately. Pieces of paper, paper clips, rubber bands, pencils, and other loose objects must be picked up as soon as they are spotted.

Whoever breaks any glass should sweep it up at once. Do not allow employees to discard loose broken glass in a waste container. Instead, tell them to wrap it in heavy paper and mark it BROKEN GLASS FOR DISPOSAL. Glass that shatters into fine pieces can be picked up with damp paper towels.

All tripping hazards, such as defective floors, rugs, or floor mats, should be reported to the maintenance department and immediately repaired. Many falls could be prevented if employees wore supportive footwear with nonslip soles; shoes with high heels are undesirable.

Chair falls. Habits that lead to chair falls must be discouraged. Supervisors should instruct employees not to scoot across the floor while sitting on a chair nor to lean sideways from the chair to pick up objects on the floor. Discourage workers from leaning back in the chair and placing their feet on the desk. It is easy to fall over backwards in this position.

People should properly seat themselves in their chairs. (See Chapter 5, Ergonomics in the Workplace, in the *Engineering and Technology* volume, for details.) They should form the habit of placing a hand behind them to make sure the chair is in place. Sitting down on the edge of the seat rather than in the center, or backing too far without looking, or kicking the chair out from under can result in a sudden fall to the floor. Standing on a chair to reach an overhead object is particularly dangerous and must be forbidden.

Filing cabinets, as discussed earlier in this chapter, are a major cause of injuries. These include bumped heads from getting up too quickly under open drawers, mashed fingers from improperly closing drawers, and hand injuries and strains from moving the cabinets.

Some precautions are necessary against these accidents:

- People should never close file drawers with their feet or any other part of their body. They should use the drawer handle to close the cabinet, making sure their fingers are not curled over the edge when the drawer closes. All file drawers should be closed immediately after use.
- Employees should open only one file drawer in the cabinet at a time to prevent the cabinet from toppling over. As previously indicated, where possible, have the file cabinets bolted together or otherwise secured to a stationary object to safeguard against this chance of human failure.
- Do not open a file drawer if someone is close by or underneath and could be injured by the drawer. Do not leave an open drawer unattended—not even for a minute. Whoever opens a file drawer should warn others working in the area so they do not turn around or straighten up quickly and bump or trip over an open drawer.
- Workers should never climb on open file drawers.
- Small stools used in filing areas are tripping hazards when left in passageways. Any person who sees one out of place should put it where it cannot cause a fall.
- Filing personnel should wear rubber finger guards to eliminate cut fingers from metal fasteners or paper edges.

Office personnel should never move desks or files; they should be moved by maintenance workers preferably using special dollies or trucks. In general, furniture should not be rearranged without authorization from office management. When desks or cabinets are moved, workers should consider whether they will obstruct floor space or aisles before making the move. If a telephone terminal box on the floor or electrical outlet box is exposed after moving furniture, the box should be marked with a tripping hazard sign until it is removed. Maintenance staff must remove the outlet and, if it is needed, relocate it; this step is far cheaper than to pay for a fall.

Do not run electric cords under rugs; they sometimes come out because of traffic movement and form tripping

hazards. They also are fire hazards. New outlets should be installed to eliminate the necessity for extension cords.

Materials storage. There are a number of precautions to be taken when storing materials. Neat storage makes it easier to find and recover materials without dropping or knocking over other materials. Supervisors must keep employees from stacking boxes, papers, and other heavy objects on file cabinets, desks, and window ledges, or from placing these materials carelessly on shelves where they could tumble down in an avalanche. If heavy objects fall toward a window, the glass might break and cause a serious accident.

Instruct workers not to place card index files, dictionaries, or other heavy objects on top of file cabinets and other high furniture. Movable objects such as flowerpots, vases, and bottles should not be allowed on windowsills or ledges. Caution workers not to throw loose razor blades, thumbtacks, and other sharp objects into their drawers but to put them in small boxes. Tools with blades and points should have the cutting or sharp end stuck in foamed-polystyrene blocks.

Lifting. Occasionally it is necessary for office personnel to lift light or heavy objects, such as files, books, boxes, and computer tapes. For these times, make sure office workers are trained in proper lifting techniques.

Other hazards. Some additional hazards are as follows: (1) never allow a spindle (spike) file in the office; (2) never store pencils in a glass on the desk with points outward; (3) never leave a knife or scissors on a desk with the point toward the user and never hand sharp-pointed objects to anyone, point first; (4) equip paper cutters with a guard that affords maximum protection (bar guards or single-rod barriers found in some cutters are not considered full protection); and (5) do not leave glass objects on the edge of desks or tables where they can easily be pushed off. Make sure that office machinery is operated only by authorized persons.

Some offices have an employee lounge or eating area with a hot plate for brewing coffee or a microwave oven for warming lunches. In these areas, spilled beverages can be a burning and a slipping hazard. Microwave ovens must be properly used to avoid injuries or damage to the oven. (See the discussion in Chapter 7 under Food Service in the *Engineering and Technology* volume of this *Manual*.)

If the sales department is part of the office staff, or if employees travel on company business, a safe driving program should be part of the company's safety program.

Supervisors should encourage employees to report all broken chairs, missing casters, stuck drawers, cracked glass, and other hazards for correction. Management should establish a policy for immediate correction of these defects and set up a formal program requiring quarterly office safety inspections.

Fire Protection

Fire hazards. To prevent spontaneous combustion fires, store all solvent-soaked or oily rags used for cleaning duplicating equipment in a metal safety container. Management should prohibit smoking within 10 ft (3 m) of where flammable solvents are used in duplicating or any other office operation. Workers should be trained in handling solvents to prevent eye injuries from splashes and wear proper protective equipment.

In recognition of the health hazards caused by smoking, many organizations do not allow smoking at any place in the facility. Other safety rules regarding smoking include (1) never allow smoking on elevators and (2) do not throw matches or cigarettes into wastebaskets; the contents usually are highly combustible. Supervisors or department heads should establish procedures so cleaning and maintenance personnel do not collect possible smoldering combustible material from ashtrays and throw it into combustible containers, such as cardboard boxes or cloth bags.

Some waste containers made from plastic or other flame-resistant material may actually be combustible if subjected to fire or intense heat. Such fires can generate dangerous toxic gases and dense smoke that can easily endanger a whole office. To control this hazard, use only metal or fire-safe tested materials designed to contain the fire.

Fire extinguishers. Portable fire extinguishers in a fully charged, operable condition must be kept in their designated places at all times when they are not used (Figure 19–4). (See Chapter 6, Fire Protection, in the *Engineering and Technology* volume of this *Manual*, for correct type of extinguishers for specific office hazard areas.)

Employees in general should know what to do in case of fire. Supervisors must train workers to operate extinguishers and fire hoses, if provided, and show them how to react in case of fire or other emergency. (Panic and confusion can be as dangerous as flame and smoke.)

When a fire is discovered, an employee should do three things: (1) turn in the alarm (no matter how small the fire is), (2) alert fellow workers, and (3) if trained to do so, use the proper firefighting equipment, but only if the employee always has a safe path of escape while fighting the fire.

Office employees should receive annual fire and other emergency training. The training should include use of portable fire extinguishers, procedures for reporting emergencies, and location of escape routes and shelters.

Emergency plan. Every office should have a written emergency plan. Supervisors should be appointed in every area to safely guide people out of the building. Every department should be assigned a specific route, and an alternate, in case the exit is blocked (See Chapter 6 and Figure 6–7). A surprise fire drill may save lives in an emergency situation in the future.

When the alarm sounds, supervisors should direct the show, but every employee must play a part. The group should move calmly along in an businesslike manner, without hurrying or pushing, and wait on a different floor or outside the building for the signal to return. In a real emergency, the officials in charge would authorize return to the building. (See Chapter 6, Emergency Preparedness.)

Ergonomics in the Office

Much care and consideration needs to be focused on the design of workstations and how equipment is used at the

Figure 19–4. Note that this fire extinguisher is placed in an easily accessible location, has nothing stored in front of it, has a sign indicating its location and other instructions regarding replacement, and carries an inspection tag. (Courtesy Signode.)

workstation. At the least, the supervisor or department head should be concerned with the selection of proper chairs, position of the individual in relation to the work surface, arrangement of work materials, and the proper use of video display terminals (VDT) at workstations (Figure 19–5).

A more in-depth ergonomic approach would include a survey of the office environment. Factors such as general lighting considerations, computer-terminal lighting, room temperature and drafts, ventilation, and noise would be included. More information about human factors engineering and ergonomics can be found in Chapter 5, Ergonomics in the Workplace, in the *Engineering and Technology* volume of this *Manual*.

SAFETY ORGANIZATION IN THE OFFICE

The supervisor is, of course, the key person in the office safety program. However, even the hardest working supervisor will have difficulty maintaining full-time interest in safety all alone. On the other hand, the office safety committee can help to maintain interest in the accident

prevention program, but it cannot substitute for good management.

Safety and Health Training

Safety training has a tendency to be overlooked in an office environment. All office workers should be provided with safety training that focuses on accident prevention, fire prevention, fire emergency response, and medical emergency response. Depending on the specific nature of the office environment, hazard communications training may be required. If an office environment contains an art department, print shop, or duplicating center, hazard communications training should be given to all employees.

To develop proper safety behavior, supervisors must provide safety instructions for all new office employees. The personnel or industrial relations department can supply an accident prevention brochure or a set of printed rules. They also should arrange for all explanations of procedures as quickly as possible during the employee's early workdays. Motivation and training were covered in Chapters 15 and 16.

Unfamiliar surroundings, new equipment, or altered work tasks increase the likelihood of accidents, even among veteran employees. Therefore, these people also should be trained when beginning a new job and given specific instructions for each piece of equipment. No one should ever be permitted to use a machine unless fully instructed in its operation and shown the location of fire equipment, how to use it, and how to summon medical aid.

Supervisors should have instruction in safe office operation because they are likely to be as unaware of accident hazards as the employees. However, prevention of accidents requires the dedicated vigilance of the supervisor throughout every working day. If he or she fails to carry out this function, the number of accidents due to unsafe practices by employees will continue undiminished.

Routine safety and health training should be part of company policy, and employees should receive continuing information on work-related hazards and safe practices. Training meetings are recommended on such topics as slip and fall prevention, proper lifting, fire safety, emergency procedures, office chemical safety, and off-the-job safety. Safe attitudes and behavior are not merely put on when an employee enters the office and taken off when the individual walks out the front door (see Chapter 15 for details).

All office employees who must enter production areas where safety hats, eye protection, and hearing protection are necessary must be provided with these items and shall be required to wear them. Every employee who visits the plant should have a card of the general safety rules that apply to the plant and should be familiar with them. The same requirement should be enforced for all visitors. Safety rules should apply to everyone if the program is to be successful.

Another critical aspect of training focuses on ergonomic and stress-related areas. Supervisors need to educate workers to recognize common physical complaints that could be caused by the ergonomic design of the worksta-

Figure 19–5. Computer workstations must be carefully designed so that the keyboard is at elbow level, the screen is at eye level and free from glare, and the terminal is at a safe distance to reduce VDT radiation emissions (see Chapter 5, Ergonomics in the Workplace, in the *Engineering and Technology* volume). Note that the lamp has an adjustable arm and can be positioned to avoid glare.

tion or how they use equipment at the workstation. Common physical complaints include eyestrain; back pain; neck pain; shoulder and arm symptoms; wrist, hand and finger symptoms; leg and foot symptoms; headache; and fatigue. Training materials should emphasize how employees can change their work habits or adapt equipment and furniture to reduce these complaints.

Office workers tend to work under a high level of stress. One definition of stress is an employee's physical and emotional reaction to change. A definite focus of training should be stress reduction and occupational wellness. Workers can learn how to recognize stressors in the work environment, detect physical symptoms of stress overload, and use various methods to alleviate stress. Stress reduction programs can include progressive relaxation, positive imagery, values clarification, proper nutrition, and exercise.

Many other programs can be developed that focus on occupational wellness. Chapter 16 of this volume contains additional information on safety training.

Finally, to repeat what has been said in previous chapters, a company safety program cannot succeed unless it has the whole-hearted backing of its top management. Supervisors must know that their accident prevention performance is watched and that good performance is appreciated. The Council's *Supervisors' Safety Manual* contains information on this topic that can guide both top management and the safety professional.

Safety and Health Committee

In planning an office safety program, management should ensure that worker representation on the safety and health committee reflects the composition of the workforce in company departments or divisions. The organization of the office safety committee can be the same as that of the company and joint safety committees discussed in Chapter 3.

The office should be on the inspection itinerary of the company's safety professional. The office supervisor should accompany the safety inspector on every inspection, along with an office safety committee member. (See Chapter 11 for inspection procedures.)

The committee's responsibilities include helping all supervisors maintain safe working conditions in office areas. The committee reports directly to the safety director or whoever is in charge of the program. Along with the department head, it can make periodic inspections of the office to look for accident or fire hazards. The committee makes

recommendations, many of them based on suggestions from supervisors and other employees. It also can help to prepare and revise company safety rules.

Often the committee is in charge of office-wide communication, training, and incentive programs. These are designed to maintain peak interest in safety, and use such means as posters, bulletins, and contests.

Accident Records

If a safety program is to succeed, the company needs to keep accurate accident records. Not only do accident investigations and analysis of records spotlight problems that must be corrected, but the records show if the company is making progress in accident prevention.

Office employees, like plant workers, should report every accident, no matter how minor the injury. The reports should be detailed and made as quickly as possible following the accident or near-accident. Unsafe conditions or procedures indicated in the reports should be corrected as quickly as possible, because near-miss accidents are warnings of worse accidents to come.

Records are the concrete foundation of the safety structure. They tell the "who, what, when, why, and how" of accidents in the office—and help supervisors and employees to prevent repeat performances. Accurate records also provide guidelines on which company insurance rates are based.

The average office will not have enough major injuries to warrant extensive investigation and analysis. However, it is urgent that records be kept to pinpoint problems and prevent future accidents. If, for instance, a large number of falls are injuring workers, supervisors can double-check possible hazards and devote special attention to the problem in meetings and other communications. Standard report forms are available. They were discussed in Chapter 13, Record-keeping and Incidence Rates. Accident investigation was discussed in Chapters 11 and 12.

SUMMARY

- Office risks often go unrecognized and unmanaged, yet the risks of serious injury are as great to office workers as they are to production workers. Therefore, a company's safety policy must include office workers in its program.
- Two studies, one by the State of California Department of Industrial Relations and one by the Equitable Life Assurance Society of the United States, revealed that on-the-job office accidents disable thousands of people and cost industry about $100 million dollars each year.
- Complacency is one of the prime causes of office accidents. Workers and management both give little thought to safety because office work is not perceived as hazardous. As a result, office workers and managers must be informed of the hazards and safe work procedures that apply to their jobs.
- Researchers have found that new surroundings and length of service increase the chances of worker accidents and injuries. On the other hand, there is little difference be-

tween men and women office workers in the number and severity of accidents the two groups suffered.

- The most common categories of major injuries are (1) falls from chairs, from slipping hazards, or from elevation; (2) strains or other injuries related to overexertion; (3) workers either being struck by objects (drawers, doors, filing cabinets) or striking them; and (4) workers being caught in or between machinery or equipment.
- Office accidents can be controlled by eliminating hazards or reducing exposure to them. Management can accomplish this goal most easily when the office is being planned, equipment purchased, or new procedures established.
- Offices should be laid out for efficiency, convenience, and safety. The health and safety professional should see to it that hazards related to stairways, exits, aisles and passageways, doors, glass windows, lighting, ventilation, electrical wiring, equipment, and storage are properly controlled or eliminated.
- Adequate lighting and proper office design can prevent glare or distracting reflections and help to reduce or eliminate eyestrain and visual fatigue. Ventilation should be carefully planned and installed by professionals to ensure that fumes, molds, smoke, and other contaminants do not accumulate in office areas.
- Management must protect office workers against electrical equipment hazards by using UL-listed equipment and by installing proper grounding and shielding equipment. Workers can be trained in the correct use of electrical machines and to place cords and wires so they do not become tripping hazards. All electrical supplies and equipment must be inspected regularly for shock hazards.
- Movable equipment must be properly attached to a desk or other surface, and heavy equipment should be braced against walls or columns. Floor surfaces should be slip-resistant, and all carpets kept in good repair or replaced when frayed or warped.
- Supervisors and managers should plan storage areas for safety and to make items easily accessible. Workers should be trained in how to stack materials in stable rows or piles and to keep passageways free. Smoking should be banned in storage areas and all flammable and combustible materials stored in proper containers.
- Management must purchase good quality office furniture that enhances both office safety and employee morale. The principles of ergonomics should be applied when ordering chairs, desks, and files, noncombustible waste containers, ladders and stands, and other items. Computers must be properly grounded and protected from fire and water damage.
- Office chemical products represent another area of potential hazard. Workers must be trained in the safe use and storage of chemicals used in copying, printing, and duplicating machines and for cleaning purposes. Where possible, nontoxic and nonflammable materials should be substituted for more hazardous chemicals.
- Heavier equipment such as office machines, offset presses, gathering and stitching machines, folding machines, and

the like must be properly guarded and designed with built-in safeguards. The safety professional can work with the purchasing department to ensure that heavy equipment is designed and manufactured for safe operation.

- The supervisor and safety professional are responsible for developing and enforcing safety procedures in office areas. Workers should receive safety training from the first day on the job to prevent accidents and to spot hazardous conditions. They must know the safety procedures for accident prevention on stairways, while carrying or lifting objects, when working with office equipment, and while handling potentially hazardous objects or materials.
- Employees should understand how to prevent fires in office work areas and what to do during a fire drill or outbreak of fire: (1) turn in the alarm, (2) alert fellow workers, and (3) use proper firefighting equipment.
- Every company should draw up written emergency plans in the event of a natural or human-caused disaster. Every department should be assigned a specific route, and each worker should know his or her particular task.
- The company should establish a formal safety program for office workers and establish training sessions for all employees. Supervisors have the major responsibility for training workers and following up on safety and health practices. Training should focus on accident prevention, fire prevention and emergency response, medical emergency response, ergonomics and stress-related disorders, and stress reduction.
- Finally, the company needs to keep accurate records on all accidents and injuries. The records will show if the company is making progress in accident prevention. Standard report forms can be used to help pinpoint major problems so that management can design effective solutions.

REFERENCES

American National Standards Institute, 11 West 42nd Street, New York, NY 10036.
Safety Performance Specifications and Methods of Test for Glazing Materials Used in Buildings, ANSI Z97.1–1984.
Practice for Office Lighting, ANSI/IES, RP1–1982.
Construction Safety Requirements for Temporary Floor and Wall Openings, Flat Roofs, Stairs, Railings, and Toeboards, ANSI A10.18-1983.

Baldwin D. Caution: office zone. *Job Safety and Health* 4(2); February 1976.

How safe is your office? *National Safety News* 112(4); October 1975.

National Fire Protection Association, Batterymarch Park, Quincy, MA 02269.
Life Safety Code, NFPA 101, 1988.
National Electrical Code, NFPA 70, 1990.

National Safety Council, 444 North Michigan Avenue, Chicago, IL 60611.
Falls on Floors, 12304–0495, 1986.
Motor Fleet Safety Manual, 3rd ed., 1986.
Starting an Office Safety Program, 1990.
Supervisors' Safety Manual, 7th ed., 1990.

Scott D. *Sitting on the Job: How to Survive the Stresses of Sitting Down to Work—A Practical Handbook.* Boston: Houghton-Mifflin, 1989.

State of California, Department of Industrial Relations, Division of Labor Statistics and Research, 525 Golden Gate Avenue, San Francisco, CA 94102.
"Disabling Work Injuries to Office Employees," 1963 and 1978 editions.
"Work Injuries and Illness in California," annually.

Statistical Abstracts of the United States, U.S. Department of Commerce, 1987.

20

Workers with Disabilities

This chapter is intended to assist companies and organizations in the safe and productive placement of disabled individuals in the workforce. Topics covered include organizations' professional and legal responsibilities and information on how to implement required policies and procedures.

Almost every worker with a physical or mental impairment can qualify for some type of job. Industry and government surveys made during the past several decades prove it is good business to hire workers with permanent disabilities. They report for work promptly, can produce as well as other workers, and their turnover rate is often lower. Unfortunately, little had been done to enforce the prohibition of discrimination against disabled people. That trend will likely change in the United States under the 1990 Americans with Disabilities Act, which provides for punitive damages against employers who discriminate.

The 1990 U.S. Americans with Disabilities Act encompasses the following areas: Title 1—Employment Provisions, Title 2—State and Local Government Provisions, Title 3—Public Accommodations and Services Operated by Private Entities, Title 4—Telecommunications, and Title 5—Miscellaneous Provisions.

Title 1 of the Act requires that employers make reasonable accommodations for the disabled. An employer is required to provide sufficient accommodation to allow qualified individuals with a disability to attain the same level of job performance as co-workers having similar skills and abilities. An employer is not required to employ an individual where to do so would pose a "direct threat" to the health or safety of others. The determination of "direct threat" must be based on the actual condition of the individual and not upon generalizations or stereotypes about the disability.

A significant aspect of Title 3 is that public accommodations and services operated by private entities are not required to permit individuals to participate in their services where the individual poses a direct threat to the health and safety of others which cannot be eliminated by modification of policies or practices or provision of auxiliary aids or services.

The law requires that companies establish affirmative action guidelines for the hiring, upgrading, promotion, award of tenure, demotion, transfer, layoff, termination, right-of-return from layoff, and rehiring of qualified disabled individuals. The law also requires employers to provide "reasonable accommodation" to modify the work environment or the job for these workers when necessary.

If an employer denies a disabled individual a specific job, the burden of proof is upon management to show that the person is unqualified because of one or more of the following reasons:

- The job would put the individual in a hazardous situation.
- Other employees would be placed in a hazardous situation if the person were on the job.
- The job requirements cannot be met by an individual with certain physical or mental limitations.
- And (for all of the above) accommodation of the job cannot reasonably be accomplished.

Developing affirmative action programs, including those for hiring workers with disabilities, is usually the responsibility of personnel other than the safety and health professional. The safety and health professional should serve as a resource person and must play a critical role in matching disabled workers to the right job and work environment. Safety evaluations for the worker must include adequate access to and exit from the workplace as well as safety on the job. Safety and health professionals must be sensitive in their use of language to describe the disabled and the tasks that the disabled may be asked to perform (Table 20–A).

Companies should not be too quick to decide that safety and health problems are insurmountable. There are many organizations available that can help management find solutions. (See Sources of Help at the end of this chapter.)

HISTORY AND THE LAW

In the 1940s, special attention was given to employing workers with disabilities by a number of large companies that realized hiring these people was smart and sound business practice. Although some companies employed such workers before the 1940s, three events occurred in that decade to stimulate these programs and encourage other companies to institute hire-the-disabled programs.

World War II. Many individuals with disabilities were hired to help fill job vacancies left by employees who joined the military.

Rehabilitation of World War II wounded veterans. For example, in the early 1940s, one company undertook an affirmative action program to help each returning disabled veteran to become an employable person. Other companies established similar programs (Figure 20–1).

The third event was a study published by the U.S. Department of Labor that debunked some myths about workers with disabilities being less productive, suffering more accidents, and losing more time from work than other workers. On the contrary, the Department of Labor study showed disabled workers were as productive as other workers, had lower frequency and severity of injury rates, and were absent from work only one day more per year than other workers. In researching over 11,000 workers with disabilities for almost two years, the study's authors did not find a single serious injury caused by a disabled worker, to himself/herself or to a fellow worker, that was a direct result of the disability (Figure 20–2).

One company in a study in 1958 and another in 1981, found that of its more than 2,700 disabled workers, 96% rated average or better on safety performance; 92% rated average or better on turnover; and 85% rated average or better on attendance. After a decade or more of direct experience in hiring the disabled worker, the personnel files of many companies contain indisputable proof of the value of these employees—and of affirmative action programs—to their companies.

Rehabilitation Act of 1973

The U.S. Vocational Rehabilitation Act of 1973 (Public Law 93-112), commonly referred to as the Rehabilitation Act, was the first major civil rights law protecting the rights of persons with disabilities. This law applies to federal contractors (Section 503) and recipients of federal assistance programs (Section 504). Therefore, all employers who do work for the federal government, or receive funds from the government, are subject to this law.

Section 503 of the Act requires employers to take "affirmative action" to recruit, hire, and advance qualified individuals with disabilities. The law applies only to employers who have federal government contracts or subcontracts of $2,500 or more. Those holding contracts or subcontracts of $50,000 or more, with at least 50 employees, must prepare and maintain (review and update annually) an affirmative action program at each establishment. The program, which sets forth the employer's policies, practices, and procedures regarding disabled workers, must be available for inspection by job applicants and employees.

Section 504 of the Act forbids acts of discrimination in employment against qualified disabled persons by employers who receive federal funds. This Section is enforced by each department or agency that provides federal funds.

In the Rehabilitation Act of 1974, amendments broadened the definition of "handicapped [disabled] individual" for purposes of Section 504. With this amended definition, it became clear that Section 504 was intended to forbid discrimination against all disabled individuals, regardless of their need for or ability to benefit from vocational rehabilitation services. Thus, Section 504 reflects a national commitment to end discrimination on the basis of disability and establishes a mandate to bring persons with disabilities into the mainstream of American life.

Other U.S. federal departments and agencies also have issued regulations similar to those of the Department of Education. For example, the U.S. Department of Labor has issued regulations (29 *CFR* 32), effective November 6, 1980, which implement Section 504 of the Act for the department. All of these regulations require federal contractors and recipients of federal funds to make reasonable accommodations to the workplace when necessary for employing people with disabilities.

All records pertaining to compliance with the Act, including employment records, and any complaints and actions taken as a result, must be retained by the employer for at least one year. If the company fails to maintain complete and accurate records or fails to update the affirmative action program each year, the government can impose "appropriate sanctions" against the employer. When complaints are brought against the employer or there is some question about affirmative action programs, the employer must allow investigators to have access during normal business hours to its place of business, its books, records, rules and regulations, and accounts pertinent to compliance with the Act.

U.S. State and Local Laws

All 50 states and many local governments have now adopted building codes or legislation requiring barrier-free design or

Table 20–A. Language Issues Regarding Disabilities

Negative Language	Positive Language
handicapped/handicap	disabled/person with disability
cripple/crippled by	person who has (whatever)
victim	person who uses (assistive device)
spastic	
paralytic	
afflicted/afflicted by	caused by . . .
deformed/deformed by	as a result of . . .
suffering from	
confined/restricted to a wheelchair	person who uses a wheelchair
wheelchair bound	wheelchair rider, user
deaf and dumb	pre-lingually deaf (at birth)
deaf mute	post-lingually deaf (after birth)
poor, unfortunate and similar words	

Normal—When used in the statistical sense or to express "average," the term is fine. However, this word should never be used to refer to people without apparent physical, emotional, or mental disabilities. Because most people are disabled at some time in their lives, the average person can be a disabled person. The disabilities may be inability to control one's temper, effects of past broken bones, strong prejudices, substance abuse, and other problems.

Wheelchair—People who are able to walk and run usually see a wheelchair as a confining device. In reality, it gives mobility to people who would otherwise be confined to bed. Wheelchairs come in many types, including some that have variable height controls, extra-narrow axle widths, or other features. It is often possible to build a wheelchair or other mobility device to suit a particular worker and his or her job duties. Employers should explore several options if a workplace cannot be safely or easily adapted to a standard wheelchair.

Deaf/hearing impaired—All deaf people are hearing impaired, but not all hearing-impaired people are deaf. This distinction is important as the needs of a deaf employee are very different from those of a hearing-impaired employee. Some deaf people use sign language, while others lipread. Some hearing-impaired people need relative quiet to understand verbal communication; others need a person with a deep voice to relay messages from a person with a higher-pitched voice. Ask what a particular hearing-impaired person needs.

Blind/visually impaired—Likewise, not all visually impaired people are blind or even legally blind. Some have tunnel vision, others have peripheral but no central vision. Some need strong light while others require dimmer light. Simply because a visually impaired person wears glasses does not mean their vision is 20/20. Ask what a particular visually impaired employee needs to be able to see comfortably.

468

Figure 20–1. After losing his leg in combat, Felix "Phil" Sitkowski returned, as a clerk, to U.S. Steel South Works in South Chicago, Illinois. Later he applied for and was promoted to field lubricating analyst. Sitkowski's prosthesis allowed him to use ladders for making inspections. He is shown at the left with two of the eight millwright helpers he supervised. (Courtesy U.S. Steel South Works.)

Figure 20–2. Studies show that workers with disabilities, such as these employed at a candy factory near Peoria, Illinois, are productive, dependable, and have low work-injury rates. (Courtesy Illinois Department of Rehabilitation Services.)

removal of barriers preventing access to the building by disabled persons. Many of these codes mandate that any public building or facility must be barrier free if the public is invited to use it for any normal purpose such as shopping, employment, recreation, lodging, or services. If accessible facilities need to be identified, the organization should use international symbol of accessibility (Figure 20–3).

WHO ARE DISABLED JOB-SEEKERS?

The law defines three types of disabled persons seeking employment—the disabled individual, the disabled veteran, and the qualified disabled individual.

"Disabled Individual"

ADA–90 defines "disability":

1. Has a physical or mental impairment that substantially limits one or more of the person's major "life activities," such as:
   - ambulation

- communication
- education
- employment
- housing
- self-care
- socialization
- transportation
- vocational training.

2. Has a record of such impairment, or
3. Is perceived as having such an impairment.

The term "substantially limits," as used above, has to do with the degree to which the disability affects the person's employability. A qualified disabled person who, because of the disability, finds it difficult to obtain an appropriate job or advance on a job would be considered substantially limited (Figure 20–4).

The term "physical or mental impairment" would include, but not be limited to, these conditions:
- diseases and infections
- orthopedic
- visual, speech, and hearing impairments
- cerebral palsy
- epilepsy
- muscular dystrophy
- multiple sclerosis
- HIV
- cancer
- heart disease
- diabetes
- mental retardation

Figure 20–3. International symbol designates access for the disabled. The symbol is in white on a blue background. (Printed with permission from American National Standards Institute.)

- emotional illness
- drug addiction
- alcoholism.

It should be noted that homosexuality and bisexuality are not defined as impairments. Compulsive gambling, illegal drug use, kleptomania, and others are also not considered to be impairments.

"Disabled Veteran"

A disabled veteran is a "special handicapped [sic] individual" who:

1. Is entitled to disability compensation under laws administered by the Veterans Administration for disability rated at 30% or more
2. Was discharged or released from active duty due to a disability incurred or aggravated in the line of duty.

A veteran with nonservice-connected disabilities is not considered a special disabled veteran but may still qualify as a disabled individual under Sections 503 and 504 of the Rehabilitation Act of 1973.

The Vietnam War had the highest proportion of disabled service personnel of any war in U.S. history. A disabled veteran of the Vietnam War is a person who was discharged or released from active duty for a service-connected disability if any part of such duty was performed between August 5, 1964, and May 7, 1975.

"Qualified Individual With a Disability"

Not every disabled person is covered by the ADA–90 Act. The crucial word is *qualified*. A person must be capable of performing the essential function of a job—with reasonable accommodation to the disability (Figure 20–5).

Nor is every disabled veteran and every Vietnam Era disabled veteran covered by the Rehabilitation Act of 1973 or the Vietnam Era Veterans' Readjustment Assistance Act of 1974. The veteran also must be qualified, that is, capable of performing a particular job, with reasonable accommoda-

tion to the disability. Organizations should be aware of the fact that the Americans with Disabilities Act requires that a certain number of deadlines be met.

REASONABLE ACCOMMODATION

According to the U.S. ADA law, an employer shall make "reasonable accommodation" to the known physical or mental limitations of an otherwise qualified disabled applicant or employee, unless the employer can demonstrate that the accommodation would impose an undue hardship. Accommodations can include modifications of equipment or facilities and alterations in processes or job descriptions. The employer may not deny any employment opportunity to a qualified disabled employee or applicant if the only basis for the denial is the need to make a reasonable accommodation.

"Undue hardship" means an action requiring significant expenses or difficulties and is determined by considering the following factors:

1. The overall size of the employer's operation with respect to number of employees, number and type of facilities, and size of budget
2. The type of operation, including the composition and structure of the workforce
3. The nature and cost of the accommodation needed.

Reasonable accommodation may include but is not limited to:

1. Making facilities used by employees readily accessible to and usable by disabled persons
2. Job restructuring, part-time or modified work schedules, acquisition or modification of equipment or devices, provision of readers or interpreters, and other similar actions.

Examples

Reasonable accommodation is demonstrated in these three examples:

A construction equipment salesperson, whose job description required him to climb onto the equipment and demonstrate its operation during sales presentations, was given a desk job after he suffered the amputation of his arm during an off-the-job motor vehicle accident. Although his prosthetic device enabled him to operate the equipment controls, the employer had considerable concern about the man's ability to climb on and off the equipment using the prosthetic device. This resulted in the job change. Upon enactment of the Rehabilitation Act of 1973, and its amendments, his employer reinstated the man as a salesperson, making a reasonable accommodation for his handicap. The accommodation consisted of providing him with a portable climbing device so that he could get on and off the equipment safely.

A forklift (powered industrial truck) mechanic became blind in one eye due to a nonindustrial health problem. The man's job description required him to test drive each forklift truck after completing maintenance or repair work on it. Because the employer's standard safety policy required that all drivers of powered industrial trucks must have binocular vision (use of both eyes), the employer at first was going to

Figure 20—4. As a result of being struck by a high-voltage highline wire, Charles Dannheim lost his legs and arms. Although he would be considered "substantially limited" and a disabled individual by definition, Dannheim works full-time as a cattle rancher with the aid of hooks and artificial legs. (Printed with permission from Beef Magazine.)

switch the man to another job, which would have lowered his earnings. Upon reviewing the requirements of the U.S. Rehabilitation Act of 1973, however, the employer provided the mechanic with a reasonable accommodation. The company altered the mechanic's specific job description to eliminate the requirement to test drive the vehicles. Instead, the job descriptions of the other mechanics were broadened to include the test driving of any vehicle repaired by the disabled mechanic.

A disabled individual working for an electrical appliance firm was provided with a reasonable accommodation to assemble small parts. It consisted of minor adjustments to the workbench to accommodate a wheelchair.

The following example was considered to be unreasonable accommodation:

A company would have to completely redesign or alter the circuitry or operating levers of a machine to accommodate a physically disabled individual.

Accommodation Is Not New

Accommodation in employment is not a novel concept. The first applications of machine guards and ventilating fans were job accommodations. Also, the first hod carrier who lacquered and reinforced his bowler as a hard hat made a job accommodation. Job placement of employees based on medical examinations, when newly hired or returning to work after an illness or injury, is again an application of accommodation. This experience is common to every employer.

Until individuals receive adequate training in the field of rehabilitation medicine, they cannot be qualified to evaluate reasonable accommodations of the workplace, its procedures, and access for the physical or mental limitations of a disabled worker. The safety and health professional without such expertise should be one member of the team consisting of the worker's physician, the occupational physician, and a rehabilitation specialist, and, in some situations, other individuals with similar disabilities. A team approach will greatly enhance the organization's commitment to an affirmative action program and nondiscrimination. It will also contribute to an affirmative action program and nondiscrimination against disabled employees and applicants in the organization.

Figure 20–5. This woman will soon be a "qualified disabled individual," and very employable, because of the IBM computer programming training she is receiving at the El Valor Rehabilitation Facility in Chicago. (Courtesy Illinois Department of Rehabilitation Services.)

Job safety analysis and safe work procedures are a means of eliminating or reducing work hazards to minimize worker risk. They directly transfer to the process of accommodation. Training in safe work procedures will be important in accommodating the person with a disability to the job.

ROLE OF THE SAFETY AND HEALTH PROFESSIONAL

Affirmative action programs required by the U.S. government usually come under the responsibility of the Equal Employment Opportunity (EEO) manager or coordinator (or labor affairs personnel). As a result, the placement of qualified disabled individuals normally is under their jurisdiction.

The safety and health professional, nonetheless, should be a resource person to those responsible for job placement of qualified disabled individuals. This professional should be consulted before workers are placed and asked to evaluate any proposed reasonable accommodation. The following is an example of some of the responsibilities of the safety and health professional in relation to disabled employees.

General Responsibilities

Maintain close liaison with the EEO manager-coordinator and with medical and personnel departments when they are placing disabled employees. Rehabilitation specialists and people with similar disabilities may also be necessary members of the placement team.

Make job safety hazard analysis of existing work based on the job responsibilities and the abilities and limitations of the disabled employee or applicant when employing, promoting, transferring, and selecting workers with disabilities.

Make recommendations for safety modifications of machine tools, established processes and procedures, and existing facilities and workplace environment when the company must make reasonable accommodations for disabled employees.

As required, cooperate with the plant or building engineer or mechanical engineer and the planning, production, and maintenance departments when disabled employee accommodations are being evaluated.

Specific Responsibilities

Review the company's affirmative action program.

Establish specific communication channels, pertaining to disabled employees, with:

1. EEO manager-coordinator. Make sure this person knows the safety and health professional is part of the team and available when a job needs a safety evaluation.
2. Medical department. Let them know the professional will be requesting their help when evaluating a job.
3. Employment department. Let them know the safety and health professional is ready when necessary to help them evaluate a job's safety and remind them of safety considerations such as:
   a. Don't place a worker with a coronary condition in a job that would aggravate that condition, which is a medical evaluation process.
   b. Don't place a person with a back problem in a job requiring heavy manual lifting, unless other considerations are given.
   c. Make certain to place a disabled employee in a job that would be safe for that person and that will not cause a hazard to others.
   NOTE: Individual judgments are based on a physician's evaluation with input from the safety and health professional.
4. Plant and mechanical engineers. Reasonable accommodation does not necessarily mean reinstalling machines; rather it could mean minor relocating of a machine's controls so a disabled employee could operate them properly and safely. Therefore, advise the engineers that the safety and health professional will evaluate all safety aspects of such an accommodation. Also advise them that the professional will be available for safety evaluations when they design reasonable accommodations into future facilities such as:
   a. ramps for wheelchairs
   b. wider door passages for wheelchairs
   c. grab bars in accessible washroom facilities

d. braille numbers (within a disabled person's reach) on elevators (Figure 20–6)
e. easy access to company facilities such as lunchrooms
f. elimination of curbed crosswalks

5. Production and maintenance departments.
   a. Because reasonable accommodations also refer to job restructuring and modifications, tell the production and maintenance departments you will help by evaluating the safety aspects of such changes. Experienced rehabilitation experts should be consulted before the company declares it is unable to make the job suitable for a particular applicant. Often a fresh look at the situation and job from an outsider's perspective will reveal new strategies that have been missed.
   b. When they delete any job specification that would arbitrarily and without justification screen out disabled individuals, you will be available if a safety evaluation is needed.

Conduct a safety evaluation of a disabled employee (in relation to the specific job or prospective job) and perform an entire job hazard analysis if needed. Consult rehabilitation specialists and others as appropriate, particularly if it appears at first that the disabled employee is simply unable to perform the job. If the consensus is that the job and employee are incompatible, the company's decision, made along with the rehabilitation and other specialists, is easier to defend.

Conduct a safety evaluation whenever a reasonable accommodation is being planned for a disabled employee.

Coordinate with both the EEO manager-coordinator and the employment department to make certain any disabled employee being considered for a new position is qualified to safely and capably perform the job. This usually requires the safety and health professional to make a safety evaluation and observe the employee during training.

Evaluate any reported harassment of a disabled employee that affects safety. For example, name calling would not normally jeopardize the employee's safety, but pranks by other employees could result in accidents and injuries. Verbal harassment of any worker for *any* reason should not be tolerated. It is particularly inexcusable for a safety and health professional to stand by, claiming that such harassment does not generate a safety hazard. Angry, defensive, or depressed workers do not make sound safety judgments on the job. Further, when verbal harassment is tolerated by management, it often escalates into physical harassment.

Failure to discipline workers who are harassing others can lead to serious consequences. These can range from workplace deaths and injuries to a decline in production to a loss of valued employees who may quit in disgust and resort to lawsuits that the employer will find difficult to counteract. The safety and health professional should work with the EEO manager-coordinator and other pertinent personnel to appropriately and effectively handle such situations as soon as they come to one's attention.

Refer all questions about interpretation of government requirements to the personnel responsible for implementing

Figure 20–6. Reasonable accommodation for the disabled includes marking elevator buttons in braille. (Courtesy Governors State University, Park Forest South, IL)

the affirmative action program (usually the EEO manager-coordinator) or to counsel.

The evaluation form (Figure 20–7) can help the safety and health professional perform safety evaluations for disabled employees, especially when reasonable accommodation is involved. A written report and supporting material (such as memos, blueprints, and photos) can be attached to the form to provide detailed information explaining why certain decisions were made.

Such evaluations should be kept for at least one year after the employee leaves the company. Records can be destroyed only after approval from the EEO manager-coordinator or other personnel responsible for the government-required affirmative action program.

INSURANCE CONSIDERATIONS

Most companies mistakenly believe that employing disabled workers will raise their workers' compensation premiums. This is not true. Rates are based on experience by the class of industry and modified in most cases by the individual plant experience. There is no indication that losses are increased when persons with disabilities are properly placed.

JOB PLACEMENT

Time and again, companies have found that disabled workers make for a first-class workforce. When properly placed and trained so they can compete on an equal basis with others,

```
┌─────────────────────────────────────────────────────────────────────────────┐
│                  HANDICAPPED EMPLOYEE SAFETY EVALUATION                        │
│                                                                               │
│   ☐ Applicant:                                                                │
│   ☐ Employee: _____  _____     │
│               (Last Name          First Name          M.I.)    (Clock No.)    │
│                                                                               │
│       Handicap: _____         │
│                                                                               │
│       Evaluation of              ☐ Current job            ☐ Prospective job   │
│                                                                               │
│       Job Title: _____         │
│                                                                               │
│       Job Description (primary duties): _____         │
│       _____          │
│                                                                               │
│       Hazards to This Employee:                                               │
│       (State "none" if none) _____          │
│       _____          │
│                                                                               │
│       Hazards to Other Employees:                                             │
│       (State "none" if none) _____          │
│       _____          │
│                                                                               │
│       Proposed "reasonable accommodation" (if any): _____         │
│       _____          │
│                                                                               │
│       CONCLUSION (Based on all known factors at this time):                   │
│                                                                               │
│       ☐ Job is safe for this employee:                                        │
│            ☐ as is     ☐ with proposed reasonable accommodation               │
│                                                                               │
│       ☐ Job is unsafe for this employee:                                      │
│            ☐ as is     ☐ with proposed reasonable accommodation               │
│                                                                               │
│       ☐ No hazard to other employees:                                         │
│            ☐ as is     ☐ only with proposed reasonable accommodation          │
│                                                                               │
│       ☐ Hazard to other employees:                                            │
│            ☐ as is     ☐ with proposed reasonable accommodation               │
│                                                                               │
│   ┌──────────────────────────────┬───────────────────────────────────┐       │
│   │ Location:                     │ Safety Supervisor (Print Name)    │       │
│   ├──────────────────────────────┼───────────────────────────────────┤       │
│   │ Date:                         │ Safety Supervisor (Signature)     │       │
│   └──────────────────────────────┴───────────────────────────────────┘       │
│                                                                               │
│   NOTE: Complete two copies of this form and give one copy to local EEO       │
│   manager-coordinator. Second copy is for the safety file.                    │
└─────────────────────────────────────────────────────────────────────────────┘
```

Figure 20–7. This evaluation form is an excellent administrative tool, especially when a written report and other supporting documentation are attached.

workers with disabilities usually equal or prove to be slightly better than other employees in production and safety. Their overall attendance and job turnover records are usually superior to those without disabilities.

General Concepts

To place a disabled worker properly, the following requirements should be observed, where applicable, after receiving a physician's evaluation of the individual.

The worker should meet the physical demands of the job. When necessary, the worker should receive the support of reasonable accommodation.

The worker should not be a hazard to himself/herself. For example, a person subject to dizzy spells should not work on a ladder or scaffold or around moving machinery, where injury or death could occur.

The worker should not be a hazard to others. For example, a person with severe vision impairment should not drive a bus or operate an overhead crane, because of the potential for personal injury or injury to others.

The task should not aggravate the known degree of disability. A person with skin disease should not be exposed to substances that may aggravate this condition. Another example might be a worker with impaired lung function.

These workers should not be exposed to substances such as smoke that will further impair lung function.

To obtain valuable input, a conference with the individual should be held before job placement is made.

Proper placement matches the right worker to the right job on the basis of the person's ability to meet the job qualifications (Figure 20–8). As a result, the impairment virtually disappears as a factor of job performance. Moreover, employers should be aware that most disabled persons have more ability than disability, because few jobs actually require all of a worker's abilities. The job-employee match forms shown in Figure 20–8 are not only used to place disabled persons; some companies use them to place all new and transferred workers.

On the other hand, employers should also remember that each impairment can impose limitations on the type and number of activities in which the disabled person can engage. The impairment will also limit the working conditions and hazards to which this person can be exposed.

Many workers with disabilities are particularly vulnerable to tobacco smoke. Quadriplegics are endangered because they cannot cough to clear their lungs; people with heart or respiratory disease should not be exposed to environmental tobacco smoke (ETS), which contains many irritants. Workers with circulatory problems are further compromised by being exposed to nicotine, a compound that constricts the blood vessels. Other workers' disabilities may also make them hypersensitive to tobacco smoke. Clearly, prohibiting smoking in the workplace is not only reasonable but can benefit all employees, not just the disabled worker.

The safety and health professional should be aware that there are regulatory standards which, although promulgated for the protection of the average employee, are not sufficient to protect employees with disabilities. Some examples are:

Standards referring to storage of flammable and combustible liquids, *CFR* 1910.106 (d)(6)(iii), include a requirement for a curb to capture spilled liquids. Disabled employees are just as vulnerable as anyone else to the hazard of chemicals spilled or escaping from a storage room. The curb can be equipped easily with a portable ramp, which not only permits a wheelchair to get in and out but allows carts to enter and exit the room. This reduces the hazard of workers transferring chemicals one by one from the cart and possibly tripping on the curb while doing so.

The "Means of Egress" standards, *CFR* 1910.37, are based on the ability of an individual to move 100 ft (30 m) in 30 seconds. Perhaps some employees with disabilities cannot move that fast.

Respirators are required by the standards, for example, *CFR* 1910.134, for certain jobs. Some individuals may have a physical impairment that can be affected by restricted breathing. If there is some indication of this problem, such employees must not wear a respirator until it is determined by a physician that it can be worn safely. This may preclude the individual from performing a specific job where a respirator is necessary.

The permissible exposure levels (PELs) listed in the OSHA "Air Contaminants" standards, *CFR* 1910.1000, are based on the susceptibility of persons with normal breathing capacities to such contaminants. Some disabled individuals do not have normal breathing capacities and therefore are susceptible to lower levels of contaminants.

The safety and health professional must consider whether existing safety standards are sufficient to protect a particular disabled worker or candidate. More stringent standards, such as less exposure to an air contaminant than the average employee can tolerate without harm, may be necessary to protect the disabled worker or applicant. Such decisions must be made with the assistance of medical and rehabilitation consultants.

ANALYSIS OF THE JOB

Each job must be evaluated to make sure the individual being considered can do it safely. The following areas should be taken into account when making the analysis.

Physical Classification

The labor market simply does not supply only "physically perfect" workers. In fact, the percentage of workers in perfect health is relatively low—the working population now includes many persons with disabilities. Advances in medical science prolong the lives of many who would have died of war injuries or such illnesses as smallpox, tuberculosis, diabetes, and heart disease. Accidents in industry, in traffic, and in the home continue to increase the number of persons with physical disabilities.

Because the company's physician conducts physical examinations of prospective employees and makes regular plant inspections, he or she often has a better understanding of various physical and mental requirements for company jobs than do noncompany physicians. The company physician's responsibility should be to provide management with clear evaluations of each applicant's fitness for a particular job. The physician's determinations must be made only on the basis of job-worker compatibility.

Although the company physician is better able to understand job requirements, he or she may need to consult with a rehabilitation specialist.

On the other hand, safety, medical, and even many rehabilitation professionals often cannot review a work situation and ask the right questions to develop a workable compromise or accommodation for a particular disabled worker or applicant. For this reason, it is wise to consult with those who are experts in a particular disability involved.

Many support groups for disabled people exist, and the members can share experiences and insights that only those living with a particular disability can provide. Such groups can help the safety, medical, and rehabilitation specialist find creative ways to accommodate a particular job to a disabled person or suggest accommodation devices that can be made or purchased.

To locate such support groups, companies should contact their national headquarters, which can supply the names and

PHYSICAL CAPACITIES FORM

Leg Amputation 5″ below knee

Artificial leg—good fitting

Name ________ Sex M Age 31 Height 5′9½″ Weight 155

PHYSICAL ACTIVITIES		WORKING CONDITIONS	
x 1 Walking	16 Throwing	51 Inside	x 66 Mechanical Hazards
0 2 Jumping	x 17 Pushing	52 Outside	x 67 Moving Objects
0 3 Running	x 18 Pulling	53 Hot	0 68 Cramped Quarters
x 4 Balancing	19 Handling	54 Cold	69 High Places
x 5 Climbing	20 Fingering	55 Sudden Temp. Changes	70 Exposure to Burns
0 6 Crawling	21 Feeling	x 56 Humid	71 Electrical Hazards
x 7 Standing	22 Talking	57 Dry	72 Explosives
8 Turning	23 Hearing	x 58 Wet	73 Radiant Energy
9 Stooping	24 Seeing	59 Dusty	74 Toxic Conditions
0 10 Crouching	25 Color Vision	60 Dirty	75 Working With Others
x 11 Kneeling	26 Depth Perception	61 Odors	76 Working Around Others
12 Sitting	27 Working Speed	62 Noisy	77 Working Alone
13 Reaching	28	63 Adequate Lighting	78
x 14 Lifting	29	64 Adequate Ventilation	79
x 15 Carrying	30	65 Vibration	80

Blank Space=Full Capacity x=Partial Capacity 0=No Capacity

May work__hours per day__days per week. (IF TB, cardiac or disability requiring limited working hours.)

May lift or carry up to ** pounds.

DETAILS OF LIMITATIONS FOR SPECIFIC PHYSICAL ACTIVITIES:
Should not be required to walk, balance, climb, stand, kneel for prolonged periods of time.
**Should not lift heavy weights continuously.
Should not carry long distances.

PHYSICAL DEMANDS FORM

Job Title Data Processing Entry Occupational Code 4-44.110

Dictionary Title Data Processing Entry

Firm Name & Address ________

Industry ________ Industrial Code ________

Branch ________ Department ________

Company Officer ________ Analyst Wetzel Date ____

PHYSICAL ACTIVITIES		WORKING CONDITIONS	
x 1 Walking	16 Throwing	x 51 Inside	66 Mechanical Hazards
2 Jumping	17 Pushing	52 Outside	67 Moving Objects
3 Running	18 Pulling	53 Hot	68 Cramped Quarters
4 Balancing	x 19 Handling	54 Cold	69 High Places
5 Climbing	x 20 Fingering	55 Sudden Temp. Changes	70 Exposure to Burns
6 Crawling	21 Feeling	56 Humid	71 Electrical Hazards
7 Standing	22 Talking	57 Dry	72 Explosives
8 Turning	23 Hearing	58 Wet	73 Radiant Energy
9 Stooping	x 24 Seeing	59 Dusty	74 Toxic Conditions
10 Crouching	25 Color Vision	60 Dirty	75 Working With Others
11 Kneeling	26 Depth Perception	61 Odors	x 76 Working Around Others
12 Sitting	27 Working Speed	62 Noisy	77 Working Alone
x 13 Reaching	28	63 Adequate Lighting	78
x 14 Lifting	29	64 Adequate Ventilation	79
x 15 Carrying	30	65 Vibration	80

DETAILS OF PHYSICAL ACTIVITIES:
Sits at computer most of the day, reads copy, and fingers keyboard to enter data. Periodically walks short distances, reaches for, lifts, and carries small stacks of billing materials. Reaches for, handles, pushes, and pulls when organizing and filing paperwork at desk level.

Figure 20–8. Example of an employment service form used when matching workers to jobs.

telephone numbers of local chapters. The service directories have Rehabilitation Services and Vocational Rehabilitation listings of companies and agencies that can provide contacts for local groups. Local hospitals and physicians are also good sources along with governmental agencies. It may be a good idea for the safety and health professional to call the groups to find out if they can provide assistance.

Many systems of employee disability classification are now in use. Generally, however, these systems use broad statements, such as "physically fit for any work"; "defect that limits applicant to certain jobs" (the defect may or may not be correctable, but may require medical supervision); and "defect that requires medical attention and is presently handicapping."

In yet another method, which approaches an ideal functional evaluation of the individual, the physician documents an employee's capacities on a form that uses identical terminology to evaluate both the physical (functional) factors and the working conditions of jobs (Figure 20–9). This effective method of presenting information from a physical examination clearly indicates the specific work capability and limitations of the individual.

Thus, the medical report is more meaningful to the placement manager because the examining physician is responsible for determining the occupational significance of physical disorders. This method, in turn, makes proper analysis of each job essential.

Job Appraisal—Job Description

Employers must know the physical requirements of jobs and the accident and health hazards involved in each one. Each job-appraisal factor can make the position unsuitable or potentially undesirable for employees with one or more types of disability. The factors to be considered in job appraisal are physical requirements, working conditions, health hazards, and accident hazards.

Physical requirements include agility, strength, exertion, vision, hearing, talking, sitting, standing, walking, running, climbing, crawling, kneeling, squatting, stooping, twisting, lifting, and handling. They should be evaluated according to quality of ability and duration of activity. For example, a job involving a considerable amount of stair climbing is unsuitable for workers with heart disease, respiratory diseases, obesity, or lower limb orthopedic disorders. Some of these

JOB ANALYSIS
FOR PHYSICAL FITNESS REQUIREMENTS

TITLE OF POSITION	GRADE

NAME AND LOCATION OF ESTABLISHMENT	AGENCY

Does establishment have medical supervision ☐ Yes ☐ No
Is there an industrial safety branch ☐ Yes ☐ No

Refer to the manual for job analyses before using this form. Check all functional and working condition factors as well as acceptable disabilities whenever appropriate.

I. FUNCTIONAL FACTORS
L - *Little*   M - *Moderate*   G - *Great*   O - *None*

Hands - Fingers	L	M	G	O	Arms	L	M	G	O	Legs - Feet	L	M	G	O	Body - Trunk	L	M	G	O
1. Reaching					8. Reaching					14. Walking or running					22. Sitting				
2. Pushing or pulling					9. Lifting					15. Standing					23. Bending				
3. Handling					10. Pushing or pulling					16. Sitting					24. Reaching				
4. Fingering					11. Carrying					17. Carrying					25. Lifting				
5. Climbing					12. Climbing					18. Climbing					26. Carrying				
6. Throwing					13. Throwing					19. Jumping					27. Jumping				
7. Touching					**Eyes**					20. Turning					28. Turning				
Voice					30. Near vision					21. Lifting									
29. Talking					31. Far vision					**Ears**									
					32. Color vision					33. Hearing									

II. WORKING CONDITION FACTORS

34. Inside			41. High humidity			48. Odors			55. Toxic conditions		
35. Outside			42. Low humidity			49. Body injuries			56. Infections		
36. High elevations			43. Wetness			50. Burns			57. Dust		
37. Cramped body positions			44. Air pressure			51. Electrical hazards			58. Silica dust		
38. High temperature			45. Noise			52. Explosives			59. Moving objects		
39. Low temperature			46. Vibration			53. Slippery surfaces			60. Working with others		
40. Sudden temperature changes			47. Oily			54. Radiant energy					

III. ACCEPTABLE DISABILITIES   *Check appropriate square if acceptable*
A - *Amputation*   D - *Disability*   Y - *Yes*   N - *No*

Hands - Fingers	A	D	Arms	A	D	Legs - Feet	A	D	Body - Trunk		D
1 or 2 on primary hand			1 Arm			1 Leg (high)			1 Hip		
1 or 2 on secondary hand			2 Arms			2 Legs (high)			2 Hips		
More than 2 on primary hand			None	☐		1 Leg (low)			1 Shoulder		
More than 2 on secondary hand						2 Legs (low)			2 Shoulders		
1 Hand						1 Foot			Back		
2 Hands						2 Feet			None	☐	
None	☐					None	☐				

Eyes	Y	N	Ears	Y	N	Cardio - Vascular	Y	N	Tuberculosis	Y	N
Blind			Deaf			Moderate tension			Minimal (healed, stable or arrested)		
Industrially blind			Hard of hearing, 1 ear			High tension			Moderate (healed, stable or arrested)		
Blind one eye			Hard of hearing, 2 ears			Organic heart disease compensated			Far advanced (healed, stable or arrested)		
Color blind			Hearing aid acceptable						Collapse therapy		
Color blind for shades											

Figure 20-9. This form is used when analyzing jobs for fitness requirements.

people can tolerate only a small amount of stair climbing.

Working conditions include indoors, outdoors, excessive heat or cold, excessive humidity or dryness, wetness, sudden temperature changes, ventilation, lighting, noise, and whether the work is performed alone, near others, with others, or as shift work or piecework. Some of these conditions could be harmful for individuals with certain disabilities. For example, work in excessive heat is generally unsuitable for persons who have had malaria, or for those with high blood pressure, heart disease, or skin disease, and for older or obese workers.

Health hazards include air pressure extremes; radiant energy (ultraviolet, infrared, radium emanations, and x-rays); silica, ETS, asbestos, dusts, and skin irritants; respiratory irritants; systemic poisons; and asphyxiants. These hazards have serious effects and can aggravate a preexisting disorder. For example, a job might involve exposure to respiratory irritants of insignificant quantities to most people; yet this condition might aggravate the disability of a person who has chronic bronchitis.

The job description needs to spell out how much lifting, how much standing, how much vision is required to do the job. Specific details will be of benefit to the employer and the applicant. (These issues were discussed in the previous section, Job Placement.)

Hazards include danger of falls from elevations; working while on moving surfaces; slipping and tripping hazards; exposure to vehicles or moving objects; objects falling from overhead; exposure to sources of foot injuries, eye injuries, cuts, abrasions, bruises, and burns; mechanical and electrical hazards; and fire and explosion hazards. These hazards could present greater dangers to workers with particular disabilities. For example, a job that may involve foot injury hazards is unsuitable for the diabetic because of his or her impaired circulation, which reduces sensation in the extremities, slows the healing of wounds and fractures, and increases susceptibility to gangrene of the foot.

Preemployment—medical evaluation. Preemployment medical evaluation must be limited to the ability of the applicant to perform job-related functions.

Post-offer preemployment medical examinations. These medical exams may be allowed after an offer has been made. They also are made to determine the nature of any accommodations that may be needed.

Drug testing is not considered a medical examination and may be required before offering employment. Other regulatory tests may also be required.

ACCESS TO FACILITIES

One of a company's primary safety considerations is providing a safe and accessible parking lot for disabled employees. If people cannot find a safe place to park in the lot, they will not make it to the building. Parking spaces need to be wider than normal; details on requirements can be obtained from the Motor Vehicles departments. Both a painted symbol on the space itself and a blue-and-white sign above

it should mark the parking space. The company must make sure that nondisabled employees, visitors, and others do not use these spaces. This may require rigorous enforcement.

People with mobility impairments are not the only ones who need such spaces. Employees with severe heart and lung disease, arthritis, or other chronic conditions that restrict movement will also need specially marked spaces. Hearing-impaired people, although able to walk, are at great risk walking through a parking lot and should be given special parking considerations. Finally, assistants of seriously disabled employees will need additional space in which to maneuver wheelchairs, walkers, and other equipment.

In some instances, employees may have new wheelchairs that can climb curbs and steps without assistance. However, this does not mean that curbs should not be eliminated or that some wheelchair users will not require assistance.

Safety considerations for hiring a disabled person begin with the first step in the employment process, and on the first day of work for that employee. For example, is there reasonable, safe access to the reception area, applicant-processing area, or workplace for the new employee? Curbs and stairs present barriers to those who use canes, crutches, walkers, and wheelchairs, and also increase their chances of falling. A wheelchair user cannot safely negotiate even one step without assistance, which can cause a slight risk to both the wheelchair user and the person assisting him. Wheelchair ramps are discussed under General Access later in this section.

Access to and from Workstation

Can disabled job applicants safely proceed to where they must go to complete an application? If already employed, can such individuals safely proceed to their workstations? Access and safety are interdependent factors that need to be reconciled when employing these people. A lack of access to the company premises has been the principal factor preventing disabled persons from seeking jobs with some organizations.

One concern within the facility is adequate means of escape for all workers in an emergency. This safety requirement frequently restricts or denies disabled persons the freedom to use premises as they would wish. However, many establishments employing workers with mobility disabilities have successfully devised safe evacuation plans.

Safety management of the disabled in many firms is well established and provides a freedom of movement that is entirely compatible with principles of general safety. These measures include supervised use of the stairs for means of escape (this is discussed later), designated staff to assist in an emergency, strategic ramping of entrances-exits, and alarm systems suitable for vision- and hearing-impaired workers.

Employees with impaired hearing and/or vision may need additional devices to perceive an alarm. Some people with impaired hearing can hear an alarm in a lower pitch than normal; others do best with a buzzer or a device that vibrates against their skin. These employees need to be interviewed separately and a workable notification systems developed. Often, all that is necessary is a "buddy" system in

which a worker who can see or hear the alarm is assigned to notify the disabled employee whenever there is an emergency or drill.

One means of safely evacuating wheelchair users and permanently or temporarily disabled persons is through use of an evacuation chair (Figure 20–10). This chair is designed to ride on the ends of stair treads so one person can easily guide the disabled worker down fire stairs without putting either person in additional jeopardy during an emergency evacuation. The evacuation chair is lightweight, folds flat, and can be safely and easily stored on a wall bracket.

Without these measures, many disabled persons would be denied access to their places of employment. It is suggested that employers discuss their safety problems with fire prevention specialists and check regulatory requirements.

There are many possible ways to enable disabled people to escape safely in an emergency.

General Access

In designing access for disabled workers, do not overlook cafeteria, washroom, and restroom facilities, width of doors, height of plumbing fixtures, electrical controls, phones, and drinking fountains. Facilities designed to be truly accessible to disabled employees enhance their feelings of dignity and independence—which, in turn, can raise the morale of the entire workforce.

With minimum expense, improvements and special considerations made for workers with disabilities also will benefit other employees. For instance:

Wheelchair ramps are safer than steps—for everyone. However, the slope should be correct, not too steep and without sharp turns, to make negotiating the ramp safe and easy for wheelchair users. Ramp surfaces should be slip-resistant and kept free of obstacles. All ramps should be cleaned of mud, snow, and ice, and railings must be in place where needed.

Clean and unobstructed aisles are necessary for safe wheelchair, cane, and crutch use. They also make the workplace safer for all employees and visitors. Good housekeeping improves traffic patterns and eliminates hazards.

WHEELCHAIR CHECKLIST

The space requirements of the average wheelchair are as follows: Most wheelchairs are 36 in. high, 26 in. wide, and 42 in. long (Figure 20–11a). They require at least 60 x 60 in. to make a 180-or 360-degree turn. However, 60 x 78 in. is preferred to make a smoother U-turn. All access aisles should be wide enough to allow a wheelchair user to make a smooth turn. "Accessibility Standards," published by the State of Illinois, contains illustrated information on maneuvering space requirements for wheelchairs (see References).

The average arm reach of people who use wheelchairs is usually 48 in. on the diagonal and 64 in. to the side. The average reach directly upward is 60 in. (Figure 20–11b). The usual maximum downward reach from the chair

Figure 20–10. Using an evacuation chair, this woman safely guides a disabled fellow worker down fire stairs without putting either of them under additional risk during an emergency. (Courtesy of Evac+Chair Corporation, New York.)

is 10 in. (Figure 20–11b). Shorter distances may be needed, however, to accomplish certain tasks (Figure 20–11c).

Access to Buildings

Parking spaces 8 feet wide, next to a 5-foot-wide access aisle, should be reserved for automobiles driven by disabled personnel and visitors (Figure 20–12a). Two accessible parking spaces, however, can share a common access aisle (Figure 20–12b). Workers need room to remove and replace their wheelchairs in an auto and also to ride the chairs between aisles.

Parking spaces for disabled persons should be marked with an upright marker; symbols painted on the ground are often difficult to see or can be covered by snow or debris. At least one accessible route and entrance to the building must be provided. The entrance width should be 32 in. (80 cm); if a ramp leads to the entrance, it should be at least 36 in. (90 cm) wide. The ramp should be a maximum of 30 ft (9 m) long with an open, level area of at least 5 ft (1.5 m) at the bottom. Employees (and visitors) using wheelchairs, crutches, or canes can then move in and out of the building completely on their own. Protective side rails must be at least 36 in. (90 cm) apart if not adjacent to a wall to prevent persons in wheelchairs running or falling off the side.

Revolving doors are inaccessible for anyone in a wheelchair and for most people using crutches, wearing a leg cast,

HUMAN FUNCTIONING DIMENSIONS

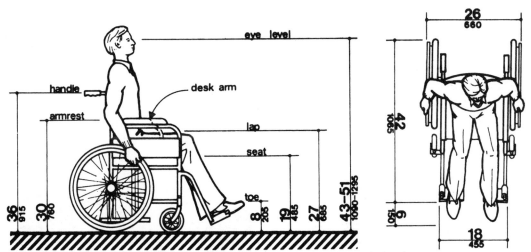

Figure 20–11a. Dimensions of adult-sized wheelchairs. Foot-rests can extend farther for very large people. Dimensions in this figure and Figures 20-11c and -11b are in both inches and millimeters. (Printed with permission from American National Standards Institute.)

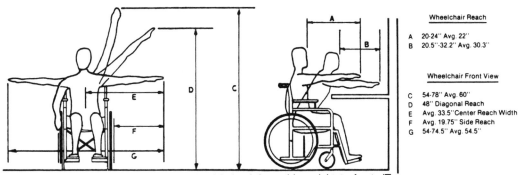

Figure 20–11b. Maximum reach from wheelchair—*left:* to sides; *right:* to front. (To convert to millimeters: 1 in. = 25.4 mm.) (Printed with permission from State Board of Barrier-Free Design, Columbia, S.C.)

Wheelchair Reach

A 20-24" Avg. 22"
B 20.5"-32.2" Avg. 30.3"

Wheelchair Front View

C 54-78" Avg. 60"
D 48" Diagonal Reach
E Avg. 33.5" Center Reach Width
F Avg. 19.75" Side Reach
G 54-74.5" Avg. 54.5"

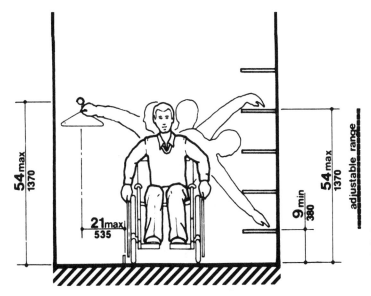

Figure 20–11c. Shorter distances are required when tasks are done. Shown here are the suggested dimensions for storage shelves and clothes racks. (Printed with permission from American National Standards Institute.)

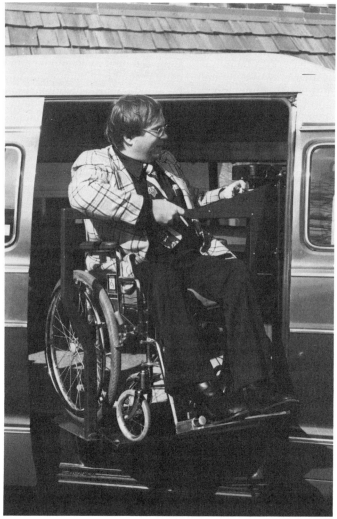
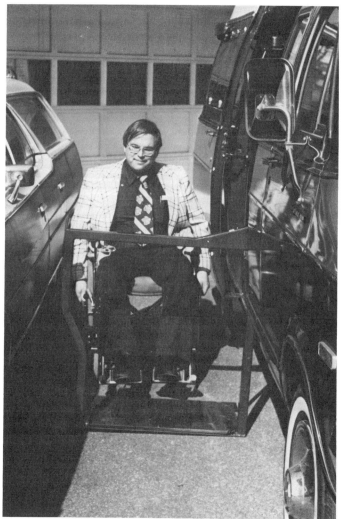

Figure 20-12a. Access aisle adjacent to a normal-width parking space is required for wheelchair clearance. Shown here is a chair lift that swings alongside of a van. For return, a locking, outside control box opens sliding doors and allows the user to control the lift. (Courtesy ABC Enterprises, Inc.)

or even carrying bulky packages. These people need entry doors that preferably open automatically. When a doorknob is necessary, it should be 36 in. (90 cm) from the floor. It is better, however, to have a vertical grab handle on the door.

Interior Access

Both entry and interior doors should be a minimum of 32 in. (80 cm) wide. Interior doors should open by a single effort and have thresholds as level with the floor as possible. To make a 180- or 360-degree turn, persons in wheelchairs need an open area of 60 x 60 in. (1.5 square m) in a typical building corridor.

Restrooms should have at least one stall wide enough for wheelchair entry. The stall should be equipped with grab bars and other fixtures no higher than 36 in. (90 cm) above floor level (Figure 20–13). The grab bars should be at 33 in. (83 cm). Stall doors should be at least 32 in. (80 cm) wide. Controls, switches, fire alarms, and other devices that might be used by a disabled individual must be within convenient reach of a wheelchair user.

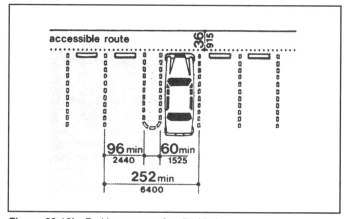

Figure 20-12b. Parking spaces for disabled persons can share a common access aisle. Aisle should be part of the accessible route to the building entrance. Overhangs from parked vehicles must not reduce the clear width of the accessible route, which must be the shortest possible distance to the entrance. (Printed with permission from American National Standards Institute.)

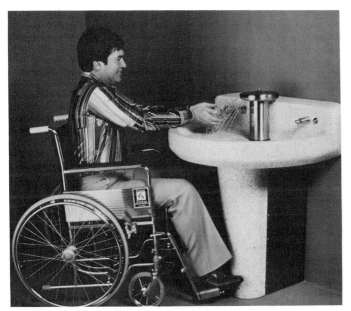

Figure 20-13. A semicircular washstand no higher than 3 ft (0.9 m) allows this wheelchair user easy access. (Courtesy Bradley Corporation.)

Restrooms should be located on each of the floors where disabled persons work.

Office Accommodations

Desk tops should be no less than 28 in. (70 cm) above the floor to accommodate wheelchairs. Metal desks usually have adjustable feet that workers can raise to the maximum. If more room is needed, the desk can be raised with additional blocks.

Chairs can be regular height, but they should be sturdy and have arms to help disabled people lift themselves up. Some individuals need a chair that will not move easily so they can stand without the chair sliding out from under them. For other disabled workers, however, casters placed on the bottom of chairs may be desirable to help them slide the chair in and out. In addition, they allow a worker to pull himself/herself from one piece of furniture to another, thereby avoiding the need to continually move in and out of the chair.

Business machines should be placed, if possible, where they will not become a barrier or obstruct traffic. This precaution is particularly important for wheelchair users.

The following accommodations may sometimes apply, depending on the individual worker and the job requirements:

File cabinets should be placed so that workers can reach drawers from both the front and side. This eliminates awkward reaches by those who must use crutches, a cane, or a wheelchair.

If books, reports, or other bulky objects must be carried from place to place, make sure a wheeled cart is kept handy so workers can load materials on the cart.

Cords for venetian blinds, window shades, and draperies should be long enough that disabled workers can reach them easily.

Floors must be free of extension cords, raised box receptacles, or any other raised items.

If a disabled employee has an assistive dog, then space, exercise facilities, and water for the dog will have to be arranged. The employee will know exactly what is needed. Other employees must understand that the dog is a working animal; they should not pet or feed it without express consent from the owner. Otherwise, the dog can become confused about its role and stop being dependable as the eyes or ears of the disabled employee.

Discrimination

Under the ADA-1990, discrimination includes—for example—limiting, segregating, or classifying a job applicant or employee in a way that adversely affects the person's opportunities or status. Discrimination also includes not making reasonable accommodations for the known physical or mental limitation of an otherwise qualified person with a disability. It also includes the denial of employment because a qualified person with a disability needs reasonable accommodation.

SUMMARY

- Almost every worker with a physical or mental impairment can qualify for some type of job because of medical advances and governmental affirmative action legislation. The Americans with Disabilities Act, 1990, identifies responsibilities involving U.S. workers. It also mandates employers to provide "reasonable accommodation" for these workers when necessary. Supervisors and other management personnel must be given training in what disabled people can and cannot do and in the proper terminology for referring to various disabilities.

- Employers who deny disabled workers a job must prove that they are unqualified because of one or more of the following reasons: (1) they would endanger themselves or others, (2) they cannot meet job requirements, and (3) the employer cannot make reasonable accommodation of the workplace or job without undue hardship to the firm.

- In the United States, the Federal Rehabilitation Act of 1973, known as the Rehabilitation Act, and Vietnam Era Veteran's Readjustment Assistance Act of 1974 protect disabled workers' rights and provide legal incentives for employers to hire and promote disabled workers and veterans. Various state and local laws also mandate affirmative action programs for the hiring and advancement of disabled persons.

- The law defines three types of disabled job seekers: disabled individual, disabled veteran, and qualified disabled individual.

- The disabled individual is anyone with a physical or mental impairment, has a record of such impairment, or is regarded as having an impairment that interferes with major life activities. The disabled veteran is anyone discharged or released from active duty because of a disability incurred in the line of duty. The qualified disabled person

is someone who is capable of performing a particular job with reasonable accommodation to the disability.

- Under regulations, employers must make "reasonable accommodation" to the known physical or mental limitations of an otherwise qualified disabled applicant or employee. This can include modifying equipment or changing job descriptions. The employer is excused from this duty only if it can show that the accommodation would impose an undue hardship.
- The safety and health professional can work with the EEO coordinator in evaluating reasonable accommodation and ensuring compliance with government affirmative action regulations.
- Management must evaluate each disabled job applicant to place the right person in the right job so the impairment is not a factor of job performance. In addition, management should realize that certain regulatory standards may not be enough to protect disabled workers. More stringent standards may need to be developed to ensure safe working conditions for these employees.
- Employers must evaluate jobs in terms of physical requirements, working conditions, health hazards, and accident hazards to eliminate or control job risks that might endanger disabled workers. Supervisors and other managers must ensure that disabled workers are not harassed or abused by other workers.
- Companies can also contact self-help and support groups; governmental agencies; and private agencies to learn more about how to accommodate a job for a disabled employee and how to work with disabled persons.
- Companies should ensure safe, convenient access to all facilities for disabled workers with mobility limitations. This includes safe parking areas, special ramps, lack of curbs, easy access to and from workstations, unobstructed means of escape in case of emergencies, and adequate space for wheelchair users in office areas, elevators, hallways, entrances, restrooms, and the like.

SOURCES OF HELP

The overall goal of all employers should be to hire qualified disabled individuals and place them in available and safe occupations. One of the goals of the safety and health professional is to assist the employer in this worthwhile endeavor.

The following list includes government and private agencies that can help achieve this goal.

Alabama Institute for the Deaf and Blind, Talladega, AL 35160.

Arthritis Foundation, 1314 Spring Street NW, Atlanta, GA 30309. (404)872-7100.

Association for Children and Adults with Learning Disabilities, 4156 Library Road, Pittsburgh, PA 15234. (412)341-1515.

Asthma and Allergy Foundation of America, 1717 Massachusetts Avenue NW, Suite 305, Washington, DC 20036.

(202)265-0265.

Cystic Fibrosis Foundation, 6931 Arlington Road, Suite 200, Bethesda, MD 20814. (800)344-4823; (301)951-4422.

Easter Seal Society, 70 East Lake Street, Chicago, IL 60601. (312)726-6200.

Epilepsy Foundation of America, 4351 Garden City Drive, Landover, MD 20785. (800)332-1000; (301)459-3700; Fax (301)577-2684.

Lupus Foundation of America, 1717 Massachusetts Avenue NW, Suite 203, Washington, DC 20036. (800)558-0121; (202)328-4550.

George Washington University, Job Development Laboratory, Rehabilitation Research and Training Center, 2300 I Street NW, Washington, DC 20052.

Mainstream, Inc., 1200 15th Street NW, Washington, DC 20005.

Multiple Sclerosis Society, 205 East 42nd Street, New York, NY 10017. (800)624-8236; (212)986-3240.

Muscular Dystrophy Association, 801 Seventh Avenue, New York, NY 10019. (212)586-0808.

Myasthenia Gravis Foundation, Inc., 53 West Jackson Boulevard, Suite 1352, Chicago, IL 60604. (800)541-5454; (312)427-6252.

National Association for Sickle Cell Disease, 4221 Wilshire Boulevard, Suite 360, Los Angeles, CA 90010. (213)936-7205.

National Association of Protection and Advocacy Systems, Client Assistance Program, 220 I Street NW, Suite 150, Washington, DC 20002. (202)546-8202.

National Council of Persons with Disabilities, P.O. Box 29113, Washington, DC 20017. (202)529-2933.

National Down Syndrome Society, 666 Broadway, New York, NY 10012. (800)221-4602; (212)460-9330.

National Head Injury Foundation, 333 Turnpike Road, Southborogh, MA 01772. (800)444-NHIF; (508)485-9950.

The National Institute for Rehabilitation Engineering, 97 Decker Road, Butler, NJ 07405.

National Kidney Foundation, 30 East 33rd Street, New York, NY 10016. (800)622-9010; (212)889-2210.

National Spinal Cord Injury Association, 600 West Cummings Park, Suite 2000, Woburn, MA 01801. (800)962-9629; (617)935-2722.

Paralyzed Veterans of America, 801 18th Street NW, Washington, DC 20006. (202)872-1300.

President's Committee on Employment of the Disabled, 1111 20th Street NW, Suite 636, Washington, DC

20036–3470.

The Rehabilitation Institute of Chicago, 345 East Superior Street, Chicago, IL 60611.

Retinitis Pigmentosa Foundation Fighting Blindness, 1401 Mt. Royal Avenue, Baltimore, MD 21217–4245. (800)638-2300; (314)225–9400.

Self Help for Hard of Hearing People, 7800 Wisconsin Avenue, Bethesda, MD 20814. (301)657–2248.

Spina Bifida Association of America, 1700 Rockville Pike, Suite 540, Rockville, MD 20852. (800)621-3141; (301)770–SBAA.

Support Dogs for the Handicapped, P.O. Box 966, St. Louis, MO 63044. (314)487–2004.

REFERENCES

American National Standards Institute, 11 West 42nd Street, New York, NY 10036. *Buildings and Facilities—Providing Accessibility and Usability for Physically Handicapped People,* A117.1–1986

Brooks WT. "Supervising Handicapped Workers for Safety."

Capital Development Board. *Accessibility Standards Illustrated.* Reprint of 1978 edition with all revisions to March 1, 1985 and Environmental Barriers Act, Public Act 84–948. Springfield: State of Illinois, 1985.

Hill N, et al. The merits of hiring disabled persons. *Business & Health.* February 1987.

———. Hiring the handicapped: Overcoming physical & psychological barriers in the job market. *Journal of American Insurance.* Third quarter, 1986.

President's Committee on Employment of the Handicapped, Washington, DC 20036.
"Affirmative Action to Employ Handicapped People—A Pocket Guide."
"Supervising Handicapped Employees."
Superintendent of Documents, U.S. Government Printing Office, Washington, DC 20402.
Occupational Safety and Health Act of 1970 (Public Law No. 91–596).
Occupational Safety and Health Act Regulations, Title 29, *CFR,* Chapter XVII, Part 1910.
Rehabilitation Act of 1973 (Public Law No. 93–112).
Rehabilitation Act Amendments of 1974 (Public Law No. 93–516).
Vietnam Era Veterans' Readjustment Assistance Act of 1974 (Public Law No. 93–508), Section 38 USC 2012.

U.S. Department of Health and Human Services, Washington, DC 20202. "Nondiscrimination on the Basis of Handicap in Programs and Activities Receiving or Benefiting from Federal Financial Assistance" (45 *CFR* 84).

U.S. Department of Labor, Employment Standards Administration, Office of Federal Contract Compliance Programs, Washington, DC. 20210. "Affirmative Action Obligations of Contractors and Subcontractors for Handicapped Workers" (41 *CFR* 60–741).

Woodward RE. Industry unlocks its doors to the handicapped. *Plant Facilities.* Vol. II, No. 2 (February 1979).

———. *Comprehensive Barrier-Free Design Standard Manual.* Columbia, SC, State Board for Barrier-Free Design, P.O. Box 11954. 1979.

21

Nonemployee Accident Prevention

Good management is evident in the conduct of a business's routine operations and in areas that directly affect relations among managers, employees, and the firm's customers. One of these areas, nonemployee accidents, also affects employees and the products or services they provide or sell. When accidents occur, operations stop (or at least slow down) no matter who is involved. With the increase in consumer product safety legislation (see Chapter 9, Product Safety Management) and greater cost of liability claims, management must pay close attention to this area of potential loss.

Today, more people are involved in serving and selling to the public than are engaged in making the products that are sold. Service businesses include specialty shops, department stores, shopping center complexes, restaurants, fast-food service operations, hotels and motels, automotive service and dealerships, hardware and building supply stores, amusement parks, banks and office buildings, and franchise businesses. The patrons, guests, or visitors of these businesses make up the category of nonemployee injuries.

Service and sales operations must constantly monitor patron activities for potential problems, even though they cannot control customers' personal activities or habits. For example, inebriated guests or patrons represent a serious risk when they improperly dispose of smoking materials or their instability causes them to slip, fall, or suffer other injuries. Children can also be involved in accidents in the business environment. Whenever possible, design the original facility for predictable use and misuse.

A company must ensure that its loss control plans include preventing nonemployee accidents. These unplanned events can be minor or catastrophic. A firm can be liable for damages or for injuries from the minute someone enters the property (including the parking lot). In addition, a firm can be liable for the actions of its employees when they are sent off company property on business should their actions result in damage or personal injury to others.

THE LEGAL SIDE OF NONOCCUPATIONAL INJURIES

Customer or product claim cases often wind up in a courtroom. The legal terminology and aspects of the law that deal with accident claims include the following definitions and explanations. These are not intended to be comprehensive but to provide a quick review of some principal considerations.

Tort. A private or civil wrong resulting in an injury—a violation of a right not arising out of a contract. It may be either (1) a direct invasion of some legal right of the individual, (2) the infraction of some public duty by which special damage accrues to the individual, or (3) the violation of some private obligation by which damage accrues to the individual. Torts result from negligence, accidents, trespass, assault, battery, seduction, deceit, conspiracy, malicious persecution, and many other wrongs or injuries.

Negligence. Failure to exercise that degree of care which an ordinarily careful and prudent person ("reasonable indi-

vidual") would exercise under similar circumstances. To establish a proper claim of negligence, however, there must be (1) a legal duty to use care, (2) a breach of that duty, and (3) injury or damage.

Degree of care. The degree of legally required attention, caution, concern, diligence, discretion, prudence, or watchfulness depends upon the circumstances. For example, a high degree of care is demanded from people who invite others onto their premises by written, verbal, or implied invitation. All sales and service enterprises must exercise a high degree of care for the safety of their patrons. As long as a business is open, it assumes a responsibility for the well-being of its customers.

Invitee. One whose presence on the premises is upon the invitation of another, such as a patron at a sports stadium or a person who visits an exhibition hall, even though no admission is charged.

Licensees. Licensees are neither "invitees" nor "trespassers." They have not been specifically invited to enter upon the property but they have a reasonable excuse (by permission or by operation of the law) for being there. These could be vendors, delivery personnel, people visiting executives, or purchasing agents for business purposes, and the like. Policemen and firefighters who enter property in the course of their duties have sometimes been held by the courts to be invitees (patrons) and sometimes licensees (nonpatrons).

Limitations of recovery. American businesses are not automatically insurers of their clients or patrons. All states apply some formula to limit the ability of insured persons to recover monies from a company through a civil lawsuit. These legal doctrines fall into two broad categories. The oldest, called *contributory negligence*, bars any financial recovery by an injured party if the victim contributed to the original accident in any way.

A newer, and perhaps more humane approach, is found in the legal doctrine called *comparative negligence*. This doctrine, adopted by most states, requires a court to limit the recovery of an injured party based upon how much their own action contributed to the original incident. Both doctrines illustrate the need for thorough and truthful investigation of every incident. It can strengthen a company's ability to reduce the overall cost of doing business and help to prevent future lawsuits.

Assumption of risk. Claimants cannot collect damages when they were aware of peril or danger yet were willing to proceed with their original intention and undertook the action. "That to which a person assents is not regarded by law as an injury." For example, a skier who falls while descending a slope legally assumes the normal risks associated with this sport, unless some special negligence in the design or maintenance of the slope and its environment is a contributing factor. If so, the skier assumes no risk when the owner or operator is negligent. However, an injury caused by a mechanical defect in a chair lift or tow rope could prove costly for the company running the operation or for the owners of the ski resort.

Hold harmless. A clause in a contract agreeing that one party will assume all liabilities, losses, or expenses involved, is a hold-harmless agreement. For example, a department store may have a hold harmless agreement with a manufacturer who supplies a particular type of merchandise. If a consumer claims an injury was caused by that product, the manufacturer will reimburse the store, should it be held liable, and will pay for legal and other expenses incurred by the store in defending itself. Even though the consumer purchased the item from the department store, a hold-harmless agreement may relieve the store of the financial effect of a tort.

Attractive nuisance. This item refers to liability associated with a dangerous condition that is generally a threat to children. It excuses trespassing and penalizes an organization for failure to keep children away from the hazard or for failure to protect or eliminate a hazard that may reasonably be expected to attract children to the premises. A swimming pool is often considered an attractive nuisance because children are drawn to it, regardless of the protective features.

Burden of proof. The injured party must prove injury or damage and its causal relation to the event or item that created the accident. A defendant is not liable if he or she is without fault. Proof must be established by facts, not opinion, suspicion, rumor, hearsay, gossip, or emotional reaction. Proof is a conclusion drawn from the evidence.

Honest and sincere witnesses often report widely different impressions about the same events. Thus, it is essential to assemble and preserve evidence quickly. Signed statements taken shortly after the accident or an all-important photograph can often make a big difference in a trial. In liability claims, the burden of proof rests upon the plaintiff (claimant).

This chapter cannot cover all possible nonemployee accidents. Instead it describes the most common ways that nonemployees can suffer an injury. Accident prevention techniques are broad and varied. It will be up to each safety and health professional to examine the company's operations in light of these guidelines. The following sections review some major problem areas and discuss typical accidents that occur in modern business facilities.

PROBLEM AREAS

When starting a nonemployee accident prevention program, the safety and health professional should look over the records of past nonemployee injuries to identify accident trouble spots. Next, choose general trouble spots, such as stairs, doors, those due to poor lighting, faulty interior design, and poor traffic flow. Change facilities and/or operations so that both employees and nonemployees are given a safe business environment.

Glazing

Modern office building, store, and plant design often uses glass extensively in doors, show windows, panels, and enclosures. Such areas can result in a confusing pattern, especially to the first-time visitor or patron.

Unmarked glass panels and doors can be the cause of severe injuries and cuts should employees or patrons walk or fall through them. Companies must follow recommended practices and standards to correct or eliminate this hazard. The safety and health professional or manager should understand regulatory standards on this issue.

Tempered glass panels, required for many uses, are given a special heat treatment during production so that, if broken, the pieces of glass will fall as small, blunt beads instead of sheet glass plates.

The improper use of glazing materials is a continuing source of legal problems for companies dealing with the public. A complete check of all glazed items is important.

Some plastic glazing offers advantages similar to glass but without the risk of damage or injury. Unfortunately, some plastics have higher impact resistance and tend to scratch or be more easily abraded than glass. As a result, workers may find it harder to break plastic windows or doors than glass if they need to escape from an area.

A few precautions will allow companies to enjoy the beauty and functionality of glass while lessening the chances for injury:

1. Make glass doors more visible to adults and children by placing decals or pressure-sensitive tape at their respective eye levels. Sand-blasted or etched designs serve the same purpose.
2. Decals or pressure tapes will also prevent glass panels from appearing to be doorways. A tall, attractive plant, placed in the center of the panel, will let people know the panel is there.
3. Installing safety bars reduces the size of the open glass areas and lessens the chance of glass breakage. The bars should be at the door-handle level on sliding doors and should be on both sides of a swinging door.
4. Keep doorways and areas that are close to glass panels free of tripping hazards; common hazards include scatter rugs and toys. Indeed, this is a good rule for all occupied areas.
5. Enforce no-play, no-running rules especially where there is a glass hazard.

Parking Lots

Most shopping or business centers have parking lots for patron use, usually self-service. Bicycles, mopeds, and motorcycles as well as automobiles, vans, and trucks require parking facilities. If these self-service parking lots are well designed and attractive, they can almost eliminate the risk of damage to any vehicle in the lot.

Parking lot problems depend to some extent on how the lots are used. Self-service parking lots for industrial plant employees, for instance, usually differ from public lots at shopping centers, theaters, stadiums, stores, schools, restaurants, or motels. Usually, public lots have a more steady flow of traffic, more children present, and shopping carts obstructing traffic ways. More importantly, the lots are used by persons with varied levels of driving skill who may be less familiar with the lot's layout.

Some public parking lots at places such as theaters, sports arenas, and schools have the combined hazards of extreme traffic fluctuation and great variations in driver skill and alertness.

An easy-to-use layout, adequate signs, and conspicuous markings help to make a parking lot safe and attractive. These steps represent the first way to reduce hazards. Of several other safety measures, none is more important than enclosing the parking lot with a curb or fence so that cars cannot enter or leave traffic unexpectedly. This step should significantly reduce accidents in the area. Parking lots should have separate, well-marked entrances and exits, placed so that they favor right-hand turns. Single-lane entrances should be at least 15 ft (4.5 m) wide; exits should be at least 10 ft (3 m) wide. Where entrance and exit must be combined, the double-lane drive should have at least 26 ft (8 m) of usable width. When the double-lane combination is necessary, it should have median curbs or strips to positively control the flow of traffic.

In general, entrances and exits should be:

- at least 50 ft (15 m) from intersections
- away from heavily traveled highways or streets
- well marked and well illuminated
- as few in number as possible.

Pedestrian traffic must be considered in parking lots. The most common injury occurs from hazards such as potholes, changes in elevation, and cracks in the walking surfaces due to uneven settling of the ground, wear of heavily loaded vehicles, and effects of cold weather. Maintenance staff should conduct periodic inspections to identify and rectify these hazards and to maintain a safe walking surface for patrons. Stairways and ramps should be constructed in accordance with ANSI standard A1264.1–1989, *Safety Requirements for Floor and Wall Openings and (nonresidential) Stair and Railing Systems.* They should be well marked, well illuminated, and guarded with handrails. The needs of the physically disabled must also be considered in the design of the lot and in any walkway to the building. See Chapter 20, Workers with Disabilities.

Parking aisles should be perpendicular to the buildings, so that pedestrians walk down the aisles rather than between parked cars. Where possible, marked lanes, islands, or raised sidewalks should be provided between parking rows, particularly where pedestrian traffic is especially heavy. These walks should be wide enough so that car bumpers overhanging them do not restrict pedestrians from walking to or from their cars.

Angle parking has both advantages and disadvantages. Although fewer cars may fit in the lot, angle parking is easier for customers and does not require as much room for sharp turns.

The area allowed per car in parking lots varies from 200 to more than 300 ft^2 (18.5 to 28 m^2) if aisles are included. (See Chapter 1 in the *Engineering and Technology* volume.)

Parking stalls. The design of parking stalls depends on (1) the size and shape of the lot, (2) the traffic pattern in the lot, and (3) the type of driver and pedestrian who use the lot.

(See Figure 20–12A and 20–12B in Chapter 20 for stall dimensions for the physically disabled.)

Bumpers on stalls have these advantages:

1. They prevent drivers from driving forward through facing stalls and proceeding in the wrong direction in one-way aisles.
2. They encourage drivers to pull forward against the bumper, thereby preventing rear overhang that might reduce aisle width.
3. They break up the huge expanse of an open lot that may tempt drivers to cut across aisles and endanger pedestrians and other drivers.
4. They block cars from accidentally rolling down inclines or running through walkway areas.

Stall bumpers can cause some problems, however:

1. They may require maintenance.
2. They may interfere with drainage or snow removal.
3. They may cause pedestrians to trip and fall. (Painting them a bright yellow or other distinctive contrasting color may help to reduce such incidents.)
4. They may reduce the flexibility of traffic flow.

Signs and lighting. Traffic signs in parking lots should conform to recommended standards. They should be like those used as street and highway signs for easy recognition.

For example, stop or yield signs should be installed at main crosswalks for pedestrians, wherever exits cross public sidewalks, and wherever exits enter main thoroughfares.

A well-lit parking lot reduces accidents and discourages crime. The amount of light recommended for parking at night usually ranges from 0.5 to 1.0 foot candles (decalux) per sq ft at a height of 36 in. (0.9 m). Lights are usually mounted 30 to 35 ft (9 to 10 m) high on poles whose bases are protected against impact by cars.

Design. Built-in bumps on lot surfaces have been used to discourage speeding. However, because they can also damage cars, their use seems questionable. An alternative device is to keep straight lanes short and to provide sharp curves that force drivers to reduce their speed.

Parking lot operators should control unauthorized use of company lots. Local police or organizational security personnel should check parking areas after hours. If lots cannot be fenced or entrances cannot be protected against unauthorized entry, prominent warning signs stating the lot is guarded will help to minimize the risks of unauthorized use and vandalism.

Those responsible for supervising lots should cooperate closely with local traffic authorities. It is also important to cooperate with police, sheriff, and fire departments. Fire trucks will need access to hydrants in a lot and to buildings served by the lot. Zoning boards are also interested in problems connected with parking lots because of the impact on traffic flow and congestion.

Shopping carts. Many companies want to keep shopping carts out of parking lots. To achieve this goal, companies should have customers drive their cars to a loading point where an attendant puts their goods into the car or wheels the cart to the car, unloads it, and brings it back.

If shopping carts are permitted in the parking lot, they should be collected frequently and temporarily stored in a designated space. Do not allow carts to accumulate in stalls in heavily trafficked aisles. Cart-collecting areas should be well marked, well illuminated and, preferably, separated from traffic by barriers or bumpers in the pavement. Encourage customers to leave carts in those areas by installing signs and access lanes. To keep carts from rolling into traffic lanes, install low curbs or a shallow depression. People who supervise the lot should make sure that children do not play with the empty carts or use them as scooters and racers.

Other vehicles. The use of bicycles, mopeds, and motorcycles as commuting vehicles is fairly commonplace. More people will use these vehicles if (1) the routes are safer and (2) secure parking is available. Current parking facilities are not generally adequate for these vehicles. Like most motorists, operators of bicycles, mopeds, or motorcycles prefer to park near their destination. In planning for such parking, two major problems arise: (1) convenience and (2) security.

Bicycle storage is fairly easy, but the problem of security is more difficult. Twelve to 15 bicycles can be stored in the space it takes to park one car. A lesser number of mopeds or motorcycles will also fit in the same area.

Three basic types of bicycle parking facilities are in current use. The common bicycle rack comes in many styles and shapes and requires chaining or clamping the bike to the rack. Unless the bicycle frame and both wheels are locked, the owner can lose part of the bike. Second, riders can use a hitching post with a chain secured to lock the bike. Finally, they can store bicycles in a key-operated locker much like baggage lockers. These are totally enclosed, protected from thieves and the weather—an advantage where weather protection is a concern, especially for all-day parking. Moped owners and motorcyclists face most of the same convenience, security, and weather problems as bicyclists.

Entrances to the Building

Entrances to department stores and office buildings must be adequate in number and size to meet all building codes. Revolving doors should have governors that limit their speed to 12 turns per minute. Maintenance should replace all worn weatherstripping and keep sidewalks and driveways cleared and in good repair to prevent tripping and falling hazards. Entrance lighting should be a minimum of 10 foot candles.

Types of Ramps

To provide passage from the sidewalk to the customer's car, it is customary in shopping centers to construct a ramp. Indented ramps are the best type, ones that actually cut into the sidewalk with an easy slope. Rails on each side can prevent people from falling into one of the recessed sides if they are sufficiently deep enough to be a hazard. Do not paint the ramp because paint seals the concrete surface and can create a slipping hazard.

Entrances to buildings are of considerable importance from a safety viewpoint. Often entrance and exit doors are

automatic and are activated when anyone steps onto a carpet. Actually, each door has two carpets: one on the sidewalk and one on the floor inside the door.

The "in" door should swing inward when patrons step on the outside carpet. The carpet on the inside of the door acts as a safety mat, i.e., should anyone step on the safety mat, it will automatically deactivate the outside carpet, preventing the door from swinging inward and striking the person on the other side. The same principle, in reverse, applies to the "out" door. The sidewalk carpet installation in front of the entrance door is usually flanked by two handrails so that someone merely walking by will not activate the door.

If the vestibules are glass-enclosed, decals should be applied to alert people against walking into glass panels, as discussed earlier.

Walking Surfaces

Slips and falls are the most likely sources of nonemployee injuries. These injuries can occur almost anywhere at any time. Few surfaces can be ignored, including everything from asphalt roads; concrete walks; wooden, tiled, or rug-covered floors; and special surfaces on stairs and conveyances (moving sidewalks, escalators, elevators), to bridges and catwalks.

The natural properties of any surface can change substantially when people track in mud, snow, dirt, and moisture. Moisture-absorbent mats, runners, or rugs can reduce such hazards. Floor maintenance requires special attention to eliminate the hazard of torn or curled-up floor coverings (Figure 21–1). Floors, stairs, and other walking surfaces should be kept nonslippery, clear, and in good repair.

Slip meters have been developed by testing agencies and insurance companies to measure the slipperiness of floors. One type of instrument (Figure 21–2), mounted on three leather "feet," is pulled across the floor by a motorized winch. The dial on top measures the intensity of the pull required to start moving it. This is converted into a "slip index," or the coefficient of friction (0.5 to 0.6 is ideal). Although important, this "slip index" is not foolproof. Floor slipperiness may change and usually increase because of moisture, oil, grease, foreign or waste materials, and incorrect cleaning or waxing.

Falls on Floors

Falls on floors occur in various ways and from various causes. A person may slip and thus lose traction, or trip over an open drawer, box in the aisle, or other object. The primary mechanical causes of falls on floors are unobserved, misplaced, or poorly designed movable equipment, fixtures, or displays; poor housekeeping; and defective equipment. The condition of a person's shoes or type of footwear soles and heels are likely to be major contributing factors in slips and falls.

Inadequate illumination also can cause falls. Light values at floor level should be uniform with no glare or shadows. Also, there should be no violent contrasts in light levels between floor areas, such as from bright sunlight outside the

Figure 21–1. Cotton and other fabric mats used in entrances absorb moisture and dust, but require constant watching to minimize tripping hazards. Someone may catch a toe or heel in the raised fold and trip or fall. *Top:* The first person's toe or heel will catch in the mat. *Bottom:* The next person will either trip on the raised portion or dislodge it further, building up the hazard. The remedy is to lay the mat flat and fasten it securely.

entrance to a dimly lit lounge or restaurant.

Other factors causing falls include a patron's age, illness, emotional disturbances, fatigue, lack of familiarity with the environment, and poor vision. Because a company cannot readily control many of these factors, it should make doubly sure that the walking surface is as safe as possible. For example, management should not place mirrors and other distracting decorations in areas visible from steps or from approaches to steps or escalators.

Types of floor surfaces. A wide variety of floor surfaces are available. In office buildings, hotels, mercantile stores, and similar establishments, the use of masonry (terrazzo, cement, or quarry tile) floors is common at entrances, in lobbies, on stairways, and sometimes throughout the ground floor and in upper floor corridors. Decorative materials such as terrazzo, marble, and ceramic tile are most often applied to interiors while concrete and granite are generally considered more practical for exterior use (Tables 21–A and 21–B).

In other public areas of these buildings, the base floor, usually of concrete or wood, is generally surfaced with one or more of the popular resilient floor-covering materials. Management generally uses carpeting in limited areas of various stores and hotels. Elsewhere, asphalt, linoleum, rubber, or plastic in either sheet or tile form, will usually be found. Obviously, safety, initial cost, durability, and maintenance costs are some factors governing the choice of floor covering. In every case, follow the recommendations of the floor material manufacturer. Companies should not improvise. Many accidents happen when managers attempt to improve the original appearance of a floor material at the expense of safety.

Table 21-A. Physical Properties of Floor Finishes

Types of Finish	Resistance to			Quality of			
	Abrasion	Impact	Indentation	Slipperiness	Warmth	Quietness	Ease of Cleaning
Portland cement concrete *in situ*	VG-P	G-P	VG	G-F	P	P	F
Portland cement concrete precast	VG-G	G-F	VG	G-F	P	P	F
High-alumina cement concrete *in situ*	VG-P	G-P	VG	G-F	P	P	F
Magnesite	G-F	G-F	G	F	F	F	G
Latex-cement	G-F	G-F	F	G	F	F	G-F
Resin emulsion cement	G-F	G-F	F	G	F	F	G-F
Bitumen emulsion cement	G-F	G-F	F-P	G	F	F	F
Pitch mastic	G-F	G-F	F-P	G-F	F	F	G
Wood block (hardwood)	VG-F	VG-F	F-P	G-F	F	F	G
Mastic asphalt	VG-F	VG-F	VG-F	VG	G	G	G-F
Wood block (softwood)	F-P	F-P	F	VG	G	G	G-P
Metal tiles	VG	VG	VG	F	P	P	G-F
Clay tiles and bricks	VG-G	VG-F	VG	G-F	P	P	VG
Epoxy resin compositions	VG	VG	VG	VG	F	F	VG

Code: VG—Very Good; G—Good; F—Fair; P—Poor; VP—Very Poor.

Figure 21–2. Slip meters, ranging from the motorized type (shown here) to a simple spring scale and heavy block pulled by hand, can be used to gauge the slipperiness of floors. (Courtesy Liberty Mutual Insurance Company.)

Most flooring materials, whether wood, masonry, or the resilient types, are reasonably slip resistant in their original, untreated condition. However, some of the masonry materials are exceptions to the rule. A highly polished marble, terrazzo, or ceramic tile, used for ornamental effect, can be slippery even when dry. When moisture is present, its slipperiness will be greatly increased. Improper surface-treating preparations and improper cleaning materials and methods also increase a floor's slipperiness. Unless a slip-resistant material is added to the compound during construction, the only preparation that should be used on such floors is a penetrating, slip-resistant sealer.

Floor coverings and mats. Reduce the possibility of slips and falls by using good carpeting, bound edging, and flush floor-level mats and runners (Figure 21–3). Over a given period of time, carpet threads tend to become loose, creating a tripping and falling hazard. This condition is more of a risk in wall-to-wall carpeting. Maintenance staff should in-spect carpeting regularly for potential hazards. When loose threads are found, have the carpet restretched or replaced to eliminate the hazard.

Whenever possible, provide a contrasting color on carpeted areas that meet and continue on treads and risers (or treads only) of stairways. If material such as an extruded metal runner is used to provide self-cleaning removal of snow, ice, or mud at entryways, it should be flush and not present a tripping hazard. Maintenance staff must pay careful attention to rubber mats, rug runners, and the like to prevent them from becoming tripping hazards. Oftentimes their edges become rumpled, corners and ends are torn or do not lie flat, and excess wear causes tears. Management should replace mats and runners at the first sign of wear or other unsafe conditions.

Eliminate slips and falls on throw, oriental, and area rugs by using skid-resistant rug pads. When using rugs over carpet, attach a skid-resistant underlay to the rugs. These underlay pads can be purchased from carpet dealers. Floor mats, run-

Table 21-B. Guide to Floor Materials and Surfacings

Floor types*	Characteristics	Use of Abrasives	Dressing Materials
Asphalt tile	Composed of blended asphaltic and/or resinous thermoplastic binders, asbestos fibers, and/or other inert filler materials and pigments.	Abrasive materials of various types may be used to reduce slipperiness of floors. Colloidal silica** can be incorporated in wax and synthetic resin floor coatings.	**Wax or wax-base products**—For most purposes, wax has several advantages. This is especially true of Carnauba wax, an ingredient generally used in so-called wax products. This wax, a Brazilian palm tree product, dries in place with a very hard and glossy finish, but with a characteristically slippery surface. Because of its many good qualities, it is widely used as a base for floor surface preparations, both in paste and emulsion forms. Other waxes, notably petroleum wax and beeswax, have their place in floor dressing formulas; they are softer and less slippery than Carnauba, but are still slippery to a degree depending on the formulation.
Linoleum	Cork dust, wood flour, or both, held together by binders consisting of linseed oil or resins and gum. Pigments are added for color.		
Rubber	Vulcanized, natural, synthetic, or combination rubber compound cured to a sufficient density to prevent creeping under heavy foot traffic.	Slip-resistant except when wet.	
Vinyls	Composed of inert, nonflammable, nontoxic resins compounded with other filler and stabilizing ingredients.	Adhesive fabric with ingrained abrasives can be used. They are patterned in strips, tiles, and cleats.	
Terrazzo	Consists of marble or granite chips mixed with a cement matrix.	Silicon carbide or aluminum oxide can be included in mix when floor is laid. Also an abrasive-reinforced plastic coating can be painted on.	**Slip-resistant sealant** will typically improve slip-resistant quality if renewed periodically.
Concrete	Made of portland cement mixed with sand, gravel, and water and then poured.		
Mastic	Like asphalt tile in composition but is heated on the job and troweled onto the floor to form a seamless flooring. Such floors are often used over concrete to give a new durable, resilient surface.	(Same as asphalt tile)	**Synthetic resins**—These preparations, known as "synthetics," "resins," or "polishes," are intended to supply the desirable characteristics of wax without producing the same degree of surface slipperiness. They include soaps, oils, resins, gums, and other ingredients, compounded in various ways to produce the desired result.
Wood	May be either soft or hard, in a variety of thicknesses and designs.	Metallic particles and artificial abrasives in varnish or paint give good nonslip qualities to various floors.	
Cork tile	Made of molded and compressed ground cork bark with natural resins of the cork to bind the mass together when heat cured under pressure.	(Same as asphalt tile)	**Other materials**—Paint products (paint, enamel, shellac, varnish, plastic) are semipermanent finishes used principally on wood and concrete floors. They do not materially increase the slipperiness of the base.
Steel	Iron containing carbon in any amount up to about 1.7 percent as an alloying constituent, and malleable when used under suitable conditions.	Surface can be touched up with an arc welding electrode so the shape of raised places on the surface resembles angle worms. Also an abrasive reinforced plastic coating can be painted on to any desired thickness, dries hard as cement, and has a sandpaper like finish. If a temporary nonskid surface is needed, two uses of mats can be employed: (1) flexible rubber mats made from old automobile tires; (2) rubber or vinyl runners.	
Clay and quarry tile	Kiln-dried clay products are similar to bricks and are extensively used in areas requiring wet cleaning.	Typically resistant to abrasives.	May be treated by etching. May be formulated as nonslip by adding carborundum or aluminum oxide when mixing the clay before kilning.

*Floors and stairways should be designed to have slip-resistant surfaces insofar as possible; adhesive carborundum strips may be used on stair treads or ramps and at critical concrete areas. Etching with mild hydrochloric (muriatic) acid solution will lessen slip problems.

**Colloidal silica is an opalescent, aqueous solution containing 30 percent amorphous silicon dioxide and a small amount of alkali as a stabilizer.

ners, and carpeting are used wherever water, oil, food, waste, and other material on the floor might make it slippery.

A company should establish clear procedures for placing, cleaning, removing, and storing mats. Those who put mats in place during inclement weather should have specific instructions about when and where mats should be put down and removed. If workers do not place the mats promptly and close enough to the door, the entranceways may become slippery; and patrons and others track in water and dirt beyond the entranceway, creating a hazard and maintenance problem in other locations. Maintenance staff should follow definite procedures for inspecting and checking the condition of mats and for maintaining them in safe condition.

Stair rails, treads, and surfaces should be designed per standards, codes, and regulations, and be kept in good repair and checked frequently for defects. Make sure nothing is

Figure 21–3. A heavyweight rubber or plastic mat with a nubby finish or raised design and beveled edges tends to lie flat and stay in place. Rotating the mat distributes the wear and minimizes "bald" spots in high-traffic locations.

stored on the stairways and landings that could contribute to falls. See also Chapter 19, Office Safety.

Merchandise Displays

Workers must construct displays to prevent breakables and other articles from falling and injuring or tripping the customers. Repair and smooth sharp or broken displays and counters so they do not cut or catch passersby.

Marketing people in some stores (like supermarkets) try to stack as much merchandise as possible on a display table to catch customers' attention. However, if these displays are stacked too high, they become a hazard for the shopper. A customer reaching for an item can set off a cascade of boxes, cans, bottles, or other items that can strike or injure the person or create a tripping hazard.

According to consumer statistics, the average woman shopper is 5 ft 4 in. (1.6 m) tall. If a company stacks heavy juice cans, glass bottles, or jars on the top tiers of shelves or displays, it is obvious why accidents from falling items occur. When a product is stacked too high, the customer may need to step onto a lower shelf to reach it. Staff should stack shelves evenly by layers, placing heavy items on the lower shelves and lighter items on the top shelves (Figure 21–4, center).

In many stores, workers may card and hang nonfood accessories such as hardware items, notions, and kitchen equipment on pegboard panels. These panels or sections should be adequately recessed to accommodate the extended J-hooks (with minimum J-radius of 1 m). Remember, because shoppers will be bending over to reach lower items, the hooks above should not extend so far as to be a potential risk to a person's eyes or face. Specially safeguarded extenders can eliminate this problem (Figure 21–4, center).

Merchandise with sharp or cutting edges should not be in open displays unless the edges are covered or otherwise protected. Protective sleeves or plastic coatings or covers serve to guard the edges from customer handling. Likewise, all electrical and mechanical display elements should have adequate protective features.

Construct display platforms using color(s) or lighting that contrasts with the floor or carpet, and do not allow them to obstruct aisles. Round or clip all corners of platforms. Displays and mannequins should be at least 6 in. (15 cm) off the floor so that customers will not tip them over accidentally. Make sure the display or the mannequin is fastened to this base.

Floor displays should be at least 3 ft (0.9 m) high so they are visible without becoming a tripping hazard. They should not be at the ends of aisles where shopping carts can dislodge them.

Keep hanging displays at a safe height. If hooks are installed at 90-degree points placed around a column, approximately 18 in. (50 cm) down from the ceiling, this then becomes part of the fixturing and allows for easy change of seasonal displays. It can also eliminate the need for taping, stapling, and other makeshift methods that are not always safe.

If a company hangs its displays from the ceiling, it should be sure that ceilings are structurally sound and that all code requirements have been met. In addition, any duplex electrical receptacles should be located 18 in. from the ceiling on certain columns with another outlet 12 to 18 in. from the floor.

Housekeeping

In mercantile establishments, it is estimated that fixtures,

Figure 21–4. Point-of-purchase display hangers must be located in such a way that the human eye cannot contact them (such as shown at left where both the upper and lower shelves project beyond the hangers). Shelf hangers must be safeguarded if they project into the aisle (center). Locating projections above eye level is another safe method (right).

displays, and other portable equipment are involved in over 40% of customer falls. It is, therefore, essential that management provide safe equipment and that the accident control program particularly emphasize safe placement and use of that equipment. Also, workers should remove dress racks and stock trucks from the sales area and return them to the stock room as soon as the items have been emptied.

All electrical wiring and extension cords hooked to store machines, displays, special decorations, and the like should not lie exposed on the floor. They can present a serious tripping hazard. Where necessary, install wires or cords in low-profile channels.

Poor housekeeping accounts for one-third of all customer falls. Each employee needs to realize that it is part of his or her responsibility to maintain good housekeeping in the work area and to promptly report unsafe floor conditions, such as tears in carpets and holes in the floors. Train workers to wipe up or barricade spills immediately until the hazard can be removed. A special warning sign can be used.

ESCALATORS, ELEVATORS, AND STAIRWAYS

Many business establishments move people between floors by escalators, elevators, stairways, and ramps. See also the discussion in Chapter 9, Hoisting and Conveying Equipment, in the *Engineering and Technology* volume.

Escalators

Escalators can be operated from a low of 70 feet per minute (fpm) to a speed of 125 fpm (0.36 to 0.64 m/s). The average recommended speed is 90 fpm (0.46 m/s). A 4-ft (1.2-m) wide escalator can move 4,000 to 8,000 passengers-per hour. If they are operated faster or slower than people anticipate, passengers can be injured, especially children or elderly patrons.

Accidents may occur when passengers are entering or leaving escalators or while they are riding them. Common problems that can result in such accidents are:

1. Unsafe floor conditions and poor housekeeping at landings
2. Sales counters, mannequins, display bases, and similar units that hamper the movement of passengers
3. Lights or spotlights facing passengers as they step on or off at landings
4. Mirrors near escalator landings causing passengers to misjudge their step and stumble
5. Merchandise signs or displays distracting riders, causing them to bump into one another or to fall
6. Failure of passengers to step on the center of a step tread, causing them to fall, possibly against others
7. Overshoes, particularly the thin plastic type, sneakers, or other objects catching in the comb-plate when pressed against a riser or catching at the side of the moving steps
8. A passenger "riding" a hand on the handrail beyond the combplate and back into the handrail return. This can cause an injury if exposed fingers run into the handrail guard
9. An emergency STOP button, unprotected against accidental contact. This hazard can cause an incident if a passenger (or object) mistakenly presses it and halts the escalator unexpectedly
10. A package, stroller, or other conveyance placed on the escalator and jamming between the balustrades or slipping from the grasp of the person trying to hold it
11. Parts of the body becoming involved with moving parts of the escalator resulting in falls, lacerations, and amputations of the toes, fingers, or other parts of the body. (This might occur if patrons are barefoot, as they may be in tropical areas)

12. A passenger walking or running up or down a moving escalator
13. Children sitting on escalator steps
14. A passenger who fails to hold onto a moving handrail and stumbles or falls.

Escalator standards and regulations. When installing or modifying escalators, check local and state ordinances. Also refer to National Safety Council Data Sheet No. 516, *Escalators.* The American National Standard (American Society of Mechanical Engineers) A17.1, *Elevators, Escalators, and Moving Walks,* is known as the Elevator Code. Some of its important points are:

- Hand and finger guards are to be protected at the point where the handrail enters the balustrade.
- Prominent caution signs should be displayed (Figure 21–5).
- Comb-plates with broken teeth should be replaced immediately.
- Strollers, carts, and the like must be prohibited on escalators.

The Elevator Code also states that balustrades must have handrails that move in the same direction and substantially at the same speed as the steps. The company should also follow city, state or provincial, and local code requirements.

Inspections. Examine all escalators from landing to landing every day. This includes riding them before the store opens to discover any defects.

Once a week, inspect the escalators:

- Replace any broken treads or fingers in step treads and comb-plates
- Examine handrails for damage
- Check balustrades for loose or missing screws and for damaged or misaligned trim.

Semiannually, inspect:

- step chain switches
- governor
- top and bottom oil pans
- skirt clearances
- step treads and risers
- steps
- machine brake
- skirt switch
- handrail brushes.

Start and stop controls. Escalators should have an emergency STOP button or switch located at the top and the bottom landing. These must stop but not start the escalator. By placing a STOP button at the base of a newel, with either recessed design or a cover, the company can prevent a person or object from pressing the button by accident. All employees should be trained in the location and use of emergency switches.

Elevators

Business establishments with elevators face other accident problems. One of the most common is a customer being struck by the door. This is particularly common in self-operating elevators.

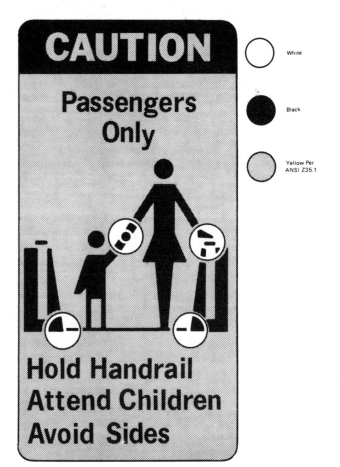

Figure 21–5. Caution sign recommended for escalators by ANSI/ASME standard A17.1d–1984. Sign should be 4 in. (102 mm) wide by 7 in. (197 mm) high.

Inspections. A logbook containing the following information should be kept:

- day, month, year, and time of inspection
- observations by mechanics or inspectors
- all breakdowns, including causes and corrective action(s). Entries should be initialed and dated.

A company must comply with all regulatory standards and should use the ANSI/ASME standard A17.1, the Elevator Code referred to under Escalators. At least every three years, all elevators should have a balance test and a contract load test. Make any required adjustments.

Once a year, hand-test the following controls:

- governors
- governor cable grip jaws
- gripping jaws of car and counterweight
- releasing carrier
- cutoff switches
- tail rope and trip rod drums
- safety rails.

Spot-check automatic elevators at the start of each day for level floor stops, brakes, and other mechanical operations. Test the alarm bell to be sure it rings and that its signal registers in the maintenance department. See also Chapter 9, Hoisting and Conveying Equipment, in the *Engineering and Technology* volume for more details.

Stairways

Because stairways are so important in emergency exits from buildings, they are covered in detail in the National Fire Protection Association's (NFPA's) publication *Life Safety Code*, NFPA 101, Chapter 5, Means of Egress. That chapter also includes ramps, exit passageways, smokeproof towers, outside stairs, fire escape stairs and ladders, illumination, exit marking, and escalators and moving walks. Only inside stairways are discussed in the following section.

The typical stair accident occurs when someone slips while descending a stairway. It has been found that most fall victims tend to look at stairway treads far less than do other people, thus they are more prone to accidents. See the U.S. Department of Commerce study "Guidelines to Stair Safety" (listed in References at the end of this chapter).

The following list represents major areas of concern, along with their corrective actions, that companies need to address.

- Because stairways are not level walking surfaces, they require different walking habits. Companies can minimize their use by posting signs directing people to the escalators and elevators. Also, never chain or otherwise lock exits to and from stairs so they cannot be used in an emergency.
- Design handrails so that employees and patrons can grasp them firmly and slide their hands along the rail without encountering obstructions. When designing handrails, keep in mind that a simple round profile permits the fingers to curl around the rail, in effect "locking" the hand in place. Two inches is the recommended maximum diameter for a typical handrail.
- Highlight treads and handrails to help riders quickly and easily distinguish them from the riser and wall surfaces. Adequate lighting is essential; NFPA specifies a minimum of one foot-candle (decalux), measured at the floor. Arrange lights so they do not create glare surfaces or temporarily blind stair users and so the failure of one unit will not leave any area in darkness. If one side of the stairwell is open to adjoining space, it is a good idea to close off that view to prevent distractions that may cause people to fall.
- The edge of the step or tread (nosing) should be easy to see. If possible, use contrasting carpeting to distinguish the approach to the steps from the stairs themselves. If not, special nosing may be installed to provide definition between the steps. Nosings must be securely fastened. Uncarpeted stairs should be edgemarked.
- A continuous handrail must be provided; see details in Chapter 19, Office Safety, in the section Aisles and Stairs. The railing preferably should be of a lighter color, because people seem to be more inclined to use a lighter-colored railing than a dark one that looks dirty and greasy. The railing should be kept clean. The handrail should extend to the top and bottom of the staircase so that it may be grasped before stepping on the first step or leaving the last step of the flight. Be sure that the railing does not extend into a passageway and become a hazard.
- Stair treads need to be stable and provide good traction.

Outdoor stairs need extra slip resistance and should have adequate water runoffs that lead away from other walking surfaces.

- Worn or defective treads or other parts should be replaced immediately.
- Stairs should be kept clear of obstructions. Make sure sharp handrail and guardrail ends are removed or covered to prevent injury. Clearly mark and protect glass areas adjacent to stair landings or at either end of the stairway to prevent people from walking through them. Fixtures should not project into the stairway.

PROTECTION AGAINST FIRE, EXPLOSION, AND SMOKE

If fire breaks out or an explosion occurs, a quick, orderly evacuation of the premises will protect all persons including visitors. Base all evacuation plans on the premise that most visitors will be on the property for the first time. They need to be directed to exits by numerous special direction signs, even though exit locations might be well-known to regular employees.

Businesses that generally attract large numbers of customers, guests, or patrons should provide well-marked exits, emergency lighting, and enough direction signs to guide people to safe exits. In a fire, dense, penetrating smoke can be as deadly as the heat or flame and can quickly sear people's lungs. Therefore, it is essential to evacuate the fire area immediately.

Enclosed stairwells provide the best fire escape routes. Doors to such stairwells must never be obstructed, locked, or propped open. A public address system, staffed by a qualified and trained person, can be used to direct the building's evacuation and issue life-saving instructions.

Usually the early detection of a fire and the use of a good evacuation plan can prevent panic and personal injuries. More details are given later in this chapter and in Chapter 6, Emergency Preparedness.

Every business should consider installing a fire-detection and suppression system for maximum safety. They can make sure architects are accountable for proper design for new construction, especially in high-rise buildings.

Fire Detection

Properly engineered fire detection systems are sound investments, but the best installation is useless if no one responds to the alarm. Systems should have a direct connection to the local fire department or security service.

The four stages of fire. Fire is a chemical combustion process created by the rapid combination of fuel, oxygen, and heat. A full discussion is found in Chapter 6, Fire Protection, in the *Engineering and Technology* volume. Most fires develop in four distinct stages: incipient, smoldering, flame, and heat. Detectors are available for each stage.

- *Incipient stage.* No visible smoke, flame, or significant heat develops but a large amount of combustion particles is generated over time. These particles, created by chemi-

cal decomposition, have weight and mass, but are too small to be visible to the human eye. They behave according to gas laws and quickly rise to the ceiling. Ionization detectors respond to these particles.

- *Smoldering stage.* As the incipient stage continues, the combustion particles increase until they become visible—a condition called "smoke." No flame or significant heat has developed. Photoelectric detectors "see" visible smoke.
- *Flame stage.* As the fire condition develops further, ignition occurs and flames start. The level of visible smoke decreases and the heat level increases. Infrared energy is given off that can be picked up by infrared detectors.
- *Heat stage.* At this point, large amounts of heat, flame, smoke, and toxic gases are produced. This stage develops very quickly, usually in seconds. Thermal detectors respond to heat energy.

Burning plastics. Some fuels such as plastic waste receptacles can produce considerable toxic smoke when they burn. Therefore, the use of nontoxic and noncombustible materials (such as metal cans) is preferable. Managers and supervisors should be aware that other materials, such as building finishes, carpentry, furniture, and office fixtures, also can give off toxic smoke. For example, polyvinyl chloride (PVC) in a single foot of one-inch sized PVC rigid nonmetallic conduit involved in a fire:

1. can produce a sufficiently heavy, dense smoke to obscure 3,500 cu ft (100 m³) of space
2. can generate enough hydrogen chloride to provide a lethal concentration of HCl in approximately 1,650 cu ft (45 m³) of space.

Engineering and control procedures. The best fire-detection system is only as good as its weakest component. Management should utilize the services of a fire protection consultant in engineering the system and establishing the control procedures. Here are four steps to consider:

1. Select the proper detector(s) for the hazard areas. For example, a computer area may involve ionization or combination detectors. A warehouse may have infrared and ionization detectors. In low-risk areas, thermal detectors or combinations of detectors may be used.
2. Determine the spacing and locations of detectors to provide the earliest possible warning.
3. Select the best control system arrangement to provide fast identification of the exact source of alarm initiation.
4. Ensure notification of responsible authorities who can immediately respond to the alarm and can take appropriate action. Every detection system must have an alarm signal transmitted to a constantly supervised point. If this cannot be ensured on the premises, the signal must be transmitted to a central station, fire department, or other reporting source.

Response Methods

Early warning systems are as important during hours of occupancy as they are when the premises are closed.

When detection pinpoints a trouble spot, immediate response by a responsible trained company representative is all

important. Shaving seconds and minutes can mean the difference between lives lost or saved, and fire confined or allowed to spread out of control.

Here are some systems that are used.

Twenty-four-hour supervisory service. If an installation has 24-hour, seven-days-a-week supervision at some point in the building, or building complex, then management should hook alarm, trouble, and zone signals into this location.

Less than 24-hour supervisory service. For periods when an installation does not have responsible personnel to respond to the alarm, a backup system should be provided. The NFPA advises connecting to a central station supervisor's service or other service.

Such systems should include a way to initiate fire and trouble signals to the central station transmitter equipment. In the case of a trouble signal, a representative can be dispatched immediately to investigate the trouble and to notify proper representatives of the property under surveillance.

Central station monitoring unavailable. In areas where no central station supervision is available, the local fire or police department may accept installation of a remote fire alarm panel at their headquarters or firehouse.

Central station or telephone leased-line tie-in unavailable. If none of the foregoing possibilities exists, then management should consider using qualified and licensed telephone answering services. Also, automatic dialing units connected to responsible officials are another alternative.

High-Rise Building Fire and Evacuation Controls

Just what is a high-rise building? These are four basic criteria to designate a high-rise structure. **First,** the size of the building makes personnel evacuation impossible or impractical. **Second,** part or most of the building is beyond the reach of fire department aerial equipment. **Third,** any fire within the building must be attacked from within because of building height. **Fourth,** the building has the potential for "stack effect."

High-rise megastructures all share one characteristic—they are intended to house people. As buildings go higher up, the population density per square foot of ground area increases, posing a whole new set of problems concerning the health, safety, and welfare of building occupants. Actually, each building is a sealed life-support system. While engineering approaches have improved heating and cooling systems, they are often extremely wasteful of energy. Because these buildings are usually airtight, occupants are at greater risk of death or injury from smoke and toxic combustion by-products.

The ladders on most fire department aerial equipment are limited to approximately 85 ft (26 m). This means that a building higher than about eight to ten stories cannot be served by this equipment. Because of the height problem and increased floor areas, fires must be fought from within the building. The principal firefighting equipment includes automatic sprinklers, hose standpipes, and portable extinguishers, as well as hose lines from the building exterior.

When a building must be evacuated, management should

EVACUATION PREPAREDNESS CHECKLIST

All questions in this checklist should be answered with "yes," "no," "NA" (not applicable), or "U" (undetermined.) For all answers that are not "yes," or "NA," the persons responsible for the specific areas in question that need correction should be noted.

Floor Diagrams
Are floor plans prominently posted on each floor?
Is each plan legible?
Does the plan indicate every emergency exit on the floor?
Does a person looking at the plan see an "X" indicating "you are here now"?
Are room number identifications for the floor as well as compass directions given?
Are directions to stairwells clearly indicated?
Are local and familiar terms used on the diagram to define directions to emergency exit stairwells? For example, are particular areas identified, such as mail room, cafeteria, personnel department, wash rooms?

Exit paths to stairwells
If color coding of pillars and doors, or stripes and markings on floors are used, are they properly explained?
Is additional clarification needed?
Are paths to exits relatively straight and clear of all obstructions?
Are proper instructions posted at changes of direction en route to an emergency exit?
Are overpressure systems and venting systems operative?

Elevators
Are signs prominently posted at and on elevators warning of the possible dangers in using elevators during fire and emergency evacuation situations?
Do these signs indicate the direction of emergency exit stairwells which are available for use?

Elderly and physically handicapped
Are there elderly or physically handicapped persons who will need assistance during a fire and emergency evacuation of premises?
What provision is made for their removal during an emergency?
Who will assist? How will the handicapped be moved?

Emergency exit doors
Are all emergency exits properly identified?
Are exit door location signs adequately and reliably illuminated?
Do exit doors open easily and swing in the proper direction (open out)?
Are any exit doors blocked, chained, locked, partially blocked, obstructed by cabinets, coat racks, umbrella stands, packages, etc.?
Note: Blockage must be removed immediately and subsequently prohibited.
Are all exit doors self-closing?
Are there complete closures of each door?
Are all exit doors kept closed, or are they occasionally propped open for convenience or to allow for ventilation?
Note: This practice must be prohibited.

Emergency stairwells
Are stair treads and risers in good condition?
Are stairwells free of mops, pails, brooms, rags, packages, barrels, or any other obstructing material?
Are all stairwells equipped with proper handrails?
Does each emergency stairwell go directly to the grade floor exit level without interruption?
Does the stairwell terminate at some interim point in the building?
If so, are there clear directions at that point which show the way to completion of exit?
Is there provision for directing occupants to refuge areas out of and away from the building when they reach the ground floor?
Are directions provided where evacuees can congregate for a "head count" during and after the evacuation has been completed?
Is there adequate lighting in the stairwell?
Are any bulbs and/or fixtures broken or missing?
Where? Describe locations.
Are exits properly identified?
Are they illuminated for day, night, and power-loss situations?
Are any confusing non-exits clearly marked for what they are?
Are floor numbers displayed prominently on both sides of exit doors?

Emergency lighting
In the event of an electrical power failure or interruption of service in the building, is automatic or manually operated emergency lighting available?
If not, what will be used?
Where are stand-by lights kept?
Who controls them?
How would they be made available during an emergency?
Is there an emergency generator in the building?
Is it operable?
Is it secured against sabotage?
Is a "fail-safe" type of emergency lighting system available for the exit stairwells that will function automatically in event of total power failure?
How long can it provide light?
Is the emergency lighting tested on a regular monthly basis with results recorded? Who maintains such records?

Communications
How should occupants of the building be notified that an emergency evacuation is necessary?
Is one or more communication systems available to each floor? (P.A. system, Muzak, stand-pipe phones, battery-operated "pagers," etc.)
If messengers must be used, have they been properly instructed?
Is the communication system(s) in good working condition?
Under what emergency conditions is it used and who operates it?
Can announcements be prerecorded by someone with a calm but authoritative voice?
Is the communications system protected from sabotage?
Do all occupants know how to contact building control to report a dangerous situation?
Is the building's emergency communications system tested monthly? By whom and to what extent?

Figure 21–6. Reprinted from National Safety Council Industrial Data Sheet 656, Evacuation System for High-Rise Buildings.

have an orderly plan in place. An evacuation checklist is given in Figure 21–6. Several sources are available to help companies devise a plan. The U.S. Occupational Safety and Health Standards Subpart E, Means of Egress (Title 29, *Code of Federal Regulations (CFR)*, Chapter XVII, Part 1910, Subpart E; Revised as of July 1, 1979; Amended by 45 *FR* 60203, September 12, 1980) mandates in Section 1910.38 (a) The emergency action plan shall be in writing. (The only exception to this regulation is for locations with ten or fewer employees.)

(a) (5) Training (ii) The employer shall review the plan with each employee covered by the plan at the following times:
(A) Initially when the plan is developed,
(B) Whenever the employee's responsibilities or designated actions under the plan change, and
(C) Whenever the plan is changed.

Although U.S. OSHA is oriented to the employees, obviously such a plan should include delegation of evacuation leaders and a series of steps to evacuate both employees and

nonemployees in an emergency.

Stack effect. Every building has its own peculiarities for creation of a "stack effect," or spread of fire, smoke, and toxic fumes from one area to another. Among the factors are structure configuration, height, number and size of openings, wind velocities, temperature extremes, number and location of mail chute openings, and elevator shafts, all of which create varying air flows that tend to accelerate and intensify an interior fire. Unprotected air-conditioning systems are an open invitation to catastrophe. If there is no automatic smoke and heat detection, no automatic fan shutdown, and no automatic fire dampers, smoke and toxic fumes can be quickly drawn into the exhaust or return air duct system and promptly distributed to all other floors and areas of the building served by the air-conditioning system.

The NFPA recognizes this potential. Its Standard No. 90A, *Installation of Air Conditioning and Ventilating Systems,* states that in systems of over 15,000 cfm (7 m³/s) capacity, smoke detectors should be installed in the main supply duct downstream of the filters. These detectors automatically shut down fans and close smoke dampers to stop the recirculation of the smoke—or they may incorporate automatic exhaust.

In planning evacuation, assume that children, elderly, and physically disabled will be involved. In addition, some people panic in a fire situation. The quantity and size of staircases will undoubtedly prohibit complete evacuation. Tests indicate that with an occupant load of 240 persons per floor, total evacuation of an 11-story building can take up to 6 minutes, while an 18-story building can take up to 7 minutes. Exits are just not designed to handle all occupants simultaneously.

Most codes do not consider elevators to be an exit component and prohibit their use during fire emergencies. But codes generally also require that one or more elevators be designated and equipped for firefighters. Key operations can transfer automatic elevators to manual operation and bring the elevators to the street floor for use by the fire service. The elevator should be situated so the fire department can find and use them easily.

Many elevators use capacitance-type call buttons, which may bring them to a stop on the fire-involved floor. Then they cannot move because smoke interrupts the light beam and keeps the doors open. Other possibilities include the inadvertent arrival of the elevator at the fire-involved floor by a passenger who does not know there is a fire, or a person who pushes the call button and then, in panic, uses the staircase for exit. With problems of this magnitude, it can be assumed that complete evacuation is impossible.

Crowd and Panic Control

Any commercial establishment may be faced with an unruly crowd because of an emergency, panic, or even a planned demonstration. Self-interest dictates that management preplan for such events to protect the facility and its employees, along with patrons and bystanders. Different measures are required depending on whether the panic occurs in the building or outside of it and who is involved.

In shopping areas, companies should have directional signs displayed at many areas in the building. Exit signs are especially important. But these alone do not reduce the higher risk potential involved when a great number of people are gathered, such as at sports events, entertainments, schools, and other places.

In almost every emergency, some panic is likely. Employees need periodic drills and practice in handling emergency situations with customers, some of whom may be confused and others who may be physically disabled. Management cannot overlook the threat of an unusual occurrence (such as a riot, bombing, and the like) and must make plans to handle such a possibility.

Although high-rise buildings pose new and special problems, other public places such as theaters and amusement and recreation facilities must also provide well-planned emergency procedures. Panic and the press of frantic, hysterical people have caused considerable loss of life in many emergencies. Oftentimes, such losses could have been prevented through the strict observance of building and fire codes to eliminate hazards and "death traps" caused by improper design and lack of firefighting and disaster-control measures.

Civil strife and sabotage are covered in Chapter 6, Emergency Preparedness.

Demonstrations. The following procedures concern controlling public demonstrations at a store, but they can easily be adapted to the needs of any establishment.

Demonstration outside of store building. Advise employees to call security and/or management. Security should telephone police, advise them of the situation, and follow their instructions.

Arrange for two or more key personnel to assume previously assigned positions at all store entrances and other key points; they should know security's telephone number to relay information and to receive instructions. They should never leave their assigned posts unless relieved or advised accordingly. Caution them to remain calm and not to interfere with the entrance or exit of customers or employees.

Those employed in portable, high-valued merchandise departments (diamonds, furs, etc.) should arrange to have such merchandise placed in an assigned secure area. Those employees working in departments selling firearms, knives, axes, straight razors, bows and arrows, and even meat cutlery, should have them removed from the selling floor to a secure area. Proceed with "business as usual" in all other departments.

Because rumors can create panic or problems among employees and customers present, all employees should be advised of two or three emergency interior telephone numbers to verify information and squelch rumors.

Demonstration moving into store. (People carrying signs, groups linking arms across aisles and taunting employees, fights between individuals.)

1. Advise all employees to avoid any comments, antagonism, or physical contact with marchers, to answer all queries courteously, and above all to keep calm.

2. In areas where demonstrations are taking place, have employees stop selling, lock their registers, remain in their areas keeping as calm as possible, and await further instructions from their supervisors.
3. In areas where "business as usual" is being maintained, arrange for frequent cash pickups.
4. Key personnel and employees should take their assigned places, as discussed above.
5. In the event such a demonstration turns into group looting or group "hit and run" stealing, employees should not attempt to make any apprehensions. Security personnel will follow previous orders for such conditions, as advised by management. Remember, personal safety is more important than property protection.

Self-Service Operations

The best-known self-service operation is found in gasoline stations. Here, customers put the gasoline in their vehicles without employee assistance. The employee merely sees that customers observe the safety rules and follow the prescribed procedure, usually posted on the pump housing or on a nearby sign. All states require that customers turn off their engines and refrain from smoking or using open flames.

Evacuation of the Physically Disabled

The evacuation of physically disabled persons from hotels, stores, and other facilities is an added problem for both management and the safety professional. Communication, especially, can be a problem. Special written instructions can be given to people with a hearing impairment; verbal instructions can be given the visually impaired.

More details on how to help the physically disabled are presented in Chapter 20, Workers with Disabilities, in this volume.

TRANSPORTATION

Some businesses by their very nature have special accident control problems. Much of their loss-control effort must be directed at protecting the nonemployee from harm.

All types of transportation—from commercial airlines, railroads, marine, and buses to local transit, taxi, and school bus operations—must be vitally concerned with preventing passenger injuries. Not only does this involve maintenance and vehicle or unit operation but also the safety of areas around vehicles, such as terminals, stations, school-bus loading areas, and the like.

Courtesy Cars

When a company provides transportation for customer courtesy and convenience (such as a hotel or motel courtesy car or bus), it should follow the same precautions used in commercial operations. These include providing the safest vehicle, maintaining it in proper working condition (meeting all regulatory requirements), and operating it with a professional driver who is trained and skilled in all facets of the vehicle's operation.

Company-Owned Vehicles

Another source of damage and injury claims arises out of the operation of company motor vehicles by employees. Because the odds are that one of every five drivers will have a collision in any given year, liability from such accidents is a constant threat and often a real dollar drain for insurance protection and claim settlement. Management must be aware of the liability resulting from an employee's use of his or her own car on company business.

PROTECT ATTRACTIVE NUISANCES

Every business can suffer losses caused by the public's curiosity about their equipment or operations. Some examples follow.

Any unattended vehicle or machine left in an operable condition is attractive to the young (and the so-called "young in heart.") "Tamper-proof" locks discourage the unauthorized use of vehicles or machines and, if necessary, security guards should be employed to watch over the equipment.

Frequently, partially finished road repairs, construction, or storms and adverse weather can present serious hazards for motor vehicles. Companies should barricade such hazards and warn motorists away from them. Refer to the U.S. Department of Transportation's *Manual on Uniform Traffic Control Devices for Streets and Highways.*

Many contractors or builders provide special, safe observation facilities for "sidewalk superintendents," that is, members of the public who like to watch construction projects. Local authorities and insurance engineers should be consulted for regulations and control measures.

Swimming Pools

Many hotels, motels, and public areas provide swimming pools for their patrons (Figure 21–7). Pool accidents usually result from inadequate protective barriers around pools, absence of lifeguards or qualified adult supervision, disregard for the rules of good pool conduct, and the failure to teach youngsters drowning-prevention techniques.

Pool precautions include:
1. Screens, fences, or other enclosures to control admittance. A tamper-proof lock and pool alarm may provide additional protection.
2. Accurately marked pool depths in feet or meters, on both pool deck and poolside.
3. No diving boards unless the pool has been constructed and staffed for diving.
4. Adequate safety measures to prevent diving accidents. The depth of water in a pool is extremely difficult for the eye to measure. This presents a hazard for persons diving. Every year numerous injuries occur when people unknowingly dive into shallow water. To prevent such accidents, the following is recommended.
   - In shallow water areas, post warning signs such as *danger, no diving allowed, shallow water.*
   - Designate specific diving areas where the water is sufficiently deep.

Figure 21–7. Basic lifesaving equipment must be available at every pool. Shown here at each lifesaving platform are ring buoys with the throwing rope attached. (Courtesy National Spa and Pool Institute.)

- Use either pool markings or plastic or wood floats strung on rope to differentiate between the diving and no-diving areas.
5. Basic lifesaving equipment, which should include a lightweight, strong pole with blunt ends at least 12 ft (3.7 m) long, or a ring buoy with long throwing rope.
6. A competent pool manager and a lifeguard on duty whenever the pool is in use.
7. A telephone nearby, such as in the bathhouse or changing room. Emergency telephone numbers should be on hand—the nearest available physician, ambulance service, hospital, police, and the fire and/or rescue unit.
8. Decks around the pool kept clear of debris and no breakable bottles allowed in the area. Make sure all cups and dishes used at poolside are unbreakable. Provide litter baskets and replace defective matting.
9. No games near the pool that could injure anyone.
10. Electrical equipment conforming to local regulations and the latest National Electrical Code requirements. Any electrical appliance used near the pool must be protected

by a ground-fault circuit interrupter (GFCI).
11. No swimming allowed in the pool during a thunderstorm.
12. All pool appliances and equipment maintained properly. Periodic safety checks should be made.
13. Sensible pool rules established, posted, and enforced.

SUMMARY

- Nonemployee accidents also affect employees, along with the products or services they provide. Preventing these accidents should be a primary goal of management not only to protect customers and patrons but to reduce a company's liability from civil lawsuits.
- Because many accidents or product claims end up in court, companies must be familiar with their legal duties and liabilities regarding nonemployee accidents and injuries. They must know such legal terms and concepts as tort, negligence, degree of care, invitee, licensee, limitations of recovery (contributory and comparative negligence), as-

sumption of risk, hold-harmless clauses, attractive nuisances, and burden of proof.

- Two of the most common sources of nonemployee accidents are (1) unmarked glass doors, panels, and windows and (2) parking lots. Companies must prevent patrons and visitors from walking or falling through plate glass by making it visible, barricading it, or replacing it with plastic. Parking lots can be made safe through proper layout, adequate signs and lighting, conspicuous markings, provision for bicycles and motorized vehicles, and use of security guards or fences.

- Building entrances can be rendered safe by ensuring the doors are adequate in number and size to meet building codes, sidewalks and driveways leading up to them are kept clear, adequate lighting is provided, and automatic doors operate properly and are safeguarded.

- Walking surfaces present many slipping, falling, and tripping hazards for nonemployees. Management should ensure that walking surfaces are kept clear and clean, adequately illuminated, made of slip-resistant materials, well maintained, and replaced when necessary. Floor coverings and mats should also be carefully chosen, regularly inspected, and well maintained or replaced to reduce or eliminate tripping and falling hazards.

- Merchandise displays must be constructed at the right height and of the right materials to prevent accidents. Otherwise, customers may be injured if the display tips over, if merchandise falls from the display, or if customers trip or catch themselves on parts of the display platform or merchandise.

- Escalators, elevators, and stairways can be a source of accidents when patrons or visitors enter, use, or leave these conveyances. Escalators and elevators should be installed or modified according to state and local ordinances and the Elevator Code. Management must establish regular inspections and make sure that all stop and start controls are guarded against accidental tampering.

- Stairway exits are important in emergency evacuations. Companies should minimize their use by nonemployees, provide proper handrails, highlight treads and handrails, supply adequate lighting in all stairwells, replace worn or defective parts immediately, and keep all stairs free from obstructions.

- In a fire, the danger to nonemployees is not only heat and flames but dense, toxic smoke as well. Every business should install adequate fire detection equipment to sound the alarm at any of the four stages of a fire: incipient, smoldering, flame, or heat. Fire detection and fire response systems must be adequate for each building's requirement and should be tied into company and municipal firefighting departments. Early fire detection and response can save lives and property.

- Many companies have evacuation plans should fire break out in the building, particularly high-rise buildings. Management and employees must be trained in and practice evacuation procedures to handle crowds and control panic among nonemployees. Company employees should be able to conduct patrons and visitors, including those physically disabled, quickly and efficiently to well-marked exits.

- Companies should also develop procedures for handling public demonstrations on their property. Employees should know how to communicate with security staff, summon police, handle demonstrators who enter the building, and guard their own safety.

- Courtesy cars and company-owned vehicles can be sources of nonemployee accidents and injuries. Companies must make sure their courtesy vehicles are in good repair and operated by a skilled driver. If employees have an accident while in a company-owned car, the firm must be sure it has adequate insurance and liability coverage to protect itself against damage claims by the victims.

- Companies can also suffer losses caused by the public's curiosity about company items or operations, known in legal terms as "attractive nuisances." Examples include unattended vehicles, construction sites, and swimming pools. Firms must follow state and local ordinances and guidelines for preventing the public from gaining access to these attractive nuisances and being injured or killed as a result.

REFERENCES

American National Standards Institute, 11 West 42nd Street, New York, NY 10036.
Buildings and Facilities—Providing Accessibility and Usability for Physically Handicapped People, A117.1–1986.
Elevators, Escalators, and Moving Walks, ANSI/ASME A17.1–1987.
Safety Requirements for Floor and Wall Openings and (nonresidential) Stair and Railing Systems, A1264.1–1989.

Guidelines to Stair Safety, Department of Commerce, U.S. Government Printing Office, Washington, DC 20402.

International Conference of Building Officials, 5360 South Workman Mill Road, Whittier, CA 90601. *Uniform Building Code*, section 3303(j).

Matwes GJ and Matwes H. *Loss Control: A Safety Guidebook for Trades and Services*. New York: Van Nostrand Reinhold Co., 1973.

National Fire Protection Association, 1 Batterymarch Park, Quincy, MA 02269.
Installation, Maintenance, and Use of Auxiliary Protective Signaling Systems, (1987) NFPA 72B.
Installation, Maintenance, and Use of Central Station Signaling Systems, (1989) NFPA 71.
Installation, Maintenance, and Use of Local Protective Signaling Systems, (1987) NFPA 72A.
Installation of Air Conditioning and Ventilating Systems, (1989) NFPA 90A.
Installation, Maintenance, and Use of Proprietary Protective Signaling Systems, (1986) NFPA 72D.

Installation, Maintenance, and Use of Remote Station Protective Signaling Systems, (1986) NFPA 72C.
Life Safety Code, (1985) NFPA 101.
National Electrical Code, (1990) NFPA 70.

National Safety Council, 444 North Michigan Avenue, Chicago, IL 60611.
Dennis L and Onion M. *Out in Front: Effective Supervision in the Workplace*. Chicago: National Safety Council, 1990.
Safety and Health Reprints.
Industrial Data Sheets.
Carbon Monoxide, (1985) 12304–0415.
Evacuation System for High-Rise Buildings, (1985) 12304–0656.

Falls on Floors, (1986) 12304–0495.
Fire Prevention in Stores, (1990) 12304–0549.
Floor Mats and Runners, (1986) 12304–0595.
Sidewalk Sheds, (1990) 12304–0368.

National Spa and Pool Institute, 2111 Eisenhower Avenue, Alexandria, VA 22314.

Superintendent of Documents, U.S. Government Printing Office, Washington, DC 20402. Commercial Practices, Title 16, *Code of Federal Regulations*, Chapter II—Consumer Product Safety Commission.

U.S. Department of Transportation. *Manual on Uniform Traffic Control Devices for Streets and Highways*, D6.1.

Appendix 1

Sources of Help

The safety professional frequently needs highly specialized or up-to-the-minute, unpublished information. The sources for obtaining this information are numerous. Professional societies and trade associations are excellent sources of help; however, their charters of responsibility are varied. As an aid in the safety professional's search for information, this chapter selects some sources and defines their functions.

On a particularly difficult problem, it may be necessary to contact a number of sources before an effective solution can be obtained. Governmental agencies can help with information about regulations or applicable standards. Insurance companies or their associations may offer assistance through their knowledge of a similar problem. The trade association in the industry may have developed materials and aids in solving the problems.

The National Safety Council, through its resources and membership, can usually provide added input to the development of an effective countermeasure.

SERVICE ORGANIZATIONS

National Safety Council
444 North Michigan Avenue
Chicago, IL 60611
after June 1992:
1121 Spring Lake Drive
Itasca, IL 60143–3201
The National Safety Council is the largest organization in the world devoting its entire efforts to the prevention of accidents and illnesses. It is nongovernmental, not-for-profit and nonpolitical. Its staff members work as a team with more than 2,000 volunteer officers, directors, and members of various divisions and committees to develop and maintain accident prevention material and programs in specific areas of safety and health. These areas include industrial, traffic, home, recreational, and public. Council headquarters' facilities include one of the largest safety libraries in the world.

At the Chicago headquarters, a staff of more than 250, about half of whom are engineers, editors, statisticians, writers, educators, data processors, librarians, and other specialists, carry out the major activities. In addition to its main office and Distribution Center near Chicago, the Council has regional offices in California and New York state, as well as a Public Policy Office and Environmental Health Center located in Washington, DC.

Industrial Division. Because recognizing industry's safety and health problems often requires specialized treatment, the Council has arranged its industrial effort into sections guided by the Industrial Division. Each section is administered by its own executive committee, elected from its own membership.

The industrial membership of the Council is organized according to the following industries. These sections are designed to provide special help for all facets of the industrial section, as shown below.

AEROSPACE

Missile and aircraft manufacture, related components

AUTOMOTIVE, TOLLING, METALWORKING, AND ASSOCIATED INDUSTRIES
Machining, fabrication, assembly, manufacturing

CEMENT, QUARRY AND MINERAL AGGREGATES
Quarrying, processing, and marketing

CHEMICAL
Manufacturing chemical compounds and substances

COAL MINING
Underground and open pit coal operations

Construction
Highway, buildings, heavy, home, specialty, and demolition operations

ELECTRONIC AND ELECTRICAL
Manufacturing and assembly of electrical apparatus

FERTILIZER AND AGRICULTURAL MATERIAL
Manufacturing, storage, transportation, retailing

FOOD AND BEVERAGE
Dairies, brewers, confectioners, distillers, canners and freezers, grain handling and processing, meat processing, restaurants, and related operation

FOREST INDUSTRIES
Logging, pulp and paper, plywood, and related products

HEALTH CARE
Hospital, patient, employee, visitor safety

INTERNATIONAL AIR TRANSPORT
Ground terminal safety, affecting personnel and equipment

MARINE
Crew, passenger, vessel safety, stevedoring, shipbuilding, repair, cargo safety

METALS
Foundries, ferrous and nonferrous manufacturing, fabricating, other metal products

MINING
Metals and minerals extraction and processing

PETROLEUM
Exploration, drilling, production, pipeline, marketing, retail

POWER PRESS AND FORGING
Metal stamping and forming, and forging

PRINTING AND PUBLISHING
Newspaper and commercial publishing

PUBLIC EMPLOYEE
Employees and governments from local to national levels and associated activities

PUBLIC UTILITIES
Communications, electric, gas, water, cables

RAILROAD
Employee, passenger freight, and public safety

RESEARCH AND DEVELOPMENT
Laboratory safety

RUBBER AND PLASTICS
Manufacturers of rubber plastic products

TEXTILE
Manufacturing and fabrication, natural and synthetic fibers, ginning and finished goods

TRADES AND SERVICES
Food service, hotels, motels, mercantile automotive, leather, financial institutions, warehouses, offices, recreational facilities.

Assignment of Council Industrial Division members to sections provides them greater interest-specific involvement with similar organizations and professionals. Many members volunteer their time and expertise to assist in accomplishing the Council's mission by serving on committees and in other leadership capacities related to their respective personal safety and health specialties and interests.

The Industrial Division meets three times a year to review current occupational safety and health problems and to determine, on a national scale, the best procedures to follow in providing increasingly beneficial programs to industry. The division is made up primarily of industrial members of the Council. It comprises section general chairmen, vice-general chairmen, and other members-at-large, drawn from business and industry member organizations, professional and trade associations, and other groups.

Improvement of Council services through new technical materials and visual aids takes a major portion of each committee's meeting time. Membership solicitation is an ongoing program. Planning of National Safety Council programs is given careful consideration. Special committees are often assigned to work on problems unique to an industry and on which there are no ready program materials.

Program materials. The following Council publications have proven to be particularly useful for industrial and off-the-job safety programs. (Unless otherwise stated, they are published monthly.)

Family Safety & Health (quarterly)
Today's Supervisor
Safety & Health
OSHA Update
Product Safety Up-to-Date
Section Newsletter (one for each of the 25 sections—six issues a year)
Safe Driver (issued in three editions—Truck, Passenger Car, and Bus)
Safe Worker
Traffic Safety (six issues a year)

In addition, the following statistical materials are also available from the Council:

Accident Facts (annually)
Section Contest Bulletins (monthly)
Work Injury and Illness Rates (annually)

Technical materials (see current Council General Materials Catalog for a complete listing):

Accident Prevention Manual for Business & Industry (this book)
   Administration & Programs volume
   Engineering & Technology volume
Aviation Ground Operations Handbook
Fundamentals of Industrial Hygiene
Industrial Data Sheets (a series; complete listing available)
Industrial Noise and Hearing Conservation
Motor Fleet Safety Manual
Occupational Health and Safety

Power Press Safety Manual
Safeguarding Concepts Illustrated
Safety Guide for Health Care Institutions
Safety Handbook for Office Supervisors

Training and motivational materials: (see Poster Catalog for a complete listing)

Banners
Booklets
Calendars
Films
Posters
Safety and health slides
Supervisory training pamphlets
Video tapes

Meetings. The National Safety Council sponsors the annual Congress and Exposition, one of the largest conventions held anywhere. Nearly 200 general or specialized sessions, workshops, and clinics are held covering the full range of safety and health topics.

The annual meetings or special business meetings of a dozen allied organizations and associations are also held concurrent with the Congress, greatly enhancing the exchange of views and information in safety and health.

Special services. Special services available through the Council to support industrial safety programs include:

Library. With a collection of more than 120,000 documents, of which 95,000 are indexed on an in-house computerized database, the National Safety Council's Library is one of the most comprehensive safety and occupational health libraries in the world. The library is available to the general public. It networks with other databases for safety and health information.

Statistics. The Council's statisticians provide a highly refined statistical capability. They are a recognized source of reliable, accurate, and authoritative data within the safety community. Equipped with data processing and research tools, they study various types of accident data in the continuing search for clues on the causes of accidents.

SAFETY TRAINING INSTITUTE

Among the strengths of the National Safety Council's wealth of safety and health advocacy programs is its renowned Safety Training Institute (STI), which was founded in 1946. STI offers training of all types for industry managers, safety professionals, and others who supervise workplace activities, including every management level from corporate senior officers to shop foremen. More than 24,000 students have completed courses. Classes are held at STI's Chicago facilities, in the Council's regional offices, and at onsite locations provided by organizations.

Course offerings cover such general subjects as Principles of Occupational Safety and Health, Safety Training Methods, and Safety Management Techniques, as well as specific topics that include industrial hygiene, hazard communication, ergonomics, product safety, chemical organizations, laboratory, hospital, advanced hospital safety, and advanced safety concepts.

The National Safety Council awards Continuing Education units (CEUs) for its STI courses. In addition to CEUs, STI students may earn the Council's Advanced Safety Certificate by completing certain course tracks.

Further information about the STI is available from the annual Safety Training Institute Course Catalog & Schedule for dates of presentation, course content, and other courses available.

The Institute also offers two Home Study Courses: Supervising for Safety, and Protecting Workers' Lives. These courses are recognized by the National Home Study Council.

Two supervisor training courses are also available for purchase from the Safety Training Institute and have been designed for presentation by in-house training personnel. These programs include the Management Development Program, which is oriented to the new supervisor, and Supervisors' Development Program for individuals with six or more months' experience. (Complete information on these and other courses can be found in the National Safety Council's General Materials Catalog.)

The Forklift Truck Operators Training Course is an eight-hour program designed to help a company comply with training regulations. The course can be presented in the standard form or can be tailored to fit an individual company's facilities and needs.

A number of these courses offered by the Safety Training Institute are also presented by local safety councils.

The Safety Training Institute can tailor courses and seminars to customer needs and present them at customer locations. Information on costs, scheduling, and specifics is available from the Safety Training Institute.

Consulting. The Council consulting services are available for loss control, motor fleet, traffic, industrial hygiene, environmental, community, product, motivation, agricultural, and research and statistical services.

The loss control consulting services can provide an analysis, recommendation, and structure for a safe and healthful working environment. Follow-up audits are also available. Information available upon request.

The Labor Division. This division and its member unions and government labor agencies represent labor and its safety and health viewpoints in many of the Council's areas of work. These are construction, public utilities, coal mining, legislation, defensive driving, and vocational education.

Specific problems requiring labor's review or consensus input are developed and approved by the Executive Committee.

Platform Speakers Group. The platform speakers group is an ideal resource for nationally known speakers who can cover a wide range of safety, health, and environmental subjects. Long or short time presentations are available. Speakers are ideal for seminars and conferences.

First Aid Institute. The Council's First Aid Institute offers programs, certification of instructors, and materials for teaching first aid and cardiopulmonary resuscitation (CPR) to employees. The courses of several levels meet requirements of governmental agencies and other organizations.

Environmental Health and Safety. To serve as a communications medium between members of the public, who demand a larger role in the management of community environmental risks, and those public and private risk managers who make the decisions, the National Safety Council established the Environmental Health Center in 1988. The goal of this special-purpose unit is to develop accurate and objective information on environmental and public health risks, improve public knowledge about these risks, and disseminate this information to the public.

Local Safety Councils. Nearly 100 local safety councils throughout the United States and Canada have been chartered by the National Safety Council. These councils work under the leadership of public-spirited citizens, commercial and industrial interests, responsible official agencies, and other important groups. They are non-profit, self-supporting organizations whose purpose is to reduce accidents and illnesses through prevention education.

Accredited councils operate under the guidance of a full-time executive and staff and receive support services from the National Safety Council. (A list of accredited safety councils can be obtained from the Safety Council Relations Department.)

These organizations give assistance to local safety engineers and others concerned with occupational safety and health. Safety professionals, in turn, render substantial service through participation in the local council's work.

Many of the local councils offer the following services:

- act as a clearinghouse of information on safety and health problems, and maintain a library of audio and visual and other communication aids
- provide a forum for exchange of experience through regularly scheduled meetings of supervisory personnel
- sponsor a variety of safety and health courses, including many presently being offered by the National Safety Council's Safety Training Institute
- conduct annual, area-wide safety and health conferences
- on request, provide assistance with safety problems and programs
- stimulate and assist in the development of accident prevention programs for all employers
- conduct safety contests, with awards for outstanding safety records.

The scope and extent of activities of each council depend, of course, upon local conditions and available resources.

American National Red Cross

17th and "D" Streets, NW
Washington, DC 20006

The American National Red Cross, through its more than 2,800 chapters, offers courses in first aid, cardiopulmonary resuscitation (CPR), swimming, lifesaving, canoeing, and sailing.

Experience in industry shows that first aid and safety training contributes to the reduction of accidents—both on and off the job—by creating an understanding of accident causes and effects and by improving attitudes toward safety. In addition, this training prepares individuals to give proper emergency care to accident victims. In some situations, immediate action may mean the difference between life and death.

Arrangements for first aid training can be made through local Red Cross chapters. Most industries prefer to select key personnel to receive training as volunteer instructors who, in turn, can conduct classes for fellow employees. Others may wish to arrange for employee training by instructors provided through the Red Cross.

Texts, instructor's manuals, charts, and other teaching materials and visual aids, such as films and posters, are available through local chapters.

Industrial Health Foundation, Inc.

34 Penn Circle W
Pittsburgh, PA 15206

The foundation, a nonprofit research association of industries, advocates industrial health programs, improved working conditions, and better human relations.

The foundation maintains a staff of physicians, chemists, engineers, biochemists, and medical technicians.

- Activities fall into three major categories:
- to give direct professional assistance to member companies in the study of industrial health hazards and their control
- to assist companies in the development of health programs as an essential part of industrial organization
- to contribute to the technical advancement of industrial medicine and hygiene by educational programs and publications.

Activities are classified as follows:

Medical:
Organization and administrative practices
Opinions on doubtful X-ray pictures
Surveys of health problems
Specific industrial medical problems
Epidemiology studies.

Chemistry, toxicology, industrial hygiene:
Field studies—plant or industry basis
Toxicity of chemicals, physical agents, or processes
Sampling and analytical procedures.

Engineering:
Ventilating systems
Exhaust hoods.

Education:
Training courses in occupational health and safety for:
    Physicians
    Nurses
    Industrial hygienists and engineers.
Symposia on special subjects of current interest in these fields.

The foundation holds an annual meeting of members, conferences of member company specialists, and special conferences on problems common in a particular industry.

The following publications are issued:
Industrial Hygiene Digest, monthly (abstracts).
Bibliographies on current interest subjects.
Technical bulletins.
Proceedings of symposia.

National Society to Prevent Blindness

500 East Remington Road
Schaumburg, IL 60173
The National Society to Prevent Blindness is the oldest national voluntary health organization that works to prevent blindness and preserve sight through community service programs, public and professional education, and research. NSPB's occupational health and safety program is guided by a professional advisory council, comprised of experts in the fields of industry, education, medicine, nursing, and accident prevention.

The National Society to Prevent Blindness:

- Sponsors the Wise Owl Program—eye-safety incentive program for industrial, military, municipal, and educational organizations. The Wise Owl Program provides information for developing eye health and safety programs, and recognition for those who saved their sight in an accident by wearing proper protective eyewear.
- Promotes state-wide eye safety for all school and college laboratory and shop students, and their teachers and visitors.
- Provides counselling to school administrators and teachers in establishing and implementing eye safety programs.
- Participates as a member of national standards organizations on committees concerning vision, eye protection, and illumination.
- Recommends the use of protective eyewear for individuals with monocular vision, and for hazardous situations in occupational and educational environments, around the home, and during sports activities.
- Develops and promotes position statements on health and safety issues including: contact lenses, fireworks, protective eyewear, etc.
- Recommends regular eye examinations for all age groups for the early detection and treatment of eye and vision disorders.
- Offers literature and audio visual materials on a wide range of eye health and safety topics.
- Sponsors the National Center for Sight, a toll-free hotline that provides information and referral services for a wide range of eye health and safety topics.

The National Safe Workplace Institute

122 South Michigan Avenue, Suite 1450
Chicago, IL 60603
The National Safe Workplace Institute was established 1987 to advance workplace safety and health, with the goal of making it a high priority for business and government. Numerous reports and publications are available.

STANDARDS AND SPECIFICATIONS GROUPS

American National Standards Institute

11 West 42nd Street
New York, NY 10036
The American National Standards Institute (ANSI) coordinates and administers the federated voluntary standardization system in the United States, which provides all segments of the economy with national consensus standards required for their operations and for protection of the consumer and industrial worker. It also represents the nation in international standardization efforts through the International Organization for Standardization (IOS), the International Electrotechnical Commission (IEC), and the Pacific Area Standards Congress (PASC).

ANSI is a federation of some 1,200 national trade, technical, professional, labor, and consumer organizations, government agencies, and individual companies. It coordinates the standards development efforts of these groups and approves the standards they produce as American National Standards when its Board of Standards Review determines that a national consensus exists in their favor. A catalog of safety standards is issued annually and is distributed without charge.

Many American National Standards, as well as other national consensus standards, have taken on additional importance since the passage of the Williams-Steiger Occupational Safety and Health Act of 1970. In promulgating standards under the act, the Occupational Safety and Health Administration has stated a definite preference for basing its regulations on consensus standards that have proved their value and practicality by use.

Under ANSI procedures, the responsibility for the management of specific standards projects is divided according to subject matter and assigned to an ANSI Safety and Health Standards Board. Standards dealing with safety fall under the jurisdiction of the Standards Management Board. Many American National Standards on safety and health are developed by ANSI-accredited Standards Committees.

American Society for Testing and Materials

1916 Race Street
Philadelphia, PA 19103
ASTM is the world's largest source of voluntary consensus standards for materials, products, systems, and services. There are currently more than 8,000 ASTM standards.

ASTM membership is drawn from a broad spectrum of individuals, agencies, and industries concerned with materials, products, and systems. The 30,000 members include engineers, scientists, researchers, educators, testing experts, companies, associations and research institutes, governmental agencies, and departments (federal, state, and municipal), educational institutions, consumers, and libraries.

ASTM standards are published for such categories as:
Nonferrous metals
Ferrous metals
Cementitious, ceramic, and masonry materials
Medical devices
Security systems
Energy
Construction
Chemicals and products
Environmental effects
Occupational safety and health
Protective equipment for sports
Electronics
Transportation systems
Business supplies

ASTM sponsors committees on geothermal resources and energy, quality control, food service equipment, and protective coatings for power generation facilities. The committee on consumer product safety has helped develop standards that will assist in protecting the public by reducing the risk of injury associated with the use of consumer products such as cigarette lighters, bathtubs, shower structures, children's furniture, trampolines, and nonpowered guns.

ASTM standards are of interest to the safety professional since they identify areas of hazards and establish guidelines for safe performance. The society also publishes standards for atmospheric sampling and analysis, fire tests of materials and construction, methods of testing building construction, nondestructive testing, fatigue testing, radiation effects, pavement skid resistance, protective equipment for electrical workers, and others.

These constitute basic reference materials for safety professionals who will frequently be confronted with ASTM standards.

FIRE PROTECTION ORGANIZATIONS

Factory Mutual Engineering Organization
1151 Boston-Providence Turnpike
Norwood, MA 02062
The Factory Mutual Engineering Organization (FMEO) is a world leader in loss control training, engineering and research. FMEO is an outgrowth of the philosophy of positive protection instead of just sharing the risk. In other words, recognition of the good risk through rate (premium) reduction is considered preferable to merely allowing the good risks to help pay for the bad.

FMEO is owned by the following companies: Allendale Mutual Insurance (Johnson, RI), Arkwright (Waltham, MA), Protection Mutual Insurance (Park Ridge, IL).

Factory Mutual Engineering and Research provides loss prevention services for policyholders. Its aim is to make properties and production facilities safe from fire, explosion, wind, water, and many other perils for which coverage is provided, including damage to boilers, pressure vessels, and machinery. Engineering services include evaluation of haz-

ards and protection through property inspections by Factory Mutual consultants located in major industrial centers. Research services involve the evaluation of fire protection devices and equipment for approval and the development of recommendations based on tests and loss experience for the prevention of loss. The Test Center in Rhode Island provides full-scale fire testing, simulating industrial conditions with regard to height, weight, and protection of major industrial storage occupancies.

Loss control training services are available to policyholders, and assistance is given to special technical problems relating to loss prevention as well as the human element aspect. The source of these services is the Factory Mutual Training Resource Center.

Training courses include Practicing Property Conservation, Boiler and Machinery Prevention Maintenance (both developed specifically for insureds), and Designing for Firesafety and Hazard Control (open to all architects and design professionals). These courses are taught by Factory Mutual engineers, research scientists, and training experts.

FMEO's publications include the *Record, Approval Guide, Handbook of Property Conservation, Loss Prevention Data Books*, and *Factory Mutual Engineering and Research Property Loss Control Catalog*. The *Record* is an internationally recognized bimonthly management magazine dealing with property conservation. The *Approval Guide* is a manual that lists industrial fire protection equipment and services that have been tested and approved by Factory Mutual Research Corporation. The *Handbook of Property Conservation* and the *Loss Prevention Data Books* cover recommended practices for protection against fire and related hazards. The *Loss Control Catalog* lists all available FMEO publications, films, training aids, and workshop kits.

Industrial Risk Insurers
85 Woodland Street
Hartford, CT 06102
The IRI is an association of more than 40 insurance companies that provides underwriting and advisory loss-control services to industry. It maintains a staff of loss prevention personnel with representation in key industrial centers.

The Loss Prevention Training Center in Hartford contains many types of fire protection equipment for examination and demonstration under working conditions. This laboratory is used for the basic and advanced training of IRI loss prevention personnel and for training plant protection personnel of IRI policyholders and of other insurance organizations who are given short courses in the proper use of fire protection devices.

National Fire Protection Association
Batterymarch Park
Quincy, MA 02269
The National Fire Protection Association is the clearinghouse for information on the subject of fire protection, fire prevention, and firefighting. It is a nonprofit technical and educational organization with a membership of some 53,000

companies and individuals.

The technical standards issued as a result of NFPA committee work are widely accepted by federal, state, and municipal governments as the basis of legislation, and widely used as the basis of good practices. More than 50 are used as OSHA regulations. Constantly revised and updated, 270 codes and standards are currently issued by NFPA, which are available in separate booklets.

Many of them supply authoritative guidance to safety engineers. Representative subjects include:
Industrial Fire Loss Prevention
Portable Fire Extinguishers
Sprinkler Systems Organization and Training of Private Fire Brigades
Flammable and Combustible Liquids Code
Hazardous Chemicals Data
Storage and Handling of Liquefied Petroleum Gases
Prevention of Dust Explosions in Industrial Plants
National Electrical Code and Handbook
Lighting Protection Code
Air Conditioning and Ventilating Systems
Life Safety Code and Handbook
Safeguarding Building Construction Operations
Protection of Records
Truck Fire Protection
Powered Industrial Trucks

The standards are also published as the *National Fire Codes* in 11 volumes totaling nearly 8,000 pages.

Other publications of interest to safety professionals concern hazardous materials, fire safety in health care facilities, and the public fire safety curriculum *Learn Not to Burn.*

The Fire Protection Handbook is also published by NFPA. It is an authoritative encyclopedia on fire and its control.

The NFPA is a sponsor of National Fire Prevention Week held every year in October in remembrance of The Great Chicago Fire.

The current NFPA catalog is available from the NFPA Customer Service Division.

Underwriters Laboratories Inc.

333 Pfingsten Road
Northbrook, IL 60062
Underwriters Laboratories, a not-for-profit organization, maintains laboratories for the examination and testing of devices, systems, and materials to determine their compliance with safety standards.

UL publishes annual directories of manufacturers whose products have met the criteria outlined in appropriate standards and whose products are covered under UL's Follow-Up Services program.

These directories are:
"Building Materials"
"Fire Protection Equipment"
"Fire Resistance"
"Recognized Component"
"Electrical Appliance and Utilization Equipment"

"General Information from Electrical Construction Materials and Hazardous Location Equipment"
"Hazardous Location Equipment"
"Marine"
"Automotive, Burglary Protection and Mechanical Equipment"
"Gas and Oil Equipment"
"Appliances, Equipment, Construction Materials, and Components Evaluated in Accordance with International Publications"

UL Follow-Up field representatives make unannounced visits to production sites where UL-Listed products are made at least four times a year to determine the effectiveness of the manufacturer's quality control program for continued compliance with UL safety standards.

Safety professionals often specify the UL Mark when they purchase fire, electrical, and other equipment that falls in categories covered under the services of Underwriters Laboratories.

Engineers should be aware, however, that UL Listings apply only within the scope of its standards for safety, and may have no bearing on performance or other factors not involved in the UL investigation. The markings and instructions provided with the products should tell a safety professional what he needs to know about the function of the device or material. UL tests are conducted under conditions of installation and use that conform to the appropriate standards of the NFPA or other applicable codes. Any departure from these standards by the user may affect the performance qualifications found by Underwriters Laboratories.

INSURANCE ASSOCIATIONS

In addition to the insurance associations listed under Fire Protection Organizations, there are a number of insurance federations with accident prevention departments that produce technical information available to safety people.

Alliance of American Insurers

1501 Woodfield Road
Schaumburg, IL 60195
The Alliance of American Insurers is a national organization of leading property-casualty insurance companies. Its membership includes more than 150 companies that safeguard the value of lives and property by providing protection against mishaps in workplaces and losses from fires, traffic accidents, and other perils. Alliance member companies have a tradition of loss prevention that dates from the organization of the first American mutual insurance company in 1752.

Through its Loss Control Department, the Alliance makes a concerted effort to reduce accidents, fires, and other loss-producing incidents. Under the guidance of its loss control advisory committee, sound safety engineering, industrial hygiene, fire protection engineering, and other loss prevention principles are promoted. Major activities include the

dissemination of information on safety subjects, conduct of specialized training courses for member company personnel, sponsorship of research, cooperation in the development of safety standards, development of visual aids, and the publication of technical and promotional safety literature. A catalog of safety materials is available without charge.

Notable among the publications issued by the Alliance are:
Safe Openings for Some Point-of-Operation Guards
Wood Working Circular Saws, Protection for Variety and Universal Types
Spreaders for Variety and Universal Saws
Nip Hazards on Paper Machines
Material Handling Manual
Handbook of Organic Industrial Solvents
Judging the Fire Risk
Tested Activities for Fire Prevention Committees
Exit Drills in the Home
Handbook of Hazardous Material
Safety Memos for Fleet Supervisors

Other Alliance departments also regularly issue a variety of bulletins, reports, research reports, and similar materials that often may relate to and support work safety and accident prevention. The communications department, for example, publishes the general-interest magazine, *Journal of American Insurance,* as well as leaflets, brochures, and other informational and educational materials. Inquiry is invited.

The Alliance cooperates extensively with trade associations, professional societies, and other organizations with similar interests, such as the National Safety Council, the National Fire Protection Association, the American National Standards Institute, Inc., the American Society of Safety Engineers, and the American Industrial Hygiene Association.

American Insurance Services Group, Inc.

Engineering and Safety Service
85 John Street
New York, NY 10038
The Engineering and Safety Service of American Insurance Services Group, Inc. is dedicated to providing information, education, and consultation to the technical and loss control personnel of its participating property-casualty insurance companies. This is accomplished through training programs, conferences, seminars, and a series of reports and bulletins.

Engineering and Safety Service's stock in trade is the latest and most beneficial information, gathered and prepared in a variety of ways for loss control professionals. The focus is always on safety and risk reduction in industrial, commercial, and residential settings within the following technical fields: occupational safety and health, pollution control, product safety, special and chemical hazards, fire protection, crime prevention, environmental science, building technology, industrial hygiene, construction hazards, commercial fleets, and boilers and machinery. In addition, Engineering and Safety Service represents the insurance industry on numerous national standards-making committees of the American National Standards Institute, National Fire Protection Association, and others.

The Engineering and Safety Service develops and publishes a wide variety of safety-related materials that it makes available to the public, in addition to those it produces for its subscriber companies.

PROFESSIONAL SOCIETIES

American Society of Safety Engineers

1800 East Oakton
Des Plaines, IL 60018
The American Society of Safety Engineers is the only organization of individual safety professionals dedicated to advancing the safety profession and to fostering the well-being and professional development of its members.

To fulfill its purpose, the Society has the following objectives:
- promote the growth and development of the profession
- establish and maintain standards for the profession
- develop and disseminate material that will carry out the purpose of the Society
- promote and develop educational programs for obtaining the knowledge required to perform the functions of a safety professional
- promote and conduct research in areas that further the purpose and objectives of the Society
- provide forums for the interchange of professional knowledge among its members
- provide for liaison with related disciplines.

An annual professional development conference is conducted for members.

The Society is actively pursuing its Professional Development Programs including the development of curricula for safety professionals, accreditation of degree programs, presentation of member education courses and publications, and definition of research needs and communications to keep safety practitioners current. In addition, the Society has increased its participation and activity in national government affairs.

The Society has established the American Society for Safety Engineers Foundation because of its concern for the need for greater research efforts in the prevention of accidents and injuries. The Society believes that increased research will improve accident and injury control techniques, thus serving the interests of its members as well as contributing to the economy of our nation and to the health and welfare of all persons.

The Society was also instrumental in establishing a separate corporation to develop a certification program for safety professionals. This corporation is known as the Board of Certified Safety Professionals of America. (See BCSP later in this chapter.)

The members of the Society receive the monthly *Professional Safety* as part of their membership. (Nonmembers may also subscribe.) Articles on new developments in the technology of accident prevention are included, as well as information on the activities of the Society, its chapters, and mem-

bers. The Society also publishes other technical or specialized information, such as "A Selected Bibliography of Reference Materials in Safety Engineering and Related Fields," a glossary of terms used in the safety profession, and a series of monographs.

Founded in 1911 as the United Association of Casualty Inspectors, it grew from the original enrollment of 35 to more than 24,000 members internationally in 1990.

Chapters engage in a number of activities designed to enhance the professional competence of their members. Most hold monthly meetings featuring speakers, demonstrations, workshops, and discussions designed to help members keep abreast of developments in their professional field. A list of chapters grouped by states and listed by name and by location can be obtained from the national office.

American Association of Occupational Health Nurses, Inc.

50 Lenox Pointe
Atlanta, GA 30324
As the national professional organization for the registered nurse working in business and industry, the association serves as an advocate for occupational health nurses, establishes health care and provides educational programs for nurses in this special field. *AAOHN Journal*, a scientific peer-reviewed journal, and *AAOHN News* are published monthly.

The annual meeting is held in conjunction with the American Occupational Health Conference, which AAOHN cosponsors with the American Occupational Medical Association. Publications that address safety and health issues in the workplace are also available.

American Board of Industrial Hygiene

4600 West Saginaw, Suite 101
Lansing, MI 48917
This specialty board is authorized to certify properly qualified industrial hygienists. The overall objectives are to encourage the study, improve the practice, elevate the standards, and issue certificates to qualified applicants.

American Chemical Society

1155 16th Street, NW
Washington, DC 20036
This society is devoted to the science of chemistry in all its branches, the promotion of research, the improvement of the qualifications and usefulness of chemists, and the distribution of chemical knowledge.

Articles on safety appear in the monthly publication *Industrial and Engineering Chemistry*, the weekly publication *Chemical and Engineering News*, and in the *Journal of Chemical Education*. Digests of papers dealing with aspects of industrial hygiene appear monthly in *Chemical Abstracts*. The environment is discussed in *Environmental Science and Technology*.

The society has a Committee on Chemical Safety and a Division of Chemical Health and Safety. It also has a Chemical Health and Safety Referral Service accessible by telephone.

American College of Surgeons

55 East Erie Street
Chicago, IL 60611
The American College of Surgeons, in addition to its role in the Joint Action Program with the National Safety Council and the American Association for the Surgery of Trauma, has now conducted Advanced Trauma Life Support Courses, through the auspices of the College's Committee on Trauma, for over 100,000 physicians. It has also initiated an educational program, developed by the Committee on Trauma, aimed at prevention of deaths caused by drinking drivers. These efforts continue the work of the College in preventing injury and in improving the care of the injured.

American Conference of Governmental Industrial Hygienists

6500 Glenway Avenue, Bldg. D-7
Cincinnati, OH 45211
ACGIH is an organization devoted to the administrative and technical aspects of worker health and safety protection. For over 50 years, ACGIH has provided leadership, educational, and investigative opportunities to professional and technical personnel representing governmental agencies or educational institutions involved in occupational health and safety activities.

ACGIH publishes a number of well-known texts, including the *Threshold Limit Values/Biological Exposure Indices*. The Conference also publishes *Applied Occupational and Environmental Hygiene*, the monthly journal that combines peer-reviewed papers with wide-ranging columns and articles to offer information and guidance to the practicing progressional.

An ACGIH Publications catalog listing over 200 texts in occupational safety and health is available upon request.

American Industrial Hygiene Association

P.O. Box 8390
345 White Pond Drive
Akron, OH 44320
AIHA is an organization of professionals dedicated to the prevention of workplace-related illness or injury that may affect the health and well-being of workers or the community. With more than 9,000 members, AIHA is the largest international association serving the needs of occupational and environmental scientists and engineers practicing industrial hygiene in industry, government, labor, academic institutions, and independent organizations. Founded in 1939, the purposes of AIHA are to promote the field of industrial hygiene, to provide education and training, to provide a forum for the exchange of ideas and information, and to represent the interests of industrial hygienists and those they serve.

AIHA offers its members and other interested professionals a wide range of products and services on a variety of topics within the occupational health and safety field. AIHA publishes an extensive library of books and

guides including *Computers in Health and Safety, Indoor Air Investigation, Emergency Response Planning Guidelines Series, Respiratory Protection Manual, Chemical Protective Clothing, Workplace Environmental Exposure Level (WEEL) Guides,* and *Noise and Hearing Conservation Manual, 4th edition.* AIHA also publishes, monthly, *The AIHA Journal,* the premier technical journal in the industrial hygiene field and *The Synergist,* a monthly informational newsletter. AIHA holds continuing education seminars on the latest topics in the field throughout the year and all over the country. AIHA offers an employment service for those professionals seeking new positions and for companies looking for employees. AIHA has four laboratory programs to assist laboratories in maintaining high quality standards, including accreditation for every qualified industrial hygiene laboratory in the world.

AIHA sponsors the American Industrial Hygiene Conference, (AIHC), the largest occupational health meeting in the world. AIHC features technical sessions, roundtable sessions, professional development courses, and a world class exhibition. Catalogs for continuing education, laboratory programs, and publications are available upon request. Information is available for all AIHA products and services.

American Institute of Chemical Engineers

345 East 47th Street
New York, NY 10017

Founded in 1908, the AIChE is a nonprofit, individual member of society dedicated to advancing the chemical engineering profession and the professional standards of 50,000 chemical engineers through publications, technical meetings, continuing education, and research. The AIChE's Safety and Health division holds annual Loss Prevention and Ammonia Safety symposia and publishes proceedings. The Institute's large continuing education program conducts short courses in many engineering safety subjects several times each year. The AIChE Design Institutes for Physical Property Data (DIPPD) and for Emergency Relief Systems (DIERS) Users Group offers corporations and government agencies the opportunity to join in the cooperative sponsorship of safety-related research that they could not afford to do individually. Symposia proceedings, DIERS and DIPPD research material and other safety publications are listed in AIChE's annually updated publications catalogue.

The Center for Chemical Process Safety established in 1985 by the American Institute of Chemical Engineers and sponsored by over 50 corporations, the nonprofit Center for Chemical Process Safety gives leadership and increased focus on measures to help improve process safety. CCPS collects, organizes, and publishes the latest in scientific and engineering practices as guidelines for preventing and mitigating major incidents that involve the release of potentially hazardous materials. Over twenty-three books in its Guidelines for Process Safety series are in print or under development. Many are referenced in government documents. A catalog of CCPS

publications that can be purchased from AIChE is available.

These books are:
Guidelines for Chemical Process Quantitative Analysis Risk
Guidelines for Risk Decision Making
Guidelines for Process Equipment Reliability Data
Guidelines for Improving Human Performance in Process Safety
Guidelines for Use of Vapor Cloud Dispersion Models
Workbook of Test Cases for Vapor Cloud Dispersion Models
Guidelines for Evaluating the Consequences of Fires and Explosion from Unconfined Vapor Clouds and BLEVES
Guidelines for the Technical Management of Chemical Process Safety
Guidelines for Technical Management of Chemical Process Safety in Chemical Plants
Guidelines for Vapor Release Mitigation
Guidelines for Post Release Mitigation Technology
Guidelines for Storage and Handling of High Toxic Hazard Materials
Guidelines for Engineering Practices to Prevent and Mitigate Fires and Explosions
Guidelines for Effective Handling of Emergency Relief Effluents
Guidelines for Processing and Handling Reactive Materials
Guidelines for Storage and Handling Reactive Materials and Warehousing
Guidelines for Safe Automation of Chemical Process Safety
Guidelines for Operations and Maintenance
Guidelines for Process Safety in Smaller Operations
Guidelines for Incident Investigation
Guidelines for Process Safety Auditing
Resources for Chemical Process Safety

CCPS sponsors international technical symposia, usually with Proceedings, where technical papers disseminate and encourage the use of safe engineering and operating practices. Teaching material, developed by CCPS, is being used at colleges to teach the concepts of Safety, Health, and Loss Prevention in Chemical Processes to undergraduate engineers. Research is also carried out to advance the state-of-the-art in measures to prevent and mitigate major events.

Both AIChE and CCPS publications, meetings, and continuing education courses are available to AIChE members and nonmembers.

American Institute of Mining, Metallurgical, and Petroleum Engineers

345 East 47th Street
New York, NY 10017

The *AIME* promotes the advancement of knowledge of the arts and sciences involved in the production and use of useful minerals, metals, energy sources, and materials and to record and disseminate developments in these areas of technology for the benefit of mankind.

Publications are *Mining Engineering, Journal of Metals, Journal of Petroleum Technology, Iron and Steelmaker,* all issued monthly and *Transactions of the Society of Mining Engineers* (quarterly), *Transactions of the Metallurgical So-*

ciety (bimonthly), and *Transactions of the Society of Petroleum Engineers of AIME* and *Society of Petroleum Engineers Journal* (quarterly).

The American Medical Association

535 North Dearborn Street
Chicago, IL 60610
The American Medical Association, with a current membership of about 200,000 physicians, has a long record of involvement in public health.

The Department of Environmental, Public, and Occupational Health (DEPOH) was organized in 1970 as a combination of AMA's Departments of Occupational Health and Environmental Health. In general terms, it is concerned with the well-being of all population groups, whether they be in the workplace or in the community. Thus the expertise of the staff covers the health effects of air, water, chemical, and physical stresses; communicable diseases; population growth; injuries; epidemiology; preventive medicine; sports medicine; and problems of the aged and the handicapped.

The department is principally an authoritative source of information for inquiries about environmental, public, and occupational health. In addition, it plans appropriate conferences and courses, prepares authoritative publications for physicians and the public, carries out special assignments and studies, and acts as the AMA liaison in federal health programs and legislation.

Since 1939, AMA has sponsored and organized annual congresses on occupational health. Intended especially to benefit the part-time occupational health practitioner, these congresses have addressed such topical subjects as: the reproductive disorders of workers, mental health in industry, group practice in occupational health, occupational pulmonary diseases, diseases and injuries of the back, and toxic chemicals in the community.

National conferences on the medical aspects of sports, which focus on the prevention and treatment of athletic injuries, are intended for team physicians, coaches, trainers, and administrators. Proceedings of the conferences are published.

The AMA continues to have an active voice in the regulatory process: for example, identifying to the National Institute of Occupational Safety and Health (NIOSH) those AMA physicians who have the expertise to review NIOSH Criteria Documents. And when the Occupational Safety and Health Administration (OSHA) proposed a rule (in July 1978) to give OSHA and NIOSH complete access to employee's medical records, department staff protested in the public hearings.

The *Journal of the American Medical Association (JAMA)* is published weekly and frequently contains articles on some aspect of occupational health.

American Nurses' Association, Inc.

2420 Pershing Road
Kansas City, MO 64108

The American Nurses' Association is the voluntary membership organization for all registered nurses. The Division on Community Health Nursing Practice is the component of the ANA that provides authoritative information about the practices of occupational health nursing practice for better employee health care.

Publications of interest are: *Standards of Community Health Nursing Practice, Nursing Case Management, Standards and Scope of Hospice Nursing Practice, Standards of College Health Nursing Practice, Community-Based Nursing Services: Innovative Models, A Practice in Correctional Facilities, Standards of School Nursing Practice, A Conceptual Model of Community Health Nursing, Guidelines for Protecting the Safety and Health of Health Care Workers.* Titles of other brochures, pamphlets, guides, and articles are included in the association's publications list which will be sent on request.

American College of Occupational Medicine

55 West Seegers Road
Arlington Heights, IL 60005
The American College of Occupational Medicine fosters the study and discussion of problems peculiar to the practice of occupational medicine, encourages the development of methods adapted to the conservation and improvement of health among workers, and promotes a more general understanding of the purpose and results of employee medical care.

Some of the association committees and sections are:
Academic Occupational Medicine Section
Alcoholism and Drug Abuse
Annual Scientific Meeting
Arts Medicine Section
Dermatology Section
Education Council
Energy Technology
Ethical Practice in Occupational Medicine
Health Achievement in Industry Award
Health Education
Maritime Medical Affairs Section
Medical Center Occupational Health Services
Medical Information Systems
Medical Practice in Small Industries
Microcomputers in Occupational Medicine Users Section
Noise and Hearing Conservation
Nuclear Industry Physicians' Association Section
Occupational and Clinical Toxicology
Occupational Medical Practice
Psychiatry and Occupational Mental Health
Railroad Medicine Section
Trauma Education.

The official publication of the association is the monthly *Journal of Occupational Medicine.*

An annual American Occupational Health Conference is held, usually in April, in collaboration with the American Association of Occupational Health Nurses.

American Psychiatric Association

1400 K Street, NW
Washington, DC 20005
This association of 33,000 members is concerned with research in all phases of mental disorders, standards of psychiatric education and treatment, medico-legal aspects of psychiatric practice, and the promotion of mental health for all citizens. Publications: *American Journal of Psychiatry*, monthly; *H&CP* (Hospital and Community Psychiatry), monthly; *Psychiatric News*, every two weeks.

American Public Health Association

1015 15th Street, NW
Washington, DC 20005
The American Public Health Association is a multidisciplinary, professional association of 30,000 health workers. Through its two monthly periodicals, *The American Journal of Public Health* and *The Nation's Health* and other publications, APHA disseminates health and safety information to those responsible for state and community health-service programs. The program area interest group is devoted to injury control and emergency health services, and to occupational health and safety.

American Society for Industrial Security

1655 North Fort Myer Drive, Suite 1200
Arlington, VA 22209
The American Society for Industrial Security, an international professional society of more than 25,000 industrial security executives in both the private and public sector, has 190 chapters worldwide. Committee activities that would be of interest to the safety professional are safeguarding proprietary information, physical security, terrorist activities, disaster management, fire prevention and safety, and investigations.

The Society publishes the magazine *Security Management* monthly, to which safety professionals may subscribe, and issues a newsletter to its members bimonthly.

The Society certifies security professionals through its Certified Protection Professional (CPP) program. Other member services include the use of the security library and job placement. The Society also produces numerous workshops and educational programs and an annual seminar open to both members and nonmembers.

American Society of Mechanical Engineers

345 East 47th Street
New York, NY 10017
This society, the professional mechanical engineers' organization, encourages research, prepares papers and publications, sponsors meetings for the dissemination of information, and develops standards and codes under the supervision of its Policy Board.

The society developed the following safety codes under the procedures meeting the criteria of the American National Standards Institute:
Safety Code for Elevators

Safety Code for Mechanical Power-Transmission Apparatus
Safety Code for Conveyors, Cableways, and Related Equipment
Safety Code for Cranes, Derricks, and Hoists
Safety Code for Manlifts
Safety Code for Powered Industrial Trucks
Safety Code for Aerial Passenger Tramways
Safety Code for Mechanical Packing
Safety Code for Garage Equipment
Safety Code for Pressure Piping
Safety Standards for Compressor Systems.

The ASME Boiler and Pressure Vessel Committee is responsible for the formation and revision of the *ASME Boiler and Pressure Vessel Code*.

The society publishes the *Transactions of the American Society of Mechanical Engineers* and the monthly publications *Mechanical Engineering & Applied Mechanical Review*.

American Society for Training and Development

Box 1433, 1630 Duke Street
Alexandria, VA 22313
This is a professional society of persons engaged in the training and development of business, industrial, and government personnel. It holds a major annual conference and has an information service for members. ASTD is publisher of the new *Technical & Skill Training* magazine.

Association of Federal Safety and Health Professionals

7549 Wilhelm Drive
Lanham, MD 20706-3737
The Association is composed of safety and health professionals who are, or have been, full-time employees of the federal government, civilian or military. Association publications (monthly newsletter and biannual Directory) are distributed to members only. The Association purposes are to promote the safety and health programs of the federal agencies and the professionalism of those who manage these programs.

Board of Certified Hazard Control Management

8009 Carita Court
Bethesda, MD 20817
Founded in 1976, the Board evaluates and certifies the capabilities of practitioners engaged primarily in the administration of safety and health programs. Levels of certification are senior and master, with master being the highest attainable status indicating that the individual possesses the skill and knowledge necessary to effectively manage comprehensive safety and health programs. The Board offers advice and assistance to those who wish to improve their status in the profession by acquiring skills in administration and combining them with technical safety abilities. It establishes curricula in conjunction with colleges, universities, and other training institutions to better prepare hazard control managers for their duties. Publications are the *Hazard Control Manager* and the *Directory*.

Board of Certified Safety Professionals

208 Burwash Avenue
Savoy, IL 61874-9510

The Board of Certified Safety Professionals (BCSP) was organized as a peer certification board in 1969 with the purpose of certifying practitioners in the safety profession. The specific functions of the Board, as outlined in its charter, are to evaluate the academic and professional experience qualifications of safety professionals, to administer examinations, and to issue certificates of qualification to those professionals who meet the Board's criteria and successfully pass its examinations. Contact the BCSP for details on the requirements for certification.

The American Board of Industrial Hygiene (ABIH)/Board of Certified Safety Professionals (BCSP) Joint Committee began a joint certification program for technologists and technicians in 1985. The purpose of this program is to certify health and safety practitioners at the technologists/technician level. For further information on certification requirements, contact the ABIH/BCSP Joint Committee.

Certified Healthcare Safety Professionals

8009 Carita Court
Bethesda, MD 20817

A program to certify health-care safety professionals has been established by the International Healthcare Safety Professional (HSP) Certification Board. The HSP program is designed to raise the competence, status, and recognition of health care safety professionals and assist in the transfer of technology and the exchange of ideas for improving safety in health-care activities.

The Healthcare Safety Professional (HSP) certification program has the following objectives:

- evaluating the qualifications of persons engaged in hazard control activities in hospital/health-care facilities
- certifying as proficient individuals who meet the level of competency for this recognition
- increasing the competence and stimulating professional development of practitioners
- providing recognition and status for those individuals who by education, experience, and achievement are considered qualified
- facilitating the exchange of ideas and technology that will improve performance.

There are three levels of certification: the Master, requiring four years' experience and a baccalaureate degree; the Senior, requiring four years' experience and an associate of arts degree; and the Affiliate, which requires assignment as a safety officer, completion of high school or equivalent, plus specified safety courses over a period of two years.

Publications include a *Directory of Diplomates* and a newsletter.

The Chlorine Institute

2001 L Street, NW, #506
Washington, DC 20036

Founded in 1924, the institute provides "a means for chlo-rine producers and firms with related interests to deal constructively with common industry problems—especially in safety, transportation, regulations and legislation, and community relations."

Results of committee deliberations are distributed worldwide to chlorine producers, consumers, and other interested groups and persons. In addition to the Chlorine Manual and audiovisual programs, some 70 engineering and design recommendations, specifications, and drawings are available in a publications catalog.

Flight Safety Foundation, Inc.

5510 Columbia Pike
Arlington, VA 22204-3194

The Flight Safety Foundation is an international membership organization dedicated solely to improving the safety of flight. Nonprofit and independent, it serves the public interest by actively supporting and participating in the development and implementation of programs, policies, and procedures affecting safety by stimulating research into ways and means of eliminating accident-inducing factors, by in-depth appraisals of actual and potential problem areas in flight and ground safety and by developing possible solutions to those problems. As a vehicle of information interchange, it cooperates with all other organizations and individuals in the field of aviation safety in educating all segments of the aviation community in the principles of accident prevention.

To further these objectives, the foundation acts as a clearinghouse for the collection, analysis, and dissemination of safety information. Its personnel participate in safety discussions and safety studies and conduct safety seminars for aviation personnel. It presents awards and otherwise encourages the growth of safety programs. It acts as a catalytic agent in drawing attention to needed improvements and changes in safety techniques.

Perhaps most well-known of the FSF's activities are its annual safety seminars, held in successive years in various parts of the world. These seminars constitute a major gathering of FSF's membership and guests to review and discuss papers presented by prominent world aviation safety experts and to share ideas for safety improvement in an informal, neutral environment.

The Corporate Aviation Safety seminar is held annually in North America. It provides a forum for discussion and review of air safety matters. The programs feature leaders of industry, operator and user experts, as well as government and university researchers.

National Association of Suggestion Systems

230 North Michigan Avenue
Chicago, IL 60601

The National Association of Suggestion Systems, incorporated as a nonprofit organization, encourages suggestion system activity in industry, commerce, finance, and government. Specific objectives are to:

- increase appreciation of the usefulness of employee suggestion systems
- encourage study of the elements necessary to successful use of employee thinking
- provide an opportunity for the personal development of those who represent member institutions
- gather and disseminate useful information through meetings, publications, factual surveys, and the like
- promote personal contacts between suggestion system administrators and the leaders of various industries.

National Safety Management Society

3871 Piedmont Avenue
Oakland, CA 94611
The National Safety Management Society, founded in 1966, is a nonprofit corporation that seeks to expand and promote the role of safety management as an integral component of total management by developing and perfecting effective methods of improving control of accident losses, be they personnel, property, or financial. Membership is open to those having management responsibilities related to loss control.

SAFE Association

25044 Peachland Avenue, Suite 205
Newhall, CA 91321
The SAFE Association, a nonprofit professional association, is dedicated to the preservation of human life. SAFE members represent the fields of engineering, psychology, medicine, physiology, management, education, industrial safety, survival training, fire and rescue, law, human factors, equipment design, and many subfields associated with the design and operation of aircraft, automobiles, buses, trucks, trains, spacecraft, and watercraft.

Regional chapters sponsor meetings and workshops to provide an exchange of ideas, information on members' activities, and presentations of new equipment and procedures encompassing governmental, private and commercial safety, and survival applications. SAFE publishes a quarterly journal, periodic newsletters, and an annual *Proceedings of the SAFE Symposium.*

System Safety Society

Technology Trading Park
Five Export Drive, Suite A
Sterling, VA 22170-4421
The System Safety Society is a nonprofit organization of professionals dedicated to safety of products and activities by the effective implementation of the system safety concept. This concept is, basically, the application of appropriate technical and managerial skills to assure that a systematic forward-looking hazard identification and control function is made an integral part of a project, program, or activity at the conceptual planning phase, continuing through design, production, testing, use, and disposal phases. The objectives include to:

- advance the state-of-the-art of system safety
- contribute to a meaningful management and technologi-

cal understanding of system safety

- disseminate newly developed knowledge to all interested groups and individuals
- further the development of the professionals engaged in system safety.

Through its local chapters, committees, executive council, publications, and meetings, the society provides many opportunities for interested members to participate in a variety of activities compatible with society objectives. In addition to its operating committees, society activities include publication of *Hazard Prevention*, the official society journal. Published five times a year, it keeps members informed of the latest developments in the field of system safety.

International System Safety Conferences are sponsored biennially and Proceedings are available.

Veterans of Safety

c/o Dr. Robert Marshall
Central Missouri State University
Humphries Building
Warrenburg, MO 64093
Membership numbers 1,700 safety engineers with 15 or more years of professional safety experience. Founded in 1941, the objective of Veterans of Safety is to promote safety in all fields. Activities include: (a) Safety Town USA, to educate preschool and elementary school children in pedestrian safety; (b) Most Precious Cargo Program, to improve school bus safety; (c) Unified Emergency Telephone Numbers Program, to establish nationwide uniformity in emergency telephone numbers to contact fire, police, and medical aid; and (d) developed and implemented the Safety and Health Hall of Fame. Gives annual awards for best technical safety papers published. Maintains placement service.

TRADE ASSOCIATIONS

American Foundrymen's Society

505 State Street
Des Plaines, IL 60016-8399
The American Foundrymen's Society is the only technical society that serves the interests of the foundry industry. It disseminates information on all phases of foundry operations, including environmental (air, water, and waste) and occupational safety and health.

Services provided include in-plant visitations by AFS environmental staff for environmental facility reviews, safety and health reviews and industrial hygiene surveys.

The AFS Lester B. Knight Analytical Laboratory is accredited by the American Industrial Hygiene Association. Analytical services include Leachate analysis of heavy metals and phenol using TCLP and industrial hygiene analysis of heavy metals, silica, and organics. Industrial hygiene pump calibration and rental is also offered.

The following Environmental/Safety publications are available through the society:

Safety Requirements for Sand Preparation, Molding and Coremaking in the Sand Foundry Industry, ANSI Z241.1
Safety Requirements for Cleaning and Finishing of Castings, ANSI Z241.2
Safety Requirements for Melting and Pouring of Metals in the Metalcasting Industry, ANSI Z241.2
Foundry Health and Safety Guides
Foundry Ventilation Manual
Industrial Noise Control
The Recirculation of Air for Energy Conservation

A complete Society Publications Catalog is available upon request.

American Gas Association

1515 Wilson Boulevard
Arlington, VA 22209
The association, through its Accident Prevention Committee, serves as a clearinghouse and in an advisory capacity to persons responsible for employee safety and to safety departments of its member companies. Its purposes are to study accident causes, recommend corrective measures, prepare manuals, and disseminate information to the gas industry that will help reduce employee injuries, motor vehicle accidents, and accidents involving the public. The committee meets several times a year and conducts a Safety and Health Workshop for Supervisors each year. It also provides speakers for regional gas associations and other gas industry meetings.

The committee has project groups that address all aspects of employee safety, including awards and statistics, distribution and utilization, education, motor vehicles, and gas transmission, and promotional and advisory.

Published material includes suggested safe practices manuals and quarterly and annual reports on the industry's accident experience and programs. A publication catalog may be obtained by contacting the A.G.A. Engineering and Operating Section.

The association also maintains laboratories in Cleveland and Los Angeles, where gas appliances are tested and design certified.

American Iron and Steel Institute

1133 15th Street, NW
Washington, DC 20005
AISI represents companies accounting for more than 80% of raw steel production in the United States.

The Occupational Health and Safety Committee investigates matters relating to the working environment of employees in the iron and steel industry to enhance the health, hygiene, and safety of steel industry personnel, both in and out of the workplace. This includes recognition and evaluation of occupational health and safety hazards with respect to existing and new technology. The Committee also develops industry information and data that may be used for a variety of purposes, and participates in the development of pertinent national and international consensus standards.

American Mining Congress

1920 N Street, NW
Washington, DC 20036
Congress membership is from coal, metal, and nonmetal mining companies. This association has a Coal Mine Safety Committee, a Metal-Nonmetal Mine Safety Committee, and an Occupational Health Committee.

The monthly *American Mining Congress Journal* regularly carries articles and news about safety and health.

American Paper Institute, Inc.

260 Madison Avenue
New York, NY 10016
The American Paper Institute, through the Safety and Health Subcommittee of its Employee Relations Committee, conducts a broad safety and health education and information service for the paper industry. The Subcommittee sponsors workshops and seminars on various safety and health subjects. Ad hoc task forces are also established to deal with specific problems requiring rapid response, specialized knowledge, or concentrated effort.

In addition, the institute's Employee Relations Department issues a quarterly "Safety and Health Report," which covers the latest developments in OSHA, NIOSH, paper industry safety issues, and other items of current interest.

The department compiles an annual "Summary of Occupational Injuries and Illnesses" and also has an industry-wide Safety Award program, which recognizes individual mills that have outstanding safety records.

American Petroleum Institute

1220 L Street, NW
Washington, DC 20005
The objective of the Committee on Safety and Fire Protection of the American Petroleum Institute is to reduce the incidence of accidental occurrences, such as injuries to employees and the public, damage to property, motor vehicle accidents, and fires. To attain this objective, the Committee on Safety and Fire Protection:

- provides statistical reports, pamphlets, data sheets, and other publications to assist the industry in the prevention of accidents and the prevention, control, and extinguishment of fires
- provides a means for the development and exchange of information on accident prevention and fire protection to be used for education and training in the industry
- provides a forum for discussion and exchange of information concerning safe practices and the science and technology of fire protection and safety engineering
- promotes research and development in the fields of accident prevention and fire protection for the benefit of the petroleum industry as a whole
- maintains contact and cooperates with association and code writing bodies such as NFPA and ANSI.

Safety and fire protection manuals have been published on such subjects as:
Protection Against Ignitions Arising Out of Static, Lightning

and Stray Currents
Safe Operation of Inland Bulk Plants
Safe Practices in Gas and Electric Cutting and Welding in Refineries, Gasoline Plants, Cycling Plants, and Petrochemical Plants
Cleaning Mobile Tanks in Flammable or Combustible Liquid Service
Cleaning Petroleum Storage Tanks
Guidelines for the Application of Water Spray Systems in the Petroleum Industry
A Guide for Controlling the Lead Hazard Associated with Tank Entry and Cleaning
Guides for Fighting Fires in and Around Petroleum Storage Tanks
Fire Hazards of Oil Spills on Waterways
Guidelines for Confined Space Work in the Petroleum Industry
Evaluation of Firefighting Foams as Fire Protection for Alcohol-Containing Fuels
Fire Protection in Refineries
Inspection for Accident Prevention
Inspection for Fire Protection
Service Station Safety
Overfill Protection for Petroleum Storage Tanks
Publications catalog is available at no cost.

American Pulpwood Association

1025 Vermont Avenue, NW, Suite 1020
Washington, DC 20005
This association fosters study, discussion, and action programs to guide and help the pulpwood industry in growing and harvesting pulpwood raw material for the pulp and paper industry. The safety and training programs of the Association are served through six regional Technical Divisions.

Available literature includes training guides, notebooks, safety alerts, and technical releases that describe items of personal protective equipment, safe working procedures, and other pertinent accident control items.

American Road and Transportation Builders

525 School Street, SW
Washington, DC 20024
ARTBA is a national federation of public and private interests concerned with transportation construction issues. Its members
are involved in highway design, construction, management, signing, lighting, and other areas related to road safety.

In 1986, ARTBA organized a national forum on work zone (road construction site) safety. A brochure summarizing the recommendations of the forum is available upon request. The organization conducts ongoing public awareness activities on this topic.

ARTBA publishes a membership newsletter 36 times per year, a quarterly membership magazine, and periodic special reports.

American Trucking Associations, Inc.

2200 Mill Road
Alexandria, VA 22314
American Trucking Associations is the national federation of the trucking industry. Through its 1,000 member Safety Management Council, programs are developed to promote safer highways and safer motor carrier operations.

The Safety Management Council is the official safety organization of the national trucking industry. It is responsible to the ATA Executive Committee through the ATA Safety and Engineering Committee. The ATA Safety Department serves as technical advisors to the Council and supports the Council's activities and programs.

Membership in the Safety Management Council is composed of professional truck fleet safety personnel in the field of motor carrier safety and/or personnel work and persons from allied fields and industries who are concerned with motor carrier safety activities. The Council's 11 technical committees provide the trucking industry with information by working on such problems as employee selection, workers compensation, training and supervision, accident investigation and reporting, transportation of hazardous materials, federal motor carrier safety regulations, drug testing, physical qualifications, and injury control. In addition to regional and national meetings, the Council also sponsors a wide range of continuing professional education programs for truck and workplace safety.

Publications, instructional videos, safety award programs, and *Trucking Safety Magazine* are available through the Council as are monthly mailings of the *Safety Bulletin*, driver letters, and safety posters.

The American Waterways Operators

1600 Wilson Boulevard
Arlington, VA 22209
The American Waterways Operators is the trade association representing the national interests of operators of towboats, tugboats, and barges engaged in domestic trade on the inland and coastal waters of the United States, as well as those of the shipyards that build and repair these vessels.

AWO has a safety committee that pursues three primary objectives:
- promotion of individual member company's safety programs
- more participation by members in the Barge and Towing Vessel Industry Safety Contest, which the association cosponsors with the National Safety Council
- preparation and distribution of AWO safety posters, which the association issues to its members each year, and for which the association conducts a safety poster contest.

The association's manual, *Basic Safety Program for the Barge and Towing Vessel Industry*, is designed either (a) to be adopted as a complete company program, or (b) to be used as a guide to develop or supplement an individual company's program. The manual covers methods and tech-

niques for accident prevention, and contains valuable guidelines on all aspects of personnel safety in the barge and towing vessel industry.

American Water Works Association
6666 West Quincy Avenue
Denver, CO 80235
The Association, through its Loss Control Committee, develops loss control programs for the water utility industry. Programs presently available include (1) collecting, analyzing, and compiling annual statistics in both employee and motor vehicle accidents; (2) a safety award program; (3) an audiovisual library; (4) one- and two-day loss control seminars for supervisors; (5) loss control audit program; and (6) a safety poster program. In addition to these programs, the Loss Control Committee has three subcommittees on accident prevention, health maintenance, and risk management that study and disseminate information that will assist the water utility industry in increasing its concepts of total loss control.

Publications available include a manual, *Safety Practices for Water Utilities* and *Safety Talks*, a compilation of 52 weekly safety talks. Slide/script and video programs available include: Work Area Protection, Cave-In Protection, Motor Vehicle Accident Prevention, Safe Handling of Water Treatment Chemicals, Entering and Working in Confined Spaces, and Safe Operation of Heavy Equipment.

Regional safety meetings are held with AWWA section safety chairmen to promote safety throughout the water utility industry. These meetings are designed to provide an exchange of ideas and experiences to enhance water utility safety.

American Welding Society
P.O. Box 351040
Miami, FL 33135
The society is devoted to the proper and safe use of welding by industry. Through its Safety and Health Committee, the society coordinates safe practices in welding by promoting new and revising existing standards.

"Safe Practices for Welding and Cutting Containers that Have Held Combustibles" is one booklet produced by this Committee and is available to industry. Some recent reports deal with noise, fumes and gases, and health effects.

The society also sponsors American National Standards Institute Committee Z49, whose publication, *Safety in Welding and Cutting*, is the authoritative standard in this field. It deals with the protection of workers from accidents, occupational diseases, and fires arising out of the installation, operation, and maintenance of electric and gas welding and cutting equipment.

Frequent articles on safety in welding appear in the official publication of the society, the *Welding Journal*.

Associated General Contractors of America, Inc.
1957 E Street, NW
Washington, DC 20006

The Associated General Contractors of America (AGC) is a construction trade association representing more than 32,500 firms, including 8,000 of America's leading general contracting companies. These companies are responsible for the employment of more than 3,500,000 employees. The member construction contractors perform more than 80% of America's contract construction of commercial buildings, highways, bridges, heavy industrial, and municipal utilities.

AGC has actively sought to improve safety performance in the construction workplace since its founding. In 1927, AGC first published its *Manual of Accident Prevention in Construction* and revised it in 1990.

Throughout the years, AGC and its Safety and Health Committee have diligently sought to keep AGC's members on the cutting edge of safety. AGC now has eleven current safety programs and publications, all designed to enhance safety in the construction industry.
- *AGC's Guidelines for a Basic Safety Program*, 1990
- AGC's On Solid Ground—A Plan for Safety Excavation and Trenching-videocassette (VHS)
- "The Choice is Yours"—An AGC Guide to Personal Protection Equipment videocassette (VHS)
- Hazard Communication "The System That Works" Training videocassette (VHS) and instructor's manual
- *Hazard Communication Compliance Guide for Construction*
- *Construction Industry Material Safety Data Sheets, Vol. 1*
- *Tool Box Talks, Vol. 1*—Safety talks
- *Construction Standard for Excavations*
- *The Drug-Free Workplace Act Compliance Guide for Contractors*
- Drug-Free Workplace videocassette (VHS).

AGC publishes a monthly magazine on construction titled *Constructor*, newsletters, and a publication and service catalogue.

Association of American Railroads
American Railroads Building
50 F Street, NW
Washington, DC 20001
All Class I railroads (those with an annual revenue in excess of $92 million) are members of the AAR. The following divisions and committees of the association are concerned with safety:
Communication and Signal Section
Engineering Division
Mechanical Division
Medical Section
Operating Rules Committee
Police and Security Section
Safety Section
State Rail Programs Division.

The Safety Section holds annual meetings; it issues a monthly newsletter, produces posters, and publishes pamphlets on railroad safety.

The Bureau for the Safe Transportation of Explosives and Other Dangerous Articals is also located at the above address.

Associations Council of the National Association of Manufacturers

1776 F Street, NW
Washington, DC 20006
The NAM safety and health activities are carried on under the aegis of its Occupational Safety and Health Committee which has a dual function: (a) promoting sound health and safety policies and programs in industry; and (b) working with the federal government to assure that present regulation of health and safety practices in industry and proposals for new regulations and legislation are realistic from industry's viewpoint.

Bituminous Coal Operators' Association

303 World Center Building
918 16th Street, NW
Washington, DC 20006
The Bituminous Coal Operators' Association has a Safety Department to furnish health and accident services to its members.

Chemical Manufacturers Association, Inc.

2501 M Street, NW
Washington, DC 20037
Formerly known as the Manufacturing Chemists Association, Inc., one of CMA's most important services is dissemination of information on the handling, transportation, and use of chemicals.

The association supports a Health and Safety Committee composed of medical directors, safety directors, and other health-related managers selected from its member companies. This committee meets monthly, supervising the activities of numerous specialized task groups and developing chemical safety and health programs and information for use by member companies, state and federal agencies, other organizations, and the public. The committee holds two open meetings each year on a broad range of topics, with frequent specialized seminars being sponsored by task groups. The committee is also charged with the development and management of the Chemical Awareness and Emergency Response (CAER) program.

Other committees of the association that include safety in their program are the Environmental Management Committee, Engineering Advisory Committee, and the Distribution Committee.

Compressed Gas Association, Inc.

1235 Jefferson Davis Highway
Arlington, VA 22202
The Compressed Gas Association is dedicated to the development and promotion of technical and safety standards and safe practices in the compressed gas industry in the United States, Canada, and other countries worldwide. More than 200 member companies work together through the

committee system to create technical specifications, safety standards, and training and educational materials; to cooperate with governmental agencies in formulating responsible regulations and standards; and to promote compliance with these regulations and standards in the workplace.

The member companies represent all facets of the industry; manufacturers, distributors, suppliers, and transporters of gases, cryogenic liquids, and related products, encompassing industrial, medical, and specialty gases in compressed or liquefied form, and a range of gas-handling equipment.

The CGA publishes more than 100 technical standards, many of which are formally recognized by U.S. government agencies, as well as safety bulletins, audiovisual safety training programs, a monthly newsletter, *Compressions,* and the *Handbook of Compressed Gases.* This handbook contains complete descriptions of the widely used industrial gases and sets forth the recognized safe methods for handling, storing, and transporting them. In addition, safety and regulatory alerts and position statements are produced when warranted. A complete catalog is available on request.

The CGA and its Canadian Division hold annual meetings each year in January and September, respectively.

Edison Electric Institute

701 Pennsylvania Avenue, NW
Washington, DC 20004
The Edison Electric Institute is the association of the nation's investor-owned electric utilities. Its members serve 99.6% of the customers serviced by the investor-owned segment of the industry.

The Safety and Industrial Health Committee is dedicated to improving working conditions through development of safe work practices. The Committee meets twice a year and compiles reports and publications on subjects of interest to safety professionals in the utility industry. Also prepared by the committee are videotape cassettes and sound-slide films.

Graphic Arts Technical Foundation

4615 Forbes Avenue
Pittsburgh, PA 15213
This is a member-supported, nonprofit, scientific, technical, and educational organization serving the international graphic communications industries since 1924. A publications and services catalog is available on all aspects of the graphic arts industry. Products include a 20-minute video titled "understanding the OSHA Hazard Communication Standard." GATF also publishes a semi-monthly technical magazine *GATFWORLD.* Sections include the Environmental Alert and Health and Safety News.

Health Physics Society

1340 Old Chain Bridge Road
McLean, VA 22101
Organized in 1955 and incorporated in 1961, the HPS has as its objectives: (1) to aid and advance health physics research and applied activities, (2) to encourage dissemination

of information between individuals in this and related fields, (3) to improve public understanding of the problems and needs in radiation protection, (4) to initiate and develop programs for training of health physicists, and (5) to promote the health physics profession. *Health Physics* is the official journal of the society.

Human Factors Society

P.O. Box 1369
Santa Monica, CA 90406

HFS is a society of psychologists, engineers, physiologists, and other related scientists who are concerned with the use of human factors in the development of systems and devices of all kinds. *Human Factors* is the official journal.

Illuminating Engineering Society of North America

345 East 47th Street
New York, NY 10017

The society is the scientific and engineering stimulus in the field of lighting. The work of the Industrial Lighting Committee and its numerous subcommittees for various specific industries should be of particular interest to industrial safety professionals. Many other projects, such as street and highway, aviation, and office lighting, may also be of interest.

Through the society, safety personnel can obtain reference material on all phases of lighting, including authoritative treatises on nomenclature, testing, and measurement procedures.

The society publishes a monthly magazine, *Lighting Design and Application;* a quarterly, *Journal of the Illuminating Engineering Society;* and the *IES Lighting Handbook*, a reference guide. In addition, the society publishes approximately 50 other publications on specific lighting areas such as mining, roadway, office, and emergency lighting.

Industrial Safety Equipment Association, Inc.

1901 North Moore Street
Arlington, VA 22209

The Industrial Safety Equipment Association (ISEA) is a nonprofit organization of manufacturers of personal protective products for industrial environments. Since its inception in 1934, ISEA has been dedicated to the safety of workers who rely on protective equipment, and to the welfare of the safety equipment industry. Member companies are located in the United States and abroad.

A results-oriented association, ISEA is primarily dedicated to fostering public interest in safety and encouraging the development and use of proper equipment to deal with industrial hazards. The organization does this primarily in three ways:

- through its member manufacturers who participate in developing consensus standards for product performance and use
- by representing the safety equipment industry before governmental agencies

- by collecting and disseminating information to the general public.

One of ISEA's major thrusts is helping to create standards that assure the high performance of protective equipment. In addition, ISEA continually evaluates existing standards to ensure industrial environments are as safe as they can feasibly be. ISEA provides manufacturer input in standards development through official representation to such organizations as the American National Standards Institute, National Fire Protection Association, National Society for the Prevention of Blindness, and the American Society for Testing and Materials.

ISEA publishes, and makes available free of charge, a "Buyers Guide" which lists manufacturers' names, addresses and telephone numbers and is divided according to product category.

Institute of Makers of Explosives

1120 19th Street, NW
Washington, DC 20036

The Institute is the safety association of the commercial explosives industry in the United States and Canada. Founded in 1913, IME is a nonprofit, incorporated association whose primary concern is safety and its application to the manufacture, transportation, storage, handling, and use of commercial explosive materials used in blasting and other essential operations. The member companies of IME produce over 85% of the commercial explosive materials consumed annually in the United States or some four billion pounds.

Safety library publications available from IME are:
Construction Guide for Storage Magazines
American Table of Distances
Suggested Code of Regulations
Warnings and Instructions for Consumers in Transporting, Storage, Handling, and Using Explosive Materials
Glossary of Commercial Explosive Industry Terms
Handbook for the Transportation and Distribution of Explosive Materials
Safety in the Transportation and Distribution of Explosive Materials
Safety Guide for the Prevention of Radio Frequency Radiation Hazards in the Use of Electric Blasting Caps
Destruction of Commercial Explosive Materials (A statement of policy—not a "how to" publication)
Recommendations for the Safe Transportation of Detonators in the Same Vehicle with Certain Other Explosive Materials
Trade Name Loading Guide for IME 22 Container

International Association of Drilling Contractors

P.O. Box 4287
Houston, TX 77210

This association works to improve oil well drilling contracting operations as a whole and to increase the value of oil well drilling as an integral part of the petroleum industry.

The association holds an annual safety clinic and has standing and special safety committees of contractor representatives to study current problems.

Safety meetings for tool pushers, drillers, crew members, and safety directors are sponsored by the association; they are conducted throughout the country in locations where these people normally reside.

A Supervisory Accident Prevention Training Program has been developed to instruct drillers and tool pushers in how to establish and maintain effective accident prevention programs. Six to eight professional safety instructors personally conduct these programs anywhere in the world where 18 to 25 people wish to enroll. Many other schools of either two-day or five-day duration are available through the association.

Safety award certificates, cards, safety hat decals, and plaques are given to member personnel and rigs that have completed one or more years without a disabling injury.

The group has produced safety manuals for the industry, inspection reports, color codes, safety signs, studies on protective clothing, and other publications. They have produced color films, film strips, and slides on specific drilling rig safety practices. The association also produces safety posters keyed to the hazards of the drilling industry.

International Association of Refrigerated Warehouses

7315 Wisconsin Avenue
Bethesda, MD 20814

The association's Safety Committee conducts a program specifically aimed at reducing accidents and injuries in refrigerated warehouses. The program includes periodic industry surveys to determine types of injuries being experienced, their causes, frequency and severity, safety bulletins, awards, and information on how to establish and operate a safety program. Members are encouraged to submit problems to the Safety Committee for study and suggested solutions.

Laser Institute of America

12424 Research Parkway, Suite 130
Orlando, FL 32826

The Laser Institute of America is a nonprofit membership society devoted to education and the advancement, promotion and safe application of laser technology. LIA is secretariat and publicity of ANSI 2136. Safe use of laser standard. The association conducts laser safety training courses and publishes a catalog of materials. The Journal of Laser Applications is published quarterly.

National Association of Manufacturers

1331 Pennsylvania Avenue, NW, Suite 1500N
Washington, DC 20004

The NAM safety and health activities are carried on under the aegis of its Risk Management Committee which has a dual function: (a) promoting sound health and safety policies and programs in industry; and (b) working with the federal government to assure that present regulation of health and safety practices in industry and proposals for new regulations and legislation are realistic from industry's viewpoint.

Metal Casting Society

455 State Street
Des Plaines, IL 60016

The Metal Casting Society is a trade association that represents gray, ferrous and nonferrous foundries in the United States and Canada. Founded in 1975 through a merger of the Gray and Ductile Iron Founders' Society and the Malleable Founders' Society, it is a nonprofit, voluntary membership organization with administrative headquarters in Des Plaines, IL. It is governed by an elected board of directors which is assisted by eleven standing committees in the formulation of programs and policy to promote the progress of its members and the industry.

A booklet developed for the Cast Metals Federation defines safe working conditions in ferrous foundries. Booklets entitled "How You Can Work Safely" are available for foundry employees in either English or Spanish language editions.

National Constructors Association

1101 15th Street, NW
Washington, DC 20005

NCA is made up of more than 50 national and regional design construction firms that build large industrial facilities for oil, steel, power, and chemicals. Its Safety and Health Committee carries out many programs to enhance the physical welfare of employees and the public. Other activities include labor and employee relations, government and international affairs, and taxes, insurance, and legal matters.

National Health Council, Inc.

350 Fifth Avenue, Room 1118
New York, NY 10018

The National Health Council, founded seven decades ago, is a private nonprofit, membership association of the nation's most respected and influential national health organizations. Its members include national voluntary health agencies, professional health associations, and federal government agencies with a strong interest in health issues.

Among a wide variety of programs, the Council produces targeted health publications, conducts ongoing Washington-based briefings and seminars on current health issues, hosts the National Health Forum, and, because of its diverse membership, provides a unique opportunity for dialogue and networking between private and public interests in the health field.

National Propane Gas Association

1600 Eisenhower Lane
Lisle, IL 60532

Founded in 1931, the association is a nonprofit, cooperative group of producers and distributors of liquefied petroleum gas (LP-gas), manufacturers of LP-gas equipment, and manufacturers and marketers of LP-gas appliances. NPGA promotes technical information and industry standards in its special field.

Its Safety Committee develops and maintains educational programs to train the public and industry in the safe han-

dling and use of LP-gas and in safe practices for the installation and maintenance of equipment and appliances.

An Educational Committee working closely with the Safety Committee arranges training schools and conferences for dealers and distributors.

The association distributes informational, technical, and legislative bulletins and publishes a weekly newsletter for members. It holds an annual meeting and sectional meetings with a definite portion of each program devoted to safety.

The Association publishes a *Safety Handbook,* training guides, audiovisual training programs, and distributes consumer education leaflets for members to help educate its customers.

National Petroleum Refiners Association

1899 L Street, NW, Suite 1000
Washington, DC 20036
The NPRA, as a national trade association of petroleum refiners and petrochemical manufacturers, gathers and disseminates industry information and statistics and provides an effective channel of communications among its member and with other associations, government, and the public.

Among the several technical meetings that are sponsored by NPRA are the National Safety Conference, which is a two-day session focusing on current safety issues. The trade group also holds one-day Fire and Accident Prevention Group meeting in each of five geographic areas. These meetings are conducted at different refinery locations and promote the exchange of information and experiences pertaining to fire and accident prevention in refining and petrochemical operations.

The association also prepares and distributes an annual summary of industry statistics dealing with occupational injuries and illnesses. In addition, the NPRA sponsors a comprehensive safety awards program in an effort to promote safety in plant operations.

National Restaurant Association

1200 17th Street, NW
Washington, DC 20036
The National Restaurant Association, through its Technical Services, Public Health, and Safety Department, carries on a program to reduce accidents and hazards that affect the safety of food service employees and patrons.

A major association activity is the preparation and distribution of educational materials to the membership. These materials include self-inspection guidelines on general safety concerns, OSHA requirements, and fire protection, as well as posters and audiovisual programs. These materials are also available for purchase through the association's Information Services Department.

The association conducts research to substantiate industry positions on DOE regulations and on OSHA.

Safety-related information appears in the association's *Washington Weekly* report and monthly magazine, *Restaurants USA.*

The Educational Foundation of the National Restaurant Association develops and markets educational and certification programs to industry members. The Educa-

tional Foundation is located at 20 North Wacker Drive, Chicago, IL 60606.

National Rural Electric Cooperative Association

1800 Massachusetts Avenue, NW
Washington, DC 20036
The association, through its Retirement Safety and Insurance Department, promotes loss control among its members. Following is a brief review of its safety programs:

Rural Electric Safety Accreditation Program. The association administers this program, which it helped to form in 1965. The Program appraises the loss control activities of its members through field audits and review of documentation relating to loss control management.

Rural Electric Accident Control Toady (REACT). The association offers a variety of safety literature for its members to make available to their consumers. The material addresses specific exposures characteristics of electric utilities.

Loss Control Resources Center. This program provides a means for members to exchange safety ideas and policies.

Loss Control Fund. Funds are provided to selected state associations to assist with their loss control activities based on their member's participation in the association sponsored property and casualty insurance program.

Publications. Loss control articles are regularly published in the association's *Rural Electrification* and *Management Quarterly* magazines.

National Sanitation Foundation

3475 Plymouth Road
Ann Arbor, MI 48106
The mission of NSF is to provide clients and the general public with objective, high quality, timely, third-party services at acceptable costs. Services include development of consensus standards, voluntary product testing and certification with policies and practices that protect the integrity of registered Marks, education and training, and research and demonstration, all relating to public health and the environmental sciences.

Listings:
- Food Service Equipment
- Plastics Piping Components
- Swimming Pool Equipment
- Drinking Water Treatment Systems
- Wastewater Treatment Devices
- Flexible Membrane Liners
- Biohazard Cabinetry
Special Categories of Equipment and Products:
- Registries
- Bottled Water
- Packaged Ice
- Drinking Water Laboratory Accreditation
- Sanitizers/Disinfectants
- Standards and Criteria
- Technical Information, Educational Materials, and Reports
- Facts About NSF
- Certification Services
- Assessment Services

National Soft Drink Association

1101 Sixteenth Street, NW
Washington, DC 20036
NSDA's Committee periodically prepares information bulletins concerning OSHA regulations and citations, educational safety procedures, and training films.

New York Shipping Association, Inc.

80 Broad Street
New York, NY 10004
New York Shipping Association is composed of American and foreign flag ocean carriers and contracting stevedores, marine terminal operators, and other employers of waterfront labor within the bi-state Port of New York and New Jersey.

Safety by NYSA is supervised by a director who is appointed by the association president. The safety director maintains contact with federal, state, and other agencies involved with industrial safety and health regulations. In this regard, the safety director disseminates relevant data to the member companies of NYSA to assist them in reducing accidents on piers and at marine facilities.

Further, the director coordinates industry activity and maintains liaison with the various stevedoring and marine terminal companies who operate their own company safety and health programs.

Portland Cement Association

5420 Old Orchard Road
Skokie, IL 60077
PCA, devoted to research, educational, and promotional activities to extend and improve the use of portland cement, is supported by more than 60 U.S., Canadian, and Mexican member companies that operate more than 130 cement manufacturing plants. Occupational safety and health have been considered important by the association since its formation in 1916.

The Washington Affairs Office works closely with an occupational safety and health committee of member company representatives to provide activities, services, and materials that are responsive to the needs of those companies. Knowledge and experience are shared through meetings and conferences.

PCA expresses the views and opinions of its member companies to governmental organizations regarding proposed legislation and regulations and other issues affecting the cement industry.

An injury/illness reporting program enables the association to accumulate data and identify significant causal and circumstantial factors. This information is used to define the nature and extent of cement industry injury/illness experience and is disseminated to member companies.

Power Tool Institute, Inc.

1095 Oceanview Drive
P.O. Box 818
Yachats, OR 97498-0818
Purposes of the Institute are:

- to promote the common business interests of the power tool industry
- to represent the industry before government
- to educate the public as to the usefulness and importance of power tools
- to encourage high standards of safety and quality control in the manufacture of power tools
- to prepare and distribute information about safe use of power tools.

Video programs:
"Safety Is Specific" Cassette
  35mm slide programs:
"Safety Is Specific" Industrial Portable Tools
"Safety Is Specific" Consumer Portable Tools
"Safety Is Specific" Stationary/Bench Top Tools
"Safety Is Specific" Lawn & Garden Tools
"Power Tool Safety, A Matter of Black & White"
"Safety Is Your Decision"
  Safety Literature:
"Safety Is Specific" illustrated brochure
On the Job Power Tool Safety Maintenance Checklist
4 page, cartooned Power Tool Safety brochure (8 ½ x 11)
"The Big Offenders" 4 page, 8 ½ x 11, for professional trades
"Safety Bank" plans for building cabinet
"Safety Is Your Decision" 16 pages, 6 x 8 ½ consumer brochure
"Safety A Matter of Black & White" 12 page pamphlet, 4 x 8 ¾
"Power Tool Safety—It's In Your Hands" 4 page brochure, 5 ½ x 8 ½
"Cordless Tools (Battery Operated)" 4 page brochure, 5½ x 8 ½

Printing Industries of America, Inc.

1730 North Lynn Street
Arlington, VA 22209
Printing Industries of America, an association of printers' organizations, actively sponsors the development of safety in the graphic arts through its affiliated local organizations and through its participation in the Graphic Arts Technical Foundation (described earlier).

Scaffolding, Shoring, and Forming Institute, Inc.

c/o Thomas Associates, Inc.
1230 Keith Building
Cleveland, OH 44115
The institute has a deep interest in safety; members try to do everything possible to improve this situation in the construction industry. Publications involving safety rules are:
Safety Rules:
Scaffolding Safety Guidelines
Suspended Powered Scaffolding Safety Rules
Flying Deck Form Safety Rules
Steel Frame Shoring Safety Rules
Horizontal Shoring Beam Safety Rules
Single Post Shore Safety Rules
Rolling Shore Bracket Safety Rules

Safety Procedures:
Guide to Scaffolding Erection and Dismantling Procedures
Recommended Steel Frame Shoring Erection Procedures
Guide to Horizontal Shoring Beam Erection Procedures for Stationary Systems
Recommended Safety Requirements for Shoring Concrete Formwork
Guide to Safety Procedures for Vertical Concrete Formwork
Standard Testing Procedures for Vertical Concrete Formwork
Safety Slide Presentations:
Forming Safety Do's and Don'ts
Scaffolding Safety Do's and Don'ts
Shoring Safety Do's and Don'ts
Suspended Powered Scaffolding Safety Do's and Don'ts

Steel Plate Fabricators Association, Inc.

2400 South Downing Avenue
Westchester, IL 60153
The association has an active safety committee that prepares publications on safety for member companies and their employees. Some of these are:
Supervisor's Accident Prevention Manual for Field Erection and Construction
Basic Safety Rules for Fabricating Shops
Basic Safety Rules for Field Erection and Construction.

The association conducts a monthly steel plate fabricators safety contest.

EMERGENCY INFORMATION BY PHONE

Although it is best to have a prepared plan of action that anticipates the emergency, sometimes this is not possible. The following source has 24-hour-a-day phone service.

CHEMTREC

Emergency information about hazardous chemicals involved in transportation accidents can be obtained 24 hours a day. It is the Chemical Transportation Emergency Center (CHEMTREC), and it can be reached by a nationwide telephone number—(800) 424-9300. The Area Code 800 WATS line permits the caller to dial the station-to-station number without charge. CHEMTREC will provide the caller with response/action information for the product or products and tell what to do in case of spills, leaks, fires, and exposures. This informs the caller of the hazards, if any, and provides sufficient information to take immediate first steps in controlling the emergency. CHEMTREC is strictly an emergency operation provided for fire, police, and other emergency services. It is not a source of general chemical information of a nonemergency nature.

ON-LINE DATABASES FOR COMPUTER ACCESS

There are more than 4,500 on-line databases available for access by users from remote computer terminals and microcomputers. These databases contain information on a wide range of subject areas that can be used to meet both general and specific needs. To help users identify databases of particular interest, the latest edition of the directories listed below can prove useful:

DataBase Directory

Knowledge Industry Publications, Inc.
701 Westchester Avenue
White Plains, NY 10604

Directory of On-line Databases

Cuadra/Elsevier
52 Vanderbilt Avenue
New York, NY 10017

Federal DataBase Finder

Information USA, Inc.
4701 Willard Avenue
Chevy Chase, MD 20815

For convenience, let's divide the databases that are available into two categories: governmental and privately held.

Governmental Information Sources

The U.S. government collects volumes of information on almost every subject imaginable. Fortunately under our Constitution, our government is, technically, subject to the control of the citizens of our country. As a result, more and more agencies are computerizing their files in order to permit public access. Frequently, one can obtain searches and cited documents free of charge; also, many of the staff are experts who can guide you as well as answer specific questions, refer you to other sources, and provide printed materials.

Some of the major governmental sources that are of interest to safety and health people are described next. Without a question, the governmental agency that provides the most databases that are of interest to safety and health professionals is the National Library of Medicine, Bethesda, MD, which has the following databases. If you want to access one or more of them, call the information desk at (800) 638-8480.

MEDLARS on-line network

The most useful databases for the safety professional who needs to know more about industrial hygiene and health subjects is MEDLARS On-line Network, which is an on-line network of approximately 20 bibliographical databases covering worldwide literature in the health sciences. Those of special value are described next. For more information, contact:
National Library of Medicine
MEDLARS Management Section
8600 Rockville Pike
Bethesda, MD 20209

AVLINE. AVLINE (Audio Visuals on-Line) contains references to 18,000 audiovisual instructional packages in the health sciences. All of these materials are professionally re-

viewed for technical quality, currency, accuracy of subject content, and educational design. AVLINE enables teachers, students, librarians, researchers, practitioners, and other health science professionals to retrieve citations which aid in evaluating audiovisual materials with maximum specificity.

CANCERLIT. CANCERLIT (Cancer Literature) is the National Cancer Institute's on-line data base dealing with all aspects of cancer. The database contains information on more than 690,000 references dealing with cancer; the sources of this database include more than 3,500 U.S. and foreign journals, as well as books and other reference sources.

CHEMLINE. CHEMLINE (Chemical Dictionary on-Line) is the National Library of Medicine's on-line, interactive chemical dictionary file created by the Specialized Information Services in collaboration with Chemical Abstracts Service (CAS). It provides a mechanism whereby more than 900,000 chemical substance names, representing nearly 100,000 unique substances, can be searched and retrieved on-line. This file contains CAS Registry Numbers; molecular formulas; preferred chemical index nomenclature; generic and trivial names derived from the CAS Registry Nomenclature File; and a locator designation which points to other files in the NLM system containing information on that particular chemical substance. For a limited number of records in the file, there are Medical Subject Headings (MeSH) terms and Wiswesser Line Notations (WLN). In addition, where applicable, each Registry Number record in CHEMLINE contains ring information including—number of rings within a ring system, ring sizes, ring elemental composition, and component line formulas.

MEDLINE. MEDLINE (Medical Literature Analysis and Retrieval System on-Line) is a database maintained by the National Library of Medicine; it contains references to over a million citations from 3,500 biomedical journals. It is designed to help health professionals find out easily and quickly what has been published recently on any specific biomedical subject. Medline is accessed from a variety of typewriter-like terminals connected to computers in Bethesda, MD, and Albany, NY, via ordinary telephone lines and nationwide communications networks.

TOXLINE. TOXLINE (Toxicology Information on-Line) is the National Library of Medicine's extensive collection of computerized toxicology information containing references to published human and animal toxicity studies, effects of environmental chemicals and pollutants, adverse drug reactions, and analytical methodology.

NIOSH's on-line network

NIOSH's on-line computerized research and reference database, called NIOSHTIC, contains bibliographic abstracts of approximately 162,000 documents in many subject areas, such as toxicology, occupational medicine, industrial hygiene, and personal protective equipment. Additional details can be obtained from:
National Institute for Occupational Safety and Health
Technical Information Branch
4646 Columbia Parkway

Cincinnati, OH 45226

RTECS

As an on-line database, RTECS is available as a real time, interactive computer database that permits the user to search the RTECS (Registry of Toxic Effects of Chemical Substances) for special data or subsets of data and to compile RTECS subfiles tailored to their particular needs. Updates to these systems are provided on a quarterly basis.

The RTECS is available online as a computer tape through the National Library of Medicine and from National Technical Information Service (NTIS), 5285 Port Royal Road, Springfield, VA 22161.

PRIVATELY OWNED INFORMATION SOURCES

Two nongovernmental organizations make their databases available to their members.

HAZARDLINE

The HAZARDLINE database provides regulatory, handling, identification, and emergency care information for over 78,000 hazardous substances. The information is gathered from regulations issued from state and federal agencies, court decisions, books, and journal articles, in order to assemble a comprehensive record for each substance.

For more information contact:
Occupational Health Services, Inc.
450 7th Avenue, Suite 2407
New York, NY 10123

SRIS

The National Safety Council's Safety Research Information Service (SRIS) includes more than 20,000 basic research documents and abstracts. SRIS is part of the Council's library database which includes more than 90,000 books and documents regarding all aspects of safety and health.

For more information, contact:
Library
National Safety Council
444 North Michigan Avenue
Chicago, IL 60611

U.S. GOVERNMENT AGENCIES

An overwhelming amount of safety information is available from the federal government concerning all aspects of safety and health, environmental problems, pollution, statistical data, and other industry problems.

Because of the constant change in government agency activities and frequent reorganizations, it is recommended that the reader consult the United States Government Organization Manual, published by the Government Printing Office,

Washington, DC 20402. It can be found in most libraries.

Information on the Occupational Safety and Health Administration, the Mine Safety and Health Administration, the National Institute for Occupational Safety and Health, the Environmental Protection Agency, the Public Health Service, and the Environmental Protection Agency can be found elsewhere in this Manual.

DEPARTMENTS AND BUREAUS IN THE STATES AND POSSESSIONS

It is important that safety professionals have a good working knowledge of the state agencies responsible for the enforcement of safety and health laws. They should, therefore, contact the proper groups in their state's labor departments or other pertinent agencies and find out how the various boards, divisions, and services function.

Some difficulty can be avoided if safety professionals are familiar with the labor legislation and safety and health codes under which they work. Codes and laws vary widely in the different states and provinces, and those persons who have safety jurisdiction in plants in a number of places must understand these differences.

In many cases, the standards set up by the code may serve only as a minimum, and safety professionals will want to compare them with American National Standards or other regulations to establish more rigid rules for their own plants. They should also know the jurisdiction rights of the factory inspectors, so that they can better understand the job inspectors have to do and how they can help them in the performance of their duties.

Labor Offices

Labor offices in the several states perform many functions, generally depending on the number and kind of labor problems.

A listing of state and provincial labor offices and the title of the chief executive of each agency or subdivision to whom inquiries should be addressed is given in the U.S. Department of Labor Bulletin "OSHA Onsite Consultation Project Directory." The bulletin, revised periodically, is available from the Occupational Safety and Health Administration, Washington, DC 20210.

For the convenience of the reader, the addresses that were published on the June 1986 list are in this chapter; also see list of regional offices in Chapter 2, Regulation History and Compliance.

Health and Hygiene Services

Departments or boards of health and industrial hygiene services are integral parts of the organization of each of the states, the District of Columbia, and the autonomous territories of the United States.

It is to these organizations that safety professionals must look for their state's specific standards and recommendations on such points as occupational health, food and health engineering, disease control, water pollution, and other fac-

ets of the overall field of industrial hygiene.

Industrial hygiene units usually function full time or, in several states, on a limited basis. In addition to the units that operate under state health departments, a number of other industrial hygiene units are run by municipalities or other local authorities.

In addition to direct industrial hygiene services, these state units are able to bring to industry a more or less complete health program by integrating their work with that of other divisions in the state government, such as sanitation and infectious disease control.

OSHA On-site Consultation Program

State and local programs coordinate their efforts with:
Directorate of Federal-State Operations
Office of Consultation Programs
Room N3476
200 Constitution Avenue, NW
Washington, DC 20210
Phone: (202) 523-8902

They also cooperate with medical societies and nurses' associations. (See descriptions earlier in this chapter.)

The names and addresses of such state, commonwealth, or territorial agencies with which the safety professional may need to communicate are as follows (as of December 1990):

State: Office/Address

Alabama
7(c)(1) Onsite Consultation Program
425 Marth Parham West
P.O. Box 870388
Tuscaloosa, ALabama 35487
(205) 348-7136

Alaska
Division of Occupational Safety and Health
Alaska Department of Labor
3301 Eagle Street, Suite 303
Pouch 7-022
Anchorage, Alaska 99510
(907) 264-2599

Arizona
Consultation and Training
Division of Occupational Safety & Health
Industrial Commission of Arizona
800 West Washington
Phoenix, Arizona 85007
(602) 255-5795

Arkansas
OSHA Consultation
Arkansas Department of Labor
10421 West Markham
Little Rock, Arkansas 72202
(501) 375-8442

California
CAL/OSHA Consultation Service
395 Oyster Point Boulevard
Wing C/3rd. Floor

South San Francisco, California 94089
(Emoryville Office)
(415) 557-2870

Colorado
Occupational Safety & Health Section
Institute of Rural Environmental Health
Colorado State University
110 Veterinary Science Building
Fort Collins, Colorado 80523
(303) 491-6151

Connecticut
Division of Occupational Safety & Health
Connecticut Department of Labor
200 Folly Brook Boulevard
Wethersfield, Connecticut 06109
(203) 566-4550

Delaware
Occupational Safety and Health
Division of Industrial Affairs
Delaware Department of Labor
820 North French Street, 6th Floor
Wilmington, Delaware 19801
(302) 571-3908

District of Columbia
Office of Occupational Safety & Health
District of Columbia Department of Employment Services
950 Upshur Street, NW
Washington, DC 20011
(202) 576-6339

Florida
7(c)(1) Onsite Consultation Program
Bureau of Industrial Safety & Health
Florida Department of Labor & Employment Security
Forrest Building, Suite 349
2728 Center View Drive
Tallahassee, Florida 32399-0663
(904) 488-3044

Georgia
7(c)(1) Onsite Consultation Program
Georgia Institute of Technology
O'Keefe Bldg.—Room 23
Atlanta, Georgia 30332
(404) 894-3806

Guam
OSHA Onsite Consultation
Government of Guam
3rd. Floor International Trade Center
P.O. Box 9970
Tamuning, Guam 96911
9-011 (671) 646-9446

Hawaii
Division of Occupational Safety & Health
830 Punchbowl Street
Honolulu, Hawaii 96813
(808) 548-2511

Idaho
Safety & Health Consultation Program

Boise State University
Department of Community & Environmental Health
Boise, Idaho 83725
(208) 385-3283

Illinois
Illinois Onsite Consultation
Industrial Services Division
Department of Commerce & Community Affairs
State of Illinois Center
100 West Randolph Street—Suite 3-400
Chicago, Illinois 60601
(312) 917-2337

Indiana
Division of Labor
Bureau of Safety, Education & Training
1013 State Office Building
Indianapolis, Indiana 46204-2287
(317) 232-2688

Iowa
7(c)(1) Consultation Program
Iowa Bureau of Labor
1000 East Grand Avenue
Des Moines, Iowa 50319
(515) 281-5352

Kansas
7(c)(1) Consultation Program
Kansas Department of Human Resources
512 West 6th Street
Topeka, Kansas 66603
(913) 296-4386

Kentucky
Consultation and Training
Kentucky OSH Program
Kentucky Labor Cabinet
U.S. Highway 127 South, Bay 4
Frankfort, Kentucky 40601
(502) 564-6895

Louisiana
7(c)(1) Consultation Program
Louisiana Department of Employment and Training
1001 North 23rd Street
P.O. Box 94094
Baton Rouge, Louisiana 70804-9094

Maine
Division of Industrial Safety
Maine Department of Labor
State Home Station 82
Hallowell Annex
Augusta, Maine 04333
(207) 289-2591

Maryland
7(c)(1) Consultation Services
Division of Labor & Industry
501 Saint Paul Place, Floor 3
Baltimore, Maryland 21202
(301) 659-4218

Massachusetts

7(c)(1) Consultation Program
Division of Industrial Safety
Massachusetts Department of Labor and Industries
100 Cambridge Street
Boston, Massachusetts 02202
(617) 727-3463

Michigan (Health)
Michigan Department of Public Health
Division of Occupational Health
3423 North Logan Street
P.O. Box 30195
Lansing, Michigan 48909
(517) 335-8250

Michigan (Safety)
Michigan Department of Labor
Bureau of Safety and Regulation
7150 Harris Drive
P.O. Box 30015
Lansing, Michigan 48909
(517) 322-1809

Minnesota
Department of Labor and Industry
Consultation Division
443 Lafayette Road
St. Paul, Minnesota 55155
(612) 297-2393

Mississippi
7(c)(1) On-site Consultation Program
Division of Occupational Safety & Health
Mississippi State Board of Health
305 West Lorenz Boulevard
Jackson, Mississippi 39219-1700
(601) 982-6315

Missouri
On-site Consultation Program
Division of Labor Standards
Department of Labor and Industrial Relations
3315 West Truman Boulevard
Jefferson City, Missouri 65102
(314) 751-3403

Montana
Department of Labor & Industry
Employment Relations Division
Safety Bureau
Arcade Building, 111 North Main
Helena, Montana 59604-8011
(406) 444-6401

Nebraska
Division of Safety Labor and Safety Standards
Nebraska Department of Labor
State Office Building
301 Centennial Mall, South
Lincoln, Nebraska 68509-5024
(402) 471-2239

Nevada
Training and Consultation
Division of Occupational Safety & Health

4600 Kietzke Lane, Building D-139
Reno, Nevada 89502
(702) 789-0546

New Hampshire
On-site Consultation Program
New Hampshire Department of Labor
19 Pillsbury Street
Concord, New Hampshire 03301
(603) 271-3170

New Jersey
Division of Workplace Standards
New Jersey Department of Labor
CN953
Trenton, New Jersey 08625-0953
(609) 292-2313

New Mexico
OSHA Consultation
Occupational Health & Safety Bureau
1190 St. Francis Drive—Room N2200
Santa Fe, New Mexico 87503
(505) 827-8949

New York
Division of Safety and Health
State Office Campus
Building 12, Room 457
Albany, New York 12240
(718) 797-7645

North Carolina
North Carolina Consultative Services
North Carolina Department of Labor
OSH Bureau of Consultative Services
413 North Salisbury Street
Raleigh, North Carolina 27603
(919) 733-4880

North Dakota
Division of Environmental Engineering
North Dakota State Department of Health
1200 Missouri Avenue, Room 304
Bismarck, North Dakota 58502-5520
(701) 224-2348

Ohio
Division of Onsite Consultation
Department of Industrial Relations
2323 West Fifth Avenue
Columbus, Ohio 43216
(800) 282-1425
(Toll-free in State)
(614) 466-7485

Oklahoma
OSHA Division
Oklahoma Department of Labor
4001 North Lincoln Boulevard
Oklahoma City, Oklahoma 73105-5212
(405) 235-0530 X240

Oregon
7(c)(1) Consultation Program
Department of Insurance & Finance/ ADP

Labor and Industries Building
Salem, Oregon 97310
(503) 378-2890

Pennsylvania
Indiana University of Pennsylvania
Safety Sciences Department
205 Uhler Hall
Indiana, Pennsylvania 15705
(800) 382-1241
(Toll-free in State)
(412) 357-2561/2396

Puerto Rico
Occupational Safety & Health Office
Puerto Rico Department of Labor and Human Resources
505 Munoz Rivera Avenue, 21st Floor
Hato Rey, Puerto Rico 00918
(809) 754-2134/2171

Rhode Island
Division of Occupational Health
Rhode Island Department of Health
206 Cannon Building
75 Davis Street
Providence, Rhode Island 02908
(401) 277-2438

South Carolina
7(c) (1) Onsite Consultation Program
Consultation and Monitoring
South Carolina Department of Labor
3600 Forest Drive
P.O. Box 11329
Columbia, South Carolina 29211
(803) 734-9599

South Dakota
S.T.A.T.E. Engineering Extension
On-site Technical Division
South Dakota State University
P.O. Box 2218
Brookings, South Dakota 57007
(605) 688-4101

Tennessee
OSHA Consultative Services
Tennessee Department of Labor
501 Union Building, 6th Floor
Nashville, Tennessee 37219
(615) 741-2793

Texas
Workers' Health & Safety Division
Texas Workers' Compensation Commission
200 East Riverside Drive
Austin, Texas 78704
(512) 458-7287

Utah
Utah Safety & Health Consultation Service
P.O. Box 510870160
East 300 South, 3rd. Floor
Salt Lake City, Utah 84151-0870
(801) 530-6868

Vermont
Division of Occupational Safety & Health
Vermont Department of Labor and Industry
118 State Street
Montpelier, Vermont 05602
(802) 828-2765

Virginia
Virginia Department of Labor and Industry
Voluntary Safety & Health
2201 West Broad Street
Richmond, Virginia 23220
(804) 786-5875

Virgin Islands
Division of Occupational Safety & Health
Virgin Islands Department of Labor
Lagoon Street
Frederiksted
Virgin Islands 00840
(809) 772-1315

Washington
Voluntary Services
Washington Department of Labor and Industries
1011 Plum Street, M/S HC-462
Olympia, Washington 98504
(206) 753-6500

West Virginia
West Virginia Department of Labor
State Capitol, Bldg. 3, Room 319
1800 East Washington Street
Charleston, West Virginia 25305
(304) 348-7890

Wisconsin (Health)
Division of Health
Wisconsin Department of Health and Human Services
1414 East Washington Avenue—Room 112
Madison, Wisconsin 53703
(608) 266-0417

Wisconsin (Safety)
Wisconsin Department of Industry, Labor and Human
  Relations
Bureau of Safety Inspections
1570 East Moreland Boulevard
Waukesha, Wisconsin 53186
(414) 521-5063

Wyoming
Occupational Health and Safety
State of Wyoming
122 West 25th, Herschler Building
2nd Floor, East
Cheyenne, Wyoming 82002
(307) 777-7786

Consultation Training Coordination
Occupational Safety and Health Administration
Training Institute
1555 Times Drive
Des Plaines, Illinois 60018
(312) 297-4810

CANADIAN DEPARTMENTS, ASSOCIATIONS, AND BOARDS

In all provinces of Canada there is a Workmen's Compensation Board or Commission. Some of these handle accident prevention directly. In other provinces there are provisions similar to Section 110 of the Quebec Workmen's Compensation Act, which stipulates "that industries included in any of the classes under Schedule I may form themselves into an Association for accident prevention and formulate rules for that purpose. Further, the Workmen's Compensation Commission, if satisfied that an Association so formed sufficiently represents the employers in the industries included in the class, may make a special grant toward the expense of any such Association."

It is under these provisions that the various safety associations were organized and are functioning. In some provinces, accident prevention is directly assumed by the board itself by establishing a safety department.

Furthermore, all provinces have legal safety requirements which are administered by the Department of Highways, Department of Labor, and the Department of Mines. These sources can be contacted by writing to the deputy minister of the department located in the capital of each province.

Governmental agencies

Federal: Director, Occupational Safety and Health Branch, Labour Canada, Ottawa, Ontario, Canada KIA OJ2.

This Branch is responsible for the implementation and administration of the Canada Labor Code Part IV and pursuant regulations which became effective January 1, 1968, and deals primarily with Occupational Safety and Health. It is also responsible for the development of Occupational Safety and Heath regulations, procedures, and standards for regulating all work places under federal jurisdiction.

Accident Prevention Associations

Canada Safety Council,
1765 Boulevard
St. Laurent, Ottawa, Ontario K1G 3V4

Provincial Associations

Alberta Safety Council
201-10526 Jasper Avenue
Edmonton T5J 1Z7
British Columbia Safety Council
8345 Winston Street
Burnaby, V5A 2H3
Nova Scotia Safety Council
5541 Russell Street
Halifax B3K 1X1
Quebec Safety League Inc.
6785 St. Jacques Ouest
Montreal H4B 1V3
Saskatchewan Safety Council
140 4th Avenue
Regina S4N 4Z4
Industrial Accident Prevention Association of Ontario

2 Bloor Street West
Toronto M4W 3N8
Included in this association are ten class associations:
Woodworkers Accident Prevention Association
Ceramics & Stone Accident Prevention Association
Metal Trades Accident Prevention Association
Chemical Industries Accident Prevention Association
Grain, Feed & Fertilizer Accident Prevention Association
Food Products Accident Prevention Association
Leather, Rubber & Tanners Accident Prevention Association
Textile & Allied Industries Accident Prevention Association
Printing Trades Accident Prevention Association
Ontario Retail Accident Prevention Association
Forest Products Accident Prevention Association
P.O. Box 270
North Bay, Ontario P1B 8H2
Ontario Pulp & Paper Makers Safety Association
91 Kelfield Street
Rexdale M9W 5A4
Ontario Safety League
21 Four Seasons Place
Etobicoke, Ontario M9B 6J8
Mines Accident Prevention Association
Box 1468
North Bay, Ontario P1B 8K6
Transportation Safety Association of Ontario (Inc.)
58-5 Whittle Road
Mississauga, Ontario L4Z 2J1
Construction Safety Association of Ontario
74 Victoria Street
Toronto M5C 2A5
Since 1929 the Construction Safety Association of Ontario (CSAO) has been dedicated to accident prevention. Unique in North America, CSAO provides health and safety information specifically designed for construction.

The association works with labor and management representatives who are directly responsible for working conditions and who can implement safety training programs.

CSAO maintains close liaison with trade associations, community colleges and various government ministries in generating a cooperative approach to construction safety standards, controls and legislation.

As part of this commitment, the Construction Safety Association has developed an international reputation for training and safety support material, from textbooks and data sheets to videos. Print productions include three state-of-the-art texts on cranes and rigging and *Construction Safety,* an extensively illustrated 16-page monthly magazine featuring valuable how-to information on a wide variety of health and safety matters.

Quebec Pulp and Paper Safety Association Inc.
1200 Avenue Germaindes-Pres, bur. 102
Ste. Foy, Quebec G1V 3M7
Quebec Logging Safety Association Inc.
1200 Avenue
Germaindes-Pres, bur. 102
Ste. Foy, Quebec G1R 5E4

INTERNATIONAL SAFETY ORGANIZATIONS

Inter-American Safety Council
(Consejo Interamericano de Seguridad)
33 Park Place
Englewood, NJ 07631

The Inter-American Safety Council was founded and incorporated in 1938, as a noncommercial, nonpolitical, and nonprofit educational association for the prevention of accidents. It is the Spanish and Portuguese language counterpart of the National Safety Council.

The Council is the first and only association of its kind rendering services to all industries and agencies in the Latin American countries and Spain. The objectives are to prevent accidents—to reduce the number and severity of accidents in every activity, both on the job and off the job. The services that the Council provides for its members are paid by membership dues and sales of the Council's monthly publications and other educational materials. All of its work is done from the headquarters in New Jersey.

Membership is open to all industries, organizations, institutions, or other groups with two or more employees, interested in accident prevention in Latin America and Spain. More than 3,200 plants or work locations in 22 countries are members or are using the materials and services of the Council. In addition, over 300 universities, technical schools, public libraries, and the like receive the monthly publications free of charge.

Among the services available to members are: monthly publications, annual contest, special awards, consultation, statistical service, reproduction and translation rights, and participation in the election of Council officers. In addition, the Council acts as a clearing house of accident prevention materials available in the United States. A catalog is available from the organization.

The Council's monthly publications include two magazines and safety posters and also a quarterly publication:
Noticias de Seguridad (Safety News)
El Supervisor (The Supervisor)
Safety posters in sizes 8 by 11 and 17 by 22 in.
Usted y su familia (You and Your Family)

In addition, the Council publishes translations of publications, films, safety slides, training programs, and other materials of the National Safety Council and other accident prevention organizations.

The Royal Society for the Prevention of Accidents

Cannon House, The Priory Queensway
Birmingham, England B4 6BS

Founded in 1916, RoSPA, Europe's largest safety organization works to prevent accidents at home, work, on the road, and at leisure. It produces safety education materials for schools and runs an occupational safety training center and extensive driver training programmes. It is frequently consulted by government departments and other organizations on technical matters.

Occupational safety and health training is offered worldwide. The RoSPA International Health and Safety Exhibition at the National Exhibition Center is the largest annual safety exhibition in Europe. There is an award scheme rewarding progressive and long term safety achievements by companies.

Three monthly journals are produced: *Occupational Safety and Health, Safety Representative*, and *RoSPA Bulletin*, cover all aspects of the workplace safety scene. The library is the largest safety library in Europe.

RoSPA is an independent charity under the patronage of HM The Queen.

International Association of Industrial Accident Boards and Commissions

1575 Aviation Center Parkway, Suite 519
Daytona Beach, FL 32114

The IAIABC founded in 1914, is a professional association of workers' compensation specialists from the private and public sector. Its purpose is the improvement of workers' compensation administration. The IAIABC active membership includes the public worker's compensation agencies in the United States, Canada, Puerto Rico, and Australia. The associate membership consists of private sector worker's compensation administrators, physicians, attorneys, labor representatives, medical and rehabilitation providers, insurers, employers, trade associations, and other groups and individuals involved with workers' compensation programs from the United States, Canada, Guam, South Africa, Australia, and Zimbabwe.

International Labor Organization

International Labor Office, CH1211
Geneva 22, Switzerland
1828 L Street, NW
Washington, DC 20036

The International Labor Organization (ILO), a specialized agency associated with the United Nations, was created by the Treaty of Versailles in 1919 as part of the League of Nations. Its purpose is to improve labor conditions, raise living standards, and promote economic and social stability as the foundation for lasting peace throughout the world. To this purpose, one of ILO's functions is "the protection of the worker against sickness, disease, and injury arising out of his employment."

The organization consists of about 140 member countries, including the United States (which joined in 1934). ILO functions through an annual conference of member states, a governing body, advisory committees, and a permanent office, the International Labor Office. ILO is distinctive from all other international agencies in that it is tripartite in character—that is, the conference, the governing body, and some of the committees are composed of representatives of governments, employers, and workers.

In the field of safety and health, the International Labor Office maintains a permanent international staff of medical doctors, engineers, and industrial hygienists. Assistance in specific fields is given by panels of consultants, drawn from all parts of the world to act in an advisory capacity and to discuss problems, draft regulations, or render help in emergencies.

The United States has several members on the panels and has been represented on all the temporary expert committees and special conferences.

The main tasks of the ILO in the field of occupational safety and health are:

- International instruments. These include conventions and recommendations, and also model safety codes and codes of practice. An *Encyclopedia of Occupational Health and Safety* has been prepared in English and French, to succeed *Occupation and Health,* which was published in 1930. This is designed to provide guidance to a wide range of people concerned with health, safety, and welfare at work. Although problems are reviewed from an international angle, special account is taken of the needs of developing countries.
- The compilation of technical studies.
- The publication of medical and technical studies.
- Direct assistance to governments, by furnishing experts, drafting regulations, supplying information, etc.
- Collaboration with other international organizations, the World Health Organization, and the International Organization for Standardization.
- Assistance to national safety and health organizations, research centers, employers' associations, trade unions, etc., in different countries.

In general, keeping in touch with the safety and health movement throughout the world and assisting the movement by all the means in its power.

Pan American Health Organization

Pan American Sanitary Bureau
525 23rd Street, NW
Washington, DC 20037

Originally established as the International Sanitary Bureau in 1902, the Pan American Health Organization serves as the regional office for the World Health Organization for the Americas. The purposes of the PAHO are to promote and coordinate the efforts of the countries of the Western Hemisphere to combat disease, lengthen life, and promote the physical and mental health of the people.

Programs encompass technical collaboration with governments in the field of public health, including such subjects as sanitary engineering and environmental sanitation, eradication or control of communicable diseases, and maternal and child health.

World Health Organization

Avenue Appia
1211 Geneva 27
Switzerland

The World Health Organization (WHO) is a specialized agency of the United Nations with primary responsibility for international health matters and public health. Through this organization, which was created in 1948, the health professions of its 166 member states exchange their knowledge and experience with the aim of making possible the attainment by all citizens of the world by the year 2000 of a level of health that will permit them to lead a socially and economically productive life.

Through technical cooperation with its member states, WHO promotes the development of comprehensive health services, the prevention and control of diseases, the improvement of environmental conditions, the development of health manpower, the coordination and development of biomedical and health services research, and the planning and implementation of health programs.

WHO also plays a major role in establishing international standards for biological substances, pesticides, and pharmaceuticals; formulating environmental health criteria; recommending international nonproprietary names for drugs; administering the International Health Regulations; revising the International Classification of Diseases, Injuries, and Causes of Death; and collecting and disseminating health statistical information.

Authoritative information on the various fields covered by WHO is to be found in its many scientific and technical publications. Of particular interest to health and safety professionals is the Environmental Health Criteria series, prepared by the International Programme on Chemical Safety, of which there are some 70 volumes so far. Each book in the series reviews all available information on a selected chemical, environmental pollutant, or method for testing toxicity and carcinogenicity, in order to provide guidance on prevention of health hazards and the setting of safe exposure limits. In 1987, WHO is launching a new series of Health and Safety Guides, based on the Criteria series and designed to enhance workers' awareness of precautions needed in the handling of individual potentially dangerous chemicals. Current titles in occupational health and safety include *Early Detection of Occupational Diseases* (1986), *Epidemiology of Occupational Health* (1986), *Recommended Health-Based Limits in Occupational Exposure to Selected Mineral Dusts* (Silica, Coal) (1986), and *Psychosocial Factors at Work and Their Relation to Health* (1987). Catalogs of new publications are available on request.

EDUCATIONAL INSTITUTIONS

Many colleges and universities offer formal courses in industrial safety. In a publication *Educational Opportunities in Occupational Safety and Health,* compiled by the American Society of Safety Engineers, accredited four-year colleges and universities are grouped by those that offer a degree program with concentration on industrial safety and those that offer one or more credit courses as an elective within engineering or education curricula. About one-half of the 1,200 four-year colleges and universities in the U.S. provided catalogues for a recent study in occupational safety and health offered by post-secondary educational institutions.

Four-year and two-year degree programs and credit courses are listed in the ASSE report. Courses in water safety, first aid and safety, firefighting, and driver or traffic education are not listed. The report shows approximately 200

four-year institutions offer one or more courses in the five main occupational categories. Write to the ASSE for a copy of this publication. (Address is listed under Professional Societies earlier in this Appendix.)

BIBLIOGRAPHY OF SAFETY AND RELATED PERIODICALS

Safety

National Safety Council
444 North Michigan Avenue
Chicago, IL 60611
See details of publications in descriptive listing earlier in this Appendix.

Accident Analysis and Prevention (quarterly)
Pergamon Press Inc.
Fairview Park
Elmsford, NJ 10523

Canadian Occupational Safety (bimonthly)
Royal Life Center
277 Lakeshore Road East
Oakville, Ontario L6J 6J3
Canada

Hazard Prevention (quarterly)
System Safety Society
5 Export Drive, Suite A
Sterling, VA 22170

Health and Safety at Work (monthly)
Drayton Bridge House
3 High Street
West Drayton, Middlesex UB7 7QT
England

Human Factors (bimonthly)
The Human Factors Society, Inc.
P.O. Box 1369
Santa Monica, CA 90406

Journal of Occupational Accidents (quarterly)
Elsevier Scientific Publishing Company
655 Avenue of the Americas
New York, NY 10010

Mine Safety and Health (bimonthly)
Department of Labor, Mine Safety and Health Administration
Superintendent of Documents
U.S. Government Printing Office
Washington, DC 20402

Nuclear Safety (bimonthly)
Superintendent of Documents
U.S. Government Printing Office
Washington, DC 20402

Occupational Hazards (monthly)
Penton Publishing Inc.
1100 Superior Avenue
Cleveland, OH 44114

Occupational Safety and Health (monthly)
Royal Society for the Prevention of Accidents

Cannon House, The Priory
Queensway
Birmingham B4 6BS
England

Professional Safety (monthly)
American Society of Safety Engineers
1800 East Oakton Street
Des Plaines, IL 60018

Protection (monthly)
Travelers Companies
One Tower Square
Hartford, CT 06183-1060

The Record, The Magazine of Property Conservation (bimonthly)
Factory Mutual System
1151 Boston-Providence Turnpike
Norwood, MA 02062

Industrial Hygiene and Medicine

Archives of Environmental Health (bimonthly)
Heldref Publications
4000 Albemarle Street, NW
Washington, DC 20016

Journal of the American Medical Association (weekly)
American Medical Association
535 North Dearborn Street
Chicago, IL 60610

American Industrial Hygiene Association Journal (monthly)
American Industrial Hygiene Association
345 White Pond Drive
P.O. Box 8390
Akron, OH 44320

American Journal of Nursing (monthly)
American Nurses Association
55 West 57th Street
New York, NY 10019

American Journal of Public Health (monthly)
American Public Health Association
1015 15th Street, NW
Washington, DC 20005

American Journal of Industrial Medicine (quarterly)
Wiley-Liss
41 East 11th Street
New York, NY 10003

Applied Occupational and Environmental Hygiene (monthly)
6500 Glenway Avenue, Bldg. D-7
Cincinnati, OH 45211-4438

Applied Ergonomics (quarterly)
Butterworth-Heinemann Ltd.
P.O. Box 63
Westbury House, Bury Street
Gilford, Surrey GU2 5BH
England

Occupational Health & Safety (13/year)
225 North New Road
Waco, TX 76710

British Journal of Industrial Medicine (monthly)

British Medical Association
Tavistock Square
London WC1H 9JR
England

CIS Abstracts (8 times a year)
International Occupational Safety and Health
Information Center (CIS)
International Labor Office
1211 Geneva 22
Switzerland

Industrial Hygiene Digest (monthly)
Industrial Health Foundation
34 Penn Circle West
Pittsburgh, PA 15206

Journal of Occupational Medicine (monthly)
Williams & Wilkins
428 Preston Street
Baltimore, MD 21202-3993

AAOHN Journal
American Association of Occupational Health Nurses
50 Lenox Pointe
Atlanta, GA 30324

Work and Stress (quarterly)
Taylor & francis Ltd.
4 John Street

London WC1N 2ET
England

Fire

Fire Command! (monthly)
Fire Journal (bimonthly)
Fire News (monthly)
Fire Technology (quarterly)
National Fire Protection Association
Batterymarch Park
Quincy, MA 02269

Fire Engineering (quarterly)
250 Fifth Avenue
New York, NY 10001

Fire Prevention (quarterly)
Fire Protection Association
140 Aldersgate Street
London EC1A 4HX
England

Fire Surveyor
Paramount Publishing Ltd.
17-21 Shenley Road
Borehanwood
Hertfordshire WD6 1RT
England

Appendix 2

Bibliography

The basic reference books listed in this section supplement the specific references appended at the end of each chapter in this *Manual*. The reference material cited in this Bibliography was selected to provide safety and health professionals with sources of information that are likely to prove most useful in coping with problems of worker health protection and hazard assessment. This compilation is not to be viewed as a comprehensive coverage of the abundant literature on this subject, nor is any endorsement implied. The reference books are listed according to the following outline:

A. General principles
B. Risk Assessment
C. Sampling methods
D. Toxicology
E. Medical
F. Dermatitis
G. Physical stresses
H. Ergonomics
I. Chemical
J. Pollution and hazardous waste
K. Control
L. Encyclopedias and handbooks
M. Safety management
N. Emergency
O. Safety training
P. Accident investigation and analysis
Q. Product safety
R. Fire
S. Loss control
T. OSHA
U. MSHA
V. Computer and applications

A. General principles

Allen RW, Ells MD, and Hart AW. *Industrial Hygiene.* Englewood Cliffs, NJ: Prentice-Hall, Inc., 1976.

American Conference of Governmental Industrial Hygienists (ACGIH). *Microcomputer Applications in Occupational Health and Safety.* Chelsea, MI: Lewis Publishers, Inc., 1986.

Annino R and Driver R. *Scientific and Engineering Applications with Personal Computers.* New York: John Wiley & Sons, Inc., 1986.

Box GER, Hunter WG, and Hunter JS. *Statistics for Experimenters.* New York: John Wiley & Sons, Inc., 1978.

Burgess, WA. *Recognition of Health Hazards in Industry.* New York: John Wiley & Sons, Inc., 1981.

Clayton, GD, and Clayton FE, eds. *Patty's Industrial Hygiene and Toxicology.* Vol. 1: *General Principles.* New York: John Wiley & Sons, Inc., 1978.

Cohen B, ed. *Human Aspects in Office Automation.* New York: Elsevier Science Publishing Co., Inc., 1984.

Computer Systems for Occupational Safety and Health Management. New York: Marcel Dekker, Inc., 1984.

Cralley LJ, and Cralley LV, eds. *Patty's Industrial Hygiene and Toxicology.* Vol. 3: *Theory and Rationale of Industrial Hygiene Practice*, 2nd ed. (in two parts: A and B), New York: John Wiley & Sons, Inc., 1985.

DeReamer R. *Modern Safety and Health Technology.* New York: John Wiley & Sons, Inc., 1980.

Esmen N, and Mehlman MA, eds. *Occupational and Industrial Hygiene: Concepts and Methods.* Princeton NJ: Princeton Scientific Publishers, Inc., 1984.

Johnson WG. *MORT Safety Assurance Systems.* New York: Marcel Dekker, Inc., 1980.

LaDou JL, ed. *Occupational Health and Safety.* Chicago: National Safety Council, 1992.

Lowry GG, and Lowry RC. *Handbook of Hazard Communication and OSHA Requirements.* Chelsea, MI: Lewis Publishers, Inc., 1985.

Nothstein GZ, et al. *The Law of Occupational Safety and Health.* New York: Free Press, Division of MacMillan Publishing Company, 1981.

O'Donnel MP and Ainsworth TH, eds. *Health Promotion in the Workplace.* Somerset, NJ: John Wiley & Sons, Inc., 1984.

Otway H and Peltu M. *Regulating Industrial Risks, Science, Hazards, and Public Protection.* Stoneham, MA: Butterworth Publishers, 1986.

Parmeggiani L, ed. *Encyclopedia of Occupational Health and Safety*, 3rd rev. ed. New York: McGraw-Hill Book Co., 1983.

B. Risk assessment—industrial hygiene

Andelman JB and Underhill DW. *Evaluation of Health Effects from Hazardous Waste Sites.* Chelsea, MI: Lewis Publishers, Inc., 1986.

Blakeslee HW and Grabowksi TM. *A Practical Guide to Plan Environmental Audits.* New York: Van Nostrand Reinhold, 1985.

Conway RA, ed. *Environmental Risk Analysis for Chemicals.* New York: Van Nostrand Reinhold, 1981.

Hallenbeck WH and Cunningham KM. *Quantitative Risk Assessment for Environmental and Occupational Health.* Chelsea, MI: Lewis Publishers, Inc. 1986.

Threshold Limit Values and Biological Exposure Indices. Cincinnati: ACGIH. Published annually.

C. Sampling methods—industrial hygiene

American Conference of Governmental Industrial Hygienists. *Air Sampling Instruments—For Evaluation of Atmospheric Contaminants*, 7th ed. Cincinnati: ACGIH, 1989.

D. Toxicology

Clayson DB, Krewski D and Munro I. *Toxicological Risk Assessment.* Boca Raton, FL: CRC Press, Inc., 1985.

Clayton GD, and Clayton FE, eds. *Patty's Industrial Hygiene and Toxicology.* Volume 2A, 2b, 2C: *Toxicology.* New York: John Wiley & Sons, Inc., 1981, 1982.

Documentation of the Threshold Limit Values for Substances in Workroom Air, rev. ed. Cincinnati: ACGIH.

Milman HA and Weisburger ED, eds. *Handbook of Carcinogen Testing.* Park Ridge, NJ: Noyes Data Corp., 1985.

Monson R. *Occupational Epidemiology.* Boca Raton, FL: CRC Press, Inc., 1980.

Neely WB and Blau, GE. *Environmental Exposure in Chemicals.* Boca Raton, FL: CRC Press, Inc., 1985.

E. Medical

Alderman MH and Hanley MJ. eds. *Clinical Medicine for the Occupational Physician.* New York: Marcel Dekker, Inc., 1982.

Alderson M. *Occupational Cancer.* Stoneham, MA: Butterworth Publishers, 1986.

Brown ML. *Occupational Health Nursing Principles and Practices.* New York: Springer Publishing Co., Inc., 1981.

Cataldo MF and Coates, TJ. *Health and Industry.* New York: John Wiley & Sons, Inc., 1986.

Preventing Illness and Injury in the Workplace. Washington, DC: Office of Technology Assessment, 1985.

Rom, WN, ed. *Environmental and Occupational Medicine.* Boston: Little, Brown & Co., 1982.

Safety Guide for Health Care Institutions, 4th ed. Chicago: National Safety Council and the American Hospital Association, 1989.

F. Dermatological

Adams RM. *Occupational Skin Disease*, 2nd ed. Orlando, FL: W. B. Saunders, 1989.

Gillin GA and Maibach HI. *Occupational Industrial Dermatology.* Chicago, IL: Year Book Medical Publishers, 1982.

Kryter KD. *The Effects of Noise on Man*, 2nd ed. San Diego, CA: Academic Press, 1985.

Pearce, B., ed. *Health Hazards of VDTs.* New York: John Wiley & Sons, Inc., 1984.

Polk C. *CRC Handbook of Biological Effects of Electromagnetic Fields.* Boca Raton, FL: CRC Press, Inc., 1986.

Sliney D and Wolbarsht ML. *Safety with Lasers and Other Optical Sources.* New York: Plenum Publishing Corp., 1980.

H. Ergonomics

Alexander DC and Babur MP. *Industrial Ergonomics: A Practitioners Guide.* Atlanta: Industrial Engineering and Management Press, 1985.

Astrand PO and Rodahl K. *Textbook of Work Physiology,* 3rd ed. New York: McGraw-Hill Book Co., 1986.

Brammer, AJ and Taylor W. *Vibration Effects on the Hand and Arm in Industry.* New York: John Wiley & Sons, Inc., 1983.

Chaffin DB and Anderson JB. *Occupational Biomechanics.* New York: John Wiley & Sons, Inc., 1984.

Eastman Kodak Company. *Ergonomic Design for People at Work,* vol. 1. Belmont, CA: Lifetime Learning Publications, 1983.

Eastman Kodak Company. *Ergonomic Design for People at Work,* vol. 2. New York: Van Nostrand Reinhold, 1986.

Grandjean, E. *Fitting the Task to the Man—An Ergonomic Approach,* 4th ed. London, England: Taylor & Francis, Ltd., 1988.

Konz S. *Work Design: Industrial Ergonomics,* 2nd ed. Worthington, OH: Publishing Horizons, 1990.

Kroemer KHE Kroemer HJ and Kroemer-Elbert KE. *Engineering Physiology: Physiologic Bases of Human Factors/Ergonomics.* New York: Elsevier Science Publishing Co., Inc., 1987.

McCormick EJ and Sander MS. *Human Factors in Engineering and Design.* 6th ed., New York: McGraw-Hill, 1986.

Salvendy G, ed. *Handbook of Human Factors.* New York: John Wiley & Sons, Inc., 1987.

Tichauer ER. "Human Factors Engineering," in 1984 *McGraw-Hill Yearbook of Science and Technology.* New York: McGraw-Hill Book Co., 1981.

I. Chemical

Alliance of America Insurers. *Handbook of Organic Industrial Solvents* (Technical Guide No. 6), 6th ed. Schaumburg, IL: AAI, 1987.

Chissick SS and Derricott R, eds. *Asbestos: Properties, Applications, and Hazards.* New York: John Wiley & Sons, Inc., 1983.

Compressed Gas Association. *Handbook of Compressed Gases,* 3rd ed. New York: Van Nostrand Reinhold, 1990.

Fawcett, HL and Wood WS. *Safety and Accident Prevention in Chemical Operations,* 2nd ed. New York: Interscience Publisher, 1982.

J. Pollution and hazardous waste

Bhatt HG, Sykes RM, and Sweeny TR.

Calvert S and England HM. *Handbook of Air Pollution Technology.* New York: John Wiley & Sons, Inc., 1984.

Dawson GW and Mercer BW. *Hazardous Waste Management.* New York: John Wiley & Sons, Inc., 1986.

Gammage RB and Kaye SV. *Indoor Air and Human Health.* Chelsea, MI: Lewis Publishers, Inc., 1985.

Godish T. *Air Quality.* Chelsea, MI: Lewis Publishers, Inc., 1985.

Levine SP and Martin WF. *Protecting Personnel at Hazardous Waste Sites.* Stoneham, MA: Butterworth Publishers, 1984.

Lioy PJ and Daisey JM, eds. *Toxic Air Pollution.* Chelsea, MI: Lewis Publishers, Inc., 1986.

Management of Toxic and Hazardous Wastes. Chelsea, MI: Lewis Publishers, Inc., 1985.

Robinson WD. *The Solid Waste Handbook, A Practical Guide.* New York: John Wiley & Sons, Inc., 1986.

K. Control

Alden JL and Kane JM. *Design of Industrial Ventilation Systems,* 5th ed. New York: Industrial Press, Inc., 1982.

Borup B. *Pollution Control for the Petrochemicals Industry.* Chelsea, MI: Lewis Publishers, Inc., 1987.

Constance JD. *Controlling In-Plant Airborne Contaminants.* New York: Marcel Dekker, Inc., 1983.

Cralley Lewis J and Cralley LV, eds. *Industrial Hygiene Aspects of Plant Operations* (in three volumes). New York: Macmillan Publishing Co., 1982, 1984, 1986.

Deisler F Jr. *Reducing the Carcinogenic Risks in Industry.* New York: Marcel Dekker, Inc., 1984.

Fullman JB. *Construction Safety, Security, and Loss Preventions.* New York: John Wiley & Sons, Inc., 1984.

Hemeon W, ed. *Plant and Process Ventilation,* 2nd ed. Ann Arbor, MI: University Microfilms International, Books-on-Demand.

Industrial Ventilation, A Manual of Recommended Practice, 20th ed. Lansing, MI: Committee on Industrial Ventilation, ACGIH, 1988.

LaDou JL, ed. *Occupational Health and Safety.* Chicago: National Safety Council, 1992.

Lees R and Smith A, eds. *Design, Construction, and Refurbishment of Laboratories.* Englewood Cliffs, NJ: Prentice-Hall, 1984.

McDermott H. *Handbook of Ventilation for Contaminant Control,* 3rd ed. Stoneham, MA: Butterworth Publishers, 1985.

Pepitone DA, ed. *Safe Storage of Laboratory Chemicals.* New

York: John Wiley & Sons, Inc., 1984.

Pitt, MJ, and Pitt E. *Handbook of Laboratory Waste Disposal*. New York: John Wiley & Sons, Inc., 1985

Rajhans G and Blackwell DS. *Practical Guide to Respirator Usage in Industry*. Stoneham, MA: Butterworth Publishers, 1985.

Schwope AD, Costas PP, Jackson JD and Witzman DJ. *Guidelines for Selection of Chemical Protective Clothing*, 2nd ed. Cincinnati: ACGIH, 1983.

L. Encyclopedias and handbooks

Agricultural Health and Safety Resource Directory. Cincinnati: ACHIG, 1984.

Calvert S and Englund HM, eds. *Handbook of Air Pollution Technology*. New York: John Wiley & Sons, Inc., 1984.

Encyclopedia of Occupational Health and Safety. Geneva, Switzerland: International Labour Office, 1983.

Hazardous Waste compliance Checklists for Supervisors. Madison, CT: Bureau of Law and Business, Inc., 1985.

King R and Hudson R. *Construction Hazard and Safety Handbook*. Stoneham, MA: Butterworth Publishers, 1985.

Lowry GG and Lowry RC. *Handbook of Hazard Communication and OSHA Requirements*. Chelsea, MI: Lewis Publishing Co., Inc., 1985.

Ridley J, ed. *Safety at Work*, 2nd ed. Stoneham, MA: Butterworth Publishers, 1986.

Sax N I. *Dangerous Properties of Industrial Materials,* 7th ed. New York: Van Nostrand Reinhold Co., 1988.

Sax N I and Lewis R J Sr, eds. *Rapid Guide to Hazardous Chemicals in the Workplace*, 2nd ed. New York: Van Nostrand Reinhold Co., 1990.

NIOSH Publications

The National Institute for Occupational Safety and Health (NISOH) has published many useful publications in the field of industrial hygiene. Consult the current "Publications Catalog," listing all NIOSH publication in print and their prices, for further information. The NIOSH publications can be obtained by requesting sing copies from:
National Institute for Occupational Safety and Health
Division of Technical Services
Publications Dissemination
4676 Columbia Parkway
Cincinnati, OH 45226

Many of these publications are also available from:
Superintendent of Documents
U.S. Government Printing Office
Washington, DC 20402

Some NIOSH publications can also be obtained from:
National Technical Information Service (NTIS)
Springfield, VA 22161

Criteria documents

The NIOSH is responsible for providing relevant data from which valid criteria for effective standards can be derived. Recommended standards for occupational exposure, which are the result of this work, are based on the health effects of exposure.

The single most comprehensive source of information on a particular material will probably be found in the NIOSH "Criteria Document" for that substance. The Table of Contents for a Criteria Document is as follows:

I. Recommendations for an Occupational Exposure Standard
Section 1—Environmental (workplace air)
Section 2—Medical
Section 3—Labeling and posting
Section 4—Personal protective equipment and clothing

M. Safety management

Asfah CR. *Intellectual Safety and Health Management*. Englewood Cliffs, NJ: Prentice-Hall, 1990.

Ferry TS. *Safety Program Administration for Engineers and Manangers*. Springfield, IL: Charles A Thiemen, 1984.

Goldsmith. *Safety Management in Construction and Industry,* New York: McGraw Hill, 1988.

Hammer. *Occupational Safety Management and Engineering*, 4th ed. Englewood Cliffs, NJ: Prentice-Hall, 1989.

Health and Safety in Small Industry. Chelsea, MI: Lewis Publishers, 1989.

Hoover et al. *Health Safety and Environmental Control*. New York: Van Hostrand Reinhold, 1989.

James, DWB. *A Safe Place to Work*. Southhampton, London: 1983.

Petersen, *Safety Management*. Goshen, NY: Aloroy, 1988.

Kavianien HR and Wentz Jr. CA *Occupational and Environment Safety Engineering and Management*. New York: Van Nostrand Reinhold, 1990.

Levy, ed. *Occupational Health: Recognizing and Preventing Work-Related Diseases*. Boston: Little, Brown & Co., 1988.

Occupational Health Service. Chicago American Medical Association, 1989.

Peterson D. *Techniques of Safety Management*, 3rd ed. Goshen, NY: ALORAY, 1989.

Peterson and Goodale J. *Readings in Industrial Accident Prevention.* New York: McGraw Hill Book Co., 1980.

Slote, ed. *Handbook of Occupational Safety & Health.* New York: John Wiley & Sons, 1987.

Tarrants WE. *The Measurement of Safety Performance.* ASSE, December 1980.

Thygersen *Safety.* 3rd ed. Englewood Cliffs, NJ: Prentice-Hall, 1986.

N. Emergency

Himmelfarb AB. *A Guide to Product Failure and Accidents.* Lancaster, PA: Technomic, 1985.

Kelly B. *Industrial Emergency Procedures.* New York: Van Nostrand Reinhold, 1989.

O. Safety training

Hendrick. *Systematic Safety Training.* New York: Marcel Dekker, Zinc., 1990.

ReVelle JB. *Safety Training Methods.* New York: Wiley, 1980.

P. Accident investigation and analysis

Ferry S. *Readings in Accident Investigation.* Springfield, IL: Charles C. Thomas, 1984.

Hendrick and Benner J. *Investigating Accident with STEP.* New York: Maracel Dekker, 1987.

Q. Product safety

Schader R and Heldman. *Product Design Liability.* New York: Practicing Law Institute, 1982.

Selden R. *Product Safety Engineering for Managers.* Englewood Cliffs, NJ: Prentice Hall, 1984.

Smith O. *Product Liability—Are You Vulnerable?* Englewood Cliffs, NJ: Prentice Hall, 1981.

R. Fire

Cote A and Linville J, eds. *Fire Protection Handbook,* 17th ed. Quincy, MA: National Fire Protection Association, 1991.

James D. *Fire Prevention Handbook.* Stoneham, MA: Butterworth Publishers, 1986.

S. Loss control

Bird FE Jr. *Management Guide to Loss Control.* Loganville, GA: Institute Press: 1980.

Bird FE Jr. and Loftus RG. *Loss Control Management.* Loganville, GA: Institute Press: 1976.

Brisbin RE. *Loss Control for the Small to Medium Size Business.* New York: Van Nostrand Reinhold, 1990.

Crowe MJ and Douglas HM. *Effective Loss Prevention.* Toronto, Canada: Industrial Accident Accident Prevention Association, 1976.

David FE. *Safety and Loss Control Management.* Burwood, Australia: The Craftsman Press Printing, Ltd., 1983.

T. OSHA

Mintz BW. *OSHA History, Law, and Policy.* Washington, DC: The Bureau of National Affairs, 1984.

OSHA Systems Safety Inspection Guide. Rockville, MD: Government Institutes, Inc., 1989.

Young DJ. *How to Comply with the OSHA Hazard Communication Standards.* New York: Van Nostrand Reinhold, 1989.

U. MSHA

U.S. Occupational Safety and Health Review commission, Washington, DC 20006. *A Guide to the Procedures of the Occupational Safety and Health Review Commission,* OSHRC Form 6 Rev. 1980.

V. Computers and applications

Microcomputer Applications in Occupational Health and Safety, Chelsea, MI, Lewis Publication, 1987.

Index

Measuring equipment, calibration of, 215–16
Mechanical engineer, in conducting inspections, 268
Medical Access Standard, 16, 45, 102
Medic Alert tags, 103
Medical examination, for fleet drivers, 226
Medical monitoring, 274
Medical personnel, in hazard control, 82
Medical records, 102–3
Medical rehabilitation, under worker's compensation, 201
Medical screening, 29
Medical treatment, distinguishing between first aid treatment and, 323
Meg, definition of, 335
Memory, definition of, 335
Merchandise displays, and nonemployee accidents, 492
Merit Program (OSHA), 30
Metal and Nonmetallic Mine Safety Act (1966), 25, 48, 50
Michigan State University (MSU) Driver performance measurement test, 226
Middle ages, safety and health awareness in, 2–3
Mine Enforcement Safety Administration (MESA), 48
Mine operators, duties of, under Mine Act, 50
Miners
    duties of, under Mine Act, 50
    rights of, under Mine Act, 49–50
    training of, under Mine Act, 50
Mine Safety and Health Act (1977), 28, 48
    administration of, 48–49
    advisory committees under, 49
    coverage of, 49
    legislative history, 48
    miners' duties, 50
    miners' rights, 49–50
    miner training, 50
Mine Safety and Health Administration (MSHA), 48, 56
    and first aid certification, 95
    periodic inspection requirements of, 248
    training requirements of, 378–81
Mine Safety and Health Review Commission, 48–49
Mine Safety and Health Standards, 50
    accident, injury, and illness reporting, 51
    citations, 52–53
    contested cases, 53
    emergency temporary standards, 50
    input from the private sector, 50
    inspection and investigation procedures, 51–52
    judicial review, 50
    penalties, 53
    variances, 50–51
    withdrawal orders, 52
Modem, definition of, 335
Monitor, definition of, 335
Monitoring
    biological, 273
    definition of, 239
    environmental, 273
    in hazard control, 73

medical, 274
personal, 273
Montreal Protocol, 108, 144
Moreell, Ben, 7
Motion pictures, for training, 401
Motivation, 347, 350
    approach-approach conflict in, 354
    approach-avoidance conflict in, 354–55
    avoidance-avoidance conflict in, 354
    complexity of, 350–52
    job satisfaction in, 352
    and learning, 360
    management theories of, 352
        job-enrichment theory, 353–54
        theory X, 352–53
        theory Y, 352–53
Mutual aid plans, 170
Mutual assistance, 170

N

Names, in safety promotions, 441
National Advisory Committee on Occupational Safety and Health (NACOSH), 28
National Ambient Air Quality Standards (NAAQS), 107
National Bureau of Standards, 6
National Contingency Plan (NCP), 110
National Council for Industrial Safety, 6
National Council on Alcoholism, 192
National Emissions Standards for Hazardous Air Pollutants (NESHAPs), 107
National Fire Protection Association (NFPA), and computer installations, 340
National Fleet Safety Contest, 422
National Foundation on Arts and Humanities Act, 25
National Institute for Occupational Safety and Health (NIOSH), 26, 56
    education and training under, 27
    employer and employee services under, 28
    research and related functions of, 27
National Institute on Alcohol Abuse and Alcoholism, 175, 192
National Institute on Drug Abuse, 175, 192
National Oceanic and Atmospheric Administration (NOAA), 171
National Pollutant Discharge Elimination System (NPDES), 109
National Safety Council
    and accident prevention, 2, 11
    birth of, 6
    defensive driving courses of, 432
    establishment of Environmental Health Center by, 16–17
    and first aid certification, 95
    First Aid Institute of, 166
    Labor Division in, 14
    and labor-management cooperation in health-safety issues, 14–15
    liaison between trade associations and, 14